Basic Mathematics
for Electricity and Electronics

Basic Mathematics
for Electricity and Electronics

8th Edition

Bertrand B. Singer
Late Assistant Chairman
Mathematics Department
Samuel Gompers High School
New York, New York

Harry Forster
Electronics Department
Miami-Dade Community College
University of Miami
Miami, Florida

Mitchel E. Schultz
Electronics Department
Western Wisconsin Technical College
La Crosse, Wisconsin

McGraw-Hill

New York, New York   Columbus, Ohio   Woodland Hills, California   Peoria, Illinois

Cover photos: (l) John Paul Endress/The Stock Market, (c) John Lund/The Stock Market, (r) Doug Martin

Photo credits appear on page 897, which is hereby made part of this copyright page.

Library of Congress Cataloging-in-Publication Data
Singer, Bertrand B.
    Basic mathematics for electricity and electronics / Bertrand B.
Singer, Harry Forster, Mitchel E. Schultz. — 8th ed.
        p. cm.
    Includes bibliographical references and index.
    ISBN 0-02-805022-3
    1. Electric engineering—Mathematics. 2.
Electronics—Mathematics. I. Forster, Harry.   II. Schultz,
Mitchel E. III. Title.
    TK153 .S54 1999
    621.3'01'51—dc21
                                            99-16614
                                               CIP

Glencoe/McGraw-Hill

A Division of The **McGraw·Hill** *Companies*

Basic Mathematics for Electricity and Electronics
Eighth Edition

Send all inquiries to:
Glencoe/McGraw-Hill
8787 Orion Place
Columbus, Ohio 43240-4027

ISBN-13: 978-0-02-805022-5

ISBN-10: 0-02-805022-3

6 7 8 9 071 09 08 07 06

Contents

Contents

The author, project editor, and publisher would like to thank the reviewers listed below. Their comments and suggestions provided the valuable input necessary to make a good book even better.

Walter Bartkiw
Former Electronics Teacher
Ontario, Canada

Brian McInerney
Delta College
University Center, Michigan

Robert Dubuc, Jr.
New England Institute of Technology
Warwick, Rhode Island

Larry R. Patterson
Orange County Electrical Training Trust
Santa Ana, California

Dave Johnson
Orange County Electrical Training Trust
Santa Ana, California

Frank E. Woodring
Pittsburgh Technical Institute
Pittsburgh, Pennsylvania

Preface

The eighth edition of *Basic Mathematics for Electricity and Electronics* continues to reflect the many new advances in technology, which have greatly affected what electricians and electronic technicians need to know on the job. Calculators and computers now play a major role in determining how problems are solved. Specifically, less emphasis is placed on the actual calculations and more emphasis is placed on what the answer means.

Electricians and electronic technicians are expected to have a wide range of knowledge. The scope of this material has grown to be so broad that it is becoming difficult to study specific isolated topics. The approach of this text is to show the applications of the many mathematical principles and concepts encountered by electricians and technicians in the field.

This edition makes reference to many industry standards such as the National Electrical Code (NEC). In some cases the standard comes from mathematical concepts. For example, Job 22-2 shows how the NEC standard 208Y/120 is derived from mathematical results. Yet the examples show that the correct "electrical answer" is not the mathematical answer but the standard that is closest to it. An understanding of the relationship between mathematics and standards helps the student to relate mathematics to the real world.

A practical application of mathematics is found in Job 16-15, Real Cables and Power Distribution. This job ties together 14 earlier jobs to show how they all relate to real cables. It proves the usefulness of the other jobs and helps users understand when to use their results and when the results can be ignored.

Another conceptual change reflected in this edition concerns the growing problems of grounding and power distribution. As power loads shift from lighting and rotating machines to electronic devices, grounding takes on a new importance. European countries are developing and imposing more stringent power-quality requirements. Some changes are being made in the NEC requirements also. The materials in this edition provide a basis for understanding these real problems.

Job 4-4, Power Distribution Line Currents, introduces what are considered to be ac concepts at the dc level. This helps the student to separate ac concepts from line current concepts and provides a basis upon which it becomes easier to introduce real ac phenomena. This is further expanded in Chapter 22, where it is shown that the mathematical techniques that should be used depend upon what needs to be demonstrated.

Many of the new power distribution problems are caused by nonlinear devices. Harmonic analysis, introduced in Job 15-9, is used to explain these devices and their related phenomena. These concepts are further explained in Job 19-5, where it is shown how they relate to power filters and ground loops.

Chapter 22 ties together the earlier developments of ideal transformers and three-phase power. This chapter gives more information about balanced-phase loads and their analysis. The mathematics is much easier, and yet it is sufficient for most purposes. The chapter also explains when more complex analysis should be used. The objective is to assist the student in finding the right tool for the job.

This edition continues to provide a solid teaching style. Checkups provide student motivators and mathematical diagnostics. Brushups review the mathematical concepts immediately necessary for progress in understanding electrical theory and application. The theory is developed in slow, simple stages and is directly and immediately applied to the solution of real problems.

Concepts are developed in short "jobs." Each job concentrates on either a mathematical or an electrical competency. Self-tests serve as a final check on the students' understanding. Each self-test is a diagnostic test that the students can perform themselves. The problems in each job range from easy to more difficult. The self-help features of this book include answers to the odd-numbered problems, which are given at the end of the book. Continual review and drill, in both mathematical and electrical concepts, increase students' understanding of the electrical theory and applications.

This edition includes several new features, among them, Learning Objectives, which are given at the beginning of each chapter; marginal items such as student success hints and calculator hints; and an end-of-chapter section containing Internet activities. These features make the text more student-friendly and will certainly enhance learning of the material.

The authors would like to thank the reviewers who took the time to examine the content and accuracy of the text. Their input was greatly appreciated. In addition, the authors would also like to thank the staff at Glencoe, McGraw-Hill, especially Jeanne Huffman and Brian Mackin, for their extra efforts in making this edition possible.

Harry Forster
Mitchel E. Schultz

Introduction to Electricity

1. Describe the basic structure of the atom.
2. List the basic particles of electric energy contained within the atom.
3. Define voltage, current, and resistance. List the units of each.
4. Describe how an ammeter is connected to measure the flow of current in a circuit.
5. Describe how a voltmeter is connected to measure the voltage in a circuit.
6. Describe how an ohmmeter is connected to measure the resistance in a circuit.
7. Explain the purpose of a fuse and its current rating.

Learning Objectives

JOB 1-1

Basic Theory of Electricity

We shall start our study of electricity with an examination of the materials from which electric energy is produced.

Elements.   Science has discovered more than 100 different kinds of material called *elements*. These cannot be made from other materials and cannot be broken up to form other materials by ordinary methods. Gold, iron, copper, oxygen, and carbon are some examples of elements.

Atoms.   An *atom* is the smallest amount of an element which has all the properties of the element. An atom may be broken down into smaller pieces, but these pieces have none of the properties of the element. It is now believed that all material is electrical in nature. All matter is made of combinations of elements, which, in turn, are made of atoms. The atoms themselves are merely combinations of different kinds of electric energy. The theory currently accepted is that an atom, because it is electrically neutral, is made of positive charges of electricity called *protons* and an equal number of negative charges called *electrons*. There are also a number of electrically neutral particles called *neutrons*. Similarly, charged particles will repel each other, and particles of opposite charge will attract each other. For example, two electrons will repel each other, and two protons will repel each other, but an electron will be attracted to a proton as shown in Fig. 1-1.

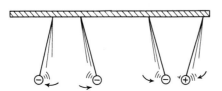

Figure 1-1 Opposite charges attract; like charges repel each other.

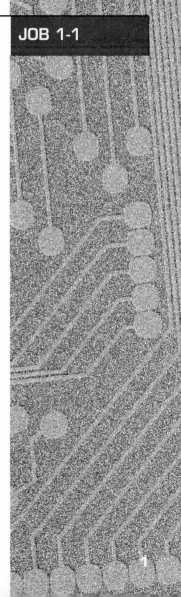

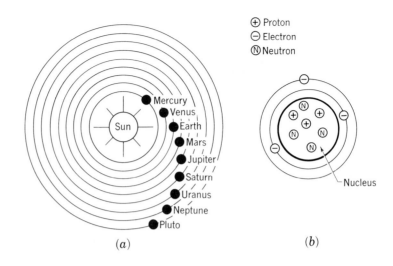

- ⊕ Proton
- ⊖ Electron
- Ⓝ Neutron

(a)

(b)

Figure 1-2 The resemblance between the solar system (a) and the theoretical diagram of an atom of lithium (b).

Structure of the atom. An atom is believed to resemble our solar system with the sun at the center and the planets revolving around it, as shown in Fig. 1-2a. An atom consists of electrons revolving around a *nucleus*, which contains the protons and neutrons, as shown in Fig. 1-2b. The number of revolving, or *planetary*, electrons is equal to the number of positively charged protons in the nucleus.

Free electrons. The distribution of planetary electrons in Fig. 1-2b determines an atom's electrical stability. Especially important is the number of electrons in the orbiting ring farthest from the nucleus. For atoms containing more than one orbiting ring of electrons, the outermost ring requires eight electrons for maximum electrical stability. (The orbiting ring of electrons closest to the nucleus can have a maximum of only two electrons, regardless of the type of atom.) The outermost ring of electrons is called the *valence ring*. Atoms containing one, two, or three valence electrons are good conductors of electricity. The reason is that with only a few valence electrons, the atom is not stable. For example, a copper atom has only one electron in its valence ring. This electron is bound only very loosely to its nucleus. Therefore, when many copper atoms are close together, such as in a copper wire, the valence electrons move easily from one atom to the next at random. Electrons that can move freely from one atom to the next are called *free electrons*. It is this movement of free electrons that provides electric current in a copper wire.

Electron flow. If you shuffle your feet over a rug on a dry day, some of these free electrons will be rubbed off the rug and will accumulate on your body. Touching a metal doorknob or some other conductor will produce a slight shock as the electrons *flow* through your body to ground. Notice that no shock will be experienced while the body is accumulating the electrons. The shock will occur only when the electrons *flow* through your body in one concerted surge. This directed flow, or movement, of the electrons is one of the most important ideas in electricity. In order to understand this better, let us compare the particles of matter with a brick wall, as in Fig. 1-3. A brick in a wall is like an atom, since it is the smallest part of the wall that has the characteristics of the wall. If an individual brick were to be powdered into dust, we could compare a single grain of dust with an electron. The grains of brick cannot do any useful work by themselves, but if a powerful pressure

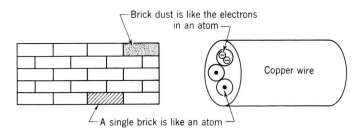

Figure 1-3 Comparison of brick dust with electrons.

like a blast of compressed air were allowed to hit the particles, as would occur in a sandblasting machine, then enormous energies would be available.

We can also consider the drops of water in a stream to be like the electrons in a piece of matter. If the drops of water move aimlessly, as in a water sprinkler, then their energy is small. But if all the drops are forced to move in the *same direction*, as in a high-pressure fire hose, then their energy is large. We can see, then, that if electrons are to do useful work, they must be *moving under pressure*, like the grains of sand or drops of water.

Electrical pressure: voltage. In order to move the drops of water or the grains of sand, mechanical pressure supplied by a water pump or an air compressor is required. Similarly, an electron pressure is required to move the electrons along a wire. This electrical pressure is called the *voltage* and is measured in units of *volts* (V). This unit of measurement was named after Count Alessandro Volta, an Italian physicist (1745–1827).

Producing electrical pressures. A working pressure is considered to exist when there is a *difference* in energies. Water in a high tank will exert a pressure because of the *difference* in the height of the water levels, as shown in Fig. 1-4. Elements differ not only in the number of electrons that make up their individual "solar systems" but also in the energies of their electrons. Therefore, if two different materials are brought together under suitable conditions, there will be a *difference* of electron energies. This difference produces the electrical pressure that we call the *voltage*.

Different combinations of materials will produce different voltages. A dry cell of carbon and zinc produces 1.5 V, a lead and sulfuric acid cell produces 2.1 V, and an Edison storage cell of nickel and iron produces 1.2 V. These combinations are called *batteries*. A voltage pressure may be produced in several other ways, which we shall learn about later. For example, the 110 to 120 V supplied by an ordinary

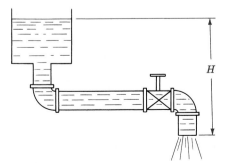

Figure 1-4 The difference in height between the water levels causes pressure.

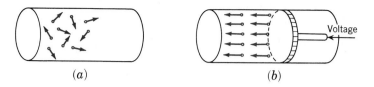

(a)   (b)

Figure 1-5 (a) The electrons in the wire move in many different directions. (b) When a voltage pushes the electrons, they all move in the *same* direction and make an electric current.

house outlet is produced by a machine called a *generator*. An automobile spark coil delivers about 1500 V. The purpose of all voltages, however produced, is to provide a force to *move* the electrons. It is appropriately termed an electron-moving, or *electromotive force*, which is abbreviated *emf*. Modern practice uses the symbol *V* to indicate emf.

Quantity of electricity. The amount of electricity represented by a single electron is very small—much too small to be used as a measure of quantity in practical electrical work. The electron has already been compared with a drop of water. It is obviously ridiculous to measure quantities of water by the number of drops. Instead, we use quantities like a gallon or a liter, each of which represents a certain number of drops. Similarly, the practical unit of electrical quantity represents a certain number of electrons and is called a *coulomb* (C). This unit was named after C. A. Coulomb, a French physicist (1736–1806). The coulomb is equal to about 6 quintillion electrons (6,000,000,000,000,000,000). The symbol for electrical quantity is *Q*.

Current. When a voltage is applied across the ends of a conductor, the electrons, which up to now have been moving in many different directions (Fig. 1-5a), are forced to move in the *same direction* along the wire (Fig. 1-5b). Individual electrons all along the path are forced to leave the atoms to which they are attached. They travel only a short distance until they find an atom that needs an electron. This motion is transmitted along the path from atom to atom, the way motion in a whip is transmitted from one end to the other. This *flow* of electrons is called an *electric current* (Fig. 1-6). The speed of this flow is very nearly equal to the speed of light, which is about 186,000 miles per second (mi/s). The actual speed of the individual electrons is much slower, but the effect of a pressure at one end of a wire is felt almost instantaneously at the other end.

The flow of water is measured as the number of gallons per minute, barrels per hour, etc. Similarly, the flow of electricity is measured by the number of electrons that pass a point in a wire in 1 s. We do not have special names for the flow of water, but we do have a special name for the flow of electrons. This name is *ampere* (A) of current. A current of 1 A represents a flow of 1 C of electricity (6 quintillion electrons) past a point in a wire in 1 s, that is, a flow of 1 C/s.

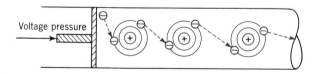

Voltage pressure

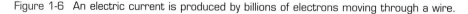

Figure 1-6 An electric current is produced by billions of electrons moving through a wire.

The symbol for current is *I*. The ampere was named after André Marie Ampère, a French physicist and chemist (1775–1836). Since an electron is so small (about 25 trillion to an inch) and since there are so many of them, it is impossible actually to count them as they go by. However, when electrons are moving as an electric current, they can do useful work, such as lighting lamps, running motors, producing heat, and plating metals. We can make use of this last ability of an electric current to measure and define it accurately. An international commission has thus defined an ampere as the number of electrons that can deposit a definite amount of silver [0.001 118 gram (g)] from a silver solution in 1 s. A 100-W 110-V lamp uses about 1 A. A 600-W 110-V electric iron uses about 5.5 A, and the current required by a television transistor may be as low as 0.001 A.

Resistance. We have learned that "free" electrons may be forced to move from atom to atom when a voltage pressure is applied. Different materials vary in their number of free electrons and in the ease with which electrons may be transferred between atoms. A *conductor* (Fig. 1-7*a*) is a material through which electrons may travel freely. Most metals are good conductors. An *insulator* (Fig. 1-7*b*) is a material that prevents the electrons from traveling through it easily. Nonmetallic materials like glass, mica, porcelain, rubber, and textiles are good insulators. No material is a perfect insulator or a perfect conductor.

The ability of a material to resist the flow of electrons is called its *resistance* and is measured in units called *ohms* (Ω). This unit of measurement was named after Georg Simon Ohm, a German physicist (1787–1854). The symbol for resistance is *R*. An international agreement defines the ohm as the resistance offered by a column of mercury of uniform cross section, 106.3 cm long (about 41.8 in), and weighing 14.45 g (about ½ oz.) For example, 1000 ft of No. 10 copper wire has a resistance of almost exactly 1 Ω; the resistance of a 40-W electric light is 300 Ω when hot; and a 150-V voltmeter has a resistance of 15,000 Ω.

There are many electric devices which make use of the fact that materials offer resistance to the flow of an electric current. Soldering irons, electric heaters, and electric light bulbs contain conductors which have a high resistance compared with the resistance of the connecting wires. Radio and television circuits contain many resistance elements, called *resistors*. They are not always made of special resistance wire. Carbon and mixtures of carbon and insulating materials are molded to form resistors.

> **Hint for Success**
>
> The resistance of a conductor is usually very small such as 1 Ω or less. An insulator, however, usually has several million ohms of resistance. For an insulator, a resistance of 1 trillion Ω is not uncommon.

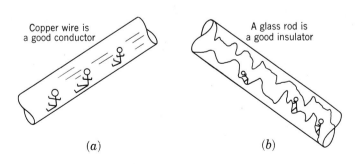

Figure 1-7 (*a*) Electrons travel freely through conductors. (*b*) Insulators prevent electrons from flowing freely.

Electrical Measurements and Circuits

The voltage, current, and resistance of an electric circuit are measured with special instruments. An *ammeter* measures the current *I* in units of amperes. A *voltmeter* measures the voltage *V* in units of volts. An *ohmmeter* measures the resistance *R* in units of ohms.

How to use the ammeter. In order to measure the flow of water in a pipe, a flowmeter is inserted into the pipe, as shown in Fig. 1-8*a*. The pump supplies the pressure to force the water through the pipe and flowmeter against the resistance of the faucet valve. Since all the water in the system must flow through the flowmeter, it must indicate the number of gallons per minute that flow through the pipe. In the same way, in order to measure the flow of current in a wire, an ammeter must be inserted *directly into the circuit* so that all the electrons will be forced through it. The ammeter has a very low resistance and thus does not stop the flow of current. When large currents are measured, certain adjustments must be made in order to prevent the ammeter from burning out (Job 10-1). In Fig. 1-8*b*, the battery supplies the electrical pressure that forces the electrons through the lamp against the resistance of the lamp. Since all the electrons in the circuit must pass through the ammeter, it will indicate the number of electrons per second, or amperes, passing through it.

How to use the voltmeter. As we have already learned, the amount of water that will flow in a pipe depends on the *difference* in the pressure between any two points. In the same way, the electric current that will flow through a resistance depends on the *difference* in electrical pressure (voltage) between the two ends, or *terminals,* of the resistance. In order to measure this difference, a voltmeter must be connected across the ends of the resistance, as shown in Fig. 1-9. Note that the current that flows through the resistance does *not* flow through the voltmeter. Since a voltmeter is designed to measure the electrical pressure, it should be placed in the circuit so that the pressure across the resistance is also across the voltmeter. An ammeter, on the other hand, is inserted *into* the circuit and receives the full current of the circuit. A voltmeter is merely attached to the ends of the part across which the voltage is to be measured.

How to use the ohmmeter. An ohmmeter is really an ammeter whose dial has been marked to read the resistance in ohms instead of the current in amperes according to a very simple relationship called *Ohm's law,* which we shall study

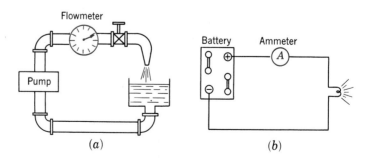

Figure 1-8 (*a*) A flowmeter is inserted in the line of flow of the water. (*b*) An ammeter is inserted in the line of flow of the electrons.

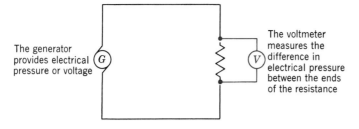

Figure 1-9 A voltmeter is always connected across the ends of a part when the voltage across it is being measured.

soon. An ohmmeter will measure the resistance of an individual part of a circuit when the leads of the instrument are connected across the ends of the part, as is done with a voltmeter. *Warning:* All power must be *off* when an ohmmeter is used.

Electric circuits.   A *circuit* is simply a *complete path* along which the electrons can move. A complete circuit must have an *unbroken* path, as shown in Fig. 1-10*a,* with the source of energy acting as an electron pump to force the electrons through the conductor (usually a copper wire) against the resistance of the device to be operated. When the switch is opened as in Fig. 1-10*b,* the electrons cannot leave one side of the switch to enter the other side and so return to their source because of the very high resistance of the air gap. This is an *open circuit,* and no current will flow through it.

Fuses.   One of the effects of the passage of an electric current is the production of heat. The larger the current, the more heat is produced. In order to prevent large currents from accidentally flowing through our expensive apparatus and burning it up, a device called a *fuse* is placed directly into the circuit, as in Fig. 1-11, so as to form a part of the circuit through which all the electrons must flow. To protect a circuit, a fuse must be a device which will *open* the circuit whenever a dangerously large current starts to flow. Accordingly, a fuse completes a circuit with a piece of special metal designed to melt quickly when heated excessively, as would occur when a large current flows. The fuse will permit currents smaller than the fuse value to flow but will melt and therefore break the circuit if a larger, dangerous current ever appears. A dangerously large current will flow, for instance, when a "short circuit" occurs. A short circuit is usually caused by an accidental connection between two points in a circuit which offer very little resistance to the flow of electrons. If the resistance is small, there will be nothing to stop the flow of the electrons, and the current will increase enormously. The resulting heat generated

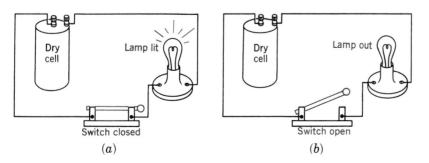

Figure 1-10   (*a*) With the switch closed, current flows through this complete circuit. (*b*) With the switch open, no current flows through this broken or open circuit.

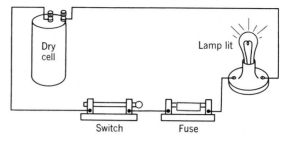

Figure 1-11 A fuse completes an electric circuit and protects it against the flow of danger-ously large currents.

might cause a fire. However, if the circuit is protected by a fuse, the heat caused by the short-circuit current will melt the fuse wire, thus breaking the circuit and reducing the current to zero.

Fuse ratings. Fuses are rated by the number of amperes of current that can flow through them before they melt and break the circuit. Thus we have 10-, 15-, 30-A, etc., fuses. We must be careful that any fuse inserted in a circuit be rated low enough to melt, or "blow," before the apparatus is damaged. For example, in a house wired to carry a current of 15 A it is best to use a fuse no larger than 15 A so that a current larger than 15 A could never flow.

Delayed fuses. In ordinary house wiring, a 15-A fuse is used to protect the line wires from overheating and producing a fire hazard. However, if a current larger than 15 A were to flow for only a few seconds, it would not harm the wires. There are many times when a wire must carry a current larger than it was designed to carry—but only for a short time. For example, a normal 5 A may be flowing in a line. If the motor of a washing machine is suddenly turned on, the current drawn may go up to 35 A for a few seconds but then drop to a normal 10 A after the motor is running. The ordinary fuse would blow, since the current is larger than the rated 15 A. However, since there is no danger, as the current drops again very quickly, we would like a fuse that could carry larger currents for a short time. This type of fuse is called a *time-delay* fuse. It is designed to carry about twice its rated current for 20 to 30 s but to blow if the rated current is exceeded for 1 min. The net effect is that the fuse will hold for *temporary* overloads but will blow on continuous, small overloads or short circuits.

Symbols. Most of our diagrams up to this point have used pictures of the various electric devices. However, not everybody can draw pictures quickly and accurately, and so we shall substitute special simple diagrams to illustrate the various parts of any circuit. Each circuit element is represented by a simple diagram called the *symbol* for the part. The standard symbols for the commonly used electrical and electronic components are given in Fig. 1-12.

Circuit diagrams. Every electric circuit must contain the following:

1. A source of electrical pressure, or voltage *V*, measured in volts.
2. An unbroken conductor through which the electrons may flow easily. The amount of electron flow per unit of time is the current *I*, measured in amperes.
3. A load, or resistance *R*, to this flow of current, measured in ohms.
4. A device to control the flow of current, or switch.

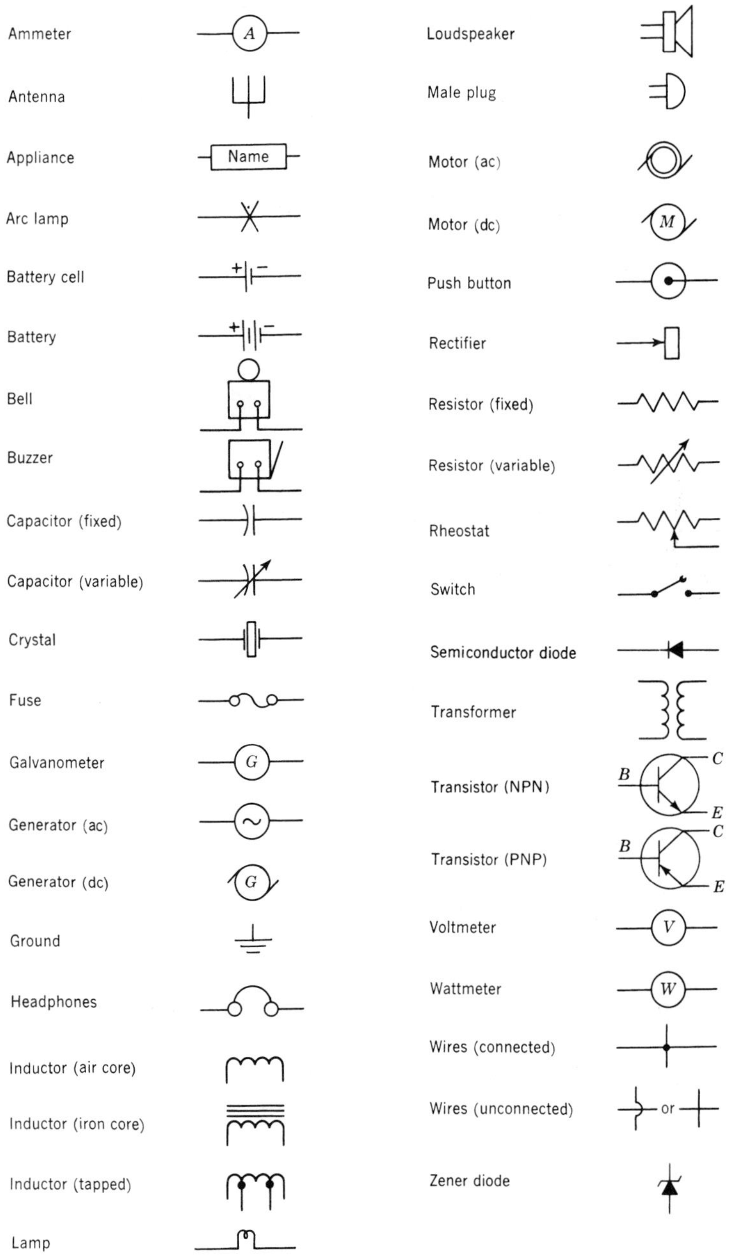

Ammeter	—(A)—	Loudspeaker	
Antenna		Male plug	
Appliance	—[Name]—	Motor (ac)	
Arc lamp		Motor (dc)	(M)
Battery cell	—+|−—	Push button	—(•)—
Battery	—+|||−—	Rectifier	—▶[
Bell		Resistor (fixed)	—◇◇◇—
Buzzer		Resistor (variable)	
Capacitor (fixed)	—|(—	Rheostat	
Capacitor (variable)		Switch	—•／•—
Crystal	—|[]|—	Semiconductor diode	—◀—
Fuse		Transformer	
Galvanometer	—(G)—	Transistor (NPN)	B (C / E)
Generator (ac)	—(∼)—	Transistor (PNP)	B (C / E)
Generator (dc)	(G)	Voltmeter	—(V)—
Ground		Wattmeter	—(W)—
Headphones		Wires (connected)	—|—
Inductor (air core)		Wires (unconnected)	—) (— or —|—
Inductor (iron core)		Zener diode	
Inductor (tapped)			
Lamp			

Figure 1-12 Standard circuit symbols.

Example 1-1

Draw a circuit containing two battery cells, an ammeter, a fuse, a single-pole switch, a resistor, and a lamp. Label each part, using V for voltage, R for resistance, and I for current.

Solution

See Fig. 1-13. *Note:* The numbers 1 and 2 under the letter R are a convenient way to indicate the first resistance R_1 and the second resistance R_2.

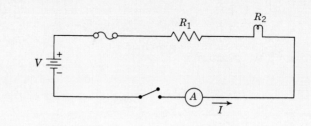

Figure 1-13

Exercises

Practice

Using the circuit symbols shown in Fig. 1-12, draw a circuit for each problem to include the elements indicated. Label the diagram completely, using the subscripts 1, 2, 3, etc., to indicate the parts of the circuit.

1. A battery of two cells, a fuse, a switch, and two resistors
2. An ac generator, a switch, an ammeter, a bell, and a buzzer
3. A battery, a switch, an ammeter, two resistors, and a lamp with a voltmeter across it
4. A battery, a switch, a bell, and a push button
5. A dc generator, a switch, a rheostat, and a dc motor
6. A dc generator, a switch, a fuse, an arc lamp, and a resistor
7. A male plug, an electric iron, a rheostat, and a switch
8. A male plug, an ac motor, a switch, and a variable resistor

Self-Test 1-2

Fill in each blank with the correct response.
1. All material is made of _____ charges.
2. Negative charges are called _____.
3. Positive charges are called _____.
4. Similar charges _____ each other, and _____ charges attract each other.
5. A _____ represents a quantity of 6 billion billion electrons.
6. An *ampere* is a current of 1 _____/s. The symbol for current is _____.
7. Current is measured by an _____, which is always inserted directly into the circuit and reads _____.
8. Electrons move because of the pressure of an electromotive force, or _____. The symbol for voltage is _____.
9. Voltage is measured by a _____, which is always connected across the _____ of the circuit element and reads _____.
10. The resistance of a circuit is the resistance offered by the circuit to the flow of _____. The symbol for resistance is _____.
11. Resistance is measured by an _____, which is always connected across the _____ of the circuit element and reads _____.

12. A *conductor* is a material that allows _____ to flow.
13. An _____ is a material that prevents current from flowing easily.
14. A fuse is a thin strip of easily _____ material. It protects a circuit from large currents by melting quickly, thereby _____ the circuit.
15. For an atom, all _____ and _____ are contained in the nucleus.
16. When measuring resistance, al power must be _____ in the circuit.
17. A copper atom has _____ valence electron(s).
18. A fuse is rated in _____.
19. The flow of electrons is called _____ _____.
20. The symbol for electrical quantity is _____.

(See CD-ROM for Test 1-1)

Assessment

1. In electrical theory, the force that is used to cause electrons to flow in a circuit is known as _____ and is measured in volts.

2. The electrical current is the quantity of electrons, in coulombs, that is moving past a point in 1 s. The unit of measure for electrical current is the _____, abbreviated _____.

3. Different materials vary in their number of free electrons and in the ease with which electrons may be transferred between atoms. A _____ is a material through which electrons may travel freely.

4. A _____ is a complete path along which electrons can move. The source of energy acts as an electron pump to force electrons through the conductor against the resistance of the device to be operated.

INTERNET
ACTIVITIES

Internet Activity A

Use the following site to find out how resistors are marked to indicate their resistance value in ohms. **http://www.proxis.com/~iguanalabs/compnets.htm**

Internet Activity B

Use the following site to find the speed of electron flow in the metal wire. **http://olbers.kent.edu/~alcomed/wwwboard/messages/2035.html**

Chapter 1 Solutions to Self-Tests and Reviews

Self-Test 1-2

1. electric
2. electrons
3. protons
4. repel, opposite
5. coulomb
6. C, *I*
7. ammeter, amperes
8. voltage, *V*
9. voltmeter, ends (or terminals), volts
10. electrons, *R*
11. ohmmeter, ends (or terminals), ohms
12. current (or electrons)
13. insulator
14. melted, breaking
15. protons, neutrons
16. off
17. one
18. amperes
19. electric current
20. *Q*

Simple Electric Circuits

chapter

Learning Objectives

1. Define the word fraction.
2. Simplify a fraction.
3. Change a mixed number to an improper fraction.
4. Change an improper fraction to a mixed number.
5. Multiply fractions and mixed numbers.
6. Apply Ohm's law to find the voltage in a circuit.
7. Convert fractions and mixed numbers to decimals.
8. Define the metric system.
9. Describe the procedure for moving the decimal point in a number when multiplying or dividing that number by a multiple of 10.
10. Convert units of measurement.
11. Add, subtract, multiply, and divide decimals.

JOB 2-1

Checkup on Fractions (Diagnostic Tests)

Before we go any further into our study of electric circuits, let's stop to check up on our knowledge of fractions. The following are some problems often encountered by electricians and electronic technicians in their daily work. They are all solved by multiplying the numbers involved in the problem. If you have difficulty solving any of these problems, see Job 2-2, which follows.

Exercises

Practice

1. $2\frac{1}{2} \times 3\frac{1}{5}$
2. $\frac{2}{5} \times 20$
3. $\frac{22}{7} \times 21$
4. $\frac{1}{2} \times \frac{1}{3}$
5. $20 \times \frac{1}{1,000}$
6. $\frac{2}{3} \times \frac{3}{4}$
7. $5\frac{1}{3} \times 8\frac{1}{2}$
8. $3\frac{1}{7} \times 5\frac{1}{11}$

Applications

9. What length of two-conductor BX cable is needed to obtain six pieces each $4\frac{1}{4}$ ft long?
10. What is the total horsepower delivered by five $\frac{3}{4}$-hp motors?
11. A voltage divider develops an emf of $\frac{1}{20}$ V for each ohm of resistance. What voltage would be measured across 1800 Ω of resistance?
12. If the resistance of one turn of a variable wirewound resistor is $2\frac{1}{3}$ Ω, what is the resistance of three turns?
13. What is the cost of $2\frac{1}{4}$ lb of magnet wire at $1.25/lb?
14. An industrial shop uses $9\frac{1}{2}$ kWh of electricity per day. How many kilowatthours are used in a month of 24 working days?
15. How many hours of work were spent on an electrical installation if five people each worked $7\frac{1}{2}$ hours (h)?

16. A neon electric sign uses $4\frac{1}{2}$ W of power per foot of tubing. How many watts are used for a sign which uses 21 ft of tubing?
17. Number 9191 Fiberduct conduit adapters weigh $1\frac{1}{2}$ lb each. Find the weight of 24 such fittings.
18. How many feet of antenna wire are there in a $1\frac{1}{8}$-lb coil if the wire runs 18 ft to the pound?

JOB 2-2

Brushup on Fractions

Meaning of a fraction. A fraction is a shorthand way of describing some part of a total amount. Figure 2-1*a* shows a whole pie. If the pie is cut into four equal pieces, as shown in Fig. 2-1*b*, then each piece is just one part out of the total of four parts. This is written as the *fraction* $\frac{1}{4}$ and means *1 part out of 4 equal parts.* The fraction $\frac{3}{4}$, as shown by the shaded portion in Fig. 2-1*c*, says that a whole was divided into 4 equal parts and that 3 of these parts were used. The fraction $\frac{1}{8}$, as shown in Fig. 2-1*d*, says that a whole was divided into 8 equal parts and that 1 of these parts was used. It is evident that we can divide the pie into any number of equal parts. Each part will be 1 part out of the total number of parts. In Fig. 2-2*a* the pie is cut into 8 equal parts. Each part is $\frac{1}{8}$ of the pie. The shaded portion is 3 of these, or $\frac{3}{8}$ of the pie. In Fig. 2-2*b* the pie is cut into 5 equal parts. Each part is $\frac{1}{5}$ of the pie. The shaded portion is 2 of these, or $\frac{2}{5}$ of the pie. In Fig. 2-2*c* the pie is again cut into 8 equal parts. Each part is $\frac{1}{8}$ of the pie. The shaded portion is 2 of these, or $\frac{2}{8}$ of the pie. If the pie had been cut into 4 equal parts, our shaded portion would then have been $\frac{1}{4}$. Thus, $\frac{2}{8} = \frac{1}{4}$.

Hint for Success

In a fraction, the numerator is the quantity written above the bar and the denominator is the quantity written below the bar.

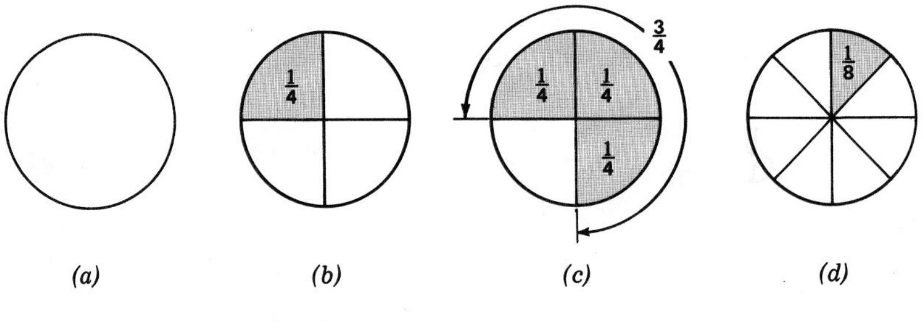

Figure 2-1 Fractional parts of a pie.

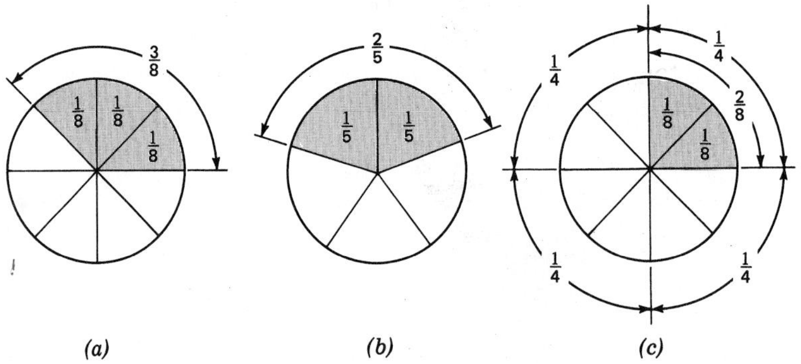

Figure 2-2

Hint for Success

The bar separating the numerator and denominator of a fraction is called the *vinculum*.

Simplifying fractions. We have seen that a portion of a total may be expressed in two different ways. That is, ¼ is the same as ⅜. The fraction ¼ is called the *simplified,* or *reduced,* form of ⅜. To simplify a fraction means to change it into another *equivalent* fraction. When the numerator and denominator of a fraction have no common divisor except 1, the fraction is said to be simplified, or reduced, to lowest terms.

RULE (Basic Principle)

If a fraction is multiplied or divided by 1, the value of the fraction is unchanged.

The number 1 may be expressed in any number of ways, such as ⅓, ¼, ⅗, ⁷⁄₇, ¹¹⁄₁₁, or ²⁵⁄₂₅. All that is required for a fraction to be equal to 1 is that the numerator and the denominator be the *same* number. Therefore, to simplify a fraction, divide it by 1, expressed as a fraction with equal numerator and denominator.

RULE   To simplify a fraction:

1. Find any number (the largest is best) that will divide evenly into both the numerator and denominator of the fraction.
2. Divide both the numerator and the denominator of the fraction by this number. (This is the same as dividing the fraction by 1.)

Example 2-1

Four inches is what fractional part of a foot?

Solution

Since there are 12 in. in 1 ft, 4 in. is ⁴⁄₁₂ of a foot. Reduce to lowest terms.

$$\frac{4}{12} \div \frac{4}{4} = \frac{4 \div 4}{12 \div 4} = \frac{1}{3} \qquad Ans.$$

Example 2-2

A dropping resistor used 24 V of the total of 96 V supplied to a circuit. Express the fraction ²⁴⁄₉₆ in lowest terms.

Solution

We can divide both the 24 and 96 by 24 to get ¼ in one step.

$$\frac{24}{96} \div \frac{24}{24} = \frac{24 \div 24}{96 \div 24} = \frac{1}{4} \qquad Ans.$$

Or we can simplify in several steps. Dividing both numerator and denominator by 8 will give

$$\frac{24}{96} \div \frac{8}{8} = \frac{24 \div 8}{96 \div 8} = \frac{3}{12}$$

and then dividing both numerator and denominator by 3 will give

$$\frac{3}{12} \div \frac{3}{3} = \frac{3 \div 3}{12 \div 3} = \frac{1}{4} \qquad \textit{Ans.}$$

Self-Test 2-3

The turns ratio of the step-up transformer shown in Fig. 2-3 is $^{36}\!/_{60}$. Express the ratio in lowest terms.

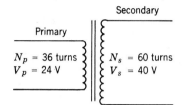

Figure 2-3   A step-up transformer.

Solution

To simplify a fraction, we must ___(1) (multiply/divide)___ both the numerator and the denominator of the fraction by the ___(2)___ number. The numbers that will divide exactly into both 36 and 60 are 2, 3, ___(3)___, ___(4)___, and ___(5)___. Suppose that we divide the 36 and the 60 by 6. Then,

6.
$$\frac{36 \div 6}{60 \div 6} = \frac{6}{?}$$

Now, to simplify further, divide both the 6 and the 10 by ___(7)___.

8.
$$\frac{6 \div 2}{10 \div 2} = \frac{?}{?}$$

Can the fraction $^{3}\!/_{5}$ be simplified any further? ___(9) (Yes/no)___.
The answer is ___(10)___.

Exercises

Practice

Express the following fractions in lowest terms:

1. $^{4}\!/_{8}$ 2. $^{3}\!/_{9}$ 3. $^{4}\!/_{16}$ 4. $^{4}\!/_{80}$
5. $^{12}\!/_{16}$ 6. $^{20}\!/_{32}$ 7. $^{5}\!/_{20}$ 8. $^{3}\!/_{12}$
9. $^{24}\!/_{72}$ 10. $^{12}\!/_{36}$ 11. $^{30}\!/_{1000}$ 12. $^{12}\!/_{48}$

13. $^{24}/_{64}$ 14. $^{40}/_{100}$ 15. $^{14}/_{32}$ 16. $^{15}/_{60}$
17. $^{15}/_{45}$ 18. $^{15}/_{75}$ 19. $^{60}/_{100}$ 20. $^{42}/_{72}$

Applications

21. The ratio of the diameter (in mils) of No. 6 wire compared with No. 12 wire is about $^{162}/_{81}$. Express the ratio in lowest terms.

22. The electric resistance of a piece of iron wire compared with the resistance of an equal length of Nichrome wire is $^{90}/_{100}$. Express the ratio in lowest terms.

23. The current gain β of a transistor is the ratio of the change in the output current compared with the change in the input current. If $\beta = ^{196}/_{4}$, express it in lowest terms.

24. The turns ratio of a step-up transformer is $^{6}/_{120}$. Express it in lowest terms.

25. The current in the parallel circuit shown in Fig. 2-4 divides in the ratio of $^{12}/_{30}$. Express it in lowest terms.

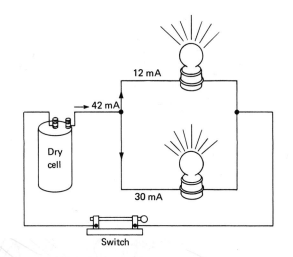

Figure 2-4   A parallel circuit provides different paths for the electrons to follow.

26. The National Electrical Code specifies that No. 14 rubber-covered wire can safely carry only 20 A. In lowest terms, what part of the wire's capacity is used by a broiler drawing 12 A?

27. The ratio of 1 horsepower (hp) to 1 kilowatt (kW) is $^{750}/_{1000}$. Express the ratio in lowest terms.

Changing mixed numbers to improper fractions.   In a *proper fraction* the top number (the numerator) is always *less* than the bottom number (the denominator). Some examples of proper fractions are ½, ¾, and $^{8}/_{11}$. A *mixed number* is made of two numbers—a whole number and a proper fraction. Some examples are 2½, 3¼, and $5^{3}/_{16}$. In an *improper fraction* the top number is always *greater* than the bottom number. Some examples are $^{5}/_{3}$, $^{8}/_{5}$, and $^{22}/_{7}$.

Example 2-4

Change 3¾ to an improper fraction.

Solution

The improper fraction will be a certain number of fourths. Each whole equals $\frac{4}{4}$. In the number 3, there are 3×4, or 12, fourths—or $\frac{12}{4}$. But in addition to the 3 wholes, we have $\frac{3}{4}$ more. Therefore,

$$3\frac{3}{4} = \frac{12}{4} + \frac{3}{4} = \frac{15}{4} \qquad \textit{Ans.}$$

RULE **To change a mixed number to an improper fraction, multiply the denominator of the fraction by the whole number and add the numerator of the fraction. Place this answer over the denominator to make the improper fraction.**

Example 2-5

Change $2\frac{3}{8}$ to an improper fraction.

Solution

The numerator is equal to 2×8, or 16, plus the 3, which equals 19. Placing this 19 over the denominator 8 gives

$$\frac{19}{8} \qquad \textit{Ans.}$$

Exercises

Practice

Change each of the following mixed numbers to improper fractions:

1. $2\frac{1}{4}$	2. $3\frac{3}{8}$	3. $2\frac{2}{9}$	4. $5\frac{3}{8}$
5. $1\frac{3}{10}$	6. $4\frac{3}{4}$	7. $4\frac{1}{6}$	8. $3\frac{7}{10}$
9. $4\frac{5}{8}$	10. $10\frac{1}{2}$	11. $3\frac{1}{3}$	12. $2\frac{1}{12}$

Changing Improper Fractions to Mixed Numbers

Example 2-6

Change $\frac{15}{2}$ to a mixed number.

Solution

$\frac{15}{2}$ means 15 divided by 2, or $15 \div 2$. This means that 2 is contained in 15 only 7 times, which is 14, leaving a remainder of 1. This is written as:

$$\frac{15}{2} = 7\frac{1}{2} \qquad \textit{Ans.}$$

RULE

To change an improper fraction to a mixed number, divide the numerator by the denominator. Any remainder is placed over the denominator. The resulting whole number and proper fraction form a mixed number.

Example 2-7

Change $^{25}/_{10}$ to a mixed number.

Solution

$25 \div 10 = 2$ with a remainder of 5, or $2^{5}/_{10}$. Any remaining fraction like $^{5}/_{10}$ should be reduced to lowest terms. Therefore,

$$2^{5}/_{10} = 2\frac{1}{2} \qquad \text{Ans.}$$

Exercises

Practice

Change the following improper fractions to mixed numbers or whole numbers:

1. $^{7}/_{4}$	2. $^{9}/_{2}$	3. $^{13}/_{3}$	4. $^{8}/_{3}$	5. $^{15}/_{4}$
6. $^{23}/_{5}$	7. $^{8}/_{5}$	8. $^{23}/_{4}$	9. $^{33}/_{10}$	10. $^{14}/_{5}$
11. $^{20}/_{2}$	12. $^{43}/_{10}$	13. $^{29}/_{3}$	14. $^{26}/_{8}$	15. $^{64}/_{4}$
16. $^{16}/_{9}$	17. $^{18}/_{5}$	18. $^{52}/_{4}$	19. $^{96}/_{8}$	20. $^{31}/_{4}$
21. $^{70}/_{10}$	22. $^{38}/_{3}$	23. $^{17}/_{10}$	24. $^{59}/_{8}$	25. $^{47}/_{9}$
26. $^{19}/_{6}$	27. $^{108}/_{9}$	28. $^{47}/_{11}$	29. $^{147}/_{12}$	30. $^{66}/_{10}$

Multiplication of Fractions

RULE

To multiply fractions, place the product of the numerators over the product of the denominators and reduce to lowest terms.

Example 2-8

The jaws of the vise shown in Fig. 2-5 will close a distance of $^{3}/_{8}$ in for each turn of the handle. How far will the jaws close in only one-half turn?

Solution

One-half turn will close the jaws $\frac{1}{2}$ of $^{3}/_{8}$ in. In this sense, the word "of" means "multiply." Thus,

$$\frac{1}{2} \text{ of } \frac{3}{8} \quad \text{means} \quad \frac{1}{2} \times \frac{3}{8} \quad \text{or} \quad \frac{1 \times 3}{2 \times 8} = \frac{3}{16} \text{ in} \qquad Ans.$$

Figure 2-5 A vise.

Cancellation. To cancel numerators and denominators means to divide both a numerator and a denominator by the same number.

Example 2-9

Multiply ⅜ × ⁴⁄₉.

Solution

Notice that the 3 on the top and the 9 on the bottom can both be divided by 3. Cross out the 3 and write 1 above it. Cross out the 9 and write 3 under it. Also, the 4 on the top and the 8 on the bottom can both be divided by 4. Cross out the 4 on the top and write 1 above it. Cross out the 8 and write 2 under it as shown below.

$$\frac{3}{8} \times \frac{4}{9} = \frac{\overset{1}{\cancel{3}}}{\underset{2}{\cancel{8}}} \times \frac{\overset{1}{\cancel{4}}}{\underset{3}{\cancel{9}}} = \frac{1 \times 1}{2 \times 3} = \frac{1}{6} \qquad \textit{Ans.}$$

Self-Test 2-10

Multiply

$$\frac{7}{15} \times \frac{25}{14}$$

Solution

The numerator 7 and the denominator ___(1)___ may both be divided by ___(2)___. Cross out the 7 and write 1 above it. Cross out the 14 and write ___(3)___ under it. The numerator 25 and the denominator ___(4)___ may be divided by ___(5)___. Cross out the 25 and write ___(6)___ above it. Cross out the 15 and write ___(7)___ under it.

8.
$$\frac{\overset{1}{\cancel{7}}}{\underset{3}{\cancel{15}}} \times \frac{\overset{5}{\cancel{25}}}{\underset{2}{\cancel{14}}} = \frac{1 \times 5}{3 \times 2} = \frac{?}{?} \qquad \textit{Ans.}$$

Exercises

Practice

Multiply the following fractions and reduce to lowest terms:

1. 8 × ½
2. ⅖ × ⅓
3. ¹⁄₁₀ × 3
4. ¼ × 3
5. 5 × ½
6. ¼ × 7
7. ⅓ × ⅕
8. ⅜ × 4
9. ⅜ × 16

Applications

10. A 40-W lamp uses about ⅜ A. How many amperes would four lamps use when connected in parallel?
11. A motor delivers only seven-eighths of the power it receives. Find the power delivered if it receives 6 hp.
12. The actual value of resistance for some inexpensive resistors may be off by as much as one-fifth of the rated value. What would be the possible error in a 230-Ω resistor of this type?
13. What total horsepower is required for the operation of three ⅙-hp single-phase capacitor motors?
14. About two-fifths of the electrolyte in a storage battery is acid. If the battery contains 5 oz of electrolyte, how many ounces of acid are in the cells of the battery?
15. Only four-fifths of the wattage of an electric range is used in computing the size of entrance wires. Find the wattage to use if the range is rated at 4000 W.
16. The load placed on a service line is 375 kilovoltamperes (kVA). If one-third of this load is for motors, three-fifths for lighting, and one-fifteenth for heating, find the kilovoltamperes used for each.
17. ⅔ × ½   18. ⁷⁄₁₀ × ⁵⁄₁₄   19. ⅐ × 5   20. ⅗ × ⁵⁄₉

(See CD-ROM for Test 2-1)

Multiplication of Mixed Numbers

Example 2-11

What length of two-conductor BX cable (Fig. 2-6) is needed to obtain six pieces, each 4¼ ft long?

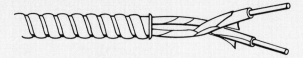

Figure 2-6   A piece of two-conductor BX cable.

Solution

$$6 \times 4\frac{1}{4} = 6 \times \frac{17}{4}$$

$$\frac{\overset{3}{\cancel{6}}}{1} \times \frac{17}{\underset{2}{\cancel{4}}} = \frac{51}{2} = 25\frac{1}{2} \text{ ft} \qquad Ans.$$

Example 2-12

If electric energy costs $6\frac{7}{8}$ cents per kilowatthour (kWh), find the cost for $5\frac{3}{5}$ kWh.

Solution

$$6\frac{7}{8} \times 5\frac{3}{5} = \frac{55}{8} \times \frac{28}{5}$$

$$\frac{\overset{11}{\cancel{55}}}{\underset{2}{\cancel{8}}} \times \frac{\overset{7}{\cancel{28}}}{\underset{1}{\cancel{5}}} = \frac{11 \times 7}{2 \times 1} = \frac{77}{2} = 38\frac{1}{2} \text{ cents} \qquad Ans.$$

Exercises

Practice

Multiply the following fractions and mixed numbers. Express the answer in lowest terms.

1. $1\frac{7}{9} \times \frac{3}{8}$
2. $\frac{1}{5} \times 4\frac{3}{8}$
3. $3\frac{1}{6} \times 3$
4. $1\frac{1}{4} \times 3\frac{2}{5}$
5. $2\frac{1}{3} \times \frac{1}{7}$
6. $3\frac{1}{3} \times 1\frac{1}{5}$
7. $1\frac{1}{3} \times 3\frac{3}{4}$
8. $2\frac{1}{4} \times 3\frac{3}{5}$
9. $\frac{1}{5} \times 3\frac{1}{3}$

Applications

10. A neon electric sign uses $11\frac{1}{2}$ W of power per meter of tubing. How many watts are used for a sign which uses 20 m of tubing?
11. Find the total voltage supplied by three $1\frac{1}{2}$-V dry cells connected in series.
12. UTC power transformers are given a surge test for insulation breakdown of $2\frac{1}{2}$ times the normally developed voltage. What is the test voltage for a transformer which delivers 510 V?
13. Electricians get time and one-half for overtime. If an electrician worked $4\frac{3}{4}$ h overtime, for how many hours would he be paid for his overtime?
14. If one-twentieth of the energy put into a motor is lost by friction, copper, and iron losses, how many kilowatts are lost if the motor receives $\frac{2}{3}$ kW?
15. If one electric outlet requires $13\frac{1}{2}$ ft of flexible conduit, how many feet of conduit are required for five such outlets?
16. A capacitor should be checked for dielectric breakdown at $1\frac{1}{2}$ times its rated working voltage. If the capacitor to be tested has a working voltage of 600 V, what test voltage should be applied?
17. A factory uses four $\frac{3}{4}$-hp motors and five $\frac{1}{4}$-hp motors. What is the total horsepower used when all motors are operating?

Self-Test 2-13

Fill in each blank with the correct response.

1. A fraction is made of two parts, a _____ and a _____.
2. The _____ is the top portion of the fraction.

3. The _____ is the bottom portion of the fraction.
4. A fraction is a mathematical way to describe some _____ of a total amount.
5. If you divide both the numerator and the denominator of the fraction $^{10}\!/_{15}$ by 5, the form of the fraction will be different but the _____ of the fraction will not be changed.
6. Consider the fraction $^{12}\!/_{18}$. The largest whole number which divides evenly into both parts of the fraction is _____.
7. When this number is divided into both parts of the fraction $^{12}\!/_{18}$, the value of the fraction will not change, and the new fraction will be _____.
8. Can you divide the parts of $^{2}\!/_{3}$ evenly by any other number except 1? ___(yes/no)___
9. When you write $^{12}\!/_{18} = {}^{2}\!/_{3}$, you have expressed the fraction in _____ terms.
10. To express a fraction in lowest terms, we must __(multiply/divide)__ both the numerator and the denominator by the largest number that divides _____ into both of them.
11. An improper fraction is one in which the numerator is __(larger/smaller)__ than the denominator.
12. Mixed numbers should be changed into _____ fractions before multiplying.
13. To change a mixed number into an improper fraction, multiply the _____ of the fraction by the whole number and add the numerator of the fraction. Place this answer over the _____ to form the improper fraction.
14. To change an improper fraction into a mixed number, __(multiply/divide)__ the numerator by the denominator. Any remainder is placed over the _____.
15. To multiply fractions, place the __(product/quotient)__ of the numerators over the __(product/quotient)__ of the denominators and express in _____ terms.
16. To cancel means to _____ any one of the numerators and any one of the denominators by the _____ number.

(See CD-ROM for Test 2-2)

Ohm's Law

In Job 1-2 we learned that the following three factors must be present in every electric circuit:

1. The *electromotive force V* expressed in *volts,* which causes the current to flow
2. The *resistance* of the circuit R expressed in *ohms,* which attempts to stop the flow of current
3. The *current I* expressed in *amperes,* which flows as a result of the voltage pressure exceeding the resistance

24   Chapter 2   Simple Electric Circuits

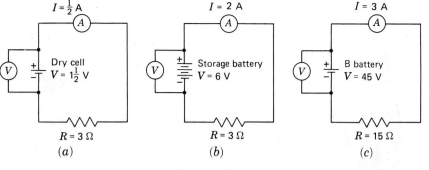

Figure 2-7   The voltage equals the current multiplied by the resistance in a simple circuit.

A definite relationship exists among these three factors, which is known as *Ohm's law*. This relationship is very important, since it is the basis for most of the calculations in electrical and electronic work. In the three circuits shown in Fig. 2-7, the voltage *V* of the battery is measured by a voltmeter. The current *I* that flows is measured by an ammeter. The resistance *R* of the resistor is indicated by the manufacturer by distinctive markings on the resistor. Let us put the information from Fig. 2-7 into a table (see Table 2-1).

Table 2-1

Figure	V	I	R	I × R
2-7*a*	1½	½	3	½ × 3 = 1½
2-7*b*	6	2	3	2 × 3 = 6
2-7*c*	45	3	15	3 × 15 = 45

In the last column of Table 2-1 we have multiplied the current *I* by the resistance *R* for each circuit. Evidently, the result of this multiplication is always equal to the voltage *V* of the circuit. This is true for all circuits and was first discovered by George S. Ohm. This simple relationship is called *Ohm's law*.

 RULE   Voltage equals current multiplied by resistance.

$$V = I \times R$$ Formula 2-1

In this formula,
   V must always be expressed in volts (V).
   I must always be expressed in amperes (A).
   R must always be expressed on ohms (Ω).

Solving Problems

1. Read the problem carefully.
2. Draw a simple diagram of the circuit.

3. Record the given information directly on the diagram. Indicate the values to be found by question marks.
4. Write the formula.
5. Substitute the given numbers for the letters in the formula. If the number for the letter is unknown, merely write the letter again. Be sure to include all mathematical signs like $\times$ or $=$.
6. Do the indicated arithmetic at the side so as not to interrupt the continued progress of the solution.
7. In the answer, indicate the letter, its numerical value, and the units of measurement.

Example 2-14

A doorbell requires ¼ A in order to ring. The resistance of the coils in the bell is 24 Ω. What voltage must be supplied in order to ring the bell?

Solution

The diagram for the circuit is shown in Fig. 2-8.

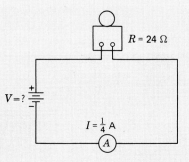

Figure 2-8

1. Write the formula.	$V = I \times R$
2. Substitute numbers.	$V = ¼ \times 24$
3. Multiply the numbers.	$V = 6$ V *Ans.*

Example 2-15

A relay used to control the large current to a motor is rated at 28 Ω resistance. What voltage is required to operate the relay if it requires a current of 0.05 A?

Solution

The diagram for the circuit is shown in Fig. 2-9.

1. Write the formula.	$V = I \times R$
2. Substitute numbers.	$V = 0.05 \times 28$
3. Multiply the numbers.	$V = 1.4$ V *Ans.*

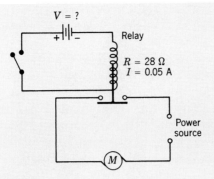

Figure 2-9 A relay controls the large current drawn by the motor.

Self-Test 2-16

An automobile battery supplies a current of 7.5 A to a headlamp with a resistance of 0.84 Ω. Find the voltage delivered by the battery.

Solution

The diagram for the circuit is shown in Fig. 2-10.

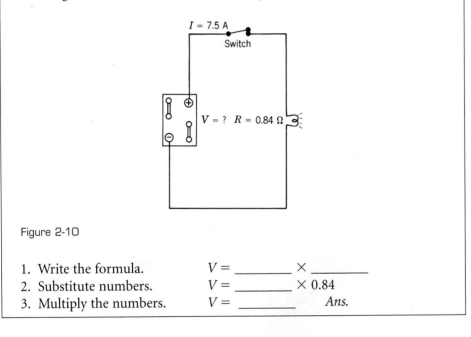

Figure 2-10

1. Write the formula. $\qquad$ $V =$ _____ $\times$ _____
2. Substitute numbers. $\qquad$ $V =$ _____ $\times 0.84$
3. Multiply the numbers. $\qquad$ $V =$ _____   *Ans.*

Exercises

Applications

1. An arc lamp with a hot resistance of 9 Ω draws 6.2 A. What voltage is required?
2. A 52-Ω electric toaster uses 2¼ A. Find the required voltage.
3. What voltage is needed to energize the field coil of a loudspeaker if its resistance is 1100 Ω and it uses 0.04 A?
4. The coils of a washing-machine motor have a resistance of 21 Ω. What voltage does it require if the motor draws 5.3 A?

Figure 2-11 Electric impact wrench.

5. What is the voltage required to operate the electric impact wrench shown in Fig. 2-11 if its resistance is 24.5 Ω and it draws 4.8 A?

6. The line from an automobile battery to a distant transmitter must not develop more than 0.25 V when the transmitter is operating. If the current from the battery is 18 A, will a 0.015-Ω line be satisfactory?

7. The winding of a transformer has a resistance of 63 Ω. What is the voltage drop in the winding when it carries a current of 0.059 A?

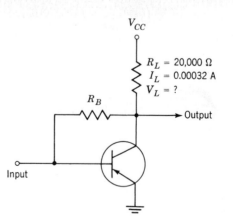

V_{CC}

$R_L = 20{,}000 \ \Omega$
$I_L = 0.00032 \ \text{A}$
$V_L = ?$

R_B

Output

Input

Figure 2-12 Self-bias transistor circuit.

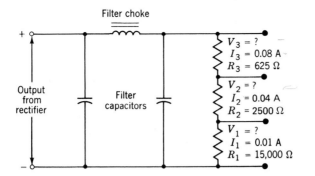

Filter choke

$V_3 = ?$
$I_3 = 0.08$ A
$R_3 = 625\ \Omega$

$V_2 = ?$
$I_2 = 0.04$ A
$R_2 = 2500\ \Omega$

$V_1 = ?$
$I_1 = 0.01$ A
$R_1 = 15{,}000\ \Omega$

Output from rectifier

Filter capacitors

Figure 2-13   A voltage divider provides different voltages from a single source.

8. An electric bell has a resistance of 25 Ω and will not operate on a current of less than 0.25 A. What is the smallest voltage that will ring the bell?

9. A 125-Ω relay coil needs 0.15 A to operate. What is the lowest voltage needed to operate the relay?

10. In the self-bias transistor circuit shown in Fig. 2-12, the load resistor $R_L = 20{,}000$ Ω and carries a current $I_L = 0.00032$ A. Find the voltage drop across the load.

11. What is the voltage across the shunt of an ammeter if the shunt has a resistance of 0.005 Ω and carries 9.99 A?

12. What voltage is registered by a voltmeter with an internal resistance of 150,000 Ω when 0.001 A flows through it?

13. In the voltage-divider circuit shown in Fig. 2-13, find the voltage across the resistors R_1, R_2, and R_3.

(See CD-ROM for Test 2-3)

JOB 2-4

Checkup on Decimals (Diagnostic Test)

Did you have any difficulty with the decimals in the last job? The following problems occur in the everyday work of the electrician and electronic technician. They all involve decimals and will help you to check up on their use. If you have

Exercises

Applications

1. Find the total current drawn by the following appliances by adding the currents: electric iron, 4.12 A; electric clock, 0.02 A, 100-W lamp, 0.91 A; and radio, 0.5 A.

2. Find the total capacitance of a parallel group of capacitors by adding these values: 0.000 25, 0.01, and 0.005 μF (microfarad).

3. How much larger in diameter is No. 10 copper wire (0.1019 in) than No. 14 wire (0.0641 in)?

4. The current drawn by a motor is 1.21 A at no load and 1.56 A at full load. What is the increase in the current?

5. A motor receives only 117.4 V when connected to a distant generator delivering 120 V. Find the voltage lost in the line wires.
6. The laminated core of a power transformer is made of 15 sheets of steel, each 0.079 in thick. Find the total thickness of the core.
7. The maximum value of an ac voltage wave is 1.414 times its ac meter reading of 46.5 V. Find the maximum value of the voltage wave.
8. If 65 ft of BX cable costs $38.95, what is the cost of 1 ft of this cable?
9. The resistance of 24.5 ft of No. 16 copper wire is 0.0982 Ω. Find the resistance of 1 ft of this wire.
10. Add 7.05, 2, and 3.5.
11. Which current is larger: 0.4 or 0.25 A?
12. Write the following decimals in words: (*a*) 4.3, (*b*) 0.359, and (*c*) 0.41.
13. Multiply ³⁄₁₀ by 0.05.
14. Write as a decimal (*a*) ¼, (*b*) ½, (*c*) ⅝, and (*d*) ¹⁄₁₆.
15. Subtract 12 from 18.24.
16. Arrange the following numbers starting with the largest: 0.050, 0.30, 0.0070, 1.1.
17. Find the difference between three-tenths and twenty-five hundredths.
18. What is the excess of 5 over 2.75?

JOB 2-5

Introduction to Decimals

The decimal system is an extremely fast and accurate system to use for most mathematical calculations. The modern electronic calculator is a marvel of speed and capability, and gives answers in extremely accurate decimal figures. Many calculators contain keys for conversions to the metric system of measurement.

A decimal is a fraction in which the denominator is not written; instead, the denominator's value is indicated by the position of a decimal point (.) in the numerator. The denominator of a decimal is always a number like 10, 100, or 1000.

The number of digits to the right of the decimal point tells us whether the decimal is to be read as tenths, hundredths, thousandths, etc.

When a decimal has *one* digit to the right of the decimal point, the decimal is read as that many *tenths*. Thus,

$$0.3 = ³⁄₁₀ \text{ and is read as "3 } tenths\text{"}$$

$$0.9 = ⁹⁄₁₀ \text{ and is read as "9 } tenths\text{"}$$

When a decimal has *two* digits to the right of the decimal point, the decimal is read as that many *hundredths*. Thus,

$$0.23 = ²³⁄₁₀₀ \text{ and is read as "23 } hundredths\text{"}$$

$$0.47 = ⁴⁷⁄₁₀₀ \text{ and is read as "47 } hundredths\text{"}$$

Now, if we wish to write ³⁄₁₀₀ as a decimal, the decimal point must be placed so that there will be *two* digits following it in order to represent *hundredths*. To make up these two digits, we must place a zero between the decimal point and the digit 3. The zero will now push the 3 into the second place, which means hundredths.

Thus

$$\frac{3}{100} = 0.03 \text{ and is read as "3 } \textit{hundredths}\text{"}$$

Note: The zero *must* be placed between the decimal point and the digit. It is wrong to place the zero after the digit 3, as in 0.30, since this number would be read as 30 hundredths, *not* 3 hundredths. Similarly,

$$\frac{7}{100} = 0.07 \text{ and is read as "7 } \textit{hundredths}\text{"}$$

$$\frac{9}{100} = 0.09 \text{ and is read as "9 } \textit{hundredths}\text{"}$$

When a decimal has *three* digits to the right of the decimal point, the decimal is read as that many *thousandths.* Thus,

$$0.123 = \frac{123}{1000} \text{ and is read as "123 } \textit{thousandths}\text{"}$$

$$0.457 = \frac{457}{1000} \text{ and is read as "457 } \textit{thousandths}\text{"}$$

If we wish to write $\frac{43}{1000}$ as a decimal, the decimal point must be placed so that there will be *three* digits following it in order to represent *thousandths.* To make up these three digits, we must place a zero between the decimal point and the 43. Thus,

$$\frac{43}{1000} = 0.043 \text{ and is read as "43 } \textit{thousandths}\text{"}$$

$$\frac{87}{1000} = 0.087 \text{ and is read as "87 } \textit{thousandths}\text{"}$$

If we wish to write $\frac{9}{1000}$ as a decimal, the decimal point must still be placed so that there will be *three* digits following it. To make up these three digits, we must now place *two* zeros between the decimal point and the 9.

$$0.009 = \frac{9}{1000} \text{ and is read as "9 } \textit{thousandths}\text{"}$$

$$0.005 = \frac{5}{1000} \text{ and is read as "5 } \textit{thousandths}\text{"}$$

Following this system, four places means ten-thousandths, five places means hundred-thousandths, six places means millionths, etc. Part of the system is shown in picture form in Fig. 2-14.

Note: Zeros placed at the *end* of a decimal do *not* change the value of the decimal. They merely describe the decimal in another way. For example: 0.5 (5 tenths) = 0.50 (50 hundredths) = 0.500 (500 thousandths).

A number such as $3\frac{13}{100}$ is written as 3.13 and is read as "3 *and* 13 hundredths." In this type of number, the decimal point is read as the word "and."

Comparing the value of decimals. When comparing the value of various decimals, we must first be certain that they have the same denominators. This means that the decimals must be written so that they have the same number of decimal places.

CALCULATOR HINT

For decimals less than 1, such as 0.3, the general practice is to write a zero to the left of the decimal point to emphasize that there is no value in the units place. On your calculator, a zero will always be displayed to the left of the decimal point for numbers less than 1.

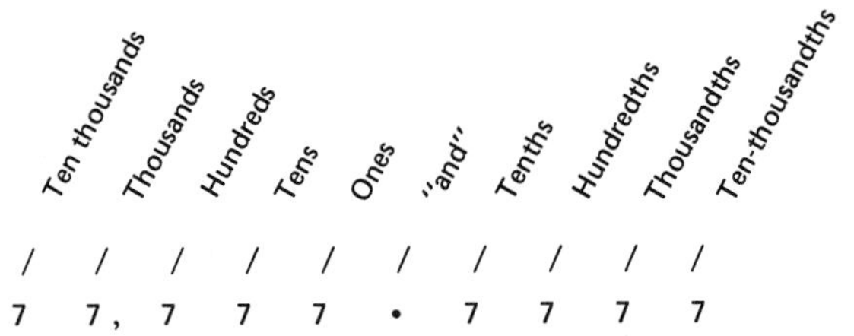

Figure 2-14 Names of the placeholders in the decimal system.

Example 2-17

Which is larger: 0.3 or 0.25?

Solution

Since 0.3 has only one decimal place and 0.25 has two decimal places, we must change 0.3 into a two-place decimal by adding a zero. This does *not* change the value but merely expresses it differently. Therefore, 0.3 or 0.30 (30 hundredths) is larger than 0.25 (25 hundredths).   *Ans.*

Exercises

Practice

Write the following fractions as decimals.

1. $\frac{7}{10}$ 2. $\frac{29}{100}$ 3. $\frac{114}{1000}$ 4. $\frac{3}{10}$

5. $\frac{6}{100}$ 6. $\frac{9}{1000}$ 7. $\frac{18}{1000}$ 8. $\frac{3}{1000}$

9. $\frac{11}{100}$ 10. $\frac{4}{10}$ 11. $\frac{13}{1000}$ 12. $\frac{74}{100}$

13. $\frac{45}{1000}$ 14. $\frac{316}{1000}$ 15. $\frac{6}{10}$ 16. $\frac{23}{100}$

Arrange the following decimals in order starting with the largest:

17. 0.007, 0.16, 0.4   18. 0.2, 0.107, 0.28   19. 0.8, 0.06, 0.040

20. 0.496, 0.8, 0.02   21. 0.5, 0.051, 0.18   22. 0.90, 0.018, 0.06

23. 0.1228, 0.236, 0.4   24. 0.006, 0.05, 0.3   25. 0.19, 0.004, 0.08

Changing mixed numbers to decimals. When a mixed number is read as a decimal, the word "and" appears as a decimal point. For example, $2\frac{7}{10}$ is read as "two *and* seven-tenths" and is written as 2.7. A whole number may be written as a decimal if a decimal point is placed at the *end* of the number. For example, the number 4 means 4.0 or 4.00 or 4.000.

Exercises

Practice

Change the following mixed numbers to decimals:

1. $2\frac{3}{10}$ 2. $18\frac{5}{100}$ 3. $3\frac{144}{1000}$ 4. $1\frac{17}{100}$

5. $2\frac{25}{1000}$ 6. $7\frac{35}{100}$ 7. $2\frac{20}{1000}$ 8. $3\frac{9}{100}$

9. $1\frac{2}{1000}$ 10. $62\frac{90}{100}$ 11. $9\frac{145}{1000}$ 12. $3\frac{27}{1000}$

Changing fractions to decimals. We shall discover that many of the answers to our electrical problems will be fractions like $\frac{1}{8}$ A, $\frac{7}{40}$ μF, and $\frac{3}{13}$ Ω. These will be perfectly correct mathematical answers, but they will be completely worthless to an electrician or electronic technician. Electric measuring instruments give values expressed as decimals and *not* as fractions. In addition, the manufacturers of electric components give the values of the parts in terms of decimals.

Suppose that we worked out a problem and found that the current in the circuit should be ⅛ A. Then, using an ammeter, we tested the circuit and found that 0.125 A flowed. Is our circuit correct? How would we know? How can we compare ⅛ and 0.125? The easiest way is to change the fraction ⅛ to its equivalent decimal and then to compare the decimals.

RULE   To change a fraction to a decimal, divide the numerator by the denominator.

Example 2-18

Change ⅛ to an equivalent decimal.

Solution

⅛ means 1 ÷ 8. To write this as a long-division example, place the numerator inside the long-division sign and the denominator outside the sign as shown below.

$$8\overline{)1}$$

We can't divide 8 into 1; but remember that every whole number may be written with a decimal point at the *end* of the number. As many zeros as we desire may be added without changing the value. Our problem now looks like this:

$$8\overline{)1.000}$$

1. Put the decimal point in the answer directly above its position in 1.000.

$$8\overline{)1.000}$$

2. Try to divide the 8 into the first digit. We cannot divide 8 into 1. Then try to divide the 8 into the first two digits. We can divide 8 into 10 one time. Place this number 1 in the answer directly above the last digit of the number into which the 8 was divided.

$$\begin{array}{r} 0.1 \\ 8\overline{)1.000} \end{array}$$

3. Multiply this 1 by the divisor 8 and place it as shown below. Draw a line and subtract.

$$\begin{array}{r} 0.1 \\ 8\overline{)1.000} \\ \underline{8} \\ 2 \end{array}$$

4. Bring down the next digit 0 and divide this new number 20 by the 8. The 8 will divide into 20 two times.

$$\begin{array}{r} 0.1 \\ 8\overline{)1.000} \\ \underline{8\downarrow} \\ 20 \end{array}$$

5. Place this 2 in the answer directly above the last digit brought down.

$$\begin{array}{r} 0.12 \\ 8\overline{)1.000} \\ \underline{8\downarrow} \\ 20 \end{array}$$

6. Multiple the 2 by the divisor 8, and continue steps 3 to 5. The answer comes out even: 0.125. This means that $\frac{1}{8} = 0.125$ *Ans.*

$$\begin{array}{r} 0.125 \\ 8\overline{)1.000} \\ \underline{8\downarrow} \\ 20 \\ \underline{16\downarrow} \\ 40 \\ \underline{40} \\ 0 \end{array}$$

Example 2-19

The current in a parallel electric circuit divides in the ratio of $\frac{6}{13}$. Express the ratio as a decimal.

Solution

The answer does not come out even:

$$\frac{6}{13} = 13\overline{)6.000} \begin{array}{r} 0.461 \\ \\ \underline{5\,2\downarrow} \\ 80 \\ \underline{78\downarrow} \\ 20 \\ \underline{13} \\ 7 \end{array}$$

We see that there is a remainder. If the remainder is more than half the divisor, we drop it and add an extra unit to the last place of the answer. Since 7 is more than half of 13,

$$0.461\frac{7}{13} \text{ becomes} \quad \begin{array}{r} 0.461 \\ \underline{+\ 1} \\ 0.462 \end{array} \quad \textit{Ans.}$$

If any remainder is less than half the divisor, drop it completely and leave the answer unchanged. For example,

$$0.236\tfrac{5}{12} = 0.236 \qquad \text{(since 5 is less than half of 12)}$$
$$0.483\tfrac{1}{4} = 0.483 \qquad \text{(since 1 is less than half of 4)}$$

An important question may have occurred to you by now: "If it doesn't come out even, how long should I continue to divide?" The answer to this depends on the use to which the answer is to be put. Some jobs require five or six decimal places, while others need only one place or none at all. For example, the capacitor in the tuning circuit of a radio receiver should be worked out to an answer like 0.000 25 microfarad (μF). The pitch diameter of a screw thread requires a measurement like 0.498 in, while a bias resistor of 203.4 Ω is just as well written as 203 Ω or even 200 Ω, because precision is not needed in this application.

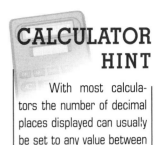

CALCULATOR HINT

With most calculators the number of decimal places displayed can usually be set to any value between 0 and 9.

Degree of accuracy. A very general rule for the number of decimal places required in an answer is given in Table 2-2.

Table 2-2

Answer	Number of Decimal Places	
Less than 1	3	0.132, 0.008
From 1 to 10	2	3.48, 6.07
From 10 to 100	1	28.3, 52.9
From 100 up	None	425, 659

Exercises

Practice

Change the following fractions to equivalent decimals:

1. $\tfrac{1}{4}$
2. $\tfrac{1}{5}$
3. $\tfrac{5}{8}$
4. $\tfrac{1}{3}$
5. $\tfrac{2}{5}$
6. $\tfrac{3}{10}$
7. $\tfrac{3}{20}$
8. $\tfrac{2}{7}$
9. $\tfrac{7}{8}$
10. $\tfrac{3}{16}$
11. $\tfrac{4}{9}$
12. $\tfrac{13}{15}$
13. $\tfrac{3}{32}$
14. $\tfrac{21}{25}$
15. $\tfrac{25}{40}$
16. $\tfrac{9}{16}$
17. $\tfrac{1}{50}$
18. $\tfrac{1}{200}$
19. $\tfrac{5}{26}$
20. $\tfrac{9}{64}$

21. A 40-W lamp uses about $\tfrac{3}{8}$ A. Express the current used as a decimal correct to three places.
22. A bell transformer reduced the primary voltage of 120 V to the 18 V delivered by the secondary. Express the voltage ratio ($\tfrac{120}{18}$) as a decimal.

(See CD-ROM for Test 2-4)

Using the decimal equivalent chart. There are some fractions that are used very often. These are the fractions which represent the parts of an inch on a ruler, like $\tfrac{1}{16}$, $\tfrac{3}{8}$, $\tfrac{5}{32}$, and $\tfrac{9}{64}$. Since they are so widely used, a table of decimal equivalents has been prepared. In order to find the decimal equivalent of a fraction of this type, refer to Table 2-3 on page 36.

Table 2-3 Decimal Equivalents

Fraction	$\frac{1}{32}$ds	$\frac{1}{64}$ths	Decimal	Fraction	$\frac{1}{32}$ds	$\frac{1}{64}$ths	Decimal
		1	0.015625			33	0.515625
	1	2	0.03125		17	34	0.53125
		3	0.046875			35	0.546875
$\frac{1}{16}$	2	4	0.0625	$\frac{9}{16}$	18	36	0.5625
		5	0.078125			37	0.578125
	3	6	0.09375		19	38	0.59375
		7	0.109375			39	0.609375
$\frac{1}{8}$	4	8	0.125	$\frac{5}{8}$	20	40	0.625
		9	0.140625			41	0.640625
	5	10	0.15625		21	42	0.65625
		11	0.171875			43	0.671875
$\frac{3}{16}$	6	12	0.1875	$\frac{11}{16}$	22	44	0.6875
		13	0.203125			45	0.703125
	7	14	0.21875		23	46	0.71875
		15	0.234375			47	0.734375
$\frac{1}{4}$	8	16	0.25	$\frac{3}{4}$	24	48	0.75
		17	0.265625			49	0.765625
	9	18	0.28125		25	50	0.78125
		19	0.296875			51	0.796875
$\frac{5}{16}$	10	20	0.3125	$\frac{13}{16}$	26	52	0.8125
		21	0.328125			53	0.828125
	11	22	0.34375		27	54	0.84375
		23	0.359375			55	0.859375
$\frac{3}{8}$	12	24	0.375	$\frac{7}{8}$	28	56	0.875
		25	0.390625			57	0.890625
	13	26	0.40625		29	58	0.90625
		27	0.421875			59	0.921875
$\frac{7}{16}$	14	28	0.4375	$\frac{15}{16}$	30	60	0.9375
		29	0.453125			61	0.953125
	15	30	0.46875		31	62	0.96875
		31	0.484375			63	0.984375
$\frac{1}{2}$	16	32	0.5		32	64	1.

Exercises

Practice

Use Table 2-3 to find the decimal equivalent of each of the following fractions and mixed numbers:

1. $\frac{5}{8}$

2. $\frac{7}{32}$

3. $\frac{9}{16}$

4. $\frac{41}{64}$

5. $2\frac{1}{4}$

6. $3\frac{11}{32}$

7. $1\frac{9}{64}$

8. $4\frac{3}{4}$

Use the table to find the fraction nearest in value to each of the following decimals:

9. 0.37   10. 0.785   11. 0.449   12. 0.88
13. 0.035   14. 0.525   15. 0.41   16. 0.615

Write each of the following decimals as a mixed number or fraction:

17. 0.75   18. 0.625   19. 0.141   20. 0.5625
21. 2.8125   22. 3.641   23. 1.844   24. 0.9375

Applications

25. A 100-W lamp requires 0.91 A, and a table radio requires ⅞ A. Which item uses more current?

26. An adjustment screw on a carburetor must be at least 0.56 in long. Will a ¹⁵⁄₃₂-in-long screw be acceptable?

(See CD-ROM for Test 2-5)

JOB 2-6

Introduction to the Metric System

The metric system is a decimal system of measurement that is the practical standard throughout most of the world. Many industries in the United States use metric measurements.

The basic unit of length in the metric system is the meter (m), which is equal to about 39.37 in.

If we divide a meter into 1000 parts, each part is called a *millimeter* (mm), since the prefix "milli" means one one-thousandth of a quantity. A millimeter is approximately equal to the thickness of a dime.

If we divide a meter into 100 parts, each part is called a *centimeter* (cm), since the prefix "centi" means one one-hundredth of a quantity. A centimeter is about ⅜ in, approximately equal to the diameter of an ordinary aspirin tablet.

If we divide a meter into 10 parts, each part is called a *decimeter* (dm), since the prefix "deci" means one-tenth of a quantity. A decimeter is almost 4 in long.

The relationship among these three quantities is shown in Fig. 2-15. For measurements larger than a meter,

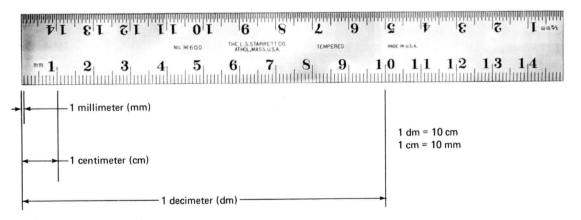

Figure 2-15   A metric scale. (*Courtesy The L.S. Starrett Company.*)

1 dekameter (dam) = 10 meters, since the prefix "deka" means 10 times as large.
1 hectometer (hm) = 100 meters, since the prefix "hecto" means 100 times as large.
1 kilometer (km) = 1000 meters, since the prefix "kilo" means 1000 times as large.

The set of lengths in the metric system is given in Table 2-4. This table is described in picture form in Fig. 2-16.

Table 2-4   Length in the Metric System

1 cm	= 10 mm	or	1 mm	= ⅒ cm	= 0.1 cm
1 dm	= 10 mm	or	1 cm	= ⅒ dm	= 0.1 dm
1 m	= 10 dm	or	1 dm	= ⅒ m	= 0.1 m
1 m	= 100 cm	or	1 cm	= ¹⁄₁₀₀ m	= 0.01 m
1 m	= 1000 mm	or	1 mm	= ¹⁄₁₀₀₀ m	= 0.001 m
1 dam	= 10 m	or	1 m	= ⅒ dam	= 0.1 dam
1 hm	= 100 m	or	1 m	= ¹⁄₁₀₀ hm	= 0.01 hm
1 km	= 1000 m	or	1 m	= ¹⁄₁₀₀₀ km	= 0.001 km

Before we try to change any of these measurements from one unit to another, let us look at a simple method for multiplying and dividing by numbers like 10, 100, and 1000.

Multiplying by 10, 100, 1000, Etc.

RULE   To multiply values by numbers like 100, 1000, and 1,000,000, move the decimal point one place to the right for every zero in the multiplier.

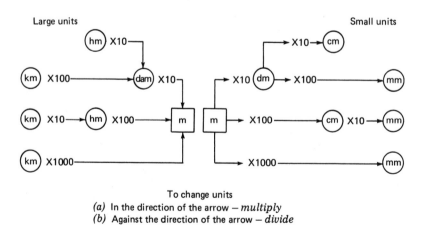

To change units
(a) In the direction of the arrow – *multiply*
(b) Against the direction of the arrow – *divide*

Figure 2-16   Rules for changing units in the metric system.

Example 2-20

1.59 × 10 = 1⊙5.9 = 15.9 (move decimal point 1 place to the right)

4.76 × 100 = 4⊙76. = 476. or 476 (move decimal point 2 places to the right)

A decimal point is ordinarily not written at the end of a whole number. However, it may be written if desired. In the following example, the decimal point must be written in together with two zeros so that we can move the decimal point past the required number of places.

$$34 \times 100 = 34 \odot 00. = 3400. \text{ or } 3400$$

If the decimal point must be moved past more places than are available, zeros are added at the end of the number to make up the required number of places. This is shown in the following example.

$$0.0259 \times 1,000,000 = \odot 025900. = 25,900$$

Self-Test 2-21

Multiply 25.2 × 100.

Solution

To multiply, the decimal point is moved to the ___(1) (left/right)___. The number 100 contains ___(2)___ zeros. The point will be moved ___(3)___ places. Zeros ___(4) (are/are not)___ needed to account for the correct number of places. The answer is 25.2 × 100 = ___(5)___.

Exercises

Practice

1. 0.0072 × 1000
2. 45.76 × 100
3. 3.09 × 1000
4. 0.0045 × 100
5. 37 × 100
6. 0.08 × 1000
7. 0.0006 × 1000
8. 0.00056 × 1,000,000
9. 27 × 100
10. 15 × 1000
11. 0.00078 × 100
12. 7.8 × 1,000,000
13. 15.4 × 1000
14. 4.5 × 10
15. 0.005 × 100
16. 6 × 10,000
17. 0.008 × 10,000
18. 0.009 × 100,000
19. 2.34 × 1000
20. 0.0008 × 1,000,000

Dividing by 10, 100, 1000, Etc.

RULE To divide values by numbers like 10, 100, and 1000, move the decimal point one place to the left for every zero in the divisor.

Example 2-22

$17.4 \div 10 = 1.7 \widehat{\odot} 4 = 1.74$ (move decimal point one place to the left)

$45 \div 100 = .45 \odot = 0.45$ (move decimal point two places to the left)

$6.5 \div 1000 = .006 \widehat{\odot} 5 = 0.0065$ (move decimal point three places to the left)

Self-Test 2-23

Divide 8.5×1000.

Solution

To divide, the decimal point is moved to the ____(1) (left/right)____. The number 1000 contains ____(2)____ zeros. The point will be moved ____(3)____ places. Zeros ____(4) (are/are not)____ needed to account for the correct number of places.

The answer is $8.5 \div 1000 = $ ____(5)____.

Exercises

Practice

1. $6500 \div 1000$
2. $7500 \div 1000$
3. $880{,}000 \div 1000$
4. $32 \div 100$
5. $6 \div 100$
6. $17.8 \div 10$
7. $835 \div 1000$
8. $550 \div 1{,}000{,}000$
9. $653.8 \div 1000$
10. $100{,}000{,}000 \div 1{,}000{,}000$
11. $0.45 \div 1000$
12. $0.08 \div 10$
13. $8.5 \div 1000$
14. $7 \div 1{,}000{,}000$
15. $28.6 \div 1000$
16. $15.6 \div 100$

17. $\dfrac{2}{1000}$
18. $\dfrac{0.08}{10}$

19. $\dfrac{398}{10{,}000}$
20. $\dfrac{600}{1000}$

(See CD-ROM for Test 2-6)

Hint for Success

Being able to convert units of measurement is very important in the field of electronics.

Changing units of measurement. There are two factors to be considered when describing any measurement: (1) how many of the measurements and (2) what *kind* of measurements. For example, $1 may be described as 2 half-dollars, 4 quarters, 10 dimes, 20 nickels, or 100 pennies. When the $1 was changed into each of the new measurements, *both* the unit of measurement as well as the number of them were changed. For example:

Since 3 ft $= 1$ yd,	2 yd	$= 2 \times 3$	$= 6$ ft
Since 2000 lb $= 1$ ton,	3 tons	$= 3 \times 2000$	$= 6000$ lb
Since 100¢ $= 1$ dollar,	4 dollars	$= 4 \times 100$	$= 400$¢

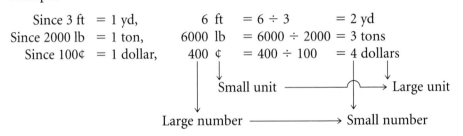

Notice that a *small number* of *large units* is always changed into a *large number* of *small units* by *multiplying* by the number showing the relationship between the units.

RULE To change from a large unit to a small unit, multiply by the number showing the relationship between the units.

Now let us reverse the process and change small units into large units. For example:

Since 3 ft $= 1$ yd,	6 ft	$= 6 \div 3$	$= 2$ yd
Since 2000 lb $= 1$ ton,	6000 lb	$= 6000 \div 2000$	$= 3$ tons
Since 100¢ $= 1$ dollar,	400 ¢	$= 400 \div 100$	$= 4$ dollars

Notice that a *large number* of *small units* is always changed into a *small number* of *large units* by *dividing* by the number showing the relationship between the units.

RULE To change from a small unit to a large unit, divide by the number showing the relationship between the units.

Figure 2-16 illustrates the use of these previous two rules for ordinary metric units of measurement.

Example 2-24

Change the following units of measurement:
1. Change 8 cm to millimeters.
    There are 10 mm in each centimeter, and since we are changing large units (cm) to small units (mm), we *multiply* by 10.

$$8 \text{ cm} = 8 \times 10 = 80 \text{ mm} \qquad Ans.$$

2. Change 42 mm to centimeters.

There are 10 mm in each centimeter, and since we are changing small units (mm) to large units (cm), we *divide* by 10.

$$42 \text{ mm} = 42 \div 10 = 4.2 \text{ cm} \qquad \textit{Ans.}$$

3. Change 2 m to decimeters.
   There are 10 dm in each meter, and since we are changing large units (m) to small units (dm), we *multiply* by 10.

$$2 \text{ m} = 2 \times 10 = 20 \text{ dm} \qquad \textit{Ans.}$$

4. Change 300 cm to meters.
   There are 100 cm in each meter, and since we are changing small units (cm) to large units (m), we *divide* by 100.

$$300 \text{ cm} = 300 \div 100 = 3 \text{ m} \qquad \textit{Ans.}$$

5. Change 5 m to millimeters.
   There are 1000 mm in each meter, and since we are changing large units (m) to small units (mm), we *multiply* by 1000.

$$5 \text{ m} = 5 \times 1000 = 5000 \text{ mm} \qquad \textit{Ans.}$$

6. Change 125 m to dekameters.
   There are 10 m in each dekameter, and since we are changing small units (m) to large units (dam), we *divide* by 10.

$$125 \text{ m} = 125 \div 10 = 12.5 \text{ dam} \qquad \textit{Ans.}$$

7. Change 6450 m to kilometers.
   There are 1000 m in each kilometer, and since we are changing small units (m) to large units (km), we *divide* by 1000.

$$6450 \text{ m} = 6450 \div 1000 = 6.45 \text{ km} \qquad \textit{Ans.}$$

Self-Test 2-25

Change the following units of measurement.

Solution

1. Change 85 mm to centimeters.
   Since there are ——————— mm in each centimeter,

   $$85 \text{ mm} = 85 \underline{\quad (\times / \div) \quad} 10 = \underline{\qquad\qquad} \text{ cm}$$

2. Change 2 m to centimeters.
   Since there are _____ cm in each meter,

   $$2 \text{ m} = 2 \underline{\quad (\times / \div) \quad} 100 = \underline{\qquad\qquad} \text{ cm}$$

3. Change 4.5 km to meters.
   Since there are _____ m in each kilometer,

   $$4.5 \text{ km} = 4.5 \underline{\quad (\times / \div) \quad} 1000 = \underline{\qquad\qquad} \text{ m}$$

Practice

Change the following units of measurement:

1. 35 cm to millimeters
2. 48 mm to centimeters
3. 2.4 dm to centimeters
4. 80 cm to decimeters
5. 36 m to decimeters
6. 3 dm to meters
7. 5 m to centimeters
8. 325 cm to meters
9. 2.45 m to millimeters
10. 6500 mm to meters
11. 125 mm to centimeters
12. 6.8 cm to millimeters
13. 4.2 hm to dekameters
14. 36 dam to hectometers
15. 75 km to hectometers
16. 8 hm to kilometers
17. 3.9 hm to meters
18. 100 m to hectometers
19. 20 km to meters
20. 3498 m to kilometers
21. 1.5 dam to decimeters
22. Arrange in order of size, starting with the largest: 0.15 m, 20 cm, and 180 mm.
23. In a vacuum test on an engine operating at 3000 ft, the vacuum gage read 34.56 cm. Express this measurement as millimeters.

(See CD-ROM for Test 2-7)

Brushup on Decimals

Addition of decimals. When adding decimals, be sure to write the numbers so that the decimal points are lined up vertically.

Example 2-26

Add 2.52 + 0.007 + 13.03 + 5 + 0.4.

Solution

The number 5 is written as 5.000 to locate the decimal point.

$$
\begin{array}{r}
2.52 \\
0.007 \\
13.03 \\
5. \\
0.4 \\
\end{array}
\quad \text{or} \quad
\begin{array}{r}
2.520 \\
0.007 \\
13.030 \\
5.000 \\
0.400 \\
\hline
20.957 \quad \textit{Ans.}
\end{array}
$$

←————— Lined-up decimal points

The empty spaces are filled in with zeros, as shown in the column at the right, to aid in keeping the numbers in the correct columns.

Example 2-27

Find the total thickness of insulation on shielded radio wire covered with resin (0.55 mm), lacquered cotton braid (0.49 mm), and copper shielding (0.225 cm).

Solution

1. Change 0.225 cm to millimeters

$$0.225 \text{ cm} = 0.225 \times 10 = 2.25 \text{ mm}$$

2. The total thickness is the sum of the individual thicknesses.

$$
\begin{array}{r}
0.55 \text{ mm} \\
0.49 \text{ mm} \\
\underline{2.25 \text{ mm}} \\
3.29 \text{ mm} \quad \textit{Ans.}
\end{array}
$$

←——————— Lined-up decimal points

Exercises

Practice

Add the following decimals:

1. 3.28 + 9.5 + 0.634 + 0.078
2. 56.09 + 14 + 4.876 + 49.007
3. 13 + 3.072 + 0.7 + 6.06
4. 54 + 0.033 + 0.713 + 8.05
5. 0.087 + 6.18 + 4 + 1.7

Applications

6. What is the cost to rewind a motor if the following materials and labor were used: top sticks at $0.42, No. 17 wire at $3.25, No. 26 wire at $2.29, armature lacquer at $1.09, and labor at $32.50?
7. Find the total drop in voltage in a distribution system if the voltage drops across the sections are 1.06, 36.4, and 8 V.
8. Find the total thickness of insulation on shielded radio wire covered with resin (0.56 mm), lacquered cotton braid (0.47 mm), and copper shielding (0.76 mm).
9. Number 14 copper wire has a diameter of 0.0641 in; No. 10 wire is 0.0378 in larger. Find the diameter of No. 10 copper wire.
10. Find the total resistance of the leads of an installation if the resistances are 0.054, 1.004, 1.2, and 1.2 Ω.
11. The emitter current of a transistor is always equal to the sum of the collector current and the base current. Find the emitter current if the collector current is 0.03 A and the base current is 0.0015 A.
12. What is the cost to repair three electric outlets if each outlet requires one loom box at $2.50, one toggle switch at $2.95, one hickey at $0.32, and 1 h of labor at $22.75?
13. A carbon brush 2.46 cm thick has a copper plating of 0.39 mm on each side. What is the total thickness?

14. The insulation to be used in a slot in a motor is as follows: fish paper, 0.0156 in; Tufflex, 0.0313 in; varnished cambric, 0.0156 in; and top stick, 0.125 in. What is the total thickness of all the insulation?
15. The following materials were charged to an electrical wiring job: conduit, $4.25; No. 8 wire, $1.75; BX cable, $18.50; conduit fittings, $3.85; outlet boxes, $6.58; switches, $5.72; and $4.25 for tape, solder, and pipe clips. What was the total amount charged for materials?

Subtracting decimals. When subtracting decimals, write the numbers in columns as for addition, lining up the decimal points in a straight vertical column.

Example 2-28

The electric-meter readings for successive months were 70.08 and 76.49. Find the difference.

Solution

$$
\begin{array}{r}
76.49 \\
- \ 70.08 \\
\hline
6.41 \quad \textit{Ans.}
\end{array}
$$

Example 2-29

Subtract 1.04 from 3.

Solution

The number after the word "from" is written on top. The number after the word "subtract" is written underneath. The number 3 is written as 3.00 to locate the decimal point correctly.

$$
\begin{array}{r}
3.00 \\
- \ 1.04 \\
\hline
1.96 \quad \textit{Ans.}
\end{array}
$$

Exercises

Practice

1. 0.26 − 0.03
2. 1.36 − 0.18
3. 0.4 − 0.06
4. 0.05 − 0.004
5. 18.92 − 11.36
6. ⅝ − 0.002
7. 0.627 − 0.31
8. 0.827 − 0.31
9. 3 − 0.08
10. 0.5 − 0.02
11. 6 − 0.1
12. 0.83 − ½
13. 2.89 − 0.5
14. 12.6 − 7
15. 0.316 − 0.054
16. 14 − 8.06
17. 5¼ − 2.63
18. 3.125 − ⅛
19. Subtract ¼ from 0.765.
20. Find the difference between 110 and 54.9.
21. Find the difference between (a) 0.316 and 0.012; (b) 3.006 and 1.9; (c) 0.5 and 0.11; (d) 7.07 and 1.32; and (e) 2 and 0.02.

about electronics

If you need to replace a motor, you will probably have to read the name plate on the existing motor. Here's an example of a metric motor. Look for the frame size (like DM250M or a longer number) and the serial number (helpful for knowing when the motor was made). A plate notes the machine's kilowatts (or horsepower) and revolutions per minute (speed). Voltage written as 220/380 means that the motor runs at 220 V or 380 V, depending on the type of connection at the terminal block. You'll see the current, A (as pulled by the motor under a full load), the phase, Ph (usually 1 or 3 depending on the required supply), and hertz, Hz (usually 50 or 60). You will possibly also see the rating and insulation class, efficiency, bearing sizes, and capacitor sizes.

22. Subtract (*a*) 0.008 from 0.80; and (*b*) 0.216 from 2.16.
23. What is the difference in the diameters of No. 1 wire (0.2893 in) and No. 7 wire (0.1447 in)?

Applications

24. The resistance of the windings of an electromagnet at room temperature is 28.69 Ω If its resistance at 175°F (degrees Fahrenheit) is 36.98 Ω, find the increase in the resistance.
25. A locating pin which should be 0.875 cm in diameter measures 9.18 mm in diameter. How much is it oversize?
26. An electric generator delivers 223.8 V. If 4.95 V is lost in the line wires, find the voltage delivered at the end of the line.
27. A circuit in a television receiver calls for a 0.0005-μF capacitor. A capacitor valued at 0.00035 μF is available. How much extra capacitance is needed if connected in parallel?
28. The intermediate frequency at the output of a converter stage is found by obtaining the difference between the oscillator frequency and the radio frequency. If the radio frequency is 1.1 MHz (megahertz) and the oscillator frequency is 1.555 MHz, what will be the intermediate frequency?

(See CD-ROM for Test 2-8)

Multiplication of decimals.   Decimals are multiplied in exactly the same way as ordinary numbers are multiplied. However, in addition to the normal multiplication, the decimal point must be correctly set in the answer.

RULE   The number of decimal places in a product is equal to the sum of the number of decimal places in the numbers being multiplied.

Example 2-30

Multiply 0.62 by 0.3.

Solution

$$
\begin{array}{ll}
0.62 & \text{(multiplicand, 2 places)} \\
\times\ 0.3 & \text{(multiplier, 1 place)} \\
\hline
0.186 & \text{(product, } 2 + 1 = 3 \text{ places)} \qquad \textit{Ans.}
\end{array}
$$

Example 2-31

Multiply 0.35 by 0.004.

Solution

$$0.35 \text{ (multiplicand, 2 places)}$$
$$\underline{\times \ 0.004 \text{ (multiplier, 3 places)}}$$
$$0.001 \ 40 \text{ (product, } 2 + 3 = 5 \text{ places)} \qquad \textit{Ans.}$$

In this problem, extra zeros must be inserted between the decimal point and the digits of the answer to make up the required number of decimal places.

Exercises

Practice

Find the product of

1. 0.005×82	2. 1.732×40	3. 1.13×0.41
4. 0.9×0.09	5. 0.866×35	6. 44.6×805
7. 7.63×0.029	8. 0.354×0.008	9. 6.2×0.003
10. 0.033×0.0025	11. 106×0.045	12. 73.8×1.09

Applications

13. If a 100-W lamp uses 0.91 A, how much current would be used by five such lamps in parallel?
14. If BX cable costs $0.875/ft, what would be the cost of 52.5 ft?
15. The number of milliamperes is found by multiplying the number of amperes by 1000. Find the number of milliamperes equal to (*a*) 0.25 A; (*b*) 0.025 A; and (*c*) 2.5 A.
16. An electrician worked 1.5 h to install a junction box. If her rate of pay is $14.25/h, how much did she earn?
17. The standard unit of resistance is measured by the resistance of a column of mercury 106.3 cm high. If 1 cm equals 0.3937 in, what is the height of the mercury column correct to the nearest thousandth of an inch?
18. The inductive reactance of a coil is found by multiplying the constant 6.28 by the frequency and the inductance. Find the inductive reactance of a coil if the frequency is 60 Hz and the inductance is 0.15 H (henry).
19. What is the capacity of a battery (expressed in ampere-hours) if it discharges at the rate of 9.6 A for 7.25 h?

Division of decimals

Example 2-32

A contractor must locate nine equally spaced electric outlets in a school corridor. The total distance from the first outlet to the last is 117.2 ft. Find the distance between two adjacent outlets.

Solution

The number of spaces between outlets is one less than the number of outlets. Therefore, the distance between any two adjacent outlets is the distance

117.2 divided by 8. This division is accomplished in the same manner as in changing fractions to decimals. See Examples 2-18 and 2-19.

$$
\begin{array}{r}
14.65 \text{ ft} \quad \textit{Ans.} \\
8\overline{)117.20} \\
\underline{8} \\
37 \\
\underline{32} \\
5\ 2 \\
\underline{4\ 8} \\
40 \\
\underline{40} \\
0
\end{array}
$$

Example 2-33

Divide 1.38 by 0.06.

Solution

When a decimal is divided, it is best to move the decimal point all the way over to the right so as to bring it to the end of the divisor.

$$
.06.\,\overline{)1.38}
$$

If this is done, the decimal point in the dividend must also be moved to the right *for the same number of places*. This is the equivalent of multiplying both dividend and divisor by 100. Now we can divide as before.

$$
\begin{array}{r}
23. \quad \textit{Ans.} \\
.06.\,\overline{)1\,.38.} \\
\underline{1\ \ 2} \\
18 \\
\underline{18} \\
0
\end{array}
$$

Example 2-34

Divide 3.6 by 0.08.

Solution

A zero must be added after the 6 to provide the two places that the decimal point must be moved to the right.

$$
\begin{array}{r}
45. \quad \textit{Ans.} \\
.08.\,\overline{)3\,.60.} \\
\underline{3\ \ 2} \\
40 \\
\underline{40} \\
0
\end{array}
$$

Example 2-35

Divide 0.0007 by 0.125.

Solution

$$
\begin{array}{r}
0.0056 \\
125. \overline{)000.7000} \\
625 \\
\hline
750 \\
750 \\
\hline
0
\end{array}
\qquad \textit{Ans.}
$$

Exercises

Practice

1. $3.9 \div 0.3$ 　　　2. $12.56 \div 0.4$ 　　　3. $80.5 \div 0.5$
4. $51 \div 0.06$ 　　　5. $38.54 \div 8.2$ 　　　6. $1591 \div 0.43$
7. $2.8296 \div 0.0036$ 　　8. $140.7 \div 0.021$ 　　9. $9.1408 \div 3.94$
10. Using the formula $I = V/R$, find I if $V = 79.5$ V and $R = 265\ \Omega$.

Applications

11. Shielded rubber-jacketed microphone cable weighs 0.075 lb/ft. How many feet of cable are there in a coil weighing 15 lb?
12. What is the smallest number of insulators, each rated at 12,000 V, that should be used to safeguard a 220,000-V transmission line?
13. The Q, or "quality," of a coil is a measure of its worth in a tuned circuit. It is found by dividing the reactance of the coil by its effective resistance (see Job 16-3). Find the Q of a coil if its reactance is 1820 Ω and its effective resistance is 30 Ω.
14. A 50-ft-long wire has a resistance of 10.35 Ω. What is the resistance of 1 ft of this wire?
15. The current-amplifying ability β of a transistor is obtained by dividing the collector current I_C by the base current I_B. Find β if $I_C = 0.0004$ A and $I_B = 0.000\ 01$ A.

(See CD-ROM for Test 2-9)

JOB 2-8

Review of Working with Decimals

1. Decimal fractions are fractions whose _____ are numbers like 10, _____, 1000, etc.
2. The denominator is shown by the number of digits to the _____ of the decimal point. Thus,

   　　　0.6 represents six-_____.
   　　　0.57 represents fifty-seven _____.

0.123 represents one hundred twenty-three _____.

3.09 represents three and nine-_____.

3. Decimals can be compared only when they have the _____ number of decimal places.

4. The word "and" in a mixed number such as five and three-hundredths is written as a _____ point. This number would be written as _____.

5. Fractions are changed to decimals by _____ the _____ by the _____.

6. A whole number always has a decimal point understood to be at the _____ of the number.
   (beginning/end)

7. When dividing decimals, if a remainder is more than _____ of the divisor, drop it and add a full unit to the last _____ of the answer. If the remainder is _____ than half, drop it completely.

8. When adding or subtracting decimals, line up the decimal points in a _____ column.

9. The product of two decimals has as many decimal places as the _____ of the places in the numbers being multiplied.

10. When you divide decimals, move the decimal point in the divisor to the _____ as many places as is necessary to bring the point behind the last digit. Then move the point in the dividend to the right for the _____ number of places.

11. $18.5 \times 100 =$ _____

    $6.28 \div 1000 =$ _____

12. To change large units into small units, we _____. To change
    (multiply/divide)
    small units into large units, we _____.
    (multiply/divide)

13. 1 cm = _____ mm

    1 km = _____ m

    1 m   = _____ cm

    1 m   = _____ mm

    1 m   = _____ dm

Applications

1. Find the dimensions A and B in the spindle shaft shown in Fig. 2-17.

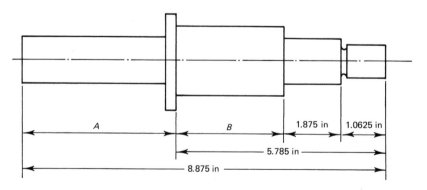

Figure 2-17

2. In an ac voltage wave, the maximum voltage is equal to 1.414 times the effective voltage as read on a meter. Find the maximum voltage of a wave whose meter reading is 117 V.

3. The power factor (pf) of an ac circuit is found by dividing its resistance R by its impedance Z. Find the power factor of a circuit if $R = 2.46 \ \Omega$ and $Z = 30 \ \Omega$.

4. The laminated core of a transformer is made of 32 thicknesses of metal. If each lamination is 7.5 mm thick, find the total thickness of the core in centimeters.

5. Multiply: (a) 0.735×1000; (b) $81.7 \times 1{,}000{,}000$; (c) 0.0035×1000; and (d) 0.004×100. Divide: (e) $0.07 \div 100$; (f) $6.28 \div 1000$; (g) $4750 \div 100$; and (h) $397 \div 1{,}000{,}000$.

6. A rebuilt alternator that cost $95.50 was sold for $147.95. (a) Find the profit. (b) What decimal part of the cost was the profit?

7. If BX cable costs $0.825/ft, what would be the cost of 52.5 ft?

8. What is the total current supplied to a parallel circuit consisting of a broiler drawing 9.25 A, a radio drawing 0.93 A, and two lamps, each drawing 0.91 A?

9. The diameter of a motor shaft bearing is 0.12 mm larger than the shaft of the motor. What is the diameter of the bearing if the shaft has a diameter of 2.25 cm?

10. Ms. Graham received an increase from $12.40 to $13.10 per hour for her work as an electrician. Find the amount of increase for a 40-h week.

11. A 60-W lamp uses about 0.625 A when connected in an ordinary house line. How many amperes would three such lamps use when connected in parallel?

12. An electrician figured the cost to rewind a motor as follows: No. 17 wire at $2.29, No. 28 wire at $1.65, No. 00 top sticks at $0.60, armalac at $1.25, and 2 h of labor at $14.25/h. Find the total cost.

13. A neon electric sign uses 4.75 W of power for each foot of tubing. Find the power consumption for a sign which uses 34.5 ft of tubing.

14. A 10.5-cm-diameter piece of work is turned down in a lathe in three cuts. The first cut takes off 40 mm, the second takes off 2.4 mm, and the third removes 1.7 mm. What is the finished diameter of the work?

15. An electrician is to wire a light switch to operate the following light bulbs on a truck: two marker lamps at 0.3 A each; three clearance lamps at 0.4 A each; two stop lights at 2.4 A each; two front parking lamps at 1.5 A each; and two headlamps at 7.6 A each. The maximum current-carrying capacities of wires are 16 gage, 6 A; 14 gage, 15 A; 12 gage, 20 A; and 10 gage, 30 A. Find the smallest gage wire that may be safely used when all lights are operating.

(See CD-ROM for Test 2-10)

Assessment

1. Seven light fixtures are to be hung on 5¾-foot centers. What is the distance (center to center) between the two outermost fixtures?
2. The windings of a step-down transformer have a ratio of 250:50. Reduce this fraction to lowest terms.
3. What is the voltage drop across R_3 in the circuit in Fig. 2-18 if the current is 100 mA?

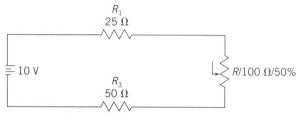

Figure 2-18

4. John bought twelve boxes of clean-up towels, and the total price of the purchase is $167.40, before tax. What is the unit cost of each box of towels? What is the total cost if a 7.75% sales tax is added to John's purchase?
5. Give the decimal equivalent for the following commonly used fractions: ⅛, ¼, ⅜, ½, ⅝, ¾, and ⅞.
6. Knowing that there are 2.54 centimeters per inch, convert 3½ feet into centimeters.
7. Sam installs a run of conduit in a straight line that is to be 100 ft between junction boxes. The wire he pulls must have 18 in. of free conductor at each box. If the #12 THHN/THWN stranded wire he pulls has a cost of $24.99 per 500-ft reel, what is the cost of wire if he pulls three hot conductors, one neutral conductor, and one ground wire?

Internet Activity A

Use the following site to find all three forms of Ohm's law.
http://www.proaxis.com/~iguanalabs/ohmslaw.htm

Internet Activity B

Use the following site to reinforce your understanding of fractions and mixed numbers. What does this site say about adding and subtracting fractions that have the same denominators? **http://tqd.advanced.org/2949/addsub.htm**

INTERNET ACTIVITIES

Chapter 2 Solutions to Self-Tests and Reviews

Self-Test 2-3

1. divide
2. same
3. 4
4. 6
5. 12
6. 10
7. 2
8. $\dfrac{3}{5}$
9. no
10. ⅗

Self-Test 2-10

1. 14
2. 7
3. 2
4. 15
5. 5
6. 5
7. 3
8. $\dfrac{5}{6}$

Self-Test 2-13

1. numerator, denominator
2. numerator
3. denominator
4. part
5. value
6. 6
7. ⅔
8. no
9. lowest
10. divide, evenly
11. larger
12. improper
13. denominator, denominator
14. divide, denominator
15. product, product, lowest
16. divide, same

Self-Test 2-16

1. *I, R*
2. 7.5
3. 6.3

Self-Test 2-21

1. right
2. 2
3. 2
4. are
5. 2520

Self-Test 2-23

1. left
2. 3
3. 3
4. are
5. 0.0085

Self-Test 2-25

1. 10, ÷ , 8.5
2. 100 × 200
3. 1000 × 4500

Job 2-8 Review

1. denominators, 100
2. right, tenths, hundredths, thousandths, hundredths
3. same
4. decimal, 5.03
5. dividing, numerator, denominator
6. end
7. half, digit, less
8. vertical
9. sum
10. right, same
11. 1850, 0.006 28
12. multiply, divide
13. 10, 1000, 100, 1000, 10

3

Formulas

1. Change rules into formulas.
2. List the order of operations for evaluating mathematical expressions.
3. Add signed numbers.
4. Multiply and divide signed numbers.
5. Describe an exponent.
6. Work with formulas containing exponents.
7. Multiply and divide by positive and negative powers of 10.
8. Express numbers greater than 1 as a lesser number times a power of 10.
9. Express numbers less than 1 as a whole number times a power of 10.
10. Multiply and divide with powers of 10.
11. Convert units of measurement.
12. Solve simple equations.

JOB 3-1

Checkup on Formulas in Electrical Work (Diagnostic Test)

In Job 2-3 we used our first *formula*. When Ohm's law is written using only the letters which represent the words of Ohm's law, it is called a *formula*. As we continue with our study of electricity, we shall meet many new formulas. Some are simple like Ohm's law, but others are more complicated. The electrician and the electronic technician find it necessary to solve problems with formulas in their daily work. If you have any difficulty with these problems, see Job 3-2, which follows.

In the applications that follow, you will be using exponents. You may want to do the diagnostic test in Job 3-4 for formulas with exponents. If you have difficulty with exponents, you may want to review Job 3-5 on exponents and Job 3-6 on powers of 10.

Many of the problems that you will work with will lead to signed numbers. If these give you any trouble, review Job 3-3 on positive and negative numbers. Most of the problems use whole numbers but the answers may be decimals. To see if these are giving you any problems, try the diagnostic test in Job 2-4.

When working with formulas, you may have problems with fractions. If so, try the diagnostic test for fractions in Job 2-1.

Exercises

Applications

1. Using Ohm's law ($V = I \times R$), find the current I drawn by a 10-Ω automobile horn R from a 6-V battery V.

2. Using Ohm's law, find the number of ohms of resistance R needed to obtain a bias voltage V of 6 V if the current I is 0.02 A.
3. Using the formula for electric power, $P = V \times I$, find the voltage V necessary to operate a 500-W electric percolator P if it draws a current I of 4.5 A.
4. Using the formula $P = V \times I$, find the current I drawn by a 440-W vacuum cleaner P from a 110-V line V.
5. Write the formula for the following rule: kilowatts (kW) equals current I multiplied by voltage V and divided by 1000.
6. Using the series-circuit formula $I = V/(R_1 + R_2)$, find I if $V = 100$V, $R_1 = 20\ \Omega$, and $R_2 = 30\ \Omega$.
7. Using the ac formula $I = V/Z$, find the impedance Z of an ac circuit if the voltage V is 50 V and the current I is 2 A.
8. Using the formula mA = A $\times$ 1000, find the number of amperes equivalent to 125 mA (milliamperes).
9. $I_T = I_1 + I_2 + I_3$ is the formula for the total current in a parallel circuit. Find the total current I_T if $I_1 = 2$ A, $I_2 = 5$ A, and $I_3 = 4$A.
10. The formula for the number of coulombs of electricity which can be placed on the plates of a capacitor is $Q = C \times V$. Find the voltage V which is necessary to place a charge Q of 0.000 000 2 C on the plates of a capacitor whose capacitance C is 0.000 000 002 F (farad).

Brushup on Formulas

Meaning.   A formula is a convenient shorthand method for expressing and writing a rule or relationship among several quantities.

Signs of operation.   The quantities involved in any simple relationship are held together by one or more of the following operations:
1. Multiplication ($\times$)
2. Division ($\div$)
3. Addition ($+$)
4. Subtraction ($-$)
5. Equality ($=$)

Each of these operations may be written in several ways.

Multiplication.   The multiplication of two quantities is often expressed as the "product of" the two quantities. This may be written as follows:
1. Using a multiplication sign ($\times$) between the numbers or letters
2. Using a dot ($\cdot$) between the numbers or letters
3. Writing nothing at all between the numbers or letters

For example, the product of 3 and 4 may be written as (1) 3×4 or (2) $3 \cdot 4$. The third method cannot be used when only numbers are involved because the meaning would not be clear. For example, 34 would mean the number thirty-four and *not* 3×4. This method of indicating multiplication by omitting all signs between the quantities can be used only for combinations of numbers and letters or combinations of letters.

The product of 6 and R may be written as (1) $6 \times R$, (2) $6 \cdot R$, or (3) $6R$. All three forms indicate that 6 is to be multiplied by the quantity called R.

The product of P, R, and T may be written as (1) $P \times R \times T$, (2) $P \cdot R \cdot T$, or (3) PRT. All three forms indicate that the quantity P is to be multiplied by the quantity R and then multiplied by the quantity T.

Division. The division of two quantities is often expressed as the "quotient of" the two quantities. This may be written as follows:

1. Using a division sign ($\div$) between the numbers or letters
2. Using a fraction bar to indicate division

For example, the quotient of 8 divided by 2 may be written as (1) $8 \div 2$ or (2) $\frac{8}{2}$.

The quotient of 12 divided by I may be written as (1) $12 \div I$ or (2) $12/I$

The quotient of V divided by R may be written as (1) $V \div R$ or (2) V/R.

Addition. The addition of two or more quantities is often expressed as the "sum of" the quantities and is indicated by a plus sign ($+$) between the quantities. For example,

The sum of 6 and 4 is written as $6 + 4$.

The sum of 3 and R is written as $3 + R$.

The sum of V_1 and V_2 is written as $V_1 + V_2$.

Subtraction. The subtraction of two quantities is often expressed as the "difference between" the quantities and is indicated by a minus sign ($-$) between them. For example,

The difference between 9 and 4 is written as $9 - 4$. This is read as (1) 9 minus 4 or (2) 4 subtracted from 9.

The difference between 20 and R is written as $20 - R$. This is read as (1) 20 minus R or (2) R subtracted from 20.

The difference between V_T and V_1 is written as $V_T - V_1$. This is read as (1) V_T minus V_1 or (2) V_1 subtracted from V_T.

Equality. An equal sign ($=$) is used to indicate that the combination of numbers and letters on one side of the equal sign has the same value as the combination of numbers and letters on the other side. For example,

$$3 \times 4 = 12$$
$$a \cdot b = ab$$
$$2R = 10$$

Changing rules into formulas. To change a rule into a formula:

1. Replace each quantity with a convenient letter.
2. Rewrite the rule. Substitute these letters for the words of the rule. Include the symbols for the signs of operation.

Note: The letter used to replace a word is usually the first letter of the word representing the quantity. However, any letter may be used. For example, if a letter has already been used to represent some quantity, it cannot be used again in the same formula to represent *another* quantity. In this event a letter which is *not* the first letter of the word would be used.

Example 3-1

Change the following rule into a formula: The area of a rectangle is equal to its length multiplied by its width.

Solution

The length is replaced by the letter L. The width is replaced by the letter W. The area is replaced by the letter A.

The area is equal to the length multiplied by the width,

$$A \qquad = \qquad L \qquad \times \qquad W$$

Thus,

$$A = L \times W \quad \text{or} \quad A = L \cdot W \quad \text{or} \quad A = LW \qquad Ans.$$

Example 3-2

Change the following rule into a formula: The voltage is equal to the current multiplied by the resistance.

Solution

The voltage is replaced by the letter V. The current is replaced by the letter I. The resistance is replaced by the letter R.

The voltage is equal to the current multiplied by the resistance.

$$V \qquad = \qquad I \qquad \times \qquad R$$

Thus,

$$V = I \times R \quad \text{or} \quad V = I \cdot R \quad \text{or} \quad V = IR \qquad Ans.$$

Example 3-3

Change the following rule into a formula: The voltage V of a series circuit of two resistors is equal to the current I multiplied by the sum of the resistances R_1 and R_2.

Solution

The word "sum" is indicated by a plus sign. This sum is actually a *single* quantity obtained by adding R_1 and R_2. This must be shown by enclosing R_1 and R_2 in a pair of parentheses. The current I will then be multiplied by the parentheses. Thus,

$$V = I \times (R_1 + R_2) \quad \text{or} \quad V = I(R_1 + R_2) \qquad Ans.$$

CALCULATOR HINT

When using a calculator to solve a formula that contains parentheses like $V = I(R_1 + R_2)$, you must add $R_1 + R_2$ before multiplying by I. When working the formula with a calculator, enter the value of I followed by the ⊠ key. Next press the left parentheses key Ⓛ. Then add the values $R_1 + R_2$ and press the right parentheses key Ⓡ. Then press the equal key to obtain your answer.

Self-Test 3-4

Change the following rule into a formula: The total resistance R_T of two resistors in parallel is equal to the product of the resistances R_1 and R_2 divided by the sum of the resistances.

Solution

The diagram for the circuit is shown in Fig. 3-1.

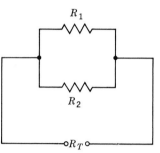

Figure 3-1   Resistances in parallel.

1. The word "product" means to __(add/multiply/divide)__.
2. The product of R_1 and R_2 is written as R_1 _____ R_2.
3. The word "sum" means to ___(add/subtract)___.
4. The sum of R_1 and R_2 is written as R_1 _____ R_2.
5. The division in our formula may be written as a fraction. The numerator of the fraction will be the ___(sum/product)___ of the resistances.
6. The denominator will be the ___(sum/product)___ of the resistances.
7. The formula will be read as:

$$R_T = \frac{R_1 \;\underline{\quad}\; R_2}{R_1 \;\underline{\quad}\; R_2} \qquad Ans.$$

Exercises

Practice

Write the formula for each of the rules given. Use the italic letters and abbreviations in parentheses to indicate each word.

1. The electric power P is equal to the current I multiplied by the voltage V.
2. The effective voltage V of an ac voltage wave is equal to 0.707 times the maximum value V_{max}.
3. The efficiency (eff) of a motor is equal to the power output P_o divided by the power input P_i.
4. The total resistance R_T of a series circuit is equal to the sum of the individual resistances R_1, R_2, and R_3.
5. The capacitive reactance X_C of a capacitor is equal to 159,000 divided by the product of the frequency f and the capacitance C.
6. The power factor (PF) of an ac circuit is equal to the total resistance R_T divided by the impedance Z.

7. The perimeter P of a rectangle is equal to the sum of twice the length L and twice the width W.
8. The current I_m in an ammeter is the difference between the line current I and the shunt current I_s.
9. The total current I_T in a series circuit of two resistors is equal to the total voltage V_T divided by the sum of the resistances R_1 and R_2.
10. The resistance R_s of an ammeter shunt is equal to the meter resistance R_m divided by 1 less than the multiplying factor N.

Substitution in formulas.   We can change a rule into a formula by substituting letters for words. Since each word or letter actually represents some number in a specific problem, we can go one step further and substitute specific numbers for the letters in any formula or expression. This is called *substitution in a formula.* The numbers are then combined according to the signs of operation shown by the formula.

Example 3-5

Find the value of $10 - 2 \times 4$.

Solution

If we do this arithmetic in order from left to right—subtract first and then multiply—we get

$$10 - 2 \times 4$$
$$8 \times 4$$
$$32 \quad \textit{Ans.}$$

But if we multiply first and then subtract, we get

$$10 - 2 \times 4$$
$$10 - 8$$
$$2 \quad \textit{Ans.}$$

We must decide which mathematical operation is performed first. If we are all expected to get the *same* answer in a particular problem, we must agree on which operation is to be done first, second, etc. Over the entire world, mathematicians have agreed on a definite *order* in which the operations of arithmetic should be performed. Failure to follow this order often gives results which do *not* represent what was actually meant.

Order of Operations for Evaluating Mathematical Expressions
1. Substitute the given numbers for the letters.
2. Find the value of all expressions in parentheses.
3. Do all multiplications and divisions in order from left to right.
4. Do all additions and subtractions in order from left to right.

Example 3-6

Find the value of $12 - 2 \times 4 + 3$.

Solution

1. Write the expression.	$12 - 2 \times 4$		$+ 3$
2. Multiply first.	$12 -$	8	$+ 3$
3. Add or subtract.		4	$+ 3$
4. Answer.			7

Example 3-7

Find the value of $10 - 2 \times 3 + \frac{8}{2}$.

Solution

1. Write the expression.	$10 - 2 \times 3$		$+ \frac{8}{2}$
2. Multiply first.	$10 -$	6	$+ \frac{8}{2}$
3. Divide next.	$10 -$	6	$+ 4$
3. Add or subtract.		4	$+ 4$
4. Answer.		8	

Example 3-8

Using the formula $P = VI$, find the power P needed to operate an electric iron using a voltage V of 110 V and a current I of 6 A. Refer to Fig. 3-2.

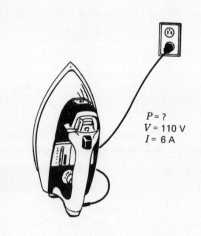

$P = ?$
$V = 110$ V
$I = 6$ A

Figure 3-2

Solution

1. Write the formula. $P = VI$
2. Substitute numbers. $P = 110 \times 6$

Note that no sign of operation between the letters means multiply.

3. Multiply the numbers. $P = 660$ W   *Ans.*

Self-Test 3-9

Using the formula $V_T = I(R_1 + R_2)$, find the total voltage V_T in the series circuit shown in Fig. 3-3.

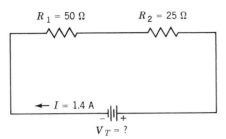

Figure 3-3

Solution

1. Write the formula. $\qquad\qquad\qquad V_T = I(R_1 + R_2)$
2. Substitute numbers. $\qquad\qquad\; V_T = 1.4(50 + 25)$
3. The arithmetic to be done first is the <u>(multiplication/addition)</u> of _____ and _____.
4. Simplifying the parentheses gives $\;\; V_T = 1.4 \times$ _____.
5. Multiply. $\qquad\qquad\qquad\qquad\;\; V_T =$ _____ V $\qquad$ *Ans.*

Exercises

Practice

Evaluate the following expressions:

1. $9 + 2 \times 4 - 3$
2. $12 - 2 \times 5$
3. $2 \times 7 + 2 \times 3$
4. $5(8 - 3) + 7$
5. $5 \times 6 - 4 \times 3$
6. $15 - 2(8 - 3)$
7. $6 + \frac{2}{8}$
8. $\dfrac{6 \times 7}{2 + 5}$
9. $3(10 - 7) - 2$
10. $21 - 2(15 - 5)$

Applications

Solve the following problems using the formulas that are given in each problem. If there is no diagram that applies to the problem, set down the given information in the space ordinarily used for the diagram.

11. Using the formula $A = LW$, find the number of square feet of area A in a rectangle if the length L is 15 ft and the width W is 9 ft.
12. Using the formula $I = V/Z$, find the number of amperes I if V is 110 V and Z is 22 Ω.
13. Using the formula $I_T = I_1 + I_2 + I_3$, find the total number of amperes I_T if $I_1 = 2$ A, $I_2 = 3$ A, and $I_3 = 5$ A.
14. Using the formula kW $= VI/1000$, find the kilowatts (kW) of power if the voltage V is 200 V and the current I is 2 A.

15. Using the formula $R_T = R_1 + R_2 + R_3$, find the total resistance R_T of a series circuit if R_1 is 18.2 Ω, R_2 is 45.8 Ω, and R_3 is 76.4 Ω.

16. The formula for the shunt current in an ammeter hookup is $I_s = I - I_m$. Find the current I_s through the shunt if the line current I is 0.045 A and the meter current I_m is 0.009 A.

17. Using the formula $R_T = (R_1 \times R_2)/(R_1 + R_2)$, find the total resistance R_T of a parallel circuit if R_1 is 30 Ω and R_2 is 60 Ω.

18. Using the formula $X_L = 6.28fL$, find the inductive reactance X_L of a coil to a frequency f of 60 Hz if the coil has an inductance L of 0.15 H.

19. Using the series-circuit formula $I = V/(R + r)$, find the current I if $V = 6$, $R = 16$, and $r = 8$.

20. Using the formula $F = \frac{9}{5}C + 32$ to change Celsius temperature to Fahrenheit temperature, find F if $C = 20°$.

21. Using the formula $R_t = R_o(1 + 0.0042t)$ for the effect of temperature on resistance, find the resistance R_t at a temperature $t = 50°$, if the resistance at $0°C$ (R_o) is 60 Ω.

(See CD-ROM for Test 3-1)

JOB 3-3

Positive and Negative Numbers

Many of the calculations, graphs, and tables used to solve problems in the jobs to follow require an understanding of positive and negative numbers. These numbers, commonly called *signed* numbers, are used to indicate *opposite* amounts, such as a *gain* or a *loss* in voltage, an *increase* or *decrease* in loudness, or currents that flow in *opposite directions.*

Writing positive and negative numbers. We often indicate opposite quantities by pairs of words such as north or south, up or down, gain or loss, and win or lose. In our electrical work, it is much easier to indicate opposite quantities by the use of the plus sign ($+$) or the minus sign ($-$). For example, 5° above zero may be written as $+5°$, and 5° below zero may be written as $-5°$. A current in one direction is written as $+10$ A but as -10 A if the flow is in the opposite direction. Numbers preceded by a minus sign are called *negative numbers.* The minus sign must always be written before a negative number. If no sign is written before a number, the quantity is understood to be a positive number. Signed numbers are usually written to agree with the following system: Gains, increases, or directions to the right or upward are written as positive ($+$), and losses, decreases, or directions to the left or downward are written as negative ($-$).

Exercises

Practice

Write the following quantities as signed numbers:

1. A loss of $3.
2. A temperature of 8° below zero.
3. 23° north latitude.
4. 70° east longitude.

5. An increase in loudness of 2 dB (decibels).
6. A drop of 10° in temperature.
7. Ten miles per hour slower.
8. Eight paces to the right.
9. Five blocks downtown.
10. Twenty feet below sea level.
11. If the voltage of the ground is considered to be 0 V, indicate a voltage of 4 V below ground.
12. Indicate a grid bias of 2 V below ground.

Addition of Signed Numbers

RULE To add two positive numbers, add the numbers and place the plus sign before the sum.

Example 3-10

"A gain of 7 plus a gain of 3 equals a gain of 10" may be indicated as

$$(+7) + (+3) = +10 \qquad Ans.$$

RULE To add two negative numbers, add the numbers and place the minus sign before the sum.

CALCULATOR HINT

To enter a negative number into your calculator, enter the number first and then press the $+ / -$ key.

Example 3-11

"A loss of 5 plus a loss of 2 equals a loss of 7" may be indicated as

$$(-5) + (-2) = -7 \qquad Ans.$$

RULE To add two numbers of *different* signs, *subtract* the numbers and place the sign of the larger before the answer.

Example 3-12

"A gain of 7 plus a loss of 3 equals a total gain of 4" may be indicated as

$$(+7) + (-3) = +4 \qquad Ans.$$

Example 3-13

"A loss of 9 plus a gain of 4 equals a total loss of 5" may be indicated as

$$(-9) + (+4) = -5 \qquad Ans.$$

Example 3-14

Combine the following signed numbers:

$$7 - 2 - 3 + 5 + 4 - 2$$

Solution

These signs are the signs of the numbers. To combine signed numbers *always* means to *add* the signed numbers using the rules for algebraic addition. Therefore, reading from left to right, we shall consider this problem to mean

$+7$ *added to* -2 is $+5$ (7 means $+7$)
$+5$ *added to* -3 is $+2$
$+2$ *added to* $+5$ is $+7$
$+7$ *added to* $+4$ is $+11$
$+11$ *added to* -2 is $+9$ *Ans.*

Self-Test 3-15

Fill in the blank spaces with the correct responses.

1. $(-4) + (-6)$ means to ___(add/subtract)___ the numbers and prefix the sum with a _____ sign. The answer is _____.
2. $(+4) + (+8)$ means to _____ the numbers and prefix the sum with a _____ sign. The answer is _____.
3. $(7) + (+4)$ means to _____ the numbers and prefix the sum with a _____ sign. The answer is _____.
4. $(+6) + (-4)$ means to _____ the numbers and prefix the difference with a _____ sign because the sign of the larger number is $+$. The answer is _____.
5. $(-12) + (-3)$ means to _____ the numbers and prefix the answer with a _____ sign. The answer is _____.
6. $(-10) + (+4)$ means to _____ the numbers and prefix the answer with a _____ sign because the sign of the larger is _____. The answer is _____.
7. $7 - 4$ means $(+7) + ($ _____ $)$. The answer is _____.
8. $(23) + (-5)$ means to _____ the numbers and prefix the answer with a _____ sign. The answer is _____.
9. $-5 - 7$ means $(-5) + ($ _____ $)$. The answer is _____.
10. $7 - 3 - 9 - 2$ means $(+7) + (-3) + ($ _____ $) + (-2)$:

$+7$ added to $-3 =$ _____
$+4$ added to $-9 =$ _____
-5 added to $-2 =$ _____   *Ans.*

Exercises

Practice

Add the signed numbers indicated in each problem.

1. $(+3) + (+9)$
2. $(-6) + (-5)$
3. $(+8) + (-2)$
4. $(-10) + (+3)$
5. $(-12) + (-5)$
6. $(+16) + (+3)$
7. $(+18) + (-11)$
8. $(-21) + (+9)$
9. $(-8) + (-9)$

10. $+18$
    $\underline{+14}$
11. -26
    $\underline{+12}$
12. $+36$
    $\underline{-14}$

13. -6
    $\underline{+22}$
14. -47
    $\underline{-23}$
15. -75
    $\underline{+23}$

16. -8.2
    $\underline{+11.6}$
17. -10.5
    $\underline{+12.4}$
18. -16.8
    $\underline{+7}$

19. $(-\frac{1}{4}) + (+\frac{1}{2})$
20. $(+\frac{5}{8}) + (-\frac{1}{4})$
21. $(-\frac{1}{2}) + (-\frac{1}{3})$

22. $-16\frac{7}{8}$
    $\underline{+8\frac{3}{4}}$
23. $+3\frac{9}{16}$
    $\underline{-9\frac{1}{4}}$
24. $-15\frac{3}{4}$
    $\underline{+28\frac{1}{2}}$

25. $+8 - 2 - 9 + 6$
26. $-6 - 3 + 11 + 4 - 5$
27. $13 - 2 - 3 + 4 - 5$
28. $-9 + 3 - 2 + 4 - 6$
29. $-3 - 5 + 2 - 26 + 7$
30. $6 - 2 - 9 - 1 + 3$
31. $1.4 - 0.2 - 0.7$
32. $-16 + 4.8 + 3.4$
33. $6.4 - 8.5 + 3.2 - 4$
34. $3.2 - 8 + 2.5 + 5.7$

(See CD-ROM for Test 3-2)

Multiplying and Dividing Signed Quantities

The following rules for signs apply to *both* multiplication and division.

RULE When quantities with the same sign are multiplied or divided, the answer is plus.

RULE When quantities with different signs are multiplied or divided, the answer is minus.

Example 3-16

Perform the indicated operation.

$$(+4) \times (+2) = +8 \qquad (-6) \times (-2) = +12$$
$$(+4) \times (-2) = -8 \qquad (-3) \times (+4) = -12$$
$$(+2) \times (+3R) = +6R \qquad (-4) \times (-2I) = +8I$$
$$(-4) \times (0) = 0 \qquad (+3) \times (0) = 0$$

$$\frac{+20}{+5} = +4 \qquad \frac{-12}{-4} = +3 \qquad \frac{+3}{+6} = +\frac{1}{2} \qquad \frac{-2}{-8} = +\frac{1}{4}$$

$$\frac{+20}{-5} = -4 \qquad \frac{-12}{+4} = -3 \qquad \frac{+3}{-6} = -\frac{1}{2} \qquad \frac{-2}{+8} = -\frac{1}{4}$$

$$\frac{-8R}{-2} = +4R \qquad \frac{-12I}{+6} = -2I \qquad \frac{+6R}{-2} = -3R \qquad \frac{-3R}{-6} = +\frac{1}{2}R$$

Exercises

Practice

Multiply:

1. $(+6)$ by $(+3)$	2. (-3) by (-4)	3. $(+4)$ by (-2)
4. (-3) by (6)	5. (-5) by (-6)	6. $(+2)$ by $(3R)$
7. $(6R)$ by $(+3)$	8. (-3) by $(4I)$	9. $(5R)$ by (-2)
10. (-3) by $(-5R)$	11. $(-2R)$ by (-5)	12. (-12) by (-6)
13. (-6) by (0)	14. $(-8T)$ by (0)	15. (8) by (-13)
16. (-25) by (-6)	17. $(-5R)$ by (17)	18. (2) by $(-13I)$
19. $(-4R)$ by (36)	20. (0.2) by (-10)	21. (-0.3) by (20)
22. (-0.2) by $(5R)$	23. (-1.2) by $(2I)$	24. $(-\frac{1}{2})$ by (-20)
25. $(-\frac{1}{4})$ by $(12R)$	26. (-1.2) by (-18)	27. (4.6) by (-3.2)
28. $(-16I)$ by (-0.2)	29. $(3.4R)$ by (-1.5)	30. (-0.6) by $(8.2R)$

Divide:

1. $(+12)$ by $(+3)$	2. (-24) by (-4)	3. $(+8)$ by (-2)
4. (-16) by $(+2)$	5. (-84) by (-7)	6. $(24R)$ by (6)
7. $(24R)$ by (-4)	8. $(-60) \div (15)$	9. $(20) \div (-40)$
10. $(-48) \div (-8)$	11. $(-16) \div (-32)$	12. $(+9) \div (-27)$
13. (0) by (-6)	14. $(0) \div (+7)$	15. $(63) \div (-9)$
16. $(-100) \div (+5)$	17. $(-60) \div (+8)$	18. $(-4) \div (20)$

19. $\dfrac{-3.6}{1.2}$	20. $\dfrac{4}{-7}$	21. $\dfrac{-1.6}{-0.5}$	22. $\dfrac{4.2}{-3}$
23. $\dfrac{-65}{-0.5}$	24. $\dfrac{0.4}{-0.1}$	25. $\dfrac{-2.6}{1.3}$	26. $\dfrac{0.4}{-3.2}$
27. $\dfrac{9.6}{-3}$	28. $\dfrac{-2.4}{-0.6}$	29. $\dfrac{-50}{0.8}$	30. $\dfrac{-10.8}{-0.3}$

(See CD-ROM for Test 3-3)

JOB 3-4

Checkup on Using Formulas with Exponents (Diagnostic Test)

The following problems involve the use of formulas which contain exponents. If you have any difficulty with these problems, see Job 3-5, which follows.

Exercises

Applications

1. Using the formula $P = V^2/R$, find the value of P if $V = 100$ and $R = 50$.
2. Using the formula $P = I^2R$, find the value of P if $I = 3$ and $R = 40$.
3. Using the formula emf $= (L \times n)/10^8$, find the value of emf if $L = 50 \times 10^6$ and $n = 40$.
4. Using the formula $R = (k \times l)/D^2$, find the value of R if $k = 10.4$, $l = 100$, and $D = 4$.
5. Using the formula F $= \mu F/10^6$, find the number of farads equal to 0.5 μF.

JOB 3-5

Brushup on Formulas Containing Exponents

Meaning of an exponent. Exponents provide a convenient shorthand method for writing and expressing many mathematical operations. For example, $2 \times 2 \times 2$ may be written as 2^3. The number 3 above and to the right of the 2 is called an *exponent*. The exponent says, "Write the number beneath it as many times as the exponent indicates, and then multiply." An exponent may be used with letters as well as numbers. Thus, R^3 means $R \times R \times R$. The exponent 2 is read as the word "square." The exponent 3 is read as the word "cube." When the exponent is any other number, it is read as "to the fourth power," "to the seventh power," etc. For example:

Hint for Success

The exponent tells how many times a number is used as a factor in a product.

$$
\begin{aligned}
3^2 \text{ (3 square)} &= 3 \times 3 = 9 \\
5^3 \text{ (5 cube)} &= 5 \times 5 \times 5 = 125 \\
2^4 \text{ (2 to the fourth power)} &= 2 \times 2 \times 2 \times 2 = 16 \\
10^1 \text{ (10 to the first power)} &= 10 \\
10^2 \text{ (10 square)} &= 10 \times 10 = 100 \\
10^3 \text{ (10 cube)} &= 10 \times 10 \times 10 = 1000 \\
10^6 \text{ (10 to the sixth power)} &= 10 \times 10 \times 10 \times 10 \times 10 \times 10 \\
&= 1,000,000
\end{aligned}
$$

CALCULATOR HINT

To square a number using your calculator, enter the number first and then press the x^2 key.

Example 3-17

The formula for the area of a circle is $A = \pi r^2$. Find the area of a circle if the radius r equals 5 in and $\pi = 3.14$.

Solution

Given: $A = \pi r^2$ Find: $A = ?$

$$r = 5 \text{ in}$$
$$\pi = 3.14$$

1. Write the formula. $A = \pi r^2$
2. Substitute numbers. $A = 3.14 \times 5^2$
3. Simplify the exponent. $A = 3.14 \times 25$
4. Multiply. $A = 78.5 \text{ in}^2$ *Ans.*

Self-Test 3-18

The resistance R of a copper wire is found by the formula $R = 10.4 \times L/D^2$, in which $D =$ the diameter of the wire in mils (1 mil = 0.001 in) and $L =$ the length of the wire in feet. Find the resistance of 5000 ft of wire whose diameter is 50 mil.

Solution

Given: $\qquad L = 5000$ ft $\qquad\qquad$ Find: $\underline{\quad(1)\quad} = ?$

$\qquad\qquad\quad D = \underline{\quad(2)\quad}$ mil

3. Write the formula. $\qquad\qquad R = \dfrac{10.4 \times L}{D^2}$

4. Substitute numbers. $\qquad\qquad R = \dfrac{10.4 \times ?}{(?)^2}$

5. Simplify the exponent. $\qquad\qquad R = \dfrac{10.4 \times 5000}{?}$

6. Cancel. $\qquad\qquad\qquad\quad R = 10.4 \times \underline{\quad\quad}$

7. Multiply. $\qquad\qquad\qquad R = \underline{\quad\quad} \ \Omega \qquad Ans.$

Self-Test 3-19

The formula for the power gain of a common-base transistor circuit is

$$PG = \alpha^2 \times \frac{R_L}{R_i}$$

Find the power gain of a transistor if the current gain $\alpha = 0.96$, the load resistance $R_L = 15{,}000 \ \Omega$, and the input resistance $Ri = 300 \ \Omega$.

Solution

Given: $\qquad \alpha = 0.96 \qquad\qquad$ Find: PG = ?

$\qquad\qquad\quad R_L = 15{,}000 \ \Omega$

$\qquad\qquad\quad Ri = 300 \ \Omega$

1. Write the formula. $\qquad\qquad PG = \alpha^2 \times \dfrac{R_L}{R_i}$

2. Substitute numbers. $\qquad\qquad PG = (0.96)^2 \times \dfrac{15{,}000}{?}$

3. Simplify the exponent. $\qquad\qquad PG = \underline{\quad\quad} \times \dfrac{15{,}000}{300}$

4. Simplify the fraction. $\qquad\qquad PG = 0.92 \times \underline{\quad\quad}$

5. Multiply. $\qquad\qquad\qquad PG = \underline{\quad\quad} \qquad Ans.$

Exercises

Applications

1. Using the formula $A = D^2$, find the area A, in circular mils, of a wire whose diameter $D = 60$ mil.
2. The volume of a sphere is given by the formula $V = 4.2R^3$, in which V is the volume in cubic inches and R is the radius of the sphere. Find the volume of a sphere whose radius is 4 in.
3. Using the formula $S = \frac{1}{2}gt^2$, find the distance S that a body will fall if the acceleration of gravity $g = 32$ feet per second per second (ft/s²) and the time $t = 8$ s.
4. Using the formula for the volume of a cylinder, $V = \pi R^2 H$, find the volume V if $\pi = \frac{22}{7}$, $R = 3.5$, and $H = 14$.
5. Using the formula $A = mA/10^3$, change 450 mA into amperes.
6. Using the formula for the horsepower rating of a gasoline engine, $H = (nD^2)/2.5$, find the horsepower rating H of an engine if the number of cylinders $n = 6$ and the diameter of each cylinder $D = 3$ in.
7. Using the formula $V = (L \times n)/10^8$, find the value of V if $L = 40,000$ and $n = 80$.
8. Using the formula $P = I^2R$, find the power used by an electric appliance if the current $I = 6$ A and the resistance $R = 20\ \Omega$.
9. Another formula for electric power is $P = V^2/R$. Find the power consumed by a 100-Ω electric iron when it is operating on a 115-V line.

(See CD-ROM for Test 3-4)

JOB 3-6

Powers of 10

In many problems in electricity and electronics, the usual units of amperes, volts, and ohms are either too large or too small. It has been found more convenient to use new units of measurement. These new units are formed by placing a special word or prefix in front of the unit. These range from exa E (1 billion billion) to atto a (1 billionth of a billionth). As we learned in Job 3-5, these numbers may be written as powers of 10.

$$10 = 10^1 = 10 \text{ to the } first \text{ power}$$
$$100 = 10^2 = 10 \text{ to the } second \text{ power}$$
$$1000 = 10^3 = 10 \text{ to the } third \text{ power}$$
$$10,000 = 10^4 = 10 \text{ to the } fourth \text{ power}$$
$$100,000 = 10^5 = 10 \text{ to the } fifth \text{ power}$$
$$1,000,000 = 10^6 = 10 \text{ to the } sixth \text{ power}$$

Hint for Success

Powers of 10 notation provides a way to express any number as a decimal number times a power of 10. This enables us to work conveniently with both very small and very large numbers.

The use of these units will clearly involve multiplying and dividing by numbers of this size and notation. Our work will be considerably simplified if we learn to apply the following rules.

Multiplying by Powers of 10

 RULE   To multiply values by numbers expressed as 10 raised to some power, move the decimal point to the right as many places as the exponent indicates.

Example 3-20

$$0.345 \times 10^2 = \odot 34.5 \text{ or } 34.5 \text{ (move point two places right)}$$
$$0.345 \times 10^3 = \odot 345. \text{ or } 345 \text{ (move point three places right)}$$
$$0.345 \times 10^6 = \odot 345000. \text{ or } 345,000 \text{ (move point six places right)}$$
$$0.0065 \times 10^3 = \odot 006.5 \text{ or } 6.5 \text{ (move point three places right)}$$

Exercises

Practice

Multiply.

1. 0.0072×1000
2. 45.76×100
3. 3.09×1000
4. 0.0045×100
5. 37×100
6. 0.08×10^2
7. 0.0006×10^3
8. $0.000\ 56 \times 10^6$
9. 27×10^2
10. 15×10^3
11. $0.000\ 78 \times 10^2$
12. 7.8×10^6
13. 15.4×10^3
14. $3 \times 15 \times 10^2$
15. $0.005 \times 2 \times 10^2$
16. 6×10^4
17. 0.008×10^4
18. 0.009×10^5
19. 2.34×10^5
20. $0.000\ 000\ 5 \times 10^8$

Dividing by Powers of 10

 RULE   To divide values by numbers expressed as 10 raised to some power, move the decimal point to the left as many places as the exponent indicates.

Example 3-21

$$467.9 \div 10^2 = 4.67\odot 9 = 4.679 \text{ (move point two places left)}$$
$$59 \div 10^2 = .59\odot = 0.59 \text{ (move point two places left)}$$
$$8.5 \div 10^3 = .008\odot 5 = 0.0085 \text{ (move point three places left)}$$
$$5{,}500{,}000 \div 10^8 = .05500000\odot = 0.055 \text{ (move point eight places left)}$$

Exercises

Practice

Divide.

1. $6500 \div 100$
2. $7500 \div 10^3$
3. $880,000 \div 1000$
4. $32 \div 10^2$
5. $6 \div 10^2$
6. $17.8 \div 10$
7. $835 \div 10^3$
8. $550 \div 10^6$
9. $653.8 \div 10^3$
10. $100,000,000 \div 10^6$
11. $0.45 \div 10^2$
12. $0.08 \div 10$
13. $8.5 \div 10^3$
14. $7 \div 10^6$

15. $\dfrac{28.6}{10^3}$

16. $\dfrac{2 \times 1000}{10^3}$

17. $0.02 \div 10^3$

18. $\dfrac{180}{2 \times 10^2}$

Negative Powers of 10

If x^5 means $x \cdot x \cdot x \cdot x \cdot x$, and x^3 means $x \cdot x \cdot x$, then

$$\frac{x^5}{x^3} = \frac{\overset{1}{\cancel{x}} \cdot \overset{1}{\cancel{x}} \cdot \overset{1}{\cancel{x}} \cdot x \cdot x}{\underset{1}{\cancel{x}} \cdot \underset{1}{\cancel{x}} \cdot \underset{1}{\cancel{x}}} = x^2 \text{ or } x^{(5-3)}$$

In general,

$$\frac{a^m}{a^n} = a^{(m-n)} \qquad \text{when } a \neq 0$$

It follows, then, that

$$\frac{a^2}{a^5} = \frac{\overset{1}{\cancel{a}} \cdot \overset{1}{\cancel{a}}}{\underset{1}{\cancel{a}} \cdot \underset{1}{\cancel{a}} \cdot a \cdot a \cdot a} = \frac{1}{a^3}$$

or $$\frac{a^2}{a^5} = a^{(2-5)} = a^{-3}$$

Thus, $$a^{-3} = \frac{1}{a^3}$$

or $$a^{-n} = \frac{1}{a^n}$$

or $$10^{-n} = \frac{1}{10^n} \qquad \text{and} \qquad \frac{1}{10^n} = 10^{-n}$$

Thus, $$50 \times \frac{1}{10^2} \qquad \text{is the same as} \qquad 50 \times 10^{-2}$$

or $$\frac{50}{10^2} \qquad \text{is the same as} \qquad 50 \times 10^{-2}$$

Therefore, if dividing by 10^2 means to move the decimal point to the left for two places, then the $-$ sign in the exponent -2 must mean to do the same thing.

Multiplying by Negative Powers of 10

RULE To multiply values by numbers expressed as 10 raised to some negative power, move the decimal point to the *left* as many places as the exponent indicates.

Example 3-22

$$50 \times 10^{-2} = .50\text{⊙} \text{ or } 0.5 \text{ (move point two places left)}$$

$$64.9 \times 10^{-2} = .64\text{⊙}9 \text{ or } 0.649 \text{ (move point two places left)}$$

$$50,000 \times 10^{-6} = .050000\text{⊙} \text{ or } 0.05 \text{ (move point six places left)}$$

$$0.2 \times 10^{-3} = .000\text{⊙}2 \text{ or } 0.0002 \text{ (move point three places left)}$$

Exercises

Practice

Multiply.

1. $25,000 \times 10^{-3}$
2. 250×10^{-2}
3. 1.5×10^{-1}
4. 0.5×10^{-2}
5. 6250×10^{-3}
6. $100,000 \times 10^{-5}$
7. $250,000 \times 10^{-8}$
8. 6×10^{-3}
9. 75.4×10^{-4}
10. 16.5×10^{-2}

Dividing by Negative Powers of 10

Since by definition,

$$\frac{1}{10^n} = 1 \times 10^{-n}$$

and

$$1 \times 10^{-n} = \frac{1}{10^n}$$

we can transfer any power of 10 from numerator to denominator, or vice versa, by simply changing the sign of the exponent.

Example 3-23

$$\frac{15}{10^{-2}} = 15 \times 10^2 = 1500$$

$$\frac{0.05}{10^{-3}} = 0.05 \times 10^3 = 50$$

$$\frac{15,000}{10^3} = 15,000 \times 10^{-3} = 15$$

$$\frac{5 \times 0.2}{10^{-4}} = 1.0 \times 10^4 = 10,000$$

Practice

Divide.

1. $\dfrac{19.2}{10^{-2}}$ 2. $\dfrac{25.6}{10^2}$

3. $\dfrac{0.85}{10^{-3}}$ 4. $\dfrac{85}{10^3}$

5. $\dfrac{0.0072}{10^{-6}}$ 6. $\dfrac{0.9}{10^{-4}}$

7. $\dfrac{0.0045}{10^2}$ 8. $\dfrac{0.96}{10^{-3}}$

9. $\dfrac{880,000}{10^{-2}}$ 10. $\dfrac{0.005}{10^{-2} \times 10^{-3}}$

11. $\dfrac{6 \times 5}{10^{-2}}$ 12. $\dfrac{0.5 \times 0.04}{10^{-3}}$

13. $\dfrac{30 \times 10^3}{10^{-2}}$ 14. $\dfrac{50 \times 10^2}{10^{-3}}$

15. $\dfrac{60 \times 10^{-2}}{10^3}$ 16. $\dfrac{64.9 \times 10^3}{10^5}$

Expressing Numbers as Powers of 10

Multiplying and dividing numbers by powers of 10 is just a set of simple mental problems. It certainly will be to our advantage, then, if we can express the numbers in any problem as powers of 10 before we multiply or divide.

Expressing Numbers Greater than 1 as a Small Number Times a Power of 10

RULE To express a large number as a smaller number times a power of 10, move the decimal point to the *left* as many places as desired. Then multiply the number obtained by 10 to a power which is equal to the number of places moved.

Example 3-24

1. $3000 = 3.000\odot$ (moved three places left)
   $3000 = 3 \times 10^3$ ←
2. $4500 = 4.500\odot$ (moved three places left)
   $4500 = 4.5 \times 10^3$ ←

3. $770,000 = 77.0000 \odot$ (moved four places left)

   $770,000 = 77 \times 10^4$

4. $800,000 = 8.00000 \odot$ (moved five places left)

   $800,000 = 8 \times 10^5$

5. Express the following number to three significant figures and express it as a number between 1 and 10 times the proper power of 10.

   $7831 = 7830$ (since only three significant figures are wanted)

   $7830 = 7.830 \odot$ (moved three places left)

   $7830 = 7.83 \times 10^3$

6. Express 62,495 using three significant figures as a number between 1 and 10 times the proper power of 10.

   $62,495 = 62,500$ (written as three significant figures)

   $62,500 = 6.2500 \odot$ (moved four places left)

   $62,500 = 6.25 \times 10^4$

Exercises

Express the following numbers to three significant figures and write them as numbers between 1 and 10 times the proper power of 10.

1. 6000	2. 5700	3. 150,000
4. 500,000	5. 235,000	6. 7,350,000
7. 4960	8. 62,500	9. 980
10. 175	11. 12.5	12. 7303
13. 48.2	14. 12,600	15. 880,000,000
16. 54,009	17. 38,270	18. 8019.7
19. 1,754,300	20. 2,395,000	21. 482,715

Expressing Numbers Less than 1 as a Whole Number Times a Power of 10

RULE To express a decimal as a whole number times a power of 10, move the decimal point to the right as many places as described. Then multiply the number obtained by 10 to a *negative* power that is equal to the number of places moved.

Example 3-25

1. $0.005 = 0 \odot 005.$ (moved three places right)

   $0.005 = 5 \times 10^{-3}$

2. $0.006\ 72 = 0 \odot 006.72$ (moved three places right)

   $0.006\ 72 = 6.72 \times 10^{-3}$

3. $0.0758 = 0\curvearrowleft07.58$ (moved two places right)

$\quad 0.0758 = 7.58 \times 10^{-2} \leftarrow$

4. $0.000\,008\,9 = 0\curvearrowleft000008.9$ (moved six places right)

$\quad 0.000\,008\,9 = 8.9 \times 10^{-6} \leftarrow$

5. Express the number 0.000 357 8 to three significant figures and then write it as a number between 1 and 10 times the proper power of 10.

$\quad 0.000\,357\,8 = 0.000\,358$ (written as three significant figures)

$\quad 0.000\,358 = 0\curvearrowleft0003.58$ (moved four places right)

$\quad 0.000\,358 = 3.58 \times 10^{-4} \leftarrow$

Exercises

Practice

Express the following numbers to three significant figures and write them as numbers between 1 and 10 times the proper power of 10.

1. 0.006	2. 0.0075
3. 0.0035	4. 0.08
5. 0.456	6. 0.0357
7. 785×10^{-2}	8. 0.000 000 12
9. 0.0965	10. 0.004 82
11. 0.5	12. 0.000 374 3
13. 0.008 147	14. 0.000 007 949
15. $0.000\,725 \times 10^{5}$	16. 0.013 33
17. 0.0006×10^{3}	18. 3200×10^{-5}
19. 360×10^{-4}	20. $0.000\,008 \times 10^{4}$

Multiplying with Powers of 10

If x^3 means $x \cdot x \cdot x$, and x^2 means $x \cdot x$, then $x^3 \cdot x^2$ means $x \cdot x \cdot x \cdot x \cdot x = x^5$ or, $x^3 \cdot x^2 = x^{(3+2)} = x^5$, which gives us the following rule.

 RULE The multiplication of two or more powers using the *same* base is equal to that base raised to the sum of the powers.

CALCULATOR HINT

Scientific calculators allow you to enter powers of 10 for arithmetic operations.

Example 3-26

1. $a^4 \times a^5 = a^{(4+5)} = a^9$
2. $10^2 \times 10^3 = 10^5$
3. Multiply $10,000 \times 1000$.
   If $10,000 = 10^4$ and $1000 = 10^3$, then

$$10,000 \times 1000 = 10^4 \times 10^3 = 10^{(4+3)} = 10^7 \quad \textit{Ans.}$$

4. Multiply $25,000 \times 4000$.
   If $25,000 = 25 \times 10^3$ and $4000 = 4 \times 10^3$, then

$$\begin{aligned}
25,000 \times 4000 &= 25 \times 10^3 \times 4 \times 10^3 \\
&= 25 \times 4 \times 10^3 \times 10^3 \\
&= 100 \times 10^6 \\
&= 10^2 \times 10^6 = 10^8 \quad \textit{Ans.}
\end{aligned}$$

5. Multiply $0.000\,05 \times 0.003$.
   If $0.000\,05 = 5 \times 10^{-5}$ and $0.003 = 3 \times 10^{-3}$, then

$$\begin{aligned}
0.000\,05 \times 0.003 &= 5 \times 10^{-5} \times 3 \times 10^{-3} \\
&= 5 \times 3 \times 10^{-5} \times 10^{-3} \\
&= 15 \times 10^{[-5 + (-3)]} \\
&= 15 \times 10^{-8} \quad \textit{Ans.}
\end{aligned}$$

6. Multiply $7000 \times 0.000\,91$.
   If $7000 = 7 \times 10^3$ and $0.000\,91 = 9.1 \times 10^{-4}$, then

$$\begin{aligned}
7000 \times 0.000\,91 &= 7 \times 10^3 \times 9.1 \times 10^{-4} \\
&= 7 \times 9.1 \times 10^3 \times 10^{-4} \\
&= 63.7 \times 10^{[3 + (-4)]} \\
&= 63.7 \times 10^{-1} \\
&= 6.37 \quad \textit{Ans.}
\end{aligned}$$

7. Multiply $0.000\,05 \times 20,000 \times 1500$.
   If $0.000\,05 = 5 \times 10^{-5}$, and $20,000 = 2 \times 10^4$, and $1500 = 1.5 \times 10^3$, then

$$\begin{aligned}
0.000\,05 \times 20,000 \times 1500 &= 5 \times 10^{-5} \times 2 \times 10^4 \times 1.5 \times 10^3 \\
&= 5 \times 2 \times 1.5 \times 10^{-5} \times 10^4 \times 10^3 \\
&= 15 \times 10^2 \\
&= 1500 \quad \textit{Ans.}
\end{aligned}$$

Exercises

Practice

Multiply the following numbers.

1. 5000×0.001
2. 850×2000
3. $16 \times 10^2 \times 4 \times 10^3$
4. $0.0004 \times 5 \times 10^2$
5. $250 \times 4000 \times 3 \times 10^{-2}$
6. $1000 \times 10^{-4} \times 0.02$
7. $3 \times 10^{-5} \times 4 \times 10^6$
8. $15 \times 10^{-4} \times 20,000 \times 0.04$
9. $200,000 \times 0.000\,005 \times 3 \times 10^{-2}$
10. $0.004 \times 0.0005 \times 5000$
11. $0.005 \times 5 \times 10^{-3} \times 0.02$
12. $6,000,000 \times 0.000\,25 \times 0.3 \times 10^{-2}$
13. $0.3 \times 10^{-2} \times 800,000 \times 400 \times 10^{-3}$
14. $(500)^2 \times 0.0002 \times 4000$
15. $500,000,000 \times 0.000\,004 \times 3.14 \times 10^2$

16. As we shall see in Job 16-2, the inductive reactance of a coil is given by the formula

$$X_L = 6.28\,fL$$

where f = frequency, Hz
   L = inductance of the circuit, H
   X_L = reactance, Ω

Find the inductive reactance when:

a. $f = 60$ Hz and $L = 0.025$ H
b. $f = 1{,}000{,}000$ Hz and $L = 0.25$ H
c. $f = 10{,}000$ Hz and $L = 0.000\ 025$ H

Division with Powers of 10

As noted in the section on Dividing by Negative Powers of 10 of this job, we can transfer any power of 10 from numerator to denominator, or vice versa, by simply changing the sign of the exponent. This will permit us to change *all* division

Example 3-27

1. $10^6 \div 10^2 = \dfrac{10^6}{10^2} = 10^6 \times 10^{-2} = 10^4$ *Ans.*

2. $\dfrac{4000}{10^2} = 4 \times 10^3 \times 10^{-2} = 4 \times 10^1 = 40$ *Ans.*

3. $\dfrac{35{,}000}{0.005} = \dfrac{35 \times 10^3}{5 \times 10^{-3}} = 7 \times 10^3 \times 10^3$
$$= 7 \times 10^6 \quad Ans.$$

4. $\dfrac{144{,}000}{12 \times 10^3} = \dfrac{\overset{1}{144 \times 10^3}}{\underset{1}{12 \times 10^3}} = 12$ *Ans.*

Note that *any* factor divided by itself is equal to 1, and it is not necessary to transfer any powers. That is, $10^3/10^3 = 10^{(3-3)} = 10^0 = 1$.

5. $\dfrac{0.000\ 75}{500} = \dfrac{75 \times 10^{-5}}{5 \times 10^2} = 15 \times 10^{-5} \times 10^{-2}$
$$= 15 \times 10^{-7} \quad Ans.$$

6. $\dfrac{60}{0.0003 \times 40{,}000} = \dfrac{60}{3 \times 10^{-4} \times 4 \times 10^4} = \dfrac{5}{10^0}$

Now, since $10^0 = 1$, $\dfrac{5}{10^0} = \dfrac{5}{1} = 5$ *Ans.*

Practice

Divide the following numbers.

1. $\dfrac{10^8}{10^3}$

2. $\dfrac{10^3}{10^5}$

3. $\dfrac{60,000}{5 \times 10^2}$

4. $\dfrac{50,000}{0.05}$

5. $\dfrac{10}{50,000}$

6. $\dfrac{20}{0.0005}$

7. $\dfrac{0.0001}{500}$

8. $\dfrac{20}{4000 \times 0.005}$

9. $\dfrac{1000 \times 0.008}{0.002 \times 500}$

10. $\dfrac{150,000}{3 \times 10^5}$

11. $\dfrac{0.000\ 15}{3 \times 10^{-2}}$

12. $\dfrac{1}{4 \times 100,000 \times 0.000\ 05}$

13. As we shall see in Job 17-4, the capacitive reactance of a capacitor is given by the formula

$$X_C = \frac{1}{2\pi f C}$$

where f = frequency, Hz
C = capacitance, F
π = 3.14
X_C = reactance, Ω

Find the capacitive reactance when:

a. $f = 60$ Hz and $C = 0.000\ 05$ F
b. $f = 1000$ Hz and $C = 0.000\ 002\ 5$ F
c. $f = 1,000,000$ Hz and $C = 0.000\ 000\ 05$ F

(See CD-ROM for Test 3-5)

JOB 3-7

Units of Measurement in Electronics

Ohm's law and other electrical formulas use the simple electrical units of volts, amperes, and ohms. However, if the measurements given or obtained in a problem were stated in kilovolts, milliamperes, or megohms, it would be necessary to change these units of measurement to the units required by the formula (Tables 3-1 to 3-3).

Table 3-1 Metric Prefixes

Prefix	Symbol	Multiplier	Prefix	Symbol	Multiplier
Exa	E	10^{18}	Milli	m	10^{-3}
Peta	P	10^{15}	Micro	μ	10^{-6}
Tera	T	10^{12}	Nano	n	10^{-9}
Giga	G	10^{9}	Pico	p	10^{-12}
Mega	M	10^{6}	Femto	f	10^{-15}
Kilo	k	10^{3}	Atto	a	10^{-18}

Changing Units of Measurement

Table 3-2 Changing Large Units to Small Units

Multiply	By	To Obtain
Tera units	10^{12}	Units
Tera units	10^{9}	Kilo units
Tera units	10^{6}	Mega units
Tera units	10^{3}	Giga units
Giga units	10^{9}	Units
Giga units	10^{6}	Kilo units
Giga units	10^{3}	Mega units
Mega units	10^{6}	Units
Mega units	10^{3}	Kilo units
Kilo units	10^{3}	Units
Units	10^{3}	Milli units
Units	10^{6}	Micro units
Units	10^{9}	Nano units
Units	10^{12}	Pico units
Milli units	10^{3}	Micro units
Milli units	10^{6}	Nano units
Milli units	10^{9}	Pico units
Micro units	10^{3}	Nano units
Micro units	10^{6}	Pico units
Nano units	10^{3}	Pico units

Hint for Success

Electronic technicians and engineers always use metric prefixes when specifying values of measurements. For example, a technician would be much more likely to refer to a 1000-Ω resistor as a 1-kΩ resistor. In a similar way, a current of 0.000001 A would normally be stated as 1 μA.

CALCULATOR
HINT

Some calculator displays show the metric prefixes as part of the displayed answer.

Example 3-28

Change:

1. 2 GV to volts: $\quad$ 2 GV $= 2 \times 10^9 = 2{,}000{,}000{,}000$ V
2. 0.25 MΩ to ohms: $\quad$ 0.25 MΩ $= 0.25 \times 10^6 = 250{,}000$ Ω
3. 0.3 kV to volts: $\quad$ 0.3 kV $= 0.3 \times 10^3 = 300$ V
4. 880 kHz to hertz: $\quad$ 880 kHz $= 880 \times 10^3 = 880{,}000$ Hz
5. 2.4 V to millivolts: $\quad$ 2.4 V $= 2.4 \times 10^3 = 2400$ mV
6. 0.0004 A to microamperes: $\quad$ 0.0004 A $= 0.0004 \times 10^6 = 400\ \mu$A
7. 0.000 05 F to microfarads: $\quad$ 0.000 05 F $= 0.000\ 05 \times 10^6 = 50\ \mu$F
8. 0.000 000 005 F to picofarads: $\quad$ 0.000 000 005 F $=$
   $\qquad$ $0.000\ 000\ 005 \times 10^{12} = 5000$ pF
9. 20 mA to microamperes: $\quad$ 20 mA $= 20 \times 10^3 = 20{,}000\ \mu$A
10. 0.000 35 μF to picofarads: $\quad$ 0.000 35 μF $= 0.000\ 35 \times 10^6 = 350$ pF
11. 0.007 μF to nanofarads: $\quad$ 0.007 μF $= 0.007 \times 10^3 = 7$ nF
12. 0.000 005 TV to megavolts: $\quad$ 0.000 005 TV $=$
    $\qquad$ $0.000\ 005 \times 10^6 = 5$ MV

Table 3-3 Changing Small Units to Large Units

Multiply	By	To Obtain
Pico units	10^{-3}	Nano units
Pico units	10^{-6}	Micro units
Pico units	10^{-9}	Milli units
Pico units	10^{-12}	Units
Nano units	10^{-3}	Micro units
Nano units	10^{-6}	Milli units
Nano units	10^{-9}	Units
Micro units	10^{-3}	Milli units
Micro units	10^{-6}	Units
Milli units	10^{-3}	Units
Units	10^{-3}	Kilo units
Units	10^{-6}	Mega units
Units	10^{-9}	Giga units
Units	10^{-12}	Tera units
Kilo units	10^{-3}	Mega units
Kilo units	10^{-6}	Giga units
Kilo units	10^{-9}	Tera units
Mega units	10^{-3}	Giga units
Mega units	10^{-6}	Tera units
Giga units	10^{-3}	Tera units

Example 3-29

Change:
1. 500,000 Ω to megohms: 500,000 Ω = 500,000 $\times$ 10^{-6} = 0.5 MΩ
2. 660 kHz to megahertz: 660 kHz = 660 $\times$ 10^{-3} = 0.66 MHz
3. 600 V to kilovolts: 600 V = 600 $\times$ 10^{-3} = 0.6 kV
4. 14.5 mA to amperes: 14.5 mA = 14.5 $\times$ 10^{-3} = 0.0145 A
5. 2.5 μF to farads: 2.5 μF = 2.5 $\times$ 10^{-6} = 0.000 002 5 F
6. 30,000,000 pF to farads: 30,000,000 pF =
$$30,000,000 \times 10^{-12} = 0.000\ 03\ F$$
7. 400 μV to millivolts: 400 μV = 400 $\times$ 10^{-3} = 0.4 mV
8. 350 pF to microfarads: 350 pF = 350 $\times$ 10^{-6} = 0.000 35 μF
9. 4000 W to kilowatts: 4000 W = 4000 $\times$ 10^{-3} = 4 kW
10. 1,010,000 Hz to kilohertz: 1,010,000 Hz =
$$1,010,000 \times 10^{-3} = 1010\ kHz$$
11. 356 mV to volts: 356 mV = 356 $\times$ 10^{-3} = 0.356 V
12. 15,000 μV to volts: 15,000 μV = 15,000 $\times$ 10^{-6} = 0.015 V
13. 0.005 TV to kilovolts: 0.005 TV = 0.005 $\times$ 10^{9} = 5 $\times$ 10^{6} kV
14. 100,000 kV to gigavolts: 100,000 kV = 100,000 $\times$ 10^{-6} = 0.1 GV
15. 500 nF to microfarads: 500 nF = 500 $\times$ 10^{-3} = 0.5 μF

Tables 3-2 and 3-3 are combined into Table 3-4. To use the table:

1. Locate the original unit in the left-hand column.
2. Read to the right until you get to the column headed by the unit desired.
3. The number and arrow at this point indicate the number of places and the direction in which the decimal point is to be moved.

Table 3-4   Metric Conversion Table

Original Unit	Desired Unit								
	Tera	Giga	Mega	Kilo	Units	Milli	Micro	Nano	Pico
Tera		3→	6→	9→	12→	15→	18→	21→	24→
Giga	← 3		3→	6→	9→	12→	15→	18→	21→
Mega	← 6	← 3		3→	6→	9→	12→	15→	18→
Kilo	← 9	← 6	← 3		3→	6→	9→	12→	15→
Units	←12	← 9	← 6	← 3		3→	6→	9→	12→
Milli	←15	←12	← 9	← 6	← 3		3→	6→	9→
Micro	←18	←15	←12	← 9	← 6	← 3		3→	6→
Nano	←21	←18	←15	←12	← 9	← 6	← 3		3→
Pico	←24	←21	←18	←15	←12	← 9	← 6	← 3	

Check Examples 3-28 and 3-29 using Table 3-4.

Practice

Change the following units of measurement.

1. 225 mA to amperes
2. 0.076 V to millivolts
3. 3.5 MΩ to ohms
4. 5 kW to watts
5. 550 kHz to hertz
6. 700,000 Hz to kilohertz
7. 70,000 Ω to megohms
8. 0.000 08 F to microfarads
9. 0.065 A to milliamperes
10. 6500 W to kilowatts
11. 75 mV to volts
12. 2.3 MHz to hertz
13. 6000 μA to amperes
14. 0.007 F to microfarads
15. 3.9 mA to amperes
16. 75,000 W to kilowatts
17. 0.005 μF to picofarads
18. ¼ A to milliamperes
19. 1000 kHz to hertz
20. 0.5 MΩ to ohms
21. 0.008 V to millivolts
22. 0.0045 W to milliwatts
23. 0.000 06 μF to picofarads
24. 0.15 A to milliamperes
25. 0.15 μF to farads
26. 125 mV to volts
27. 8000 W to kilowatts
28. 4.16 kW to watts
29. 0.000 004 a to microamperes
30. 0.6 MHz to hertz
31. 0.5 THz to megahertz
32. 0.000 05 nF to picofarads
33. 15 MV to gigavolts
34. 0.000 25 μF to nanofarads

(See CD-ROM for Test 3-6)

JOB 3-8

Using Electronic Units of Measurement in Simple Circuits

All the formulas for Ohm's law, series circuits, parallel circuits, and power demand that the measurements be given in units of amperes, volts, and ohms only. If a certain problem gives the measurements in units other than these, we must change all the measurements to amperes, volts, and ohms before we can use any of these formulas.

Example 3-30

Find the voltage that will force 28.6 μA of current through a 70-kΩ resistor in the base circuit of a transistor.

Solution

Given: $I = 28.6 \ \mu A$ Find: $V = ?$
                        $R = 70 \ k\Omega$

1. Change μA to A. $28.6 \ \mu A = 28.6 \times 10^{-6} \ A$
2. Change 70 kΩ to Ω. $70 \ k\Omega = 70 \times 10^3 \ \Omega$
3. Find the voltage V. $V = I \times R$ (2-1)
                             $V = 28.6 \times 10^{-6} \times 70 \times 10^3$
                             $V = 28.6 \times 7 \times 10^{-2}$
                             $V = 200.2 \times 10^{-2} = 2 \ V$ *Ans.*

Self-Test 3-31

A bias voltage of 300 mV is developed across a 2-MΩ resistor. Find the current flowing using the formula $I = V/R$.

Solution

Given: $V = 300$ mV   Find: $I = ?$
          $R = \underline{\quad(1)\quad}$

2. Change 300 mV to volts. 300 mV $= 300 \times \underline{\hspace{2cm}}$ V
3. Change 2 mΩ to ohms. 2 M$\Omega = 2 \times \underline{\hspace{2cm}}$ Ω

4. Find the current I. $I = \dfrac{V}{R}$

5. $= \dfrac{300 \times 10^{-3}}{2 \times 10^{6}}$

                              $= 150 \times 10^{-?}$ A

6. Change amperes to microamperes.

$$150 \times 10^{-9} \text{ A} = 150 \times 10^{-9} \times 10^{6} \ \mu\text{A}$$
$$= 150 \times 10^{-?} \ \mu\text{A}$$
$$= \underline{\hspace{2cm}} \ \mu\text{A}$$

Exercises

Applications

1. If 2 μA of current flow in an antenna whose resistance is 50 Ω, find the voltage drop in the antenna.
2. Find the voltage drop across the 5-kΩ load of a transistor if the collector current is 2 mA.
3. Using the formula Watts $= VI$, find the number of kilowatts used by a 24-Ω device drawing 5 A from a 120-V source.
4. Using the formula $I = V/R$, find how many milliamperes of current will flow through a 100-Ω resistor if the voltage across its ends is 20 mV.
5. Using the formula $I = V/R$, find the number of microamperes flowing through a 2-MΩ grid leak if the voltage drop across it is 1000 mV.
6. Using the formula for total current in a parallel circuit, $I_T = I_1 + I_2 + I_3$, find I_T if $I_1 = 40$ mA, $I_2 = 6000$ μA, and $I_3 = 0.013$ A.
7. Using the formula for total resistance in series, $R_T = R_1 + R_2 + R_3$, find the total resistance R_T if $R_1 = 0.2$ MΩ, $R_2 = 5$ kΩ, and $R_3 = 10,000$ Ω.
8. Using the formula $R = V/I$, find the resistance R of a coil if an emf of 200 μV sends 10 mA of current through it.
9. Using the formula for capacitances in parallel, $C_T = C_1 + C_2 + C_3$, find the total capacity C_T of a 0.0025-μF and a 125-pF capacitor in parallel.
10. Using the formula $R_T = (R_1 \times R_2)/(R_1 + R_2)$, find the total resistance R_T of a 1000-Ω and a 4-kΩ resistor in parallel.
11. The time constant of an RC circuit is equal to the product of the resistance (ohms) and the capacitance (farads). Find the time constant of a circuit if $R = 10$ kΩ and $C = 0.004$ μF.

12. The radiation resistance of a shortwave antenna is 100 Ω. Using the formula Watts = I^2R, find the number of watts radiated if the transmitter delivers 900 mA to the antenna.

13. A 470-kΩ resistor in the base circuit of a 2N2924 phase-shift oscillator circuit carries a current of 30 μA. Find the voltage drop across the resistor.

14. A photoelectric-cell circuit contains a resistance of 0.12 MΩ and carries a current of 50 μA. Find the voltage drop in the resistor.

(See CD-ROM for Test 3-7)

JOB 3-9

Solving the Ohm's Law Formula for Current or Resistance

Formulas are equations. As you may have noticed, every formula contains the sign of equality. The statement that a combination of quantities is *equal* to another combination of quantities is called an *equation*. In this sense, every formula is an equation. Examples of some equations are

$$3 \times 4 = 12$$
$$2 \times I = 10$$
$$12 = 3 \times R$$
$$V = IR$$

Thus for a statement to be termed an equation, the value on the left side of the equality sign must be *truly equal* to the value on the right side.

Working with equations. Many mathematical operations may be performed on an equation. However, the basic equality of the statements on each side of the equality sign must be maintained if the equation is to remain an equation. This is accomplished by applying the following basic principle.

BASIC PRINCIPLE

Any mathematical operation performed on one side of an equality sign must also be performed on the other side.

For example, let us subject the equation $3 \times 4 = 12$ to different mathematical operations.

RULE 1
The same number may be added to both sides of an equality sign without destroying the equality.

1. Write the equation.	$3 \times 4 = 12$
2. Add 3 to both sides.	$(3 \times 4) + 3 = 12 + 3$
3. Do the arithmetic.	$12 + 3 = 12 + 3$
or	$15 = 15$

which is a true equation, since the left side is still equal to the right side.

RULE 2

The same number may be subtracted from both sides of an equality sign without destroying the equality.

1. Write the equation.	$3 \times 4 = 12$
2. Subtract 3 from both sides.	$(3 \times 4) - 3 = 12 - 3$
3. Do the arithmetic.	$12 - 3 = 12 - 3$
or	$9 = 9$

which is a true equation, since the left side is still equal to the right side.

RULE 3

Both sides of an equality sign may be multiplied by the same number without destroying the equality.

1. Write the equation.	$3 \times 4 = 12$
2. Multiply both sides by 3.	$(3 \times 4) \times 3 = 12 \times 3$
3. Do the arithmetic.	$12 \times 3 = 12 \times 3$
or	$36 = 36$

which is a true equation, since the left side is still equal to the right side.

RULE 4

Both sides of an equality sign may be divided by the same number without destroying the equality.

1. Write the equation.	$3 \times 4 = 12$
2. Divide both sides by 3.	$\dfrac{(3 \times 4)}{3} = \dfrac{12}{3}$
3. Do the arithmetic.	$\dfrac{12}{3} = \dfrac{12}{3}$
or	$4 = 4$

which is a true equation, since the left side is still equal to the right side.

These examples show that if the same mathematical operation (other than division by zero) is performed on *both* sides of an equation, it does not destroy the basic equality of the equation.

Solving equations.   Consider the equation $2R = 10$. To solve this equation means to find the value of the unknown letter R in the equation. This value is found *when the letter stands all alone on one side of the equality sign.* When this occurs, the equation has the form

$$R = \text{some number}$$

This number will obviously be the value of the letter R, and the equation will be solved.

How do we get the letter all alone?   In the equation $2R = 10$, the letter will be alone on the left side of the equality sign if we can somehow eliminate the number 2 on that side. The number 2 will actually be eliminated if we can change it to a 1, since $1R$ means the same as R. This can be done by applying the basic principle that permits us to perform the same operation on both sides of the equality sign. But which mathematical operation will eliminate the number 2? In general, the operation to be performed on both sides of the equality sign will be the *opposite* of that used in the equation. Since $2R$ means 2 *multiplied* by R, we shall use rule 4 above, and *divide both sides* of the equation by that same number 2.

1. Write the equation. $2R = 10$

2. Divide both sides by 2 $\dfrac{2R}{2} = \dfrac{10}{2}$

3. Simplify each side separately. $\dfrac{\overset{1}{\cancel{2}}R}{\underset{1}{\cancel{2}}} = \dfrac{\overset{5}{\cancel{10}}}{\underset{1}{\cancel{2}}}$

or $1R = 5$
or $R\ = 5$

Only those numbers that appear on the same side of the equality sign may be canceled. Do *not* cancel across the equality sign.

Suppose the unknown letter appears on the right side of the equality sign as in the equation $12 = 3R$. In this situation, we proceed exactly as before. To solve the equation means to get the letter *all alone* on one side of the equality sign. It is not important which side is chosen, as long as the letter is *all alone* on that side. We can get the letter R all alone on the right side by changing the $3R$ to $1R$ by applying the *opposite* operation of *dividing both sides* of the equation by the number 3.

1. Write the equation. $12 = 3R$

2. Divide both sides by 3. $\dfrac{12}{3} = \dfrac{3R}{3}$

3. Simplify each side separately. $\dfrac{\overset{4}{\cancel{12}}}{\underset{1}{\cancel{3}}} = \dfrac{\overset{1}{\cancel{3}}R}{\underset{1}{\cancel{3}}}$

or $4 = 1R$ or $4 = R$ or $R = 4$ *Ans.*

Notice that the number that is multiplied by the unknown letter will be canceled out *only* if we divide both sides of the equality sign *by that same number*. Dividing both sides by any other number will *not* eliminate this number.

RULE 5   To eliminate the number which is multiplied by the unknown letter, divide both sides of the equality sign by the multiplier of the letter.

Example 3-32

Solve the following equations for the values of the unknown letters.

$$2R = 10 \quad \text{and} \quad 14 = 7V$$

Solution

1. Write the equations.

   $$2R = 10 \qquad\qquad 14 = 7V$$

2. Divide both sides of each equation by the multiplier of the letter.

   $$\frac{\cancel{2}R}{\cancel{2}} = \frac{10}{2} \qquad\qquad \frac{14}{7} = \frac{\cancel{7}V}{\cancel{7}}$$

3. Cancel out the multiplier of the letter.

   $$R = \frac{10}{2} \qquad\qquad \frac{14}{7} = V$$

4. Divide

   $$R = 5 \quad \textit{Ans.} \qquad\qquad 2 = V \quad \textit{Ans.}$$

Notice that in each example, the effect of dividing both sides of the equality sign by the multiplier of the letter has been to *move* the multiplier *across the equality sign* into the position shown in step 3. Since this will always occur, we can eliminate step 2 and proceed as shown in Example 3-33.

Example 3-33

Solve the following equations for the values of the unknown letters:

$$3R = 12 \quad \text{and} \quad 15 = 5V$$

Solution

1. Write the equations.

   $$3R = 12 \qquad\qquad 15 = 5V$$

2. Divide the quantity all alone on one side of the equality sign by the multiplier of the letter.

   $$R = \frac{12}{3} \qquad\qquad \frac{15}{5} = V$$

3. Divide.

   $$R = 4 \quad \textit{Ans.} \qquad\qquad 3 = V \quad \textit{Ans.}$$

RULE   To solve a simple equation of the form "a number multiplied by a letter equals a number," divide the number all alone on one side of the equality sign by the multiplier of the letter.

Example 3-34

Solve the equation $18 = 0.3Z$ for the value of Z.

Solution

$$18 = 0.3Z$$

$$\frac{18}{0.3} = Z$$

$$60 = Z$$

or $$Z = 60 \qquad Ans.$$

Self-Test 3-35

In a common-emitter transistor circuit, the relationship between the base current I_B and the collector current I_C is given as $\beta I_B = I_C$. Find the values of βI_B if $\beta = 50$ and $I_C = 0.002$ A.

Solution

1. Write the formula. $\beta I_B = I_C$

2. Substitute numbers. $50 I_B = \underline{\qquad}$

3. Solve for I_B. $I_B = \dfrac{?}{?}$

4. Divide the numbers. $I_B = \underline{\qquad}$ A $= \underline{\qquad}$ mA

Exercises

Practice

Solve the following equations for the value of the unknown letter:

1. $3I = 15$	2. $5R = 20$	3. $2V = 12$
4. $48 = 8R$	5. $7I = 63$	6. $4L = 21$
7. $3R = 41$	8. $16R = 4$	9. $20I = 117$
10. $19 = 2V$	11. $\frac{1}{2}W = 20$	12. $20 = 100R$
13. $0.3R = 120$	14. $0.04Z = 60$	15. $40 = 0.2Z$
16. $8 = 0.4Z$	17. $0.15R = 120$	18. $0.003R = 78$
19. $117 = 0.3Z$	20. $\frac{3}{5}T = 12$	21. $8V = \frac{1}{2}$

Solving the Formula for Ohm's law.   The formula for Ohm's law is actually an equation. By applying rule 6, we can solve Ohm's law for any unknown value of current or resistance.

Example 3-36

The total resistance of a relay coil is 50 Ω. What current will it draw from a 20-V source?

Solution

The diagram for the circuit is shown in Fig. 3-4.

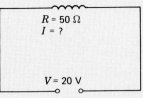

$R = 50\ \Omega$
$I = ?$

$V = 20\ V$

Figure 3-4

1. Write the formula. $V = IR$
2. Substitute numbers. $20 = I \times 50$
3. Solve for I. $\dfrac{20}{50} = I$
4. Divide the numbers. $0.4 = I$
   or $I = 0.4\ A$ *Ans.*

Self-Test 3-37

Find the total resistance of a telegraph coil if it draws 0.015 A from a 6.6-V source.

Solution

The diagram is shown in Fig. 3-5.

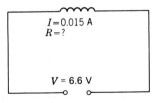

$I = 0.015\ A$
$R = ?$

$V = 6.6\ V$

Figure 3-5

1. Write the formula. $V = I \times \underline{\hspace{1.5cm}}$

2. Substitute numbers. $\underline{\hspace{1.5cm}} = 0.015 \times R$

3. Solve for R. $\dfrac{?}{?} = R$

4. Divide the numbers. $\underline{\hspace{1.5cm}} = R$

   or $R = \underline{\hspace{1.5cm}}\ \Omega$ *Ans.*

Applications

1. What is the hot resistance of an arc lamp if it draws 15 A from a 30-V line?
2. The resistance of the motor windings of an electric vacuum cleaner is 20 Ω. If the voltage is 120 V, find the current drawn.
3. An electric enameling kiln draws 9 A from a 117-V line. Find the resistance of the coils.
4. The field magnet of a loudspeaker carries 40 mA when connected to a 40-V supply. Find its resistance.
5. How much current is drawn from a 12-V battery when operating an automobile horn of 8-Ω resistance?
6. What is the hot resistance of a tungsten lamp if it draws 250 mA from a 110-V line?
7. What current would flow in a 0.3-Ω short circuit of a 6-V automobile ignition system?
8. A 2N525 transistor is used as a transistor switch to control a 20-V source across a 100-Ω load. Find the current that flows when the switch is conducting.
9. Find the resistance of an electric iron if it draws 4.8 A from a 120-V line.
10. A dry cell indicates a terminal voltage of 1.2 V when a wire of 0.2-Ω resistance is connected across it. What current flows in the wire?
11. Find the resistance of an automobile starting motor if it draws 90 A from the 12-V battery.
12. A meter registers 0.2 mA when the voltage across it is 3 V. Find the total resistance of the meter circuit.
13. What is the resistance of a telephone receiver if there is a voltage drop of 24 V across it when the current is 20 mA?
14. What current is drawn by a 5-kΩ electric clock when operated from a 110-V line?
15. Find the current drawn by a 52-Ω toaster from a 117-V line.
16. Find the resistance of an electric furnace drawing 41 A from a 230-V line.
17. The resistance of the field coils of a shunt motor is 60 Ω. What is the field current when the voltage across the coils is 220 V?
18. The resistance of a common Christmas tree lamp is about 50 Ω. What is the current through it if the voltage across the lamp is 14 V?
19. The large copper leads on switchboards are called *bus bars*. What is the resistance of a bus bar carrying 400 A if the voltage across its ends is 0.6 V?
20. If a radio receiver draws 0.85 A from a 110-V line, what is the total resistance (impedance) of the receiver?
21. A 32-candela (cd) lamp in a truck headlight draws 3.4 A from the 6-V battery. What is the resistance of the lamp?
22. If the resistance of the air gap in an automobile spark plug is 2.5 kΩ, what voltage is needed to force 0.16 A through it?
23. A voltage of 5 V appears across the 2.5-kΩ load resistor in a self-biased transistor circuit. Find the current in the resistor.
24. What current flows through an automobile headlight lamp of 1.2-Ω resistance if it is operated from the 6-V battery?
25. What is the resistance of a buzzer if it draws 0.14 A from a 3-V source?

Estimating Answers

In many applications it is important only to estimate the answer. It may only be necessary to know the order of magnitude of the answer and perhaps one digit of the answer. For example, you must know whether the voltage is 10, 100, or 1000 V in order to adjust a meter to the proper scale. Or you may be measuring resistance and need to know whether the range is $G\Omega$, $k\Omega$, or $M\Omega$.

The scientific and exponential notations that have been used in this chapter will make approximations very easy. In scientific notation, all numbers are converted to a number between 1 and 10 with the appropriate power of 10. By looking only at the powers of 10, you usually can determine the power of 10 in the answer. Since all the other numbers are between 1 and 10, the actual arithmetic is easy. This method is particularly useful for problems with which you are unfamiliar.

Example 3-38

A large launching rocket was moved 0.0301 mi in 2.2 h. Find the speed in feet per minute (ft/min). Is the speed closer to 1, 10, or 100 ft/min?

Solution

1. Change units.

$$0.0301 \text{ m} \times 5280 = \text{ft}$$
$$2.2 \text{ h} \times 60 \text{ min/h} = \text{min}$$

2. Divide feet by minutes.

$$\frac{0.0301 \times 5280}{2.2 \times 60}$$

3. Write the problem in scientific notation.

$$\frac{3.0 \times 10^{-2} \times 5.28 \times 10^{3}}{2.2 \times 6 \times 10}$$

4. Look only at the powers of 10.

$$\frac{10^{-2} \times 10^{3}}{10} = 10^{0} = 1$$

5. Now estimate the numbers.

$$\frac{3 \times 5}{2 \times 6} = \frac{15}{12} = 1.25$$

Therefore, the answer is approximately $1.25 \times 1 = 1.25$ ft/min. Since the actual answer is 1.204 ft/min, the error is only 0.046 ft/min, or about 3.8 percent.

Use the following procedure to estimate an answer.

1. Change the measurements to the proper units.
2. Write out the problem.
3. Convert the numbers to scientific notation.
4. Solve the powers of 10.
5. Estimate the numbers.

Self-Test 3-39

Find the voltage drop across the 2000-Ω load of a transistor if the collector current is 0.015 A.

Solution

Given: $R = 2000$ Find: $V = ?$
           $I = 0.105$ A

1. $V = I \times R$
2. $V = \underline{\hspace{1.5cm}} \times 2000$
3. $V = 1.5 \times \underline{\hspace{1.5cm}} \times 2 \times 10^3$
4. $V \times 1.5 \times 2 \times \underline{\hspace{1.5cm}}$
5. $V - \underline{\hspace{1.5cm}} \times 10 = 30$ V   *Ans.*

Exercises

Practice

Solve the applications in Job 3-8 by estimation. Compare the estimates with the actual answers to see whether the actual answers are reasonable.

JOB 3-11

Review of Ohm's Law

In any electric circuit,

1. The voltage forces the _____ through a conductor against its resistance.
2. The _____ tries to stop the current from flowing.
3. The current that flows in a circuit depends on the _____ and the resistance.

   The relationship among these three quantities is described by Ohm's law. Ohm's law applies to an entire circuit or to any component part of a circuit.

$$V = IR \qquad \text{Formula 2-1}$$

where V = voltage, V
        I = current, A
        R = resistance, Ω

The formula for Ohm's law may be used to find the value of any one of the quantities in the formula. It is of equal important that the student be able to determine the relative values of each quantity as one of the other quantities is changed in amount.

In Figure 3-6, if the resistance $R = 10\ \Omega$ remains unchanged,

4. When the voltage $V = 10$ V, the current $I = $ _____.
5. When the voltage $V = 20$ V, the current $I = $ _____.
6. When the voltage $V = 50$ V, the current $I = $ _____.
7. When the voltage $V = 100$ V, the current $I = $ _____.

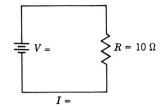

Figure 3-6 When the resistance remains constant, the larger the voltage the larger the current.

   As you can see, when the resistance remains constant,
8. The larger the voltage, the larger the current.
   The smaller the voltage, the _____ the current.

   In Fig. 3-7, if the voltage $V = 100$ V remains unchanged,
9. When the resistance $R = 1\ \Omega$, the current $I =$ _____.
10. When the resistance $R = 10\ \Omega$, the current $I =$ _____.
11. When the resistance $R = 50\ \Omega$, the current $I =$ _____.
12. When the resistance $R = 100\ \Omega$, the current $I =$ _____.

   As you can see, when the voltage remains constant,
13. The larger the resistance, the smaller the current.
The smaller the resistance, the _____ the current.

Formulas in electrical work. A formula is a shorthand method for writing a rule. Each letter in a formula represents a number which may be substituted for it. The signs of operation tells us what to do with these numbers.

Steps in solving problems
1. Read the problem carefully.
2. Draw a simple diagram of the circuit.
3. Record the given information directly on the diagram. Indicate the values to be found by question marks.
4. Write the formula.
5. Substitute the given numbers for the letters in the formula. If the number for the letter is unknown, merely write the letter again. Include all mathematical signs.
6. Do the indicated arithmetic. If after substitution the unknown letter is multiplied by some number, divide the number all alone on one side of the equality sign by the multiplier of the unknown letter.
7. In the answer, indicate the letter, its numerical value, and the units of measurement.

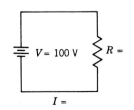

Figure 3-7 When the voltage remains constant, the larger the resistance the smaller the current.

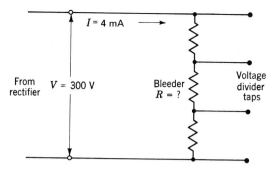

Figure 3-8 The bleeder resistor is made of three resistors in series.

Self-Test 3-40

In Fig. 3-8, the bleeder resistance draws 4 mA from the 300-V power supply when no loads are connected to the various voltage-divider terminals. Find the total bleeder resistance.

Solution

1. The diagram for the circuit is shown in Fig. 3-8.
2. Write the formula. $\qquad V = I \times \underline{\qquad}$
3. Substitute numbers. $\qquad 300 = \underline{\qquad} \times R$
4. Solve the equation. $\qquad \dfrac{300}{?} = R$
5. Divide the numbers. $\qquad \underline{\qquad} = R$

or $\qquad\qquad\qquad\qquad R = 75,000\ \Omega = 75 \underline{\qquad}$ *Ans.*

Exercises

Applications

1. The resistance of an electric percolator is 22 Ω. If it draws 5 A, what is the operating voltage?
2. An electric heater whose coil is wound with No. 18 iron wire is connected across 110 V. If it draws a current of 10 A, what is the value of its resistance?
3. According to the National Electrical Code, No. 14 asbestos-covered type A wire should never carry more than 30 A. Is this wire safe to use to carry power to a 10-Ω 230-V motor?
4. A washing-machine motor has a total resistance of 39 Ω and operates on 117 V. Find the current taken by the motor.
5. What is the voltage drop across an Allied model BK relay of 12 kΩ resistance if it carries 1.5 mA?
6. What is the voltage across a telephone receiver of 800 Ω resistance if the current flowing is 30 mA?
7. A spot welder delivers 7000 A when the voltage is 4.9 V. What is the resistance of the piece being welded?
8. A 600-W soldering iron is used on a 120-V line and has a resistance of 24 Ω. Find the current drawn.

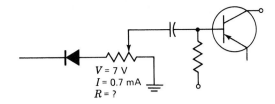

Figure 3-9 Volume control in a simple transistor radio.

9. A volume control similar to that used in the Panasonic RF 738 transistor radio is shown in Fig. 3-9. How many ohms of resistance are engaged in the potentiometer if the voltage drop is 7 V and it passes a current of 0.7 mA?

10. If the full-scale reading of an ammeter is 10 A, what is its resistance if this current causes a voltage drop of 0.05 V?

11. What is the resistance of a bus bar carrying 300 A if the voltage drop across it is 1.2 V?

12. A sensitive dc meter takes 9 mA from a line when the voltage is 108 V. What is the resistance of the meter?

13. An electromagnet draws 5 A from a 110-V line. What current will it draw from a 220-V line?

14. The resistance of the series field coils of a compound motor is 0.24 Ω, and they carry a current of 72A. Find the voltage drop across these coils.

15. A 5-kΩ resistor in a voltage divider reduces the voltage across it by 150 V. What current flows through the resistor?

16. A series resistor is used to reduce the voltage to a motor by 45 V. What must be the resistance of the resistor if the motor draws 0.52 A?

17. A voltmeter has a resistance of 27 kΩ. What current will flow through the meter when it is placed across a 220-V line?

18. A series of insulators leak 30 μA at 9 kV. Find the resistance of the insulator string.

(See CD-ROM for Test 3-8)

Assessment

chapter 3

1. Using Ohm's law for power ($P = E^2/R$), determine the resistance of a 500-W lamp that is rated for 120 V.

2. Write the formula for the following: The total impedance Z in a series LCR circuit is equal to square root ($\sqrt{}$) of the quantities of resistance R squared plus net reactance $(X_L - X_C)$ squared.

3. Knowing that there are 27 ft³ in one cubic yard, determine the number of cubic yards of concrete that will be needed to fill a form if it is 6.5 feet deep by 36 in wide and 650 ft long.

4. Using Ohm's law for power ($P = E^2/R$), if the resistance of a heating element is 10 Ω and the rated voltage is 120 V, determine the power drawn by the heater when it is energized.

5. A gas compressor has a cylinder with a bore dimension of 4.725 in and a stroke dimension of 9.265 in. Using the formula $V = R^2H$, determine the volume of air, in cubic inches, that is pushed out of the cylinder after each stroke.

6. A 117-V rated drill motor has a circuit resistance at full load of 14.625 Ω. What is the current drawn by the drill at full load?

7. What is the voltage drop across a 270-kΩ resistor when there is 0.37037 mA of current flowing through it?

8. Multiply the following numbers using scientific notation: 5,000,000,000,000 and 0.000000000000825.

9. Change 625 pF to nanofarads.

10. A heating plate used to melt ski wax draws 1200 W of power. If the unit is rated for 117 V, what is the circuit resistance?

INTERNET ACTIVITIES

Internet Activity A

Use the following site to find the definitions for (whole numbers and integers). List examples of whole numbers and integers.
http://instructor.iwcc.cc.ia.us/bmay/lesson1.htm

Internet Activity B

Use the following site to determine the two different meanings of the minus (−) symbol used in math. **http://www.idbsu.edu/people/jbrennan/algebra /numbers/addition_and_subtraction_of_real numbers.html**

Chapter 3 Solutions to Self-Tests and Reviews

Self-Test 3-4

1. multiply
2. ×
3. add
4. +
5. product
6. sum
7. ×, +

Self-Test 3-9

3. addition, 50, 25
4. 75
5. 105

Self-Test 3-15

1. add, minus, −10
2. add, plus, +12
3. add, plus, +11
4. subtract, plus, +2
5. add, minus, −15
6. subtract, minus, minus, −6

7. −4, +3
8. subtract, plus, +18
9. −7, −12
10. −9, +4, −5, −7

Self-Test 3-18

1. R
2. 50
4. 5000, 50
5. 2500
6. 2
7. 20.8

Self-Test 3-19

2. 300
3. 0.92
4. 50
5. 46

Self-Test 3-31

1. 2 MΩ
2. 10^{-3}

3. 10^6
5. 9
6. 3, 0.15

Self-Test 3-35

2. 0.002
3. $\dfrac{0.002}{50}$
4. 0.000 04, 0.04

Self-Test 3-37

1. R
2. 6.6
3. $\dfrac{6.6}{0.015}$
4. 440, 440

Self-Test 3-39

2. 0.105
3. 10^{-2}
4. 10
5. 3

Self-Test 3-40

2. R
3. 0.004
4. 0.004
5. 75,000, kΩ

JOB 3-11 Review

1. current
2. resistance
3. voltage
4. 1 A
5. 2 A
6. 5 A
7. 10 A
8. smaller
9. 100 A
10. 10 A
11. 2 A
12. 1 A
13. larger

1. Recognize a series circuit.
2. Explain why the current is the same in all parts of a series circuit.
3. Determine the total current I_T in a series circuit.
4. Determine the total voltage V_T in a series circuit.
5. Determine the total resistance R_T in a series circuit.
6. Apply Ohm's law to solve for the unknown values in a series circuit.
7. Describe the physiological effects of current passing through the human body.
8. Determine power distribution line currents with both balanced and unbalanced loads.
9. Explain the purpose of a ground-fault circuit interrupter (GFCI) device.
10. Solve for the value of the unknown letter in equations containing terms that are added and subtracted.

In this chapter you will learn how to develop your own formulas for series circuits. You will then solve these formulas to determine values of voltage, current, and resistance in series circuits. If you have any difficulties with the solution of these formulas, you may want to review Chap. 3.

JOB 4-1

Voltage, Current, and Resistance in Series Circuits

Wiring a series circuit. A series circuit is one in which all the component parts are connected in succession from plus (+) to minus (−), as shown in Fig. 4-1. In a series circuit there is *only one path* through which the electrons may flow. The flow of current in a series circuit may be compared with the flow of water in a series-connected water system. In Fig. 4-2a, the pump forces the water through the three valves in succession. In Fig. 4-2b, the electron-moving pump—the battery—forces the electrons through the three resistors in succession. In both series circuits there is *only one path* that the water or the electrons may travel. If the water circuit is broken at any point, by either closing a valve or breaking a pipe, the flow of water

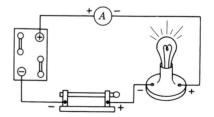

Figure 4-1   Simple series circuit. The parts are connected so that the current can flow in only one path.

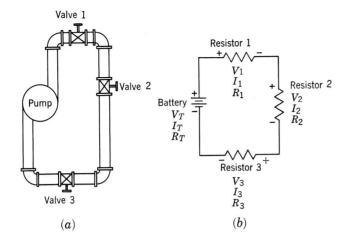

Figure 4-2 (a) Valves in series in a water-supply system. (b) Resistors in series in an electric circuit.

around the system will stop. If the electric series circuit is broken at any point, no energy will be available to any part of the circuit, since there will be no return path for the electrons to follow. For example, in Fig. 4-3, a push button is usually placed in series with the bell it controls. As long as the button is held up by the force of the spring inside it, the circuit is broken. Since no current can flow in a broken, or "open," circuit, the bell will not ring. When the button is depressed, the wires make contact, completing the circuit. The current then flows through the bell, and the bell rings.

Symbols for series circuits. Numbers or letters written underneath other numbers or letters are called *subscripts*. Numbers like 1, 2, or 3 are written under the letters V, I, or R to indicate these quantities in the first, second, or third part of the circuit. For example, in Fig. 4-2b:

V_1 represents the voltage across the first resistor.
I_2 represents the current through the second resistor.
R_3 represents the resistance of the third resistor.
V_T represents the total voltage in the circuit.
I_T represents the total current in the circuit.
R_T represents the total resistance in the circuit.

Total current in a series circuit. In Fig. 4-2a, the water was forced through each valve in turn because it had no other place to go. Whatever quantity of water flowed through the first valve had to flow through the second and third valve also. In Fig. 4-2b, the electrons that were forced through the first resistor R_1 also had to

> # Hint for Success
>
> Imagine a string of Christmas tree lights connected in series. If any one of the bulbs burns out, none of the bulbs will light. This is because the only available path for current flow has been interrupted.

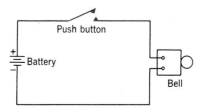

Figure 4-3 A push button in series controls the flow of current from the battery to the bell by opening or closing the circuit.

flow through the second resistor R_2 and through the third resistor R_3 because there was no other place for them to go. It follows, then, that any current entering the circuit must flow *unchanged* through all the other parts of the circuit.

 RULE The total current in a series circuit is equal to the current in any other part of the circuit.

$$I_T = I_1 = I_2 = I_3 = \text{etc.}$$

Formula 4-1

where I_T = total current
I_1 = current in first part
I_2 = current in second part
I_3 = current in third part
etc.

Total voltage in a series circuit.   In Fig. 4-4a, the total force required to lift the weights must be equal to the *sum* of the forces required to lift the individual weights. In Fig. 4-4b, the total electrical pressure supplied by the battery must be equal to the *sum* of the pressures required by each lamp. Since the available voltage pressure decreases, or "drops" as the current is forced through each resistance, the voltages V_1 and V_2 are called voltage *drops*.

 RULE The total voltage in a series circuit is equal to the sum of the voltage drops across all the parts of the circuit.

$$V_T = V_1 + V_2 + V_3 + \text{etc.}$$

Formula 4-2

where V_T = total voltage
V_1 = voltage across first part
V_2 = voltage across second part
V_3 = voltage across third part
etc.

Total resistance in a series circuit.   In Fig. 4-4a, the resistance that must be overcome by the body is equal to the *sum* of the weights. In Fig. 4-4b, the total electric resistance of the circuit is equal to the *sum* of the resistances of all the lamps.

 RULE The total resistance of a series circuit is equal to the sum of the resistances of all the parts of the circuit.

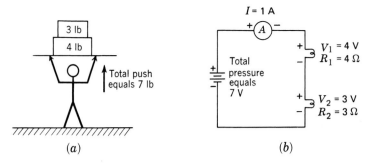

Figure 4-4 Similarity between (a) a force system and (b) an electric circuit.

$$R_T = R_1 + R_2 + R_3 + \text{etc.} \qquad \text{Formula 4-3}$$

where
R_T = total resistance
R_1 = resistance of first part
R_2 = resistance of second part
R_3 = resistance of third part
etc.

Example 4-1

A 6-V 20-Ω filament and a 12-V 40-Ω filament are connected in series with a 20-Ω limiting resistor using 6 V and 0.3 A. Find (a) the total voltage, (b) the total current, and (c) the total resistance.

Solution

Draw the circuit diagram as shown in Fig. 4-5

$$
\begin{array}{lll}
V_1 = 6\text{ V} & V_2 = 12\text{ V} & V_3 = 6\text{ V} \\
I_1 = ? & I_2 = ? & I_3 = 0.3\text{ A} \\
R_1 = 20\ \Omega & R_2 = 40\ \Omega & R_3 = 20\ \Omega
\end{array}
$$

$$
\begin{array}{l}
V_T = ? \\
I_T = ? \\
R_T = ?
\end{array}
$$

Figure 4-5

a. Find the total voltage.
   1. Write the formula. $V_T = V_1 + V_2 + V_3$ (4-2)
   2. Substitute numbers. $V_T = 6 + 12 + 6 = 24\text{ V}$ *Ans.*
b. Find the total current.
   1. Write the formula. $I_T = I_1 = I_2 = I_3$ (4-1)
   2. Substitute numbers. $I_T = I_1 = I_2 = 0.3\text{ A}$ *Ans.*
c. Find the total resistance.
   1. Write the formula. $R_T = R_1 + R_2 + R_3$ (4-3)
   2. Substitute numbers. $R_T = 20 + 40 + 20 = 80\ \Omega$ *Ans.*

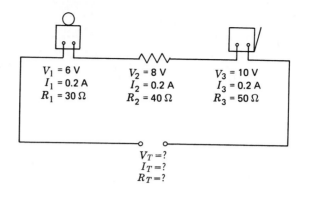

$V_1 = 6$ V
$I_1 = 0.2$ A
$R_1 = 30\ \Omega$

$V_2 = 8$ V
$I_2 = 0.2$ A
$R_2 = 40\ \Omega$

$V_3 = 10$ V
$I_3 = 0.2$ A
$R_3 = 50\ \Omega$

$V_T = ?$
$I_T = ?$
$R_T = ?$

Figure 4-6

Exercises

Applications

1. In the circuit shown in Fig. 4-6, find (*a*) the total voltage, (*b*) the total current, and (*c*) the total resistance.
2. In an antique car, a 3-V 1.5-Ω dash light and a 3-V 1.5-Ω taillight are connected in series to a battery delivering 2 A as shown in Fig. 4-7. Find (*a*) the total voltage and (*b*) the total resistance.
3. Three resistances are connected in series. $V_1 = 6.3$ V, $I_1 = 0.3$ A, $R_1 = 21\ \Omega$, $V_2 = 12.6$ V, $R_2 = 42\ \Omega$, $V_3 = 24$ V, and $R_3 = 80\ \Omega$. Find (*a*) the total voltage, (*b*) the total current, and (*c*) the total resistance.
4. The receiver, transmitter, and line coil of a telephone circuit are connected in series. For the receiver: $V = 2.5$ V, $I = ?$, and $R = 12.5\ \Omega$. For the transmitter: $V = 18.6$ V, $I = 0.2$ A, and $R = 93\ \Omega$. For the line coil: $V = 6.7$ V, $I = ?$, and $R = 33.5\ \Omega$. Find (*a*) the total voltage, (*b*) the total current, and (*c*) the total resistance.

JOB 4-2

Using Ohm's Law in Series Circuits

Ohm's law may be used for the individual parts of a series circuit. When it is used on a particular part of a circuit, great care must be taken to use *only* the voltage,

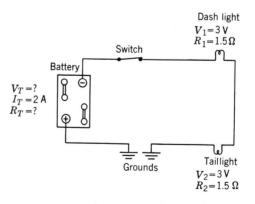

Dash light
$V_1 = 3$ V
$R_1 = 1.5\ \Omega$

Switch

Battery

$V_T = ?$
$I_T = 2$ A
$R_T = ?$

Grounds

Taillight
$V_2 = 3$ V
$R_2 = 1.5\ \Omega$

Figure 4-7 Series-connected automobile dash and taillight for Prob. 2.

current, and resistance of that particular part. That is, the *voltage of a part* is equal to the *current in that part* multiplied by the *resistance of that part*. This may be easily remembered by using the correct subscripts when writing the Ohm's law formula for a particular part.

For the first part:

$$V_1 = I_1 \times R_1 \qquad\qquad 4\text{-}4$$

For the second part:

$$V_2 = I_2 \times R_2 \qquad\qquad 4\text{-}5$$

For the third part:

$$V_3 = I_3 \times R_3 \qquad\qquad 4\text{-}6$$

Example 4-2

A solid-state relay is used to control a lamp. The output circuit shown in Fig. 4-8 shows the lamp, relay, and wiring resistance. Find all missing values of (*a*) current, (*b*) voltage, and (*c*) resistance.

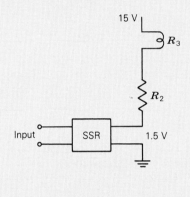

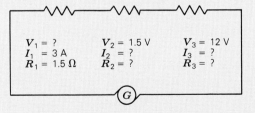

$$
\begin{aligned}
V_1 &= ? & V_2 &= 1.5\text{ V} & V_3 &= 12\text{ V}\\
I_1 &= 3\text{ A} & I_2 &= ? & I_3 &= ?\\
R_1 &= 1.5\ \Omega & R_2 &= ? & R_3 &= ?
\end{aligned}
$$

Figure 4-8

Solution

We can find the total values by the following formulas:

$$I_T = I_1 = I_2 = I_3 \qquad (4\text{-}1)$$
$$V_T = V_1 + V_2 + V_3 \qquad (4\text{-}2)$$
$$R_T = R_1 + R_2 + R_3 \qquad (4\text{-}3)$$

However, in order to use these formulas, we must know the individual values for each part of the circuit. These values may be found by using the Ohm's law formulas for each part.

$$V_1 = I_1 \times R_1 \qquad \qquad \text{(4-4)}$$
$$V_2 = I_2 \times R_2 \qquad \qquad \text{(4-5)}$$
$$V_3 = I_3 \times R_3 \qquad \qquad \text{(4-6)}$$

a. Since the current has the same value at every point in a series circuit, it is easiest to find the current first.

    1. Write the formula. $I_T = I_1 = I_2 = I_3$ (4-1)

    2. Substitute numbers. $I_T = 3 = I_2 = I_3$

    3. The current value is $I_T = I_1 = I_2 = I_3 = 3\ A$ *Ans.*

b. Find V_1. Use $I_1 = 3\ A$ from step *a*.

    1. Write the formula. $V_1 = I_1 \times R_1$ (4-4)

    2. Substitute numbers. $V_1 = 3 \times \tfrac{1}{2} = 1.5\ V$ *Ans.*

Find V_T. Use $V_1 = 1.5\ V$ from step *b*.

    1. Write the formula. $V_T = V_1 + V_2 + V_3$ (4-2)

    2. Substitute numbers. $V_T = 1.5 + 1.5 + 12 = 15\ V$ *Ans.*

c. Find R_2 and R_3. Use $I_2 = I_3 = 2\ A$ from step *a*.

    1. Write the formula. $V_2 = I_2 \times R_2$ $V_3 = I_3 \times R_3$

    2. Substitute numbers. $1.5 = 3 \times R_2$ $12 = 3 \times R_3$

    3. Solve. $\tfrac{1.5}{3} = R_2$ $\tfrac{12}{3} = R_3$

                                $R_2 = \tfrac{1}{2}\ \Omega$ *Ans.* $R_3 = 4\ \Omega$ *Ans.*

Find R_T. Use $R_2 = \tfrac{1}{2}\ \Omega$ and $R_3 = 4\ \Omega$ from step *c*.

    1. Write the formula. $R_T = R_1 + R_2 + R_3$ (4-3)

    2. Substitute numbers. $R_T = \tfrac{1}{2} + \tfrac{1}{2} + 4 = 5\ \Omega$ *Ans.*

Self-Test 4-3

Part of the first stage of a two-stage transistorized amplifier is shown in Fig. 4-9. Solve the circuit for all missing values of (*a*) current, (*b*) voltage, and (*c*) resistance.

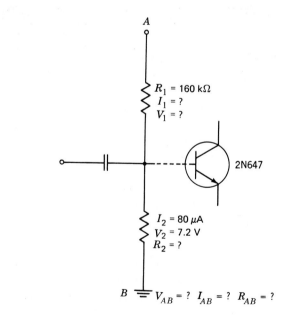

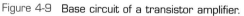

Figure 4-9   Base circuit of a transistor amplifier.

Solution

a. Find the total current.

    1. Write the formula. $I_{AB} = I_1 =$ _____ $=$ _____ μA *Ans.*

b. Find the voltages. Find V_1. Use $I_1 =$ _____ μA from step *a*.

    1. Write the formula. $V_1 = I_1 \times$ _____   (4-4)

    2. Substitute numbers. $V_1 = 80 \times 10^{-?} \times 160 \times 10^?$

                          $V_1 = 12{,}800 \times 10^{-?} =$ _____ V *Ans.*

Find E_{AB}. Use $V_1 = 12.8$ V from step *b*.

    1. Write the formula. $V_{AB} = V_1 +$ _____   (4-2)

    2. Substitute numbers. $V_{AB} = 12.8 +$ _____ $=$ _____ V *Ans.*

c. Find the resistances. Find R_2. Use $I_2 =$ _____ μA

    1. Write the formula. $V_2 =$ _____ $\times R_2$ (4-5)

    2. Substitute numbers. $7.2 =$ _____ $\times R_2$

    3. Solve. $R_2 = \dfrac{7.2}{80 \times 10^{-6}} = 0.09 \times 10^?$

                          $R_2 =$ _____ $\Omega =$ _____ $k\Omega$

d. Find R_{AB}. Use $R_2 =$ _____ $k\Omega$ from step *c*.

    1. Write the formula. $R_{AB} = R_1 +$ _____

    2. Substitute numbers. $R_{AB} = 160\ k\Omega + 90\ k\Omega =$ _____ $k\Omega$ *Ans.*

Exercises

Applications

1. A lamp using 10 V, a 10-Ω resistor drawing 4 A, and a 24-V motor are connected in series. Find (*a*) the total current, (*b*) the total voltage, and (*c*) the total resistance.

2. A small arc lamp designed to operate on a current of 6 A has a resistance of 14 Ω. It is used in series with a limiting resistor of 6 Ω. Find (*a*) the total current, (*b*) the total voltage, and (*c*) the total resistance.

3. A series-connected automobile dash light and taillight circuit similar to that shown in Fig. 4-7 operates from a 6-V battery. The dash light operates on 2 V and 0.8 A. The taillight requires 4 V. Find (*a*) the resistance of each light, (*b*) the total resistance, and (*c*) the total current.

4. A lamp, a resistor, and a soldering iron are connected in series. The lamp has a voltage of 16 V across it. The voltage across the resistor is 12.8 V. The iron has a resistance of 6 Ω and carries a current of 3.2 A. Find (*a*) the total current, (*b*) the total voltage, and (*c*) the total resistance.

5. A 17,000-Ω 150-V scale voltmeter is combined in series with a 51-kΩ resistor in order to increase the range of the meter to read 600 V. Find the current in the meter when it reads 600 V.

6. Using the circuit shown in Fig. 4-10 on page 106, find the total voltage between points *A* and *B*.

7. The first video IF amplifier in the RCA TV chassis KCS 176 uses a base circuit similar to that shown in Fig. 4-11 on page 106. Find (*a*) V_1, (*b*) V_2, and (*c*) V_T.

(See CD-ROM for Test 4-1)

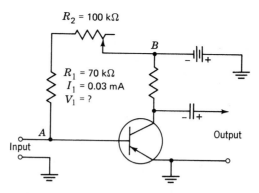

Figure 4-10   Class A audio amplifier.

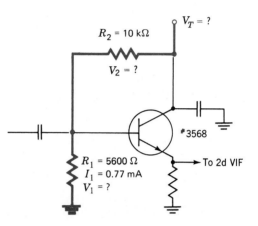

Figure 4-11   First video amplifier circuit in a modern TV chassis.

Using Ohm's Law for Total Values in Series Circuits

Ohm's law was used in the last job to find the voltage, current, and resistance of the individual parts of a series circuit. Ohm's law may also be used to find the *total values* of voltage, current, and resistance in a series circuit.

 RULE   In a series circuit, the total voltage is equal to the total current multiplied by the total resistance.

$$V_T = I_T \times R_T \qquad \text{Formula 4-7}$$

where V_T = total voltage, V
I_T = total current, A
R_T = total resistance, Ω

Finding the Total Voltage

Example 4-4

A resistor of 45 Ω, a bell of 60 Ω, and a buzzer of 50 Ω are connected in series as shown in Fig. 4-12. The current in each is 0.2 A. What is the total voltage?

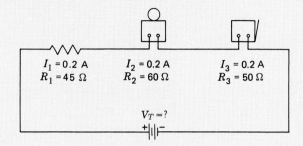

$I_1 = 0.2$ A
$R_1 = 45\ \Omega$

$I_2 = 0.2$ A
$R_2 = 60\ \Omega$

$I_3 = 0.2$ A
$R_3 = 50\ \Omega$

$V_T = ?$

Figure 4-12

Solution

In order to use Ohm's law to find the total voltage, we must first obtain the values for the total current and the total resistance.

These values may be found by the following formulas:

$$I_T = I_1 = I_2 = I_3 \qquad (4\text{-}1)$$
$$R_T = R_1 + R_2 + R_3 \qquad (4\text{-}3)$$

1. Find the total current.

$$I_T = I_1 = I_2 = I_3 = 0.2 \text{ A} \qquad Ans. \qquad (4\text{-}1)$$

2. Find the total resistance.

$$R_T = R_1 + R_2 + R_3 \qquad (4\text{-}3)$$
$$R_T = 45 + 60 + 50 = 155\ \Omega \qquad Ans.$$

3. Find the total voltage.

$$V_T = I_T \times R_T \qquad (4\text{-}7)$$
$$V_T = 0.2 \times 155 = 31 \text{ V} \qquad Ans.$$

about—electronics

The unit of measure for resistance (the ohm) is named for German physicist Georg Simon Ohm. You'll recognize his famous law: voltage = current $\times$ resistance.

Finding the Total Current

Example 4-5

A portable spotlight of 3 Ω resistance is connected to a 6-V power pack with two wires, each of 0.4 Ω resistance. Find the total current drawn from the power pack.

Solution

The diagram for the circuit is shown in Fig. 4-13. In order to use Ohm's law to find the total current, we must first obtain the values for the total voltage and the total resistance.

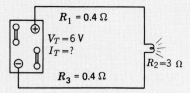

Figure 4-13

1. The total voltage is known: $V_T = 6$ V.
2. Find the total resistance. $\qquad R_T = R_1 + R_2 + R_3 \qquad\qquad$ (4-3)
$$R_T = 0.4 + 3 + 0.4 = 3.8\ \Omega \qquad Ans.$$
3. Find the total current. $\qquad V_T = I_T \times R_T \qquad\qquad$ (4-7)
$$6 = I_T \times 3.8$$
$$\frac{6}{3.8} = I_T$$
or $\qquad\qquad\qquad\qquad\qquad\quad I_T = 1.58\ \text{A} \qquad Ans.$

Finding the Total Resistance

Example 4-6

A motor, a lamp, and a rheostat are connected in series. The motor uses 80 V, the lamp takes 10 V and 2 A, and the rheostat uses 30 V. Find the total resistance of the circuit.

Solution

The diagram for the circuit is shown in Fig. 4-14. In order to use Ohm's law to find the total resistance, we must first obtain the values for the total voltage and the total current.

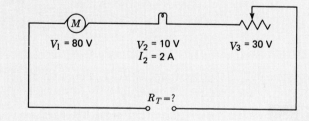

$V_1 = 80$ V $\qquad\qquad V_2 = 10$ V $\qquad\qquad V_3 = 30$ V
$\qquad\qquad\qquad\qquad\quad I_2 = 2$ A

$R_T = ?$

Figure 4-14

1. Find the total voltage. $\qquad V_T = V_1 + V_2 + V_3 \qquad\qquad$ (4-2)
$$V_T = 80 + 10 + 30 = 120\ \text{V} \qquad Ans.$$
2. Find the total current. $\qquad I_T = I_1 = I_2 = I_3 = 2\ \text{A} \qquad Ans. \qquad$ (4-1)
3. Find the total resistance. $\qquad V_T = I_T \times R_T \qquad\qquad$ (4-7)
$$120 = 2 \times R_T$$

$$\frac{120}{2} = R_T$$
$$R_T = 60\ \Omega \qquad \textit{Ans.}$$

Shock Safety

The problem with being a repairperson is that you are called upon only when things are not right. The same can be true for new installations when there are problems in the assembly of the system. Sometimes malfunctions or assembly errors inside a device can cause a safety hazard. This section will begin the study of the effects of shock hazards.

There are two important elements of a shock hazard: the magnitude of the *current* and the effect this current has on the body. When a shock situation is analyzed, the body is considered to be an equivalent resistor. There is no exact value for this resistor. There are only some approximate ranges that the resistor could have. The value in any particular case depends upon whether the skin is wet or dry, what parts of the body are involved, the condition of the skin, etc. It is also believed that body resistance may be a function of time and voltage.

The values given in this job are for 60-Hz systems. The values for direct current are about the same. Researchers have arrived at the following guidelines for body resistance:

Conditions	Body Resistance
Wet skin, outdoors	500 Ω
Dry skin, indoors	1500 Ω
Lowest resistance between two limbs, excluding skin resistance	500–1000 Ω
Across the chest	100 Ω

The physiological effects of current are shown in Fig. 4-15. On the right of the graph is a scale of the physiological effects and the approximate current levels at which they occur. The Underwriters Laboratories set a 0.5-mA level for their leakage tests. This value is also shown on the scale. It is above the threshold of sensation but well below that causing any serious effects to the body.

To the left of the scale of effects are curves that show that the level at which an effect is observed is different for men and for women. The current level at which an effect occurs is different for any particular individual. This produces a statistical distribution of the current level at which the effect is observed.

The let-go current is the current at which a victim's muscles are taken over by the current. The victim can be conscious and can attempt to let go of the current source but will not be able to. On the scale the let-go current is 10 mA. To the left are two let-go curves. One curve is the current at which 99.5% of the victims will

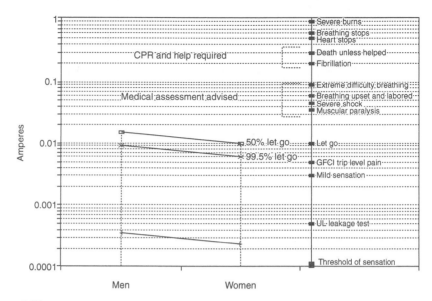

Figure 4-15

be able to let go. The other curve is the current level at which only 50% of the victims will be able to let go. The 10-mA value is a representative value selected between these two curves. Notice also that for each of these curves, the current levels are lower for women.

With these two pieces of information, you can now convert a physical situation into an equivalent circuit by replacing the body with a resistor. The equivalent circuit can then be analyzed using conventional techniques. You can solve for the current flowing through the equivalent body resistor and then look up the physiological effect that will come from the current using Fig. 4-15.

Figure 4-16a shows the situation in which a 90-V line is shorted through 7500 Ω to the housing of some machinery. A person who is standing indoors with

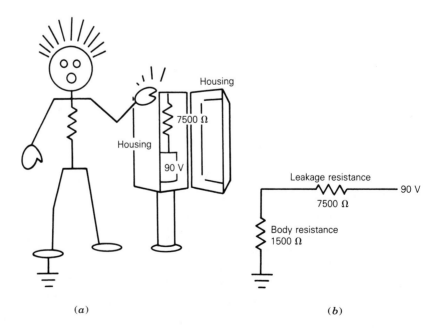

(a)   (b)

Figure 4-16   (a) Leakage from unit to housing. (b) Equivalent resistance model.

dry skin on ground accidentally touches the housing. What is the effect of the current flowing through the victim?

Self-Test 4-7

The equivalent circuit for the situation is shown in Fig. 4-16b. In order to use Ohm's law to find the ____(1)____ through the victim was must find the total ____(2)____.

Find the body resistance.

From the table, on page 109, the total resistance for dry skin indoors is ____(3)____ Ω.

Find the total resistance.

The total resistance of the circuit is 1500 Ω plus ____(4)____ = ____(5)____ Ω.

Find the current in the circuit.

The total current is ____(6)____ V/____(7)____ Ω = ____(8)____ mA.

The current through the victim is greater than the "let-go current," which means that it may be necessary to turn the voltage off in order to get to the victim.

Exercises

Applications

1. Five lamps are connected in series for use in a subway lamp bank. Each lamp has a resistance of 100 Ω. What is the subway-circuit total voltage if the total current drawn is 1 A?

2. A railroad signal lamp and the coil of a semaphore are connected in series. The resistance of the lamp is 16 Ω, and that of the coil is 5 Ω. The current is 2 A. What is the total voltage?

3. In a telephone circuit, the receiver, transmitter, and line coil are connected in series. The receiver resistance is 2.2 Ω; the transmitter resistance is 6.2 Ω; the coil resistance is 1.2 Ω. If the current in the receiver is 0.5 A, what is the total voltage of the circuit?

4. A motor, a lamp, and a rheostat are connected in series. The voltage across the motor is 96 V, across the lamp 24 V, and across the rheostat 40 V. If the current in the motor is 2 A, find (a) the total current, (b) the total voltage, and (c) the total resistance.

5. A current of 0.003 A flows through a resistor that is connected to a 1.5-V dry cell. If three additional 1.5-V cells are connected in series to the first cell, find the current now flowing through the resistor.

6. A series circuit consists of a heating coil, an ultraviolet lamp, and a motor for an electric clothes drier. The current in the motor is 5 A. The voltages are 50 V for the lamp, 80 V for the motor, and 100 V for the heating coil. Find (a) the total current, (b) the total voltage, and (c) the total resistance.

7. Four lamps are wired in series. The resistance of each lamp is 0.5 Ω. The voltage across each lamp is 0.4 V. Find (a) the total current, (b) the total voltage, and (c) the total resistance.

8. Three arc lights are connected in series with a resistance of 21 Ω. The resistance of each arc light when hot is 10 Ω. The voltage across each arc light is 40 V, and the voltage across the resistance is 84 V. Find (*a*) the total voltage, (*b*) the total current, and (*c*) the total resistance.
9. A voltage divider in a television receiver consists of a 6-kΩ, a 3-kΩ, and a 1500-Ω resistor in series. If the total current is 15 mA, find the total voltage drop.

JOB 4-4

Power Distribution Line Currents

This job will develop some basic concepts in power distribution systems. They will be analyzed as simple dc circuits; later the same circuits will be analyzed as ac circuits. In this example the power comes from 12-V batteries. In the ac circuits the power comes from 120-V transformers. The ideas are the same.

The basic dc circuit is shown in Fig. 4-17. A 12-V battery provides power to a load represented by R_1. Current is supplied to the load through line L_1. The current returns through the line N, which is called the *neutral line*. The circuit is not different from Fig. 4-3 or Fig. 4-13.

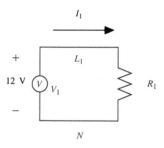

Figure 4-17

Assume that two circuits like those in Fig. 4-17 are stacked one on top of the other. The resulting circuit is shown in Fig. 4-18. The new circuit provides current to load R_2 through line L_2. The current returns through the neutral line. It is apparent that the name *neutral* comes from the fact that the one line serves two circuits; it is neutral to both circuits.

The two batteries represent two phases. Line L_1 provides a positive voltage relative to the neutral line, and line L_2 provides a negative voltage relative to the neutral line. Another important observation is that the current returned from line L_1 is in the opposite direction from the current returned from line L_2. This means that the currents will subtract to provide the total current in line N.

Unbalanced Load

Example 4-8

Assume that in Fig. 4-18, $R_1=6$ Ω and $R_2=3$ Ω. Find the currents in lines 1 and 2 and in the neutral wire.

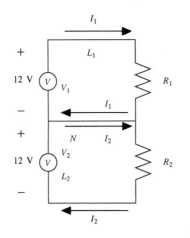

Figure 4-18

Solution

1. Find the current in each of the loads.

$$I_1 = \frac{V_1}{R_1} = \frac{12}{3} = 2 \text{ A} \qquad Ans.$$

$$I_2 = \frac{V_2}{R_2} = \frac{12}{3} = 4 \text{ A} \qquad Ans.$$

2. The line currents are

$$I_{L1} = I_1 = 2 \text{ A} \qquad Ans.$$
$$I_{L2} = I_2 = 4 \text{ A} \qquad Ans.$$

3. The current flowing out of the batteries in the neutral line is

$$I_N = I_2 - I_1 = 4 - 2 = 2 \text{ A} \qquad Ans.$$

Notice that the sum of the currents flowing out of the sources through lines 1, 2, and N is zero. This is because all of the current that goes out of a power source must come back.

Balanced Load

Example 4-9

A balanced load is when the load on each phase of the power supply is the same. In this case $R_1 = R_2$. It is assumed that the voltages in each phase are equal. Find the current in the neutral line under a balanced load condition.

Solution

1. The line currents under balanced conditions are

$$I_1 = \frac{V_1}{R_1} = \frac{V_2}{R_2} = I_2 \qquad Ans.$$

2. The current in the neutral line is

$$I_N = I_1 - I_2 = 0 \qquad Ans.$$

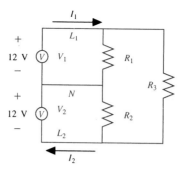

Figure 4-19

Adding a Third Series Circuit

An additional, double-voltage circuit is available in a two-phase system such as the preceding circuit. A 24-V supply can be obtained by connecting a load between line 1 and line 2. The resulting circuit is shown in Fig. 4-19.

The new load, R_3, draws current from line 1 and returns it to line 2. This load current increases both I_1 and I_2. No current flows in the neutral line. There is no neutral on this circuit. There are just two "hot" lines.

The following example shows how two 12-V supplies can provide both 12 V and 24 V. The feeder circuit has just three wires, as just described. These three wires provide power on three separate circuits, two 12-V circuits and a 24-V circuit. The principle used is exactly the same as that with the 120/240 ac circuit. The concepts for calculating line currents are also the same.

Two-Phase 12(120) V/24(240) V Power Supply

Example 4-10

The new load on the 24-V system is shown in Fig. 4-19. Assume that $R_1 = 6 \ \Omega$ and $R_2 = 3 \ \Omega$, as in Example 4-8, and let $R_3 = 8 \ \Omega$. The currents in loads 1 and 2 are the same as in the previous example.

Solution

1. Find the load currents.

$$I_{R_1} = 2 \text{ A} \qquad Ans.$$
$$I_{R_2} = 4 \text{ A} \qquad Ans.$$

The current in the new load is

$$I_{R_3} = \frac{24 \text{ V}}{8 \ \Omega} = 3 \text{ A} \qquad Ans.$$

2. Find the line currents.

$$I_1 = I_{R_1} + I_{R_3}$$
$$= 2 + 3 = 5 \text{ A} \qquad Ans.$$
$$I_2 = I_{R_2} + I_{R_3}$$
$$= 4 + 3 = 7 \text{ A} \qquad Ans.$$
$$I_N = I_2 - I_1$$
$$= 7 - 5 = 2 \text{ A} \qquad Ans.$$

Notice that there is no change in the neutral current. The current in the neutral line is the difference between the phase 1 and phase 2 currents. The new high-voltage phase uses only lines 1 and 2.

Summary

This job looked at a different concept of series circuits. Each of the elementary circuits considered was a simple series circuit. Two of these circuits were stacked so that they interact. The junction was the neutral line.

A special case occurs when the two power sources are out of phase relative to the neutral line. This means that one line is positive relative to the neutral and the other line is negative. When the two individual circuits are out of phase, the currents in the neutral line oppose each other.

Another special case is when the currents in each load are identical. This is called a *balanced* load. In this special situation the current in the neutral line is zero. What would happen if the neutral line is removed under balanced conditions?

An important concept to learn here is the difference between "phase currents" and "line currents." The phase currents are the currents that flow in each of the loads. The line currents are the currents that flow in each of the lines. In Example 4-10 the line currents and the phase currents were all different.

JOB 4-5

Ground-Fault Protection

In this job we look at the concepts of GFCI (ground-fault circuit interrupter). We will develop the ideas on a dc basis. Just as with Job 4-4, the basic ideas are the same whether we work with direct or alternating current.

Job 4-3 showed the dangers of electrical shock. It would be nice if there were a way to determine if there is a "ground fault" and turn the circuit off if there is. This problem is recognized in NEC Article 210-8, titled "Ground-Fault Circuit-Interrupter Protection for Personnel" and in many other articles of the NEC.

Figure 4-17 is the basic model for a single-phase distribution circuit. The current comes from the power source through line L_1 to the load R_1. The same current returns to the power supply through the neutral line N. The current flowing to the right through line L_1 is exactly the same as the current flowing to the left in line N. If you subtract the two currents, the result is zero.

The same type of circuit is shown in Fig. 4-20. Figure 4-20a shows how a fault from the load to the case or equipment housing can create a hazard for workers. If a person touches the case, a circuit is created that goes through the fault, the person, the ground, and back to the power source.

A diagram of the circuits is shown in Fig. 4-20b. This figure shows how the fault circuit goes around the GFCI device. The current in line L_1 is the sum of the other two currents:

$$I_1 = I_2 + I_3$$

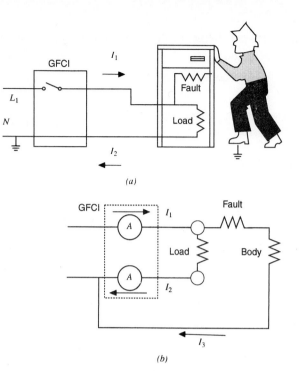

 (a)

(b)

Figure 4-20

The fault current I_3 creates an imbalance between I_1 and I_2. This imbalance is sensed by two current-measuring devices. When the imbalance gets large enough, the GFCI opens the switch and "interrupts" the circuit. The level of imbalance is usually about 5 mA.

Figure 4-15 shows this value in relation to the physiological effects of shock current. The level of 5 mA is the level at which you would get a mild sensation but probably nothing more severe.

JOB 4-6

Intermediate Review of Series Circuits

A series circuit is one in which the electrons may flow in only ____(1)____ path.

Finding the total voltage. The total voltage in a series circuit may be found by use of the following formulas:

$$V_T = V_1 + V_2 + \underline{\qquad(2)\qquad} \qquad \text{4-2}$$
$$V_T = \underline{\qquad(3)\qquad} \times R_T \qquad \text{4-7}$$

Finding the total resistance. The total resistance in a series circuit may be found by use of the following formulas:

$$R_T = R_1 + R_2 + \underline{\qquad(4)\qquad} \qquad \text{4-3}$$
$$V_T = I_T \times \underline{\qquad(5)\qquad} \qquad \text{4-7}$$

Finding the total current.   The total current in a series circuit may be found by use of the following formulas:

$$I_T = I_1 = \underline{\quad (6) \quad} = I_3 \qquad\qquad 4\text{-}1$$

$$\underline{\quad (7) \quad} = I_T \times R_T \qquad\qquad 4\text{-}7$$

Self-Test 4-11

A 12-Ω spotlight in a theater is connected in series with a dimming resistor of 31 Ω. If the voltage drop across the light is 31.2 V, find (*a*) the current in the light, (*b*) the current in the dimmer, (*c*) the total current, (*d*) the voltage drop across the dimmer, and (*e*) the total voltage and total resistance.

Solution

The diagram for the circuit is shown in Fig. 4-21.

a. Find the light current.

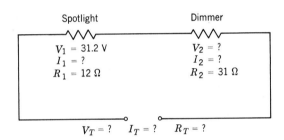

Figure 4-21

$$V_1 = I_1 \times \underline{\quad (1) \quad} \qquad\qquad (4\text{-}4)$$
$$31.2 = \underline{\quad (2) \quad} \times 12$$
$$\frac{31.2}{(3)} = I_1$$
$$I_1 = \underline{\quad (4) \quad} \text{ A} \qquad Ans.$$

b. Find the current in the dimmer.

$$I_T = I_1 = \underline{\quad (5) \quad} \qquad\qquad (4\text{-}1)$$
$$I_T = \underline{\quad (6) \quad} = I_2$$
$$I_2 = \underline{\quad (7) \quad} \text{ A} \qquad Ans.$$

c. The total current $= I_T = \underline{\quad (8) \quad}$ A   *Ans.*

d. Find the voltage drop across the dimmer.

$$V_2 = \underline{\quad (9) \quad} \times R_2 \qquad\qquad (4\text{-}5)$$
$$V_2 = \underline{\quad (10) \quad} \times 31 = \underline{\quad (11) \quad} \text{ V} \qquad Ans.$$

e. Find the total voltage.

$$V_T = V_1 + \underline{\quad (12) \quad} \qquad\qquad (4\text{-}2)$$
$$V_T = \underline{\quad (13) \quad} + 80.6$$
$$V_T = \underline{\quad (14) \quad} \text{ V} \qquad Ans.$$

Find the total resistance.

$$R_T = R_1 + \frac{(15)}{}$$ (4-3)
$$R_T = \frac{(16)}{} + 31$$
$$R_T = \frac{(17)}{} \ \Omega \quad Ans.$$

Check: If we have been correct in our calculations, our values for the total voltage, total current, and total resistance should satisfy the Ohm's law formula for total values.

$$V_T = \frac{(18)}{} \times R_T$$ (4-7)
$$\frac{(19)}{} = 2.6 \times 43$$
$$111.8 = \frac{(20)}{}$$

The equation ___(does/does not)___ check.

Exercises

Applications

1. Three resistors are connected in series. R_1 has a resistance of 40 Ω. The second resistor causes a 6-V drop, and the third resistor has a resistance of 120 Ω. Find the total voltage across the circuit if the series current is 0.3 A.

2. A 40-Ω resistor is in series with a 25-Ω resistor. Find the total voltage if the series current is 0.5 A.

3. Three resistors are in series, $I_1 = 0.5$ A, $R_1 = 2$ Ω, $R_2 = 3$ Ω, and $V_3 = 3.5$ V. Find the total voltage.

4. Two resistors and a motor are connected in series. The motor current is 3 A, and its resistance is 25 Ω. The voltage across each resistor is 37.5 V. Find (*a*) the total current, (*b*) the total voltage, and (*c*) the total resistance.

5. A voltage divider consists of a 3-, a 5-, and a 10-kΩ resistor in series. The series current is 0.015 A. Find (*a*) the voltage drop across each resistor, (*b*) the total voltage, and (*c*) the total resistance.

6. A lamp and two bells are connected in series in a burglar-alarm circuit. The lamp draws a current of 0.25 A and has a resistance of 160 Ω. Each bell uses 35 V. Find (*a*) the total current, (*b*) the total voltage, and (*c*) the total resistance.

7. Three resistors are in series: $V_1 = 31.2$ V, $V_2 = 48$ V, $I_2 = 1.2$ A, and $R_3 = 44$ Ω. Find (*a*) the total current, (*b*) the total voltage, and (*c*) the total resistance of the circuit.

8. In a series circuit, $I_1 = 0.5$ A, $R_1 = 35$ Ω, $V_2 = 91$ V, $R_3 = 23$ Ω, and $V_4 = 40$ V. Find the total voltage.

9. Three resistances are in series. They use 8, 10, and 14 V, respectively. The total resistance of the circuit is 80 Ω. Find all missing values of current, voltage, and resistance, including the total values.

10. Three resistances of 30, 40, and 50 Ω are in series. The total voltage impressed across the circuit is 24 V. Find all missing values of current, voltage, and resistance, including the total values.

(See CD-ROM for Test 4-2)

Checkup on Formulas Involving Addition and
Subtraction (Diagnostic Test)

In our next electrical job we shall be required to solve some formulas that are
slightly different from any we have solved up to this point. The following
10 problems are of this type. If you have any difficulty with them, see Job 4-8,
which follows.

Exercises

Practice

1. Using the formulas $V_T = V_1 + V_2 + V_3$, find V_2 if $V_T = 78$, $V_1 = 17$,
   and $V_3 = 32$.
2. Using the formula $P = 2L = 2W$, find L if $P = 80$ and $W = 14$.
3. Using the formula $I_T = I_1 + I_2 + I_3$, find I_3 if $I_T = 5$ A, $I_1 = 1.6$ A, and
   $I_2 = 2.3$ A.
4. Find R in the equation $7 + R = 5.4 + 19$.
5. Find V in the equation $3V + 8 = 29$.
6. Find I in the equation $I - 5 = 40$.
7. Find I in the equation $3I - 4 = 32$.
8. Find R in the formula $R + r = V/I$ if $V = 60$, $I = 5$, and $r = 9$.

Applications

9. What resistance must be placed in series with six 15-V Christmas tree lights
   in order to operate them on a 110-V circuit? Each light requires 0.8 A.
10. What series resistor is necessary to operate a circuit requiring 79 V at 0.3 A if
    it is to be operated from a 110-V line?

Brushup on Formulas Involving Addition and Subtraction

Solving a formula means finding the value of the unknown letter in the formula.
In Job 3-9, after substituting the numbers for the letters, we obtained statements of
equality such as $2 \times R = 10$ or $12 = 3 \times R$. In each instance, a number *multiplied*
by a letter was equal to another number. We obtained the value of the unknown
letter by *eliminating* the number multiplied by it. This was accomplished by *dividing both sides* of the equality sign by that *same* number. In general, to solve any formula or equation, we must eliminate all numbers and letters which appear on the
same side of the equality sign as the *unknown* letter. This is done by applying the
basic principle given in Job 3-9.

BASIC PRINCIPLE

Any mathematical operation performed on one side of an equality sign
must also be performed on the other side.

In simple language this says, "whatever we do to one side of an equality sign must also be done to the other side."

Formulas Involving Addition

 RULE The same number may be subtracted from both sides of an equality sign without destroying the equality.

Self-Test 4-12

Solve the equation $x + 3 = 7$.

Solution

Solving an equation means to get the letter all alone on one side of the equality sign, with its value on the other side. Therefore, all we have to do to solve the equation is to get the x alone. This will be accomplished if we can get rid of the number _____(1)_____. We can do this by applying the *opposite* operation of _____(2)_____ 3 *from both sides* of the equality sign.

$$
\begin{array}{rcl}
x + 3 &=& 7 \\
-\,3 & & -3 \\
\hline
x + 0 &=& \underline{\quad(3)\quad} \\
x &=& 4 \qquad Ans.
\end{array}
$$

Self-Test 4-13

Solve the equation $y + 4 = 9$.

Solution

We can get y alone on the left side of the equality sign if we can eliminate the number $+4$ from that side. This can be done by applying the *opposite* operation of *subtracting* _____(1)_____ *from both sides*.

$$
\begin{array}{rcl}
y + \quad 4 &=& 9 \\
-\quad 4 & & -4 \\
\hline
y + \underline{\quad(2)\quad} &=& 5 \\
y &=& \underline{\quad(3)\quad} \qquad Ans.
\end{array}
$$

Example 4-14

Using the formula $R_T = R_1 + R_2 + R_3$, find the resistance R_1 if $R_T = 100\ \Omega$, $R_2 = 20\ \Omega$, and $R_3 = 40\ \Omega$.

Solution

1. Write the formula. $R_T = R_1 + R_2 + R_3$ (4-3)
2. Substitute numbers. $100 = R_1 + 20 + 40$

3. Simplify (add 20 + 40). $100 = R_1 + 60$

We can get R_1 alone on the right side of the equality sign if we can eliminate the number +60 from that side. This can be accomplished by applying the *opposite* operation of *subtracting 60 from both sides* of the equality sign.

4. Subtract 60 from both sides.

$$\begin{array}{r} 100 = R_1 + 60 \\ -\ 60 \qquad -\ 60 \\ \hline \end{array}$$

5. Subtract.

$$40 = R_1 \ + 0$$

6. The resistance is

$$R_1 = 40\ \Omega \quad Ans.$$

Exercises

Practice

Solve each of the following equations for the value of the unknown letter:

1. $V + 3 = 9$
2. $2 + R = 10$
3. $10 + P = 80$
4. $110 = V + 60$
5. $V + 18 = 70$
6. $I + 2\frac{1}{2} = 3\frac{1}{2}$
7. $V + 40 + 30 + 25 = 120$
8. $I + \frac{1}{2} = 5$
9. $I + 3.5 = 10.8$
10. $20 + 10 + V = 110$
11. $V + 6.8 = 35$
12. $I + 2\frac{1}{2} = 15\frac{1}{2}$
13. $R + \frac{3}{4} = 6$
14. $V + 12.9 = 75.6$
15. $I + 0.2 + 1.3 = 7.6$
16. $12.6 + 6.3 + V = 120$
17. $110 = 35 + 12.6 + 6.3 + V$
18. $I + 2\frac{1}{2} = 8.4$
19. Using the formula $R_T = R_1 + R_2 + R_3$, find the resistance R_1 if $R_T = 175$, $R_2 = 40$, and $R_3 = 15$.
20. Using the formula $P_T = P_1 + P_2 + P_3$, find the power P_3 if $P_T = 1100$, $P_1 = 300$, and $P_2 = 275$.
21. Using the formula $I_T = I_1 + I_2$, find the current I_2 if $I_T = 99$ and $I_1 = 26$.
22. Using the formula $I_s + I_m = I$, find the shunt current I_s if the line current $I = 1.64$ and the meter current $I_m = 0.014$.
23. Using the formula $C_1 + C_2 + C_3 = C_T$, find C_1 if $C_T = 0.000\ 25$, $C_2 = 0.000\ 12$, and $C_3 = 0.000\ 05$.
24. Using the formula $V_1 + (I_2 \times R_2) = V_T$, find V_1 if $I_2 = 2$, $R_2 = 25$, and $V_T = 117$.
25. Using the formula $I_T = I_1 + (V_2/R_2)$, find I_1 if $I_T = 1.5$, $V_2 = 6.3$, and $R_2 = 126$.

Formulas Involving Subtraction

 RULE   The same number may be added to both sides of an equality sign without destroying the equality.

Self-Test 4-15

Solve the equation $x - 3 = 7$.

Solution

Solving an equation means to get the letter all alone on one side of the equality sign and its value on the other side. We can get x alone on the left side of the equality sign if we can get rid of the number ____(1)____. We can do this by applying the *opposite* operation of ____(2)____ 3 to both sides of the equality sign.

$$
\begin{array}{ccccc}
x & - & 3 & = & 7 \\
 & + & 3 & + & 3 \\
\hline
x & + & \underline{\quad(3)\quad} & = & 10 \\
 & & x & = & \underline{\quad(4)\quad} \quad \textit{Ans.}
\end{array}
$$

Self-Test 4-16

Using the series-circuit formula $V - Ir = IR$, find the voltage V in the circuit shown in Fig. 4-22.

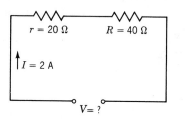

Figure 4-22

Solution

1. Write the formula. $V - Ir = IR$
2. Substitute numbers. $V - 2 \times 20 = 2 \times$ _____
3. Multiply numbers. $V -$ _____ $=$ 80
4. Add 40 to both sides. $V - 40 =$ 80
   $$+\,40 = +\,40$$
5. Add. $V + 0 =$ _____
6. The voltage is $V =$ _____ V *Ans.*

Practice

Solve each of the following equations for the value of the unknown letter:

1. $R - 5 = 12$
2. $12 = V - 3$
3. $I - 4 = 18$
4. $30 + R - 10 = 50$
5. $I - \frac{1}{2} = 4\frac{1}{2}$
6. $P - 3.2 = 8.3$
7. $9 = I - 3.5$
8. $V - 2.2 = 6.3$
9. $V - 17 = 62$
10. $82 = R - 14$
11. $I - 7\frac{3}{4} = 3$
12. $I - 1\frac{1}{2} = 6\frac{3}{4}$
13. $12 = R - 3\frac{1}{4}$
14. $72\frac{1}{2} = R - 5\frac{1}{4}$
15. $120 = R + 70 - 30$
16. $I - 0.045 = 0.85$
17. $C - 0.08 = 0.019$
18. $0.000\,25 = C - 0.000\,05$
19. Using the formula $C = S - P$, find S if $C = \$18.23$ and $P = \$3.60$.
20. Using the formula $I_1 = I_T - I_2$, find I_T if $I_1 = 8$ and $I_2 = 6$.
21. Using the formula $V_T - V_h = V_R$, find V_T if $V_R = 110$ and $V_h = 62$.
22. Using the formula $V_T - (I_1 \times R_1) = V_2$, find V_T if $I_1 = 2$, $R_1 = 30$, and $V_2 = 57$.
23. Using the formula $S - 5L = A$, find S if $L = 3.2$ and $A = 1.05$.
24. Using the formula $V - R = RM$, find V if $R = 8$ and $M = 2.5$.
25. Using the formula $R_1 = R_2 - V/I$, find R_2 if $R_1 = 9$, $V = 2.5$, and $I = 0.025$.

(See CD-ROM for Test 4-3)

Transposition. The method of eliminating a number from one side of an equality sign by adding or subtracting the same number may be shortened by the method know as *transposition*. Let us investigate the following four examples in an effort to determine the rule for the transposition of quantities.

Example 4-17

Find R in the equation $R + 2 = 10$.

Solution

1. Write the equation. $\qquad R + 2 = 10$
2. Subtract 2 from both sides. $\quad R + 2 - 2 = 10 - 2$
3. Subtract. $\qquad\qquad\qquad R + 0 = 10 - 2$
$$R = 8 \qquad Ans.$$

> **Hint for Success**
>
> The expressions that are joined by the equals sign (=) to form an equation are called the *members of the equation*. Thus, in the equation $R + 2 = 10$, the left member is $R + 2$ and 10 is the right member.

Example 4-18

Find V in the equation $8 = V + 3$.

Solution

1. Write the equation.	$8 = V + 3$
2. Subtract 3 from both sides.	$8 - 3 = V + 3 - 3$
3. Subtract.	$8 - 3 = V + 0$
	$5 = V$
or	$V = 5$ *Ans.*

Example 4-19

Find S in the equation $S - 4 = 10$.

Solution

1. Write the equation.	$S - 4 = 10$
2. Add 4 to both sides.	$S - 4 + 4 = 10 + 4$
3. Add.	$S + 0 = 10 + 4$
	$S = 14$ *Ans.*

Example 4-20

Find I in the equation $9 = I - 5$.

Solution

1. Write the equation.	$9 = I - 5$
2. Add 5 to both sides.	$9 + 5 = I - 5 + 5$
3. Add.	$9 + 5 = I + 0$
	$14 = I$
or	$I = 14$ *Ans.*

Notice that in each of the last four examples, the number to be eliminated—the number in red—seems to have *moved* from one side of the equality sign to the *other side*. Also, the sign in front of the number changed from $+$ to $-$ or from $-$ to $+$. This gives us the following rule for transposing.

 RULE   Plus or minus quantities may be moved from one side of an equality sign to the other if the sign of the quantity is changed from $+$ to $-$ or from $-$ to $+$.

Example 4-21

Find V in the equation $V + 8 = 12$.

Solution

1. Write the equation. $V + 8 = 12$
2. Transpose the 8. $V = 12 - 8$
3. Subtract. $V = 4$ *Ans.*

Example 4-22

Find R in the equation $9 = R + 3$.

Solution

1. Write the equation. $9 = R + 3$
2. Transpose the 3. $9 - 3 = R$
3. Subtract. $6 = R$
or $R = 6$ *Ans.*

Example 4-23

Find S in the equation $S - 5 = 16$.

Solution

1. Write the equation. $S - 5 = 16$
2. Transpose the 5. $S = 16 + 5$
3. Add. $S = 21$ *Ans.*

Example 4-24

Find I in the equation $4 = I - 9$.

Solution

1. Write the equation. $4 = I - 9$
2. Transpose the 9. $4 + 9 = I$
3. Add. $13 = I$
or $I = 13$ *Ans.*

Self-Test 4-25

Using the series-circuit formula $R + r = V/R$, find the resistance r in the circuit shown in Fig. 4-23.

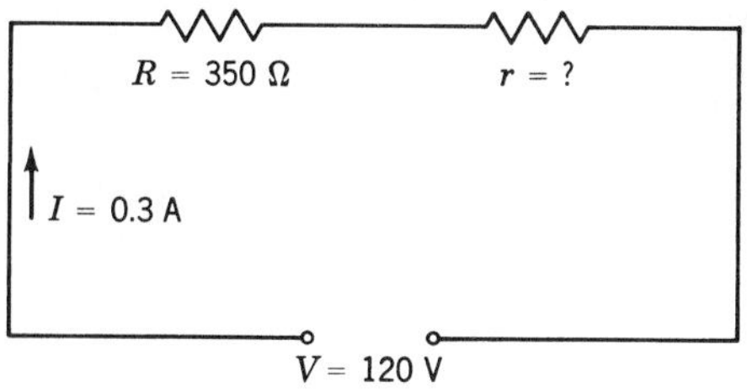

R = 350 Ω   r = ?

I = 0.3 A

V = 120 V

Figure 4-23

Solution

1. Write the formula.

$$R + r = \frac{V}{I}$$

2. Substitute numbers.

$$350 + \underline{\hspace{2cm}} = \frac{120}{0.3}$$

3. Divide numbers.

$$350 + r = \underline{\hspace{2cm}}$$

4. Transpose the 350. (A + sign is understood to be present in front of the 350.)

$$r = \underline{\hspace{2cm}} - 350$$

5. Subtract.

$$r = \underline{\hspace{2cm}} \, \Omega \qquad Ans.$$

Exercises

Practice

Solve the following equations by transposition:

1. $V + 3 = 14$
2. $16 = V + 3$
3. $2 + R = 8$
4. $13 = I + 2$
5. $6 + 4 + V = 20$
6. $3 + R + 2 = 15$
7. $21 = 4 + 5 + I$
8. $6 + I = 15$
9. $P - 2 = 6$
10. $14 = R - 4$
11. $0.06 + I = 1.6$
12. $20 = V - 2$
13. $18 - 3 + R = 25$
14. $16 = R - 4 - 5$
15. $R + 6.7 = 18.2$
16. $I + 0.045 = 0.09$
17. $0.025 = C + 0.004$
18. $X + 2.6 + 1.05 + 3.0 = 9.4$
19. $R + 7 - 2.40 + 3.60 = 19.68$
20. $I + 0.07 = 3.4$
21. $X + 3\frac{1}{2} = 5$
22. $I + 0.03 = 6.3$
23. Using the formula $P_T = P_1 + P_2 + P_3$, find P_1 if $P_T = 600$, $P_2 = 120$, and $P_3 = 300$.
24. Using the formula $I_1 = I_T - I_2$, find I_T if $I_1 = 8$ and $I_2 = 3$.

25. Using the formula $V_T = V_1 + V_2 + V_3$, find V_3 if $V_T = 110$, $V_1 = 30$, and $V_2 = 50$.
26. Using the formula $R_T = R_1 + R_2 + R_3$, find R_2 if $R_T = 245$, $R_1 = 85$, and $R_3 = 90$.
27. Using the formula $I_s = I_L - I_m$, find I_L if $I_s = 75$ and $I_m = 1.5$.
28. Using the formula $V_T - V_h = V_R$, find V_T if $V_R = 97.2$ and $V_h = 6.3$.
29. For $C_T = C_1 + C_2$, find C_1 if $C_T = 0.000\ 35$ and $C_2 = 0.0001$.
30. For $C_T = C_1 + C_2$, find C_2 if $C_T = 0.004$ and $C_1 = 0.000\ 15$.

(See CD-ROM for Test 4-4)

JOB 4-9

Control of Current in a Series Circuit

In Job 3-9 we learned that resistance affects the current in an electric circuit. The greater the resistance, the smaller the current; the smaller the resistance, the greater the current. There are many times when a certain current must be maintained in a circuit. We can accomplish this by adjusting the resistance of the circuit so as to obtain any required current.

Example 4-26

How much resistance must be added to the circuit shown in Fig. 4-24 in order to allow only the rated current of 2 A to flow?

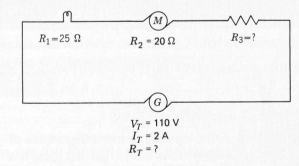

$R_1 = 25\ \Omega$ $R_2 = 20\ \Omega$ $R_3 = ?$

$V_T = 110$ V
$I_T = 2$ A
$R_T = ?$

Figure 4-24

Solution

The diagram for the circuit is shown in Fig. 4-24. If we can find the total resistance required to limit the current to 2 A, then we can use formula (4-3) to find the missing extra resistance.

Find R_T.

1. Write the formula. $V_T = I_T \times R_T$ (4-7)

2. Substitute numbers. $110 = 2 \times R_T$

3. Solve for R_T. $\dfrac{110}{2} = R_T$

4. Divide. $R_T = 55 \ \Omega$ *Ans.*

Find R_3, the resistance to be added.

1. Write the formula. $R_T = R_1 + R_2 + R_3$ (4-3)

2. Substitute numbers. $55 = 25 + 20 + R_3$

3. Simplify (add 25 + 20). $55 = 45 + R_3$

4. Transpose the 45. $55 - 45 = R_3$

5. Subtract. $10 = R_3$

or $R_3 = 10 \ \Omega$ *Ans.*

Example 4-27

What resistance must be added in series with a lamp rated at 12 V and 0.3 A in order to operate it from a 24-V source?

Solution

The diagram for the circuit is shown in Fig. 4-25. If we can discover how many volts are available in excess of that required by the lamp, then we can find the resistance which will use up that extra voltage at the rated 0.3 A.

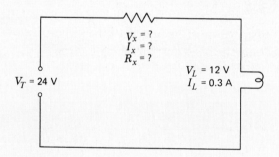

Figure 4-25

Find the extra voltage supplied V_x.

1. Write the formula. $V_T = V_x + V_L$ (4-2)

2. Substitute numbers. $24 = V_x + 12$

3. Transpose the 12. $24 - 12 = V_x$

4. Subtract. $12 = V_x$

or $V_x = 12 \text{ V}$ *Ans.*

Find the current I_x in the unknown resistor.

1. Write the formula. $I_T = I_x = I_L$ (4-1)

2. Substitute numbers. $I_T = I_x = 0.3$

                                   $I_x = 0.3 \text{ A}$ *Ans.*

Find the resistance R_x which will use up the extra 12 V at the rated series current of 0.3 A.

1. Write the formula.	$V_x = I_x \times R_x$		(2-1)
2. Substitute numbers.	$12 = 0.3 \times R_x$		
3. Solve for R_x.	$\dfrac{12}{0.3} = R_x$		
4. Divide.	$40 = R_x$		
or	$R_x = 40\ \Omega$	*Ans.*	

Self-Test 4-28

In the voltage-divider circuit shown in Fig. 4-26, the total load resistance between points A and B also serves to "bleed off" any charge on the filter capacitors after the rectifier is turned off. If the bleeder current I_b is to be limited to 0.015 A, find the resistance of R_1.

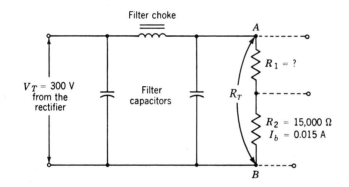

Figure 4-26 Voltage divider with no loads attached.

Solution

If we can find the total resistance of the load which will limit the bleeder current to 0.015 A, then we can use the formula for total ____(1)____ to find R_1.

Find R_T.

a. Write the formula. $V_T = \underline{\quad(2)\quad} \times R_T$ (4-7)

b. Substitute numbers. $\underline{\quad(3)\quad} = 0.015 \times R_T$

c. Solve for R_T. $\dfrac{300}{(4)} = R_T$

d. Divide. $R_T = \underline{\quad(5)\quad}\ \Omega$ *Ans.*

Find R_1, the resistance to be added to R_2 to make a total of $\underline{\quad(6)\quad}\ \Omega$.

a. Write the formula. $R_T = R_1 + R_2$ (4-3)

b. Substitute numbers. $20{,}000 = R_1 + \underline{\quad(7)\quad}$

c. Transpose the 15,000. $20{,}000 - 15{,}000 = \underline{\quad(8)\quad}$

d. Subtract. $\underline{\quad(9)\quad} = R_1$

or $R_1 = \underline{\quad(10)\quad}\ \Omega$ *Ans.*

Applications

1. How much resistance must be connected in series with a 30-Ω lamp rated at 2 A if it is to be used on a 110-V line?
2. A 10-Ω lamp rated at 5 A is to be operated from a 120-V line. How much resistance must be added in series to reduce the current to the desired value?
3. A motor of 22 Ω resistance and a signal lamp of 25 Ω resistance are connected in series across 110 V. Find the series resistor that must be added to limit the current to 2 A.
4. A boy wants to illuminate five houses on his model railroad, using five 20-Ω lamps in series. What resistance must be connected in series in order to limit the current to the 0.5 A needed by the lamps if the circuit is to be operated from the ordinary 110-V house line?
5. What resistance must be placed in series with a 12-Ω bell if it is to draw exactly ¼ A from a 24-V source?
6. Five ordinary 0.90-A 110-V lamps must be used in series when operated from the 550-V subway system. If only one of these lamps were used, what resistance would have to be placed in series with it to operate it from the 550-V line?
7. A 3-V airplane instrument lamp is to be operated from the 12-V electrical system. What resistance should be inserted in series so that the lamp will receive its rated current of 0.1 A?
8. The Panasonic AM-FM RE 7329 receiver uses an output stage similar to that shown in Fig. 4-27. If the effective $R = 100$ Ω, the collector current $I_C = 5$ mA, and the collector supply voltage $V_{CC} = 9$ V, find the collector voltage V_C.
9. An 80-A motor is connected to a 250-V generator through leads which have a resistance of 0.3 Ω for each lead. What is the voltage available at the motor?
10. The wiring in a house has a resistance of 0.4 Ω. What is the voltage available at an electric range using 12 A if the voltage at the meter is 117 V?
11. A floodlamp of 10 Ω resistance is connected in series with a variable dimming resistor of 0 to 45 Ω and is operated from a 110-V line. What are the maximum and minimum currents that may be supplied to the lamp?
12. The voltage drop in the line cord of a portable transmitter must not exceed 0.25 V. If the transmitter draws 20 A from a 12-V battery, find the maximum resistance of the line cord.

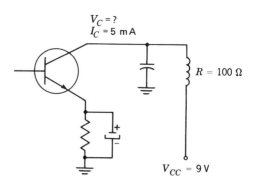

$V_C = ?$
$I_C = 5$ mA

$R = 100$ Ω

$V_{CC} = 9$ V

Figure 4-27   Output stage of a transistor radio.

13. A portable transmitter of 0.575 Ω resistance is to operate at 11.5 V. What is the maximum resistance of the line cord connecting the transmitter to a 12-V battery?

14. A relay has a coil resistance of 550 Ω and is designed to operate on 150 mA. If the relay is used on a 120-V source, what value of resistance should be connected in series with the coil?

Review of Series Circuits

Definition. A series circuit is a circuit in which the electrons can flow in only one path.

RULE 1

The total current is equal to the current in any part of the circuit.

$$I_T = I_1 = I_2 = I_3 \qquad\qquad \text{4-1}$$

RULE 2

The total voltage is equal to the sum of the voltages across all the parts of the circuit.

$$V_T = V_1 + V_2 + V_3 \qquad\qquad \text{4-2}$$

RULE 3

The total resistance is equal to the sum of the resistances of all the parts of the circuit.

$$R_T = R_1 + R_2 + R_3 \qquad\qquad \text{4-3}$$

RULE 4

Ohm's law may be used for any part of a series circuit.

$$
\begin{aligned}
V_1 &= I_1 \times R_1 & &\text{4-4} \\
V_2 &= I_2 \times R_2 & &\text{4-5} \\
V_3 &= I_3 \times R_3 & &\text{4-6}
\end{aligned}
$$

RULE 5   Ohm's law may be used for total values in a series circuit.

$$V_T = I_T \times R_T$$ 4-7

Self-Test 4-29

In Fig. 4-28 the type 2N247 transistor has a collector current I_C of 0.01 A and a collector voltage V_C of 9.2 V. Find the total supply voltage V_{CC}.

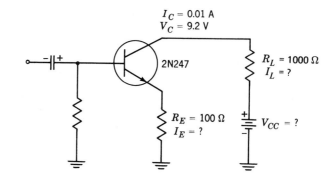

I_C = 0.01 A
V_C = 9.2 V

2N247

R_L = 1000 Ω
I_L = ?

R_E = 100 Ω
I_E = ?

V_{CC} = ?

Figure 4-28   The total supply voltage is equal to the sum of the voltages around the circuit.

Solution

Since the electrons complete the circuit flowing from the emitter to the collector of the transistor, it is a ___(1) (series/parallel)___ circuit.

Find the current in each part of the circuit.

$$I_T = I_E = \underline{\quad (2) \quad} = I_L \qquad (4\text{-}1)$$
$$I_T = I_E = 0.01 = I_L$$

Find the voltage across each part of the circuit.

$$V_{RE} = I_E \times \underline{\quad (3) \quad} \qquad (4\text{-}4)$$
$$V_{RE} = 0.01 \times \underline{\quad (4) \quad} = \underline{\quad (5) \quad} \text{ V}$$
$$V_L = \underline{\quad (6) \quad} \times R_L$$
$$V_L = 0.01 \times \underline{\quad (7) \quad} = \underline{\quad (8) \quad} \text{ V} \qquad (4\text{-}5)$$

Find the total supply voltage V_{CC}.

$$V_{CC} = V_{RE} + V_C + \underline{\quad (9) \quad} \qquad (4\text{-}2)$$
$$V_{CC} = \underline{\quad (10) \quad} + 9.2 + 10 = \underline{\quad (11) \quad} \text{ V} \qquad Ans.$$

Exercises

Applications

1. Three resistors are connected in series. $V_1 = 24$ V, $I_2 = 2$ A, $R_2 = 30$ Ω, and $R_3 = 13$ Ω. Find all missing values of voltage, current, and resistance, including the total values.

132   Chapter 4   Series Circuits

2. The two field coils of a generator have a resistance of 55 Ω each and are connected in series across the brushes, which deliver 110 V. What is the current in the field coils?

3. An electric heater whose resistance is 10 Ω is to be used on a 220-V line. The maximum current permitted through it is 10 A. What resistance must be added in series with the heater in order to hold the current to 10 A?

4. In the universal transistor circuit shown in Fig. 4-29, the resistances between points A and B are in series. Find (a) the total current and (b) the voltage drop across each resistance.

5. The field coils of a motor draw 4 A from a 112-V line. What is the resistance of the coils? If a 14-Ω resistor is added in series with the coils, find the current in the coils and the voltage across the coils.

6. What value of resistance must be placed in series with two 50-Ω lamps, each taking 50 V, if they are to be operated from a 220-V line?

7. Three resistors are connected in series across 220 V. $R_1 = 15\ \Omega$, $R_2 = 25\ \Omega$, and $R_3 = 60\ \Omega$. (a) Find the total current. (b) If the maximum permissible current is 2 A, how much resistance must be added in series to keep the current at this value?

8. In order to dim a bank of stage lights, a rheostat may be connected in series with the lights to reduce the current and therefore the brightness of the lights. What value of resistance must be connected in series with a lamp bank drawing 20 A from a 120-V line in order to reduce the total current drawn to 5 A?

9. Two 25-V 25-Ω incandescent lamps are connected in series with a demonstration motor requiring 30 V at 1 A. What extra resistance should be added in series in order to draw the required current from a 110-V line?

10. When a voltmeter indicates its maximum rated voltage of 150 V, it is drawing a current of 0.01 A. What value of resistance is in series with the moving coil of the voltmeter if its resistance is 20 Ω?

(See CD-ROM for Test 4-5)

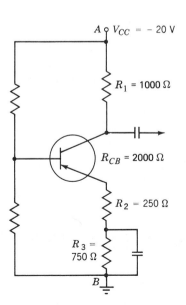

Figure 4-29   Universal transistor circuit.

Assessment

chapter **4**

1. There are three resistors in series, and the voltage drop across each resistor is 16 V, 22 V and 12 V, respectively. What is the total voltage applied to the circuit?

2. In a series circuit you find that the voltage drop across two of three resistors is 28 V each when 100 V is applied. What is the voltage drop across the third resistor?

3. You find an old-fashioned string of Christmas lights at a garage sale. The string of lights is rated for 115 V and contains 24 individual lamps that are all connected in series. Assuming that each lamp has the same resistance, and thus the same wattage rating, what is the voltage drop across each lamp when the lamps are all lit?

4. Refer to the voltage divider circuit in Fig. 4-30.

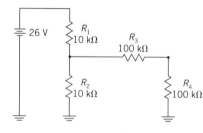

Figure 4-30

What is the voltage drop across R_4 if the voltage drop across R_3 is 6.3414 V?

5. How much current flows through the lamp in Fig. 4-31 when the switch is closed and the circuit completed?

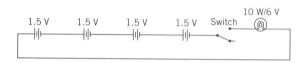

Figure 4-31

6. A current of 3.625 mA is flowing in a series circuit containing a 15-, a 23-, and a 44-Ω resistor. What is the total voltage drop across the three resistors?

7. Given the formula $V_T + 2.6 + 3.9 + 5.2 = 31.7$, what is the value of V_T?

8. How much resistance must be placed in series with a 12-V, 25-W fire alarm panel buzzer if the supply voltage in the circuit is 16 V?

9. Two 120-V 100-W lamps are connected in series. If the lamps are connected to a 240-V supply, what value of resistance will have to be placed in the line to limit the current to the required value?

INTERNET
ACTIVITIES

Internet Activity A

Use the following site to find the procedure for testing a ground fault interrupting duplex outlet, commonly found in a modern home.
http://www.nuaire.com/custsupport/manuals/accessories/gfiinfo.shtml

Internet Activity B

Use the following site to define the term *effective resistance* as it relates to series circuits.
http://www.unb.Ca/Physics/1940/resistance.fm.htm

Chapter 4 Solutions to Self-Tests and Reviews

Self-Test 4-3

a.1. I_2, 80

b. 80

b.1. R_1

b.2. 6, 3, 3, 12.8

1. V_2

2. 7.2, 20

c. 80

c.1. I_2

c.2. 80×10^{-6}

3. 6, 90,000, 90

d. 90

1. R_2

2. 250

Self-Test 4-7

1. current
2. resistance
3. 1500
4. 7500 Ω
5. 9000
6. 90
7. 9000
8. 10

Job 4-6 Review

1. one

---

2. V_3
3. I_T
4. R_3
5. R_T
6. I_2
7. V_T

Self-Test 4-11

1. R_1
2. I_1
3. 12
4. 2.6
5. I_2
6. 2.6
7. 2.6
8. 2.6
9. I_2
10. 2.6
11. 80.6
12. V_2
13. 31.2
14. 111.8
15. R_2
16. 12
17. 43
18. I_T
19. 111.8

---

20. 111.8
21. does

Self-Test 4-12

1. +3
2. subtracting
3. 4

Self-Test 4-13

1. 4
2. 0
3. 5

Self-Test 4-15

1. −3
2. adding
3. 0
4. 10

Self-Test 4-16

2. 40
3. 40
5. 120
6. 120

Self-Test 4-25

2. r

---

3. 400
4. 400
5. 50

Self-Test 4-28

1. resistance
2. I_T
3. 300
4. 0.015
5. 20,000
6. 20,000
7. 15,000
8. R_1
9. 5000
10. 5000

Self-Test 4-29

1. series
2. I_C
3. R_E
4. 100
5 1
6. I_L
7. 1000
8. 10
9. V_L
10. 1
11. 20.2

Parallel Circuits

1. Recognize a parallel circuit.
2. Explain why the voltage is the same across all branches in a parallel circuit.
3. Determine the total voltage V_T across the branches of a parallel circuit.
4. Calculate the total current I_T in a parallel circuit.
5. Calculate the total resistance R_T of a parallel circuit.
6. Apply Ohm's law to solve for the unknown values in a parallel circuit.
7. Add and subtract fractions having the same denominator.
8. Determine the least common denominator (LCD) for fractions containing different denominators.
9. Add and subtract fractions with different denominators.
10. Solve simple fractional equations for the value of the unknown letter.
11. Determine the division of current in a parallel circuit when the total current I_T and individual branch resistances are known.

Chapter 5 is similar to Chap. 4 except that it relates to parallel circuits. You will learn how to develop formulas to solve for values in parallel circuits. If you have difficulties with the mathematics, you may want to review Chap. 3 on formulas.

Total Voltage, Total Current, and Total Resistance in Parallel Circuits

Recognizing a parallel circuit.   A parallel circuit is a circuit connected in such a manner that the current flowing into it may *divide* and flow in *more than one path.* In the parallel circuit shown in Fig. 5-1, the current divides at point A, part of the current flowing through the lamp in path 1 and the rest of the current flowing through the motor in path 2.

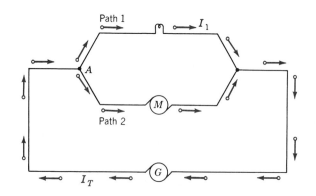

Figure 5-1   The current in a parallel circuit flows in more than one path.

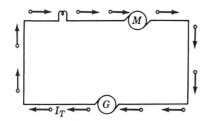

Figure 5-2   The current in a series circuit can flow in only one path.

The various paths of a parallel circuit are called the *branches* of the circuit. Notice how different this is from a series circuit shown in Fig. 5-2, in which the current can flow in *only one path*.

Total voltage in a parallel circuit.   Figure 5-3 shows a water-distribution system that might be called a parallel system because the water can flow in more than one pipe. A similar electrical system is shown in Fig. 5-4. In the circuit, the 110-V outlet acts as an electric pump and supplies an *equal* pressure to all the branches in parallel.

 RULE   The voltage across any branch in parallel is equal to the voltage across any other branch and is also equal to the total voltage.

$$V_T = V_1 = V_2 = V_3 = \text{etc.} \qquad\qquad \text{Formula 5-1}$$

Total current in a parallel circuit.   In Fig. 5-3 we can see that the total number of gallons per minute flowing in the main pipe equals the number of gallons per minute discharged by all the pipes together, or 12 gal/min. In the same way, the total current that flows in the circuit of Fig. 5-4 is equal to $2 + 4 + 6 = 12$ A.

RULE   The total current in a parallel circuit is equal to the sum of the currents in all the branches of the circuit.

$$I_T = I_1 + I_2 + I_3 + \text{etc.} \qquad\qquad \text{Formula 5-2}$$

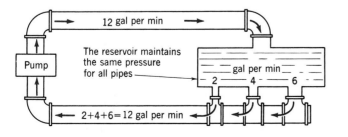

Figure 5-3   Water-distribution system in "parallel."

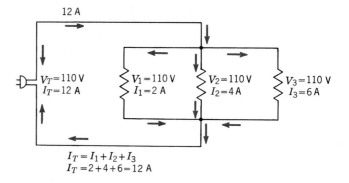

$$I_T = I_1 + I_2 + I_3$$
$$I_T = 2 + 4 + 6 = 12 \text{ A}$$

Figure 5-4 Three resistors in parallel. The voltages across all the branches are equal.

Distribution of current in parallel.   The current in a parallel circuit might be distributed as shown in Fig. 5-5. A total of 12 A is drawn from the line. At point A, 2 A is drawn off through R_1, leaving only 10 A to flow along to point B. At point B, 4 A is drawn off through R_2, leaving 6 A to flow through the rest of the circuit R_3 around to point C. At point C, this 6 A combines with 4 A from R_2 to give 10 A, which flows along to point D. Here it combines with the 2 A from R_1 to make up the total of 12 A available originally.

Line drop.   The distribution of the current as indicated above is based on the assumption that the connecting wires have no resistance, which is never true. In most instances, however, the resistance of the connecting wires is so small that we may neglect it completely. If the line wires are so long that their resistance is large, its effect must be included. This case will be taken up later in Job 6-4 under Line Drop.

Total Resistance in a Parallel Circuit

RULE   The total resistance in a parallel circuit is found by applying Ohm's law to the total values of the circuit.

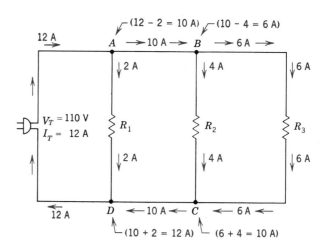

Figure 5-5 Distribution of current in a simple parallel circuit.

Formula $$V_T = I_T \times R_T \qquad (4\text{-}7)$$

Example 5-1

A toaster, a waffle iron, and a hot plate are connected in parallel across a house line delivering 110 V. The current through the toaster is 2 A, through the waffle iron 6 A, and through the hot plate 3 A. Find (*a*) the total current drawn from the line, (*b*) the voltage across each device, and (*c*) the total resistance of the circuit.

Solution

The diagram for the circuit is shown in Fig. 5-6.

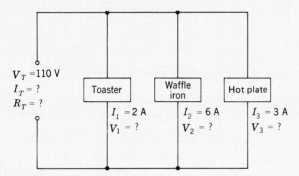

Figure 5-6

a. Find the total current I_T.
   1. Write the formula. $\qquad I_T = I_1 + I_2 + I_3 \qquad (5\text{-}2)$
   2. Substitute numbers. $\qquad I_T = 2 + 6 + 3 = 11 \text{ A} \qquad Ans.$
b. Find the voltage across each device.
   1. Write the formula. $\qquad V_T = V_1 = V_2 = V_3 \qquad (5\text{-}1)$
   2. Substitute numbers. $\qquad V_T = V_1 = V_2 = V_3 = 110 \text{ V} \qquad Ans.$
c. Find the total resistance R_T.
   1. Write the formula. $\qquad V_T = I_T \times R_T \qquad (4\text{-}7)$
   2. Substitute numbers. $\qquad 110 = 11 \times R_T$
   3. Solve for R_T. $\qquad \dfrac{110}{11} = R_T$
   4. Divide. $\qquad R_T = 10 \ \Omega \qquad Ans.$

Self-Test 5-2

An ammeter carrying 0.06 A is in parallel with a shunt resistor carrying 1.84 A, as shown in Fig. 5-7. If the voltage drop across the combination is 3.8 V, find (*a*) the total current and (*b*) the total resistance of the combination.

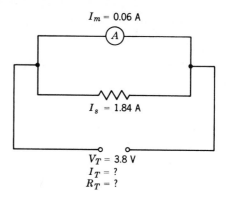

Figure 5-7 The shunt prevents large currents from damaging the meter.

Solution

a. Find the total current I_T.

 1. Write the formula. $\quad I_T = I_m +$ _____

 2. Substitute numbers. $\quad I_T =$ _____ $+ 1.84 =$ _____ A $\qquad$ *Ans.*

b. Find the total resistance R_T.

 1. Write the formula. _____ $= I_T \times R_T$ $\qquad\qquad$ (4-7)

 2. Substitute numbers. $\quad 3.8 =$ _____ $\times R_T$

 3. Solve for R_T. $\qquad\qquad \dfrac{3.8}{?} = R_T$

 4. Divide. $\qquad\qquad\qquad R_T =$ _____ Ω $\qquad$ *Ans.*

Applications

1. Two lamps each drawing 2 A and a third lamp drawing 1 A are connected in parallel across a 110-V line. Find (*a*) the total current drawn from the line, (*b*) the voltage across each lamp, and (*c*) the total resistance of the circuit.
2. The ignition coil and the starting motor of an automobile are connected in parallel across a 12-V battery through the ignition switch as shown in Fig. 5-8. Find (*a*) the total current drawn from the battery, (*b*) the voltage across the coil and the motor, and (*c*) the total resistance of the circuit.
3. Find the total current drawn by eight trailer-truck warning lights in parallel if each takes 0.5 A.

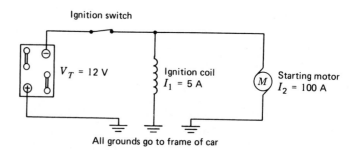

All grounds go to frame of car

Figure 5-8 The ignition coil and the starting motor of an automobile are in parallel.

4. A motor, a heating coil, and an ultraviolet lamp of a modern clothes drier are connected in parallel across 117 V. The lamp current is 1 A, the coil current is 4 A, and the motor current is 3 A. Find (*a*) the total current drawn and (*b*) the total resistance of the circuit.

5. In Table 13-1 the National Electrical Code specifies that No. 14 rubber-covered wire can safely carry only 20 A. How many 0.5-A lamps could be safely operated at a time on a line using this wire? How many lamps could be safely operated at one time if a 5-A electric iron were connected in the circuit?

6. A toaster drawing 2 A, a coffee percolator drawing 3.5 A, and a refrigerator motor drawing 4.5 A are connected in parallel across a 110-V line. Find (*a*) the total current drawn, (*b*) the voltage across each device, and (*c*) the total resistance.

7. A bank of ten 110-V 100-W lamps is connected in parallel in a stage lighting circuit. Each lamp uses 0.91 A. Find the total current drawn. Will a fuse rated at 10 A safely carry this load?

8. A 40-W lamp drawing 0.36 A, a 60-W lamp drawing 0.54 A, and a 100-W lamp drawing 0.9 A are connected in parallel across 110 V. Find (*a*) the total current, (*b*) the voltage across each lamp, and (*c*) the total resistance of the circuit.

9. Two 32-cd headlight lamps each drawing 3.9 A and two taillight lamps (4 cd, 0.85 A each) are wired in parallel to the 12-V storage battery. Find the total current drawn and the total resistance of the circuit.

10. A washing machine drawing 7.5 A, an electric fan drawing 0.85 A, and an electric clock drawing 0.02 A are in parallel with a 110-V line. Find the total current and the total resistance.

Using Ohm's Law in Parallel Circuits

JOB 5-2

In some instances, it may be impossible to find the total current by adding the individual currents because the individual currents may be unknown. Therefore, the current in each branch must be found before we can find the total current.

Example 5-3

The circuit of an electric clothes drier is shown in Fig. 5-9. The ultraviolet lamp R_1 is 120 Ω and draws 1 A. The heating coil R_2 is 30 Ω, and the motor R_3 draws a current of 4 A. Find (*a*) the voltage for each part and the total voltage, (*b*) the current in each part and the total current, and (*c*) the resistance of each part and the total resistance.

Hint for Success

An open in one branch of a parallel circuit results in zero current in that branch. However, all other branch currents remain the same. This is because the total voltage V_T still exists across the other branches even though one branch may be open.

$$V_T = ?$$
$$I_T = ?$$
$$R_T = ?$$

$$V_1 = ?$$
$$I_1 = 1\ A$$
$$R_1 = 120\ \Omega$$

$$V_2 = ?$$
$$I_2 = ?$$
$$R_2 = 30\ \Omega$$

$$V_3 = ?$$
$$I_3 = 4\ A$$
$$R_3 = ?$$

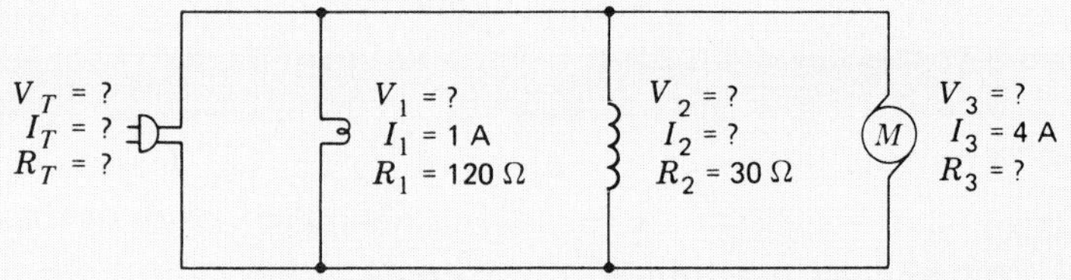

Figure 5-9

Solution

a. Find V_1.
    1. Write the formula. $V_1 = I_1 \times R_1$ (4-4)
    2. Substitute numbers. $V_1 = 1 \times 120 = 120$ V   *Ans.*

Find V_T and the voltage across each part of the circuit.
    1. Write the formula. $V_T = V_1 = V_2 = V_3$ (5-1)
    2. Substitute numbers. $V_T = V_1 = V_2 = V_3 = 120$ V

b. Find the current I_2.
    1. Write the formula. $V_2 = I_2 \times R_2$ (4-5)
    2. Substitute numbers. $120 = I_2 \times 30$
    3. Solve for I_2. $I_2 = \dfrac{120}{30} = 4$ A   *Ans.*

Find the total current I_T.
    1. Write the formula. $I_T = I_1 + I_2 + I_3$ (5-2)
    2. Substitute numbers. $I_T = 1 + 4 + 4 = 9$ A   *Ans.*

c. Find the resistance R_3.
    1. Write the formula. $V_3 = I_3 \times R_3$ (4-6)
    2. Substitute numbers. $120 = 4 \times R_3$
    3. Solve for R_3. $R_3 = \dfrac{120}{4} = 30$ Ω *Ans.*

Find the total resistance R_T.
    1. Write the formula. $V_T = I_T \times R_T$ (4-7)
    2. Substitute numbers. $120 = 9 \times R_T$
    3. Solve for R_T. $R_T = \dfrac{120}{9} = 13.3$ Ω *Ans.*

In Example 5-3, R_2 and R_3 are both equal to 30 Ω. They both draw 4 A of current. This gives us the following rules.

 RULE If the branches of a parallel circuit have the same resistance, then each will draw the same current.

 RULE If the branches of a parallel circuit have different resistances, then each will draw a different current. The larger the resistance, the smaller the current drawn.

Exercises

Applications

1. Two resistors of 3 and 6 Ω are connected in parallel across 18 V. Find (*a*) the voltage across each resistor, (*b*) the current in each resistor and the total current, and (*c*) the total resistance.

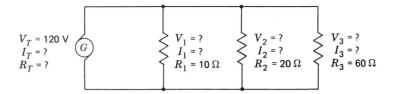

Figure 5-10

2. Find all missing values of voltage, current, and resistance in the circuit shown in Fig. 5-10.

3. A parallel circuit has three branches of 12, 6, and 4 Ω resistance. If the current in the 6-Ω branch is 4 A, what current will flow in each of the other branches? What is the total current?

4. The secondary of a power transformer is connected across the motors of three toy trains in parallel. Motor 1 has a resistance of 50 Ω and draws 0.4 A. Motor 2 has a resistance of 40 Ω. Motor 3 draws a current of 0.3 A. Find (a) the total voltage, (b) the total current, and (c) the total resistance.

5. Three resistors are in parallel. $I_1 = 12$ A, $V_2 = 114$ V, $R_2 = 19$ Ω, and $R_3 = 57$ Ω. Find (a) the voltage across each resistor and the total voltage, (b) the current in each resistor and the total current, and (c) the resistance of each resistor and the total resistance.

6. Three buzzers are wired in parallel. They draw currents of 0.2, 0.4, and 0.6 A, respectively. If their total resistance is 6 Ω, find (a) the resistance of each buzzer and (b) the voltage across each buzzer.

7. Four 1½-V lamps are wired in parallel. Three of these lamps draw a current of 0.05 A each. The fourth draws a current of 0.1 A. (a) What is the resistance of each filament? (b) What is the total current drawn? (c) What is the total voltage required? (d) What is the total resistance of the four filaments in parallel?

8. Solve the circuit shown in Fig. 5-11 for the values indicated.

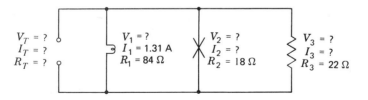

Figure 5-11

9. The following GE panel lights are to be wired in parallel from a 6.8-V source: two No. 40 lamps (0.15 A) and three No. 44 lamps (0.25 A). Find the total current, total voltage, and total resistance of the circuit.

10. What is the total current drawn from the 12-V automobile battery by two 4-Ω headlights and a 12-Ω taillight if they are all connected in parallel?

(See CD-ROM for Test 5-1)

Checkup on Addition and Subtraction of
Fractions (Diagnostic Test)

Our next electrical job considers problems which are solved by the addition of
fractions. In order to prepare for them, let us try the following problems, which are
often met by electricians and electronic technicians in their work. If you have any
difficulty with any of these, see Job 5-4, which follows.

Exercises

Practice

1. What is the total horsepower delivered by a $\frac{1}{3}$-, a $\frac{1}{4}$-, and a $\frac{1}{8}$-hp motor in parallel?
2. Add the following conductances in a circuit to obtain the total conductance: $\frac{1}{60}$, $\frac{1}{20}$, and $\frac{1}{10}$ S (siemens).
3. In rewinding the armature of a motor, the following thicknesses of insulation were used: $\frac{1}{16}$, $\frac{1}{32}$, and $\frac{3}{64}$ in. What was the total thickness of the insulation?
4. What length of a bolt is covered by a lock washer ($\frac{1}{32}$ in), a washer ($\frac{1}{8}$ in), and a nut ($\frac{5}{16}$ in)?
5. What is the total weight of three coils of wire weighing $16\frac{1}{4}$, $4\frac{1}{8}$, and $2\frac{1}{2}$ lb?
6. Add $\frac{1}{5}$, $\frac{1}{3}$, and $\frac{1}{10}$.
7. Add $\frac{3}{4}$, $\frac{7}{8}$, and $\frac{3}{16}$.
8. Add $\frac{3}{4}$, $\frac{5}{6}$, and $\frac{7}{8}$.
9. Add $\frac{5}{6}$, $\frac{1}{9}$, and $\frac{2}{3}$.
10. Add $\frac{1}{2}$, $\frac{5}{7}$, and $\frac{3}{4}$.
11. Subtract $1\frac{1}{2}$ from $3\frac{5}{8}$.
12. What is the difference between $9\frac{3}{8}$ and $4\frac{7}{16}$?
13. Subtract: $8 - 3\frac{3}{4}$.
14. Subtract: $5\frac{1}{4} - 2\frac{9}{16}$.
15. Subtract: $\frac{1}{24} - \frac{1}{40}$.

Brushup on Addition of Fractions

Adding Fractions with the Same Denominator

Example 5-4

1. 1 apple + 3 apples + 2 apples = _____ apples.
2. $1\,\Omega + 3\,\Omega + 2\,\Omega =$ _____ Ω.
3. $1\,A + 3\,A + 2\,A = 6$ _____.
4. $1\,V + 3\,V + 2\,V = 6$ _____.

Apparently, in order to add the same *kind* of thing, it is only necessary to add
the numbers involved. Thus

5. One-*eighth* + three-*eighths* + two-*eighths* = six-_____

6. or $\frac{1}{8} + \frac{3}{8} + \frac{2}{8} =$ ________

And $\frac{6}{8}$ may be simplified to $\frac{3}{4}$. *Ans.*

RULE **To add fractions with the same denominator, add the numerators and place the sum over the same denominator.**

Example 5-5

Add $\frac{3}{16}$, $\frac{5}{16}$, and $\frac{7}{16}$.

Solution

7. $$\frac{3}{16} + \frac{5}{16} + \frac{7}{16} = \frac{3 + 5 + 7}{16} = \frac{15}{?}$$

Adding Fractions with Different Denominators

Procedure

1. Find the *least common denominator.* (See Examples 5-7 and 5-8.)
2. Change the fractions to *equivalent* fractions using this new denominator.
3. Add these fractions with the same denominator.

Example 5-6

Add $\frac{1}{4} + \frac{3}{8}$.

Solution

The least common denominator (LCD) is the smallest number into which all the other denominators will divide evenly. In this example the denominators are 4 and 8. To find the LCD, we first find all the numbers that "make up" each of the denominators. These numbers are the "factors."

$$4 = 2 \times 2$$
$$8 = 2 \times 2 \times 2$$

The factors are all the numbers that cannot be made up from numbers other than themselves and 1. In this example all of the factors are 2. Four is made up of a 2 and a 2. Eight is made up of 2×4, but the 4 has factors 2 and 2. Therefore, the factors of 8 are 2, 2, 2.

We now find the LCD by taking all the factors that are required to make each of the denominators without any extras. The 4 requires two 2's, but this is not enough to make 8. Thus the LCD is $2 \times 2 \times 2 = 8$. In Fig. 5-12, since $\frac{1}{4} = \frac{2}{8}$, we have

$$\frac{2}{8} + \frac{3}{8} = \frac{5}{8} \quad \textit{Ans.}$$

Hint for Success

Another method for finding the LCD for fractions is to list the multiples of each denominator. For example, suppose it is necessary to add the following fractions: $\frac{2}{3} + \frac{1}{4} + \frac{2}{6}$. The multiples of 3 are 3, 6, 9, 12, 15, 18, 21, 24, etc. The multiplies of 4 are 4, 8, 12 16, 20, 24, 28, etc. The multiples of 6 are 6, 12, 18, 24, 30, 36, etc. Examining each of the multiples shows that 12 is the lowest multiple common to each of the denominators, thus making 12 the LCD. Therefore, $\frac{2}{3} + \frac{1}{4} + \frac{2}{6} = \frac{8}{12} + \frac{3}{12} + \frac{4}{12} = \frac{15}{12} = 1\frac{3}{12} = 1\frac{1}{4}$

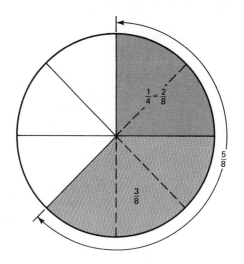

Figure 5-12

In this problem, it was very easy to change all the fractions into equivalent fractions with the same denominator. Not all problems are so simple. One of the difficulties will be to decide on what the new denominator will be.

How to Find the Least Common Denominator

Example 5-7

Find the LCD for ½ + ⅓ + ⅛.

Solution

In this example the denominators are 2, 3, and 8. Their factors are:

$$2 = 2 \times 1$$
$$3 = 3 \times 1$$
$$8 = 2 \times 2 \times 2$$

The numbers in the set of factors are 2 and 3. We need three 2's to make an 8. One of those 2's can be used for the denominator 2; therefore, the LCD is

$$\text{LCD} = 2 \times 2 \times 2 \times 3 = 24 \qquad \textit{Ans.}$$

We can start to see that factors are numbers that cannot be divided evenly by any other number except themselves and 1. They are 1, 2, 3, 5, 7, 11, 13, 17, 19, etc.

Example 5-8

Add $\dfrac{1}{7} + \dfrac{1}{2} + \dfrac{1}{4}$.

1. Find the LCD. In this example the denominators are 7, 2, and 4. Their factors are

$$7 = 7 \times 1$$
$$2 = 2 \times 1$$
$$4 = 2 \times 2$$

The factors are 7 and 2, but we need two 2's to make a 4. The LCD is

$$\text{LCD} = 7 \times 2 \times 2 = 28$$

2. Change each fraction into an equivalent fraction whose denominator is 28. We can do this by multiplying the fraction by 1, written as a fraction LCD/LCD.

$$\frac{1}{7} \times \frac{7 \times 2 \times 2}{7 \times 2 \times 2} = \frac{4}{28}$$

$$\frac{1}{2} \times \frac{7 \times 2 \times 2}{7 \times 2 \times 2} = \frac{14}{28}$$

$$\frac{1}{4} \times \frac{7 \times 2 \times 2}{7 \times 2 \times 2} = \frac{7}{28}$$

3. Add the numerators of the equivalent fractions and place the result over the LCD. If necessary, reduce the fraction to lowest terms.

$$\frac{4 + 14 + 7}{28} = \frac{25}{28} \qquad Ans.$$

Self-Test 5-9

What length of cable is needed to install the lighting outlets shown in Fig. 5-13?

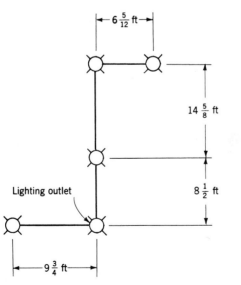

Figure 5-13

Solution

1. The total length required is equal to the _____ of all the lengths. The total length = $9\frac{3}{4} + 8\frac{1}{2} + 14\frac{5}{8} + 6\frac{5}{12}$.

2. Find the LCD. The denominators are 4, 2, 8, and 12. The factors are ________ and ________. Three 2's are required for the 8; therefore, the LCD is ________ = 24.
3. Now that you know the LCD you must find the sum of the parts of the cable by adding the ________ and the fractional parts and then ________ these two results. The whole numbers are ________, ________, ________, and ________. The sum of the whole numbers is ________.

4. *a.* For the ¾: 4 into 24 is 6. Therefore,

$$\frac{3}{4} \times \frac{24}{24} = \frac{?}{?}$$

b. For the ½: 2 into 24 is 12. Therefore,

$$\frac{1}{2} \times \frac{24}{24} = \frac{?}{?}$$

c. For the ⅝: 8 into 24 is 3. Therefore,

$$\frac{5}{8} \times \frac{24}{24} = \frac{?}{?}$$

d. For the 5⁄12: 12 into 24 is 2. Therefore,

$$\frac{5}{12} \times \frac{24}{24} = \frac{?}{?}$$

5. The sum of the fractional parts is ________.
6. Change 55⁄24 to a mixed number by ________ 24 into 55. This will give 2 and ________. Add the whole number 2 to the 37 to give the final answer which is ________ 7⁄24 ft. *Ans.*

Exercises

Practice

Add the following fractions and express in lowest terms.

1. ⅛ + ⅜ + ⅞
2. ¾ + ⅚ + 9⁄16
3. ¼ + ⅚ + ⅜
4. ⅚ + 1⁄9 + ⅔
5. 1⁄60 + 1⁄20 + 1⁄10
6. ⅓ + ⅙ + 1⁄9
7. ¼ + 1⁄12 + 1⁄15
8. ⅕ + 1⁄7 + 1⁄9
9. 1⁄200 + 1⁄300
10. 1⁄50 + 1⁄100
11. 3⁄16 + 5⁄64 + ⅝
12. 5⁄32 + ⅜ + 9⁄16
13. 1⁄10 + 1⁄30 + 1⁄60
14. 1⁄500 + 1⁄200 + 1⁄400
15. ¾ + 9⁄32 + 5⁄16
16. 1⅜ + 3½ + 2 3⁄16
17. 1¾ + 8 5⁄16 + 5⅜
18. 1⅝ + 3⅔
19. A carbon brush 29⁄32 in. thick is coated with copper to a thickness of 1⁄64 in. on each side. Find the total thickness.
20. What is the total thickness of insulation made by 1⁄64-in. fish paper, 1⁄32-in. tufflex, 3⁄64-in. varnished cambric, and ⅛-in. top stick?

Applications

21. In Fig. 5-14, the reciprocal of the resistance is known as the conductance, whose symbol is *G*. If the total conductance of a parallel circuit is the sum of the conductances, find the total conductance G_T, expressed in siemens (S).

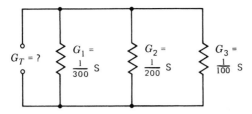

Figure 5-14 The conductance G is the reciprocal of the resistance R. A resistance $R_1 = 300 \ \Omega = 1/300$ siemens (S) has a conductance G_1.

22. Find the total resistance of a series circuit containing a $2\frac{1}{2}$-, a $1\frac{3}{10}$-, and a $\frac{9}{100}$-MΩ resistor.
23. In a layout similar to that shown in Fig. 5-13, find the total length of cable needed if the measurements between outlets are: $6\frac{1}{4}$, $8\frac{7}{12}$, $10\frac{5}{8}$, and $4\frac{5}{10}$ ft.

(See CD-ROM for Test 5-2)

Brushup on Subtraction of Fractions

Subtracting Fractions with the Same Denominator

RULE To subtract fractions with the same denominator, subtract the numerators and place the difference over the same denominator.

Example 5-10

Subtract $\frac{3}{8}$ from $\frac{5}{8}$.

Solution

$$\frac{5}{8} - \frac{3}{8} = \frac{5-3}{8} = \frac{2}{8} = \frac{1}{4} \qquad Ans.$$

Subtracting Fractions with Different Denominators

Procedure

1. Find the *least common denominator.*
2. Change the fractions to *equivalent* fractions using this new denominator.
3. Subtract these fractions with the same denominator.

Example 5-11

Subtract $\frac{3}{16}$ from $\frac{3}{4}$.

Solution

1. Find the LCD, which is 16.
2. Set up the problem as shown below and change the fractions to equivalent fractions as described in Examples 5-8 and 5-9.

$$\frac{3}{4} \times \frac{16}{16} = \frac{12}{16}$$

$$\frac{3}{16} \times \frac{16}{16} = \frac{3}{16}$$

3. Subtract these fractions with the same denominator and reduce to lowest terms.

$$\frac{12}{16} - \frac{3}{16} = \frac{9}{16} \qquad Ans.$$

Example 5-12

Subtract 3¼ from 9.

Solution

1. Set up the problem as shown below.

$$\begin{array}{r} 9 \\ -3\frac{1}{4} \end{array}$$

Since ¼ cannot be subtracted from nothing, *borrow* 1 from the 9, leaving 8, and replace this 1 as the fraction ⁴⁄₄. The 1 that is borrowed is always replaced as a fraction with identical numerator and denominator, each being equal to the denominator of the fraction being subtracted.

$$\begin{array}{rcl} 9 & = & \overset{8}{\cancel{9}}\frac{4}{4} \\ -3\frac{1}{4} & & -3\frac{1}{4} \end{array}$$

2. Subtract the fractions (⁴⁄₄ − ¼ = ¾). Subtract the whole numbers (8 − 3 = 5).
3. Write the answer.

$$5\frac{3}{4} \qquad Ans.$$

Example 5-13

Subtract 2⅝ from 8³⁄₁₆.

Solution

1. Set up the problem as shown below.

$$\begin{array}{r} 8\frac{3}{16} \\ -2\frac{5}{8} \end{array}$$

2. Find the LCD, which is 16.
3. Change the fractions to equivalent fractions using this new denominator.

$$\frac{3}{16} \times \frac{16}{16} = \frac{3}{16}$$

$$\frac{5}{8} \times \frac{16}{16} = \frac{10}{16}$$

4. As you can see, $^{10}\!/_{16}$ cannot be subtracted from only $^{3}\!/_{16}$. Therefore, we shall *borrow* one unit from the 8 as in Example 5-12, and replace it as $^{16}\!/_{16}$ as shown below.

$$8\frac{3}{16} = 7\frac{3}{16} + \frac{16}{16}$$

5. Combine the $^{3}\!/_{16} + {}^{16}\!/_{16}$ into $^{19}\!/_{16}$ and subtract the $^{10}\!/_{16}$ from the $^{19}\!/_{16}$ as shown below.

$$\begin{array}{r} 7^{19}\!/_{16} \\ -\ 2^{10}\!/_{16} \\ \hline 5^{9}\!/_{16} \quad \textit{Ans.} \end{array}$$

Self-Test 5-14

What resistance must be placed in series with a $12\frac{3}{4}$-Ω bell in order to provide a total resistance of $84\frac{3}{5}$ Ω?

Solution

1. The required resistance may be found by ——————— the $12\frac{3}{4}$ Ω from the total of $84\frac{3}{5}$ Ω. The number after the word "from" is written on the (top/bottom) of the subtraction problem. The LCD for the denominators 4 and 5 is ——————.

2. Set up the problem as shown below.

$$\begin{array}{r} 84\frac{3}{5} \\ -\ 12\frac{3}{4} \\ \hline \end{array}$$

3. Change the fractions to equivalent fractions using ——————— as the new denominator.

$$\frac{3}{5} \times \frac{20}{20} = \frac{12}{20}$$

$$\frac{3}{4} \times \frac{20}{20} = \frac{15}{20}$$

4. As you can see, $^{15}\!/_{20}$ cannot be subtracted from only $^{12}\!/_{20}$. Therefore, we must ——————— one unit from the 84 and replace it as ——————— as shown below.

$$84\frac{12}{20} = 83\frac{12}{20} + \frac{20}{20}$$

5. Combine the $^{12}\!/_{20}$ and the $^{20}\!/_{20}$ into ——————— and subtract the $^{15}\!/_{20}$ from it. Complete the problem as shown below.

$$\begin{array}{r} 83^{32}\!/_{20} \\ -\ 12^{15}\!/_{20} \\ \hline 71^{17}\!/_{20}\ \Omega \quad \textit{Ans.} \end{array}$$

Practice

Subtract the following fractions and express in lowest terms.

1. $\frac{3}{4} - \frac{1}{4}$
2. $\frac{7}{8} - \frac{3}{8}$
3. $\frac{11}{16} - \frac{5}{16}$
4. $5\frac{5}{8} - 1\frac{3}{8}$
5. $\frac{9}{16} - \frac{1}{4}$
6. $\frac{5}{6} - \frac{1}{3}$
7. $\frac{3}{4} - \frac{2}{3}$
8. $5\frac{5}{6} - 2\frac{2}{3}$
9. $\frac{1}{2} - \frac{1}{3}$
10. $\frac{1}{50} - \frac{1}{100}$
11. $\frac{1}{10} - \frac{1}{60}$
12. $\frac{1}{90} - \frac{1}{100}$
13. $\frac{1}{40} - \frac{1}{80}$
14. $\frac{1}{12} - \frac{1}{60}$
15. $\frac{1}{250} - \frac{1}{500}$
16. $\frac{1}{480} - \frac{1}{2400}$
17. $7 - 2\frac{1}{4}$
18. $13 - 5\frac{3}{8}$
19. $8 - \frac{2}{3}$
20. $5\frac{1}{4} - 2\frac{5}{8}$
21. $8\frac{3}{16} - 2\frac{3}{4}$
22. $5\frac{2}{3} - 3\frac{5}{8}$
23. $\frac{1}{200} - \frac{1}{600}$
24. $\frac{1}{1000} - \frac{1}{2500}$

25. How many feet of push-back wire are left from a 100-ft roll if a radio mechanic used $23\frac{1}{4}$ ft?

26. An electrician had a $16\frac{1}{4}$-ft-long piece of conduit. If she used $11\frac{5}{12}$ ft, how many feet were left?

Applications

27. The length of threading on a BX box connector is $\frac{7}{16}$ in. If $\frac{7}{32}$ in of thread is covered with washers, lock nuts, and metal, how much thread is left for attaching a threaded bushing?

28. Find the inside diameter of an electric conduit shown in Fig. 5-15 if its outside diameter is $2\frac{1}{4}$ in and the thickness of the conduit is $\frac{3}{16}$ in.

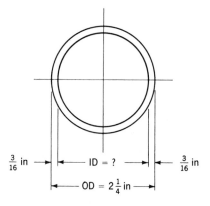

Figure 5-15   Cross section of an electric conduit.

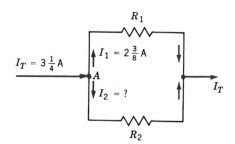

Figure 5-16   The current that enters point A is equal to the sum of the currents that leave the point.

29. Kirchhoff's law states that the current that enters a point in a circuit is equal to the sum of the currents that leave the point. In the circuit shown in Fig. 5-16, find the current I_2.

(See CD-ROM for Test 5-3)

Checkup on Solution of Formulas Involving Fractions (Diagnostic Test)

In addition to adding fractions, our next electrical job will require that we be able to solve formulas which contain fractions. Following are some problems of this type. If you have any difficulty solving any of them, see Job 5-7 which follows.

Exercises

Practice

1. Using the formula $f = PS/120$, find the speed S of an alternator if the frequency f is 60 Hz and the number of poles P is 4.
2. $R_1/R_2 = L_1/L_2$ is a formula used to compare the resistances of different lengths of wire. Find R_1 if $R_2 = 100$, $L_1 = 500$, $L_2 = 800$.
3. $V_1/V_2 = I_2/I_1$ is the formula for the relation between the voltages and the currents in the primary and secondary windings of a transformer. Find I_1 if $V_1 = 110$ V, $V_2 = 22$ V, and $I_2 = 10$ A.
4. The fundamental equation for the Wheatstone bridge (a resistance-measuring device) is $R_1/R_2 = R_3/R_x$. Find R_x if $R_1 = 1000$, $R_2 = 10,000$, and $R_3 = 84.3$.
5. In series circuits, the voltage drops are proportional to the resistances. This is stated mathematically as $V_1/V_2 = R_1/R_2$. Find V_1 if $V_2 = 8$ V, $R_1 = 3$ Ω, and $R_2 = 4$ Ω.

Brushup on Solution of Fractional Equations

Solving simple fractional equations. Many electrical formulas are stated in the form of a fraction. After substituting the given numbers for the letters of the formula, we might get an equation like $V/4 = 5$. To find the value of V, we must get the letter V all by itself on one side of the equality sign. In other words, we must eliminate the number 4. We can do this by applying the general rule (Job 3-8), which says that we may do anything to one side of an equality sign provided we do the same thing to the other side. In general, the operation to be performed on both sides of the equality sign will be the *opposite* of that used in the equation. Since our equation involves division, the opposite operation of multiplication should be performed on both sides of the equality sign.

RULE   Both sides of an equality sign may be multiplied by the same number without destroying the equality.

Self-Test 5-15

Find the value of V in the equation

$$\frac{V}{4} = 5$$

Solution

1. Solving an equation means to get the letter all alone on one side of the equality sign. We can get V alone on the left side of the equality sign if we can get rid of the denominator 4. Since the equation says that V is *divided* by 4, we can eliminate the denominator 4 by doing the opposite operation of _____ both sides of the equality sign by that same number 4.

2. Write the equation.
$$\frac{V}{4} = 5$$

3. Multiply both sides by 4.
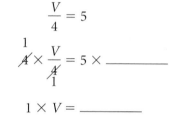

4. Multiply each side separately. $1 \times V =$ _____
or $V =$ _____   *Ans.*

Self-Test 5-16

Using the formula $I = \dfrac{V}{R}$, find V if $I = 2$ and $R = 6$.

Solution

1. Write the formula.
$$I = \frac{V}{R}$$

2. Substitute numbers.
$$2 = \frac{V}{?}$$

3. Step 2 asks the question, "What number divided by 6 gives 2 as an answer?" We can find E if we can eliminate the denominator 6. This may easily be done by _____ both sides of the equality sign by _____.

4. Multiply both sides by 6.
$$6 \times 2 = \frac{V}{\cancel{6}} \times \cancel{6}^{1}$$

5. Multiply both sides separately.   _____ $= V \times 1$
or $V =$ _____   *Ans.*

Notice that in each of the last three examples we eliminated the number in the denominator by *multiplying both sides by that same number.* If we were to multiply both sides by any other number, we would still be left with a number in the denominator and the letter would not stand alone. Therefore, to eliminate a number in the denominator of a fraction, we use the following rule:

RULE

To eliminate a number divided into a letter, multiply both sides of the equality sign by that same number.

Exercises

Practice

Find the value of the unknown letter in each problem.

1. $\dfrac{V}{3} = 8$

2. $4 = \dfrac{V}{10}$

3. $\dfrac{P}{6} = 2$

4. $0.5 = \dfrac{V}{3}$

5. $\dfrac{M}{0.2} = 5$

6. $\dfrac{A}{10} = 0.35$

7. $\dfrac{V}{3} = \dfrac{2}{3}$

8. $0.4 = \dfrac{M}{0.8}$

9. $\dfrac{P}{\frac{1}{2}} = 16$

10. $\dfrac{V}{4} = 2\frac{1}{2}$

11. $150 = \dfrac{x}{45}$

12. $117 = \dfrac{P}{4.7}$

13. In the formula $I = V/R$, find V if $I = 3$ and $R = 18$.
14. In the formula $V = P/I$, find P if $V = 110$ and $I = 5$.
15. In the formula $Q = X_L/R$, find X_L if $Q = 100$ and $R = 40$.
16. In the formula $\text{Eff} = O/I$, find O if $I = 36$ and $\text{Eff} = 0.85$.
17. In the formula $Z = V/I$, find V if $Z = 2000$ and $I = 0.015$.
18. In the formula $\cos\theta = I_R/I_T$, find I_R if $I_T = 10$ and $\cos\theta = 0.866$.
19. In the formula $\sin\theta = i/I_{max}$, find i if $I_{max} = 100$ and $\sin\theta = 0.9397$.
20. In the formula $R_1 + R_2 = V_T/I_T$, find V_T if $R_1 = 25$, $R_2 = 85$, and $I_T = 2$.

Using the Least Common Denominator to Solve Fractional Equations

Example 5-17

Find V in the equation

$$\frac{2V}{9} = \frac{4}{3}$$

Solution

When fractions appear on both sides of the equality sign, the solution will be simplified if we eliminate *all* denominators first. We could do this one denominator at a time by the method used in Examples 5-15 and 5-16, but it is faster if we eliminate both denominators at the same time. In order to eliminate *both* the 9 and the 3 at the same time, we must multiply both sides of the equality sign by some number so that *both* denominators will be canceled out. This must be a number into which *both* denominators will evenly divide—which is the *least common denominator*. In this problem, the LCD of 9 and 3 is 9.

1. Write the equation.
$$\frac{2V}{9} = \frac{4}{3}$$

2. Multiply both sides by the LCD (9).

$$\overset{1}{\cancel{9}} \times \frac{2V}{\cancel{9}} = \frac{4}{\cancel{3}} \times \cancel{9}$$

3. Multiply each side separately.

$$2V = 12$$

4. Solve for V.

$$V = \frac{12}{2} = 6 \qquad Ans.$$

Example 5-18

Find L in the equation

$$\frac{6}{L} = \frac{2}{3}$$

Solution

1. Write the equation.

$$\frac{6}{L} = \frac{2}{3}$$

2. Multiply both sides by the LCD. In general, although it may not be the *least* common denominator, a workable common denominator will be the value obtained by multiplying all the denominators. Thus, in this problem, the LCD is $3L$.

$$\overset{1}{3\cancel{L}} \times \frac{6}{\cancel{L}} = \frac{2}{\cancel{3}} \times \overset{1}{\cancel{3}L}$$

Note that L cancels L on the left side, and 3 cancels 3 on the right side.

3. Multiply each side separately. $18 = 2L$

4. Solve for L.

$$L = \frac{18}{2} = 9 \qquad Ans.$$

RULE   When fractions appear on both sides of the equality sign, eliminate the denominators by multiplying both sides by the least common denominator.

Self-Test 5-19

$R/K = L/A$ is a formula that may be used to calculate the length of a wire of particular material and cross section that will provide a definite resistance. Find the length L if $K = 60$ for pure iron wire, $A = 25$ cmil, and $R = 24$ Ω.

Solution

1. Write the formula.

$$\frac{R}{K} = \frac{L}{A}$$

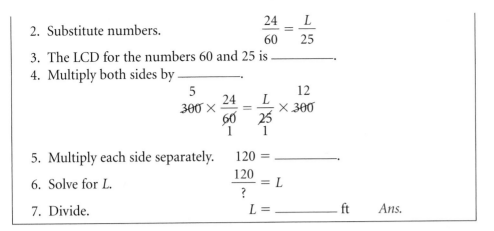

2. Substitute numbers.

$$\frac{24}{60} = \frac{L}{25}$$

3. The LCD for the numbers 60 and 25 is _____.
4. Multiply both sides by _____.

$$\overset{5}{\cancel{300}} \times \frac{24}{\underset{1}{\cancel{60}}} = \frac{L}{\underset{1}{\cancel{25}}} \times \cancel{300}\overset{12}{}$$

5. Multiply each side separately. $120 = $ _____.

6. Solve for L.

$$\frac{120}{?} = L$$

7. Divide.

$$L = \underline{\quad\quad} \text{ ft} \quad\quad Ans.$$

Exercises

Practice

Solve each equation for the value of the unknown letter.

1. $\dfrac{R}{3} = \dfrac{2}{3}$

2. $\dfrac{V}{8} = \dfrac{3}{4}$

3. $\dfrac{2}{3} = \dfrac{P}{6}$

4. $\dfrac{2V}{3} = \dfrac{4}{9}$

5. $\dfrac{M}{5} = \dfrac{2}{3}$

6. $\dfrac{4}{5} = \dfrac{T}{2}$

7. $\dfrac{2V}{5} = \dfrac{3}{0.5}$

8. $\dfrac{2B}{5} = \dfrac{7}{10}$

9. $\dfrac{N}{20} = \dfrac{3}{4}$

10. $\dfrac{4}{7} = \dfrac{S}{2}$

11. $\dfrac{0.3}{V} = \dfrac{1}{4}$

12. $\dfrac{3}{5} = \dfrac{2}{R}$

13. $\dfrac{2}{L} = \dfrac{6}{30}$

14. $\dfrac{10}{C} = \dfrac{150}{200}$

15. $\dfrac{22}{7} = \dfrac{11}{R}$

16. $\dfrac{100}{250} = \dfrac{0.1}{R}$

17. $\dfrac{N}{0.4} = \dfrac{3.9}{0.2}$

18. $\dfrac{9}{L} = \dfrac{0.3}{8.9}$

19. $\dfrac{21}{0.3} = \dfrac{2V}{9}$

20. $\dfrac{37}{250} = \dfrac{74}{R}$

21. $N_1/N_2 = V_1/V_2$ is a formula used in transformer calculations. Find N_2 if $N_1 = 40$ turns, $V_1 = 6$ V, and $V_2 = 18$ V.
22. Using the formula of Prob. 21, find V_1 if $N_1 = 30$ turns, $N_2 = 70$ turns, and $V_2 = 21$ V.
23. $A_2/A_1 = R_1/R_2$ is a formula used to calculate the sizes of wires in electrical installations. Find A_2 if $A_1 = 100$, $R_1 = 1000$, $R_2 = 3000$.
24. $I_1/I_2 = R_2/R_1$ is a formula used for calculating the way the current divides in a parallel circuit. Find I_1 if $I_2 = 2$ A, $R_2 = 100$ Ω, and $R_1 = 25$ Ω.
25. Using the same formula as in Prob. 24, find R_2 if $I_1 = 3$ A, $I_2 = 5$ A, and $R_1 = 100$ Ω.

Applications

26. $R_1/R_2 = R_3/R_x$ is the formula used for calculations in the Wheatstone-bridge method for measuring resistance. Find R_x if $R_1 = 1000$, $R_3 = 26.9$, and $R_2 = 10{,}000$.

27. $E/R = A_1/A_2$ is a formula used to determine the effort required to raise a large weight in a hydraulic lift. Find the effort E required to raise a weight R of 200,000 lb if the area of the small piston A_1 is 4 in² and the area of the large piston A_2 is 1600 in².

28. $R_1/R_2 = L_2/L_1$ is a formula used to locate the position of a break in an underground cable. Find the length to the break in the cable L_2 if the length of the unbroken cable L_1 is 4000 ft, R_1 is 10 Ω, and R_2 is 250 Ω.

29. $R_s/R_m = I_m/I_s$ is the formula for finding the shunt resistor needed to extend the range of an ammeter. Find R_s if $I_m = 0.001$, $R_m = 50$, and $I_s = 0.049$.

30. $R_1/R_2 = C_x/C_1$ is a formula used to measure the capacitance of an unknown capacitor. Find C_x if $R_1 = 100\ \Omega$, $R_2 = 425\ \Omega$, and $C_1 = 0.5\ \mu\text{F}$.

(See CD-ROM for Test 5-4)

Cross multiplication.   When a fractional equation contains *only two fractions* equal to each other, the equation may be solved by a very simple method known as cross multiplication. This method automatically multiplies both sides of the equality sign by the LCD and therefore eliminates one step in the solution. This method is most useful when the unknown letter is in the denominator of one of the fractions.

RULE   To cross-multiply, set the product of the numerator of the first fraction and the denominator of the second equal to the product of the numerator of the second fraction and the denominator of the first.

It is easier to understand this rule by putting it in picture form. The multiplication of the numbers along one diagonal line is equal to the multiplication of the numbers along the other diagonal line.

$$2 \times 6 = 4 \times 3$$

or

$$3 \times 4 = 6 \times 2$$

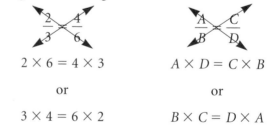

$$A \times D = C \times B$$

or

$$B \times C = D \times A$$

Example 5-20

Find the value of R in the equation

$$\frac{3}{R} = \frac{2}{5}$$

Solution

1. Write the equation.

$$\frac{3}{R} \diagtimes \frac{2}{5}$$

2. Cross-multiply. $2 \times R = 3 \times 5$
3. Simplify each side. $2R = 15$
4. Solve for R. $R = \frac{15}{2} = 7\frac{1}{2}$ *Ans.*

Exercises

Practice

Find the value of the unknown letter in each equation.

1. $\dfrac{V}{10} = \dfrac{3}{5}$

2. $\dfrac{8}{3} = \dfrac{R}{6}$

3. $\dfrac{60}{A} = \dfrac{3}{4}$

4. $\dfrac{9}{V} = \dfrac{15}{40}$

5. $\dfrac{84}{V} = \dfrac{28}{17}$

6. $\dfrac{84}{28} = \dfrac{66}{R}$

7. $\dfrac{V}{18} = \dfrac{3}{2}$

8. $9 = \dfrac{54}{R}$

9. $\dfrac{50}{T} = 2$

10. $\dfrac{2R}{3} = \dfrac{10}{5}$

11. $\dfrac{24}{2} = \dfrac{6M}{5}$

12. $\dfrac{3P}{0.4} = 6$

13. $\dfrac{3.6}{5} = \dfrac{A}{2}$

14. $0.88 = \dfrac{R}{3}$

15. $\dfrac{3}{12} = \dfrac{4}{3T}$

16. $5 = \dfrac{1}{0.2V}$

17. $\dfrac{1}{R} = 10$

18. $80 = \dfrac{1}{R}$

19. $\dfrac{1}{R} = \dfrac{13}{24}$

20. $\dfrac{10}{R} = 50$

(See CD-ROM for Test 5-5)

about electronics

If your car needed a battery boost from another car's battery, would you know how to minimize risk? The cars shouldn't be touching, and no one should be smoking. The engines should be turned off, and emergency brakes set. The two batteries must be connected properly, to avoid sparks leading to an explosion from hydrogen gas. It can help to place a damp cloth over each battery. The dead battery to be boosted should be connected in parallel to the other, good battery; so first you should connect the red cable from the positive post on the good car's battery to the positive post on your car's dead battery. Then you should connect the black cable to the negative post on the good car's battery. Next, choose a solid "ground" on the dead car, like an unpainted metal surface; that's where the last cable end should go, far from the battery or moving engine parts. Start the good car up, idling fast for several minutes; then start the dead car. When the car starts, disconnect the cables in the reverse order from the way you put them on.

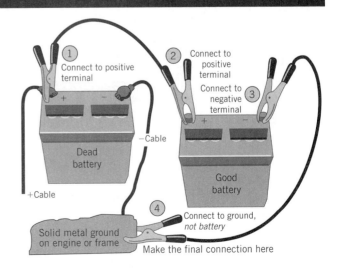

Total Resistance in a Parallel Circuit

When resistances are connected in parallel, the total resistance is always *less* than the resistance of any branch. When resistances are added to a circuit in parallel, they merely provide extra paths for the current to follow. Each extra path will draw its own current from the voltage source, *increasing* the total current drawn. If the addition of a resistance *increases* the current, then this resistance must have *reduced* the total resistance. For example, in the circuit of Fig. 5-17*a*, the current that flows in the resistor is 1 A and the total resistance R_T is 50 Ω. Now let us add an extra 50-Ω resistor in parallel as shown in Fig. 5-17*b*. Since equal resistors carry equal currents, the total current equals $I_1 + I_2 = 2$ A. The total resistance will be

$$V_T = I_T \times R_T \qquad (4\text{-}7)$$
$$50 = 2 \times R_T$$
$$R_T = \frac{50}{2} = 25 \ \Omega$$

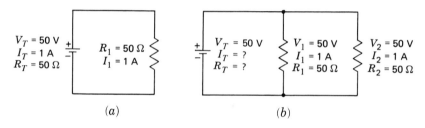

(a) (b)

Figure 5-17 Adding resistances in parallel *reduces* the total resistance.

Thus we see that the net effect of adding another resistance in parallel *reduced* the total resistance from 50 to 25 Ω. Also, the current drawn by the new circuit *increased* from 1 to 2 A.

The total resistance in parallel is given by

$$\frac{1}{R_T} = \frac{1}{R_1} + \frac{1}{R_2} + \frac{1}{R_3} + \text{etc.} \qquad \text{Formula 5-3}$$

where R_T is the total resistance in parallel, and R_1, R_2, and R_3 are the branch resistances.

Hint for Success

Formula 5-3 can be written as:
$$R_T = \frac{1}{1/R_1 + 1/R_2 + 1/R_3 + etc.}$$
This formula shows that R_T equals the reciprocal of the sum of the reciprocals.

Example 5-21

Find the total resistance of a 3-, a 4-, and an 8-Ω resistor in parallel.

Solution

The diagram for the circuit is shown in Fig. 5-18.

1. Write the formula. $$\frac{1}{R_T} = \frac{1}{R_1} + \frac{1}{R_2} + \frac{1}{R_3} \qquad (5\text{-}3)$$

2. Substitute numbers. $$\frac{1}{R_T} = \frac{1}{3} + \frac{1}{4} = \frac{1}{8}$$

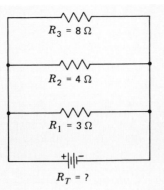

$R_3 = 8\,\Omega$

$R_2 = 4\,\Omega$

$R_1 = 3\,\Omega$

$R_T = ?$

Figure 5-18

3. Add fractions. $\dfrac{1}{R_T} = \dfrac{17}{24}$

4. Cross-multiply. $17 \times R_T = 1 \times 24$

5. Simplify. $17 R_T = 24$

6. Solve for R_T. $R_T = \dfrac{24}{17} = 1.4\,\Omega$ *Ans.*

Example 5-22

What resistance must be connected in parallel with a 40-Ω resistance in order to provide a total resistance of 24 Ω?

Solution

Given: $R_T = 24\,\Omega$ Find: $R_2 = ?$
  $R_1 = 40\,\Omega$

1. Write the formula. $\dfrac{1}{R_T} = \dfrac{1}{R_1} + \dfrac{1}{R_2}$ (5-3)

2. Substitute numbers. $\dfrac{1}{24} = \dfrac{1}{40} + \dfrac{1}{R_2}$

3. Transpose the ¹⁄₄₀ to the left side of the equality sign.

$$\dfrac{1}{24} - \dfrac{1}{40} = \dfrac{1}{R_2}$$

4. Subtract fractions. $\dfrac{1}{60} = \dfrac{1}{R_2}$

5. Cross-multiply. $R_2 = 60\,\Omega$ *Ans.*

Self-Test 5-23

A 40-, a 70-, and a 150-Ω resistor are connected in parallel. Find the total resistance.

Solution

Given: $R_1 = 40 \ \Omega$ Find: $R_T = ?$
               $R_2 = 70 \ \Omega$
               $R_3 = 150 \ \Omega$

1. Write the formula. $\dfrac{1}{?} = \dfrac{1}{R_1} + \dfrac{1}{R_2} + \dfrac{1}{R_3}$

2. Substitute numbers. $\dfrac{1}{R_T} = \dfrac{1}{40} + \dfrac{1}{?} + \dfrac{1}{150}$

The next step would be to add these fractions. However, it is very difficult to find the LCD when the denominators are as large as these. In this situation, it is best to find the decimal equivalent for each fraction and use these decimals instead of the fractions.

$$\frac{1}{40} = 40 \overline{\smash{)}\begin{array}{r} 0.025 \\ 1.000 \\ \underline{80} \\ 200 \\ \underline{200} \end{array}}$$

$$\frac{1}{70} = 70 \overline{\smash{)}\begin{array}{r} 0.014 \\ 1.000 \\ \underline{70} \\ 300 \\ \underline{280} \\ 20 \end{array}}$$

$$\frac{1}{150} = 150 \overline{\smash{)}\begin{array}{r} 0.006\frac{2}{3} = 0.007 \\ 1.000 \\ \underline{900} \\ 100 \end{array}}$$

3. Substitute the decimals for the fractions.

$$\frac{1}{R_T} = 0.025 + 0.014 + \underline{\qquad\qquad}$$

4. Add the decimals. $\dfrac{1}{R_T} = \underline{\qquad\qquad}$

5. Cross-multiply. $0.046 R_T = \underline{\qquad\qquad}$

6. Solve for R_T. $R_T = \dfrac{1}{?}$

7. Divide. $R_T = \underline{\qquad\qquad} \ \Omega$ *Ans.*

CALCULATOR HINT

With a calculator, the total resistance R_T is very easy to calculate using Formula 5-3. Simply enter the resistance value followed by the ⅟ₓ key. Next, press the ± key and enter the next resistance value followed by the ⅟ₓ key. Once you have added all the reciprocal values, press the ⅟ₓ key to obtain your answer.

Exercises

Practice

1. Find the total resistance of a 3-, a 4-, and a 12-Ω resistor in parallel.
2. Find the total resistance of the circuit shown in Fig. 5-19.

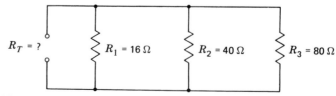

$R_T = ?$ $R_1 = 16 \ \Omega$ $R_2 = 40 \ \Omega$ $R_3 = 80 \ \Omega$

Figure 5-19

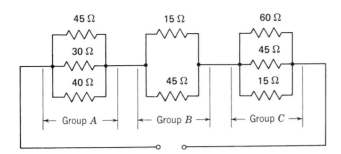

Figure 5-20

3. Three resistances of 3, 6, and 9 kΩ are in parallel. What is the combined resistance?
4. Find the total resistance of a 25-Ω coffee percolator and a 30-Ω toaster in parallel.
5. Find the total resistance of 4, 8, 12, and 15 Ω in parallel.
6. Find the total resistance of 6, 8, and 4.8 Ω in parallel.
7. Find the total resistance of 15, 7.5, and 5 Ω in parallel.
8. Find the total resistance of 100, 250, and 500 Ω when connected in parallel.

Applications

9. Find the resistance of each group of resistors in the circuit shown in Fig. 5-20.
10. Find the total resistance of 1, 1.5, and 2 kΩ when connected in parallel.
11. What resistance must be connected in parallel with a 20-Ω resistance in order to provide a total resistance of 15 Ω?
12. What resistance must be connected in parallel with a 100-Ω resistance in order to provide a total resistance of 90 Ω?
13. What resistance must be connected in parallel with a 600-Ω resistance in order to provide a total resistance of 400 Ω?
14. A voltage of 120 V is applied across a parallel combination of a 100-Ω resistor and an unknown resistor. If the total current is 1.5 A, find the value of the unknown resistor.
15. A spotlight of unknown resistance is placed in parallel with an automobile cigarette lighter of 80 Ω resistance. If a current of 0.75 A flows when a voltage of 12 V is applied, find the resistance of the spotlight.

Total Resistance of a Number of Equal Branches

RULE

The total resistance of a number of equal resistors in parallel is equal to the resistance of one resistor divided by the number of resistors.

$$R_T = \frac{R}{N}$$ Formula 5-4

where R_T = total resistance of equal resistors in parallel
R = resistance of one of the equal resistors
N = number of equal resistors

Example 5-24

Three lamps, each having a resistance of 60 Ω, are connected in parallel. Find the total resistance of the combination.

Solution

Given: $R_1 = R_2 = R_3 = 60 \ \Omega$ Find: $R_T = ?$

$$R_T = \frac{R}{N} \tag{5-4}$$

$$R_T = \frac{60}{3} = 20 \ \Omega \qquad \textit{Ans.}$$

Exercises

Applications

1. Find the total resistance of two 100-Ω lamps in parallel.
2. Find the total resistance of three 48-Ω bells in parallel.
3. Find the total resistance of two 1.5-Ω headlight lamps in parallel.
4. If the total resistance of two identical bells in parallel is 20 Ω, what is the resistance of each bell?
5. A toaster, an electric iron, and a coffee percolator, all of 22 Ω resistance, are connected in parallel across a 110-V line. Find (a) the total resistance and (b) the total current.

Two resistors in parallel. When only two resistors are in parallel, the total resistance may be calculated by a simple rule.

RULE   To find the total resistance of only two resistors in parallel, multiply the resistances and then divide the product by the sum of the resistors.

$$R_T = \frac{R_1 \times R_2}{R_1 + R_2} \qquad \text{Formula 5-5}$$

where R_T is the total resistance in parallel, and R_1 and R_2 are the two resistors in parallel.

Example 5-25

Find the total resistance of a 4- and a 12-Ω resistor in parallel.

Solution

Given: $R_1 = 4 \ \Omega$ Find: $R_T = ?$
        $R_2 = 12 \ \Omega$

$$R_T = \frac{R_1 \times R_2}{R_1 + R_2} \qquad\qquad (5\text{-}5)$$

$$R_T = \frac{4 \times 12}{4 + 12}$$

$$R_T = \frac{48}{16} = 3\ \Omega \qquad Ans.$$

Exercises

Applications

1. Find the total resistance of two 40-Ω coils in parallel.
2. Find the total resistance of a 20- and a 60-Ω motor in parallel.
3. An ammeter has a coil whose resistance is 56 Ω and is shunted by a 42-Ω resistor in parallel. Find the equivalent resistance of the combination.
4. Find the total resistance of 90-Ω galvanometer in parallel with a 10-Ω shunt resistor.
5. A section of the picture-control circuit of a television receiver uses a 10,000- and a 25,000-Ω resistor in parallel. Find the total resistance of the combination.
6. A 24- and a 48-Ω solenoid used in semaphore signals of an H-O electric train set are connected in parallel. Find the total resistance.
7. Find the total resistance of a 5- and a 12-kΩ resistor in parallel.
8. Find the total resistance of 9 kΩ and 2000 Ω when connected in parallel.
9. A 40-Ω soldering iron and a 100-Ω lamp are connected in parallel. Find the total resistance.
10. Find the total resistance of 20 and 27 kΩ when connected in parallel.

(See CD-ROM for Test 5-6)

Total Voltage in a Parallel Circuit

Example 5-26

What voltage is needed to send 3 A through a parallel combination of a 3-, a 4-, and a 12-Ω resistance?

Solution

The diagram for the circuit is shown in Fig. 5-21. In order to find the total voltage V_T, we must know the value of the total current I_T and the total resistance R_T. If either value is unknown, it must be found first.

1. Find the total resistance R_T.

$$\frac{1}{R_T} = \frac{1}{R_1} + \frac{1}{R_2} + \frac{1}{R_3} \qquad\qquad (5\text{-}3)$$

$$\frac{1}{R_T} = \frac{1}{3} + \frac{1}{4} + \frac{1}{12}$$

$$\frac{1}{R_T} = \frac{2}{3}$$

$$2R_T = 3$$

$$R_T = \frac{3}{2} = 1.5 \ \Omega \qquad Ans.$$

2. Find the total voltage V_T.

$$V_T = I_T \times R_T \qquad\qquad (4\text{-}7)$$
$$V_T = 3 \times 1.5 = 4.5 \ \text{V} \qquad Ans.$$

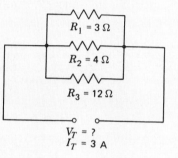

Figure 5-21

Exercises

Applications

1. Find the voltage needed to send 2 A through a parallel combination of three 60-Ω resistors.
2. Find the total voltage needed to send 9 A through a parallel circuit consisting of a 25-Ω percolator and a 30-Ω refrigerator motor.
3. Find the voltage needed to send 3 A through a parallel combination of a 2- and an 8-Ω resistor.
4. Find the voltage needed to send 2 A through a parallel combination of a 3- and a 6-Ω resistor.
5. Find the voltage required to send a current of 2.4 A through a 10-Ω coil, a 20-Ω coil, and a 60-Ω motor if they are wired in parallel.
6. In Fig. 5-22, a 25-Ω galvanometer is shunted with a 4-Ω resistor when indicating a current of 3 mA. What is the voltage across the galvanometer?

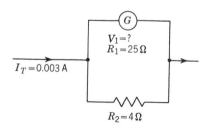

Figure 5-22   A shunt across the galvanometer carries most of the line current.

7. Find the voltage required to send 2 A through a parallel combination of a 20-, a 30-, and a 40-Ω resistance.
8. Find the total voltage required to send 0.15 A through three coils in parallel if their effective resistances are 200, 400, and 800 Ω.
9. A faulty resistor R_L in the self-bias circuit shown in Fig. 5-23a was replaced with a 6000- and a 4000-Ω resistor in parallel as shown in Fig. 5-23b. If $I_L = 2$ mA, find the voltage across the parallel combination.

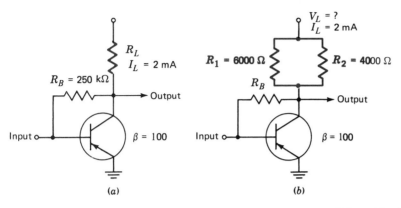

Figure 5-23 (a) Self-bias circuit. (b) R_L is replaced with an equivalent parallel combination.

Division of Current in a Parallel Circuit

In our study of Ohm's law we learned that the resistance of a circuit affected the current in the circuit. The greater the resistance, the smaller the current. The smaller the resistance, the greater the current. When the current in a parallel circuit reaches a point at which it may divide and flow in more than one path, the largest current will flow in that portion of the circuit which offers the smallest resistance, and the smallest current will flow through the largest resistance. The way a current will divide in a parallel circuit is shown in the following example.

Example 5-27

Find the current that flows in each branch of the parallel circuit shown in Fig. 5-24.

Solution

1. Find the total resistance R_T.

$$\frac{1}{R_T} = \frac{1}{R_1} + \frac{1}{R_2} + \frac{1}{R_3} \tag{5-3}$$

$$\frac{1}{R_T} = \frac{1}{16} + \frac{1}{48} + \frac{1}{24}$$

$$\frac{1}{R_T} = \frac{1}{8}$$

$$R_T = 8 \ \Omega$$

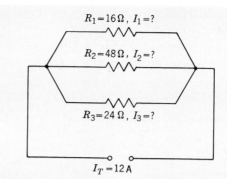

Figure 5-24

2. Find the total voltage V_T.

$$V_T = I_T \times R_T \qquad (4\text{-}7)$$
$$V_T = 12 \times 8 = 96 \text{ V}$$

3. Find the branch voltages.

$$V_T = V_1 = V_2 = V_3 = 96 \text{ V} \qquad (5\text{-}1)$$

4. Find the branch currents.

$$V_1 = I_1 \times R_1 \qquad\qquad\qquad V_2 = I_2 \times R_2$$
$$96 = I_1 \times 16 \qquad\qquad\qquad 96 = I_2 \times 48$$

$$I_1 = \frac{96}{16} = 6 \text{ A} \qquad Ans. \qquad I_2 = \frac{96}{48} = 2 \text{ A} \qquad Ans.$$

$$V_3 = I_3 \times R_3$$
$$96 = I_3 \times 24$$

$$I_3 = \frac{96}{24} = 4 \text{ A} \qquad Ans.$$

5. *Check*:

$$I_T = I_1 + I_2 + I_3 \qquad (5\text{-}2)$$
$$12 = 6 + 2 + 4$$
$$12 = 12 \qquad Check$$

Example 5-28

This example will illustrate the problem of attempting to use discrete transistors in parallel to handle higher current loads. Because each transistor is manufactured separately, the parameter variations cause an unequal distribution of the load current. Two 2N2652 transistors with input impedances of $R_1 = 2 \text{ k}\Omega$ and $R_2 = 8 \text{ k}\Omega$ are used in the circuit shown in Fig. 5-25a. The equivalent input circuit is shown in Fig. 5-25b. Find the current into the base of each transistor.

Solution

1. Find the resistance of the parallel combination.

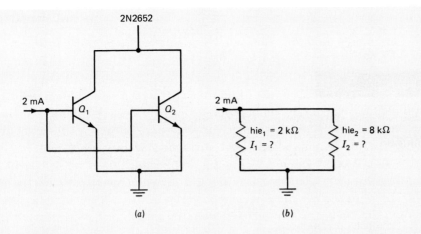

Figure 5-25

$$R_T = \frac{R_1 \times R_2}{R_1 + R_2} \qquad (5\text{-}5)$$

$$R_T = \frac{2\ k\Omega \times 8\ k\Omega}{0.002\ k\Omega + 1600\ k\Omega}$$

$$R_T = \frac{16}{10}\ k\Omega = 1.6\ k\Omega$$

2. Find the total voltage.

$$V_T = I_T \times R_T \qquad (4\text{-}7)$$
$$V_T = 2\ mA \times 1.6\ k\Omega = 3.2\ V$$

3. Find the branch voltages.

$$V_T = V_1 = V_2 = 3.2\ V \qquad (5\text{-}1)$$

4. Find the branch currents.

Transistor No. 1

$$V_1 = I_1 \times R_1$$
$$3.2 = I_1 \times 2\ k\Omega$$
$$I_1 = \frac{3.2}{2\ k\Omega}$$
$$I_1 = 1.6\ mA \qquad Ans.$$

Transistor No. 2

$$V_2 = I_2 \times R_2$$
$$3.2 = I_2 \times 8\ k\Omega$$
$$I_2 = \frac{3.2}{8\ k\Omega}$$
$$I_2 = 0.4\ mA \qquad Ans.$$

Division of Current in Two Branches in Parallel

RULE   When only two branches are involved, the current in one branch will be only some fraction of the total current. This fraction is the quotient of the second resistance divided by the sum of the resistances.

$$I_1 = \frac{R_2}{R_1 + R_2} \times I_T \qquad \text{Formula 5-6}$$

$$I_2 = \frac{R_1}{R_1 + R_2} \times I_T \qquad \text{Formula 5-7}$$

Hint for Success

Another formula for calculating the division of currents in a parallel circuit is $I_R = \frac{R_T}{R} \times I_T$, where R represents the resistance R of a given branch resistance and I_R represents the branch current. This formula works for any number of branch resistances.

Example 5-29

We can now solve Example 5-28 by this method.

Solution

Given: $R_1 = 2 \text{ k}\Omega$ Find: $I_1 = ?$
         $R_2 = 8 \text{ k}\Omega$ $I_2 = ?$
         $I_T = 2 \text{ mA}$

1. Find the current I_1.

$$I_1 = \frac{R_2}{R_1 + R_2} \times I_T \qquad (5\text{-}6)$$

$$I_1 = \frac{8 \text{ k}\Omega}{2 \text{ k}\Omega + 8 \text{ k}\Omega} \times 2 \text{ mA} = \frac{8}{10} \times 2 \text{ mA}$$

$$I_1 = 0.8 \times 2 \text{ mA} = 1.6 \text{ mA} \qquad Ans.$$

2. Find the current I_2.

$$I_2 = \frac{R_1}{R_1 + R_2} \times I_T \qquad (5\text{-}7)$$

$$I_2 = \frac{2 \text{ k}\Omega}{2 \text{ k}\Omega + 8 \text{ k}\Omega} \times 2 \text{ mA} = \frac{2}{10} \times 2 \text{ mA}$$

$$I_2 = 0.2 \times 2 \text{ mA} = 0.4 \text{ mA} \qquad Ans.$$

Self-Test 5-30

A 2N405 in a two-transistor receiver receives 36 mA from a parallel combination of a 40-kΩ resistor and a 5-kΩ headphone, as shown in Fig. 5-26. Find the current flowing in each part.

Solution

1. Find the current I_2 in the headphone.
Write the formula.

$$I_2 = \frac{?}{R_1 + R_2} \times I_T$$

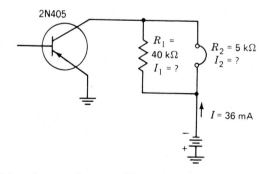

Figure 5-26   Division of current in a parallel circuit.

$$= \frac{40 \text{ k}\Omega}{40 \text{ k}\Omega + 5 \text{ k}\Omega} \times 36 \text{ mA}$$

$$= \frac{40 \times \cancel{10^3}}{45 \times \cancel{10^3}} \times 36 \times \underline{\hspace{2cm}}$$

$$= 32 \times 10^{-3} \underline{\text{(A/mA)}}$$

$$= \underline{\hspace{2cm}} \text{ mA} \qquad Ans.$$

2. Find the current I_1 in the resistor.
Write the formula.

$$I_1 = \frac{?}{R_1 + R_2} \times I_T$$

$$= \frac{5 \text{ k}\Omega}{40 \text{ k}\Omega + 5 \text{ k}\Omega} \times 36 \text{ mA}$$

$$= \frac{5 \times \cancel{10^3}}{45 \times \cancel{10^3}} \times 36 \times 10^{-3}$$

$$= \underline{\hspace{2cm}} \times 10^{-3} \underline{\text{(mA/A)}}$$

$$= \underline{\hspace{2cm}} \text{ mA} \qquad Ans.$$

Exercises

Applications

1. Two resistances are connected in parallel. $R_1 = 48 \ \Omega$, $R_2 = 48 \ \Omega$, and $I_T = 8$ A. Find the current flowing in each branch. On the basis of your answers, state a rule about the division of current between equal resistors in parallel.
2. Find the current in each branch of a parallel circuit consisting of a 20-Ω percolator and a 30-Ω toaster if the total current is 9 A.
3. Two 1.5-Ω automobile headlight lamps in parallel draw a total of 8 A. Find the total voltage supplied and the current drawn by each lamp.
4. A galvanometer with a resistance of 48 Ω and a parallel shunt of 2 Ω draw a total current of 0.2 A. Find the current through the galvanometer and through the shunt.
5. A generator supplies a current of 19.5 A to three small electroplating tanks arranged in parallel. The resistances of the tanks are 8, 12, and 16 Ω. What current does each tank draw?
6. A generator supplies a current of 26 A to three motors arranged in parallel. The resistances of the motors are 24, 36, and 48 Ω. What current does each motor draw?
7. Two resistances are arranged in parallel as shown in Fig. 5-27. Find the current in each resistance.

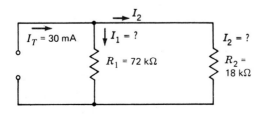

Figure 5-27

8. The resistance of an ammeter is 2.8 Ω. A shunt of 0.02 Ω is connected in parallel with it. If the combination is inserted into a line carrying 10 A, how much current actually flows through the ammeter?

9. In a circuit similar to that shown in Fig. 5-21, $R_1 = 5\ \Omega$, $R_2 = 7\ \Omega$, $R_3 = 8\ \Omega$, and $I_T = 13.1$ A. Find (a) the total resistance of the circuit, (b) the total voltage of the circuit, (c) the voltage across each branch, and (d) the current in each branch.

10. A low-power flasher circuit contains a resistor $R_1 = 16$ kΩ in parallel with another resistor $R_2 = 4$ kΩ. If the current entering the combination is 3 mA, find the current in each resistor.

JOB 5-11

Review of Parallel Circuits

Definition. A parallel circuit is a circuit in which the current may divide so as to flow in ____(a)____ than one path.

Rules and formulas

1. The total voltage across a parallel circuit is _____ to the voltage across any branch of the circuit.

$$V_T = V_1 = \text{_____} = \text{_____}$$

5-1

2. The total current in a parallel circuit is equal to the _____ of the currents in all the branches of the circuit.

$$I_T = I_1 + \text{_____} + \text{_____}$$

5-2

3. The total resistance in a parallel circuit may be found by applying Ohm's law to the _____ values of the circuit.

$$V_T = I_T \times \text{_____}$$

4-7

4. The total resistance may also be found as follows:
   For any number of resistors:

$$\frac{1}{?} = \frac{1}{R_1} + \frac{1}{R_2} + \frac{1}{R_3}$$

5-3

   For just two resistors:

$$R_T = \frac{R_1\ ?\ R_2}{R_1\ ?\ R_2}$$

5-5

   For any number N of equal resistors of $R\ \Omega$ each:

$$R_T = \frac{R}{?}$$

5-4

5. Ohm's law may be used on any branch of a parallel circuit.

$$V_1 = I_1 \times \text{_____}$$

4-4

$$V_2 = \text{_____} \times R_2$$

4-5

$$\text{_____} = I_3 \times R_3$$

4-6

6. The total resistance in parallel is always __(more/less)__ than the resistance of any branch.

7. Division of current between two branches in parallel:

$$I_1 = \frac{?}{R_1 + R_2} \times I_T \qquad \text{5-6}$$

$$I_2 = \frac{?}{R_1 + R_2} \times I_T \qquad \text{5-7}$$

8. If the branches of a parallel circuit have the same resistance, then each will draw the _____ current. If the branches of a parallel circuit have different resistances, then each will draw a _____ current. The larger the resistance, the _____ the current drawn.

9. Adding or subtracting fractions:
   In order to add or subtract fractions, all the fractions must have the _____ denominator.
   The LCD is the _____ number into which _____ the denominators will evenly divide.

10. Solving fractional equations:
    Simplify fractions whenever possible.
    When a fractional equation contains only _____ fractions equal to each other, the equation may be solved by cross _____.
    To cross-multiply, the product of the numerator of the first fraction and the denominator of the second is set _____ to the product of the numerator of the second fraction and the denominator of the first. For example, if

$$\frac{C}{D} = \frac{E}{F}$$

then $C \times$ _____ $= E \times D$

or $E \times D =$ _____ $\times F$

Exercises

Applications

1. An electric iron, a radio, and an electric clock are connected to a three-way 110-V kitchen outlet which puts the appliances in parallel. The iron draws 5 A, the radio draws 0.5 A, and the clock draws 0.25 A. Find (*a*) the total current drawn from the line, (*b*) the voltage across each device, and (*c*) the total resistance of the circuit.

2. A semaphore signal, a floodlight tower, and a coal loader of a model railroad are connected in parallel across the 12-V winding of the power transformer. The signal draws 0.1 A, and the coal loader draws 0.2 A. The floodlight tower has a resistance of 48 Ω. Find (*a*) the voltage across each device, (*b*) the resistances of the semaphore and the coal loader, (*c*) the total current, and (*d*) the total resistance of the circuit.

3. Three motors are wired in parallel across 440 V. Motor 1 draws a current of 10 A, and motor 3 draws 15 A. Motor 2 has a resistance of 20 Ω. Solve the circuit for all missing values of current, voltage, and resistance.

4. Find all missing values of voltage, current, and resistance in the circuit shown in Fig. 5-28.

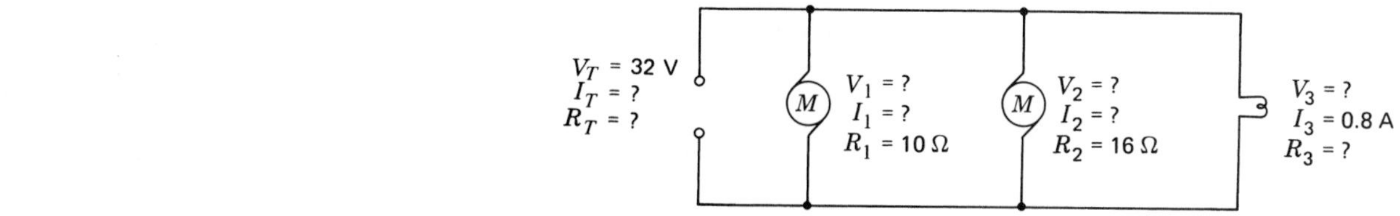

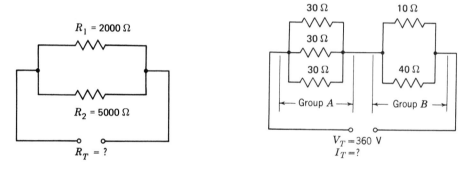

Figure 5-28

5. Find the total resistance of a 60-, an 80-, and a 120-Ω resistor in parallel.
6. A 2000- and a 5000-Ω resistor are connected in parallel as shown in Fig. 5-29. Find the total resistance of the combination.
7. In the circuit shown in Fig. 5-30, find (*a*) the total resistance R_A of group *A* and (*b*) the total resistance R_B of group *B*. (*c*) Draw a new circuit using a single resistor (R_A and R_B) in place of the groups they represent. (*d*) Is the new circuit a series or a parallel circuit? (*e*) What is the total resistance of the new circuit? (*f*) Find the total current in the new circuit.

Figure 5-29

Figure 5-30

8. What is the combined resistance of a 480-Ω galvanometer and its parallel 20-Ω shunt?
9. What resistance must be connected in parallel with a 2400-Ω resistor in order to provide a total resistance of 480 Ω?
10. What resistance must be connected in parallel with a 20- and a 60-Ω resistor in parallel in order to provide a total resistance of 10 Ω?
11. Find the voltage needed to send 2 A through a parallel combination of a 75- and a 100-Ω resistor.
12. A 30-Ω resistor is connected in parallel with an unknown resistor across a 120-V source. If the total current is 5 A, find the value of the unknown resistor.
13. In a circuit similar to that shown in Fig. 5-27, R_2 equals 3 kΩ, R_1 equals 17 kΩ, and I_T equals 40 mA. Find (*a*) the total resistance and (*b*) the total voltage.
14. A generator supplies 26 A to three motors in parallel. The resistances of the motors are 12, 18, and 24Ω. What current does each motor draw?

(See CD-ROM for Test 5-7)

1. What is the total resistance of the circuit shown in Fig. 5-31?

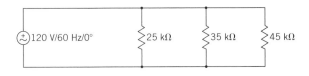

120 V/60 Hz/0°   25 kΩ   35 kΩ   45 kΩ

Figure 5-31

2. Fourteen 40-W cool-white fluorescent lamps are connected to a 120-V source. What is the total current drawn in this circuit as well as the branch current to each lamp?

3. Which is the larger number, $\frac{1}{25}$ or $\frac{1}{10}$?

4. If four mounting holes are to be drilled in a straight line in a panel, and each hole is $1\frac{7}{8}$ in on center, what is the distance between the two outside holes?

5. If you have 42 cartons on a pallet, each weighing $6\frac{3}{4}$ lb, what is the remaining weight if half of the cartons are taken away?

6. If $X_C = 45\ \Omega$, $C = 0.0005F$, and $\pi = 3.141592$, use the formula $X_C = 1/2\pi fC$ to determine the frequency (in hertz).

7. What is the value for V_T if $23.5 + 17.2 = V_T/S$?

8. What is the voltage drop across R_3 in Fig. 5-32?

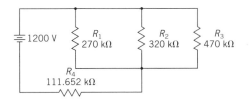

1200 V R_1 270 kΩ R_2 320 kΩ R_3 470 kΩ

R_4 111.652 kΩ

Figure 5-32

9. If each of five parallel resistors of 120 kΩ has 2.5 mA of current flowing through it, what is the source voltage?

10. You test a circuit and find that three parallel branches have 110 mA, 230 mA, and 335 mA flowing in their respective branches when 100 V is applied. What is the total resistance of the circuit?

11. Find the total resistance for the circuit in Fig. 5-33.

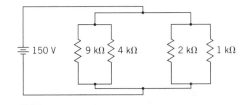

150 V   9 kΩ   4 kΩ   2 kΩ   1 kΩ

Figure 5-33

INTERNET
ACTIVITIES

Internet Activity A

The following site lists the steps for finding the greatest common factor (GCF) of two or more numbers. (1) List the three steps. (2) Using the steps outlined in (1), find the GCF of the numbers 24 and 36.
http://tqd.advanced.org/2949/math.htm

Internet Activity B

Use the following site to find the meaning of the term *node* as it relates to parallel circuits.
http://www.physics.uoguelph.ca/tutorials/ohm/Q.ohm.intro.parallel.html

Chapter 5 Solutions to Self-Tests and Reviews

Self-Test 5-2

a. 1. I_s
   2. 0.06, 1.9
b. 1. V_T
   2. 1.9
   3. 1.9
   4. 2

Job 5-4 Review

1. 6
2. 6
3. A
4. V
5. Eighths
6. $\dfrac{6}{8}$
7. 16

Self-Test 5-9

1. sum or addition
2. 2, 3, $2 \times 2 \times 2 \times 3$
3. whole numbers, adding, 9, 8, 14, 6, 37

4. a. $\dfrac{18}{24}$

   b. $\dfrac{12}{24}$

   c. $\dfrac{15}{24}$

   d. $\dfrac{10}{24}$

5. $\dfrac{55}{24}$

6. dividing, $^7/_{24}$, 39

Self-Test 5-14

1. subtracting, top, 20
3. 20
4. borrow, $^{20}/_{20}$
5. $^{32}/_{20}$

Self-Test 5-15

1. multiplying
3. 4
4. 20, 20

Self-Test 5-16

2. 6
3. multiplying, 6
5. 12, 12

Self-Test 5-19

3. 300
4. 300
5. 12L
6. 12
7. 10

Self-Test 5-23

1. R_T
2. 70
3. 0.007
4. 0.046
5. 1
6. 0.046
7. 21.7

Self-Test 5-30

1. R_1, 10^{-3}, A, 32
2. R_2, 4, A, 4

Job 5-11 Review

   a. more
1. equal, V_2, V_3
2. sum, I_2, I_3
3. total, R_T
4. R_T, $\times$, +, N
5. R_1, I_2, V_3
6. less
7. R_2, R_1
8. same, different, smaller
9. same, smallest, *all*
10. two, multiplication, equal, *F*, *C*

Combination Circuits

1. List the advantages and disadvantages of a series circuit.
2. List the advantages and disadvantages of a parallel circuit.
3. Solve for all unknown values of voltage, current, and resistance in a basic parallel-series circuit.
4. Solve for all unknown values of voltage, current, and resistance in a basic series-parallel circuit.
5. Explain the meaning of the term *line drop* in power distribution systems.
6. Find the line drop in a circuit for which the line current and line resistance are known.
7. Explain the effects caused by the line drop in power distribution systems.
8. Find the load voltages when the generator voltage and line drops are known.

Combination circuits are circuits that have both series and parallel components. You will learn how to develop formulas to solve for values in these more complicated circuits. If you have difficulty with these, you may have problems with the concepts of series and/or parallel circuits.

To review the concepts of series circuits go to Chap. 4, and then to Chap. 5 for a review of the concepts of parallel circuits. If you feel that you understand the concepts but have troubles with the formulas, review Chap. 3 on formulas and applications.

Introduction to Combination Circuits

Each simple circuit has its own advantages and disadvantages.

Series circuits.   An advantage of a series circuit is that it may be used to connect small voltages to obtain high voltages. Also, high voltages may be reduced by connecting resistances in series. Series circuits provide a means for reducing and controlling the current by connecting resistances in series. However, this series current remains unchanged throughout the circuit. This is a serious disadvantage. Since the current is constant, we are forced to use only those devices which require the same current. Thus it would be impossible to use any two household appliances at the same time because their current requirements range from 0.25 to 20 A. In addition, if any part of a series circuit should burn out, it would cause an open circuit and put the entire circuit out of operation. Series circuits are used where different voltage drops and a constant current are needed.

Parallel circuits.   If a break should occur in any branch of a parallel circuit, it would not affect the other branch circuits. Houses are wired in parallel so that any device may be operated independently of any other device. This is both an advantage and a disadvantage. In a parallel circuit, more branches may be added at any time. Each new load draws current from the line. So much current may be drawn that the original line wires may not be able to carry the new current and the fuse will "blow." In this event, it is necessary to rewire the circuits completely, using wire capable of carrying the larger currents, or to put in extra independent circuits to supply the installation. Parallel circuits are used wherever a constant voltage and a large supply of current are required.

Combination circuits. If we combine series circuits with parallel circuits, we produce a combination circuit which makes use of the best features of each. A combination circuit makes it possible to obtain the different voltages of a series circuit and the different currents of a parallel circuit. This is the condition most generally required, particularly when the different voltages and currents must be supplied from the same source of power. The different voltage and current needs of the different circuits of a typical AM/FM radio are all obtained from combination circuits drawing power from a single power supply.

Simple combination circuits are of two types:

1. A *parallel-series* circuit (Fig. 6-1) is a circuit in which one or more *groups* of resistances in series are connected in parallel.
2. A *series-parallel* circuit (Fig. 6-12) is a circuit in which one or more *groups* of resistances in parallel are connected in series.

General method for solving combination circuits

1. A *group* of resistances is a simple combination of two or more resistances which are arranged in either a *simple* series or a *simple* parallel circuit. Identify these groups.
2. Every group must be removed from the circuit as a unit and *replaced* by a single resistor which offers the identical resistance. This equivalent resistance is the total resistance of the group.
3. Redraw the circuit, using the equivalent resistance in place of each group.
4. Solve the resulting simple circuit for all missing values.
5. Go back to the *original* circuit to find the voltage, current, and resistance for each resistance in the circuit.

JOB 6-2

Solving Parallel-Series Circuits

Example 6-1

Solve the circuit shown in Fig. 6-1 for all missing values of voltage, current, and resistance.

Solution

1. Identify groups *A* and *B* as simple series circuits.
2. Find the equivalent resistance of each group. This means that the circuit is broken at points *X* and *Y*; resistors R_1 and R_2 are removed and replaced by a single resistance R_A. This single resistance will do the work of the combination of R_1 and R_2. Similarly, break the circuit at points *W* and *Z*;

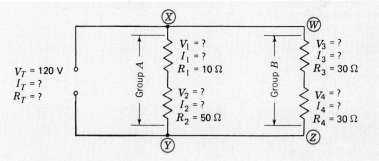

Figure 6-1 Parallell-series circuit.

resistors R_3 and R_4 are removed and replaced by a *single* resistance R_B. This single resistance will do the work of the combination of R_3 and R_4.

3. Since the resistors of groups A and B are in *series*,

$$R_A = R_1 + R_2 \qquad\qquad (4\text{-}3)$$
$$R_A = 10 + 50 = 60\ \Omega \qquad Ans.$$
$$R_B = R_3 + R_4 \qquad\qquad (4\text{-}3)$$
$$R_B = 30 + 30 = 60\ \Omega \qquad Ans.$$

4. Redraw the circuit, using these 60-Ω resistors in place of the series groups as shown in Fig. 6-2.

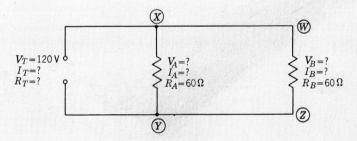

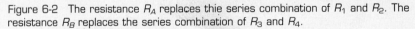
Figure 6-2 The resistance R_A replaces thie series combination of R_1 and R_2. The resistance R_B replaces the series combination of R_3 and R_4.

5. Solve the new *parallel* circuit.

Find the voltage for each group.

$$V_T = V_A = V_B = 120\ V \qquad Ans. \qquad (5\text{-}1)$$

Find the current in each group.

$$V_A = I_A \times R_A \qquad\qquad\qquad V_B = I_B \times R_B$$
$$120 = I_A \times 60 \qquad\qquad\qquad 120 = I_B \times 60$$
$$I_A = \frac{120}{60} = 2\ A \quad Ans. \qquad I_B = \frac{120}{60} = 2\ A \quad Ans.$$

Find the total current I_T.

$$I_T = I_A + I_B \qquad\qquad (5\text{-}2)$$
$$I_T = 2 + 2 = 4\ A \qquad Ans.$$

Hint for Success

When a parallel branch contains series resistors, as in Fig. 6-1, both resistors have the same current but the individual resistor voltage drops are less than the voltage applied across the entire branch. However, the individual resistor voltages in that branch add up to equal the voltage applied across that branch.

Find the total resistance R_T.

$$V_T = I_T \times R_T \qquad (4\text{-}7)$$
$$120 = 4 \times R_T$$
$$R_T = \frac{120}{4} = 30 \ \Omega \qquad Ans.$$

6. Go back to the original circuit to find the voltage and current for each resistor.

Find the current in each resistor.

$$I_A = I_1 = I_2 = 2 \text{ A} \qquad Ans. \qquad (4\text{-}1)$$
$$I_B = I_3 = I_4 = 2 \text{ A} \qquad Ans. \qquad (4\text{-}1)$$

Find the voltage drop across each resistor.

$$V_1 = I_1 \times R_1 \qquad\qquad V_3 = I_3 \times R_3$$
$$V_1 = 2 \times 10 \qquad\qquad V_3 = 2 \times 30$$
$$V_1 = 20 \text{ V} \quad Ans. \qquad V_3 = 60 \text{ V} \quad Ans.$$

$$V_2 = I_2 \times R_2 \qquad\qquad V_4 = I_4 \times R_4$$
$$V_2 = 2 \times 50 \qquad\qquad V_4 = 2 \times 30$$
$$V_2 = 100 \text{ V} \quad Ans. \qquad V_4 = 60 \text{ V} \quad Ans.$$

Self-Test 6-2

In Fig. 6-3, find the resistance R_3 that must be connected in parallel with the resistances of group A in order to obtain a total resistance of 15 Ω.

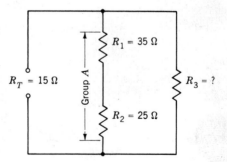

Figure 6-3 The series combination of R_1 and R_2 is in parallel with R_3.

Solution

1. Identify group A as a simple _____ circuit.
2. Find the resistance of group A.

$$R_A = R_1 + \text{_____} \qquad (4\text{-}3)$$
$$R_A = 35 + \text{_____} = \text{_____} \ \Omega$$

3. Redraw the circuit, using $R_A = 60 \ \Omega$ in place of the series group as shown in Fig. 6-4.

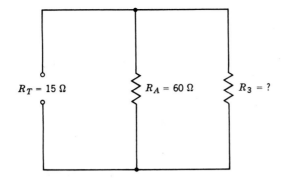

$R_T = 15 \, \Omega$ $R_A = 60 \, \Omega$ $R_3 = ?$

Figure 6-4 The resistance R_A replaces the series combination of R_1 and R_2 in Fig. 6-3.

4. Find R_3.

$$\frac{1}{R_T} = \frac{1}{?} + \frac{1}{R_3} \qquad\qquad (5\text{-}3)$$

$$\frac{1}{15} = \frac{1}{60} + \frac{1}{R_3}$$

$$\frac{1}{15} - \underline{\quad\quad} = \frac{1}{R_3}$$

$$\underline{\quad\quad} = \frac{1}{R_3}$$

$$R_3 = \underline{\quad\quad} \, \Omega \qquad Ans.$$

Exercises

Applications

1. Find all missing values in the circuit shown in Fig. 6-5.

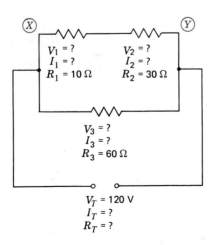

$V_1 = ?$ $V_2 = ?$
$I_1 = ?$ $I_2 = ?$
$R_1 = 10 \, \Omega$ $R_2 = 30 \, \Omega$

$V_3 = ?$
$I_3 = ?$
$R_3 = 60 \, \Omega$

$V_T = 120 \, V$
$I_T = ?$
$R_T = ?$

Figure 6-5

2. Find the total resistance and the total current for the circuit shown in Fig. 6-6.

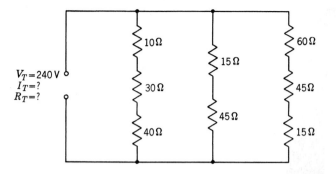

Figure 6-6

3. Find the total resistance and the total current for the circuit shown in Fig. 6-7.

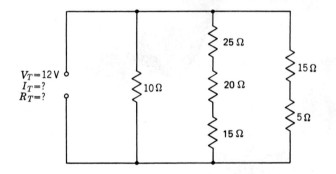

Figure 6-7

4. Find all missing values in the circuit shown in Fig. 6-8.

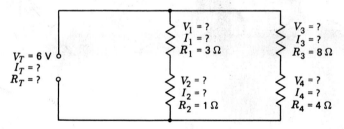

Figure 6-8

5. In the circuit shown in Fig. 6-9, find (a) the total resistance, (b) the total current, (c) the voltage V_1, (d) the current I_1, (e) the current I_3, (f) the voltage V_2, and (g) the voltage V_3.

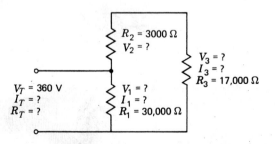

Figure 6-9

6. Part of the picture-control circuit of a television receiver is shown in Fig. 6-10. Find the total resistance of the circuit when the variable resistor R_3 has engaged (*a*) 6000 Ω, (*b*) 10,000 Ω, and (*c*) 5000 Ω.

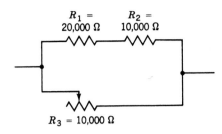

Figure 6-10

7. In a circuit similar to that shown in Fig. 6-3, find R_3 if $R_1 = 6\ \Omega$, $R_2 = 18\ \Omega$, and $R_T = 6\ \Omega$.
8. Find R_T in the circuit shown in Fig. 6-11. *Hint:* Find the equivalent resistance of R_3 and R_4 in parallel. Then proceed as in a normal parallel-series circuit. See also Example 6-3 below.

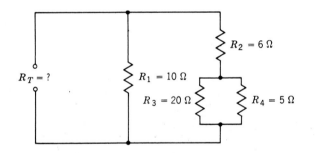

Figure 6-11

(See CD-ROM for Test 6-1)

Solving Series-Parallel Circuits

Example 6-3

Solve the circuit shown in Fig. 6-12 for all values of voltage, current, and resistance.

Solution

1. In Fig. 6-12, the resistors R_2 and R_3 form the *parallel* group A. Find the total resistance R_A of group A. This means that the circuit is broken at points X and Y; resistors R_2 and R_3 are removed and replaced by a *single*

resistance R_A. This single resistance will do the work of the combination of R_2 and R_3.

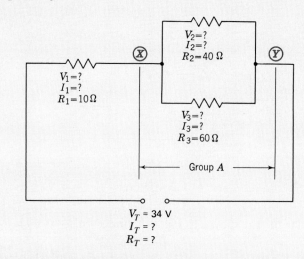

Figure 6-12   Series-parallel circuit.

2. Since R_2 and R_3 are in parallel,

$$R_A = \frac{R_2 \times R_3}{R_2 + R_3} \qquad (5\text{-}5)$$

$$R_A = \frac{40 \times 60}{40 + 60} = \frac{2400}{100} = 24 \ \Omega \qquad Ans.$$

3. Redraw the circuit using this 24-Ω resistor R_A in place of the combination as shown in Fig. 6-13.

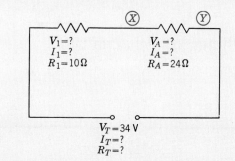

Figure 6-13   The resistance R_A replaces the parallel combination of R_2 and R_3 of Fig. 6-12.

4. Solve the new circuit. Notice that we have a simple *series* circuit. Find the total resistance R_T.

$$R_T = R_1 + R_4 \qquad (4\text{-}3)$$
$$R_T = 10 + 24 = 34 \ \Omega \qquad Ans.$$

Find the total current I_T.

$$V_T = I_T \times R_T \qquad (4\text{-}7)$$
$$34 = I_T \times 34$$
$$I_T = \frac{34}{34} = 1 \ A \qquad Ans.$$

Find the current in each part of the series circuit.

$$I_T = I_1 = I_A = 1 \text{ A} \qquad Ans. \qquad (4\text{-}1)$$

Find the voltage in each part of the series circuit.

$$V_1 = I_1 \times R_1 \qquad\qquad V_A = I_A \times R_A$$
$$V_1 = 1 \times 10 = 10 \text{ V} \quad Ans. \qquad V_A = 1 \times 24 = 24 \text{ V} \qquad Ans.$$

5. Go back to group A in Fig. 6-12 to find V, I, and R for each resistance in group A as shown in Fig. 6-14.

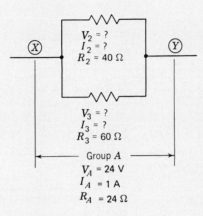

Figure 6-14

Since V_A represents the total voltage of the parallel group A,

$$V_A = V_2 = V_3 = 24 \text{ V} \qquad Ans. \qquad (5\text{-}1)$$

Find the current in each resistor of group A.

$$V_2 = I_2 \times R_2 \qquad\qquad V_3 = I_3 \times R_3$$
$$24 = I_2 \times 40 \qquad\qquad 24 = I_3 \times 60$$
$$I_2 = \frac{24}{40} = 0.6 \text{ A} \qquad\qquad I_3 = \frac{24}{60} = 0.4 \text{ A} \qquad Ans.$$

6. *Check*:

$$I_T = I_2 + I_3 \qquad\qquad (5\text{-}2)$$
$$I_T = 0.6 + 0.4$$
$$1 = 1 \qquad Check$$

Example 6-4

Solve the circuit shown in Fig. 6-15 on page 186 for all values of voltage, current, and resistance.

Solution

1. Find the total resistance. It is usually very helpful if the circuit is redrawn so as to put it in standard form with easily recognizable series or parallel subcircuits. Therefore, redraw Fig. 6-15 to appear as shown in Fig. 6-16.

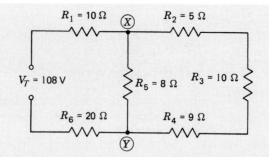

Figure 6-15

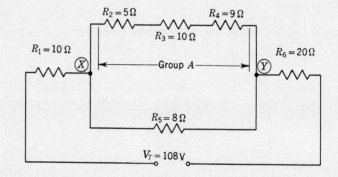

Figure 6-16   The circuit of Fig. 6-15 is redrawn in standard form.

a. In Fig. 6-16, the resistors R_2, R_3, and R_4 form the *series* group A. Find the total resistance of this group and replace the group with the equivalent resistance R_A as shown in Fig. 6-17.

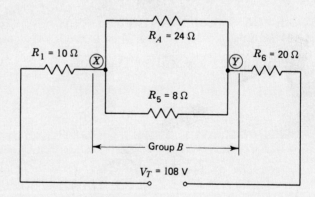

Figure 6-17   The resistance of R_A replaces the series combination of R_2, R_3, and R_4 of Fig. 6-16.

$$R_A = R_2 + R_3 + R_4 \qquad (4\text{-}3)$$
$$R_A = 5 + 10 + 9 = 24 \ \Omega \qquad \textit{Ans.}$$

b. In Fig. 6-17, the resistances R_A and R_5 form the *parallel* group B. Find the total resistance of group B and replace the group with the equivalent resistance R_B as shown in Fig. 6-18,

$$R_B = \frac{R_A \times R_5}{R_A + R_5} = \frac{24 \times 8}{24 + 8} = \frac{192}{32} = 6 \ \Omega \qquad Ans.$$

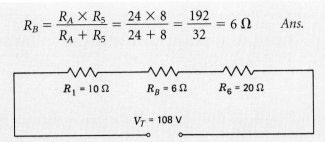

Figure 6-18 The resistance R_B replaces the parallel combination of R_A and R_5 of Fig. 6-17.

 c. In Fig. 6-18, the resistances R_1, R_B, and R_6 form a simple *series* circuit. Find the total resistance of the entire circuit.

$$R_T = R_1 + R_B + R_6 \qquad (4\text{-}3)$$
$$R_T = 10 + 6 + 20 = 36 \ \Omega \qquad Ans.$$

2. Find the total current I_T.

$$V_T = I_T \times R_T \qquad (4\text{-}7)$$
$$108 = I_T \times 36$$
$$I_T = \frac{108}{36} = 3 \ A \qquad Ans.$$

3. Find the currents and voltages in each part. Start with the *simplest* circuit obtained in the calculation of the total resistance. This would be Fig. 6-18. In this figure, R_1, R_B, and R_6 are in *series*. Therefore,

$$I_T = I_1 = I_B = I_6 = 3 \ A \qquad (4\text{-}1)$$

Find the voltage across each part by Ohm's law.

$$
\begin{array}{lll}
V_1 = I_1 \times R_1 & V_B = I_B \times R_B & V_6 = I_6 \times R_6 \\
V_1 = 3 \times 10 & V_B = 3 \times 6 & V_6 = 3 \times 20 \\
V_1 = 30 \ V & V_B = 18 \ V & V_6 = 60 \ V
\end{array}
$$

These values should be entered on the figure so that it will appear as in Fig. 6-19.

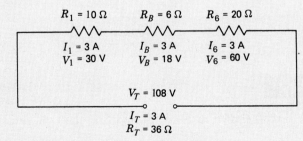

Figure 6-19 Values of voltage and current are entered on the circuit of Fig. 6-18.

4. We are now ready to find the voltages and currents for group A and resistor R_5. In Fig. 6-17, R_A and R_5 are in parallel. Therefore,

$$V_B = V_A = V_5 = 18 \ V \qquad (5\text{-}1)$$

Find the current in R_A and R_5 by Ohm's law.

$$V_A = I_A \times R_A \qquad\qquad V_5 = I_5 \times R_5$$
$$18 = I_A \times 24 \qquad\qquad 18 = I_5 \times 8$$
$$I_A = \frac{18}{24} = 0.75 \text{ A} \qquad\qquad I_5 = \frac{18}{8} = 2.25 \text{ A}$$

Enter these values on Fig. 6-17 so that it will appear as shown in Fig. 6-20.

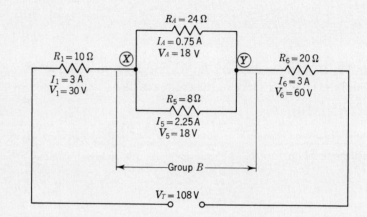

Figure 6-20   Values of voltage and current are entered on the circuit of Fig. 6-17.

5. We are now ready to find the currents and voltages for the individual resistors of group A. In Fig. 6-16, R_2, R_3, and R_4 are in series. Therefore,

$$I_A = I_2 = I_3 = I_4 = 0.75 \text{ A} \tag{4-1}$$

Find the voltage across these resistors by Ohm's law.

$$V_2 = I_2 \times R_2 \qquad V_3 = I_3 \times R_3 \qquad V_4 = I_4 \times R_4$$
$$V_2 = 0.75 \times 5 \qquad V_3 = 0.75 \times 10 \qquad V_4 = 0.75 \times 9$$
$$V_2 = 3.75 \text{ V} \qquad V_3 = 7.5 \text{ V} \qquad V_4 = 6.75 \text{ V}$$

6. *Check*: The voltage across V_2, V_3, and V_4 should equal V_A.

$$V_A = V_2 + V_3 + V_4 = 18 \text{ V} \tag{4-2}$$
$$18 = 3.75 + 7.50 + 6.75$$
$$18 = 18 \qquad Check$$

Example 6-5

Simplify the circuit shown in Fig. 6-21 and find the total resistance of the circuit.

Solution

1. Redraw the circuit in standard form as shown in Fig. 6-22.
2. In Fig. 6-22 the resistors R_3, R_4, and R_5 form the *series* group A. Find the total resistance of the group and replace the group with the equivalent resistance R_A as shown in Fig. 6-23.

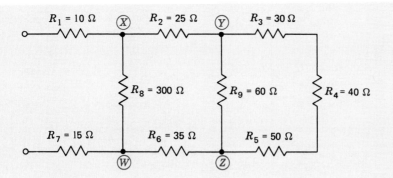

Figure 6-21

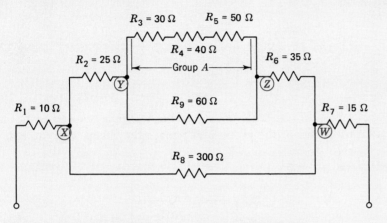

Figure 6-22 The circuit of Fig. 6-21 is redrawn in standard form.

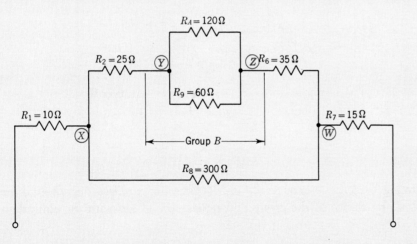

Figure 6-23 The resistance R_A replaces the series combination of R_2, R_4, and R_5 of Fig. 6-22.

$$R_A = R_3 + R_4 + R_5 \qquad (4\text{-}3)$$
$$R_A = 30 + 40 + 50 = 120\ \Omega \qquad Ans.$$

3. In Fig. 6-23 the resistances R_A and R_9 form the *parallel* group B. Find the total resistance of the group and replace the group with the equivalent resistance R_B as shown in Fig. 6-24.

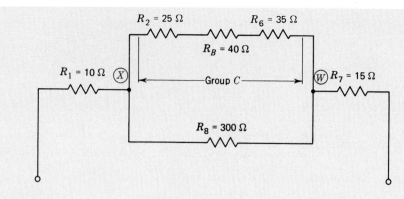

Figure 6-24 The resistance R_B replaces the parallel combination of R_A and R_9 of Fig. 6-23.

$$R_B = \frac{R_A \times R_9}{R_A + R_9} \qquad (5\text{-}5)$$

$$R_B = \frac{120 \times 60}{120 + 60} = \frac{7200}{180} = 40\ \Omega \qquad Ans.$$

4. In Fig. 6-24 the resistances R_2, R_B, and R_6 form the *series* group C. Find the total resistance of the group and replace the group with the equivalent resistance R_C as shown in Fig. 6-25.

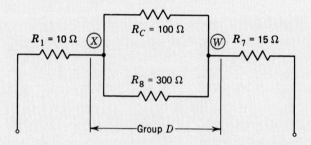

Figure 6-25 The resistance R_C replaces the series combination of R_2, R_B, and R_6 of Fig. 6-24.

$$R_C = R_2 + R_B + R_6 \qquad (4\text{-}3)$$
$$R_C = 25 + 40 + 35 = 100\ \Omega \qquad Ans.$$

5. In Fig. 6-25 the resistances R_C and R_8 form the *parallel* group D. Find the total resistance of the group and replace the group with the equivalent resistance R_D as shown in Fig. 6-26.

$$R_D = \frac{R_C \times R_8}{R_C + R_8} \qquad (5\text{-}5)$$

$$R_D = \frac{100 \times 300}{100 + 300} = \frac{30,000}{400} = 75\ \Omega \qquad Ans.$$

6. In Fig. 6-26 the resistances R_1, R_D, and R_7 form a *series* circuit. Find the resistance of the entire circuit.

$$R_T = R_1 + R_D + R_7 \qquad (4\text{-}3)$$
$$R_T = 10 + 75 + 15 = 100\ \Omega \qquad Ans.$$

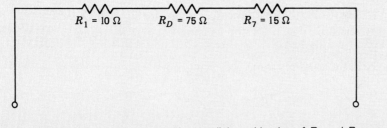

Figure 6-26 The resistance R_D replaces the parallel combination of R_C and R_B of Fig. 6-25.

Self-Test 6-6

Find the total resistance of the circuit shown in Fig. 6-27.

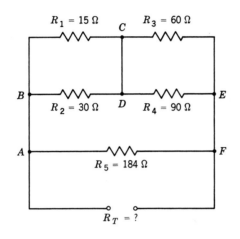

Figure 6-27

Solution

1. Redraw the circuit in standard form as shown in Fig. 6-28.

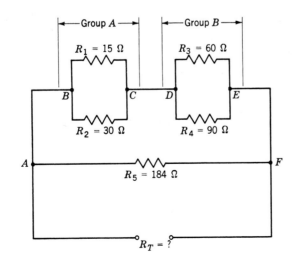

Figure 6-28 Figure 6-27 redrawn in standard form.

2. In Fig. 6-28, the resistors R_1 and R_2 are in ___(series/parallel)___, forming group A. In Fig. 6-28, the resistors R_3 and R_4 are in ___(series/parallel)___, forming group B.

3. Find the total resistance of each group.

$$R_A = \frac{R_1 \times R_2}{?}$$

$$R_A = \frac{15 \times 30}{? + 30}$$

$$R_A = \frac{450}{45} = \underline{\hspace{1.5cm}} \ \Omega \qquad Ans.$$

$$R_B = \frac{60 \times ?}{60 + 90}$$

$$R_B = \frac{5400}{150} = \underline{\hspace{1.5cm}} \ \Omega \qquad Ans.$$

4. Redraw the circuit, replacing each group with its equivalent resistance R_A and R_B as shown in Fig. 6-29.

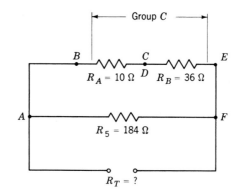

Figure 6-29 The resistance R_A replaces the parallel combination of R_1 and R_2; R_B replaces the parallel combination of R_3 and R_4 in Fig. 6-28.

5. In Fig. 6-29, the resistances R_A and R_B form the ___(series/parallel)___ group C. Find the total resistance of the group C.

$$R_C = \underline{\hspace{1.5cm}} \qquad\qquad (4\text{-}3)$$
$$R_C = 10 + \underline{\hspace{1.5cm}} = \underline{\hspace{1.5cm}} \ \Omega \qquad Ans.$$

6. Redraw the circuit, replacing R_A and R_B with their equivalent resistance R_C as shown in Fig. 6-30.

7. In Fig. 6-30, the resistances R_C and R_5 are in ___(series/parallel)___. Find R_T.

$$R_T = \frac{?}{?}$$

$$R_T = \frac{46 \times ?}{46 + ?}$$

$$R_T = \frac{8464}{230} = \underline{\hspace{1.5cm}} \ \Omega \qquad Ans.$$

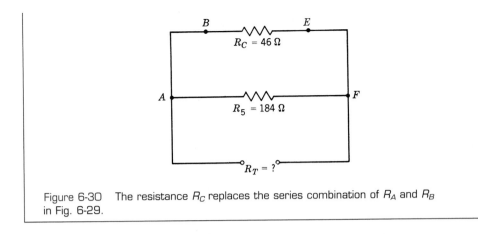

Figure 6-30   The resistance R_C replaces the series combination of R_A and R_B in Fig. 6-29.

Exercises

Applications

1. In a circuit similar to that shown in Fig. 6-12, $V_T = 130$ V, $R_1 = 10\ \Omega$, $R_2 = 4\ \Omega$, and $R_3 = 12\ \Omega$. Find all missing values of voltage, current, and resistance.
2. Find all missing values in the circuit shown in Fig. 6-31.

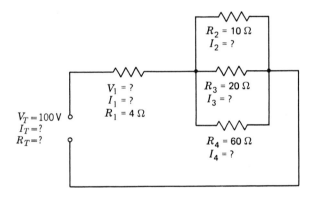

Figure 6-31

3. In a circuit similar to that shown in Fig. 6-31, find the total current if $V_T = 8.5$ V, $R_1 = 12\ \Omega$, $R_2 = 10\ \Omega$, $R_3 = 15\ \Omega$, and $R_4 = 30\ \Omega$.
4. Use a circuit similar to that shown in Fig. 6-15. $V_T = 120$ V, $R_1 = 5\ \Omega$, $R_2 = 1\ \Omega$, $R_3 = 3\ \Omega$, $R_4 = 6\ \Omega$, $R_5 = 10\ \Omega$, and $R_6 = 10\ \Omega$. Find (a) the total resistance, (b) the total current, (c) the current in each resistor, and (d) the voltage drop across each resistor.
5. In Fig. 4-16 the problem of accidental shock was discussed. Suppose that you attempt to grasp the victim and remove him from the housing. What current would go through you and what would be the effect?
6. Use a circuit similar to that shown in Fig. 6-21. $V_T = 400$ V, $R_1 = 50\ \Omega$, $R_2 = 25\ \Omega$, $R_3 = 8\ \Omega$, $R_4 = 2\ \Omega$, $R_5 = 10\ \Omega$, $R_6 = 15\ \Omega$, $R_7 = 105\ \Omega$, $R_8 = 450\ \Omega$, and $R_9 = 20\ \Omega$. Find (a) the total resistance, (b) the total current, (c) the current in each resistor, and (d) the voltage drop across each resistor.
7. Find the total resistance of the circuit shown in Fig. 6-32.

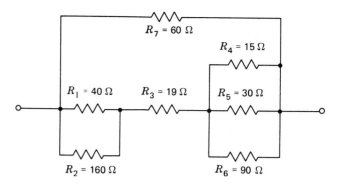

Figure 6-32

8. Find the total resistance of the circuit shown in Fig. 6-27 if the 60-Ω resistor were to burn out and open.

9. In Fig. 6-33, (*a*) draw the equivalent resistance network, (*b*) find the total resistance between the point *P* and the ground *G*, and (*c*) find the resistance that must be connected between *P* and *G* (in parallel with the entire circuit) to reduce the total resistance to 36 kΩ.

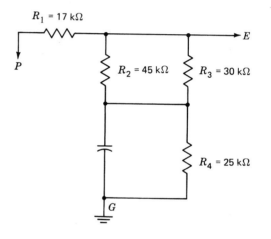

Figure 6-33

10. In Fig. 6-34, find the total resistance from *A* to *B*.

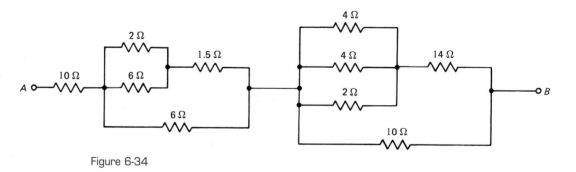

Figure 6-34

11. Find the total resistance of the circuit shown in Fig. 6-35.

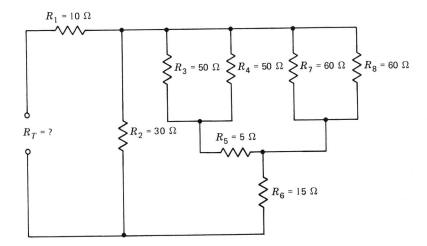

Figure 6-35

12. Find the total resistance of the circuit shown in Fig. 6-36.

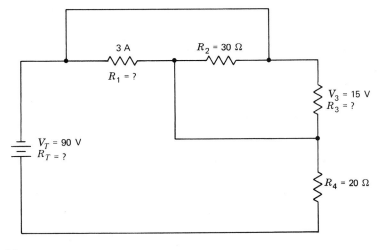

Figure 6-36

Self-Test 6-7

Finding the total resistance

1. Redraw the circuit in _____ form if necessary.
2. First simplification.
   a. Locate the combination of resistors which forms a simple _____ or parallel circuit.
   b. Indicate the beginning and _____ of the group with a dimension line and name it "group _____."
   c. Calculate the resistance of this group and label it _____.
3. Redraw the circuit, substituting _____ in place of the group.

4. Second simplification: Repeat steps 2a, 2b, and 2c, using the name "group _____" and R_B in the circuit of step 3.
5. Repeat steps 2 and 3 using the names "group C," "group D," etc., until the circuit is reduced to the total resistance named _____.

Solving combination circuits

1. Find the total resistance R_T by repeated simplification of the original circuit.
2. Find the total current I_T using the formula

$$V_T = \underline{\quad\quad} \times R_T \qquad (4\text{-}7)$$

3. Find the current and voltage in each resistor.
   a. Start with the __(simplest/most complicated)__ circuit obtained in the calculations for R_T.
   b. The value of current in this circuit will be the _____ current. Enter this value of I_T on this circuit diagram.
   c. Calculate the voltage drops across all resistors of this circuit by the formula

   _____

   d. Enter these values on the next simplest circuit obtained in the original calculation for _____.
   e. Solve this circuit for all missing values of current and _____.
   f. Repeat steps d and e on successive circuits found in the _____ simplification until all parts have been found.
(See CD-ROM for Test 6-2)

JOB 6-4

Line Drop

Meaning of line drop. In our last two jobs, we worked with circuits in which the connecting wires were very short. The resistances of these short lengths were so very small that we did not bother to include them in our calculations. In home and factory installations, however, where long lines of wire (feeders) are used, the resistance of these long lengths must be included in all calculations.

In Fig. 6-37, the voltage that is needed to force the current through the resistance of the line wires is called the *line drop*. For example, if a generator delivers 120 V but the voltage available at a motor some distance away is only 116 V, then

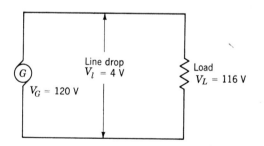

Figure 6-37   Voltage is lost in the line wires between a generator and a load.

there has been a "drop" in voltage of 4 V. The connecting wires apparently had enough resistance to use up 4 V of electrical pressure.

We must be very careful about the kind and size of wires used in any installation. If the wires are poorly chosen, then the *line drop* may be very large and the voltage available to the electric apparatus will be too low for proper operation. The national and city electrical codes permit definite amounts of line drop for specific installations. This limitation of the line drop is accomplished by specifying definite sizes of wire to be used in specific installations. We shall study this in detail in Chap. 13.

You may have noticed the effect of line drop in your home. The lights may suddenly get dim when the refrigerator motor starts. The large current drain required to start the motor increases the line drop in the house wiring so that the voltage left over for the lights is less than normal. Since the running current is much less than the starting current, as soon as the motor has started the current drain decreases—the line drop decreases—and the lights once again come up to full brilliance.

Definitions

1. Generator voltage (V_G): The total voltage supplied to the circuit from the source of voltage
2. Load voltage (V_L): The voltage that is available to operate the devices or loads
3. Line drop (V_l): The voltage that is lost in sending the current through the line wires

Rules and Formulas

1. By Ohm's law, the total line drop is equal to the line current multiplied by the total line resistance.

$$V_{\text{line}} = I_{\text{line}} \times R_{\text{line}}$$

or

$$\boldsymbol{V_l = I_l \times R_l} \qquad \text{Formula 6-1}$$

2. The generator voltage is the total voltage of the series circuit, made up of the line drop and the load voltage.

$$V_T = V_1 + V_2 \qquad (4\text{-}2)$$

or

$$V_{\text{generator}} = V_{\text{line}} + V_{\text{load}}$$

$$\boldsymbol{V_G = V_l + V_L} \qquad \text{Formula 6-2}$$

3. The load voltage is equal to the generator voltage minus the line drop. From

$$V_l + V_L = V_G \qquad (6\text{-}2)$$

we get, by transposing the V_l,

$$\boldsymbol{V_L = V_G - V_l} \qquad \text{Formula 6-3}$$

and, by transposing the V_L,

$$\boldsymbol{V_l = V_G - V_L} \qquad \text{Formula 6-4}$$

When using these formulas, be sure that

1. The line current I_l is the current flowing in the line wires and *not* the current in the load.
2. The line drop V_l is the voltage lost in the line wires only.
3. The line resistance is the resistance of the connecting wires only. The resistances of *both* lead and return wires must be considered when finding the total resistance of the line.

Example 6-8

A lamp bank consisting of three lamps, each drawing 2 A, is connected to a 120-V source. Each line wire has a resistance of 0.2 Ω. Find the drop in voltage in the line and the voltage available at the load.

Solution

The diagram for the circuit is shown in Fig. 6-38.

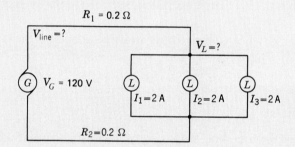

Figure 6-38

1. Find the current drawn by the load.

$$I_L = I_1 + I_2 + I_3 \qquad (5\text{-}2)$$
$$I_L = 2 + 2 + 2 = 6\,\text{A} \qquad Ans.$$

2. Find the line current. Since the line wires are in series with the load,

$$I_l = I_L = 6\,\text{A} \qquad Ans. \qquad (4\text{-}1)$$

3. Find the resistance of the line wires. Since the line wires are in series,

$$R_l = R_1 + R_2 \qquad (4\text{-}3)$$
$$R_l = 0.2 + 0.2 = 0.4\,\Omega \qquad Ans.$$

4. Find the total line drop.

$$V_l = I_l \times R_l \qquad (6\text{-}1)$$
$$V_l = 6 \times 0.4 = 2.4\,\text{V} \qquad Ans.$$

5. Find the voltage available at the load.

$$V_L = V_G - V_l \qquad (6\text{-}3)$$
$$V_L = 120 - 2.4 = 117.6\,\text{V} \qquad Ans.$$

Example 6-9

Find the generator voltage required for the circuit shown in Fig. 6-39.

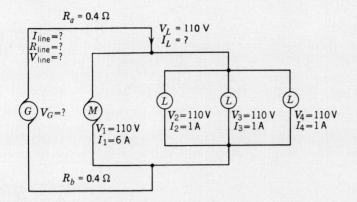

Figure 6-39

Solution

1. Find the current drawn by the load.

$$I_L = I_1 + I_2 + I_3 + I_4 \qquad (5\text{-}2)$$
$$I_L = 6 + 1 + 1 + 1 = 9\,\text{A} \qquad Ans.$$

2. Find the line current.

$$I_l = I_L = 9\,\text{A} \qquad Ans. \qquad (4\text{-}1)$$

3. Find the resistance of the line wires.

$$R_l = R_a + R_b \qquad (4\text{-}3)$$
$$R_l = 0.4 + 0.4 = 0.8\,\Omega \qquad Ans.$$

4. Find the total line drop.

$$V_l = I_l \times R_l \qquad (6\text{-}1)$$
$$V_l = 9 \times 0.8 = 7.2\,\text{V} \qquad Ans.$$

5. Find the generator voltage.

$$V_G = V_l + V_L \qquad (6\text{-}2)$$
$$V_G = 7.2 + 110 = 117.2\,\text{V} \qquad Ans.$$

Self-Test 6-10

Find the resistance of each line wire in the circuit shown in Fig. 6-40.

Solution

In order to find the line resistance we must know the voltage drop in the line and the current in the line. Find the current distribution by investigating group A first.

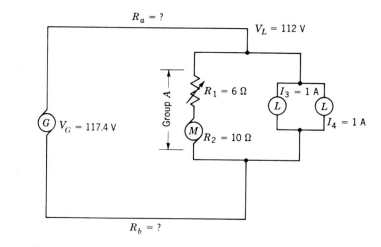

Figure 6-40

1. Find the total resistance of group A.

$$R_A = R_1 + R_2 \qquad \text{(4-3)}$$
$$R_A = 6 + 10 = \underline{\hspace{2cm}} \ \Omega$$

2. Since V_A is in parallel with the lamp load and the load voltage V_L,

$$V_A = V_L = \underline{\hspace{2cm}} \text{ V} \qquad \text{(5-1)}$$

3. Find the current in group A.

$$V_A = I_A \times R_A \qquad \text{(4-7)}$$
$$112 = I_A \times \underline{\hspace{2cm}}$$
$$I_A = \frac{112}{16} = \underline{\hspace{2cm}} \text{ A}$$

4. Find the total load current I_L.

$$I_L = I_A + I_3 + I_4 \qquad \text{(5-2)}$$
$$I_L = \underline{\hspace{2cm}} + 1 + 1 = \underline{\hspace{2cm}} \text{ A}$$

5. Since the line wires are in series with the load,

$$I_l = I_L = \underline{\hspace{2cm}} \text{ A} \qquad \text{(4-1)}$$

6. Find the voltage drop in the line wires.

$$V_l = V_G - V_L \qquad \text{(6-4)}$$
$$V_l = 117.4 - 112 = \underline{\hspace{2cm}} \text{ V}$$

7. Find the resistance of the line wires.

$$V_l = I_l \times R_l \qquad \text{(6-1)}$$
$$5.4 = \underline{\hspace{2cm}} \times R_l$$
$$R_l = \frac{5.4}{9} = \underline{\hspace{2cm}} \ \Omega$$

8. Find the resistance of each line wire.

$$R_a = R_b = \frac{1}{2} \times R_l$$

$$R_a = R_b = \frac{1}{2} \times 0.6 = \underline{\hspace{2cm}} \ \Omega \qquad Ans.$$

Exercises

Applications

1. In a circuit similar to that shown in Fig. 6-38, the generator voltage is 117 V. Each line wire has a resistance of 0.4 Ω, and each lamp draws 1 A. Find the line drop and the voltage available at the lamps.
2. A motor is connected by two wires of 0.15 Ω each to a generator. The motor takes 30 A at 211 V. What must be the generator voltage?
3. If the voltage at a load drawing 6 A is 117 V while the generator voltage is 120 V, what is the resistance of each line wire? *Hint*: Find the line drop, then the line resistance, and finally the resistance of each wire.
4. Home wiring is often done with No. 16 wire, which has a resistance of 0.401 Ω for a 100-ft length. What is the loss in voltage from the house meter to an electric broiler using 12 A and located 100 ft from the meter?
5. What would be the voltage drop if No. 14 wire (0.252 Ω/100 ft) were used in Prob. 4? Which size of wire is better for wiring homes?
6. In a circuit similar to that shown in Fig. 6-38, the generator voltage is 117 V. Each line wire has a resistance of 0.45 Ω, and the lamps draw currents of 0.9, 1.4, and 1.8 A. Find the line drop and the voltage available at the lamps.
7. In a circuit similar to that shown in Fig. 6-39, the motor draws 8.2 A and each of the three lamps draws 0.92 A. Each line wire has a resistance of 0.15 Ω. Find the generator voltage if the load voltage must be 110 V.
8. In a circuit similar to that shown in Fig. 6-38, the generator voltage is 117 V, the resistance of each line wire is 0.2 Ω, and the total resistance of the lamp bank is 16.1 Ω. Find (a) the total resistance of the circuit, (b) the current delivered to the lamp bank, and (c) the voltage across the lamp bank.

Distribution Systems

In order to distribute current throughout an installation, the various loads are connected in parallel across the feeder lines. The feeder lines form various combination circuits with the loads. To solve circuits like these, we must first break down the combination into simple series or parallel circuits. It is best to follow a definite system like the following:

1. Find the current distribution. Start with the section *farthest* from the generator. Find the current in this section and work backward toward the generator, finding the current in the different parts of the circuit.
2. Name the sections. Call the section nearest to the generator section *A*, the next section *B*, etc.

3. Find the resistance of each *pair* of line wires for each section using Eq. (4-3).
4. Find the line drop for each section of the circuit using Eq. (6-1). The line wires connecting the generator to section *A* are called line 1, the next line-wire pair is called line 2, etc.
5. Find the voltage across each section. Start where the voltage is known, and apply Eqs. (6-2) and (6-3).

Example 6-11

Find the voltage across the motor and across the lamp bank of the circuit shown in Fig. 6-41.

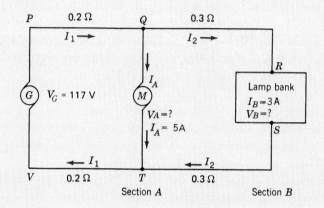

Figure 6-41

1. Find the current distribution. Start with the section *B* farthest from the generator. $I_B = 3$ A. Line 2, from *Q* to *R* and from *S* to *T*, must carry this 3 A. Therefore, $I_2 = 3$ A. Since $I_A = 5$ A, the wire from *Q* to *T* must carry this 5 A. At point *Q*, we have the beginning of a parallel circuit made of the motor (section *A*) and the lamp bank (section *B*).

$$I_1 = I_A + I_2 \qquad (5\text{-}2)$$
$$I_1 = 5 + 3 = 8 \text{ A}$$

Thus, the current in line 1 from *P* to *Q* equals 8 A. Similarly, at *T*, the current I_A from the motor and the current I_2 from the lamp bank will combine.

$$I_1 = I_A + I_2 \qquad (5\text{-}2)$$
$$I_1 = 5 + 3 = 8 \text{ A}$$

Thus, the current in line 1 from *T* to *V* equals 8 A.

2. Find the resistance of the *pairs* of line wires.

$$R_{l_1} = 0.2 + 0.2 = 0.4 \ \Omega \qquad R_{l_2} = 0.3 + 0.3 = 0.6 \ \Omega \quad (4\text{-}3)$$

3. Find the line drop for each section.

$$V_{l_1} = I_1 \times R_{l_1} \qquad\qquad V_{l_2} = I_2 \times R_{l_2} \qquad (6\text{-}1)$$
$$V_{l_1} = 8 \times 0.4 = 3.2 \text{ V} \qquad V_{l_2} = 3 \times 0.6 = 1.8 \text{ V}$$

4. Find the load voltages.

$$V_A = V_G - V_{l_1}$$ (6-3)
$$V_A = 117 - 3.2 = 113.8 \text{ V} \quad Ans.$$
$$V_B = V_A - V_{l_2}$$ (6-3)
$$V_B = 113.8 - 1.8 = 112 \text{ V} \quad Ans.$$

Example 6-12

Change to a mixed number. Find the generator voltage and the voltage across the motor in the circuit shown in Fig. 6-42.

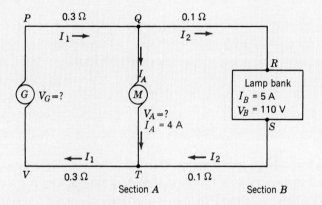

Figure 6-42

Solution

1. Find the current distribution. Start with the section B farthest from the generator. $I_B = 5$ A. Line 2, from Q to R and from S to T, must carry this 5 A. Therefore, $I_2 = 5$ A.
   Since $I_A = 4$ A, the wire from Q to T must carry this 4 A.

At point Q:	At point T:	
$I_1 = I_A + I_2$	$I_1 = I_A + I_2$	(5-2)
$I_1 = 4 + 5 = 9$ A	$I_1 = 4 + 5 = 9$ A	

2. Find the resistance of the *pairs* of line wires.

$$R_{l_1} = 0.3 + 0.3 = 0.6 \ \Omega \qquad R_{l_2} = 0.1 + 0.1 = 0.2 \ \Omega \quad (4\text{-}3)$$

3. Find the line drop for each section.

$$V_{l_1} = I_1 \times R_{l_1} \qquad\qquad V_{l_2} = I_2 \times R_{l_2} \qquad (6\text{-}1)$$
$$V_{l_1} = 9 \times 0.6 = 5.4 \text{ V} \qquad V_{l_2} = 5 \times 0.2 = 1 \text{ V}$$

4. Find the load and generator voltages.

$$V_A = V_{l_2} + V_B \qquad (6\text{-}2)$$
$$V_A = 1 + 110 = 111 \text{ V} \quad Ans.$$
$$V_G = V_{l_1} + V_A$$
$$V_G = 5.4 + 111 = 116.4 \text{ V} \quad Ans.$$

Self-Test 6-13

Find the voltage across (*a*) the lamp bank, (*b*) motor 1, and (*c*) motor 2 in the circuit shown in Fig. 6-43.

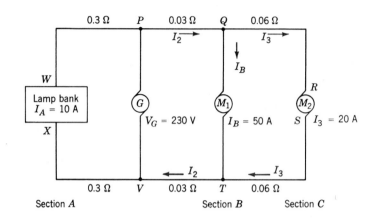

Figure 6-43

Solution

1. Find the current distribution. Start with section ————. I_3 from Q to R and from S to $T = $ ———— A. At point Q, the incoming current I_2 divides into two branches, ———— and ————.

$$I_2 = I_3 + \text{————}$$
$$I_2 = 20 + \text{————} = \text{————} A$$

Thus, the current from P to $Q = $ ———— A. At point T, I_3 and I_B combine to form ————. Thus, the current from T to $V = $ ———— A. The current I_A from W to P and from V to X equals ———— A.

2. Find the resistance of the *pairs* of line wires.

$$R_{I_A} = 0.3 + 0.3 = \text{————} \ \Omega$$
$$R_{I_2} = 0.03 + \text{————} = \text{————} \ \Omega$$
$$R_{I_3} = \text{————} + \text{————} = 0.12 \ \Omega$$

3. Find the line drop for each section.

$$V_{I_A} = I_A \times \text{————}$$
$$V_{I_A} = 10 \times \text{————} = \text{————} \ V$$
$$V_{I_2} = \text{————} \times R_{I_2}$$
$$V_{I_2} = 70 \times \text{————} = \text{————} \ V$$
$$V_{I_3} = I_3 \times \text{————}$$
$$V_{I_3} = \text{————} \times 0.12 = \text{————} \ V$$

4. Find the load voltages. The voltage across the lamp bank is called V_A.

$$V_A = V_G - \text{————} \qquad\qquad (6\text{-}3)$$
$$V_A = \text{————} - 6 = \text{————} \ V \qquad Ans.$$

The voltage across motor 1 is called V_B.

$$V_B = \underline{\hspace{2cm}} - V_{l_2} \qquad\qquad (6\text{-}3)$$
$$V_B = 230 - \underline{\hspace{2cm}} = 225.8 \text{ V} \qquad Ans.$$

The voltage across motor 2 is called \underline{\hspace{2cm}}.

$$V_C = V_B - \underline{\hspace{2cm}} \qquad\qquad (6\text{-}3)$$
$$V_C = 225.8 - 2.4 = \underline{\hspace{2cm}} \text{ V} \qquad Ans.$$

Exercises

Applications

1. In the circuit shown in Fig. 6-44, the motor takes 30 A and the lamp bank takes 8 A. The generator voltage is 117 V. Find (*a*) the line drop in each section, (*b*) the voltage across the motor, and (*c*) the voltage across the lamp bank.

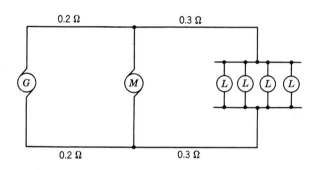

Figure 6-44

2. Each lamp in the circuit shown in Fig. 6-44 takes 1.5 A. The motor takes 12 A. The generator voltage is 115 V. Find (*a*) the line drop in each section, (*b*) the voltage across the motor, and (*c*) the voltage across the lamp bank.
3. Each lamp in the diagram shown in Fig. 6-45 takes 0.5 A. Find V_A and V_B.

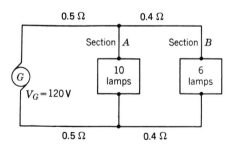

Figure 6-45

4. In the circuit shown in Fig. 6-46, the resistance of the wire *AB* is 0.3 Ω, *BC* is 0.5 Ω, *EF* is 0.5 Ω, and *DE* is 0.3 Ω. If each lamp takes 1 A, what is the terminal voltage of the generator? The voltage across *CF* is 105 V.

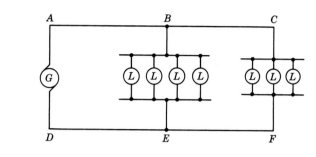

Figure 6-46

5. In the circuit shown in Fig. 6-47, each lamp takes 1.2 A. V_A = 113 V, and I_A = 12 A. Find V_G and V_B.

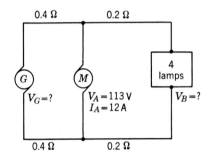

Figure 6-47

6. In the circuit shown in Fig. 6-48, each lamp takes 1 A. V_A = 114 V, and V_G = 117 V. Find V_B and I_A.

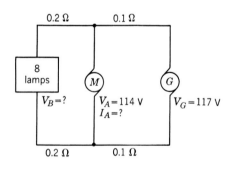

Figure 6-48

7. Find the generator voltage in the circuit shown in Fig. 6-49.
8. Each line wire of a two-wire distribution system has a resistance of 0.2 Ω. A motor drawing 10 A is connected by two of these wires to a generator delivering 121 V. Two more of these line wires continue from the motor and carry current to a lamp bank drawing 4.5 A. Find (a) the voltage at the motor and (b) the voltage at the lamp bank.

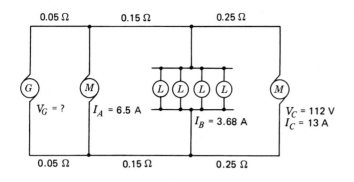

Figure 6-49

9. In Fig. 6-50, find (*a*) the voltage across motor 1 and (*b*) the voltage across motor 2. *Hint:* $V_{M_1} = V_G - (V_{l_1} + V_{l_2} + V_{l_4})$, and $V_{M_2} = V_G - (V_{l_1} + V_{l_3} + V_{l_2})$.

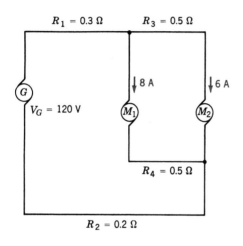

Figure 6-50

(See CD-ROM for Test 6-3)

Analysis with Ideal Circuit Elements

Sometimes a circuit looks just like a series-parallel circuit, but it is not analyzed that way. Job 4-6 is just such a situation. The circuit of Job 4-6 is reproduced in Fig. 6-51 on page 208. Using the definitions of series and parallel connections, there is no doubt that resistor R_1 is in series with resistor R_2. Both these resistors are in parallel with resistor R_3.

In spite of these obvious properties, the circuit was analyzed as three simple series circuits in Job 4-6. A very similar situation will be seen later in 120/240 distribution systems and again in three-phase ac transformer circuits. In all these cases, the reason for the apparent conflict in analysis is the same. *The circuits are*

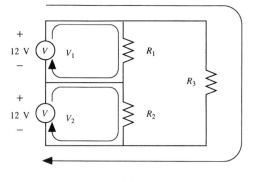

(a)

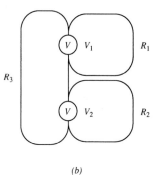

(b)

Figure 6-51   (a) Circuit model. (b) Loop currents.

ideal. Ideal circuits simplify calculations. As long as the results adequately represent the real world, then they are valid.

The ideal elements of these problems are the connecting wires and the batteries. It is assumed that there is no resistance in any of these components. Because of these assumptions, the voltages are connected directly across the loads. Each load resistor has a voltage source connected directly across it. By ignoring the resistances, there is no mechanism by which these loads can affect each other.

This separation can be seen more easily if you look at the loop currents in Fig. 6-51a. If these loop circuits are redrawn, as in Fig. 6-51b, it is more obvious that there is no interaction among the three separate simple series circuits. The current that goes through R_1 is completely determined by V_1. That current comes out of source V_1 and returns to V_1. The current through R_2 is determined in a similar manner.

The current through R_3 is determined by both V_1 and V_2. The current through R_3 flows through both V_1 and V_2. Because it is assumed that V_1 and V_2 have no resistance, it is not possible for the currents in these sources to interact.

Figure 6-52 shows two more common ideal-circuit configurations that appear to be complex but are no more than a collection of simple series circuits. The upper figure is a "delta" circuit, and the lower is a four-wire "Y," or "wye," circuit. In each case load R_1 is connected directly across voltage source V_1, R_2 is connected across V_2, and R_3 is connected across V_3. These circuits have ac counterparts.

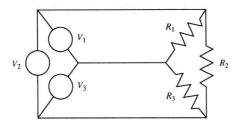

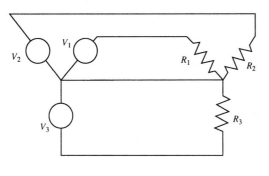

(a)

(b)

Figure 6-52   (a) Delta circuit. (b) Y circuit.

RULE   If all connecting lines are considered ideal—i.e., they have no voltage drop—and if the voltage source is considered ideal—i.e., it has no internal resistance, then all loads connected directly across the ideal voltage source are considered as independent simple series circuits.

Assessment

chapter 6

1. Find the missing values in the circuit in Fig. 6-53.

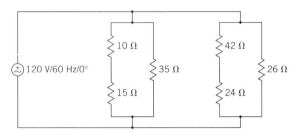

Figure 6-53

2. What are the total resistance and total current in the voltage divider network in Fig. 6-54?

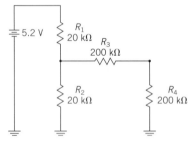

Figure 6-54

3. In the circuit in Fig. 6-55, determine the total current, the voltage that is supplied to the load, and the total line voltage drop. Is this acceptable in terms of the National Electrical Code recommendation of no more than 3% drop?

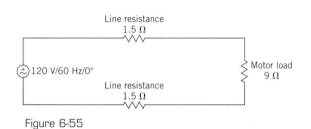

Figure 6-55

4. In your own words, explain why understanding how to determine line drop is important and how an installation electrician might make use of this information.

INTERNET ACTIVITIES

Internet Activity A

Use the following site to obtain the definition for the term *combination circuit*.
http://webhome.idirect.com/~jadams/electronics/circuit.htm

Internet Activity B

Using the Web site listed in Activity A identify (1) which loads are in the parallel path and (2) which load is in the series path, for the combination circuit shown.

Chapter 6 Solutions to Self-Tests and Reviews

Self-Test 6-2

1. series
2. R_2, 25, 60
4. R_A, $\dfrac{1}{60}$, $\dfrac{1}{20}$, 20

Self-Test 6-6

2. parallel, parallel
3. $R_1 + R_2$, 15, 10, 90, 36
5. series, $R_A + R_B$, 36, 46
7. parallel, $R_C \times R_5$, $R_C + R_5$, 184, 184, 36.8

Self-Test 6-7

Finding the total resistance
1. standard
2. a. series
   b. end, A
   c. R_A
3. R_A
4. B
5. R_T

Solving combination circuits
2. I_T
3. a. simplest

b. total
c. $V = I \times R$
d. R_T
e. voltage
f. resistance

Self-Test 6-10

1. 16
2. 112
3. 16, 7
4. 7, 9
5. 9
6. 5.4

7. 9, 0.6
8. 0.3

Self-Test 6-13

1. C, 20, I_3, I_B, I_B, 50, 70, 70, I_2, 70, 10
2. 0.6, 0.03, 0.06, 0.06, 0.06
3. R_{I_A}, 0.6, 6, I_2, 0.06, 4.2, R_{I3}, 20, 2.4
4. V_{I_A}, 230, 224, V_G, 4.2, V_G, V_{I_3}, 223.4

7

c h a p t e r

Electric Power

Learning Objectives

1. Describe the meaning of *electric power*.
2. Determine the power when the voltage and current are known.
3. Calculate the total power used by an electric circuit.
4. Calculate the voltage when power and current are known.
5. Calculate the current when power and voltage are known.
6. Square powers of 10.
7. Calculate power using the exponential power formulas $P = I^2R$ and $P = V^2/R$.
8. Determine the square root of a number.
9. Determine the square root of a power of 10.
10. Solve for voltage or resistance when using the formula $P = V^2/R$.
11. Solve for current or resistance when using the formula $P = I^2R$.

This chapter develops formulas for power in dc circuits. You will learn the concepts of different aspects of power and how to develop formulas for them. There are two areas of possible computational difficulty: formulas, and squares and square roots. If you have problems with formulas, see Chap. 3. If you have difficulty with squares, you may want to do the diagnostic test in Job 3-4 and Review Job 7-5 on squaring powers of 10. If you have difficulties with square roots, try the diagnostic test in Job 7-7.

JOB 7-1

Electric Power in Simple Circuits

Meaning of electric power.   Did you know that it is possible to push a candle through a wooden board? You may not be successful when you try it because you may omit an essential element necessary to do this. This missing quantity is *speed*. Power depends on *how fast* a certain amount of work is done. Thus, if you *shoot* the candle from a shotgun, it will have the necessary speed to penetrate the board.

More than speed, however, is necessary to increase the power. In recognition of this fact, the Roman gladiators of long ago wrapped strips of lead around their fists to increase the power of their blows. Our modern boxers are not permitted to increase the weight of their gloves, but they increase their hitting power by punching with short, *fast* blows rather than with looping, slow swings.

Power, then, depends on two quantities—the work done (or the weight moved) and the speed of doing it.

$$\text{Power} = \text{work done} \times \text{speed}$$

In electrical work,

$$\text{Work done} = Q \times V$$

$$\text{Speed} = \frac{1}{t}$$

Therefore,
$$P = Q \times V \times \frac{1}{t}$$

or
$$P = \frac{Q \times V}{t} = \frac{Q}{t} \times V$$

but since $Q/t = I$

$$P = I \times V$$

RULE The electric power in any part of a circuit is equal to the current in that part multiplied by the voltage across that part of the circuit.

$$\boldsymbol{P = I \times V}$$
Formula 7-1

where P = power, W
   I = current, A
   V = voltage, V

A *watt* of electric power is the power used when one volt causes one ampere of current to flow in a circuit. A *kilowatt* of power is equal to 1000 W.

Example 7-1

A variable resistor in the volume control of a code practice oscillator passes 0.05 A with a voltage drop of 30 V. Find the power consumed.

Solution

Given: $V = 30$ V   Find: $P = ?$
   $I = 0.05$ A

$$P = I \times V \qquad (7\text{-}1)$$
$$P = 0.05 \times 30$$
$$P = 1.5 \text{ W} \qquad Ans.$$

Example 7-2

A 1-kΩ resistor in a power-supply filter circuit carries 0.06 A at 60 V. How many watts of power are developed in the resistor? What must be the wattage rating of the resistor in order to dissipate this power safely as heat?

Solution

Given: $I = 0.06$ A   Find: $P = ?$
   $V = 60$ V   Wattage rating $= ?$

$$P = I \times V \qquad (7\text{-}1)$$
$$P = 0.06 \times 60 = 3.6 \text{ W} \qquad Ans.$$

Hint for Success

Every electrical appliance in your home is given both a wattage rating and a voltage rating. For example, a typical toaster might have a wattage rating of 1050 W, with a voltage rating of 120 V.

The wattage rating of a resistor describes its ability to dissipate the heat produced in it by the passage of an electric current without itself overheating. For example, a 2-W resistor could dissipate 2 W of heat energy without overheating. However, if 3 W of power were to be developed in it, it would overheat because of the 1 W of power that it could not dissipate. Owing to the lack of ventilation in the close quarters of most television receivers, the wattage ratings of these resistors is usually at least twice the wattage developed in them.

$$\text{Wattage rating} = 2 \times P$$
$$\text{Wattage rating} = 2 \times 3.6 = 7.2 \text{ W} \qquad \textit{Ans.}$$

Self-Test 7-3

A 56-Ω resistor is shunted with a 168-Ω resistor in a 0.3-A circuit as shown in Fig. 7-1. Find the wattage rating of the 168-Ω resistor.

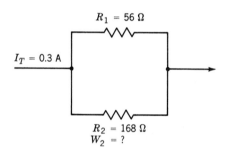

Figure 7-1   The wattage rating of a resistor is equal to at least twice the wattage developed in it.

Solution

1. In order to find the wattage developed in R_2 we must know the value of V_2 and _____.

$$I_2 = \frac{?}{R_1 + R_2} \times I_T \qquad (5\text{-}7)$$

$$I_2 = \frac{56}{56 + ?} \times 0.3$$

$$I_2 = \frac{56}{224} \times 0.3 = \underline{\hspace{1.5cm}} \text{ A}$$

2. Find V_2.

$$V_2 = I_2 \times \underline{\hspace{1.5cm}} \qquad (4\text{-}5)$$

$$V_2 = \underline{\hspace{1.5cm}} \times 168 = \underline{\hspace{1.5cm}} \text{ V}$$

3. Find the wattage developed in R_2.

$$P = I \times V \qquad (7\text{-}1)$$

$$P = 0.075 \times \underline{\hspace{1.5cm}} = \underline{\hspace{1.5cm}} \text{ W}$$

4. Find the wattage rating.

Wattage rating = ———— × wattage

Wattage rating = 2 × 0.945 = ———— W

or   Wattage rating = ———— W   *Ans.*

Exercises

Applications

1. An automobile starting motor draws 80 A at 6 V. How much power is drawn from the battery?
2. A 20-hp motor takes 74 A at 230 V when operating at full load. Find the power used in kilowatts.
3. What is the wattage dissipated as heat by a 550-Ω resistor operating at 110 V and 0.2 A?
4. What is the power consumed by an automobile headlight if it takes 2.8 A at 6 V?
5. Find the power used by a 3.4-A soldering iron at 110 V.
6. The high contact resistance of a poorly wired electric toaster plug reduced the current by 1 A on a 110-V line. Find the power wasted in the plug.
7. How much power is consumed by an electric clock using 0.02 A at 110 V?
8. If the voltage drop across a spark plug air gap is 30 V, find the power consumed in sending 2 mA across the gap.
9. Find the power dissipated by the collector of a transistor that passes 0.25 A at 9.2 V.
10. A power supply delivers 0.16 A at 250 V to a public address amplifier. Find the watts of power delivered.
11. A window air conditioner is rated at 7.5 A and is operated on a 120-V line. Find the wattage used by the conditioner.
12. An electric oven uses 36.3 A at 117 V. Find the power used in kilowatts.
13. An emitter bias resistor carries 45 mA at 10 V. What must be its wattage rating?
14. An electric enameling kiln takes 9.2 A from a 117-V line. Find the power used.
15. How much power is used by a ¾-ton air conditioner drawing 11.4 A from a 220-V line?
16. Fifteen lamps in parallel each take 1.67 A when connected to a 120-V line. Find the wattage of each lamp and the total wattage used.

JOB 7-2

Total Power in an Electric Circuit

When using Ohm's law, we found that it could be used for total values in a circuit as well as for the individual parts of the circuit. In the same way, the formula for power may be used for total values.

$$P_T = I_T \times V_T \qquad \text{Formula 7-2}$$

where P_T = total power, W
         I_T = total current, A
         V_T = total voltage, V

Example 7-4

Two 0.62-A lamps, each drawing 120 V, are connected in series. Find the total power used.

Solution

Given: $I_1 = I_2 = 0.62$ A   Find: $P_T = ?$
$V_1 = V_2 = 120$ V

1. Find the total voltage.

$$V_T = V_1 + V_2 = 120 + 120 = 240 \text{ V} \qquad (4\text{-}2)$$

2. Find the total current.

$$I_T = I_1 = I_2 = 0.62 \text{ A} \qquad (4\text{-}1)$$

3. Find the total power.

$$P_T = I_T \times V_T \qquad (7\text{-}2)$$
$$P_T = 0.62 \times 240 = 148.8 \text{ W} \qquad Ans.$$

Example 7-5

If the same lamps were connected in parallel across 120 V, find the total power used and compare it with the power used when the lamps were connected in series as in Example 7-4.

Solution

Given: $I_1 = I_2 = 0.62$ A   Find: $P_T = ?$
$V_1 = V_2 = 120$ V

1. Find the total voltage. $V_T = 120$ V   (given)
2. Find the total current.

$$I_T = I_1 + I_2 = 0.62 + 0.62 = 1.24 \text{ A} \qquad (5\text{-}2)$$

3. Find the total power. $P_T = I_T \times V_T$ $\qquad (7\text{-}2)$
$$P_T = 1.24 \times 120 = 148.8 \text{ W} \qquad Ans.$$

The total power in parallel is the same as the total power in series.

If the power used by the parts of a circuit is known, the total power may be found by the following

$$\boldsymbol{P_T = P_1 + P_2 + P_3} \qquad \text{Formula 7-3}$$

where $\boldsymbol{P_T}$ = total power, W
$\boldsymbol{P_1, P_2, P_3}$ = power used by the parts of the circuit, W

Self-Test 7-6

Two resistors in series form the base-bias voltage divider for an audio amplifier. The voltage drops across them are 2.4 V and 6.6 V, respectively, in the 1.5-mA circuit. Find the power used by the circuit.

Solution

Given: $V_1 = 2.4$ V
$V_2 = \underline{\quad (a) \quad}$ V
$I_T = I_1 = I_2 = 1.5$ mA $= \underline{\quad (b) \quad}$ A   Find: $P_T = ?$

1. Find P_1. $P_1 = I_1 \times V_1$ (7-1)
                               $P_1 = 0.0015 \times \underline{\quad (c) \quad} = \underline{\quad (d) \quad}$ W

2. Find P_2. $P_2 = I_2 \times V_2$
                               $P_2 = 0.0015 \times 6.6 = \underline{\quad (e) \quad}$ W

3. Find the total power. $P_T = P_1 + P_2$ (7-3)
                               $P_T = 0.0036 + 0.0099 = \underline{\quad (f) \quad}$ W   Ans.

The problem can be solved by another method.
1. Find the total voltage used by both resistors. Since they are in series,

$$V_T = V_1 + \underline{\quad (g) \quad}$$ (4-2)
$$V_T = 2.4 + \underline{\quad (h) \quad} = \underline{\quad (i) \quad}$$ V

2. Find the total power.

$$P_T = I_T \times \underline{\quad (j) \quad}$$ (7-2)
$$P_T = \underline{\quad (k) \quad} \times 9 = \underline{\quad (l) \quad}$$ W

Exercises

Applications

1. Seven Christmas tree lamps are connected in series. Each lamp requires 16 V and 0.1 A. Find the total power used.
2. If the same lamps were connected in parallel across a 16-V source, what would be the power taken?
3. A certain transistor must pass a maximum of 0.3 A. If it is rated at 0.15 W maximum, will it be able to withstand a 0.45-V collector-emitter voltage?
4. What is the total power used by a 4.5-A electric iron, a 0.85-A fan, and a 2.2-A refrigerator motor if they are all connected in parallel across a 115-V line?
5. Find the power drawn from a 6-V battery by a parallel circuit of two headlights (4 A each) and two taillights (0.9 A each).
6. In Fig. 6-47, the generator voltage is 120 V. The motor takes 12.4 A, and each lamp takes 0.92 A. Neglecting the line drop, find the total power used by the devices.
7. An electric percolator drawing 9 A and an electric toaster drawing 10.2 A are connected in parallel to the 117-V house line. Find the total power consumed.
8. In a circuit similar to that shown in Fig. 6-1, $V_T = 200$ V, $R_1 = 10\ \Omega$, $R_2 = 15\ \Omega$, $R_3 = 40\ \Omega$, and $R_4 = 60\ \Omega$. Find (a) the total current, (b) the power used by each resistor, and (c) the total power.

9. In a circuit similar to that shown in Fig. 6-15, $V_T = 13$ V, $R_1 = 2\ \Omega$, $R_2 = 1\ \Omega$, $R_3 = 6\ \Omega$, $R_4 = 3\ \Omega$, $R_5 = 40\ \Omega$, and $R_6 = 3\ \Omega$. Find (a) the total current, (b) the power used by each resistor, and (c) the total power.

JOB 7-3

Solving the Power Formula for Current or Voltage

In Job 3-9 we learned how to solve Ohm's law for the current or the resistance. The power formula is the same general type of equation, and we can use the methods learned in Job 3-9 to solve it for values of current or voltage.

Hint for Success

Note this interesting point about the power relations. For a given power rating, a higher voltage will require less current to obtain the desired power in watts. For example, a 1.5-kW electric heater draws 12.5 A of current from the 120-V power line; $I = 1.5$ kW/120 V = 12.5 A. However, the same electric heater will only need 6.25 A of current if it is able to operate from the 240-V power line; $I = 1.5$ kW/240 V = 6.25 A. With less current in the wires carrying current to the heater, there is less line drop, which is an advantage.

Example 7-7

What is the operating voltage of an electric toaster rated at 600 W if it draws 5 A?

Solution

Given: $P = 600$ W   Find: $V = ?$
        $I = 5$A

$$P = I \times V \qquad\qquad (7\text{-}1)$$
$$600 = 5 \times V$$
$$V = \frac{600}{5} = 120 \text{ V} \qquad Ans.$$

Self-Test 7-8

A stabilizing resistor in the emitter circuit of a 2N109 output transistor develops a voltage drop of 1.2 V while consuming 0.06 W of power. Find the current through the resistor.

Solution

Given: $P = 0.06$ W   Find: ___(1)___ = ?
        $V =$ ___(2)___ V

$$P = I \times V \qquad\qquad (7\text{-}1)$$
$$\underline{\quad(3)\quad} = I \times 1.2$$
$$I = \frac{0.06}{(4)}$$
$$I = \underline{\quad(5)\quad} \text{ A} \qquad Ans.$$

Exercises

Applications

1. A 550-W neon sign operates on a 110-V line. Find the current drawn.
2. What current is drawn by a 480-W soldering iron from a 120-V line?

3. Find the current drawn by a 1200-W aircraft system from a 24-V source.
4. A resistor is capable of dissipating 10 W of power. If the current is 0.3 A, what is the maximum voltage drop permitted across the resistor?
5. The quiescent point of a transistor is at a collector current of 0.06 A and a collector voltage of 20 V. Find the transistor power dissipation with no signal applied.
6. What current is drawn by a 1500-W electric ironing machine from a 120-V line?
7. What current is drawn by a 55-cd 110-V lamp if it uses 1 W per candela?
8. Two resistors, each dissipating 2 W, are connected in series with a 40-V source. What is the total current drawn? What is the current in each resistor?
9. Twenty 60-W lamps are connected in parallel to light a stage. Find the current drawn from a 220-V source.
10. The nameplate on a resistance welder reads 25 kVA at 440 V. What is its rated primary current?
11. An electric broiler rated at 1550 W operates from a 117-V line. The available fuses are rated at 20, 30, 50, and 60 A. Which fuse should be used in the circuit to protect the broiler?

Intermediate Review of Power

In some problems, the power cannot be found because either the voltage or the current is unknown. In these instances, the unknown value is found by Ohm's law.

Example 7-9

Find the power taken by a soldering iron of 60 Ω resistance if it draws a current of 2 A.

Solution

Given: $R = 60\ \Omega$ Find: $P = ?$
        $I = 2$ A

1. Find the voltage. $V = I \times R$ (2-1)
                       $V = 2 \times 60 = 120$ V

2. Find the power. $P = I \times V$ (7-1)
                       $P = 2 \times 120 = 240$ W   *Ans.*

Self-Test 7-10

Find the power used by the 11-Ω resistance element of an electric furnace if the voltage is 110 V.

Solution

Given: $R = 11\ \Omega$ Find: $P = ?$
              $V = \underline{\quad(1)\quad}\ V$

1. Find the current. $V = I \times R$ (2-1)
                        $110 = I \times \underline{\quad(2)\quad}$
                              $I = \dfrac{110}{11} = \underline{\quad(3)\quad}\ A$

2. Find the power. $P = \underline{\quad(4)\quad} \times V$ (6-1)
                        $P = 10 \times 110 = \underline{\quad(5)\quad}\ W$ *Ans.*

Exercises

Applications

1. A 20-Ω neon sign operates on a 120-V line. Find the power used by the sign.
2. Find the power used by a 55-Ω electric light which draws 2 A.
3. What is the wattage dissipated by a 10,000-Ω voltage divider if the voltage across it is 250 V? What is its wattage rating?
4. What is the maximum power obtainable from a Grenet cell of 2 V that has an internal resistance of 0.02 Ω?
5. A 240-Ω resistor in the emitter circuit of the output stage of a transistor radio carries 0.005 A. Find the wattage developed in the resistor.
6. The resistance of an ammeter is 0.025 Ω. Find the power used by the meter when it reads 4A.
7. What is the power consumed by a 90-Ω subway-car heater if the operating voltage is 550 V?
8. A 40-Ω pilot light is to be operated in a 0.15-A circuit. How many watts are developed in the lamp?
9. A 20-Ω toaster operates on a 115-V line. Find the power used.
10. A voltmeter has an internal resistance of 220,000 Ω. How much power does it use when the meter reads 110 V?
11. In the circuit shown in Fig. 7-2, find (*a*) the total resistance, (*b*) the total current, and (*c*) the total power.

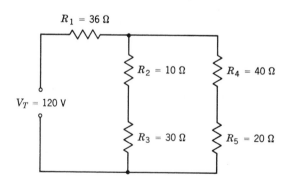

Figure 7-2

12. In a circuit similar to that shown in Fig. 6-15, $V_T = 188$ V, $R_1 = 22$ Ω, $R_2 = 18$ Ω, $R_3 = 70$ Ω, $R_4 = 80$ Ω, $R_5 = 56$ Ω, and $R_6 = 30$ Ω. Find (a) the total current, (b) the power used by each resistor, and (c) the total power.
13. In Fig. 7-3, find (a) the total resistance, (b) the total current, and (c) the total power.

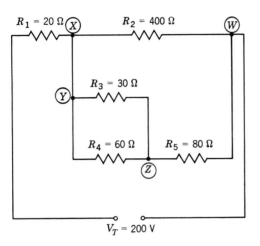

Figure 7-3

(See CD-ROM for Test 7-1)

Squaring Powers of 10

The next job, as well as many others, will require us to find the square of very large or very small numbers. The calculations will be simplified if we express these numbers as powers of 10.

Example 7-11

Find the value of $(10^3)^2$.

Solution

An exponent means that a base should be multiplied by itself as many times as the exponent indicates. If the base happens to be 10^3, then

$$(10^3)^2 \text{ means } 10^3 \times 10^3$$

and, since exponents are added when multiplying (see page 75),

$$(10^3)^2 = 10^3 \times 10^3 = 10^6 \qquad \textit{Ans.}$$

The answer may be obtained more directly by applying the following rule.

 RULE   To find the power of a power, express the base to a power obtained by multiplying the exponents.

For example:

1. $(10^3)^2 = (10)^{(3 \times 2)} = 10^6$
2. $(10^4)^2 = (10)^{(4 \times 2)} = 10^8$
3. $(10^{-3})^2 = (10)^{(-3 \times 2)} = 10^{-6}$
4. $(10^{-6})^2 = 10^{-12}$

Example 7-12

Express the number $(12,000)^2$ as a power of 10.

Solution

1. Express 12,000 as a power of 10.

$$12,000 = 12 \times 10^3$$

2. Square this number.

$$(12,000)^2 = (12 \times 10^3)^2$$
$$= (12)^2 \times (10^3)^2$$

Notice that both the 12 and the 10^3 are squared because each is a part of the base (12×10^3) that is being squared.

$$(12,000)^2 = 144 \times 10^6 \quad Ans.$$

Example 7-13

Express $(0.00008)^2$ as a power of 10.

Solution

$$0.00008 = 8 \times 10^{-5}$$

Therefore, $(0.00008)^2 = (8 \times 10^{-5})^2$
$$= 8^2 \times (10^{-5})^2$$
$$= 64 \times 10^{-10} \quad Ans.$$

Example 7-14

Find the value of $(2.5 \times 10^2)^2$.

Solution

$$(2.5 \times 10^2)^2 = (2.5)^2 \times (10^2)^2$$
$$= 6.25 \times 10^4 = 62,500 \quad Ans.$$

Exercises

Practice

Perform the indicated operations.

1. $(10^5)^2$
2. $(10^{-4})^2$
3. $(3 \times 10^3)^2$
4. $(5 \times 10^{-3})^2$
5. $(10^2 \times 10^3)^2$
6. $(10^6 \times 10^{-2})^2$
7. $(8000)^2$
8. $(15,000)^2$
9. $(1100)^2$
10. $(0.007)^2$
11. $(0.00025)^2$
12. $(0.015)^2$
13. $(1.5 \times 10^3)^2$
14. $(5.1 \times 10^{-4})^2$
15. $(1.2 \times 10^{-3})^2$

The Exponential Power Formulas

In Job 7-4, either the voltage or the current was unknown. These values were found by Ohm's law and then used to find the power. The two steps involved in solving problems of this type may be combined into a single formula. The power is given by

$$P = I \times V \qquad (7\text{-}1)$$

But by Ohm's law,

$$V = I \times R \qquad (2\text{-}1)$$

Therefore, we may substitute the quantity $I \times R$ for V. This gives

$$P = I \times I \times R$$

$$\boldsymbol{P = I^2 R} \qquad \text{Formula 7-4}$$

Also by Ohm's law, $I = V/R$. Therefore, we may substitute the quantity V/R for I. Thus,

$$P = I \times V \qquad (7\text{-}1)$$

Substituting,

$$P = \frac{V}{R} \times V$$

$$P = \frac{V^2}{R} \qquad \text{Formula 7-5}$$

Example 7-15

A 4-kΩ base-bias resistor in a silicon NPN amplifier transistor carries a base current of 5 mA. Find the power consumed and the wattage rating required by the resistor.

Solution

The diagram for the circuit is shown in Fig. 7-4.

> **Hint for Success**
>
> When electric power companies transmit electric power, they typically use very high voltages such as 160 kV or more. With such a high voltage, the current carried by the power lines can be kept low to transmit the desired power from one location to another. The reduction in I considerably reduces the I^2R power losses in the power-line conductors.

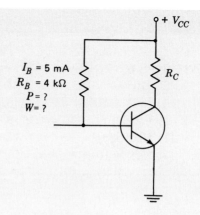

Figure 7-4

$$P = I^2R \qquad (7\text{-}4)$$
$$P = (5 \times 10^{-3})^2 \times 4 \times 10^3$$
$$P = 25 \times 10^{-6} \times 4 \times 10^3$$
$$P = 100 \times 10^{-3} = 0.1 \text{ W} \qquad Ans.$$
$$\text{Wattage rating} = 2 \times P$$
$$= 2 \times 0.1 = 0.2 \text{ W} \qquad Ans.$$

Example 7-16

A motor has a total resistance of 20 Ω and operates on a 120-V line. Find the power used.

Solution

Given: $R = 20 \ \Omega$ Find: $P = ?$
             $V = 120 \text{ V}$

$$P = \frac{V^2}{R} \qquad (7\text{-}5)$$

$$P = \frac{(120)^2}{20} = \frac{120 \times 120}{20} = 720 \text{ W} \qquad Ans.$$

Self-Test 7-17

Find the wattage rating of R_1 and R_2 in the voltage-divider circuit shown in Fig. 7-5.

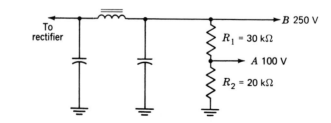

Figure 7-5   Simple voltage-divider circuit.

Solution

The voltage drop across R_1 is equal to the difference between the voltage V at B (250) and the voltage V at A (100). Therefore, $V_1 = $ ___(a)___ V.

$$P_1 = \frac{V_1^2}{R_1}$$

$$P_1 = \frac{(150)^2}{(b)}$$

$$P_1 = \frac{225 \times 10^2}{(c)}$$

$$P_1 = 75 \times \frac{(d)}{} = \underline{\quad(e)\quad} \text{ W}$$

$$\text{Wattage rating} = \frac{(f)}{} \times P_1$$

$$= 2 \times 0.75 = 1.5 \text{ W}$$

However, since the 1.5-W size is not usually stocked, we must use a 2-W 30-kΩ resistor.   *Ans.*

For R_2, the voltage drop $V_2 = $ ___(g)___ V

$$P_2 = \frac{(100)^2}{2 \times 10^4} = \frac{1 \times 10^4}{2 \times 10^4} = \underline{\quad(h)\quad} \text{ W}$$

Therefore, wattage rating = ___(i)___ W   *Ans.*

Exercises

Applications

1. A 90-Ω device draws 5 A. Find the power used in kilowatts.
2. If the voltage drop across a 10,000-Ω voltage divider is 90 V, find the power used.
3. Find the power consumed by a 100-Ω electric iron operating on a 115-V line.
4. A poorly soldered joint has a contact resistance of 100 Ω. What is the power lost in the joint if the current is 0.5 A?
5. Two 2N406 transistors operating in push-pull deliver an average current of 0.05 A through a 2600-Ω load resistance. Find the power developed.
6. Find the power used by a 15-Ω neon sign on a 110-V line.
7. The motor shown in Fig. 7-6 takes 1400 W at 220 V. Find the current drawn. How many watts are consumed in the line wires if each has a resistance of 0.2 Ω?

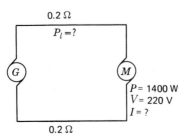

Figure 7-6

8. How much power is dissipated in the form of heat in a ballast resistor of 60 Ω if the current is 0.3 A?

9. A 100- and a 260-Ω resistor are connected in series to a 120-V source. Find the power used by each resistor.

10. Find the total voltage across the circuit shown in Fig. 7-7. What is the voltage across each resistor? What is the power taken by each resistor?

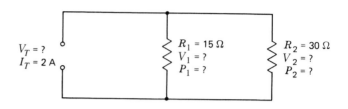

$V_T = ?$
$I_T = 2 A$

$R_1 = 15 Ω$
$V_1 = ?$
$P_1 = ?$

$R_2 = 30 Ω$
$V_2 = ?$
$P_2 = ?$

Figure 7-7

(See CD-ROM for Test 7-2)

<hr>

JOB 7-7

Checkup on Square Root (Diagnostic Test)

In the last job we learned two new formulas which use the exponent 2. This exponent is read as the word *square*. We can use these formulas to find the current, voltage, or resistance if we know the power. To do this, we must be able to do the opposite of "squaring" a number. This is called *finding the square root*. We shall also use "square root" when we get to the study of ac circuits. Can you do the following problems? If you have any difficulty with them, turn to Job 7-8, which follows.

Exercises

Practice

Find the square root of each of the following numbers:

1. 64	2. 169	3. 17.64	4. 3481
5. 12,544	6. 57.76	7. 14	8. 567.9
9. 76,432	10. 30	11. 0.652	12. 870,000

<hr>

JOB 7-8

Brushup on Square Root

The *square root* of a number is that number which must be multiplied by *itself* in order to obtain the original number. The symbol for the square root is $\sqrt{}$. When a number appears under this symbol, it means that we are to find a number which can be multiplied by itself to give the number under the square-root symbol. For example, the square root of 9 ($\sqrt{9}$) must be the number 3 because only $3 \times 3 = 9$. Also,

$$\sqrt{16} = 4 \text{ because only } 4 \times 4 \text{ will equal } 16$$

$$\sqrt{25} = 5 \text{ because only } 5 \times 5 \text{ will equal } 25$$

$$\sqrt{36} = 6 \text{ because only } 6 \times 6 \text{ will equal } 36$$

These numbers under the square-root sign are called *perfect squares* because the square root of each is a whole number.

We can continue to find the square roots of numbers in this manner until our knowledge of the multiplication table is insufficient to keep up with the large numbers involved. For example,

$$\sqrt{49} = 7 \qquad\qquad \sqrt{121} = 11$$

$$\sqrt{64} = 8 \qquad\qquad \sqrt{144} = 12$$

$$\sqrt{81} = 9 \qquad\qquad \sqrt{169} = 13$$

$$\sqrt{100} = 10$$

Somewhere along here we begin to forget whether 14×14 is 196 or not or whether 16×16 is 256 or not. You can see that we will not get very far if we rely on just our knowledge of the multiplication table. Besides, what about all those smaller numbers between the perfect squares, like

$$\sqrt{7} = ?$$

$$\sqrt{15} = ?$$

$$\sqrt{32} = ?$$

There doesn't seem to be any number that we can multiply by itself to get 7 or 15 or 32. No, there aren't any *whole* numbers which are the square roots of these numbers, but there are decimal numbers which will satisfy.

In the following system, it will help if we mark off the numbers into groups of two digits to a group. We start marking off the groups *at the decimal point*. If there is no decimal point indicated, it may be assumed to be at the *end* of the number.

Example 7-18

Group the digits in the number 3456.

Solution

The decimal point is at the end of the number. Starting at the decimal point, the numbers are grouped two to a group as we move to the left.

34 56.

Example 7-19

Group the digits in the number 54,819.8.

Solution

Starting at the decimal point and proceeding to the left, we find that the digit 5 is left over. In situations like this, the single digit at the extreme left is considered to be a group. Now, return to the decimal point and group the digits to the *right* of the point—two to a group. A zero must be added after the 8 at the right to complete the group.

$$5 \; 48 \; 19 \; . \; 80$$

Example 7-20

Find the square root of 5776.

Solution

1. Locate the decimal point. In a whole number the decimal point is at the end of the number. Place the point in the answer directly above its position in the number.

$$\sqrt{5 \; 7 \; 7 \; 6 \; .}$$

2. Separate the digits into groups—two digits to a group. Start at the decimal point, and group the digits to the left.

$$\sqrt{57 \; 76 \; .}$$

3. Start with the first group (57). Find a number which when multiplied by itself will give an answer close to or equal to but not larger than 57. For example, 8 squared is 64, but that is too large; 7 squared is 49, which is just right. Place the 7 over the first group in the answer, and the 49 under the 57. Draw a line and subtract the 49 from the 57, leaving a remainder of 8.

$$\begin{array}{r} 7 . \\ \sqrt{57 \; 76 \; .} \\ -49 \\ \hline 8 \end{array}$$

4. Bring down the next *group. Never* bring down a single number. Make a little box to the left of this new number 876. *Double* the answer at this point (the 7), and place it in this box. This will be the number 14.

$$\begin{array}{r} 7 . \\ \sqrt{57 \; 76 \; .} \\ -49 \downarrow \\ \hline 14 \; | \; 8 \; 76 \end{array}$$

5. Place your finger over the *last digit* in the number 876. The number there will now appear to be 87. Divide the number in the box (14) into this 87. It will go about 6 times. Place this 6 in the answer above the second group, *and also place it next to the 14 in the box.*

$$
\begin{array}{r}
76\\
\sqrt{5776}\\
-49\downarrow\\
\hline
146876
\end{array}
$$

6. Multiply the 6 by the number just formed (the 146). If the product is larger than 876, we shall be forced to change the 6 to a smaller number. However, $6 \times 146 = 876$. Write this 876 under the 876 already there and subtract. Since there is no remainder, 76 is the exact square root of 5776. *Check*: $76 \times 76 = 5776$.

$$
\begin{array}{r}
76\\
\sqrt{5776}\\
-49\downarrow\\
\hline
146876\\
876\\
\hline
0
\end{array}
$$

Example 7-21

Find the square root of 930.25.

Solution

1. Locate the decimal point. Place the point in the answer directly above its position in the number.

$$\sqrt{930.25}$$

2. Separate the digits into groups—two digits to a group. Start at the decimal point, and group the digits to the left. Return to the decimal point, and group the digits to the right.

$$\sqrt{930\,.\,25}$$

3. Start with the first group (9). Find a number which when multiplied by itself will give an answer close to or equal to but not larger than 9. The number 3 squared is 9, which is exactly right. Place the 3 over the first group in the answer, and the 9 under the 9 in the number. Draw a line and subtract, leaving a remainder of 0.

$$
\begin{array}{r}
3.\\
\sqrt{930\,.\,25}\\
-9\\
\hline
\end{array}
$$

4. Bring down the next *group* (30). *Never* bring down a single digit. Make a little box to the left of this number. *Double* the answer at this point (the 3), and place it in this box. This will be the number 6.

$$\begin{array}{r} 3 \quad . \\ \sqrt{9\ 30\ .\ 25} \\ -9 \downarrow \\ \hline 6 \boxed{}\ 30 \end{array}$$

5. Place your finger over the *last digit* in the number 30. The number there will now appear to be 3. Divide the number in the box (6) into this 3. It will go 0 times. Place this 0 in the answer above the second group, *and also place it next to the 6 in the box.*

$$\begin{array}{r} 3\ 0\ . \\ \sqrt{9\ 30\ .\ 25} \\ -9 \downarrow \\ \hline 6\ 0 \boxed{}\ 30 \end{array}$$

6. Multiply the 0 by the number just formed (the 60). The answer is 00. Write this 00 under the 30 and subtract, leaving a remainder of 30.

$$\begin{array}{r} 3\ 0\ . \\ \sqrt{9\ 30\ .\ 25} \\ -9 \downarrow \\ 6\ 0 \boxed{}\ 30 \\ -00 \\ \hline 30 \end{array}$$

7. Bring down the next group (25). Make a little box to the left of this new number (3025). *Double* the answer up to this point (the 30), and place the product (60) in this box.

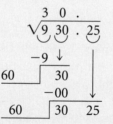

8. Place your finger over the *last digit* in the number 3025. The number there will now appear to be 302. Divide the number in the box (60) into this 302. It will go 5 times. Place this 5 in the answer above the third group, *and also place it next to the 60 in the box.*

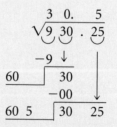

9. Multiply the 5 by the number just formed (the 605). The answer is 3025. Write this 3025 under the 3025 already there and subtract. Since there is

no remainder, 30.5 is the exact square root of 930.25. *Check:*
$30.5 \times 30.5 = 930.25$.

$$
\begin{array}{r}
3\ \ 0.\ \ \ 5 \\
\sqrt{9\ \ 30\ .\ 25} \\
\end{array}
$$

$$
\begin{array}{r}
-9 \downarrow \\
60\ \ |\ \ 30 \\
-00 \downarrow \\
605\ \ |\ \ 30\ \ 25 \\
-30\ \ 25 \\
\hline
0
\end{array}
$$

Example 7-22

Find the square root of 12.

Solution

1. Locate the decimal point. In a whole number the decimal point is at the end of the number. Place the point in the answer directly above its position in the number.

$$\overset{\textstyle .}{\sqrt{12\ .}}$$

2. Separate the digits into groups—two digits to a group. Start at the decimal point, and group the digits to the left.

$$\overset{\textstyle .}{\sqrt{1\ 2\ .}}$$

3. Start with the first group (12). Find a number which when multiplied by itself will give an answer close to or equal to but not larger than 12. The number 4 squared is 16, but that is too large; 3 squared is 9, which is less than 12 and so is just right. Place the 3 over the first group in the answer, and the 9 under the 12 in the number. Draw a line and subtract, leaving a remainder of 3.

$$
\begin{array}{r}
3\ . \\
\sqrt{1\ 2\ .} \\
-9 \\
\hline
3
\end{array}
$$

4. Since there is a remainder, the square root will be a decimal. Add two *pairs* of zeros after the decimal point. Bring down the next group (00). Make a little box to the left of this number (300). *Double* the answer up to this point (the 3), and place the product (6) in this box.

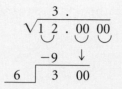

5. Place your finger over the *last digit* in the number 300. The number there will now appear to be 30. Divide the number in the box (6) into this 30. It will go 5 times. Place this 5 in the answer above the second group, *and also place it next to the 6 in the box.* Multiply the 5 by the number just formed (the 65), and place the product (325) under the 300 already there. *But 325 is larger than 300, and we have evidently made an error in using the 5.* Since 5 was too large, use 4 instead of 5.

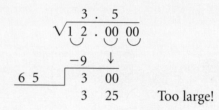

6. Be sure to change *both* the 5 in the answer and the 5 in the 65 in the box to the number 4. Now multiply the 4 in the answer by the 64 in the box. Place the product (256) under the 300 already there and subtract, leaving a remainder of 44.

$$
\begin{array}{r}
3\ .\ 4 \\
\sqrt{1\ 2\ .\ 00\ 00} \\
-9\ \ \downarrow \\
\hline
6\ 4\ \overline{)\ \ 3\ \ 00} \\
-2\ \ 56 \\
\hline
44
\end{array}
$$

7. Bring down the next group (00). Make a little box to the left of this new number (4400). Double the answer up to this point (the 34), and place the product (68) in this box. Always disregard the decimal point in this step.

$$
\begin{array}{r}
3\ .\ 4 \\
\sqrt{1\ 2\ .\ 00\ 00} \\
-9\ \ \downarrow \\
\hline
6\ 4\ \overline{)\ \ 3\ \ 00} \\
-2\ \ 56 \\
\hline
68\ \overline{)\ \ 44\ \ 00}
\end{array}
$$

8. Place your finger over the *last digit* in the number 4400. The number there will now appear to be 440. Divide the number in the box (68) into this 440. It will go 6 times. Place this 6 in the answer over the third group, *and also place it next to the 68 in the box.*

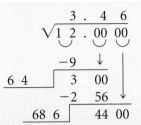

9. Multiply the 6 by the number just formed (the 686), and place the product (4116) under the 4400 already there. Draw a line and subtract, leaving a remainder of 284. This remainder may be disregarded, as we shall rarely need an answer more accurate than two decimal places. The problem may be worked out to any number of decimal places and then "rounded off" to suit.

$$
\begin{array}{r}
3\ .\ 4\ \ 6 \\
\sqrt{1\ 2\ .\ 00\ \ 00} \\
\end{array}
$$

$$
\begin{array}{r}
-9 \quad \downarrow \\
6\ 4 \quad \overline{3\ \ 00} \\
-2\ \ 56 \quad \downarrow \\
68\ 6 \quad \overline{\quad\ 44\ \ 00} \\
-41\ \ 16 \\
\hline
2\ \ 84 \\
\end{array}
$$

Check: $3.46 \times 3.46 = 11.97$, or practically 12

Exercises

Practice

Find the square roots of the following numbers:

1. 3481	2. 17.64	3. 15.21	4. 12,544	5. 18,769
6. 151.29	7. 40	8. 267	9. 65	10. 53.87

JOB 7-9

The Square Root of a Power of 10

 RULE   To find the square root of a power, express the base to a power obtained by dividing the exponent by 2.

For example:
1. $\sqrt{10^6} = 10^{(6 \div 2)} = 10^3$
2. $\sqrt{10^{10}} = 10^{(10 \div 2)} = 10^5$
3. $\sqrt{10^{-12}} = 10^{(-12 \div 2)} = 10^{-6}$
4. $\sqrt{10^{-8}} = 10^{(-8 \div 2)} = 10^{-4}$

Example 7-23

Express $\sqrt{250,000}$ as a power of 10.

Solution

$$250,000 = 25 \times 10^4$$

Find the square root of this number. Be careful to find the square root of *both* the 25 and the 10^4.

$$\sqrt{250,000} = \sqrt{25 \times 10^4} = \sqrt{25} \times \sqrt{10^4}$$
$$= 5 \times 10^{(4 \div 2)}$$
$$= 5 \times 10^2 \qquad Ans.$$

Example 7-24

Express $\sqrt{0.0004}$ as a power of 10.

Solution

$$\sqrt{0.0004} = \sqrt{4 \times 10^{-4}} = \sqrt{4} \times \sqrt{10^{-4}}$$
$$= 2 \times 10^{(-4 \div 2)}$$
$$= 2 \times 10^{-2} \qquad Ans.$$

Now let's see what happens if we get an exponent which is *not* exactly divisible by 2. For example, $\sqrt{10^3} = 10^{1.5}$. This decimal exponent, although it has wide use in all the sciences, is very inconvenient, and should be avoided if at all possible. Let's see how this may be done.

Example 7-25

$\sqrt{500,000}$ as a power of 10.

Solution

When 500,000 is expressed as a power of 10, it *must* be written so that the power of 10 is an *even* number that can be divided by 2. Thus, 500,000 written as 5×10^5 is not good because the exponent 5 is not evenly divisible by 2. And 500,000 written as 500×10^3 is wrong for the same reason. Therefore, we must write 500,000 as a number $\times$ 10 to an *even* exponent.

$$\sqrt{500,000} = \sqrt{50 \times 10^4} = \sqrt{50} \times \sqrt{10^4}$$
$$= 7.07 \times 10^2 \qquad Ans.$$

Example 7-26

Express $\sqrt{0.000\ 64}$ as a power of 10.

Solution

$$\sqrt{0.000\ 64} = \sqrt{6.4 \times 10^{-4}}$$
$$= \sqrt{6.4} \times \sqrt{10^{-4}}$$
$$= 2.53 \times 10^{-2} \quad \textit{Ans.}$$

Example 7-27

$$\sqrt{0.9 \times 10^3} = \sqrt{9 \times 10^2}$$
$$= 3 \times 10 = 30 \quad \textit{Ans.}$$

Example 7-28

$$\sqrt{0.0016 \times 0.0004} = \sqrt{16 \times 10^{-4} \times 4 \times 10^{-4}}$$
$$= 4 \times 10^{-2} \times 2 \times 10^{-2}$$
$$= 8 \times 10^{-4} \quad \textit{Ans.}$$

Example 7-29

$$\sqrt{8000 \times 400} = \sqrt{8 \times 10^3 \times 4 \times 10^2}$$
$$= \sqrt{32 \times 10^5}$$
$$= \sqrt{320 \times 10^4}$$
$$= 17.9 \times 10^2 \quad \textit{Ans.}$$

Example 7-30

This example will be particularly important when we study the impedance of an ac circuit. Make note of the fact that quantities involving *different* powers of 10 may *not* be added or subtracted. If a problem involves different powers of 10, we must change the power in one quantity to agree with the other before addition or subtraction.

Find the value of $\sqrt{4 \times 10^4 + 41 \times 10^2}$.

Solution

Since the powers of 10 are different, we change 4×10^4 to 400×10^2 so that both quantities contain the same power 10^2.

$$\sqrt{4 \times 10^4 + 41 \times 10^2} = \sqrt{400 \times 10^2 + 41 \times 10^2}$$
$$= \sqrt{(400 + 41) \times 10^2}$$
$$= \sqrt{441 \times 10^2}$$
$$= 21 \times 10 = 210 \quad \textit{Ans.}$$

Practice

Perform the indicated operation.

1. $\sqrt{10^8}$
2. $\sqrt{10^{-6}}$
3. $\sqrt{16 \times 10^4}$
4. $\sqrt{9 \times 10^{-6}}$
5. $\sqrt{160,000}$
6. $\sqrt{0.0064}$
7. $\sqrt{0.000\ 025}$
8. $\sqrt{20 \times 80}$
9. $\sqrt{120 \times 30}$
10. $\sqrt{200,000}$
11. $\sqrt{800 \times 500}$
12. $\sqrt{0.0006}$
13. $\sqrt{0.016 \times 10^3}$
14. $\sqrt{0.25 \times 10^8}$
15. $\sqrt{0.4 \times 10^5}$
16. $\sqrt{0.0049 \times 0.0009}$
17. $\sqrt{6 \times 10^3 \times 500}$
18. $\sqrt{8 \times 10^{-3} \times 0.08}$
19. $\sqrt{14.4 \times 10^7}$
20. $\sqrt{0.081 \times 10^9}$
21. $\sqrt{0.225 \times 10^{-3}}$
22. $\sqrt{0.625 \times 10^{-11}}$
23. $\sqrt{6 \times 10^4 + 25 \times 10^2}$
24. $\sqrt{4 \times 10^3 + 900}$

(See CD-ROM for Test 7-3)

JOB 7-10

Applications of the Exponential Power Formula

The formula $P = I^2R$ may be used to find the current I or the resistance R. The formula $P = V^2/R$ may be used to find the voltage V or the resistance R.

> **Hint for Success**
>
> When the power P and resistance R are both known, the equation $P = I^2R$ can be arranged as $I = \sqrt{P/R}$ or $I^2 = P/R$ to solve for I. Similarly, when the power P and current I are both known, the equation $P = I^2R$ can be arranged as $R = P/I^2$ to solve for R.

Example 7-31

A 2N1479 transistor delivers an output power of 4 W at an average current of 0.2 A. What is the value of the load resistance?

Solution

Given: $P = 4$ W   Find: $R = ?$

                $I = 0.2$ A

$$P = I^2R \tag{7-4}$$
$$4 = (0.2)^2 \times R$$
$$4 = 0.04 \times R$$
$$R = \frac{4}{0.04} = 100 \ \Omega \qquad Ans.$$

Example 7-32

What is the maximum current-carrying capacity of a resistor marked 1000 Ω and 10 W?

Solution

Given: $R = 1000\ \Omega$ Find: $I = ?$
$P = 10\ W$

$$P = I^2 R \qquad (7\text{-}4)$$
$$10 = I^2 R$$
$$10 = I^2 \times 1000$$
$$I^2 = \frac{10}{1000} = 0.01\ A$$

Now $I^2 = 0.01$ means that some number I multiplied by itself will equal 0.01. Another way to say this is "What number multiplied by itself will equal 0.01?" This can be written as

$$I = \sqrt{0.01}$$

Actually, we have transformed the equation $I^2 = 0.01$ into $I = \sqrt{0.01}$ by taking the square root of both sides of the equality sign as shown in the following step. Since $\sqrt{I^2} = I$,

$$\sqrt{I^2} = \sqrt{0.01}$$
$$I = \sqrt{0.01}$$
$$I = 0.1\ A \qquad Ans.$$

CALCULATOR HINT

To use your calculator for a problem like Example 7-31, proceed as follows. Enter the digit ④ followed by the ÷ key. Next, enter ⓪·② followed by the x̄ key. Next, press the = key to display the answer of 100 Ω.

Example 7-33

The total resistance of the field coils of a 240-W motor is 60 Ω. Find the voltage needed to operate the motor at its rated power.

Solution

Given: $R = 60\ \Omega$ Find: $V = ?$
$P = 240\ W$

1. Write the formula.

$$P = \frac{V^2}{R} \qquad (7\text{-}5)$$

2. Substitute numbers.

$$\frac{240}{1} = \frac{V^2}{60}$$

3. Cross-multiply.

$$V^2 = 240 \times 60$$

4. Multiply.

$$V^2 = 14{,}400$$

5. Take the square root of both sides.

$$\sqrt{V^2} = \sqrt{14{,}400}$$

6. Since $\sqrt{V^2} = V$,

$$V = \sqrt{14{,}400}$$

Hint for Success

When the power P and resistance R are both known, the equation $P = V^2/R$ can be arranged as $V^2 = P \times R$ or $V = \sqrt{PR}$ to solve for V. Similarly, when the power P and voltage V are both known, the equation $P = V^2/R$ can be arranged as $R = V^2/P$ to solve for R.

7. The voltage is

$$V = 120 \text{ V} \qquad Ans.$$

Self-Test 7-34

What is the maximum current-carrying capacity of a resistor marked 10 kΩ and 4 W?

Solution

Given: $P = \underline{\quad (1) \quad}$ W   Find: $I = ?$
$R = \underline{\quad (2) \quad}$ Ω

$$P = I^2 \times \underline{\quad (3) \quad}$$
$$4 = I^2 \times \underline{\quad (4) \quad}$$

$$I^2 = \frac{4}{(5)}$$

$$I^2 = \underline{\quad (6) \quad}$$
$$\sqrt{I^2} = \underline{\quad (7) \quad}$$
$$I = \underline{\quad (8) \quad} \text{ A} = \underline{\quad (9) \quad} \text{ mA} \qquad Ans.$$

Exercises

Applications

1. What current flows through a line supplying 1500 W of power to an electric range of 15 Ω resistance?
2. What voltage is necessary to operate an 18-W automobile headlight bulb of 2 Ω resistance?
3. What is the current flowing through a 50-Ω electromagnet drawing 200 W?
4. What is the maximum current-carrying capacity of a resistor marked 500 Ω and 10 W?
5. Find the voltage drop across a corroded connection if its contact resistance is 100 Ω and it uses 4 W of power.
6. Find the internal resistance of a 2-W electric clock which operates on a 110-V line.
7. A 60- and a 40-W lamp are in parallel across 120 V. Find the combined resistance of the lamps.
8. What is the voltage necessary to operate a 600-W neon sign that has a resistance of 20 Ω?
9. An ammeter shunt has a resistance of 0.01 Ω and is rated at 15 W. Find the maximum safe current it can carry.
10. What is the maximum current-carrying capacity of a resistor marked 4 kΩ and 10 W?

Review of Electric Power

The power used by any part of a circuit is equal to the ____(1)____ in that part multiplied by the ____(2)____ across that part.

The formula for power is

$$P = I \times \underline{\quad(3)\quad}$$

where $P = \underline{\quad(4)\quad}$, measured in ____(5)____
$I = \underline{\quad(6)\quad}$, measured in ____(7)____
$V = \underline{\quad(8)\quad}$, measured in ____(9)____

This formula may be used to find

P if ____(10)____ and V are known
I if ____(11)____ and V are known
V if P and ____(12)____ are known

The wattage rating of a resistor is equal to ____(13)____ times the wattage developed in the resistor.

The total power in a circuit may be found by the formula

$$P_T = \underline{\quad(14)\quad} \times V_T$$

or

$$P_T = P_1 + \underline{\quad(15)\quad} + \underline{\quad(16)\quad}$$

In the expression 10^3, the exponent is the number ____(17)____.

5^3 means $5 \times 5 \times$ ____(18)____ or ____(19)____.

$10^3 = $ ____(20)____ and $10^6 = $ ____(21)____.

Some other formulas for power are:

$$P = \underline{\quad(22)\quad} \times R$$

and

$$P = \frac{V^2}{(23)}$$

The square root of a number is that number which must be multiplied by ____(24)____ to get the original number.

The symbol for square root is ____(25)____.

$$\sqrt{16} = \underline{\quad(26)\quad}$$
$$\sqrt{V^2} = \underline{\quad(27)\quad}$$
$$\sqrt{I^2} = \underline{\quad(28)\quad}$$

When finding the square root of a number, the digits should be marked off, ____(29)____ digits to a group, starting at the ____(30)____ point.

The decimal point in a whole number is at the ____(31)____ of the number.

Group the digits in the following numbers preparatory to finding the square root.

a. 469.4 ____(32)____
b. 8062 ____(33)____
c. 12,345 ____(34)____
d. 0.012 ____(35)____

e. 0.002 _____(36)_____

f. 6.05 _____(37)_____

g. 0.000 06 _____(38)_____

To find the power of a power, the base is expressed to a power obtained by ___(39) (adding/multiplying)___ the powers.

To find the square root of a power, the base is expressed to a power obtained by ___(40) (multiplying/dividing)___ the exponent by 2.

Exercises

Applications

1. Find the power used by an electric toaster if it draws 6 A from a 110-V line.
2. Find the power used by a 22-Ω motor if the current is 10 A.
3. Find the total power drawn by the four lamps shown in Fig. 7-8.

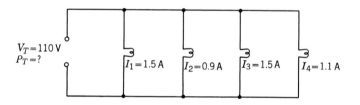

Figure 7-8

4. What current is drawn by a 250-W electric vacuum cleaner when it is operated on 110 V?
5. Three 18-V 0.8-A bells are in series. Find the total power.
6. How many watts of power are dissipated in a 100-kΩ voltage divider if the voltage across it is 300 V?
7. What voltage is required to operate a 25-W automobile headlight bulb properly if the current drawn is 4 A?
8. Find the resistance of a 1000-W electric ironing machine if it uses 5 A.
9. A toy electric-train semaphore is made of a 28-Ω lamp in parallel with a solenoid coil with an effective resistance of 42 Ω. If the total current drawn is 0.6 A, find (*a*) the total resistance and (*b*) the total power used.
10. The combined resistance of a coffee percolator and toaster in parallel is 22 Ω. Find the total power used if the line voltage is 110 V.
11. What power is dissipated in the form of heat in a 130-Ω ballast resistor designed to use up 40 V of excess voltage?
12. What is the voltage needed to operate a 10-W electric train accessory that has a resistance of 15 Ω?
13. A 100- and a 60-W lamp are connected in parallel across 120 V. Find the combined resistance.
14. What is the maximum current-carrying capacity of a resistor marked 5000 Ω and 20 W?
15. A number of incandescent lamps in parallel are supplied by a generator delivering 112 V at its brushes. The resistance of each of the two leads carrying cur-

rent to the lamps is 0.05 Ω and causes a total voltage drop of 2 V. If each lamp draws 50 W, how many lamps are lit? *Hint:*

a. Find the line current.

$$V_l = I_l \times R_l \qquad\qquad (6\text{-}1)$$

b. Find the voltage at the load.

$$V_L = V_G - V_l \qquad\qquad (6\text{-}3)$$

c. Find the power supplied to the load.

$$P_L = I_l \times V_L \qquad\qquad (7\text{-}1)$$

d. The number of lamps $= P_L \div$ wattage per lamp.

16. A dc generator supplies a 5-A lamp bank and a 1.5-A motor as shown in Fig. 7-9. The resistor R reduces the voltage to that required by the motor. Find (*a*) the terminal voltage of the generator, (*b*) the wattage dissipated by the resistor R, (*c*) the wattage lost in all the line wires, and (*d*) the total power supplied to the circuit.

17. In a circuit similar to that shown in Fig. 6-21, $V_T = 100$ V, $R_1 = 50$ Ω, $R_2 = 25$ Ω, $R_3 = 8$ Ω, $R_4 = 2$ Ω, $R_5 = 10$ Ω, $R_6 = 15$ Ω, $R_7 = 105$ Ω, $R_8 = 450$ Ω, and $R_9 = 20$ Ω. Find (*a*) the total resistance of the circuit, (*b*) the total current in the circuit, and (*c*) the total power used by the circuit.

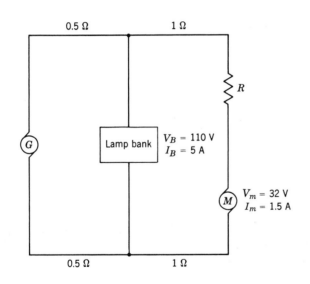

Figure 7-9

(See CD-ROM for Test 7-4)

Assessment

chapter 7

1. A two-slice toaster uses 925 W of energy and is designed for 120-V operation. Determine the current flow.
2. You have three appliances that use 950, 1225, and 1440 W, respectively. If the total current is 30.125 A, what is the supply voltage?
3. You are going to use a 500-Ω, 10-W resistor in a circuit. What is the maximum voltage you can supply to this resistor, and why?
4. Find the value of $(5.638 \times 10^{-6})^2$.
5. A wire feeding a 480-V, 3-θ compressor draws 80 A. One of the feed conductors is loose on the load side of the circuit breaker. When the resistance across the connection is checked with an ohmmeter, the resistance is found to be 46.875 mΩ. What is the power in heat dissipated by this loose connection? Could it be enough to cause the breaker to trip from thermal load?
6. You have discovered that a capacitor that you have bought is actually of a capacitance value that is equal to the square of the value that you needed. If the capacitor you purchased was a 7.25×10^{-9} farad device, what was the rating you had intended to buy?
7. An 1800-W soldering iron has an element resistance of 16 Ω. What is the total current drawn when it is operating?

INTERNET ACTIVITIES

Internet Activity A

View the following site to learn how to calculate the cost of running any appliance in your home. Using the information available at this site, determine how much it will cost to operate a 1500-W quartz heater for 12 hours if the average kilowatthour cost equals 7 cents. **http://www.cipco.org/brochure/cost.htm**

Internet Activity B

View the following site to learn more about square roots. After viewing the information at this site, list the square roots for each of the following positive numbers: 9, 16, and 25. Define what is meant by the *principal root of a number*. **http://www.boisestate.edu/people/jbrennan/Algebra/radicals/square_roots. html**

Chapter 7 Solutions to Self-Tests and Reviews

Self-Test 7-3

1. I_2, R_1, 168, 0.075
2. R_2, 0.075, 12.6
3. 12.6, 0.945
4. 2, 1.89, 2

Self-Test 7-6

a. 6.6
b. 0.0015
c. 2.4
d. 0.0036
e. 0.0099
f. 0.0135
g. V_2
h. 6.6
i. 9
j. V_T
k. 0.0015
l. 0.0135

Self-Test 7-8

1. I
2. 1.2
3. 0.06

4. 1.2
5. 0.05

Self-Test 7-10

1. 110
2. 11
3. 10
4. I
5. 1100

Self-Test 7-17

a. 150
b. 30,000
c. 3×10^4
d. 10^{-2}
e. 0.75
f. 2
g. 100
h. 0.5
i. 1

Self-Test 7-34

1. 4
2. 10 k
3. R

4. 10^4
5. 10^4
6. 4×10^{-4}
7. $\sqrt{4 \times 10^{-4}}$
8. 2×10^{-2}
9. 20

Review 7-11

1. current
2. voltage
3. V
4. power
5. watts
6. current
7. amperes
8. voltage
9. volts
10. I
11. P
12. I
13. 2
14. I_T
15. P_2
16. P_3
17. 3

18. 5
19. 125
20. 1000
21. 1,000,000
22. I^2
23. R
24. itself
25. $\sqrt{}$
26. 4
27. V
28. I
29. 2
30. decimal
31. end
32. 4 69. 40
33. 80 62.
34. 1 23 45.
35. 0. 01 20
36. 0. 00 20
37. 6. 05
38. 0.00 00 60
39. multiplying
40. dividing

8
chapter

Algebra for Complex Electric Circuits

Learning Objectives

1. Combine like and unlike terms.
2. Solve a simple algebraic equation for the value of the unknown letter.
3. Use the methods of cross multiplication and transposition to solve an equation for the value of the unknown letter.
4. Combine unlike terms involving signed numbers.
5. Solve equations having unknowns and numbers on both sides of the equation.
6. Solve equations having unknowns that are negative.
7. Remove parentheses and collect like terms.
8. Solve equations containing parentheses for the value of the unknown letter.
9. Solve equations containing fractions.
10. Solve simultaneous equations using the methods of addition and substitution.

Many circuits cannot be solved by the methods used in Chap. 6. These extremely complicated circuits must be solved by the application of Kirchhoff's laws, which will be discussed in the next chapter. However, the solution of these circuits by Kirchhoff's laws requires an extension of our knowledge of algebra.

JOB 8-1

Combining Like Terms

Different quantities of the same item may be added or subtracted. Thus, 2 apples plus 3 apples will equal 5 apples. Similarly, 3 pencils subtracted from 5 pencils will equal 2 pencils. If we use the letter a for apples and p for pencils, these statements are shortened to read

$$2a + 3a = 5a \qquad \text{and} \qquad 5p - 3p = 2p$$

Quantities involving the *same* letter or letter combinations are called *like terms*. The process of adding or subtracting these like terms is called *combining* terms.

RULE To combine like terms, combine the numerical quantities and place the result before the letter.

Example 8-1

Combine the following like terms:

$$3x + 4x = 7x$$
$$9y - 2y = 7y$$
$$4I_1 + 5I_1 = 9I_1$$

Note: A letter that stands alone such as x, R, or T means $1x$, $1R$, or $1T$. Thus, $4x + x$ means $4x + 1x$ or $5x$.

$$6R + 3R - R = 8R \qquad 1.2x + 3.4x = 4.6x$$
$$4.7R + 2R = 6.7R \qquad 7.8x - 4x = 3.8x$$

Self-Test 8-2

Combine the following like terms:

$$2\text{ mA} + 8\text{ mA} = \underline{\quad(1)\quad}\text{ mA}$$
$$8R + R + 0.2R = 9.2\underline{\quad(2)\quad}$$
$$5K - 2K + 4K = \underline{\quad(3)\quad}$$
$$8I_2 + 4I_2 - 3I_2 = \underline{\quad(4)\quad}$$

Exercises

Practice

Combine the following like terms:

1. $3x + 5x$
2. $6y - 4y$
3. $8R - 2R$
4. $2I + 5I$
5. $4x + 7x + x$
6. $8y - y + 4y$
7. $4x + 3x - 2x$
8. $2.5R + 1.2R$
9. $5.6R - 1.2R$
10. $2I + 3.7I$
11. $7.5x - 4x$
12. $\frac{1}{2}y + \frac{1}{4}y$
13. $\frac{1}{2}T + \frac{1}{3}T$
14. $x - 0.2x$
15. $1.2R + 5.4R + 2.4R$
16. $3I + 5I + 0.2I$
17. $5.6x + 3.7x - 1.8x$
18. $x + 5x - 0.6x$
19. $7x - 3.5x + x$
20. $y - 0.4y$
21. $5.2R + 1.8R - 0.4R$

Using Fig. 8-1, express the total length D of the block in terms of x if A, B, and C have the values given in Exercises 22 to 24.

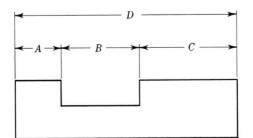

Figure 8-1

22. $A = x$, $B = 2x$, and $C = 3x$
23. $A = \frac{3}{4}x$, $B = 1\frac{1}{2}x$, and $C = \frac{5}{8}x$
24. $A = 0.05x$, $B = 1.25x$, and $C = 0.4x$

Using Fig. 8-2, find the measurement B if A, C, and D have the values given in Exercises 25 to 27.

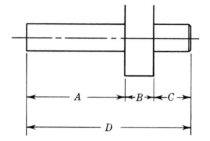

Figure 8-2

25. $A = 3x$, $C = x$, and $D = 7x$
26. $A = 2\frac{1}{2}y$, $C = \frac{3}{4}y$, and $D = 5\frac{7}{8}y$
27. $A = 1.6x$, $C = 0.28x$, and $D = 3x$

JOB 8-2

Combining Unlike Terms

Unlike terms are those in which the letters are *different*. $4R$, $3X$, $3Y$, and $4I$ are all unlike terms. Similarly, $4X$ and the number 6 are unlike terms because the number 6 has no letter.

RULE To combine several quantities involving unlike terms, combine each group of like terms separately.

Example 8-3

Combine terms.

$$2x + 4y + 5x + 7y$$

Solution

1. Combine the like x terms. $2x + 5x = 7x$
2. Combine the like y terms. $4y + 7y = 11y$
3. Since unlike terms may *not* be combined, state the answer.

$$7x + 11y \qquad Ans.$$

Example 8-4

Combine terms.

$$2R + 4I + 7 + 5R - I - 3$$

Solution

1. Combine the like R terms. $2R + 5R = 7R$
2. Combine the like I terms. $4I - I = 3I$
3. Combine the like "number" terms. $7 - 3 = 4$
4. State the answer. $7R + 3I + 4$ *Ans.*

Example 8-5

Combine terms.

$$8I_1 + 5I_2 + 7I_3 + 2I_1 - I_3 - 3I_2$$

Solution

1. Combine the like I_1 terms. $8I_1 + 2I_1 = 10I_1$
2. Combine the like I_2 terms. $5I_2 - 3I_2 = 2I_2$
3. Combine the like I_3 terms. $7I_3 - I_3 = 6I_3$
4. State the answer. $10I_1 + 2I_2 + 6I_3$ *Ans.*

Self-Test 8-6

Combine terms.

$$3I + 10 + 6I - 2 = 9I + \underline{\quad(1)\quad}$$
$$5R + 3I + 4I - R = \underline{\quad(2)\quad} R + \underline{\quad(3)\quad}$$
$$12 - 2R - 3 + 7R = \underline{\quad(4)\quad} + \underline{\quad(5)\quad}$$
$$4I_1 + 2I_2 - I_1 + 3I_2 = 3 \underline{\quad(6)\quad} + \underline{\quad(7)\quad} I_2$$
$$3I_1 + 4I_2 - 3I_1 + 5I_2 + 7 = \underline{\quad(8)\quad}$$

Exercises

Practice

Combine the terms in the following algebraic expressions:

1. $2R + 4I + 3I + 8R$
2. $3R + 2I - R + 5I$
3. $3x + 7 - x + 2$
4. $6x + 2y - y + x$
5. $4R + 3R + 6 - R + 2$
6. $1.5I + 7 + 0.5I - 3$
7. $40 + 30 + 3R - 10$
8. $9I + 16 - 4 + 2I - I$
9. $4.2x + 6y + 1.8x - y$
10. $6R + R - 0.5R + 2I$
11. $1.5I + 0.7R - 0.6I + 1.2R$
12. $2I + 60 - 20 + 3I - 5$

13. $3I_1 + 4I_2 + 5I_1 + 6I_2$

14. $5I_2 + I_2 + 3I_1 - 0.5I_1$

15. $5x + 2y + 7 - x + 3y + 3$

16. $2.5I_1 + 40 - 0.4I_1 + 20$

17. $2.5I_2 + 10 - 1.3I_2 - 8$

18. $8.6R + 7.4 + 2.6 - 1.3R$

19. $3.1 + 4I - 0.7 - 0.6I$

20. $8R - 0.3R + 26 - 9.2$

Find the total voltage V_T of the circuit shown in Fig. 8-3 if A, B, C, and D have the values given in Exercises 21 to 23.

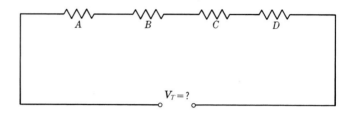

Figure 8-3

21. $A = 2x$ V, $B = 40$ V, $C = 4x$ V, and $D = 15$ V

22. $A = 12.6$ V, $B = 2.5y$ V, $C = y$ V, and $D = 35$ V

23. $A = 0.15R$ V, $B = 0.15R$ V, $C = 25$ V, and $D = 0.15R$ V

In Fig. 8-4, find the sum of the voltage drops by tracing around the circuit named in the following exercises.

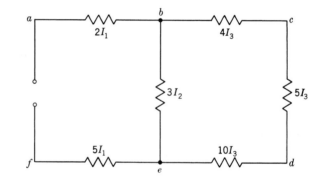

Figure 8-4

24. Circuit *abef*

25. Circuit *bcdeb*

26. Circuit *abcdef*

(See CD-ROM for Test 8-1)

Solving Simple Algebraic Equations

In Job 3-9 we learned to solve the standard equation of the type $3x = 12$. In Job 4-8 we learned to solve equations of the type $R + 2 = 10$ and $x - 3 = 7$. In Job 5-7 we learned to solve equations of the type $V/4 = 5$.

These are the three basic types of equations. The more difficult equations are solved by simplifying them step by step into one or the other of these basic types. The concept of combining like terms should be used whenever possible, as this combination will often simplify the equation into one of the basic types which are readily solved.

Example 8-7

Solve the equation

$$8x - 2x - 4x = 10$$

Solution

1. Write the equation.	$8x - 2x - 4x = 10$
2. Combine like terms.	$2x = 10$
3. Solve the basic equation.	$x = 5$ *Ans.*

> **Hint for Success**
>
> The equation in Example 8-7 is called a *conditional equation* because it is not satisfied by any value of x except $x = 5$. The value of 5 for x is sometimes called the *root of the equation*.

Example 8-8

Solve the equation

$$21 = 0.7x - 0.4x$$

Solution

1. Write the equation.	$21 = 0.7x - 0.4x$
2. Combine like terms.	$21 = 0.3x$
3. Solve the basic equation.	$\dfrac{21}{0.3} = x$
4. Write the answer.	$70 = x$ or $x = 70$ *Ans.*

Example 8-9

Solve the equation

$$3x + 5 = 50$$

Solution

In this equation, as in all equations, our main objective is to get the unknown letter *all alone on one side of the equality sign*. The number that remains on the other side will be the answer. The letter x will remain alone on the left side of the equality sign if we can eliminate the $+5$ and the 3 from that side. Since transposing is a very simple operation, your *first* step will *always* be to transpose any single $+$ or $-$ quantity which is not part of a group. Thus, our first step in this problem will be to transpose the $+5$ from the left side of the equal sign to the right side as -5.

1. Write the equation.	$3x + 5 = 50$
2. Transpose the $+5$.	$3x = 50 - 5$
3. Combine like terms.	$3x = 45$
4. Solve the basic equation.	$x = 15$ *Ans.*

Example 8-10

Solve the equation

$$3I - 4 = 20$$

Solution

1. Write the equation. $\qquad\qquad\qquad 3I - 4 = 20$
2. Transpose the -4. $\qquad\qquad\qquad 3I = 20 + 4$
3. Combine like terms. $\qquad\qquad\qquad 3I = 24$
4. Solve the basic equation. $\qquad\qquad\quad I = 8 \qquad$ *Ans.*

Example 8-11

Solve the equation

$$25 = 4x - 3$$

Solution

1. Write the equation. $\qquad\qquad\qquad 25 = 4x - 3$
2. Transpose the -3 to the *left* side of the equality sign.

$$25 + 3 = 4x$$

3. Combine like terms. $\qquad\qquad\qquad 28 = 4x$
4. Solve the basic equation. $\qquad\quad 7 = x \quad$ or $\quad x = 7 \qquad$ *Ans.*

Self-Test 8-12

Solve the equation $3x + 2 + x = 14$.

Solution

1. Write the equation. $\qquad\qquad 3x + 2 + x = 14$
2. Combine like terms. $\qquad\underline{\quad(1)\quad} + 2 = 14$
3. Transpose the $+2$ $\qquad\qquad\quad 4x = 14 \,\underline{\quad(2)\quad}$
4. Combine like terms. $\qquad\qquad\quad 4x = \underline{\quad(3)\quad}$
5. Solve the basic equation. $\qquad\quad\; x = \underline{\quad(4)\quad} \qquad$ *Ans.*

Self-Test 8-13

Solve the equation $4x - 6.2 - x = 1.3$.

Solution

1. Write the equation. $\qquad\; 4x - 6.2 - x = 1.3$
2. Combine like terms. $\qquad\underline{\quad(1)\quad} - 6.2 = 1.3$
3. Transpose the -6.2. $\qquad\qquad 3x = 1.3 \,\underline{\quad(2)\quad}$
4. Combine like terms $\qquad\qquad\; 3x = \underline{\quad(3)\quad}$
5. Solve for x. $\qquad\qquad\qquad\; x = \underline{\quad(4)\quad} \qquad$ *Ans.*

Self-Test 8-14

An architect's plan calls for five lights laid out as shown in Fig. 8-5. If the total length of electrical conduit used was 58 ft, find the length from B to C.

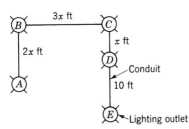

Figure 8-5

Solution

1. The total length is equal to the _____ of the individual lengths.
2. Write an equation to describe this.

$$2x + 3x + \text{_____} + 10 = 58$$

3. Combine like terms. $\text{_____} + 10 = 58$
4. Transpose the $+10$. $6x = 58 - \text{_____}$
5. Combine like terms. $6x = \text{_____}$
6. Solve the basic equation. $x = \text{_____}$
7. The length from B to C is _____ ft.
8. By substituting our answer 8 for x, we get the length from B to $C = 3 \times$ _____ = _____ ft   *Ans.*

Self-Test 8-15

Solve the circuit shown in Fig. 8-6 for the value of R_1.

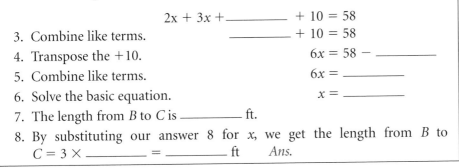

Figure 8-6

Solution

We can obtain the value of V_1 by using the formula

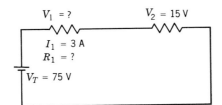

$$V_1 = I_1 \times \underline{\quad (1) \quad}$$ $(4\text{-}4)$
$$V_1 = \underline{\quad (2) \quad} \times R_1$$
$$V_1 = 3R_1$$

1. In this series circuit, the total voltage is equal to the ____(3)____ of all the voltages.

$$V_T = V_1 + \underline{(4)}$$

2. Substitute numbers. $75 = \underline{(5)} + 15$

3. Transpose the +15. $75 - 15 = \underline{(6)}$

4. Combine like terms. $\underline{(7)} = 3R_1$

Exercises

Practice

Solve the following equations:

1. $3x = 21$
2. $14 = 2R$
3. $2a + 3a = 25$
4. $3I + I = 24$
5. $40 = 3R + 5R$
6. $\dfrac{I}{5} = 6$
7. $\dfrac{2R}{5} = \dfrac{8}{4}$
8. $x - 4 = 6$
9. $\dfrac{2x}{3} = 8$
10. $\dfrac{2x}{3} = 24$
11. $20 = 6x - 2x$
12. $x + 3 = 9$
13. $3x = 16$
14. $10R = 2$
15. $0.2x = 40$
16. $14 = R - 4$
17. $9 = 0.3x$
18. $9R - 5R = 38$
19. $0.3a + 0.2a = 35$
20. $26 = 1.1R + 0.2R$
21. $2.4x - 0.4x = 10$
22. $0.4R + 0.3R = 4.9$
23. $\dfrac{3y}{4} = 0.6$
24. $R - 5.6 = 14$
25. $5x + 2x + x = 40$
26. $3I + 7I - 2I = 56$
27. $26 = 8x + 6x - 2$
28. $0.12x = 0.06$
29. $\dfrac{2R}{7} = 7$
30. $5a - 6 = 4$
31. $25 = 3R + 4$
32. $6 = \dfrac{3x}{5}$
33. $1.9x - 0.4x = 4.5$
34. $3x + 2 = 11$
35. $42 = 6x - x + 2x$
36. $34 = 7R - 8$
37. $29 = 3x + 11$
38. $2R + 3 = 3$
39. $8x + 1 = 3$
40. $2x - 0.3 = 1.1$
41. $4x + 2.6 = 5.4$
42. $4R - R + 2 = 20$
43. $3R - 9 = 11$
44. $59 = 5x - 6$
45. $5y - 17 = 37$
46. $12x - 1 = 8$
47. $5x + 0.2x = 10.4$
48. $x + 1.2x = 4.4$
49. $x + 1.2x + 9 = 97$
50. $2I + 7I + 30 - I = 70$

Applications

51. Find the value of each dimension shown in Fig. 8-7.

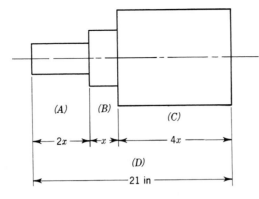

Figure 8-7

52. Using Fig. 8-1, find the value of each dimension if $A = x$, $B = 3x$, $C = 4x$, and $D = 32$
53. $A = 1.3x$, $B = 0.04x$, $C = 2.16x$, and $D = 35$
54. $A = x + 4$, $B = 5$, $C = 2x$, and $D = 21$
55. $A = 1.2y + 8$, $B = 0.8y$, $C = 0.4y$, and $D = 56$
56. The current in a transistor circuit divides according to the formula $I_E = I_B + \beta I_B$. Find the current gain β of the transistor shown in Fig. 8-8.

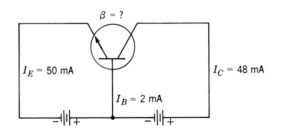

Figure 8-8   Division of current in an NPN transistor circuit.

57. Two equal resistors are connected in series with a 150-Ω resistor to make a total resistance of 200 Ω as shown in Fig. 8-9. (*a*) Form an equation which can be used to solve for R. (*b*) Solve for R.

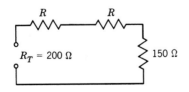

Figure 8-9

58. Two holes, 2.25 in apart on center, are to be drilled equidistant from the ends of a steel plate 4.75 in long as shown in Fig. 8-10 on the next page. How far from the ends should the centers of the holes be marked?

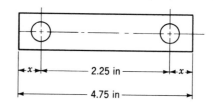

Figure 8-10

59. A motor, a lamp, and a soldering iron are in parallel. The motor draws twice as much current as the lamp, and the soldering iron draws 3½ times as much current as the lamp. Find the current drawn by each item if the total current drawn in 13 A.

60. A voltage divider is made in three sections connected in series. If R_2 is three times R_1, and R_3 is twice R_2, find the resistance of each section when the total resistance is 100,000 Ω.

61. In a series circuit, the voltage across V_2 is twice that across V_1, and the voltage V_3 is one-half of V_2. If the total voltage across the circuit is 120 V, find the voltage across each resistor.

(See CD-ROM for Test 8-2)

JOB 8-4

Solving Equations by Transpositions and Cross Multiplication

Hint for Success

When you are learning how to solve an equation like the one shown in Example 8-16, it is important to show each step of the solution until you become very proficient with the process. Only then is it all right to take a shortcut or two.

Example 8-16

Solve the equation

$$\frac{2x}{5} + 3 = 7$$

Solution

1. Write the equation. $\dfrac{2x}{5} + 3 = 7$

2. Transpose the +3. $\dfrac{2x}{5} = 7 - 3$

3. Combine like terms. $\dfrac{2x}{5} = 4$

4. Cross-multiply. $2x = 20$
5. Solve the basic equation. $x = 10$ *Ans.*

Example 8-17

Solve the equation

$$23 = \frac{2R}{3} + 7$$

Solution

1. Write the equation.

$$23 = \frac{2R}{3} + 7$$

2. Transpose the + 7.

$$23 - 7 = \frac{2R}{3}$$

3. Combine like terms.

$$16 = \frac{2R}{3}$$

4. Cross-multiply.

$$2R = 48$$

5. Solve the basic equation.

$$R = 24 \quad \textit{Ans.}$$

Self-Test 8-18

Solve the equation

$$1\frac{1}{4}x + 1\frac{1}{2}x + 3 = 14$$

Solution

1. Write the equation.

$$1\frac{1}{4}x + 1\frac{1}{2}x + 3 = 14$$

2. Combine like terms.

$$\underline{\hspace{2cm}} + 3 = 14$$

3. Change the mixed number to an improper fraction.

$$\underline{\hspace{2cm}} + 3 = 14$$

4. Transpose the +3.

$$\frac{11x}{4} = 14 \,\underline{\hspace{2cm}}$$

5. Combine like terms.

$$\frac{11x}{4} = \underline{\hspace{2cm}}$$

6. Cross-multiply.

$$11x = \underline{\hspace{2cm}}$$

7. Solve the basic equation.

$$x = \underline{\hspace{2cm}} \quad \textit{Ans.}$$

Exercises

Practice

Solve the following equations.

1. $\dfrac{x}{2} + 4 = 9$

2. $\dfrac{x}{3} + 7 = 12$

3. $\dfrac{R}{3} - 2 = 8$

4. $\dfrac{y}{5} - 6 = 3$

5. $9 = \dfrac{x}{2} + 1$

6. $5 = \dfrac{y}{3} - 2$

7. $\dfrac{2R}{3} - 4 = 6$

8. $\dfrac{3R}{5} + 4 = 10$

9. $\dfrac{x}{2} - 1\dfrac{1}{2} = 3\dfrac{1}{2}$

10. $5 = \dfrac{3x}{2} - 7$

11. $\dfrac{x}{3} - 1.2 = 3.2$

12. $\dfrac{x}{5} + 0.2 = 1.8$

13. $\dfrac{5x}{3} + 4 = 9$

14. $\dfrac{4x}{5} - 18 = 22$

15. $32 = \dfrac{3x}{2} - 10$

16. $\dfrac{x}{3} + \dfrac{1}{4} = 1\dfrac{1}{4}$

17. $1\dfrac{1}{4}x - 3 = 12$

18. $2\dfrac{1}{3}x + 4 = 18$

Applications

19. In Fig. 8-5, $AB = 2x$, $BC = 3x$, $CD = x$, and $DE = 10$. If the total length of conduit is 87 ft, find the length of CD.

20. A lamp, a toaster, and a coffee maker are connected in parallel and draw a total of 12 A. The lamp draws one-sixth as much current as the toaster, and the coffee maker draws 1 A less than the toaster. Find the current drawn by each appliance.

Intermediate Review

Self-Test 8-19

Find the distance x in Fig. 8-11.

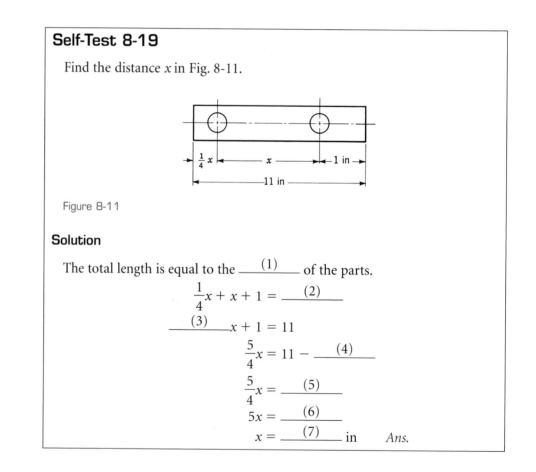

Figure 8-11

Solution

The total length is equal to the ____(1)____ of the parts.

$$\frac{1}{4}x + x + 1 = \underline{\quad(2)\quad}$$

$$\underline{\quad(3)\quad}x + 1 = 11$$

$$\frac{5}{4}x = 11 - \underline{\quad(4)\quad}$$

$$\frac{5}{4}x = \underline{\quad(5)\quad}$$

$$5x = \underline{\quad(6)\quad}$$

$$x = \underline{\quad(7)\quad} \text{ in} \qquad Ans.$$

Self-Test 8-20

Three identical resistors are in series with a 100- and a 300-Ω resistor. If the total resistance of the circuit is 1600 Ω, find the resistance of each unknown resistor.

Solution

In a series circuit, the total resistance is equal to the _____(1)_____ of the individual resistances.

$$R_T = R_1 + R_2 + R_3 + \text{etc.} \qquad (4\text{-}3)$$
$$1600 = R + R + \underline{\quad(2)\quad} + 100 + 300$$
$$1600 = \underline{\quad(3)\quad} + 400$$
$$1600 - \underline{\quad(4)\quad} = 3R$$
$$\underline{\quad(5)\quad} = 3R$$
$$R = \underline{\quad(6)\quad} \; \Omega \qquad Ans.$$

(See CD-ROM for Test 8-3)

Combining Unlike Terms Involving Signed Numbers

The procedure is exactly the same as that outlined in Job 8-2 except that the *signs* of the quantities must now be taken into account in the addition.

Example 8-21

Combine the following terms:

$$40 - 55 + 2x - 10 - 5x$$

Solution

1. Combine the like "number" terms. $+40 - 55 - 10 = -25$
2. Combine the like x terms. $+2x - 5x = -3x$
3. State the answer. $-25 - 3x$ *Ans.*

Example 8-22

Combine the following terms:

$$5x - 2y - 5y - 7x + 14 - y - 6$$

Solution

1. Combine the like x terms. $+5x - 7x = -2x$
2. Combine the like y terms. $-2y - 5y - y = -8y$
3. Combine the like "number" terms. $+14 - 6 = +8$
4. State the answer. $-2x - 8y + 8$ *Ans.*

Example 8-23

Combine the following terms:

$$2I_1 - 6I_2 - 4I_1 - 30 - 4I_1 + 2I_2$$

Solution

1. Combine the like I_1 terms. $+2I_1 - 4I_1 - 4I_1 = -6I_1$
2. Combine the like I_2 terms. $-6I_2 + 2I_2 = -4I_2$
3. State the answer. $-6I_1 - 4I_2 - 30$ *Ans.*

Self-Test 8-24

An equation that might result from an application of Kirchhoff's laws (which we shall study in the next chapter) is

$$-10 - 2x + 2y - 3x - 5y = 0$$

Combine the terms on the left side of the equation.

Solution

1. Combine the like x terms. $-2x - 3x =$ _____
2. Combine the like y terms. $+2y - 5y =$ _____
3. State the simplified equation. $-5x$ _____ $-10 = 0$ *Ans.*

Exercises

Practice

Combine the following terms:

1. $3x + 4 - 5x + 3$
2. $16 - 4y - 2y - 7$
3. $2I_1 + 3I_2 - 5I_2 - 6I_1$
4. $3 - 8 + 4x - 2 - x$
5. $2x - 3y - 2y - 4x + y$
6. $8 - 2I_2 - 3 - 4I_2 - 3I_2$
7. $4x - 13 - 5x + 6 - 2x$
8. $2x - y - 3x - 4y + 7 - x$
9. $20 - 35 + 4x - 5 - 6x$
10. $-x - y - 4 + 4x + 2 - 3y$

JOB 8-6   Solving Equations Which Have Unknowns and Numbers on Both Sides of the Equality Sign

As noted in Job 8-3, our main objective is to get the unknown letter all alone on one side of the equality sign. If the unknowns appear on *both sides* of the equality sign, we must gather them together as our first step. This will be accomplished by the normal process of transposing. In general then, transpose *all* letters to one side of the equality sign and transpose *all* numbers to the other side. Letters are collected on the left side or on the right side, the side chosen depending on the whim of the solver.

Example 8-25

Solve the equation

$$7R - 84 = 4R$$

Solution

1. Write the equation. $\qquad 7R - 84 = 4R$
2. Transpose the $4R$ to the left side, and the -84 to the right side.

$$7R - 4R = 84$$

3. Combine like terms. $\qquad\qquad 3R = 84$
4. Solve the basic equation. $\qquad\quad R = 28 \qquad$ *Ans.*

Example 8-26

Solve the equation

$$8x - 7 = 3x + 8$$

Solution

1. Write the equation. $\qquad\qquad 8x - 7 = 3x + 8$
2. Transpose the $3x$ to the left side and the -7 to the right.

$$8x - 3x = 8 + 7$$

3. Combine like terms. $\qquad\qquad 5x = 15$
4. Solve the basic equation. $\qquad\quad x = 3 \qquad$ *Ans.*

Example 8-27

Solve the equation

$$3x - 3 = 8x - 18$$

Solution

1. Write the equation. $\qquad\qquad 3x - 3 = 8x - 18$
2. Since the greater number of x's appear on the right side of the equality sign, we shall transpose all x's to this right side and all numbers to the left side.

$$-3 + 18 = 8x - 3x$$

3. Combine like terms. $\qquad\qquad 15 = 5x$
4. Solve the basic equation. $\qquad\quad 3 = x \quad$ or $\quad x = 3 \qquad$ *Ans.*

Exercises

Practice

Solve the following equations:

1. $7R = 2R + 10$
2. $8x = 10 + 6x$
3. $4x = 27 + x$
4. $5R = -2R + 14$
5. $3y + 24 = 9y$
6. $28 - 2I = 5I$
7. $5x - 20 = 3x$
8. $7R - 48 = 3R$
9. $2x = 35 - 5x$
10. $-x - 19 = -2x$
11. $-2y - 21 = -5y$
12. $6T = 4 - 2T$
13. $3 - R = 8R$
14. $2x = 5x - 39$
15. $y = 9y - 40$
16. $x = 3.9 - 2x$
17. $5R + 1 = 3R + 9$
18. $6x - 4 = 2x + 28$
19. $26 + x = 4x - 1$
20. $7x - 25 = 4x + 23$
21. $R + 10 = 45 - 4R$
22. $2x - 8 = 9x - 50$
23. $7x - x + 8 = 2x + 40$
24. $12R - 5 = 6R - R + 23$
25. $8x - 3 + 9 = 2x + x + 31$
26. $12 + 2x = 8x - 60$
27. $8I - 1 = 3 - 4I$
28. $1 - 3R = 7R - 4$
29. $5T = 3T + 4.2$
30. $1.4x - 12 = -0.6x$

(See CD-ROM for Test 8-4)

JOB 8-7

Solving Equations with Possible Negative Answers

Consider the equation

$$2R = -10$$

Solving,

$$R = \frac{-10}{+2}$$

or

$$R = -5 \quad Ans.$$

The signs in this problem might appear as shown below.

$$-2R = 10 \qquad\qquad -2R = -10$$

Solving,

$$R = \frac{+10}{-2} \qquad\qquad R = \frac{-10}{-2}$$

or

$$R = -5 \quad Ans. \qquad\qquad R = +5 \quad Ans.$$

The answers above are not incorrect. It is quite possible for an answer to be a negative number. In the next chapter on complex circuits and in our future study of ac electricity, we shall meet these frequently. The basic equation is solved in a normal manner, but care must be used in dividing the signed numbers. Another method for handling these signs is illustrated in the next example.

Example 8-28

Solve the equation

$$-2R = -10$$

Solution

Since all parts of an equation may be multiplied by the same number without destroying the equality, we can eliminate the cumbersome negative sign of the unknown by multiplying the entire equation by -1. Therefore,

$$-1(-2R = -10)$$

becomes $\qquad 2R = 10$

or $\qquad R = 5 \qquad$ *Ans.*

This operation may be described in the following simple rule.

RULE All the signs of the individual parts of an equation may be changed without destroying the equality.

Example 8-29

Solve the following equations.

$$-R = -6 \qquad\qquad -I = 25 \qquad\qquad 7 = -R$$
$$R = 6 \qquad\qquad I = -25 \qquad\qquad -7 = R \qquad \textit{Ans.}$$

Example 8-30

Solve the equation

$$4x - 5 = 5x + 1$$

Solution

1. Write the equation. $\qquad\qquad\qquad 4x - 5 = 5x + 1$
2. Transpose. $\qquad\qquad\qquad\qquad 4x - 5x = 1 + 5$
3. Combine like terms. $\qquad\qquad\qquad -x = 6$
4. Change signs throughout. $\qquad\qquad x = -6 \qquad \textit{Ans.}$

Exercises

Practice

Solve the following equations:

1. $-x = +8$
2. $12 = -R$
3. $3x = -15$
4. $-21 = 3T$

5. $8x = -4$

6. $-5 = 10x$

7. $-2x = 16$

8. $40 = -4x$

9. $-2x = -14$

10. $-24 = -3R$

11. $3x + 17 = 5$

12. $6 - 3x = 18$

13. $9 = 8 - R$

14. $y = 5y + 28$

15. $3R + 10 = R$

16. $5x + 9 = 4x + 1$

17. $5R + 7 = 6R + 18$

18. $3I + 22 = 9I - 20$

19. $-4R - 11 = 6R + 19$

20. $0.4x = -20$

21. $-1.2I = 2.4$

22. $-39 = 0.3T$

23. $1.6R + 10 = 0.6R + 3$

24. $-2I_2 = 10 - 8.58$

JOB 8-8

Removing Parentheses

In order to solve the equations in the next job, we must be able to simplify an equation by removing any parentheses in it. Parentheses are used to indicate that the quantities within them represent a single idea. For example, if a transformer costs $30 and the tax is $1, then the total actual cost is represented as the quantity $(30 + 1)$. Of course, this would be written as $31 because we would naturally combine the like terms. However, if the cost were unknown, we would represent it as x dollars. If the tax remains constant, then the only way to represent the total cost would be as the quantity $x + 1$ or $(x + 1)$.

Now, if we bought six transformers, the total cost would be six times the cost of one transformer of $6 \times (x + 1)$. This is usually written as $6(x + 1)$. This means that the 6 is multiplied by the x *and also* by the number 1. Thus,

$$6(x + 1) = 6x + 6$$

We can check the accuracy of this method by using the actual cost.

$$6(30 + 1) = 6(31) = 186$$
or
$$6(30 + 1) = 6 \times 30 + 6 \times 1 = 180 + 6 = 186$$

RULE   To multiply a parenthesis by a single quantity:

1. Multiply each term of the parenthesis by the multiplier.
2. Combine like terms.

Example 8-31

Multiply $+3(2R - 7)$.

Solution

1. Multiply each part of the parenthesis by $+3$.

$$(+3) \times (2R) = +6R$$
$$(+3) \times (-7) = -21$$

2. State the answer.

$$6R - 21 \quad \textit{Ans.}$$

Example 8-32

Remove parentheses and collect like terms.

a. $8x - 3(2 + x)$

   $8x - 6 - 3x$

   $5x - 6 \quad \textit{Ans.}$

b. $11 + 2(3x - 9)$

   $11 + 6x = 18$

   $6x - 7 \quad \textit{Ans.}$

c. $-3(I + 4) - 5$

   $-3I - 12 - 5$

   $-3I - 17 \quad \textit{Ans.}$

Example 8-33

Remove parentheses and collect like terms.

$$5x + (3 - 8x)$$

Solution

When the parenthesis is preceded by just a plus sign or a minus sign, the number 1 is understood to be present.

1. Write the problem.

$$5x + (3 - 8x)$$

2. Insert the number 1 after the sign.

$$5x + 1(3 - 8x)$$

3. Multiply the parenthesis by $+1$.

$$5x + 3 - 8x$$

4. Collect like terms.

$$-3x + 3 \quad \text{or} \quad 3 - 3x \quad \textit{Ans.}$$

Certain problems can lead to errors. Be careful to note the difference between the following examples.

Example 8-34

Remove parentheses and collect like terms.

$-9(4x - 3)$

$-36x + 27 \quad \textit{Ans.}$

$-9 - (4x - 3)$

$-9 - 1(4x - 3)$

$-9 - 4x + 3$

$-6 - 4x \quad \textit{Ans.}$

Exercises

Practice

Remove parentheses and collect like terms.

1. $2(3 - 4x)$
2. $3(5x - 6)$
3. $-3(x - 4)$
4. $4(2x - 3) - 3x$
5. $7 + 3(x - 4)$
6. $2y - (y - 3)$
7. $9I - 3(I + 8)$
8. $6 + (R - 7)$
9. $+(8 - 2R) - 7$
10. $-(2 + 3x) + 5$
11. $14 - 3(x - 5)$
12. $60 - 3(I_1 - 5)$
13. $20 + 2(I_1 + I_2)$
14. $30 - 4(6 - I_2)$
15. $5R - 2(10 - 4 - R)$
16. $(x - 2) + 2(x - 4)$
17. $-3(-2 + R) + (R - 1)$
18. $7x - 3x - 2(x + 4)$
19. $-(I_1 - I_2) + 4(I_1 - 6)$
20. $3x - 7 - 2(x - 3) + 20$

(See CD-ROM for Test 8-5)

JOB 8-9

Solving Equations Containing Parentheses

Example 8-36

Solve the equation

$$6x - 2(x - 4) = 32$$

Solution

In order to solve an equation, we collect *all* the unknowns on one side of the equality sign and *all* the numbers on the other side. We can't do this in this equation until we release the x and the -4 from the parentheses. Once this has been done, the solution proceeds normally.

1. Write the equation. $\qquad 6x - 2(x - 4) = 32$
2. Remove parentheses. $\qquad 6x - 2x + 8 = 32$
3. Transpose the $+8$. $\qquad\qquad 6x - 2x = 32 - 8$

4. Combine like terms. $4x = 24$
5. Solve the basic equation. $x = 6$ *Ans.*

Self-Test 8-37

Find R_2 in the circuit shown in Fig. 8-12.

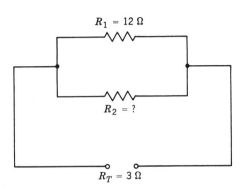

Figure 8-12

Solution

1. This is a ————— circuit.

2. Write the formula for R_T. $R_T = \dfrac{(R_1 \times R_2)}{(R_1 + ?)}$

3. Substitute numbers. $\dfrac{3}{1} = \dfrac{(12 \times ?)}{(? + R_2)}$

4. Cross-multiply. $3(12 + R_2) = $ ————

5. Remove parentheses. $36 + $ ———— $= 12 R_2$

6. Transpose the $3R_2$. $36 = 12R_2 - $ ————

7. Combine like terms. $36 = $ ————

8. Solve the basic equation. $R_2 = $ ———— Ω *Ans.*

Self-Test 8-38

An electric iron uses 90 W more than twice the power used by an electric toaster. Three times the toaster power equals 460 W less than twice the iron wattage. Find the power used by each.

Solution

In order to describe the problem in the language of algebra, we need algebraic names for the power used by each appliance.

Let x = the power used by the toaster. Then twice this power = ———(1)———, and

$$90 \text{ W more than this} = (\text{———(2)———})$$

or $(2x + 90)$ = the power used by the ____(3)____.
Now to make the equation.

3 times the toaster power = twice the iron power − 460
$$3 \times \underline{\quad (4) \quad} = 2(\underline{\quad (5) \quad}) - 460$$

Solve this equation.

$$3x = 2(2x + 90) - 460$$

1. Remove parentheses.

$$3x = \underline{\quad (6) \quad} + 180 - 460$$

2. Transpose the $4x$.

$$3x - \underline{\quad (7) \quad} = 180 - 460$$

3. Combine like terms.

$$-x = \underline{\quad (8) \quad}$$

4. Solve for x.

$$x = \underline{\quad (9) \quad}$$

5. State the answer. Since x was our algebraic name for the toaster power,

$$\text{Toaster power} = \underline{\quad (10) \quad} \text{ W} \qquad Ans.$$

Since $(2x + 90)$ was our algebraic name for the iron power, by substituting 280 for x, we get

$$\text{Iron power} = (2 \times 280) + 90$$
$$= \underline{\quad (11) \quad} + 90$$
$$= \underline{\quad (12) \quad} \text{ W} \qquad Ans.$$

Exercises

Practice

Solve the following equations:

1. $2(x + 3) = 8$
2. $3(R - 4) = 3$
3. $4(R + 3) = 4$
4. $2(3 - R) = 4$
5. $5(2x + 3) = 35$
6. $7(3y - 2) = 28$
7. $3(2x - 2) = -24$
8. $3(3x - 1) - 4 = 11$
9. $5(3R - 2) + 8 = 43$
10. $3(a - 4) = 2(a + 1)$
11. $7(x + 1) = 5(x + 1)$
12. $2(I - 4) = 5(I + 2)$
13. $3(3x - 2) + 5 = 26$
14. $2(y + 4) + y = 23$
15. $4(x - 3) - 2x = -20$
16. $6x - (x - 3) = 28$
17. $4x - 5(x - 2) = 3$
18. $8R - (R - 4) = 11$
19. $6x - 2(x + 6) = x$
20. $10x = 38 + (4x - 2)$
21. $2R = 15 + 3(R + 2)$
22. $6y = 20 - (y - 8)$
23. $8y = 1 - 2(y - 2)$
24. $5x - 2(x - 3) = 17$
25. $2y + (y - 3) = 18 - 2(y + 3)$
26. $4(R - 2) - 7 = 9 - (R + 4)$

27. $0.4(x - 2) = 1.4$

28. $0.3(5R - 6) = 2.7$

29. $4(x - 0.2) = 2(x + 0.7)$

30. $2(x - 1.2) = 0.2(x - 3)$

Applications

31. In a circuit similar to that shown in Fig. 8-8, $I_E = 52$ mA and $I_B = 2$ mA. Find β using the transistor formula $I_E = I_B (\beta + 1)$.

32. In a circuit similar to that shown in Fig. 8-12, find R_1 if $R_2 = 60$ Ω and $R_T = 15$ Ω. Use the formula given in Example 8-37.

33. In a parallel circuit of two resistances, $R_1 = 16$ Ω, $I_2 = 2$ A, $I_T = 8$ A. Find R_2 using the formula

$$I_2 = \frac{R_1}{(R_1 + R_2)} \times I_T$$

34. Using the formula to change Fahrenheit temperature to Celsius temperature, $C = 5(F - 32)/9$, find the number of degrees Fahrenheit F that is equivalent to $C = 20°$.

35. Each heating element in a cafeteria grill uses 130 W of power less than three times the wattage used by a toaster. When two toasters and three heating elements are in use, they require a total of 2470 W of power. Find the wattage used by one heating element.

36. Using the transistor formula $\beta = \alpha/(1 - \alpha)$, find α if $\beta = 49$.

(See CD-ROM for Test 8-6)

JOB 8-10

Solving Equations Containing Fractions

In Job 5-7 we learned how to solve some simple fractional equations. These equations were limited to those containing only two fractions which were equal to each other. They were solved by cross multiplication, which is merely a short cut to be used only in the situation where two fractions equal each other. If there are *more* than two fractions, we are forced to use the least common denominator method discussed in Examples 5-17 to 5-19 of Job 5-7.

RULE If an equation contains more than two fractions, eliminate the denominators by multiplying all parts of the equation by the least common denominator.

Example 8-39

Solve the equation

$$\frac{x}{2} - \frac{x}{3} = \frac{1}{2}$$

Solution

1. Write the equation.

$$\frac{x}{2} - \frac{x}{3} = \frac{1}{2}$$

2. Multiply all terms by the LCD, which is 6.

$$\frac{\overset{3}{\cancel{(6)}}\,x}{\underset{1}{\cancel{2}}} - \frac{\overset{2}{\cancel{(6)}}\,x}{\underset{1}{\cancel{3}}} = \frac{\overset{3}{\cancel{(6)}}\,1}{\underset{1}{\cancel{2}}}$$

3. Multiply.

$$3x - 2x = 3$$

4. Combine like terms.

$$x = 3 \qquad Ans.$$

Example 8-40

Solve the equation

$$\frac{x}{4} + \frac{x}{8} = 6$$

Solution

1. Write the equation

$$\frac{x}{4} + \frac{x}{8} = 6$$

2. Complete all fractions by placing whole numbers or decimals over 1.

$$\frac{x}{4} + \frac{x}{8} = \frac{6}{1}$$

3. Multiply all terms by the LCD, which is 8.

$$\frac{\overset{2}{\cancel{(8)}}\,x}{\underset{1}{\cancel{4}}} + \frac{\overset{1}{\cancel{(8)}}\,x}{\underset{1}{\cancel{8}}} = \frac{\overset{8}{\cancel{(8)}}\,6}{\underset{1}{\cancel{1}}}$$

4. Multiply.

$$2x + x = 48$$

$$3x = 48$$

$$x = 16 \qquad Ans.$$

Example 8-41

Solve the equation

$$\frac{2R}{5} + 3 = \frac{R}{4}$$

Solution

1. Write the equation.

$$\frac{2R}{5} + 3 = \frac{R}{4}$$

2. Complete all fractions by placing whole numbers or decimals over 1.

$$\frac{2R}{5} + \frac{3}{1} = \frac{R}{4}$$

3. Multiply all terms by the LCD, which is 20.

$$\overset{4}{(\cancel{20})} \frac{2R}{\underset{1}{\cancel{5}}} + \overset{20}{(\cancel{20})} \frac{3}{\underset{1}{\cancel{1}}} = \overset{5}{(\cancel{20})} \frac{R}{\underset{1}{\cancel{4}}}$$

4. Multiply.

$$8R + 60 = 5R$$

5. Transpose.

$$8R - 5R = -60$$
$$3R = -60$$
$$R = -20 \quad Ans.$$

Exercises

Practice

Solve the following equations.

1. $\dfrac{x}{2} - \dfrac{x}{3} = \dfrac{1}{3}$

2. $\dfrac{x}{2} - \dfrac{x}{4} = \dfrac{1}{2}$

3. $\dfrac{x}{4} - \dfrac{x}{5} = 2$

4. $\dfrac{3R}{4} - \dfrac{2R}{3} = 1$

5. $\dfrac{x}{3} - \dfrac{x}{6} = 5$

6. $\dfrac{x}{2} + \dfrac{x}{3} = 5$

7. $\dfrac{V}{2} + \dfrac{V}{8} = 5$

8. $\dfrac{V}{5} + \dfrac{V}{3} = \dfrac{8}{15}$

9. $\dfrac{10}{3} = \dfrac{x}{3} + \dfrac{x}{7}$

10. $\dfrac{R}{4} + \dfrac{2R}{5} = 13$

11. $\dfrac{5x}{3} - 7 = \dfrac{x}{2}$

12. $\dfrac{V}{30} - 2 = \dfrac{V}{40}$

13. $\dfrac{V}{20} = \dfrac{V}{30} = 5$

14. $\dfrac{V}{20} + \dfrac{V}{10} + \dfrac{V}{60} = 5$

15. $\dfrac{3V}{9} - \dfrac{3V}{10} = \dfrac{1}{3}$

16. $\dfrac{R}{10} + \dfrac{R}{30} + \dfrac{R}{60} = \dfrac{9}{10}$

17. $\dfrac{6x}{50} + \dfrac{3x}{50} = \dfrac{9}{25}$

18. $\dfrac{V}{200} - \dfrac{V}{500} = \dfrac{3}{2}$

19. $\dfrac{R}{8} - \dfrac{R}{15} = 1$

20. $\dfrac{5x}{6} + \dfrac{x}{9} - \dfrac{2x}{3} = \dfrac{5}{9}$

21. $\dfrac{V}{6} + \dfrac{V}{9} + \dfrac{V}{12} = 39$

22. $\dfrac{V}{10} + \dfrac{V}{8} + \dfrac{V}{6} = 18.8$

(See CD-ROM for Test 8-7)

JOB 8-11

Solving Simultaneous Equations by Addition

Simultaneous equations are those that arise at the same time (simultaneously) from the same problem. In simple problems, an equation can usually be found to describe the conditions of the problem. The solution of the equation finds the unknown. In the complex circuits to be discussed in the next chapter, there will be *two* unknowns to find. Suppose that the two unknown currents are represented as x and y and an equation connecting these currents is found to be

$$x + y = 12$$

There is an infinite variety of answers which will fit this equation. If $x = 2$, then $y = 10$. If $x = 7$, then $y = 5$. If $y = 3$, then $x = 9$, etc. Obviously, then, an equation which contains two unknown quantities *cannot* be solved. However, if a *second* equation connecting the *same* two quantities can be found, we might be able to find a solution by working the two equations simultaneously. In general, *two* unknowns will require *two* equations. *Three* unknowns will require *three* equations, etc.

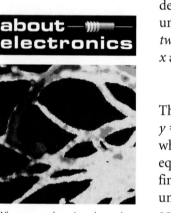
Example 8-42

Solve the pair of simultaneous equations given below for the value of x and y.

Solution

1. The equations are:

$$x + y = 12 \qquad (1)$$
$$x - y = 4 \qquad (2)$$

2. Since both equations contain two unknowns and therefore cannot be solved separately, let us *add* the equations. Since only like terms may be added, we must be sure to have the equations arranged so that the like terms are one under the other. If they are not so arranged, we must transpose one equation so as to agree with the other. Adding,

$$2x = 16$$

Note that the sum of $+y$ and $-y$ equals zero, and that the letter y disappears. The resulting equation contains *only one unknown*, and is easily solved. Solving,

$$x = 8 \qquad Ans.$$

3. Substitute this value of x in *either* of the original equations to find the value of y. Using equation (1),

$$x + y = 12 \tag{1}$$

Substituting,	$8 + y = 12$	
Transposing,	$y = 12 - 8$	
Combining like terms.	$y = 4$	*Ans.*

We were very lucky in the last problem. The simple act of adding the two equations *eliminated* one letter and provided a *third* equation which contained only *one* unknown which was easily solved. Will this occur all the time, or are certain conditions necessary for the elimination of one unknown by addition?

Consider the following additions: $(+7) + (-7) = 0$; $(+4) + (-4) = 0$; $(-5) + (+5) = 0$; $(+19) + (-19) = 0$. Apparently, the total is zero whenever we add *identical* quantities with *opposite* signs. This is what happened when we added $+y$ and $-y$ in the last problem. The addition totaled zero, which therefore eliminated the unknown y.

The number portion of an algebraic expression is called its *numerical coefficient*. For example,

2 is the numerical coefficient of the expression $2x$.

7 is the numerical coefficient of the expression $7xy$.

1 is the numerical coefficient of the expression ab.

RULE To eliminate an unknown by addition, the numerical coefficients of the unknown must be identical, but with opposite signs.

> **Hint for Success**
>
> Simultaneous equations can also be solved by subtracting the equations from each other. For an unknown to be eliminated by subtraction, the numerical coefficient of the unknown must be identical.

If necessary, one or *both* equations may be multiplied by the proper number and sign to create identical coefficients with opposite signs. Let's see how this works out in the next example.

Example 8-43

Solve the pair of simultaneous equations given below for the value of x and y.

Solution

1. The equations are:

$$x + 2y = 10 \tag{1}$$
$$x - y = 1 \tag{2}$$

You can see that neither x nor y will be eliminated by adding these two equations. (x added to x gives $2x$ and *not* zero. $+2y$ added to $-y$ gives $+y$ and *not* zero.) However, if the $-y$ in equation (2) were $-2y$, then the total of $+2y$ in equation (1) and $-2y$ in equation (2) *would* total zero. We shall therefore *change* the $-y$ in equation (2) into $-2y$ by multiplying the *entire* equation by the quantity which will change $-y$ into $-2y$. This multiplier must be $+2$.

2. Rewrite the problem as shown.

$$x + 2y = 10$$
$$+2(x - y = 1)$$

3. Multiplication gives

$$x + 2y = 10$$
$$2x - 2y = 2$$

4. Add the equations.

$$3x = 12$$
$$x = 4 \quad \textit{Ans.}$$

5. Substitute this value for x in *either* original equation.

$$x + 2y = 10 \qquad (1)$$
$$4 + 2y = 10$$
$$2y = 10 - 4$$
$$2y = 6$$
$$y = 3 \quad \textit{Ans.}$$

Example 8-44

Solve the pair of simultaneous equations given below for the value of a and b.

Solution

1. The equations are:

$$a + b = 5 \qquad (1)$$
$$2a + b = 7 \qquad (2)$$

In order to eliminate the unknown b, we must multiply equation (1) by -1.

$$-1(a + b = 5)$$
$$2a + b = 7$$

2. Multiplication gives

$$-a - b = -5$$
$$2a + b = 7$$

3. Add the equations.

$$a = 2 \quad \textit{Ans.}$$

4. Substitute 2 for a in equation (1).

$$a + b = 5 \qquad\qquad (1)$$
$$2 + b = 5$$
$$b = 5 - 2$$
$$b = 3 \qquad Ans.$$

Example 8-45

Solve the pair of simultaneous equations given below for the value of x and y.

Solution

1. The equations are:

$$5x + 2y = 4 \qquad\qquad (1)$$
$$4x - 3y = 17 \qquad\qquad (2)$$

The elimination of *either* unknown will solve the problem. However, since the signs of the unknown y are exactly what they should be, we shall attempt to eliminate y. It is very difficult to change a 2 into a 3 or a 3 into a 2 by multiplication. However, it is not necessary that the coefficients of y remain what they were in the original problem.

All that is necessary is that the coefficients be the same number, but with opposite signs.

We shall therefore change $+2y$ in equation (1) into $+6y$ by multiplying it by 3, and the $-3y$ in equation (2) into $-6y$ by multiplying it by 2.

$$3(5x + 2y = 4)$$
$$2(4x - 3y = 17)$$

2. Multiplication gives

$$15x + 6y = 12$$
$$8x - 6y = 34$$

3. Add the equations.

$$23x = 46$$

4. Solve for x.

$$x = 2 \qquad Ans.$$

5. Substitute this value for x in equation (1).

$$5x + 2y = 4 \qquad\qquad (1)$$
$$5(2) + 2y = 4$$
$$10 + 2y = 4$$
$$2y = 4 - 10$$
$$2y = -6$$
$$y = -3 \qquad Ans.$$

Self-Test 8-46

Solve the pair of simultaneous equations given below for the value of x and y.

Solution

1. The equations are:

$$2x + 3y = -8 \qquad (1)$$
$$7x + 4y = 11 \qquad (2)$$

2. We plan to eliminate one letter by _____ the equations.
3. The letter will disappear if the addition of the coefficients of that letter is equal to _____.
4. The addition of these coefficients will equal zero if
   a. The coefficients are _____ in value.
   b. The signs are _____.
5. The coefficients of x ___(do/do not)___ satisfy these conditions. The coefficients of y ___(do/do not)___ satisfy these conditions.
6. Yet, if we are going to solve the equations, the coefficients of one of these letters *must* be _____ and with _____ signs.
7. Let us try to attain this condition for the letter y.
8. The smallest number into which we can change both 3 and 4 by multiplication is _____.
9. If we change the $3y$ of equation (1) into $12y$, then we must change the $4y$ of equation (2) into _____.
10. To get $12y$, we must multiply equation (1) by _____ because only $4 \times 3y = 12y$. To get $-12y$, we must multiply equation (2) by _____ because only $-3 \times 4y = -12y$.
11. Indicate the plan by placing the entire equation in parentheses with the proper multiplier in front.

$$\underline{\qquad}(2x + 3y = -8)$$
$$\underline{\qquad}(7x + 4y = 11)$$

12. Multiply each equation by its multiplier to give

$$8x + 12y = \underline{\qquad}$$
$$\underline{\qquad} - 12y = -33$$

13. Add the equations.

$$-13x = \underline{\qquad}$$

14. Solve this equation.

$$x = \underline{\qquad} \quad Ans.$$

15. Substitute this value for x in equation (1).

$$2x + 3y = -8$$
$$2(\underline{\qquad}) + 3y = -8$$
$$\underline{\qquad} + 3y = -8$$
$$3y = -8 \underline{\qquad}$$
$$3y = \underline{\qquad}$$
$$y = \underline{\qquad} \quad Ans.$$

16. *Check*: If we are correct, then the substitution of these values in the original equations should make true statements. Substitute $x = 5$ and $y = -6$ in equation (1).

$$2x + 3y = -8 \qquad (1)$$
$$2(5) + 3(\underline{\quad(16)\quad}) = -8$$
$$\underline{\quad(17)\quad} - 18 = -8$$
$$\underline{\quad(18)\quad} = -8 \qquad Check$$

Substitute $x = 5$ and $y = -6$ in equation (2).

$$7x + 4y = 11 \qquad (2)$$
$$7(\underline{\quad(19)\quad}) + 4(-6) = 11$$
$$35 - \underline{\quad(20)\quad} = 11$$
$$\underline{\quad(21)\quad} = 11 \qquad Check$$

Exercises

Practice

Solve the following sets of equations:

1. $2x + y = 8$
   $x - y = 1$

2. $x + y = 6$
   $x - y = 4$

3. $2x + y = 12$
   $x + y = 7$

4. $x + y = 1$
   $x - y = 5$

5. $I_1 + I_2 = 8$
   $I_1 - 2I_2 = 2$

6. $x + y = 8$
   $x - y = 0$

7. $V - 4R = -1$
   $V - 2R = 3$

8. $x + 2y = 7$
   $x - 2y = 3$

9. $2I + 3V = 8$
   $4I - 3V = 34$

10. $7x + 4y = 26$
    $2x - 4y = -8$

11. $5a + 4b = -7$
    $a - 2b = 7$

12. $3I_1 - I_2 = 5$
    $2I_1 + 3I_2 = 18$

13. $2x - 3y = 7$
    $3x + 4y = 19$

14. $3I_1 + 7I_2 = 5$
    $2I_1 + 3I_2 = 5$

15. $3a + 4b = 58$
    $5a - 2b = 10$

16. $12I_2 + 3I_1 = 75$
    $4I_2 - I_1 = 7$

17. $4I_2 + I_3 = 25$
    $-3I_2 + 8I_3 = -45$

18. $20x - 7y = -36$
    $-4x - 2y = -20$

19. $6I_2 + 5I_3 = 40$
    $-4I_2 + 7I_3 = -6$

20. $5I_1 + 2I_2 = 36$
    $3I_1 - 4I_2 = -20$

(See CD-ROM for Test 8-8)

Solving Simultaneous Equations by Substitution

There are some pairs of equations which may be solved more efficiently by this method of substitution than by the addition method discussed in Job 8-11.

Example 8-47

Solve the pair of simultaneous equations given below for the value of x and y.

Solution

1. The equations are:

$$x = 4 \qquad (1)$$
$$2x + y = 11 \qquad (2)$$

2. Since x is given equal to 4, we may *substitute* this value for x in equation (2).

   a. Write the equation. $\qquad 2x + y = 11 \qquad (2)$

   b. Substitute 4 for x. $\qquad 2(4) + y = 11$

   c. Multiply. $\qquad 8 + y = 11$

   d. Transpose. $\qquad y = 11 - 8$

   e. Combine like terms. $\qquad y = 3 \qquad$ *Ans.*

Example 8-48

Solve the pair of simultaneous equations given below for the value of x and y.

Solution

1. The equations are:

$$y = 2x \qquad (1)$$
$$x + 2y = 20 \qquad (2)$$

2. Since y is given equal to $2x$, we may *substitute* this value for y in equation (2).

   a. Write the equation. $\qquad x + 2y = 20 \qquad (2)$

   b. Substitute $2x$ for y. $\qquad x + 2(2x) = 20$

   c. Multiply. $\qquad x + 4x = 20$

   d. Combine like terms. $\qquad 5x = 20$

   e. Solve the basic equation. $\qquad x = 4 \qquad$ *Ans.*

3. Substitute this value for x in equation (1) to find y.

   a. Write the equation. $\qquad y = 2x \qquad (1)$

   b. Substitute 4 for x. $\qquad y = 2(4)$

   c. Multiply. $\qquad y = 8 \qquad$ *Ans.*

Example 8-49

Solve the pair of simultaneous equations given below for the value of x and y.

Solution

1. The equations are:

$$3x + y = 6 \qquad (1)$$
$$y = 2x + 1 \qquad (2)$$

2. In equation (2), y is given equal to the *quantity* $2x + 1$, which is best written as $(2x + 1)$. We can substitute *this* value for y in equation (1).
   - *a.* Write the equation. $3x + y = 6$ (1)
   - *b.* Substitute $(2x + 1)$ for y. $3x + (2x + 1) = 6$
   - *c.* Remove parentheses. $3x + 2x + 1 = 6$
   - *d.* Transpose. $3x + 2x = 6 - 1$
   - *e.* Combine like terms. $5x = 5$
   - *f.* Solve the basic equation. $x = 1$ *Ans.*
3. Substitute this value for x in equation (2) to find y.
   - *a.* Write the equation. $y = 2x + 1$ (2)
   - *b.* Substitute 1 for x. $y = 2(1) + 1$
   - *c.* Multiply. $y = 2 + 1$
   - *d.* Combine like terms. $y = 3$ *Ans.*

Example 8-50

Solve the pair of simultaneous equations given below for the value of x and y.

Solution

1. The equations are:

$$2x + y = 6 \qquad (1)$$
$$3x + 0.5y = 5 \qquad (2)$$

2. Solve equation (1) so as to obtain a value for y expressed in terms of x.
   - *a.* Write the equation.

$$2x + y = 6 \qquad (1)$$

   - *b.* Transpose *all* quantities *except* y to the other side.

$$y = 6 - 2x$$

3. Since y is now expressed as the *quantity* $(6 - 2x)$, we may substitute *this* value for y in equation (2).
   - *a.* Write the equation. $3x + 0.5y = 5$ (2)
   - *b.* Substitute $(6 - 2x)$ for y. $3x + 0.5(6 - 2x) = 5$
   - *c.* Remove parentheses. $3x + 3 - 1x = 5$
   - *d.* Transpose. $3x - x = 5 - 3$

CALCULATOR HINT

A graphing calculator is able to show the solution to a simultaneous equation by drawing a graph of each of the two equations. The graph of each equation is a straight line. The point at which the graphs intersect represents one common set of values for the unknown variables.

e. Combine like terms. $2x = 2$

f. Solve the basic equation. $x = 1$ *Ans.*

4. Substitute this value for x in the equation found in step 2.

    a. Write the equation. $y = 6 - 2x$

    b. Substitute 1 for x. $y = 6 - 2(1)$

    c. Multiply. $y = 6 - 2$

    d. Combine like terms. $y = 4$ *Ans.*

Summary of Method

1. Write the equation.
2. Solve either equation for one of the unknowns in terms of the other.
3. Substitute this value in the other equation.
4. Solve the resulting equation.
5. Substitute the value found in step 4 in the equation found in step 2.
6. Solve the resulting equation for the value of the second unknown.
7. Check.

Self-Test 8-51

Solve the pair of simultaneous equations given below for the value of x and y.

Solution

The equations are:

$$5x + 3y = 6.5 \tag{1}$$
$$\underline{2x - y = 7} \tag{2}$$

1. Solve one of these equations for one letter in terms of the other. The better equation to use for this purpose is equation (———) in which we shall solve the equation for the letter ——— because the coefficient of y is already 1.

2. *a.* Write the equation.

$$2x - y = 7 \tag{2}$$

    b. Transpose to get $+y$ even if it is on the right side.

$$2x - \text{———} = y$$
or
$$y = \text{———}$$

3. Since y is now expressed as the quantity ———, we may now ——— this value for y in equation (1).

    a. Write the equation. $5x + 3y = 6.5$ (1)

    b. Substitute. $5x + 3(\text{———}) = 6.5$

    c. Remove parentheses.

$$5x + \text{———} - \text{———} = 6.5$$

    d. Transpose. $5x + 6x = 6.5 + \text{———}$

    e. Combine like terms. $11x = \text{———}$

f. Solve the basic equation. $x = \dfrac{27.5}{?}$

g. Divide. $x =$ _____ Ans.

4. Substitute this value for x in the equation found in step 2.

    a. Write the equation. $y = 2x - 7$

    b. Substitute. $y = 2(\underline{\hspace{1.5cm}}) - 7$

    c. Multiply. $y =$ _____ $- 7$

    d. Combine like terms. $y =$ _____ Ans.

5. *Check*: Substitute these values for x and y in the original equations (1) and (2).

Substitute $x =$ _____ and $y =$ _____ in equation (1).

$$5x + 3y = 6.5 \tag{1}$$

$$5(\underline{\hspace{1.5cm}}) + 3(\underline{\hspace{1.5cm}}) = 6.5$$

$$\underline{\hspace{1.5cm}} - 6 = 6.5$$

$$\underline{\hspace{1.5cm}} = 6.5 \quad \textit{Check}$$

Substitute $x = 2.5$ and $y = -2$ in equation (2).

$$2x - y = 7 \tag{2}$$

$$2(\underline{\hspace{1.5cm}}) - (\underline{\hspace{1.5cm}}) = 7$$

$$5\underline{\hspace{1.5cm}} = 7$$

$$\underline{\hspace{1.5cm}} = 7 \quad \textit{Check}$$

Exercises

Practice

Solve the following sets of equations:

1. $y = 3$
    $4x + y = 23$

2. $4y - x = 9$
    $x = 3$

3. $x = 5$
    $4x - y = 18$

4. $2x + y = 10$
    $y = 3x$

5. $y = 3x + 1$
    $2x + y = 11$

6. $3x + 2y = 17$
    $y = x - 4$

7. $2y - x = 1$
    $x = y + 2$

8. $2x + 5y = 31$
    $x + y = 8$

9. $y - x = 3$
    $2x + y = 6$

10. $2x - y = 7$
    $x - y = 3$

11. $x + 2y = 9$
    $x + 3y = 13$

12. $11x - y = 18$
    $5x + 3y = 22$

Review of Algebra

1. *Like* terms are those which contain the _____ letters.
2. *Unlike* terms are those in which the letters are _____.

3. Adding signed numbers.
   a. To add two signed numbers of the *same* sign, _____ the numbers and use the common sign.
   b. To add two signed numbers of *different* signs, _____ the smaller from the larger and use the sign of the _____ number.
4. To combine a string of signed numbers means to _____ the individual signed numbers using the rules for _____.
5. Multiplying and dividing signed numbers.
   a. Two signed numbers with the *same* sign yield a __(plus/minus)__ answer.
   b. Two signed numbers with *different* signs yield a __(plus/minus)__ answer.
6. A parenthesis is removed by _____ the signed number in front of the parenthesis by each and every quantity within the parentheses.
7. To combine several quantities involving *unlike* terms, combine each group of _____ terms separately.
8. Solving equations.
   a. Fractional equations containing one fraction.
      (1) Transpose any other number to the other side of the equality sign.
      (2) Combine like terms.
      (3) Make two fractions by placing any whole numbers or decimals over the number _____.
      (4) Cross-multiply.
      (5) Solve the basic equation.
   b. Fractional equations containing two fractions.
      (1) Cross-_____.
      (2) Solve the basic equation.
   c. Fractional equations with more than two fractions.
      (1) Find the least common _____.
      (2) Multiply each term of the equation by this _____.
      (3) Transpose *all* unknowns to one side of the _____ sign and *all* numbers to the other side.
      (4) Combine the _____ terms on each side separately.
      (5) Solve the basic equation.
   d. Equations with parentheses.
      (1) Remove parentheses.
      (2) _____ *all* unknowns to one side of the equality sign, and *all* numbers to the other side.
      (3) _____ the like terms on each side separately.
      (4) Solve the basic equation.
   e. Simultaneous equations solved by addition.
      (1) In each equation, transpose *all* unknowns to one side of the equality sign, and *all* _____ to the other side.
      (2) Combine the _____ terms on each side separately.
      (3) Arrange the _____ letters under each other in both equations.
      (4) Eliminate one letter by _____ one or both equations by the proper number and sign so as to obtain the _____ coefficient with _____ signs.
      (5) _____ the two equations to eliminate the letter.
      (6) Solve the resulting equation, which will now contain only _____ letter.

(7) Substitute this value in *either* of the two _____ equations.

(8) Solve this equation to obtain the value of the second letter.

f. Simultaneous equations solved by substitution.

(1) Solve either equation for one of the unknowns in terms of the _____ unknown.

(2) Substitute this expression for the letter in the other _____.

(3) Solve the resulting equation.

(4) Substitute the value found in step 3 in the equation found in step _____.

(5) Solve the resulting equation for the value of the _____ unknown.

(See CD-ROM for Test 8-9)

Assessment

1. Combine the following like terms:
   $0.4Xy + 16.2Xy + 23.02Xy + 10.52Xy$.
2. Combine the terms in the following expression:
   $4R + 3aB + 5T + 8T + 3.2R + 2aB$.
3. Solve the following equation:
   $b + (3b + 2) = 34$.
4. Solve the following equation:
   $[(5\ T)/3] + 17 = 77$.
5. Combine the following terms:
   $-3x + 14y - 22z - 4x + 2Y + 12z$.
6. Solve the following equation:
   $6W + 49 = -11 - 13W$.

7. Solve the following equation:
   $16K + 42 = 20K + 7$.
8. Remove the parentheses and collect like terms:
   $3(5R + 3) + 12(4 + 2.5R)$.
9. Solve the following equation:
   $0.75(X) + 0.25(2.25) = 1.50$.
10. Solve the following equation:
    $x/5 + x/8 = 26$.
11. Solve the following two sets of equations:
    $4P + 3B = 23.75$
    $2P + B\ = 9.75$

INTERNET
ACTIVITIES

Internet Activity A

View the following site to learn about graphing simultaneous equations.
http://www.scas.bcit.bc.ca/scas/math/examples/ary_7_2/backgd3b.htm.
From the information provided at the site, answer the following questions:
When the two equations $1x + 1y = 4$ and $2x - 3y = 6$ are graphed, what are the x and y values at the point at which the two lines intersect?

Internet Activity B

Using the Web site above, explain what is significant about the point at which the lines cross.

Chapter 8 Solutions to Self-Tests and Reviews

Self-Test 8-2

1. 10
2. R
3. $7K$
4. $9I_2$

Self-Test 8-6

1. 8
2. 4
3. $7I$
4. 9
5. $5R$
6. I_1
7. 5
8. $9I_2 + 7$

Self-Test 8-12

1. $4x$
2. -2
3. 12
4. 3

Self-Test 8-13

1. $3x$
2. $+6.2$
3. 7.5
4. 2.5

Self-Test 8-14

1. sum
2. x
3. $6x$
4. 10
5. 48
6. 8
7. $3x$
8. 8, 24

Self-Test 8-15

1. R_1
2. 3
3. sum
4. V_2
5. $3R_1$
6. $3R_1$

7. 60

Self-Test 8-18

2. $2\frac{3}{4}x$
3. $\frac{11x}{4}$
4. -3
5. 11
6. 44
7. 4

Self-Test 8-19

1. sum
2. 11
3. $1\frac{1}{4}$
4. 1
5. 10
6. 40
7. 8

Self-Test 8-20

1. sum
2. R
3. $3R$
4. 400
5. 1200
6. 400

Self-Test 8-24

1. $-5x$
2. $-3y$
3. $-3y$

Self-Test 8-35

1. $+2y$
2. $5y$

Self-Test 8-37

1. parallel
2. R_2
3. R_2, 12
4. $12R_2$
5. $3R_2$

6. $3R_2$
7. $9R_2$
8. 4

Self-Test 8-38

1. $2x$
2. $2x + 90$
3. iron
4. x
5. $2x + 90$
6. $4x$
7. $4x$
8. -280
9. 280
10. 280
11. 560
12. 650

Self-Test 8-46

2. adding
3. zero
4. a. equal
   b. opposite
5. do not, do not
6. equal, opposite
8. 12
9. $-12y$
10. 4, -3
11. 4, -3
12. -32, $-21x$
13. -65
14. $+5$
15. 5, 10, -10, -18, -6
16. -6
17. 10
18. -8
19. 5
20. 24
21. 11

Self-Test 8-51

1. 2, y
2. b. 7, $2x - 7$
3. $(2x - 7)$, substitute
   b. $2x - 7$

c. $6x$, 21
d. 21
e. 27.5
f. 11
g. 2.5
4. b. 2.5
   c. 5
   d. -2
5. 2.5, -2, 2.5, -2, 12.5, 6.5 (equation 1), 2.5, -2, $+2$, 7 (equation 2)

Review 8-13

1. same
2. different
3. a. add
   b. subtract, larger
4. add, addition
5. a. plus
   b. minus
6. multiplying
7. like
8. a. (3) 1
   b. (1) multiply
   c. (1) denominator
      (2) LCD
      (3) equality
      (4) like
   d. (2) Transpose
      (3) Combine
   e. (1) numbers
      (2) like
      (3) same
      (4) multiplying, same, different or opposite
      (5) Add
      (6) one
      (7) original
   f. (1) other
      (2) equation
      (4) 1
      (5) second

9 Kirchhoff's Laws

chapter

Learning Objectives

1. In your own words, state Kirchhoff's current law.
2. In your own words, state Kirchhoff's voltage law.
3. Apply Kirchhoff's voltage law to find the unknown values of voltage and current in a series circuit.
4. Apply Kirchhoff's current law to find the unknown values of voltage, current, and resistance in a parallel circuit.
5. Apply Kirchhoff's laws to solve for the unknown values of voltage and current in complex electric circuits.
6. Simplify the analysis of an unbalanced bridge circuit by converting a delta circuit to a star circuit.
7. Apply Thevenin's theorem in solving for an unknown value of voltage or current.
8. Apply Norton's theorem in solving for an unknown value of voltage or current.

JOB 9-1

Kirchhoff's Current Law

RULE At any point in an electric circuit, the sum of the currents that enter a junction is equal to the sum of the currents that leave it.

$$\Sigma I_e = \Sigma I_l \qquad \text{Formula 9-1}$$

where ΣI_e = summation of all currents *entering* a point
ΣI_l = summation of all currents *leaving* a point

In Fig. 9-1, three resistors meet at point X. The figure shows three ways in which this condition may be drawn. In each, the current I_3 that *enters* point X is equal to

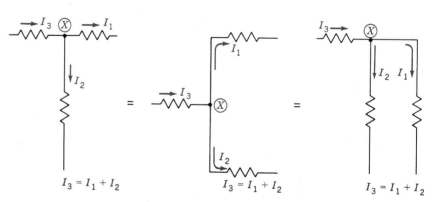

Figure 9-1 The current that enters a point is equal to the current that leaves it.

the *sum* of the currents I_1 and I_2 that *leave* the junction. This fact may be expressed as the formula $I_3 = I_1 + I_2$. You will probably recognize this as the formula for the total current in a parallel circuit.

Finding the Current in Branch Wires

Example 9-1

Find the current in R_2 of the circuit shown in Fig. 9-2.

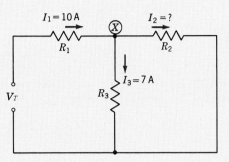

Figure 9-2

Solution

At point X,

$$\Sigma I_e = \Sigma I_l \qquad (9\text{-}1)$$

$$I_1 = I_2 + I_3$$
$$10 = I_2 + 7$$
$$10 - 7 = I_2$$
$$3 = I_2 \quad \text{or} \quad I_2 = 3 \text{ A} \qquad Ans.$$

Example 9-2

Find the current I_2 in the circuit shown in Fig. 9-3.

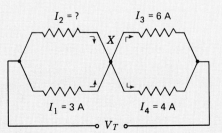

Figure 9-3

Solution

The current that enters point X is equal to the current that leaves it. Since I_1 and I_2 enter point X, and I_3 and I_4 leave point X, we can apply Eq. (9-1).

$$\Sigma I_e = \Sigma I_l \qquad\qquad (9\text{-}1)$$

$$I_1 + I_2 = I_3 + I_4$$
$$3 + I_2 = 6 + 4$$
$$I_2 = 10 - 3 = 7 \text{ A} \qquad Ans.$$

JOB 9-2

Kirchhoff's Voltage Law

 RULE The voltage supplied to a circuit is always equal to the sum of the voltage drops across the individual parts of the circuit.

This fact was used when we studied series circuits and was expressed as the formula $V_T = V_1 + V_2 + V_3$. Kirchhoff's voltage law explains that the sum of the voltage *gains* supplied to a circuit from a generator, battery, etc., is always equal to the sum of the voltage *drops* (V_1, V_2, V_3, etc.) in the circuit. Since the gains are equal to the losses, the total sum of the gains plus the losses must equal zero. The quantities involved must be added *algebraically* in order to include the *direction* of the various voltages. Kirchhoff's voltage law may now be expressed as follows.

 RULE The *algebraic* sum of all the voltages in any complete electric circuit is equal to zero.

$$V_1 + V_2 + V_3 + \text{etc.} = 0 \qquad \text{Formula 9-2}$$

Direction of voltages. If two equal voltages oppose each other, they may be expressed as $+V_1$ and $-V_2$, where the minus sign in $-V_2$ represents the opposition to $+V_1$. Their algebraic sum would be indicated as

$$+V_1 + (-V_2)$$
or
$$V_1 - V_2$$

Therefore, as we go around a circuit, adding the voltages, we must be very careful to indicate voltage *gains* as *plus* values and voltage *losses* as *minus* values. The following procedure will determine the sign of the voltage.

1. Note the direction of the electron flow around the circuit. Electrons *leave* the negative terminal of the voltage source and *enter* the positive terminal. This is illustrated in Fig. 9-4a.

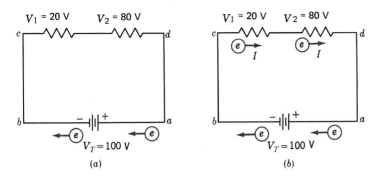

Figure 9-4 Electrons *leave* the negative terminal of a voltage source and *enter* the positive terminal.

2. Mark the circuit to indicate the electron flow through *each* resistance. This is shown in Fig. 9-4b. In more complicated circuits, the direction may be unknown. In this event, we shall *assume* a direction. If our assumption is incorrect, the error will be indicated by a minus sign when the resulting equation is solved. A minus current therefore tells us that the true direction is *opposite* to the direction originally assumed. The value of the current will not be affected.

3. We intend to move around a circuit to combine all the voltages encountered. We shall start at some convenient point and move around the *complete* circuit until we return to the original starting point. The direction of this movement is usually in the direction of the electron flow.

4. *Source voltages* are *positive* (+) if we move through them from the positive (+) to the negative (−) terminal. *Source voltages* are *negative* (−) if we move through them from the negative (−) to the positive (+) terminal. Therefore, in Fig. 9-4b, if we start at point *a* and move around the circuit in the direction *abcd*, we shall go through V_T from + to − and V_T will equal +100 V. However, if we start at point *a* and move around the circuit in the direction *adcb*, we shall go through V_T from − to + and V_T will equal −100 V.

5. The voltage across any resistance will be *negative* if we go through it *in the direction of the assumed electron flow*. Thus, in Fig. 9-4b, if we go through the circuit in the direction *abcd*, $V_1 = -20$ V and $V_2 = -80$ V.

6. The voltage in any resistance will be *positive* if we go through it in a direction *opposite* to the assumed electron flow. Thus, in Fig. 9-4b, if we go through the circuit in the direction *adcb*, $V_2 = +80$ V and $V_1 = +20$ V.

Example 9-3

Find the signs of the voltages encountered when tracing the circuits shown in Fig. 9-5a on page 288.

Solution

1. Determine the direction of the electron flow around the circuit. Since electrons leave the negative terminal, the direction of flow will be *up* to point *c* from both V_1 and V_4. At point *c*, since the electrons that enter point *c* must also leave it, the electrons must flow *out* of point *c up* to point *g* and continue around to point *f*. At *f*, the current divides, part going to point *a* and the rest to point *e*. This is indicated in Fig. 9-5b.

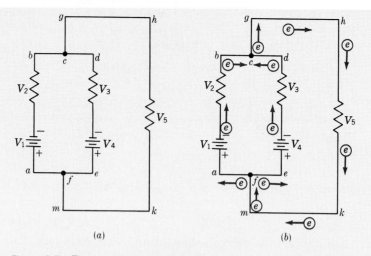

Figure 9-5 Electrons leave the negative terminal of a voltage source.

2. Determine the direction of the voltages around the circuit *abcdef.*
   a. V_1 is +, since we go through it from + to −.
   b. V_2 is −, since we go through it in the direction of the electron flow.
   c. V_3 is +, since we go through it in a direction *opposite* to the electron flow.
   d. V_4 is −, since we go through it from − to +.
3. Determine the direction of the voltages around the circuit *abcghkmf.*
   a. V_1 is +, since we go through it from + to −.
   b. V_2 is −, since we go through it in the direction of the electron flow.
   c. V_5 is −, since we go through it in the direction of the electron flow.
4. Determine the direction of the voltages around the circuit *fedcghkm.*
   a. V_4 is +, since we go through it from + to −.
   b. V_3 is −, since we go through it in the direction of the electron flow.
   c. V_5 is −, since we go through it in the direction of the electron flow.

Self-Test 9-4

Find the signs of the voltages encountered when tracing the circuit *abcdefa* shown in Fig. 9-6.

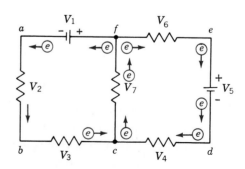

Figure 9-6 The direction of the electron flow is assumed.

Solution

1. Assume that the direction of the electron flow is that shown on Fig. 9-6.
2. Determine the signs of the voltages. The direction of our trace is in a counterclockwise direction starting from point a.

   a. V_2 and V_3 are both —————, since we are going through them both in the —————(same/opposite)————— direction as the electron flow.

   b. V_4 and V_6 are both —————, since we are going through them both in the —————(same/opposite)————— direction as the electron flow.

   c. V_5 is —————, since we go through it from ————— to —————.

   d. V_1 is —————, since we go through it from ————— to —————

Using Kirchhoff's Laws in Series Circuits

Example 9-5

Find the current in the series circuit shown in Fig. 9-7a.

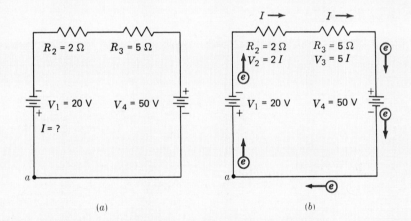

(a)   (b)

Figure 9-7   The sources are connected so as to aid each other.

Solution

1. Determine the direction of the electron flow. V_1 and V_4 are connected so that the electron flow comes out of one source and continues through the other in the same direction. This connection of sources aiding each other causes an electron flow moving *clockwise* around the circuit.
2. Mark this direction on the circuit as shown in Fig. 9-7b.
3. By Ohm's law, the voltages in R_2 and R_3 will equal $2I$ and $5I$, respectively.
4. Determine the direction of the voltages. Trace the circuit in a clockwise direction starting at point a.

   a. V_1 and V_4 are both $+$, since we go through them from $+$ to $-$.

   b. V_2 and V_3 are both $-$, since we go through them in the direction of the electron flow.

> ### Hint for Success
>
> Kirchhoff's current and voltage laws can be applied to all types of electronic circuits, not just those containing dc voltage sources and resistors. For example, Kirchhoff's laws can be applied when analyzing circuits containing electronic devices such as diodes and transistors.

5. Apply Kirchhoff's voltage law. The *algebraic* sum of all the voltages around a circuit will equal zero. Add the voltages starting at point *a*.

$$V_1 + V_2 + V_3 + V_4 = 0$$
$$+20 - 2I - 5I + 50 = 0$$
$$-2I - 5I = -20 - 50$$
$$-7I = -70$$
$$7I = 70$$
$$I = 10 \text{ A} \qquad Ans.$$

6. *Check*: If we trace the circuit in a counterclockwise direction starting from point *a*, the voltages through the sources are now negative and the voltages through the resistors are positive, since we are now going through them in a direction *opposite* to the electron flow.

$$V_4 + V_3 + V_2 + V_1 = 0$$
$$-50 + 5I + 2I - 20 = 0$$
$$5I + 2I = 50 + 20$$
$$7I = 70$$
$$I = 10 \text{ A} \qquad Check$$

Example 9-6

Find the current in the series circuit shown in Fig. 9-8a.

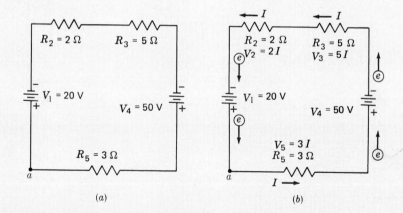

Figure 9-8   The sources are connected so as to oppose each other.

Solution

1. Determine the direction of the electron flow. Voltage V_1 and V_4 are connected so that they *oppose* each other. Voltage V_4 (50 V) is stronger than V_1 (20 V), and the electron flow is therefore governed by V_4. This will cause the electron flow to be in a *counterclockwise* direction.
2. Mark this direction on the circuit as shown in Fig. 9-8b.
3. By Ohm's law, the voltages in R_2, R_3, and R_5 will equal $2I$, $5I$, and $3I$, respectively.

4. Determine the direction of the voltages. Trace the circuit in a counter-clockwise direction starting at point a. V_5, V_3, and V_2 are all $-$, since we go through them in the direction of the electron flow. V_4 is $+$, since we go through it from $+$ to $-$. $V1$ is $-$, since we go through it from $-$ to $+$.
5. Apply Kirchhoff's voltage law.

$$V_5 + V_4 + V_3 + V_2 + V_1 = 0$$
$$-3I + 50 - 5I - 2I - 20 = 0$$
$$-3I - 5I - 2I = 20 - 50$$
$$-10I = -30$$
$$10I = 30$$
$$I = 3 \text{ A} \qquad Ans.$$

Example 9-7

Use the circuit shown in Fig. 9-9a. Find (a) the total current and (b) the voltage drop across each resistance.

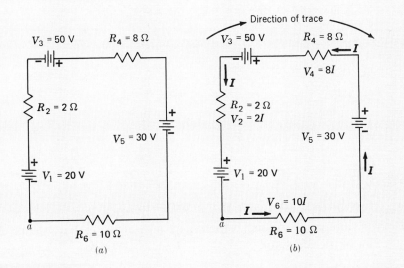

Figure 9-9

Solution

a. Find the total current.

1. Determine the direction of the electron flow. Voltages V_1 and V_3 aid each other, and since their total (70 V) is greater than the opposing V_5 (30 V), the flow will be in a counterclockwise direction.
2. Mark this direction on the circuit as shown in Fig. 9-9b.
3. By Ohm's law, the voltages in R_2, R_4, and R_6 will equal $2I$, $8I$, and $10I$, respectively.
4. Determine the direction of the voltages. Trace the circuit in a *clockwise* direction starting at point a.

   V_1 and V_3 are $-$, since we go through them from $-$ to $+$.
   V_5 is $+$, since we go through it from $+$ to $-$.

V_2, V_4, and V_6 are all $+$, since we go through them in a direction *opposite* to the electron flow.

5. Apply Kirchhoff's voltage law.

$$V_1 + V_2 + V_3 + V_4 + V_5 + V_6 = 0$$
$$-20 + 2I - 50 + 8I + 30 + 10I = 0$$
$$2I + 8I + 10I = 20 + 50 - 30$$
$$20I = 40$$
$$I = 2 \text{ A} \qquad Ans.$$

b. Find the voltage drop across each resistance.

$V_2 = IR_2$	$V_4 = IR_4$	$V_6 = IR_6$
$V_2 = 2 \times 2$	$V_4 = 2 \times 8$	$V_6 = 2 \times 10$
$V_2 = 4 \text{ V}$	$V_4 = 16 \text{ V}$	$V_6 = 20 \text{ V} \qquad Ans.$

Check: The sum of the voltages around the complete circuit should equal zero. Add the voltages in a counterclockwise direction starting from point *a*.

$$V_6 + V_5 + V_4 + V_3 + V_2 + V_1 = 0$$
$$-20 - 30 - 16 + 50 - 4 + 20 = 0$$
$$0 = 0 \qquad Check$$

Self-Test 9-8

In Fig. 9-10, R_2 represents the internal resistance of the generator. Find (*a*) the total current in the circuit, (*b*) the voltage drop across each resistance, and (*c*) the emf of the generator.

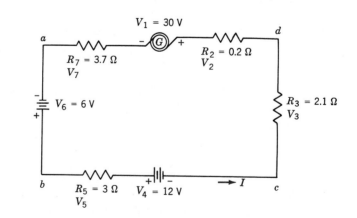

Figure 9-10 R_2 represents the internal resistance of the generator.

Solution

a. Find the total current.

1. Determine the direction of the electron flow. V_1 and V_4 _(aid/oppose)_ each other, and since their total $(30 + 12 = 42)$ is greater than V_6 (6 V), which _(aids/opposes)_ them, the flow will be in a _(clockwise/counterclockwise)_ direction.

2. Mark this direction on the circuit diagram.

3. By Ohm's law, the voltages in the resistances are

$$V_2 = 0.2I \qquad\qquad V_3 = \underline{\hspace{2cm}}$$
$$V_5 = \underline{\hspace{2cm}} \qquad\qquad V_7 = \underline{\hspace{2cm}}$$

4. Determine the direction of the voltages. Trace the circuit in the same direction as the electron flow, which will be _____. Start at point a.

   V_6 is _____, since we go through it from _____ to _____.

   V_4 and V_1 are both _____, since we go through them both from _____ to _____.

   V_5, V_3, V_2, and V_7 are all _____, since we go through them all in the _(same/opposite)_ direction as the electron flow.

5. Apply Kirchhoff's voltage law.

$$V_6 + V_5 + V_4 + V_3 + V_2 + V_1 + V_7 = \underline{\hspace{2cm}}$$
$$-6 - 3I + 12 - 2.1I - 0.2I + 30 - \underline{\hspace{1.5cm}} = 0$$
$$36 - \underline{\hspace{1.5cm}} = 0$$
$$36 = \underline{\hspace{1.5cm}}$$
$$I = \underline{\hspace{1.5cm}} \text{ A} \qquad Ans.$$

b. Find the voltage drop across each resistance.

$$V_2 = IR_2 \qquad\qquad\qquad V_3 = IR_3$$
$$V_2 = 4 \times 0.2 = 0.8\text{ V} \qquad V_3 = 4 \times 2.1 = \underline{\hspace{1cm}(1)\hspace{1cm}}\text{ V}$$
$$V_5 = IR_5 \qquad\qquad\qquad V_7 = IR_7$$
$$V_5 = 4 \times 3 = 12\text{ V} \qquad V_7 = 4 \times \underline{\hspace{0.8cm}(2)\hspace{0.8cm}} = \underline{\hspace{0.8cm}(3)\hspace{0.8cm}}\text{ V}$$

Check: The sum of the voltages around the circuit should equal _(4)_. Add the voltages in a clockwise direction starting from point a. In this direction, R_7, R_2, R_3, and R_5 will all be _(5)_ because we go through them in a direction _(6)_ to the flow of the electrons.

V_1 and V_4 will both be _(7)_ because we go through them from _(8)_ to _(9)_.

V_6 will be _(10)_ because we go through it from _(11)_ to _(12)_.

$$V_7 + V_1 + V_2 + V_3 + V_4 + V_5 + V_6 = 0$$
$$14.8 - 30\underline{\hspace{0.8cm}(13)\hspace{0.8cm}} + 8.4 - 12 + 12 + 6 = 0$$
$$0 = 0 \qquad Check$$

c. Find the emf of the generator.

$$\text{emf} = V_1 + V_2 = 30 + 0.8 = \underline{\hspace{1.5cm}}\text{ V} \qquad Ans.$$

Practice

1. In the circuit shown in Fig. 9-11, find (*a*) the total current and (*b*) the voltage drop across each resistor.

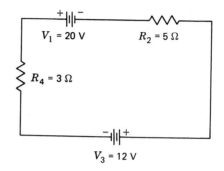

Figure 9-11

2. In the circuit shown in Fig. 9-12, find (*a*) the total current and (*b*) the voltage drop across each resistor.

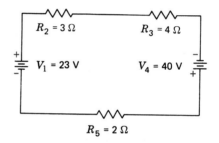

Figure 9-12

3. Use a circuit similar to that shown in Fig. 9-12. $V_1 = 20$ V, $R_2 = 6$ Ω, $R_3 = 4$ Ω, $V_4 = 16$ V, and $R_5 = 5$ Ω.
4. Use a circuit similar to that shown in Fig. 9-12. $V_1 = 45$ V, $R_2 = 2$ Ω, $R_3 = 4$ Ω, $R_5 = 6$ Ω, and $V_4 = 9$ V with the polarity reversed.
5. In the circuit shown in Fig. 9-13, find (*a*) the total current and (*b*) the voltage drop across each resistor.

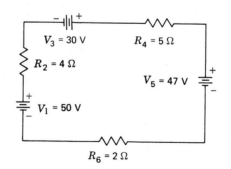

Figure 9-13

6. In the circuit shown in Fig. 9-14, find (*a*) the total current and (*b*) the voltage drop across each resistor.

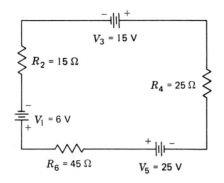

$V_3 = 15$ V

$R_2 = 15\ \Omega$

$R_4 = 25\ \Omega$

$V_1 = 6$ V

$R_6 = 45\ \Omega$

$V_5 = 25$ V

Figure 9-14

7. Find the total current flowing in the circuit shown in Fig. 9-15. Be sure to include the internal resistance of the batteries and generator.

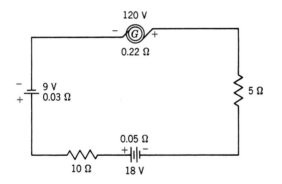

120 V

0.22 Ω

9 V
0.03 Ω

5 Ω

0.05 Ω

10 Ω   18 V

Figure 9-15   The circuit includes the internal resistances of the voltage sources.

8. A current of 4 A flows in the circuit shown in Fig. 9-16. Find the value of *R*.

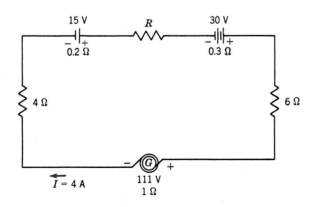

15 V   *R*   30 V

0.2 Ω   0.3 Ω

4 Ω   6 Ω

$I = 4$ A   111 V
               1 Ω

Figure 9-16

Using Kirchhoff's Laws in Parallel Circuits

Example 9-9

Find all missing values of voltage, current, and resistance in the circuit shown in Fig. 9-17.

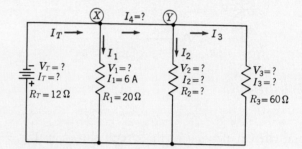

Figure 9-17   Using Kirchhoff's first law in parallel circuits.

Solution

1. Start at the point where the greatest amount of information is available. This will be branch R_1. Find V_1.

$$V_1 = I_1 \times R_1$$
$$V_1 = 6 \times 20 = 120 \text{ V} \qquad Ans.$$

2. Find the other voltages. Since this is a parallel circuit,

$$V_T = V_1 = V_2 = V_3 = 120 \text{ V} \qquad Ans.$$

3. Find I_T and I_3 by Ohm's law.

$$V_T = I_T \times R_T$$
$$120 = I_T \times 12$$
$$I_T = 10 \text{ A}$$
$$V_3 = I_3 \times R_3$$
$$120 = I_3 \times 60$$
$$I_3 = 2 \text{ A}$$

4. Find the current I_2 by applying Kirchhoff's current law.
   a. Assume the direction of the currents at point X to be as shown on the diagram. $I_T = 10$ A *entering* point X, $I_1 = 6$ A *leaving* point X, and $I_4 = ?$ *leaving* point X.

$$\Sigma I_e = \Sigma I_l \qquad\qquad (9\text{-}1)$$
$$I_T = I_4 + I_1$$
$$10 = I_4 + 6$$
$$10 - 6 = I_4$$
$$I_4 = 4 \text{ A } leaving \text{ point } X$$

b. At point Y, I_4 enters the point, and I_2 and I_3 leave the point.

$$\Sigma I_e = \Sigma I_l \qquad (9\text{-}1)$$
$$I_4 = I_2 + I_3$$
$$4 = I_2 + 2$$
$$I_2 = 2 \text{ A } \textit{leaving} \text{ point Y} \qquad \textit{Ans.}$$

5. Find R_2 by Ohm's law.

$$V_2 = I_2 \times R_2$$
$$120 = 2 \times R_2$$
$$R_2 = 60 \ \Omega \qquad \textit{Ans.}$$

Example 9-10

Find the total voltage of the circuit shown in Fig. 9-18.

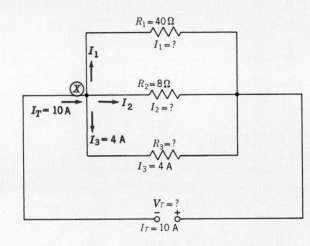

Figure 9-18

Solution

The total voltage V_T in this parallel circuit will be equal to any of the branch voltages. V_1 or V_2 may be found if we can determine the value of either I_1 or I_2.

1. Find I_2. Assume the direction of the currents to be as shown on the diagram. At point X, by Kirchhoff's current law,

$$\Sigma I_l = \Sigma I_e \qquad (9\text{-}1)$$
$$4 + I_1 + I_2 = 10$$
$$I_2 = 10 - 4 - I_1$$
or $$I_2 = 6 - I_1 \qquad (1)$$

2. Apply Kirchhoff's voltage law. Trace around the circuit through R_1 in a clockwise direction starting at V_T.

$$+V_T - 40I_1 = 0$$
or $$V_T = 40I_1 \qquad (2)$$

Trace the circuit again, but this time through R_2.

$$+V_T - 8I_2 = 0$$
or
$$V_T = 8I_2 \qquad (3)$$

3. In equation (1), we found that $I_2 = 6 - I_1$. Substitute this value for I_2 in equation (3).

$$V_T = 8I_2 \qquad (3)$$
$$V_T = 8(6 - I_1)$$
$$V_T = 48 - 8I_1 \qquad (4)$$

4. Set the values for V_T found in equations (2) and (4) equal to each other.

$$40I_1 = 48 - 8I_1$$
$$40I_1 + 8I_1 = 48$$
$$48I_1 = 48$$
$$I_1 = 1 \text{ A} \qquad Ans.$$

5. Find V_1 by Ohm's law.

$$V_1 = I_1 \times R_1 = 1 \times 40 = 40 \text{ V}$$

6. Find V_T. Since this is a parallel circuit,

$$V_T = V_1 = V_2 = V_3 = 40 \text{ V} \qquad Ans.$$

Self-Test 9-11

A voltage source supplies 12 A to a parallel combination of 60 Ω and 20 Ω, as shown in Fig. 9-19. Find I_1, I_2, and V_T.

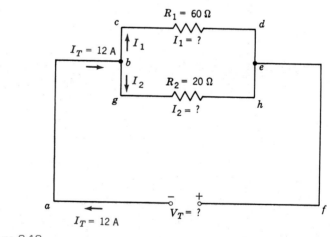

Figure 9-19

Solution

1. Apply Kirchhoff's voltage law. Trace around the circuit in the direction *fabcdef*.

$$V_T - 60I_1 = 0$$
$$V_T = \underline{\hspace{2cm}} \tag{1}$$

Trace around the circuit in the direction *fabghef*.

$$\underline{\hspace{3cm}} - 20I_2 = 0$$
$$V_T = \underline{\hspace{2cm}} \tag{2}$$

These two expressions for V_T may be set $\underline{\hspace{2cm}}$ to each other, which gives

$$20I_2 = \underline{\hspace{2cm}}$$
$$I_2 = \frac{60I_1}{?}$$
$$I_2 = \underline{\hspace{2cm}} I_1 \tag{3}$$

2. Now apply Kirchhoff's current law.

$$\Sigma I_l = \Sigma I_e \tag{9-1}$$

At point *b*,

$$I_1 + I_2 = \underline{\hspace{2cm}} \tag{4}$$

Now, according to equation (3), we can substitute $3I_1$ for I_2 in equation (4). This will give

$$I_1 + \underline{\hspace{2cm}} = 12$$
$$\underline{\hspace{2cm}} = 12$$
$$I_1 = \underline{\hspace{2cm}} A \qquad Ans.$$

Now we can go back to equation (3) to find I_2.

$$I_2 = 3I_1$$
$$I_2 = 3 \times \underline{\hspace{2cm}} = \underline{\hspace{2cm}} A \qquad Ans.$$

3. Find V_2. By Ohm's law,

$$V_2 = I_2 \times R_2$$
$$V_2 = 9 \times \underline{\hspace{2cm}} = \underline{\hspace{2cm}} V$$

But in a parallel circuit,

$$V_2 = V_T \tag{5-1}$$
$$V_T = \underline{\hspace{2cm}} V \qquad Ans.$$

Note: You have already learned a much simpler method for solving this problem. Can you recall it? If not, see Job 5-9.

Exercises

Practice

1. In a circuit similar to that shown in Fig. 9-17, $R_T = 6\,\Omega$, $I_1 = 9$ A, $R_1 = 10\,\Omega$, and $R_3 = 60\,\Omega$. Find all missing values of voltage, current, and resistance.

2. In a circuit similar to that shown in Fig. 9-17, $R_T = 8\ \Omega$, $I_1 = 5$ A, $I_2 = 1$ A, and $R_2 = 80\ \Omega$. Find all missing values of voltage, current, and resistance.
3. In a circuit similar to that shown in Fig. 9-17, $R_T = 15\ \Omega$, $I_1 = 10$ A, $R_1 = 30\ \Omega$, and $R_3 = 40\ \Omega$. Find all missing values of voltage, current, and resistance.
4. In a circuit similar to that shown in Fig. 9-18, $I_T = 40$ A, $R_1 = 4\ \Omega$, $R_2 = 12\ \Omega$, and $I_3 = 20$ A. Find the total voltage V_T and R_3.
5. In a circuit similar to that shown in Fig. 9-18, $I_T = 39$ A, $R_1 = 12\ \Omega$, $R_2 = 9\ \Omega$, and $I_3 = 18$ A. find the total voltage V_T and R_3.
6. In a circuit similar to that shown in Fig. 9-18, $I_T = 15$ A, $I_1 = 9$ A, $R_2 = 10\ \Omega$, and $R_3 = 30\ \Omega$. Find the total voltage V_T and R_1.
7. In a circuit similar to that shown in Fig. 9-19, $R_1 = 4\ \Omega$, $R_2 = 12\ \Omega$, and $I_T = 2.6$ A. Find I_1, I_2, and V_T.
8. In a circuit similar to that shown in Fig. 9-19, $R_1 = 100\ \Omega$, $R_2 = 45\ \Omega$, and $I_T = 7.975$ A. Find I_1, I_2, and V_T.
9. Check Prob. 7 using the method outlined in Job 5-9.
10. Check Prob. 8 using the method outlined in Job 5-9.
11. A generator supplies 12 A to three resistors of 16 Ω, 48 Ω, and 24 Ω connected in parallel. Find (a) the current in each resistor and (b) the generator voltage.
12. In Fig. 9-20, find (a) I_1, (b) I_2, and (c) V_T. Check this problem using the method outlined in Job 6-3.

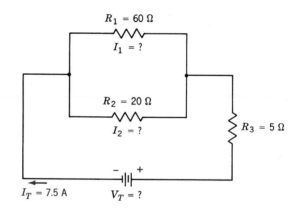

Figure 9-20

(See CD-ROM for Test 9-1)

(See CD-ROM for Test 9-1)

JOB 9-5

Using Kirchhoff's Laws in Complex Circuits

As the circuits under consideration become more complicated, it is generally easier to designate the currents in the various branches as x, y, and z, rather than as I_1, I_2, or I_3.

Example 9-12

Solve the circuit shown in Fig. 9-21 for the values of the currents x, y, and z.

Solution

1. Assume the direction of the electron flow to be that indicated in Fig. 9-21.

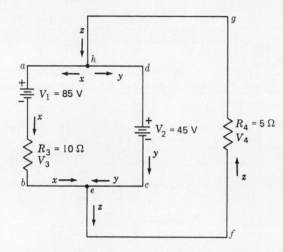

Figure 9-21   In complex circuits, the branch currents are designated as x, y, and z A.

2. Apply Kirchhoff's voltage law. Trace around the circuit in the direction *abcd*.

$$V_1 + V_3 + V_2 = 0$$
$$85 - 10x - 45 = 0$$
$$-10x = 45 - 85$$
$$-10x = -40$$
$$x = 4 \text{ A} \qquad Ans.$$

Trace around the circuit in the direction *abefgh*.

$$V_1 + V_3 + V_4 = 0$$
$$85 - 10x - 5z = 0$$

Substitute 4 for x.

$$85 - 10(4) - 5z = 0$$
$$85 - 40 - 5z = 0$$
$$-5z = 40 - 85$$
$$-5z = -45$$
$$z = 9 \text{ A} \qquad Ans.$$

3. Find the current y by applying Kirchhoff's current law. At point e,

$$\Sigma I_e = \Sigma I_l \qquad (9\text{-}1)$$
$$x + y = z$$
$$4 + y = 9$$
$$y = 9 - 4 = 5 \text{ A} \qquad Ans.$$

Hint for Success

Many types of circuits have components that are not in series, in parallel, parallel-series, or series-parallel. A good example is the circuit shown in Fig. 9-21. Where the rules of series and parallel circuits cannot be applied, more general methods of analysis must be used. These include Kirchhoff's laws, Thevenin's theorem, and Norton's theorem, all of which are covered in this chapter. Without these laws or theorems, these circuits would be impossible to solve.

4. Check by tracing the circuit *dcefgh*, using Kirchhoff's voltage law.

$$V_2 + V_4 = 0 \qquad (9\text{-}2)$$
$$45 - 5z = 0$$
$$45 - 5(9) = 0$$
$$45 - 45 = 0$$
$$0 = 0 \qquad Check$$

Example 9-13

Solve the circuit shown in Fig. 9-22 for the values of the currents *x*, *y*, and *z*.

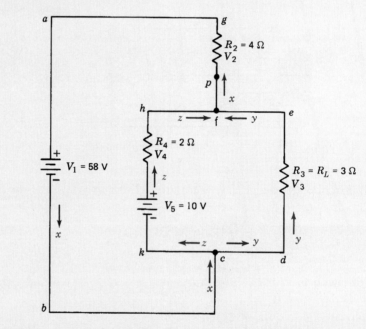

Figure 9-22 Since V_1 is stronger than V_5, the electron flow is *in* to point *c* from the source V_1.

Solution

1. Since V_1 is greater than V_5, the electron flow *x* is assumed to be *in* to point *c* from V_1. At point *c*, the flow is assumed to be as shown on the diagram. If we are incorrect in these assumptions, a minus sign for a calculated current will advise us to change the direction.
2. Express the current *z* in terms of currents *x* and *y*. At point *c*,

$$\Sigma I_l = \Sigma I_e \qquad (9\text{-}1)$$
$$y + z = x$$
$$z = x - y \qquad (a)$$

3. Apply Kirchhoff's voltage law. Trace around the circuit in the direction *abcdefg*.

$$58 - 3y - 4x = 0$$
$$-3y - 4x = -58$$
or $\qquad 3y + 4x = 58 \qquad\qquad (1)$

Trace around the circuit in the direction *khfedck*.

$$-10 - 2z + 3y = 0$$

Substitute $(x - y)$ for z from equation (a).

$$-10 - 2(x - y) + 3y = 0$$
$$-10 - 2x + 2y + 3y = 0$$
$$5y - 2x - 10 = 0$$
or $\qquad 5y - 2x = 10 \qquad\qquad (2)$

4. Solve equations (1) and (2) simultaneously.

$$3y + 4x = 58 \qquad\qquad (1)$$
$$5y - 2x = 10 \qquad\qquad (2)$$

Multiply equation (2) by 2.

$$3y + 4x = 58$$
$$\underline{10y - 4x = 20}$$
Add: $\qquad 13y \qquad\;\; = 78$
$$y = 6 \text{ A} \qquad Ans.$$

5. Substitute $y = 6$ in equation (2) to find x.

$$5y - 2x = 10 \qquad\qquad (2)$$
$$5(6) - 2x = 10$$
$$30 - 2x = 10$$
$$-2x = 10 - 30$$
$$-2x = -20$$
$$x = 10 \text{ A} \qquad Ans.$$

6. Substitute the values for x and y in equation (a) to find the current z.

$$z = x - y \qquad\qquad (a)$$
$$z = 10 - 6 = 4 \text{ A} \qquad Ans.$$

7. Find the voltage drop in each resistor.

$V_2 = 4x$	$V_3 = 3y$	$V_4 = 2z$
$V_2 = 4(10)$	$V_3 = 3(6)$	$V_4 = 2(4)$
$V_2 = 40$ V	$V_3 = 18$ V	$V_4 = 8$ V

8. Check by tracing the circuit *abckhfg*, using Kirchhoff's voltage law.

$$V_1 + V_5 + V_4 + V_2 = 0$$
$$58 - 10 - 8 - 40 = 0$$
$$58 - 58 = 0$$
$$0 = 0 \qquad Check$$

In order to illustrate the technique of handling problems in which the assumed direction of electron flow proves to be incorrect, let us investigate the following example.

Example 9-14

Solve the circuit shown in Fig. 9-23 for the values of the currents x, y, and z.

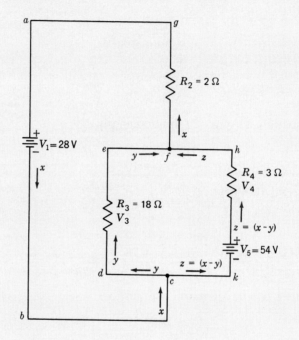

Figure 9-23

Solution

1. Assume the direction of the electron flow to be that shown on the diagram.
2. Express the current z in terms of currents x and y. At point c,

$$\Sigma I_l = \Sigma I_e \qquad (9\text{-}1)$$
$$y + z = x$$
$$z = x - y \qquad (a)$$

3. Apply Kirchhoff's voltage law. Trace around the circuit in the direction *abcdefg*.

$$28 - 18y - 2x = 0$$
$$-2x - 18y = -28$$

Divide the equation by -2 in order to simplify. This gives

$$x + 9y = 14 \qquad (1)$$

Trace around the circuit in the direction *khfedc*.

$$-54 - 3z + 18y = 0$$

Substitute $(x - y)$ for z from equation (a).

$$-54 - 3(x - y) + 18y = 0$$
$$-54 - 3x + 3y + 18y = 0$$
$$-3x + 21y = 54$$

Divide the equation by 3 in order to simplify. This gives

$$-x + 7y = 18 \qquad (2)$$

4. Solve equations (1) and (2) simultaneously.

$$x + 9y = 14 \qquad (1)$$
$$-x + 7y = 18 \qquad (2)$$

Add: $\qquad \overline{\qquad 16y = 32}$

$$y = 2 \text{ A} \qquad Ans.$$

5. Substitute $y = 2$ in equation (1) to find x.

$$x + 9y = 14 \qquad (1)$$
$$x + 9(2) = 14$$
$$x + 18 = 14$$
$$x = 14 - 18 = -4 \text{ A} \qquad Ans.$$

The fact that the current x is *minus* 4 A means that current x does *not* flow *into* point c but *does* flow *out* of this point. Similarly, the current x does *not* flow *out* of point f but *does* flow *into* this point.

6. Substitute the values for x and y in equation (a) to find the current z.

$$z = x - y \qquad (a)$$
$$z = -4 - 2 = -6 \text{ A} \qquad Ans.$$

Since z is *minus* 6 A, the actual direction of flow of current z is *into* point c and *out* of point f.

7. Find the voltage drop in each resistor.

$V_2 = 2x$	$V_3 = 18y$	$V_4 = 3z$
$V_2 = 2(-4)$	$V_3 = 18(2)$	$V_4 = 3(-6)$
$V_2 = -8$ V	$V_3 = 36$ V	$V_4 = -18$ V

8. Check by tracing the circuit *abckhfg*, using Kirchhoff's voltage law.

$$V_1 + V_5 + V_4 + V_2 = 0$$

Note:

$V_5 = -54$ V because we go through it from $-$ to $+$.
$V_4 = -(-18)$ V because we go through R_4 in the direction of the assumed current flow.
$V_2 = -(-8)$ V because we go through R_2 in the direction of the assumed current flow.

$$28 - 54 - (-18) - (-8) = 0$$
$$28 - 54 + 18 + 8 = 0$$
$$54 - 54 = 0$$
$$0 = 0 \qquad Check$$

Example 9-15

Solve the circuit shown in Fig. 9-24 for the values of the currents x, y, and z.

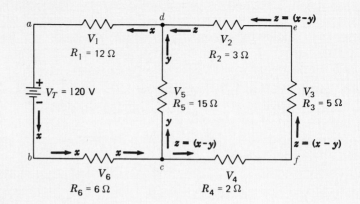

Figure 9-24

Solution

1. Assume the direction of the electron flow to be that shown in Fig. 9-24. Remember that these directions could be assumed to point in *other* directions without affecting the validity of the calculations. As we have seen, the truth of our assumptions will be determined by the signs of the calculated currents.

2. Express current z in terms of currents x and y. By Kirchhoff's current law, operating at point c,

$$y + z = x$$

    or
    $$z = x - y \qquad\qquad (a)$$

3. Apply Kirchhoff's voltage law. Trace around the circuit in the direction *cdef*. Note that in tracing from point d around to point c, we shall be tracing a *series* circuit made of the resistors R_2, R_3, and R_4. We can therefore simplify our calculations by considering these three resistors to be one resistor of $3 + 5 + 2$, or $10\ \Omega$.

$$-15y + 10z = 0$$

    Substitute $(x - y)$ for z from equation (a).

$$-15y + 10(x - y) = 0$$
$$-15y + 10x - 10y = 0$$
$$-25y + 10x = 0$$
$$-25y = -10x$$
$$y = \frac{-10}{-25}x = \frac{2}{5}x \qquad\qquad (1)$$

4. Trace around the circuit in the direction *abcd*.

$$120 - 6x - 15y - 12x = 0$$
$$-18x - 15y = -120$$

Divide the equation by -3 in order to simplify. This gives

$$6x + 5y = 40 \qquad (2)$$

5. Substitute the value of y from equation (1) in equation (2).

$$6x + 5y = 40 \qquad (2)$$
$$6x + 5\left(\frac{2}{5}x\right) = 40$$
$$6x + 2x = 40$$
$$8x = 40$$
$$x = 5 \text{ A} \qquad \textit{Ans.}$$

6. Substitute $x = 5$ in equation (1).

$$y = \frac{2}{5}x \qquad (1)$$

$$y = \frac{2}{5}(5) = 2 \text{ A} \qquad \textit{Ans.}$$

7. Substitute the values for x and y in equation (a) to find the current z.

$$z = x - y \qquad (a)$$
$$z = 5 - 2 = 3 \text{ A} \qquad \textit{Ans.}$$

8. Find the voltage drop in each resistor.

$V_1 = 12x$	$V_2 = 3z$	$V_3 = 5z$
$V_1 = 12(5)$	$V_2 = 3(3)$	$V_3 = 5(3)$
$V_1 = 60$ V	$V_2 = 9$ V	$V_3 = 15$ V
$V_4 = 2z$	$V_5 = 15y$	$V_6 = 6x$
$V_4 = 2(3)$	$V_5 = 15(2)$	$V_6 = 6(5)$
$V_4 = 6$ V	$V_5 = 30$ V	$V_6 = 30$ V

9. Check by tracing the circuit $abcfeda$, using Kirchhoff's voltage law.

$$V_T + V_6 + V_4 + V_3 + V_2 + V_1 = 0$$
$$120 - 30 - 6 - 15 - 9 - 60 = 0$$
$$120 - 120 = 0$$
$$0 = 0 \qquad \textit{Check}$$

Example 9-16

Solve the circuit shown in Fig. 9-25 on page 308 for the values of the currents $w, x, y,$

Solution

and z.

1. Assume the direction of the electron flow to be that shown in Fig. 9-25.
2. Express current w in terms of currents x and y. By Kirchhoff's current law, operating at point c,

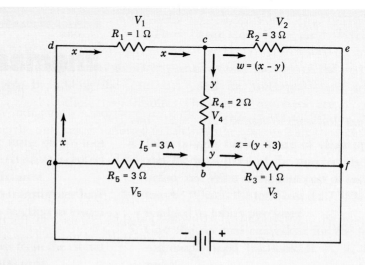

Figure 9-25

$$w + y = x$$

or
$$w = x - y \qquad (a)$$

At point b,
$$z = y + 3 \qquad (b)$$

3. Apply Kirchhoff's voltage law. Trace around the circuit in the direction *abcd*.

$$-3(3) + 2y + 1(x) = 0$$
$$-9 + 2y + x = 0$$
or
$$x + 2y = +9 \qquad (1)$$

4. Trace around the circuit in the direction *bcef*.

$$+2y - 3w + 1z = 0$$

Substitute $(x - y)$ for w from equation (a) and $(y + 3)$ for z from equation (b).

$$+2y - 3(x - y) + 1(y + 3) = 0$$
$$+2y - 3x + 3y + y + 3 = 0$$
$$-3x + 6y = -3$$

Divide the equation by 3 in order to simplify. This gives

$$-x + 2y = -1 \qquad (2)$$

5. Solve equations (1) and (2) simultaneously.

$$x + 2y = 9 \qquad (1)$$
$$\underline{-x + 2y = -1} \qquad (2)$$
Add:
$$4y = 8$$
$$y = 2 \text{ A} \qquad Ans.$$

6. Substitute this value for y in equation (1) to find x.

$$x + 2y = 9$$
$$x + 2(2) = 9$$

$$x + 4 = 9$$
$$x = 9 - 4 = 5 \text{ A} \qquad Ans.$$

7. Substitute these values for x and y in equation (a) to find w.

$$w = x - y \qquad\qquad\qquad (a)$$
$$w = 5 - 2 = 3 \text{ A} \qquad Ans.$$

8. Substitute $y = 2$ in equation (b) to find z.

$$z = y + 3 \qquad\qquad\qquad (b)$$
$$z = 2 + 3 = 5 \text{ A} \qquad Ans.$$

9. Find the voltage drop in each resistor.

$$
\begin{array}{lll}
V_1 = 1x & V_2 = 3w & V_3 = 1z \\
V_1 = 1(5) & V_2 = 3(3) & V_3 = 1(5) \\
V_1 = 5 \text{ V} & V_2 = 9 \text{ V} & V_3 = 5 \text{ V}
\end{array}
$$

$$
\begin{array}{ll}
V_4 = 2y & V_5 = 3(3) \\
V_4 = 2(2) = 4 \text{ V} & V_5 = 9 \text{ V}
\end{array}
$$

10. Check by tracing the circuit *abfecd* using Kirchhoff's voltage law.

$$V_5 + V_3 + V_2 + V_1 = 0$$
$$-9 - 5 + 9 + 5 = 0$$
$$0 = 0 \qquad Check$$

Example 9-17

Solve the unbalanced bridge circuit shown in Fig. 9-26 on page 310 for the values of the currents x, y, v, and w and the total voltage.

Solution

1. Assume the direction of the electron flow to be that shown on the diagram.
2. Express current y in terms of x and the total current. By Kirchhoff's current law, operating at point a,

$$y + x = 6$$

or
$$y = 6 - x \qquad\qquad (a)$$

At point c:
$$w = x + z \qquad\qquad (b)$$

At point b:
$$v + z = y$$

or
$$v = y - z$$

Substituting $(6 - x)$ for y from equation (a),

$$v = 6 - x - z \qquad\qquad (c)$$

3. Apply Kirchhoff's voltage law. Trace around the circuit in the direction *abca*.

$$-3y - 6z + 9x = 0$$

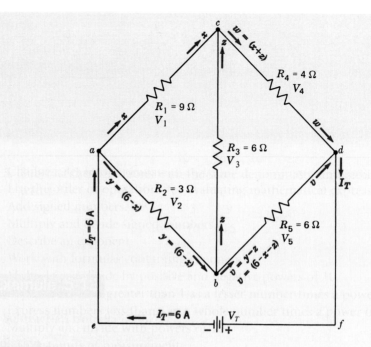

Figure 9-26　Unbalanced bridge circuit.

Substitute $(6 - x)$ for y from equation (a).

$$-3(6 - x) - 6z + 9x = 0$$
$$-18 + 3x - 6z + 9x = 0$$
$$12x - 6z = 18$$

Divide the equation by 6 in order to simplify. This gives

$$2x - z = 3 \qquad (1)$$

4. Trace around the circuit in the direction *bcdb*.

$$-6z - 4w + 6v = 0$$

Substitute $(x + z)$ for w from equation (b) and $(6 - x - z)$ for v from equation (c).

$$-6z - 4(x + z) + 6(6 - x - z) = 0$$
$$-6z - 4x - 4z + 36 - 6x - 6z = 0$$
$$-10x - 16z = -36$$

Divide the equation by -2 in order to simplify. This gives

$$5x + 8z = 18 \qquad (2)$$

5. Solve equations (1) and (2) simultaneously.

$$2x - z = 3 \qquad (1)$$
$$\underline{5x + 8z = 18} \qquad (2)$$

Multiply equation (1) by 8.

$$16x - 8z = 24 \qquad (1)$$
$$5x + 8z = 18 \qquad (2)$$

Add:
$$21x \quad = 42$$
$$x = 2 \text{ A} \qquad Ans.$$

6. Substitute this value for x in equation (1) to find z.

$$2x - z = 3 \qquad (1)$$
$$2(2) - z = 3$$
$$4 - z = 3$$
$$-z = 3 - 4$$
$$-z = -1$$
$$z = 1 \text{ A} \qquad Ans.$$

7. Substitute $x = 2$ in equation (a) to find y.

$$y = 6 - x$$
$$y = 6 - 2 = 4 \text{ A} \qquad Ans.$$

8. Substitute $x = 2$ and $z = 1$ in equation (b) to find w.

$$w = x + z \qquad (b)$$
$$w = 2 + 1 = 3 \text{ A} \qquad Ans.$$

9. Substitute $x = 2$ and $z = 1$ in equation (c) to find v.

$$v = 6 - x - z \qquad (c)$$
$$v = 6 - 2 - 1 = 3 \text{ A} \qquad Ans.$$

10. Find the voltage drop in each resistor.

$$V_1 = 9x \qquad\qquad V_2 = 3y \qquad\qquad V_3 = 6z$$
$$V_1 = 9(2) = 18 \text{ V} \qquad V_2 = 3(4) = 12 \text{ V} \qquad V_3 = 6(1) = 6 \text{ V}$$

$$V_4 = 4w \qquad\qquad V_5 = 6v$$
$$V_4 = 4(3) = 12 \text{ V} \qquad V_5 = 6(3) = 18 \text{ V}$$

11. Check by tracing the circuit in the direction *abdca*, using Kirchhoff's voltage law.

$$V_2 + V_5 + V_4 + V_1 = 0$$
$$-12 - 18 + 12 + 18 = 0$$
$$0 = 0 \qquad Check$$

12. Find the total voltage. Trace the circuit in the direction *feabd*.

$$V_T + V_2 + V_5 = 0$$
$$V_T - 12 - 18 = 0$$
$$V_T = 12 + 18 = 30 \text{ V} \qquad Ans.$$

Self-Test 9-18

In Fig. 9-27 on page 312, find (a) the currents x, y, and z and (b) the voltage drops across all the resistors. (c) Check the problem.

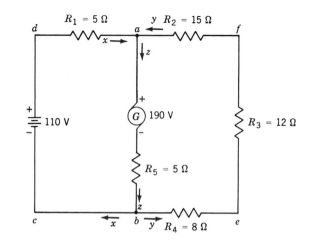

Figure 9-27   Opposing voltage sources.

Solution

1. Since the generator voltage (190 V) is greater than the battery voltage (110 V), it is assumed that the current z will flow (up/down) from point a to point b. Assume the remaining currents to be directed as shown.

2. Apply Kirchhoff's current law to point a.

$$x + \underline{\hspace{2cm}} = \underline{\hspace{2cm}}$$

or

$$y = z - \underline{\hspace{2cm}}$$

3. Apply Kirchhoff's voltage law. Trace the circuit $abcda$.

$$190 - 5z - 110 - 5 \underline{\hspace{2cm}} = 0$$
$$-5z - 5x = \underline{\hspace{2cm}}$$

Divide the equation by -5 in order to simplify. This gives

$$z + x = \underline{\hspace{2cm}} \tag{1}$$

4. Trace the circuit $abefa$. In this path, R_4, R_3, and R_2 are in $\underline{\hspace{2cm}}$ and may be considered to be one resistance of $\underline{\hspace{2cm}}$ Ω.

$$190 - 5z - \underline{\hspace{2cm}} = 0$$
$$-5z - 35y = \underline{\hspace{2cm}}$$

Divide the equation by $\underline{\hspace{2cm}}$ in order to simplify. This gives

$$z + 7y = \underline{\hspace{2cm}} \tag{2}$$

5. At this point we would solve equations (1) and (2) simultaneously. However, these two equations contain three letters at this point—z, x, and y. One of them must be expressed in terms of the other two letters. We shall express y in terms of z and x by using equation(a), and substituting $z–x$ for y in equation (2),

$$z + 7y = 38 \tag{2}$$
$$z + 7(\underline{\hspace{2cm}}) = 38$$
$$z + 7z - \underline{\hspace{2cm}} = 38$$
$$\underline{\hspace{2cm}} - 7x = 38 \tag{3}$$

6. Now, since they contain the same letters, z and x, we can solve equations (———) and (———) simultaneously.

$$z + x = 16 \qquad\qquad (1)$$
$$\underline{8z - 7x = 38} \qquad\qquad (3)$$

Multiplying equation (1) by ———,

$$7z + 7x = \underline{\hspace{2cm}}$$
$$\underline{8z - 7x = 38}$$

Adding, $\qquad\qquad 15z \qquad = \underline{\hspace{2cm}}$

$$z = \underline{\hspace{2cm}} A \qquad Ans.$$

7. Substitute this value for z in equation (1) to find x.

$$z + x = 16 \qquad\qquad (1)$$
$$\underline{\hspace{2cm}} + x = 16$$
$$x = 16 - \underline{\hspace{2cm}}$$
$$x = \underline{\hspace{2cm}} A \qquad Ans.$$

8. Substitute these values for x and z in equation (a) to find y.

$$y = z - x \qquad\qquad (a)$$
$$y = 10 - 6 = \underline{\hspace{2cm}} A \qquad Ans.$$

9. Find the voltage drops across the resistors.

$V_1 = x \times R_1$ $\qquad\qquad$ $V_5 = z \times R_5$
$V_1 = \underline{\hspace{1.5cm}} \times 5$ $\qquad$ $V_5 = \underline{\hspace{1.5cm}} \times 5$
$V_1 = 30$ V $\quad$ *Ans.* $\qquad$ $V_5 = \underline{\hspace{1.5cm}}$ V $\quad$ *Ans.*

$V_2 = y \times R_2$ $\qquad$ $V_3 = y \times R_3$ $\qquad$ $V_4 = y \times R_4$
$V_2 = \underline{\hspace{1.2cm}} \times 15$ $\quad$ $V_3 = \underline{\hspace{1.2cm}} \times 12$ $\quad$ $V_4 = \underline{\hspace{1.2cm}} \times 8$
$V_2 = 60$ V $\;$ *Ans.* $\;$ $V_3 = 48$ V $\;$ *Ans.* $\;$ $V_4 = \underline{\hspace{1.2cm}}$ V $\;$ *Ans.*

10. *Check*: Trace the circuit *abcda*.

$$190 - 50 - 110 - \underline{\hspace{2cm}} = 0$$
$$\underline{\hspace{2cm}} = 0 \qquad Check$$

11. Trace the circuit *abefa*.

$$190 - 50 - 32 - \underline{\hspace{2cm}} - 60 = 0$$
$$\underline{\hspace{2cm}} = 0 \qquad Check$$

Exercises

Practice

Set No. 1

1. In a circuit similar to that shown in Fig. 9-21, $V_1 = 22$ V, $V_2 = 20$ V, $R_3 = 1\ \Omega$, $R_4 = 4\ \Omega$. Find the currents x, y, and z, and check your answer.

2. In a circuit similar to that shown in Fig. 9-22, $V_1 = 75$ V, $R_2 = 3$ Ω, $R_3 = 12$ Ω, $R_4 = 4$ Ω, and $V_5 = 28$ V. Find the currents x, y, and z, and check your answer.
3. In a circuit similar to that shown in Fig. 9-24, $V_T = 25$ V, $R_1 = 1$ Ω, $R_2 = 3$ Ω, $R_3 = 6$ Ω, $R_4 = 1$ Ω, $R_5 = 5$ Ω, and $R_6 = 4$ Ω. Find the currents x, y, and z, and check your answer.
4. In a circuit similar to that shown in Fig. 9-25, $R_1 = 2$ Ω, $R_2 = 7$ Ω, $R_3 = 9$ Ω, $R_4 = 4$ Ω, $R_5 = 5$ Ω, and $I_5 = 4$ A. Find the currents w, x, y, and z, and check your answer.
5. In a circuit similar to that shown in Fig. 9-26, $R_1 = 10$ Ω, $R_2 = 8$ Ω, $R_3 = 15$ Ω, $R_4 = 18$ Ω, $R_5 = 2$ Ω, and $I_T = 15$ A. Find the currents v, w, x, y, and z and the total voltage.
6. Solve the circuit shown in Fig. 9-28 for the values of the current x, y, and z.

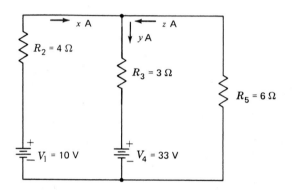

Figure 9-28

7. Solve the circuit shown in Fig. 9-29 for the values of the currents x, y, and z.

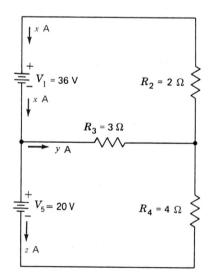

Figure 9-29

8. Solve the circuit shown in Fig. 9-30 for the values of the currents x, y, and z.

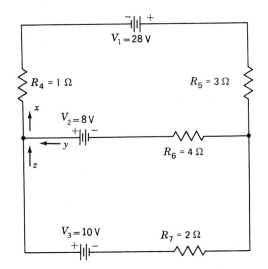

Figure 9-30

Set No. 2

1. In a circuit similar to that shown in Fig. 9-21, $V_1 = 32$ V, $V_2 = 12$ V, $R_3 = 5\,\Omega$, and $R_4 = 2\,\Omega$. Find the currents x, y, and z, and check your answer.
2. In a circuit similar to that shown in Fig. 9-22, $V_1 = 28$ V, $R_2 = 1\,\Omega$, $R_3 = 15\,\Omega$, $R_4 = 5\,\Omega$, and $V_5 = 12$ V. Find the currents x, y, and z, and check your answer.
3. In a circuit similar to that shown in Fig. 9-24, $V_T = 11$ V, $R_1 = 2\,\Omega$, $R_2 = 3\,\Omega$, $R_3 = 1\,\Omega$, $R_4 = 6\,\Omega$, $R_5 = 10\,\Omega$, and $R_6 = 4\,\Omega$. Find the currents x, y, and z, and check your answer.
4. In a circuit similar to that shown in Fig. 9-25, $R_1 = 10\,\Omega$, $R_2 = 18\,\Omega$, $R_3 = 2\,\Omega$, $R_4 = 15\,\Omega$, $R_5 = 8\,\Omega$, and $I_5 = 10$ A. Find the currents w, x, y, and z, and check your answer.
5. In a circuit similar to that shown in Fig. 9-26, $R_1 = 30\,\Omega$, $R_2 = 10\,\Omega$, $R_3 = 5\,\Omega$, $R_4 = 16\,\Omega$, $R_5 = 15\,\Omega$, and $I_T = 11$ A. Find the currents v, w, x, y, and z, and the total voltage.
6. In a circuit similar to that shown in Fig. 9-28, $V_1 = 22$ V, $R_2 = 2\,\Omega$, $R_3 = 4\,\Omega$, $V_4 = 60$ V, and $R_5 = 16\,\Omega$. Find the currents x, y, and z.
7. In a circuit similar to that shown in Fig. 9-29, $V_1 = 76$ V, $R_2 = 10\,\Omega$, $R_3 = 2\,\Omega$, $R_4 = 15\,\Omega$, and $V_5 = 54$ V. Find the currents x, y, and z.
8. In a circuit similar to that shown in Fig. 9-30, $V_1 = 23$ V, $V_2 = 11$ V, $V_3 = 8$ V, $R_4 = 5\,\Omega$, $R_5 = 2\,\Omega$, $R_6 = 2\,\Omega$, and $R_7 = 3\,\Omega$. Find the currents x, y, and z.

Set No. 3

1. In a circuit similar to that shown in Fig. 9-21, $V_1 = 10$ V, $V_2 = 5$ V, $R_3 = 10\,\Omega$, and $R_4 = 2.5\,\Omega$. Find the currents x, y, and z, and check your answer.
2. In a circuit similar to that shown in Fig. 9-22, $V_1 = 12$ V, $R_2 = 2.5\,\Omega$, $R_3 = 3.4\,\Omega$, $R_4 = 3\,\Omega$, and $V_5 = 38$ V. Find the currents x, y, and z, and check your answer.
3. In a circuit similar to that shown in Fig. 9-24, $V_T = 10$ V, $R_1 = 5\,\Omega$, $R_2 = 6\,\Omega$, $R_3 = 8\,\Omega$, $R_4 = 4\,\Omega$, $R_5 = 6\,\Omega$, and $R_6 = 3\,\Omega$. Find the currents x, y, and z, and check your answer.

4. In a circuit similar to that shown in Fig. 9-25, $R_1 = 5\ \Omega$, $R_2 = 40\ \Omega$, $R_3 = 10\ \Omega$, $R_4 = 10\ \Omega$, $R_5 = 3\ \Omega$, and $I_5 = 2$ A. Find the currents w, x, y, and z, and check your answer.
5. In a circuit similar to that shown in Fig. 9-26, $R_1 = 32\ \Omega$, $R_2 = 2\ \Omega$, $R_3 = 5\ \Omega$, $R_4 = 4\ \Omega$, $R_5 = 20\ \Omega$, and $I_T = 3.5$ A. Find the currents v, w, x, y, and z and the total voltage.
6. In a circuit similar to that shown in Fig. 9-28, $V_1 = 12$ V, $R_2 = 10\ \Omega$, $R_3 = 30\ \Omega$, $V_4 = 41$ V, and $R_5 = 20\ \Omega$. Find the currents x, y, and z.
7. In a circuit similar to that shown in Fig. 9-29, $V_1 = 31$ V, $R_2 = 2\ \Omega$, $R_3 = 10\ \Omega$, $R_4 = 4\ \Omega$, and $V_5 = 6$ V *with polarity opposite to that shown in Fig. 9-29.* Find the currents x, y, and z.
8. In a circuit similar to that shown in Fig. 9-29, $V_1 = 17$ V, $R_2 = 20\ \Omega$, $R_3 = 10\ \Omega$, $R_4 = 40\ \Omega$, and $V_5 = 13$ V. Find the currents x, y, and z.

JOB 9-6

Review of Kirchhoff's Laws

1. Kirchhoff's current law.

RULE At any point in an electric circuit, the sum of the currents that _____ a junction is equal to the sum of the currents that _____ it.

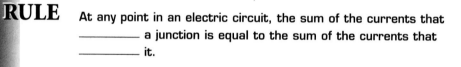

$$\Sigma I_e = \Sigma I_l \qquad \text{Formula 9-1}$$

2. Direction of electron flow.
   a. If the total known electron flow *in* to a point is *greater* than the total known flow *out* of the point, then the direction of any unknown flow must be $\underset{\text{(into/out of)}}{\underline{\hspace{2cm}}}$ the point.
   b. If the total known electron flow *into* a point is *less* than the total known flow *out* of the point, then the direction of any unknown flow must be $\underset{\text{(into/out of)}}{\underline{\hspace{2cm}}}$ the point.
3. Kirchhoff's voltage law.

RULE The _____ sum of all the voltages in any complete electric circuit is equal to _____.

$$V_1 + V_2 + V_3 + \text{etc.} = 0 \qquad \text{Formula 9-2}$$

4. Direction of voltages.
   a. *Source* voltages are positive ($+$) if we move through them from the _____ (_____) terminal to the negative ($-$) terminal.

b. *Source* voltages are ————— (—————) if we move through them from the negative (−) terminal to the positive (+) terminal.

c. A voltage across a resistance is *negative* (−) if we go through it in the __(same/opposite)__ direction as that of the assumed electron flow.

d. A voltage across a resistance is ————— (—————) if we go through it in a direction *opposite* to the assumed electron flow.

5. Using Kirchhoff's laws.

a. Determine the ————— of the electron flow. If insufficient data are given, *assume* directions of flow in each branch of the circuit. A minus sign for the solution indicates that the assumed direction was __(correct/incorrect)__. The value of the current found __(is/is not)__ affected by an incorrect choice of direction.

b. Mark the diagram to indicate all ————— of the electron flow and all polarities of sources of voltage.

c. If possible, express the currents in the various branches in terms of the ————— in other branches.

d. Trace around a ————— circuit, totaling the ————— sum of *all* voltages encountered. Set the sum equal to —————. Be careful to note the correct sign for each voltage.

e. Trace around a *second complete* circuit, obtaining a second equation.

f. Solve the set of ————— obtained in steps *d* and *e* simultaneously.

g. Find all voltage drops in all resistances by ————— law.

h. Check the solution by tracing around a *complete third* circuit, setting the algebraic sum of the voltages equal to —————.

Self-Test 9-19

In Fig. 9-31, find the generator voltage V which will reduce the current through R_3 to zero.

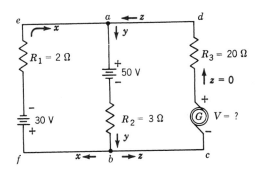

Figure 9-31 Find the generator voltage that will reduce the current through R_3 to zero.

Solution

1. Since the two batteries __(aid/oppose)__ each other, it is assumed that the current y will flow __(up/down)__ from point a to point b. Assume the currents x and z to be directed as shown.

2. Apply Kirchhoff's current law to point b.

$$y = x + \underline{\hspace{2cm}}$$
$$y = x + \underline{\hspace{2cm}}$$
or $\quad\quad\quad\quad\quad\quad\quad y = x \quad\quad\quad\quad\quad\quad\quad\quad\quad\quad (a)$

3. Apply Kirchhoff's voltage law to the circuit *efbae*.

$$+2x - 30\underline{}^{(+/-)} \ 3y - 50 = 0$$
$$2x + 3y = \underline{\hspace{2cm}}$$

Now, since $y = \underline{\hspace{2cm}}$ from equation (a),

$$2x + 3\underline{\hspace{2cm}} = 80$$
$$\underline{\hspace{2cm}} = 80$$
$$x = \underline{\hspace{2cm}} \text{ A}$$

And, since $y = x$, $y = \underline{\hspace{2cm}}$ A.

4. Trace the circuit *abcda*.

$$50 - 3y - \underline{\hspace{2cm}} - 20z = 0$$

Substitute 0 for z as given.

$$50 - 3y - V - 20(0) = 0$$
$$50 - 3y - V = 0$$

Substitute 16 for y in this equation.

$$50 - 3(16) - V = 0$$
$$50 - 48 = \underline{\hspace{2cm}}$$
$$2 = V \quad \text{or} \quad V = 2 \text{ V} \quad\quad Ans.$$

5. *Check*: Trace the circuit *efbcdae*.

$$2x - 30 - V - 20z = 0$$
$$2(16) - 30 - \underline{\hspace{1.5cm}} - 20(\underline{\hspace{1.5cm}}) = 0$$
$$32 - 30 - 2 - \underline{\hspace{1.5cm}} = 0$$
$$\underline{\hspace{2cm}} = 0 \quad\quad Check$$

Now try just one more problem.

6. In Fig. 9-32, find the power consumed in R_5.
(See CD-ROM for Test 9-2)

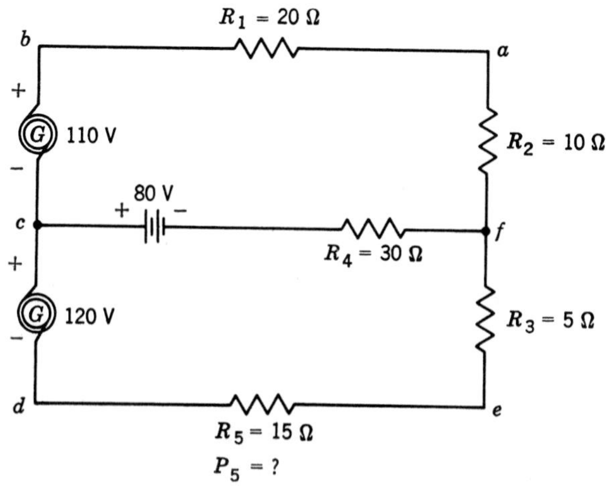

Figure 9-32

Equivalent Delta and Star Circuits

Example 9-17 of the last job can be greatly simplified by the application of the theory of delta and star circuits. We shall solve this problem again at the conclusion of this job.

The triangle formed by the three resistors R_2, R_1, and R_3 of Fig. 9-26 has been turned through one-fourth turn and is shown in Fig. 9-33a. The resistors are said to be connected in *delta* because the connection resembles the Greek letter delta (Δ). The resistors R_a, R_b, and R_c in Fig. 9-33b are connected in *star*, or **Y**, formation. Most delta circuits will be much easier to solve if we can change them into *equivalent* star circuits.

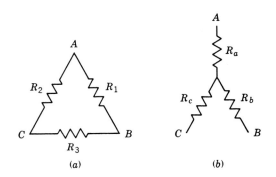

(a)   (b)

Figure 9-33 (a) R_1, R_2, and R_3 are connected in delta formation. (b) R_a, R_b, and R_c are connected in *star* or Y formation.

The word *equivalent* means that the circuits must do the same job as they did before. That is, the resistance between A and B, B and C, and C and A must be the same in each circuit.

The resistance from A to B in Fig. 9-33a is found by redrawing the circuit as shown in Fig. 9-34a. This reduces to a circuit in which R_1 is in parallel with the series combination of $R_2 + R_3$, as in Fig. 9-34b. Thus,

$$R_{AB} = \frac{R_1(R_2 + R_3)}{R_1 + R_2 + R_3} \qquad (1)$$

In Fig. 9-33b, the resistance from A to B is simply the series combination of $R_a + R_b$, or

$$R_{AB} = R_a + R_b \qquad (2)$$

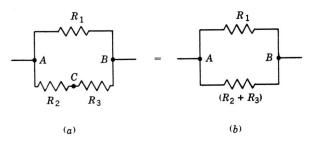

(a)   (b)

Figure 9-34 (a) Figure 9-33a redrawn in standard form. (b) R_2 and R_3 are added since they are in series.

Since the two resistances must be equal, we can set equation (2) equal to equation (1) and get

$$R_a + R_b = \frac{R_1 R_2 + R_1 R_3}{R_1 + R_2 + R_3}$$

If we use the symbol ΣR_Δ (read as "the summation of resistances in delta") to mean $R_1 + R_2 + R_3$, we get

$$R_a + R_b = \frac{R_1 R_2 + R_1 R_3}{\Sigma R_\Delta} \qquad (3)$$

In a similar manner,

$$R_b + R_c = \frac{R_1 R_3 + R_2 R_3}{\Sigma R_\Delta} \qquad (4)$$

$$R_a + R_c = \frac{R_1 R_2 + R_2 R_3}{\Sigma R_\Delta} \qquad (5)$$

When equations (3) to (5) are solved simultaneously, we get the following

$$\boldsymbol{R_a = \frac{R_1 R_2}{\Sigma R_\Delta}} \qquad \text{Formula 9-3}$$

$$\boldsymbol{R_b = \frac{R_1 R_3}{\Sigma R_\Delta}} \qquad \text{Formula 9-4}$$

$$\boldsymbol{R_c = \frac{R_2 R_3}{\Sigma R_\Delta}} \qquad \text{Formula 9-5}$$

Do not attempt to memorize these formulas. They depend on the positions of the letters A, B, and C and the positions of R_1, R_2, and R_3.

An easy way to develop these formulas is shown in Fig. 9-35a. Each equivalent Y resistance is obtained by multiplying the two adjacent delta resistances and then dividing by the sum of the delta resistances (ΣR_Δ). For example, the resistors R_2 and R_3 are adjacent (next) to R_c. Therefore,

$$R_c = \frac{R_2 \times R_3}{\Sigma R_\Delta}$$

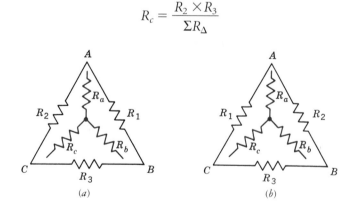

Figure 9-35 (a) R_1 and R_2 are adjacent to R_a; R_1 and R_3 are adjacent to R_b; R_2 and R_3 are adjacent to R_c. (b) R_1 and R_2 are adjacent to R_a; R_2 and R_3 are adjacent to R_b; R_1 and R_3 are adjacent to R_c.

But, in Fig. 9-35b, in which the positions of R_1 and R_2 have been interchanged, R_1 and R_3 are adjacent to R_c. Therefore,

$$R_c = \frac{R_1 R_3}{\Sigma R_\Delta}$$

Example 9-20

In Fig. 9-33a, $R_1 = 10\ \Omega$, $R_2 = 8\ \Omega$, and $R_3 = 2\ \Omega$. Find the resistances of the equivalent Y circuit of Fig. 9-33b.

Solution

1. $\Sigma R_\Delta = R_1 + R_2 + R_3$
   $\Sigma R_\Delta = 10 + 8 + 2 = 20\ \Omega$

2. $R_a = \dfrac{R_1 R_2}{\Sigma R_\Delta}$ (9-3) $R_b = \dfrac{R_1 R_3}{\Sigma R_\Delta}$ (9-4) $R_c = \dfrac{R_2 R_3}{\Sigma R_\Delta}$ (9-5)

   $R_a = \dfrac{10 \times 8}{20}$ $R_b = \dfrac{10 \times 2}{20}$ $R_c = \dfrac{8 \times 2}{20}$

   $R_a = 4\ \Omega$ $R_b = 1\ \Omega$ $R_c = 0.8\ \Omega$ *Ans.*

Self-Test 9-21

Find the total resistance between points A and D in the circuit shown in Fig. 9-36a.

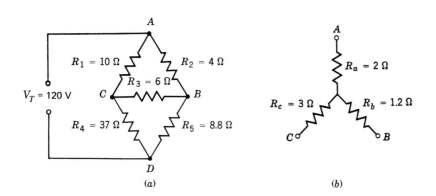

Figure 9-36   (a) An unbalanced bridge circuit. (b) The Y equivalent of the delta circuit *ABC*.

Solution

1. Change the delta circuit *ABC* into its equivalent Y circuit as shown in Fig. 9-36b. Note that R_1 and R_2 are *not* in the same position as they were in Fig. 9-33a. Develop the correct formulas for R_a, R_b, and R_c as shown above.

$$\Sigma R_\Delta = R_1 + R_2 + R_3$$
$$\Sigma R_\Delta = 10 + 4 + 6 = \underline{\hspace{1.5cm}}\ \Omega$$

$$R_a = \frac{R_1 R_2}{\Sigma R_\Delta} \qquad R_b = \frac{R_2 R_3}{\Sigma R_\Delta} \qquad R_c = \frac{R_1 R_3}{\Sigma R_\Delta}$$

$$R_a = \frac{10 \times \text{?}}{20} \qquad\qquad R_b = \frac{4 \times \text{?}}{20} \qquad\qquad R_c = \frac{10 \times \text{?}}{20}$$

$R_a =$ _____ Ω $\qquad\qquad$ $R_b =$ _____ Ω $\qquad\qquad$ $R_c =$ _____ Ω

2. Redraw the delta circuit as a Y circuit and connect it to the remainder of the original circuit as shown in Fig. 9-37a.

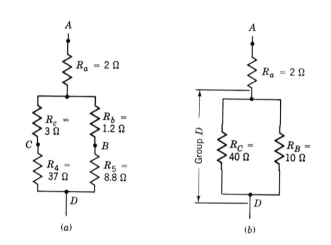

Figure 9-37 (a) The Y equivalent of delta *ABC* is attached to the remainder of the bridge circuit *BDC*. (b) R_C is the series equivalent of $R_c + R_4$: R_b is the series equivalent of $R_b + R_5$.

3. In Fig. 9-37b, R_c represents the series resistance of R_c plus R_4, which equals 3 + 37 = _____ Ω. Also, R_B represents the series resistance of _____ plus R_5, which equals 1.2 + 8.8 = _____ Ω.

4. In Fig. 9-37b, R_C and R_B are connected in _____. The total resistance for this group is

$$R_D = \frac{R_C \times R_B}{R_C + R_B} \qquad\qquad (5\text{-}5)$$

$$R_D = \frac{40 \times 10}{40 + 10} = \text{_____} \ \Omega$$

5. In Fig. 9-37b, the total resistance for the circuit from *A* to *D* is made of the $\underset{\text{(series/parallel)}}{\text{_____}}$ combination of R_a and R_D.

$$R_T = R_a + \text{_____}$$
$$R_T = 2 + 8 = \text{_____} \ \Omega \qquad Ans.$$

6. Since the voltage across points *A* and *D* equals 120 V, we can find the total current by _____ law.

$$E_T = I_T \times R_T \qquad\qquad (4\text{-}7)$$
$$120 = I_T \times \text{_____}$$
$$I_T = \text{_____} \ A \qquad Ans.$$

Self-Test 9-22

Solve the unbalanced bridge circuit given in Example 9-17 for the values of the currents x, y, v, z, and w and the total voltage.

Solution

The circuit is redrawn for your convenience as Fig. 9-38.

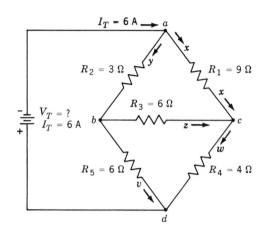

Figure 9-38   Figure 9-26 of Example 9-17 is redrawn.

1. Find the total resistance between points a and d.
    a. Change the delta circuit abc into its equivalent Y circuit as shown in Fig. 9-39a.

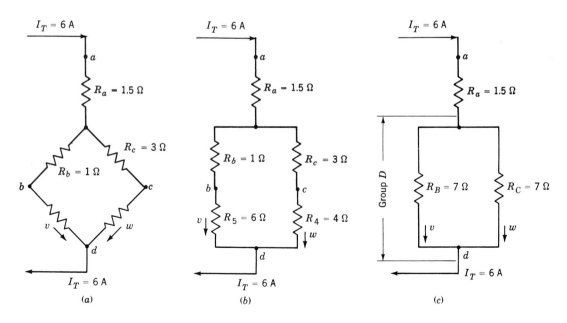

Figure 9-39   (a) Delta abc converted to Y formation. (b) Y formation of abc connected to bdc in standard form. (c) R_B is the series equivalent of $R_b + R_5$: R_c is the series equivalent of $R_c + R_4$.

$$\Sigma R_\Delta = R_1 + R_2 + R_3$$
$$\Sigma R_\Delta = 9 + \underline{\hspace{1.5cm}} + 6 = \underline{\hspace{1.5cm}} \Omega$$

$$R_a = \frac{3 \times ?}{?} \qquad\qquad R_b = \frac{? \times 6}{18} \qquad\qquad R_c = \frac{6 \times ?}{18}$$

$$R_a = \underline{\hspace{1.5cm}} \Omega \qquad R_b = \underline{\hspace{1.5cm}} \Omega \qquad R_c = \underline{\hspace{1.5cm}} \Omega$$

b. Connect the equivalent Y circuit to the remainder of the circuit as shown in Fig. 9-39b. Label all the parts with their respective values.

c. In Fig. 9-39c, R_C represents the $\underline{\text{(series/parallel)}}$ resistance of R_c and R_4, which equals $\underline{\hspace{1.5cm}} \Omega$. Also, R_B represents the series resistance of $\underline{\hspace{1.5cm}}$ and R_5, which equals $\underline{\hspace{1.5cm}} \Omega$.

d. In Fig. 9-39c, R_C and R_B are connected in $\underline{\hspace{1.5cm}}$.

$$R_D = \underline{\hspace{1.5cm}} \Omega$$

e. The total resistance for the circuit from a to d equals $R_a + R_D = 1.5 + 3.5 = \underline{\hspace{1.5cm}} \Omega$.

2. Find the total voltage V_T.

$$V_T = I_T \times R_T = 6 \times 5 = \underline{\hspace{1.5cm}} \text{V} \qquad Ans.$$

3. Find the currents v and w.

In Fig. 9-39c, since $R_C = R_B = 7\ \Omega$, the total current $I_T = 6$ A will divide so that

$$v = w = \underline{\hspace{1.5cm}} \text{A} \qquad Ans.$$

4. Find V_5 and V_4. In Fig. 9-39b,

$$V_5 = v \times R_5 \qquad\qquad\qquad V_4 = w \times R_4$$
$$V_5 = 3 \times 6 = \underline{\hspace{1.5cm}} \text{V} \qquad\qquad V_4 = 3 \times 4 = \underline{\hspace{1.5cm}} \text{V}$$

5. Trace the circuit cbd in Fig. 9-38.

$$6z - V_5 + V_4 = 0$$
$$6z - \underline{\hspace{1.5cm}} + 12 = 0$$
$$6z = \underline{\hspace{1.5cm}}$$
$$z = \underline{\hspace{1.5cm}} \text{A} \qquad Ans.$$

6. Find the currents x and y in Fig. 9-38.

Apply Kirchhoff's current law to point c.

$$x + z = \underline{\hspace{1.5cm}}$$
$$x + 1 = 3 \text{ (from step 3)}$$
$$x = \underline{\hspace{1.5cm}} \text{A} \qquad Ans.$$

Apply Kirchhoff's current law to point a.

$$y + x = \underline{\hspace{1.5cm}}$$
$$y + \underline{\hspace{1.5cm}} = 6$$
$$y = \underline{\hspace{1.5cm}} \text{A} \qquad Ans.$$

7. *Check:*

$$V_1 = 9x = 9(\underline{\hspace{2cm}}) = 18 \text{ V}$$
$$V_2 = 3 \underline{\hspace{2cm}} = 3(4) = 12 \text{ V}$$
$$V_3 = \underline{\hspace{2cm}} z = 6(1) - 6 \text{ V}$$
$$V_4 = (\underline{\hspace{1.5cm}})(\underline{\hspace{1.5cm}}) = 4(3) = 12 \text{ V}$$
$$V_5 = 6(\underline{\hspace{2cm}}) = 6(3) = 18 \text{ V}$$

8. Trace the circuit *abdca*.

$$-V_2 - V_5 + V_4 + V_1 = 0$$

Exercises

Practice

1. In the circuit shown in Fig. 9-33a, $R_1 = R_2 = R_3 = 60\ \Omega$. Find the resistances of the equivalent Y circuit.
2. In the circuit shown in Fig. 9-33a, $R_1 = 50\ \Omega$, $R_2 = 40\ \Omega$, and $R_3 = 10\ \Omega$. Find the resistances of the equivalent Y circuit.
3. In the circuit shown in Fig. 9-33a, $R_1 = 20\ \Omega$, $R_2 = 12\ \Omega$, and $R_3 = 8\ \Omega$. Find the resistances of the equivalent Y circuit.
4. In the circuit shown in Fig. 9-33a, $R_1 = 20\ \Omega$, $R_2 = 10\ \Omega$, and $R_3 = 15\ \Omega$. Find the resistances of the equivalent Y circuit.
5. In a circuit similar to that shown in Fig. 9-36a, $R_1 = 2\ \Omega$, $R_2 = 4\ \Omega$, $R_3 = 6\ \Omega$, $R_4 = 5\ \Omega$, and $R_5 = 4\ \Omega$. Find the total resistance from A to D.
6. In a circuit similar to that shown in Fig. 9-36a, $R_1 = 12\ \Omega$, $R_2 = 18\ \Omega$, $R_3 = 10\ \Omega$, $R_4 = 1\ \Omega$, and $R_5 = 1.5\ \Omega$. Find the total resistance from A to D.
7. In a circuit similar to that shown in Fig. 9-36a, $R_1 = 30\ \Omega$, $R2 = 12\ \Omega$, $R_3 = 8\ \Omega$, $R_4 = 10.2\ \Omega$, and $R_5 = 3.08\ \Omega$. Find the total resistance from A to D.
8. In a circuit similar to that shown in Fig. 9-36a, $R_1 = 10\ \Omega$, $R_2 = 30\ \Omega$, $R_3 = 5\ \Omega$, $R_4 = 15\ \Omega$, and $R_5 = 16\ \Omega$. Find the total resistance from A to D.
9. In a circuit similar to that shown in Fig. 9-38, $R_1 = 10\ \Omega$, $R_2 = 8\ \Omega$, $R_3 = 2\ \Omega$, $R_4 = 1\ \Omega$, and $R_5 = 1.2\ \Omega$. If $I_T = 10$ A, find (*a*) the total resistance of the circuit, (*b*) the currents *v, w, x, y*, and *z*, and (*c*) the total voltage.
10. In a circuit similar to that shown in Fig. 9-38, $R_1 = 10\ \Omega$, $R_2 = 10\ \Omega$, $R_3 = 20\ \Omega$, $R_4 = 19\ \Omega$, and $R_5 = 3\ \Omega$. If $I_T = 16$ A, find (*a*) the total resistance of the circuit, (*b*) the currents *v, w, x, y*, and *z*, and (*c*) the total voltage.
11. In a circuit similar to that shown in Fig. 9-38, $R_1 = 10\ \Omega$, $R_2 = 50\ \Omega$, $R_3 = 15\ \Omega$, $R_4 = 8\ \Omega$, and $R_5 = 5\ \Omega$. If $I_T = 10$ A, find (*a*) the total resistance of the circuit, (*b*) the currents *v, w, x, y*, and *z*, and (*c*) the total voltage.
12. Figure 9-40 on page 326 shows the original circuit and the transformations necessary to find the total resistance. Find (*a*) the total resistance, (*b*) the total current, (*c*) the current in R_5, and (*d*) the current in R_6.

JOB 9-8

Thevenin's Theorem

The main objective in the solution of a circuit is usually the calculation of the values for the load voltage and the load current which can usually involve many

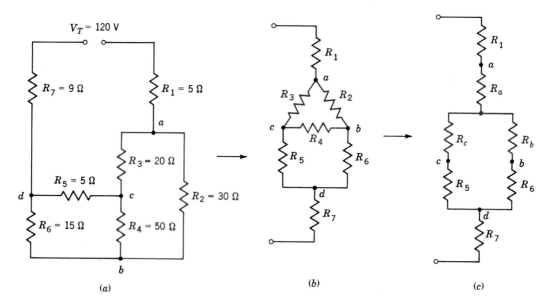

Figure 9-40 (a) Original circuit. (b) R_2, R_3, and R_4 in delta formation. (c) Delta abc in Y formation and connected to the remainder of the circuit.

intermediate steps. Thevenin's theorem will eliminate many of these intermediate steps and permit us to calculate the load values directly.

Figure 9-41a is the same as Fig. 9-24 of Example 9-15 except that the resistances R_2, R_3, and R_4 have been combined into the load resistance $R_L = 10\ \Omega$. In the solution of Example 9-15, we found

$$z = I_L = 3\ \text{A}$$
$$V_2 + V_3 + V_4 = V_L$$
$$9 + 15 + 6 = V_L = 30\ \text{V}$$

Now if this same load $R_L = 10\ \Omega$ were connected by a switching arrangement to become the load of circuit Fig. 9-41b, then $R_T = 10 + 15 = 25\ \Omega$, and

$$I_L = \frac{V_T}{R_T} = \frac{75}{25} = 3\ \text{A}$$

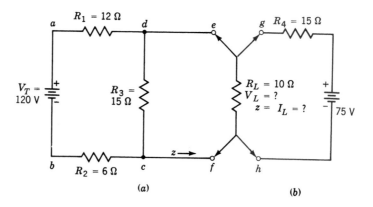

Figure 9-41 As far as the load R_L is concerned, circuit (a) is equivalent to circuit (b).

and
$$V_L = I_L \times R_L$$
$$V_L = 3 \times 10 = 30 \text{ V}$$

Thus, the *same* values for V_L and I_L were obtained when R_L was connected as part of circuit (*a*) or when R_L was connected as part of circuit (*b*). Therefore, as far as R_L is concerned, circuit (*b*) could be substituted for circuit (*a*) and the same results obtained. Since the two circuits produce the same results, they may be said to be equivalent. Obviously, if we had a choice, we would rather analyze circuit (*b*) than circuit (*a*).

Thevenin's theorem is a method for changing a complex circuit into a simple *equivalent* circuit which may be solved with a minimum of effort.

Thevenin's theorem. Any linear network of voltage sources and resistances, if viewed from any two points in the network, can be replaced by an equivalent resistance R_{TH} in series with an equivalent source V_{TH}. The problem consists of three steps.

1. Find the equivalent voltage V_{TH}.
2. Find the equivalent resistance R_{TH}.
3. Place R_{TH} in series with the load, and solve the simple series circuit.

Method

1. Disconnect the part of the circuit which is considered as the load.
2. Find the voltage that would appear across the load terminals when the load is disconnected. In Fig. 9-41*a* this would be the voltage across terminals *e* and *f*. This open-circuit voltage is called the *Thevenin voltage* (V_{TH}), as shown in Fig. 9-42*a*.

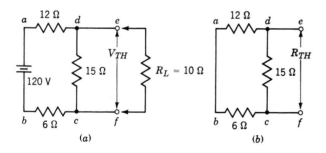

Figure 9-42 (*a*) When the load is removed, the equivalent voltage V_{TH} equals the voltage across the terminals *ef* or *dc*. (*b*) The equivalent resistance R_{TH} equals the total resistance with all sources of emf shorted and the load removed.

3. Replace each voltage source with a short, reducing the voltage to zero, and remove the load as shown in Fig. 9-42*b*.
4. Find the total resistance that would appear across the load terminals. This will be the *Thevenin resistance* (R_{TH}), as shown in Fig. 9-42*b*.
5. Draw the equivalent circuit consisting of R_{TH} in series with R_L and connected across the equivalent voltage V_{TH}, as shown in Fig. 9-43.
6. Solve for the load current and the load voltage.

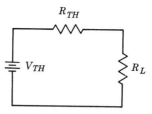

Figure 9-43 Thevenin equivalent circuit supplying a load R_L.

Example 9-23

Solve the circuit shown in Fig. 9-41a for the values of I_L and V_L.

Solution

1. Disconnect the load R_L as shown in Fig. 9-42a.
2. Find the voltage across the terminals *ef*. This is equal to the voltage across terminals *dc*. In the circuit *abcd*,

$$R_t = 6 + 15 + 12 = 33 \ \Omega \tag{4-3}$$

$$I_t = \frac{120}{33} = 3.63 \text{ A}$$

Therefore, $V_{dc} = 3.63 \times 15 = 54.45$ V

or $V_{TH} = 54.45$ V

3. Replace each voltage source with a short, and remove the load as shown in Fig. 9-42b.
4. Find the value of R_{TH}. In Fig. 9-42b, the 12 Ω is in series with the 6 Ω for a total of 18 Ω. This 18 Ω is in parallel with the 15 Ω.

$$R_{TH} = \frac{18 \times 15}{18 + 15} = 8.18 \ \Omega \tag{5-5}$$

5. Draw the equivalent circuit as shown in Fig. 9-44.

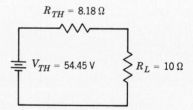

Figure 9-44 V_{TH} is in series with R_{TH} and R_L.

6. Solve for I_L.

$$Rt = 8.18 + 10 = 18.18 \ \Omega$$

$$I_L = \frac{V_{TH}}{R_t} = \frac{54.45}{18.18} = 3 \text{ A} \qquad Ans.$$

7. Solve for V_L.

$$V_L = I_L \times R_L$$
$$V_L = 3 \times 10 = 30 \text{ V} \qquad Ans.$$

The true value of Thevenin's theorem is apparent when we compare the small effort involved here with the lengthy calculations needed to solve the same problem of Example 9-15.

Example 9-24

Solve the circuit of Example 9-13 by Thevenin's theorem.

Solution

Figure 9-22 is repeated below for your convenience.

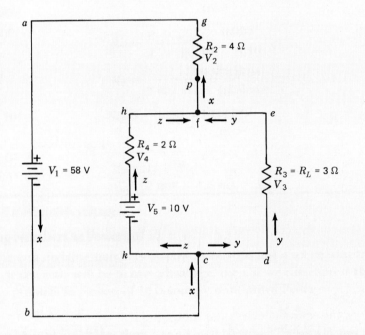

1. We must redraw the figure so that the load may be separated from the rest of the circuit at only two terminals. See Fig. 9-45. Consider the load to be R_3.

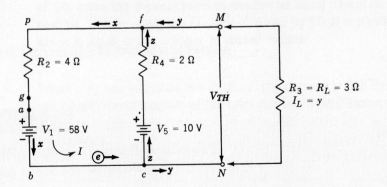

Figure 9-45   The circuit of Fig. 9-22 is redrawn to present only two terminals to the load.

2. Find V_{TH} with the load removed. This will be the voltage across the terminals M and N which is equal to the voltage across the terminals f and c.
   a. In the series circuit $pbcf$, the circulating current I is

$$I = \frac{V}{R} = \frac{58 - 10}{4 + 2} = \frac{48}{6} = 8 \text{ A}$$

   b. Find the voltage from f to c.

$$V_{fc} = IR_4 + V_5$$
$$V_{fc} = 8(2) + 10 = 26 \text{ V}$$

Therefore, $\qquad\qquad\qquad V_{TH} = 26 \text{ V}$

3. Replace each voltage source with a short, and remove the load as shown in
   Fig. 9-46.

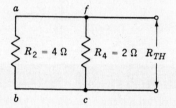

Figure 9-46 The equivalent resistance R_{TH} consists of R_2 and R_4 in parallel.

4. Find R_{TH}.

$$R_{TH} = \frac{4 \times 2}{4 + 2} = 1.33 \text{ }\Omega$$

5. Draw the equivalent circuit as shown in Fig. 9-47.

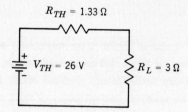

Figure 9-47 Thevenin equivalent circuit. V_{TH} is in series with R_{TH} and R_L.

6. Solve for I_L.

$$R_t = 1.33 + 3 = 4.33 \text{ }\Omega$$

$$I_L = \frac{V_{TH}}{R_t} = \frac{26}{4.33} = 6 \text{ A} \qquad Ans.$$

Solve for V_L.

$$V_L = I_L \times R_L$$
$$V_L = 6 \times 3 = 18 \text{ V} \qquad \textit{Ans.}$$

7. Find the currents x and z. In Fig. 9-45, since $V_{pb} = V_{fc} = V_L = 18$ V (all in parallel),

$$V_{pb} = -4x + 58 = 18$$
$$-4x = -40$$
$$x = 10 \text{ A} \qquad \textit{Ans.}$$

Also, since $V_{fc} = 18$ V,

$$2z + 10 = 10$$
$$2z = 8$$
$$z = 4 \text{ A} \qquad \textit{Ans.}$$

Example 9-25

Find the load current and the load voltage in the circuit shown in Fig. 9-48a.

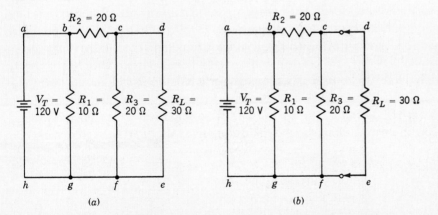

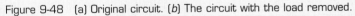

Figure 9-48 (a) Original circuit. (b) The circuit with the load removed.

Solution

1. Remove the load from the circuit as shown in Fig. 9-48b.
2. Find V_{TH} with the load removed. This will be the voltage across R_3.
   a. The resistors R_2 and R_3 between points b and f are in series. Therefore, $R_{bf} = 20 + 20 = 40 \ \Omega$.
   b. The path from b to f is in parallel with the 120-V source. Therefore,

$$I_{bf} = \frac{120}{40} = 3 \text{ A}$$

$$V_3 = V_{TH} = I_3 \times R_3$$
$$V_{TH} = 3 \times 20 = 60 \text{ V}$$

3. Replace each voltage source with a short and remove the load as shown in Fig. 9-49a. Be careful to note that *when the battery is replaced with a short, it also shorts out the resistor R_1,* which results in the circuit shown in Fig. 9-49b.

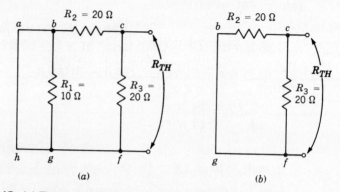

Figure 9-49 (a) The equivalent R_{TH} equals the total resistance with the source shorted and the load removed. (b) The short across *ab* of Fig. 9-49a also shorts out R_1; R_{TH} is found from this circuit.

4. Find R_{TH}. Since R_2 and R_3 are now in parallel,

$$R_{TH} = \frac{R_2 \times R_3}{R_2 + R_3} \qquad (5\text{-}5)$$

$$R_{TH} = \frac{20 \times 20}{20 + 20} = 10 \ \Omega$$

5. Draw the equivalent circuit as shown in Fig. 9-50.

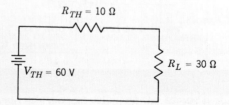

Figure 9-50 Thevenin equivalent circuit. V_{TH} is in series with R_{TH} and R_L.

6. Solve for I_L.

$$R_t = 10 + 30 = 40 \ \Omega$$

$$I_L = \frac{V_{TH}}{R_t} = \frac{60}{40} = 1.5 \ \text{A} \qquad Ans.$$

Solve for V_L.

$$V_L = I_L \times R_L$$
$$V_L = 1.5 \times 30 = 45 \ \text{V} \qquad Ans.$$

Example 9-26

Solve the circuit shown in Fig. 9-29 for the load current y and the load voltage V_3 by Thevenin's theorem.

Figure 9-29 is repeated here for your convenience.

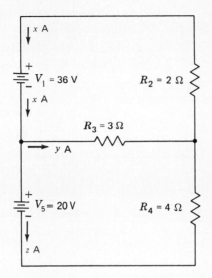

Solution

1. Remove the load R_3 from the circuit as shown in Fig. 9-51a.

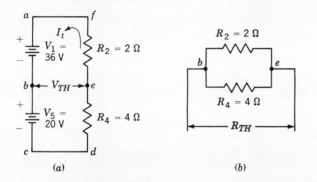

Figure 9-51 (a) V_{TH} is the voltage across the terminals from which the load has been removed. (b) The equivalent R_{TH} is the *total* resistance with all sources of emf shorted and the load removed.

2. Find V_{TH} with the load removed. This will be the voltage between points e and b.

   a. The circuit $abcdef$ is a simple series circuit in which

$$V_t = 36 + 20 = 56 \text{ V}$$
$$R_t = 4 + 2 = 6 \ \Omega$$
$$I_t = \frac{56}{6} = 9.33 \text{ A}$$

b. The voltage from e to b around the path $efab = V_{TH}$.

$$V_{TH} = -I_t R_2 + V_1$$
$$V_{TH} = -(9.33 \times 2) + 36$$
$$V_{TH} = -18.66 + 36 = 17.34 \text{ V}$$

3. Replace each voltage source with a short and remove the load as shown in Fig. 9-51b. In this parallel circuit,

$$R_{TH} = \frac{2 \times 4}{2 + 4} = 1.33 \ \Omega$$

4. Draw the equivalent circuit as shown in Fig. 9-52.

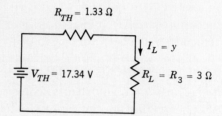

Figure 9-52 Thevenin equivalent circuit. V_{TH} is in series with R_{TH} and R_3.

5. Solve for the load current y.

$$R_t = 1.33 + 3 = 4.33 \ \Omega$$
$$y = I_L = \frac{V_{TH}}{R_t} = \frac{17.34}{4.33} = 4 \text{ A} \qquad Ans.$$

Solve for the load voltage V_3.

$$V_3 = y \times R_3 = 4 \times 3 = 12 \text{ V} \qquad Ans.$$

Self-Test 9-27

Find the load current and the load voltage in the circuit shown in Fig. 9-53.

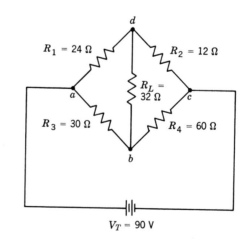

Figure 9-53

Solution

1. Remove the load from the circuit as shown in Fig. 9-54. Move $V_T = 90$ V to the inside of the figure for greater clarity.

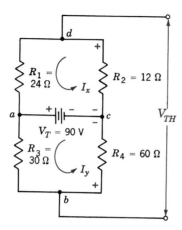

Figure 9-54 V_{TH} is the voltage across the terminals from which the load has been removed.

2. Find V_{TH} with the load removed. This will be the voltage across terminals _____ and _____.

    a. In the circuit acd, R_1 and R_2 are in _____.

    $$R_x = 24 + \underline{\qquad} = \underline{\qquad} \ \Omega$$

    $$I_x = \frac{V_T}{R_x} = \frac{90}{?} = 2.5 \text{ A}$$

    b. In circuit abc, R_3 and R_4 are in _____.

    $$R_y = \underline{\qquad} + 60 = \underline{\qquad} \ \Omega$$

    $$I_y = \frac{?}{90} = 1 \text{ A}$$

    c. Find V_2 and V_4.

$V_2 = \underline{\qquad} \times R_2$	$V_4 = \underline{\qquad} \times R_4$
$V_2 = \underline{\qquad} \times 12$	$V_4 = \underline{\qquad} \times 60$
$V_2 = 30$ V	$V_4 = 60$ V

    d. V_{TH} is equal to the *algebraic* sum of the voltages around path bcd. Be careful to note the polarities as indicated on the figure.

    $$V_{bcd} = V_{TH} = V_4 \xrightarrow{(+/-)} V_2$$
    $$V_{TH} = 60 - 30 = \underline{\qquad} \text{ V}$$

3. Replace each voltage source with a _____ and remove the _____ as shown in Fig. 9-55a. The circuit should now be redrawn in standard form as shown in Fig. 9-55b.

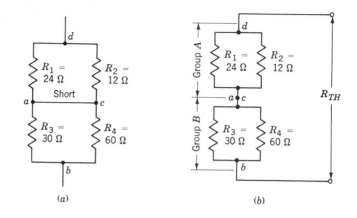

Figure 9-55 (a) Figure 9-53 with the load removed and the battery shorted. (b) Figure 9-55a redrawn in standard form.

4. Find R_{TH}. In Fig. 9-55b,
   R_1 and R_2 form a ____(series/parallel)____ group A.
   R_3 and R_4 form a ____(series/parallel)____ group B.
   Group A is in ____(series/parallel)____ with group B.

$$R_{TH} = \frac{R_1 \times R_2}{R_1 + R_2} + \underline{\hspace{2cm}}$$

$$R_{TH} = \frac{24 \times 12}{24 + 12} + \frac{30 \times 60}{30 + 60}$$

$$R_{TH} = \underline{\hspace{1.5cm}} + \underline{\hspace{1.5cm}}$$

$$R_{TH} = \underline{\hspace{1.5cm}} \, \Omega$$

5. Complete the drawing of the equivalent circuit of Fig. 9-56.

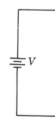

Figure 9-56 Incomplete Thevenin circuit.

6. Solve for I_L.

$$R_t = \underline{\hspace{1.5cm}} + \underline{\hspace{1.5cm}} = 60 \, \Omega$$

$$I_L = \frac{?}{R_t}$$

$$I_L = \frac{?}{60} = 0.5 \, A \qquad Ans.$$

7. Solve for V_L.

$$V_L = \underline{\hspace{1.5cm}} \times \underline{\hspace{1.5cm}}$$

$$V_L = 0.5 \times \underline{\hspace{1.5cm}} = \underline{\hspace{1.5cm}} \, V \qquad Ans.$$

Practice

Solve the following problems by applying Thevenin's theorem to the circuit.

1. In a circuit similar to that shown in Fig. 9-14a, $V_T = 25$ V, $R_1 = 1\ \Omega$, $R_2 = 4\ \Omega$, $R_3 = 5\ \Omega$, and $R_L = 10\ \Omega$. Find I_L and V_L.

2. In a circuit similar to that shown in Fig. 9-22, $V_1 = 75$ V, $R_2 = 3\ \Omega$, $R_4 = 4\ \Omega$, and $V_5 = 28$ V. If the load $R_3 = 12\ \Omega$, find the currents x, y, and z and the load voltage V_3.

3. In a circuit similar to that shown in Fig. 9-48a, $V_T = 120$ V, $R_1 = 2\ \Omega$, $R_2 = 4\ \Omega$, $R_3 = 6\ \Omega$, and $R_L = 3.6\ \Omega$. Find I_L and V_L.

4. In a circuit similar to that shown in Fig. 9-29, $V_1 = 76$ V, $R_2 = 10\ \Omega$, $R_3 = 2\ \Omega$, $R_4 = 15\ \Omega$, and $V_5 = 54$ V. Find the current y and the voltage across R_3.

5. In a circuit similar to that shown in Fig. 9-53, $R_1 = 30\ \Omega$, $R_2 = 70\ \Omega$, $R_3 = 10\ \Omega$, $R_4 = 15\ \Omega$, and $V_T = 100$ V. If $R_L = 23\ \Omega$, find I_L and V_L.

6. Solve Example 9-14 by Thevenin's theorem.

7. Using Fig. 9-28, find I_5 and V_5 by Thevenin's theorem.

8. Using Fig. 9-25, find the current w and V_2 by Thevenin's theorem.

9. Solve the circuit shown in Fig. 9-57 for the values of I_L and V_L.

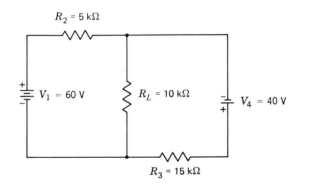

Figure 9-57

10. Solve the circuit shown in Fig. 9-58 for the values of I_L and V_L.

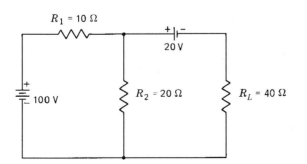

Figure 9-58

Norton's Theorem

Norton's theorem is another theorem like Thevenin's which is used to simplify circuit analysis. The two theorems give identical results, and the two equivalent circuits are very easily related. The difference is that Norton's theorem uses a *current* source instead of the voltage source used by Thevenin. This is more convenient for many of the newer semiconductor devices which are current-operated rather than voltage-operated.

The form of the Norton equivalent is shown in Fig. 9-59.

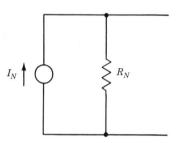

Figure 9-59   Norton equivalent circuit.

The R_N of the Norton equivalent is the same as the R_{TH} of the Thevenin equivalent. When a Thevenin equivalent is established for a given pair of points in a network, the voltage at the two points is the V_{TH}. For a Norton equivalent, the two points are shorted together and the current through them is the Norton equivalent current I_N. The relationships between the parameters of the two equivalents is shown in Fig. 9-60.

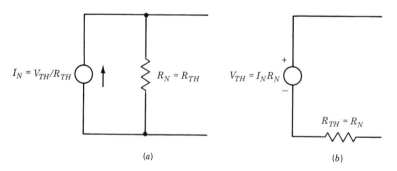

Figure 9-60   Relationship between the (a) Norton and (b) Thevenin equivalent circuits.

Norton's Theorem

Any linear network of voltage sources and resistances, if viewed from any two points in the network, can be replaced by an equivalent resistance R_N in parallel with an equivalent current source I_N.

Procedure

1. Find the equivalent current source I_N.
2. Find the equivalent resistance R_N.
3. Place the R_N in parallel with the load, and solve the simple parallel circuit.

Example 9-28

Convert the Thevenin equivalent circuit shown in Fig. 9-43 to a Norton equivalent.

Solution

1. Disconnect the load R_L from the circuit.
2. Short the open ends of the circuit together and determine the current flowing through the terminals.

$$I = \frac{V_{TH}}{R_{TH}} = I_N$$

3. Short the voltage sources and determine the resistance between the two open-circuit points.

$$R = R_{TH} = R_N$$

These equations are exactly the same as those shown in Fig. 9-60. The final circuit is shown on the left of that figure.

Hint for Success

Although valuable in circuit analysis, Norton's theorem is used less frequently than Thevenin's theorem. With Norton's theorem, a complex circuit can be reduced to a simple parallel circuit.

Example 9-29

Solve the circuit in Example 9-24, by Norton's theorem.

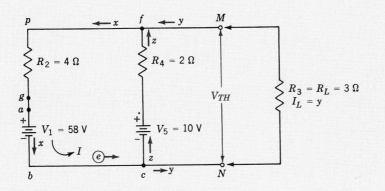

Solution

The original problem was redrawn as in Fig. 9-45, which is repeated above for your convenience.

1. Find the current through terminals M and N when they are shorted. By Kirchhoff's law, the Norton current I_N will be the sum of the currents in R_2 and R_4.

$$I_N = \frac{V_5}{R_4} + \frac{V_1}{R_2}$$

$$= \frac{10}{2} + \frac{58}{4}$$

$$= 5 + 14.5 = 19.5 \text{ A}$$

2. Find R_N. This value is determined exactly as is the R_{TH} found in step 4 on page 330. $R_N = 1.33$.
3. Draw the equivalent circuit as shown in Fig. 9-61.

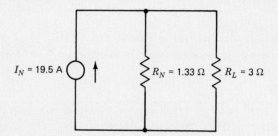

Figure 9-61   Norton equivalent circuit for Example 9-29.

4. Solve for I_L. Use the current divider Eqs. (5-6) and (5-7).

$$I_1 = \frac{R_2}{R_1 + R_2} \times I_T \tag{5-6}$$

$$I_L = \frac{R_N}{R_N + R_L} \times I_N$$

$$I_L = \frac{1.33}{1.33 + 3} \times 19.5$$

$$I_L = 6 \text{ A} \qquad Ans.$$

This is exactly the same answer as that obtained with Thevenin's theorem, and it was obtained much more easily. The reason is that once the equivalent circuit shown in Fig. 9-45 was obtained, it was very easy to see what the Norton equivalent current would be. This may not be true for all circuits, which requires that we understand both theorems.

Self-Test 9-30

Solve the circuit of Example 9-25, using Norton's theorem. Use exactly the same configuration as shown in Fig. 9-48, which is repeated on the next page for your convenience.

Solution

1. Find the Norton equivalent current.
   a. Disconnect the load R_L from the circuit.

b. Short the open ends *c* and *f* together and determine the current flowing through the terminals. This will be the current through _____.

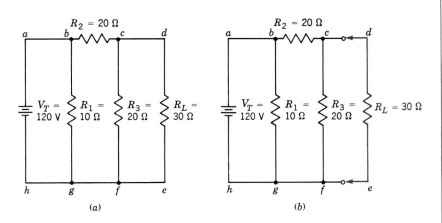

Figure 9-48 (repeated for reference)

$$I_N = \frac{V_T}{R_2} = \frac{120}{20} = 6 \text{ A}$$

2. Find R_N. This is exactly the same as R_{TH}, as shown in step 4 on page 332. $R_{TH} = 10 \ \Omega$, and $R_N = $ _____ Ω.
3. Draw the equivalent circuit as shown in Fig. 9-62.

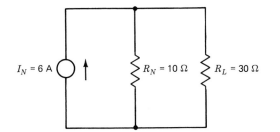

Figure 9-62 Norton equivalent circuit for Self-Test 9-30.

4. Solve for I_L. Use the current divider equations.

$$I_L = \frac{?}{R_N + R_L} \times I_N$$

$$= \frac{10}{10 + 30} \times 6 = \text{_____ A} \qquad Ans.$$

Exercises

Practice

Repeat the problems given in Job 9-8, using Norton's theorem. All the answers are the same. Compare the two methods and determine when one is better than the other.

Review

In Fig. 9-38, circuit *abc* is in ___(1) (delta/Y)___ formation.

$$\Sigma R_\Delta = \underline{\quad(2)\quad} \qquad\qquad R_b = \frac{(4)}{\Sigma R_\Delta}$$

$$R_a = \frac{(3)}{\Sigma R_\Delta} \qquad\qquad R_c = \frac{(5)}{\Sigma R_\Delta}$$

Thevenin's Theorem

Any two-terminal network can be replaced by an ___(6)___ simple series circuit of one resistance and one voltage.

1. The equivalent voltage is the terminal voltage of the complex circuit when the ___(7)___ is removed. Its symbol is ___(8)___.
2. The equivalent resistance is the resistance between the terminals of the complex circuit when the load is ___(9)___ and all sources of emf are ___(10)___. Its symbol is ___(11)___.

Procedure

1. Determine the load whose current I_L and whose voltage (___(12)___) are to be found. To simplify the problem, a voltage source ___(13) (may/may not)___ be included in the load.
2. Disconnect the ___(14)___ from the circuit at two points, and label these points as Thevenin terminals.
3. Find the open-circuit output voltage that would appear across these terminals. The symbol for this voltage is ___(15)___.
4. Replace each voltage source with a ___(16)___ and ___(17)___ the load.
5. Redraw the remaining circuit in standard form.
6. Find the total resistance that would appear across the Thevenin terminals. This is the symbol ___(18)___.
7. Draw the equivalent circuit consisting of V_{TH} in ___(19)___ with R_{TH} and the ___(20)___.
8. If the "load" has included any voltage source, it must be ___(21) (included/ excluded)___ in the equivalent Thevenin circuit with its original polarity.
9. Solve the circuit for the load ___(22)___ and the load ___(23)___. The total resistance of this circuit is the sum of R_{TH} and the ___(24)___ resistance.

Norton's Theorem

1. The Norton equivalent current is the current through the circuit terminals when the output is ___(25)___.
2. The symbol for the Norton equivalent current is ___(26)___.
3. The symbol for the Norton equivalent resistance is ___(27)___.
4. The Norton equivalent resistance is calculated in exactly the same manner as the ___(28)___.
5. The Norton equivalent is made by putting the R_N in ___(29)___ with ___(30)___.
6. The V_{TH} and the I_N are related by the equation ___(31)___.

(See CD-ROM for Test 9-3)

1. In Fig. 9-63 show the total current and the voltage drop across each resistor.

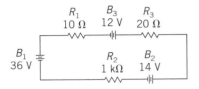

Figure 9-63

2. In Fig. 9-14 if B_1 is replaced with a 12-V battery of same polarity and R_4 is replaced with a 50-Ω resistor, what will be the total current?

3. In Fig. 9-64 determine total current, direction of current in each branch, and voltage drop across each resistor.

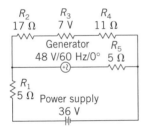

Figure 9-64

4. In Fig. 9-65 solve for the currents V, W, X, Y, and Z and find the total source voltage.

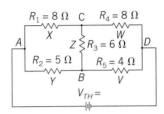

Figure 9-65

5. In Fig. 9-66 find the Thevenin voltage and resistance for part a and draw them in the correct place on part b.

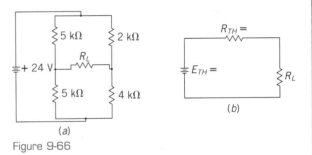

Figure 9-66

6. A circuit has a Thevenin voltage of 15 V and a Thevenin resistance of 3.5 kΩ. Draw the Norton circuit.

INTERNET
ACTIVITIES

Internet Activity A

View the following site to learn more about earth ground connections in electronic circuitry. From the information provided at this site, does the voltage at earth ground always have to be zero volts?
http://ee.elen.utah.edu/~cfurse/ee108.dir/Ex1.html

Internet Activity B

At the site given above, assume $V_a = 12$ V and $V_b = 24$ V, $R_1 = 12\ \Omega$, $R_2 = 10\ \Omega$, and $R_3 = 15\ \Omega$. Solve for V_{eq}, V_1, I_1, I_2, I_3, and V_3.

Chapter 9 Solutions to Self-Tests and Reviews

Self-Test 9-4

2. a. $-$, same
   b. $+$, opposite
   c. $-$, $-$, $+$
   d. $+$, $+$, $-$

Self-Test 9-8

a. 1. aid, opposes, counterclockwise
   3. $2.1I$, $3I$, $3.7I$
   4. counterclockwise, $-$, $-$, $+$, $+$, $+$, $-$, $-$, same
   5. 0, $3.7I$, $9I$, $9I$, 4

b. 1. 8.4
   2. 3.7
   3. 14.8
   4. zero
   5. $+$
   6. opposite
   7. $-$
   8. $-$
   9. $+$
   10. $+$
   11. $+$
   12. $-$
   13. $+0.8$

c. 30.8

Self-Test 9-11

1. $60I_1$, V_T, $20I_2$, equal, $60I_1$, 20, 3
2. 12, $3I_1$, $4I_1$, 3, 3, 9
3. 20, 180, 180

Self-Test 9-18

1. down
2. y, z, x
3. x, -80, 16

4. series, 35, $35y$, -190, -5, 38
5. z–x, $7x$, $8z$
6. 1, 3, 7, 112, 150, 10
7. 10, 10, 6
8. 4
9. 6, 10, 50, 4, 4, 4, 32
10. 30, 0
11. 48, 0

Review 9-6

1. enter, leave
2. a. out of
   b. into
3. algebraic, zero
4. a. positive, $+$
   b. negative, $-$
   c. same
   d. positive, $+$
5. a. direction, incorrect, is not
   b. directions
   c. currents
   d. complete, algebraic, zero
   f. equations
   g. Ohm's
   h. zero

Self-Test 9-19

1. aid, down
2. z, 0
3. $+$, 80, x, x, $5x$, 16, 16
4. V, V
5. 2, 0, 0, 0
6. 223.5 W

Self-Test 9-21

1. 20, 4, 6, 6, 2, 1.2, 3
3. 40, R_b, 10

4. parallel, 8
5. series, R_D, 10
6. Ohm's, 10, 12

Self-Test 9-22

1. a. 3, 18, 9, 3, 9, 18, 1.5, 1, 3
   c. series, 7, R_b, 7
   d. parallel, 3.5
   e. 5
2. 30
3. 3
4. 18, 12
5. 18, 6, 1
6. w, 2, 6, 2, 4
7. 2, y, 6, 4, w, v
8. 0

Self-Test 9-27

2. a. b, d, series, 12, 36
   b. series, 30, 90, 90
   c. I_x, I_y, 2.5, I
   d. $-$, 30
3. short, load
4. parallel, parallel, series, $\dfrac{R_3 \times R_4}{R_3 + R_4}$, 8, 20, 28
5. See Fig. 9-67.

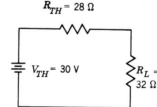

Figure 9-67 Thevenin equivalent circuit.

6. 28, 32, V_{TH}, 30
7. I_L, R_L, 32, 16

Self-Test 9-30

1. b. R_2
2. 10
4. R_N, 1.5

Review 9-10

1. delta
2. $R_1 + R_2 + R_3$
3. $R_1 \times R_2$
4. $R_2 \times R_3$
5. $R_1 \times R_3$
6. equivalent
7. load
8. V_{TH}
9. removed
10. shorted
11. R_{TH}
12. V_L
13. may
14. load
15. V_{TH}
16. short
17. remove
18. R_{TH}
19. series
20. load
21. included
22. current
23. voltage
24. load
25. shorted
26. I_N
27. R_N
28. R_{TH}
29. Parallel
30. R_L
31. $V_{TH} = I_N \times R_N$

Applications of Series and Parallel Circuits

Learning Objectives

1. Calculate the value of shunt resistance required to extend the range of a basic ammeter.
2. Calculate the value of multiplier resistance required to make a basic ammeter capable of measuring voltage.
3. Calculate the required resistances of a voltage divider.
4. Understand the voltage-comparison method of resistance measurement.
5. Understand the operation of a Wheatstone bridge.
6. Apply Thevenin's and Norton's theorems to circuit analysis.
7. Understand basic transistor operation.
8. Calculate dc voltage and currents in transistor circuits.
9. Draw dc equivalent circuits for fixed-bias transistor circuits.
10. Draw the *H*-parameter model of a transistor.
11. Describe the purpose of an attenuator pad in an electric circuit.

An *ammeter* is a device that is used to measure the current in an electric circuit. To do this, it must always be inserted directly into the line whose current is to be measured. An ammeter can be used to measure many ranges of currents by taking advantage of the fact that current in a parallel circuit will divide—the large current flowing through the small resistance and the small current flowing through the large resistance.

JOB 10-1

Extending the Range of an Ammeter

An *ammeter* is essentially a coil of very fine wire that turns in proportion to the current flowing through it. A *pointer* attached to the coil moves over a dial and indicates the value of the current. The weight of the coil must be very small so that it can react to the current in it. This weight factor necessitates the use of very fine wire that can carry only about 0.05 A at best. Yet we can use this ammeter to measure much larger currents by taking advantage of the fact that current in a parallel circuit will divide.

Range of an ammeter. The *range* of an ammeter indicates the value of current required to cause the pointer to swing over the entire scale. This is called *full-scale deflection*. Ranges are indicated as "0–1 mA," "0–100 mA," etc. Thus:

0–1 mA requires 1 mA for full-scale deflection
0–100 mA requires 100 mA for full-scale deflection
0–10 A requires 10 A for full-scale deflection

Extending the range of an ammeter. Suppose we had a milliammeter with a range of 0–1 mA. It is desired to extend its range to read up to 50 mA. This can be accomplished by connecting a low-resistance resistor called a *shunt* in parallel with the meter as shown in Fig. 10-1. The object is to use a shunt of such a value that the current I in the line will divide at point A so as to keep the current flowing through the coil unchanged. Thus, using a 0–1 mA milliammeter, if the range is to be extended to read 0–50 mA, the shunt should carry 49 mA and the coil should carry its original 1 mA. If the range is to be extended to read 100 mA, then the shunt should carry 99 mA and the coil of the meter should carry its original 1 mA. In general, with *any* shunt, the current through the coil for full-scale deflection is the current necessary for full-scale deflection *before* using the shunt.

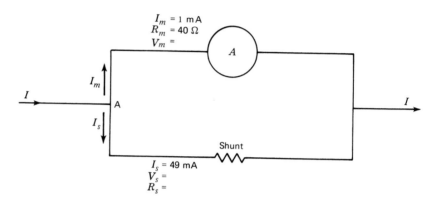

Figure 10-1   A shunt is placed in parallel with an ammeter. I = line current to be measured; V_m = voltage drop across meter; I_m = current through meter; R_m = resistance of meter; V_s = voltage across shunt; I_s = current through shunt; R_s = resistance of shunt.

In Fig. 10-1, to extend the meter to 50 mA, we require a shunt R_s of such a value that the original 1 mA will flow through the meter and the remaining 49 mA will flow through the shunt. In effect, the pointer will cover the full scale (owing to the 1 mA), but it will be indicating the full 50 mA.

$$V_m = I_m \times R_m = 0.001 \times 40 = 0.04 \text{ V}$$

but since the shunt is in parallel with the meter,

$$V_s = V_m = 0.04 \text{ V}$$
$$R_s = \frac{V_s}{I_s} = \frac{0.04}{0.049} = 0.816 \ \Omega \qquad \textit{Ans.}$$

These calculations may be combined to produce the

$$R_s = \frac{I_m \times R_m}{I_s} \qquad \qquad \text{Formula 10-1}$$

And since
$$I = I_m + I_s \qquad \qquad (5\text{-}2)$$

by transposing,
$$I_s = I - I_m \qquad \qquad 10\text{-}2$$

By substituting this value of I_s in formula (10-1), we obtain the

$$R_s = \frac{I_m \times R_m}{I - I_m} \qquad \qquad \text{Formula 10-3}$$

In this formula, the resistance of the meter must be known, and it can usually be obtained from the manufacturer.

Example 10-1

A Weston model 430 0–1-mA milliammeter has a resistance of 92 Ω. What shunt is necessary to extend its range up to 50 mA?

Solution

Given: $I_m = 1$ mA $= 0.001$ A   Find: $R_s = ?$

$R_m = 92\ \Omega$

$I = 50$ mA $= 0.05$ A

$$R_s = \frac{I_m \times R_m}{I - I_m} = \frac{0.001 \times 92}{0.05 - 0.001} = \frac{0.092}{0.049} = 1.88\ \Omega \qquad Ans.$$

A simpler formula to extend the range of ammeters may be found by a series of algebraic manipulations on formula (10-3).

$$R_s = \frac{I_m \times R_m}{I - I_m} \tag{10-3}$$

or

$$R_s = R_m \frac{I_m}{I - I_m}$$

Dividing by 1 expressed as I_m/I_m

$$R_s = R_m \frac{I_m \div I_m}{(I - I_m) \div I_m}$$

$$R_s = R_m \frac{1}{\left(\dfrac{I}{I_m} - 1\right)} \tag{a}$$

The fraction I/I_m is the ratio of the maximum *new* range as compared with the maximum *old* range. As such, it represents the amount by which the range of the meter has been multiplied. If N is used to represent this multiplying factor,

$$N = \frac{I}{I_m}$$

$$N = \frac{\textbf{max new range}}{\textbf{max old range}} \qquad \text{Formula 10-4}$$

By substituting N for I/I_m in equation (a) above, we obtain

$$R_s = \frac{R_m}{N - 1} \qquad \text{Formula 10-5}$$

where $R_s =$ shunt resistor
$R_m =$ resistance of the meter
$N =$ multiplying factor

Self-Test 10-2

A Weston model 433 0–100-mA milliammeter has a resistance of 49 Ω. Find the shunt resistor necessary to extend the range to give maximum deflection at 750 mA.

Solution

Given:
1. Old range = —————— mA   Find R_s = ?
2. New range = —————— mA
3. R_m = —————— Ω

4. Multiplying factor $N = \dfrac{\text{max ? range}}{\text{max ? range}}$

5. $N = \dfrac{750}{?} = 7.5$ (10-4)

6. $R_s = \dfrac{R_m}{N - 1}$ (10-5)

7. $R_s = \dfrac{?}{7.5 - 1}$

8. $R_s = $ —————— Ω *Ans.*

Exercises

Applications

1. A D'Arsonval meter movement has a resistance of 50 Ω and requires 1 mA of current for full-scale deflection. Find the shunt resistor needed to convert this movement to a milliammeter with a range of 1–10 mA.
2. A 0–1 mA milliammeter has a resistance of 100 Ω. If the meter is to be converted to read 100 mA, which of the following shunts should be used: (*a*) 0.1 Ω, (*b*) 10 Ω, (*c*) 1.01 Ω, (*d*) 1010 Ω?
3. Find the shunt necessary to extend the range of a 0–1-mA milliammeter with a resistance of 27 Ω to read 0–10 mA.
4. A 0–1-mA milliammeter has a resistance of 5 Ω. What shunt resistor is necessary to give a full-scale reading of 51 mA?
5. Find the shunt needed for the milliammeter of Prob. 4 that will permit full-scale deflection at 0.1 A.
6. A Weston model 45, 0–100-mA milliammeter has a resistance of 0.5 Ω. Find the shunt needed to extend its range to read 0–300 mA.
7. A Weston model 1, 0–15-mA milliammeter has a resistance of 2.24 Ω. Find the shunt needed to extend its range to read 0–60 mA.
8. Full-scale deflection results when 25 mA flows through an ammeter of 25 Ω resistance. Find the shunt resistance needed to extend its range to read (*a*) 250 mA, (*b*) 500 mA, and (*c*) 1 A.
9. A Weston model 45 0–3-A ammeter has a resistance of 5 Ω. Find the shunt needed to extend its range to read 0–75 A.
10. When the coil current in a meter is 10 mA, the pointer deflects past a full scale of 100 divisions. If the coil resistance is 15 Ω, find the shunts needed to extend its range to read (*a*) 0–1 A and (*b*) 0–5 A.

Reading Ammeters of Extended Range

Example 10-3

The range of a 0–100-mA milliammeter was increased to read 0 to 1 A. Find the true current flowing when the meter reads 60 mA.

Solution

Given: Old range = 100 mA   Find: $I = ?$
                      New range = 1 A = 1000 mA
                            I_m = 60 mA

The increase in range was from 100 to 1000 mA.

$$N = \frac{\text{max new range}}{\text{max old range}} = \frac{1000}{100} = 10 \qquad (10\text{-}4)$$

Therefore all readings will be increased 10 times, or a reading of 60 mA really means

$$60 \times 10 = 600 \text{ mA} \qquad \text{or} \qquad 0.6 \text{ A} \qquad \textit{Ans.}$$

> **Hint for Success**
>
> When the insertion of a current meter reduces the circuit current below that which exists without the meter present, the effect is called *current meter loading*.

Example 10-4

A model 221-T Triplett 0–1-mA milliammeter has a resistance of 55 Ω. When it is shunted with a 2-Ω resistor, the meter reads 0.5 mA. Find the true current.

Solution

Given: I_m = 0.5 mA   Find: $I = ?$
            R_m = 55 Ω
            R_s = 2 Ω

In this problem we do not know the increase in range and cannot use the method shown in the last example. However, we can find the shunt current by formula (10-1). The line current I can then be found by formula (5-2).

$$R_s = \frac{I_m \times R_m}{I_s} \qquad (10\text{-}1)$$

$$\frac{2}{1} = \frac{0.5 \times 55}{I_s}$$

$$2 \times I_s = 0.5 \times 55$$

$$I_s = \frac{27.5}{2} = 13.8 \text{ mA}$$

The line current can now be found.

$$I = I_m + I_s = 0.5 + 13.8 = 14.3 \text{ mA} \qquad \textit{Ans.} \qquad (5\text{-}2)$$

Note: Since the shunt usually carries most of the line current, the shunt current is very often used as the line current.

Exercises

Applications

1. The range of 0–1-mA milliammeter was increased to read 0–50 mA. Find the true current when the meter reads 0.3 mA.
2. The range of a 0–0.1-A ammeter was increased to read 0–1 A. Find the true current when the meter reads 0.07 A.
3. A 0–1-mA milliammeter of 25 Ω resistance is used with a 0.25-Ω shunt. What is the true current when the meter reads 0.4 mA?
4. A Weston model 45 0–3-A ammeter of 5 Ω resistance is used with a 0.1-Ω shunt. What is the true current when the meter reads 0.6 A?
5. A 0–1-A ammeter of 10 Ω resistance is used with a 0.1-Ω shunt. What is the true current when the meter reads 0.7 A?
6. A 0–10-mA milliammeter of 10 Ω resistance is used with a 0.2-Ω shunt. What is the true current when the meter reads 2 mA?

JOB 10-3

Extending the Range of a Voltmeter

A voltmeter is a device that is used to measure the difference in electrical pressure across two points in a circuit. To do this, the instrument must be placed in parallel with the portion of the circuit being tested as shown in Fig. 10-2. If the resistance of the voltmeter is low, the current in the line will divide at A and the large current will flow through the moving coil of the meter and burn it out. To avoid this, the resistance of the moving coil is increased by adding a *high* resistance in *series* with the coil. These resistors are called *multipliers*. Essentially, a voltmeter is really an ammeter with a series resistor. This extra resistance holds down the current to the value required for full-scale deflection when the full-scale voltage is applied to it. It is possible to use one meter to read many different voltage ranges if the correct resistances can be placed in the instrument with an arrangement for disconnecting one resistance and connecting another in series. Commercial meters are available which do this by a variety of switching arrangements.

<table>
<tr><td>

Hint for Success

The sensitivity or ohms-per-volt rating of a voltmeter can be determined by taking the reciprocal of I_M, which is the current required through the meter movement to produce full-scale deflection of the needle.

</td></tr>
</table>

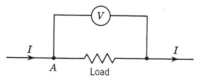

Figure 10-2 A voltmeter is always connected in parallel across the terminals of the part being tested.

Ohms per volt.   A perfect meter should measure the current but should not use any of the current in the circuit for itself. This is actually impossible, since some current, however little, is needed to operate the meter. The best, or most sensitive, meters are those which use as little current as possible. The more resistance that can be placed in a meter and still have it read 1 V, the less the current that is drawn. This means that the meter will use very little current for itself and will have a high sensitivity. Sensitivity is usually expressed as the number of ohms needed to read 1 V, or the ohms per volt R/v.

$$R/v = \frac{R_T}{V_T} \qquad \text{Formula 10-6}$$

where R/v = sensitivity, Ω/V
        R_T = total meter resistance, Ω
        V_T = maximum scale voltage, V

Example 10-5

A 50,000-Ω voltmeter reads 50 V at full scale. What is its sensitivity in ohms per volt?

Solution

Given: $R_T = 50,000 \ \Omega$ Find: $R/v = ?$
         $V_T = 50 \ V$

$$R/v = \frac{R_T}{V_T} = \frac{50,000}{50} = 1000 \ \Omega/V \qquad Ans. \qquad (10\text{-}6)$$

Example 10-6

A 1000-Ω/V voltmeter has a range of 0–10 V. What is the total resistance of the meter?

Solution

Given: $R/v = 1000 \ \Omega/V$ Find: $R_T = ?$
         $V_T = 10 \ V$

$$R/v = \frac{R_T}{V_T} \qquad (10\text{-}6)$$

$$1000 = \frac{R_T}{10}$$

$$R_T = 1000 \times 10$$
$$R_T = 10,000 \ \Omega \qquad Ans.$$

Hint for Success

Remember when dividing tens that if both the divisor and the dividend end in zeros, you can drop an equal number of zeros from both of them. For example, divide 50,000 by 50. Since 50 ends with one zero, you can drop only one zero from 50,000. So 50,000 ÷ 50 becomes 5000 ÷ 5 = 1000

Extending the range of a voltmeter.   A 0–150-V voltmeter has a 150,000-Ω resistor in series with its moving coil. How can it be altered so as to read up to 750 V?

Solution.   Since 750 V is five times as much as 150 V, we shall require five times as much resistance as is in the meter for the 150-V range. The resistance already in the meter may then be subtracted from the total required resistance to find the needed *additional* resistance to increase the range to 750 V. The increase in voltage (the five times) represents the multiplying power we desire. Therefore, the series resistor R_s which must be added to extend the range of an existing voltmeter is given by

$$R_s = (\text{multiplying power} \times R_{\text{meter}}) - R_{\text{meter}}$$

Another way to write this is as follows.

$$R_s = (MP - 1) \times R_{meter} \qquad \text{Formula 10-7}$$

where $\quad R_s$ = series resistor to be added, Ω
$\qquad R_{meter}$ = resistance of meter, Ω

and $\qquad\qquad\qquad MP = \dfrac{\textbf{max new voltage}}{\textbf{max old voltage}} \qquad \text{Formula 10-8}$

Let us use this formula to solve the problem above.

$$MP = \dfrac{\text{max new voltage}}{\text{max old voltage}} = \dfrac{750}{150} = 5 \qquad (10\text{-}8)$$

$$R_s = (MP - 1) \times R_m \qquad (10\text{-}7)$$
$$= (5 - 1) \times 150{,}000 = 4 \times 150{,}000 = 600{,}000 \; \Omega \qquad \textit{Ans.}$$

Note: The resistance of the meter must be known in order to use formula (10-7). If it is not given, it may be found by connecting the meter in series with a milliammeter across some voltage within its range as shown in Fig. 10-3. The resistance of the meter can then be found by the

$$R_m = \dfrac{\textbf{voltmeter reading (V)}}{\textbf{ammeter reading (A)}} \qquad \text{Formula 10-9}$$

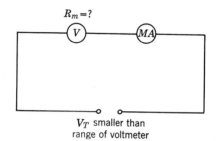

V_T smaller than
range of voltmeter

Figure 10-3 Circuit for finding the resistance of a voltmeter.

Example 10-7

A 0–10-V 10,000-Ω voltmeter is to be extended to read 100 V. What additional series resistance is needed?

Solution

Given: Max new voltage = 100 V   Find: R_s = ?
Max old voltage = 10 V
R_{meter} = 10,000 Ω

$$MP = \dfrac{\text{max new voltage}}{\text{max old voltage}} = \dfrac{100}{10} = 10 \qquad (10\text{-}8)$$

$$R_s = (MP - 1) \times R_m \qquad (10\text{-}7)$$
$$= (10 - 1) \times 10{,}000 = 9 \times 10{,}000 = 90{,}000 \; \Omega \qquad \textit{Ans.}$$

Example 10-8

A 1000-Ω/V 0–10-V voltmeter is to be extended to read 0–100 V. What is the needed series resistance?

Solution

Given: $R/v = 1000\ \Omega$ Find: $R_s = ?$
   Max new voltage $= 100$ V
   Max old voltage $= 10$ V

Before we can find the series resistance, we must know the resistance of the meter and the multiplying power.

$$R/v = \frac{R_T}{V_T} \qquad\qquad (10\text{-}6)$$

$$1000 = \frac{R_T}{10}$$

$$R_T = 10 \times 1000 = 10{,}000\ \Omega$$
$$R_m = 10{,}000\ \Omega \qquad Ans.$$

$$MP = \frac{\text{max new voltage}}{\text{max old voltage}} = \frac{100}{10} = 10 \qquad Ans. \qquad (10\text{-}8)$$

$$R_s = (MP - 1) \times R_m \qquad\qquad\qquad (10\text{-}7)$$
$$= (10 - 1) \times 10{,}000 = 9 \times 10{,}000 = 90{,}000\ \Omega \qquad Ans.$$

Self-Test 10-9

A 0–10-V voltmeter of 1000 Ω resistance is extended to read 100 V. Find the true voltage when the meter reads 5.5 V.

Solution

Given: 1. $R_m = $ _____ Ω Find: True voltage $= ?$
   2. Max new voltage $= $ _____ V
   3. Max old voltage $= $ _____ V
   4. Meter reading $= 5.5$ V

$$5.\ MP = \frac{\text{max new voltage}}{\text{max old voltage}} = \frac{?}{10} = 10$$

Therefore, all readings will be increased 10 times, so a reading of 5.5 V means

6. $5.5 \times$ _____ $= 55$ V   Ans.

Exercises

Applications

1. A 1000-Ω/V 0–10-V voltmeter is to be extended to read 300 V. Find (a) the series resistor needed and (b) the true voltage when the meter reads 5.6 V.
2. A 0–7.5-mV millivoltmeter whose resistance is 1 Ω is to be extended to read 15 V. Find (a) the series resistor needed and (b) the true voltage when the meter reads 6 mV.

A volt-ohm-multimeter (VOM) measures voltage, current, or resistance. The one pictured has an analog pointer reading and combines a function selector and range switch. In a VOM, the dc voltmeter R_V changes with range. The zero-ohms adjustment is changed for each range. Ohms range up to $R \times 10,000\ \Omega$ as a multiplying factor. A VOM is relatively inexpensive, simple, compact, and portable.

3. A 0–50-mV millivoltmeter of 5 Ω resistance is to be extended to read up to 30 V. Find (a) the series resistor needed and (b) the true voltage when the meter reads 36 mV.

4. A 0–100-V voltmeter of 12,000 Ω resistance is to read up to 300 V. Find the multiplying resistor needed.

5. A voltmeter with a full-scale deflection of 150 V has a resistance of 17,000 Ω. Find the external series resistance necessary to extend it to read (a) 300 V and (b) 600 V.

6. A 1000-Ω/V 0–1-V voltmeter is to be extended to read up to 100 V. Find (a) the series resistor needed and (b) the true voltage when the meter reads 0.7 V.

7. If the current taken by a 5000-Ω meter is 1 mA, find the series resistor needed to extend its range to read up to (a) 50 V, (b) 100 V, and (c) 150 V. *Hint:* Find V of the meter by $V = I \times R$.

8. Find the series resistor to be used with a 15,000-Ω voltmeter in order to have its readings multiplied by 5.

JOB 10-4

Review of Meters

Extending the Range of an Ammeter

Ammeters may be used to measure currents larger than their full-scale deflections by connecting appropriate shunts in parallel with the meter. The value of this shunt resistor may be found by the following formulas:

$$R_s = \frac{I_m \times R_m}{I_s} \qquad \text{10-1}$$

$$R_s = \frac{I_m \times R_m}{I - I_m} \qquad \text{10-3}$$

$$R_s = \frac{R_m}{N - 1} \qquad \text{10-5}$$

where
R_s = resistance of the ___(a)___, Ω
I_s = current in the ___(b)___, A
R_m = resistance of the ___(c)___, Ω
I_m = current in the ___(d)___, A
I = line current to be measured, A
N = multiplying factor which indicates the number of times that the range of the meter is increased

$$N = \frac{\text{max} \underline{\quad(e)\quad} \text{range}}{\text{max} \underline{\quad(f)\quad} \text{range}} \qquad \text{10-4}$$

Reading ammeters of extended range

1. If the multiplying factor N is known, the true current is equal to the actual reading multiplied by ————.

$$\textbf{True reading} = \textbf{actual reading} \times \textbf{N}$$

2. If N is unknown,
   a. Find the shunt current by

$$R_s = \frac{I_m \times R_m}{?} \qquad\qquad 10\text{-}1$$

   b. Find the true current by

$$I = I_m + ? \qquad\qquad 5\text{-}2$$

Extending the Range of a Voltmeter

The range of a voltmeter may be extended by placing it in series with the appropriate series multiplier resistor.

$$R_s = (MP - 1) \times \underline{\quad(1)\quad} \qquad\qquad 10\text{-}7$$

where R_s = series resistor to be added, Ω
$\quad\quad\;\; R_{\text{meter}}$ = resistance of meter, Ω

$$MP = \frac{\text{max} \underline{\;(2)\;} \text{ voltage}}{\text{max} \underline{\;(3)\;} \text{ voltage}} \qquad\qquad 10\text{-}8$$

Meter sensitivity.   The sensitivity of a meter is expressed as its resistance in ohms for every volt measured on its scale. For the same reading of 1 V, the meter with the larger resistance will draw ____(4) (more/less)____ current for itself and therefore give more accurate measurements.

$$R/v = \frac{R_T}{V_T} \qquad\qquad 10\text{-}6$$

where $R/v = \underline{\quad(5)\quad}, \Omega/V$
$\quad\quad\;\; R_T$ = total meter resistance, Ω
$\quad\quad\;\; V_T$ = maximum scale voltage, V

Finding the total resistance of the meter

1. Use formula (10-6) to find R_T of the meter, or
2. Connect the voltmeter in series with an ammeter across some voltage within its range. The meter resistance is given by

$$R_m = \frac{\textbf{voltmeter reading (V)}}{\textbf{ammeter reading (A)}} \qquad\qquad 10\text{-}9$$

Exercises

Applications

1. A 0–2-mA milliammeter has a resistance of 18 Ω. Find the parallel shunt resistor which will permit measurement of currents up to 200 mA.

2. The range of a 12-Ω 0–5-mA milliammeter was increased to read 0–50 mA. Find the true current when the meter reads 4.5 mA.

3. A 0–1-mA milliammeter of 20 Ω resistance is used with a 0.2-Ω shunt in parallel. Find the true current when the meter reads 0.6 mA.

4. A 10,000-Ω voltmeter reads 100 V at full-scale deflection. Find its sensitivity in ohms per volt.

5. A D'Arsonval meter movement with a resistance of 250 Ω requires 3 mA for full-scale deflection. Find its sensitivity in ohms per volt.

6. The range of a 15-V 1000-Ω/V voltmeter is to be extended to read 150 V. What value of resistance should be added?

7. Which of the following represents the full-scale sensitivity of a 20,000-Ω/V meter: (a) 20 μA, (b) 50 μA, (c) 2 mA, (d) 5 mA?

8. A 500-Ω meter movement has a full-scale current rating of 400×10^{-6} A. If this movement is used in a 200-V voltmeter with a multiplier, find the ohms per volt rating of the voltmeter.

9. A Weston model 5 dc voltmeter has a sensitivity of 100 Ω/V and a range of 0–15 V. Find the total resistance of the meter.

10. A 0–30-V voltmeter is to be extended to read 150 V. If the resistance of the meter is 3000 Ω, find (a) the sensitivity, (b) the series resistor needed, and (c) the true voltage when the meter reads 12 V.

11. A 100-Ω/V 0–7.5-V voltmeter is to be extended to read 0–30 V. Find (a) the series multiplier needed and (b) the true voltage when the meter reads 3.2 V.

12. If the current taken by a 1000-Ω meter is 15 mA, find (a) the voltage indicated by the full-scale deflection, (b) the sensitivity in ohms per volt, and (c) the series multiplier resistor needed to extend the range to read (1) 120 V and (2) 180 V.

13. The resistance of a 0–15-mA milliammeter is 3.2 Ω. Find the parallel shunt required to extend the range to 0–100 mA.

14. The range of an 18-Ω 0–3-mA milliammeter was increased to read up to 50 mA. Find the true current when the meter reads 1.7 mA.

(See CD-ROM for Test 10-1)

Voltage Dividers

The filter circuit of a power supply is designed to smooth out the dc voltage delivered from the rectifier. This voltage may be 30 V. However, although some circuits in the receiver may require these 30 V, other circuits may need only 18 or only 9 V. These different voltages may be obtained by placing a large resistor across the filter output as shown in Fig. 10-4. This resistor, known as a *bleeder* resistor, serves as a fixed load on the power supply. Also, any charge remaining on the filter capacitors when the power supply is turned off bleeds through this resistor and eliminates the danger of shock. Some current (about 11 percent of the total current) will always flow through the bleeder as long as the rectifier is operating. The different voltages required may be obtained by tapping the bleeder at various points, which changes the bleeder into a *voltage divider*.

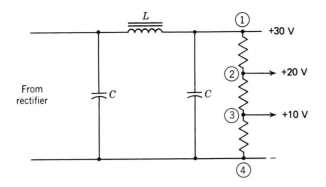

Figure 10-4   Power-supply voltage divider.

The total voltage of 30 V appears across points 1 and 4. If the resistor is divided into three equal parts by taps at points 2 and 3, at point 2, which is one-third of the resistor, one-third of the total voltage ($\frac{1}{3} \times 30 = 10$) has been used up. Therefore only

$$30 - 10 = 20 \text{ V}$$

will be available at point 2. At point 3, which is two-thirds of the resistor, two-thirds of the 30 V ($\frac{2}{3} \times 30 = 20$) has been used up, and at point 3 only $30 - 20 = 10$ V will be available. This will all be true only if the current in all parts of the voltage divider is the same current. However, the different voltages must be available at different currents, and so the calculations for a working voltage divider are a bit more complicated.

Example 10-10

Calculate the resistances for the sections of a voltage divider which is to deliver 40 mA at 300 V, 30 mA at 180 V, and 20 mA at 90 V.

Solution
The diagram for the circuit is shown in Fig. 10-5.

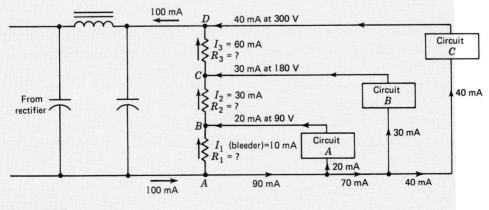

Figure 10-5

1. Determine the current distribution. The total current required equals $20 + 30 + 40 = 90$ mA. The bleeder current = 11 percent of 90 mA = $0.11 \times 90 = 10$ mA (approximately).

At A: $I_1 = I$ from rectifier $-$ current to loads
             $I_1 = 100 - 90 = 10$ mA

At B: $I_2 = I_1 +$ current from load
             $I_2 = 10 + 20 = 30$ mA

At C: $I_3 = I_2 +$ current from load
             $I_3 = 30 + 30 = 60$ mA

At D: $I_T = I_3 +$ current from load
             $I_T = 60 + 40 = 100$ mA

2. Find the sectional resistances.

For R_3: The voltage drop from D to $C = 300 - 180 = 120$ V.

$$V_3 = I_3 \times R_3 \tag{4-6}$$
$$120 = 0.06 \times R_3$$

$$R_3 = \frac{120}{0.06} = 2000 \ \Omega \qquad \textit{Ans.}$$

For R_2: The voltage drop from C to $B = 180 - 90 = 90$ V.

$$V_2 = I_2 \times R_2 \tag{4-5}$$
$$90 = 0.03 \times R_2$$

$$R_2 = \frac{90}{0.03} = 3000 \ \Omega \qquad \textit{Ans.}$$

For R_1: The voltage drop from B to $A = 90 - 0 = 90$ V.

$$V_1 = I_1 \times R_1 \tag{4-4}$$
$$90 = 0.01 \times R_1$$

$$R_1 = \frac{90}{0.01} = 9000 \ \Omega \qquad \textit{Ans.}$$

Self-Test 10-11

Find the resistances of the sections R_1, R_2, R_3, and R_4 of the voltage divider shown in Fig. 10-6. The -50-V bias terminal draws no current, and the bleeder current is 10 percent of the total load current.

Solution

1. Determine the current distribution

Total load current $= 50 + 30 +$ _____ mA
$$I_L = \text{_____} \ \text{mA}$$
Bleeder current $= 10$ percent $\times 100$
$$= \text{_____} \times 100$$
$$I_b = \text{_____} \ \text{mA}$$

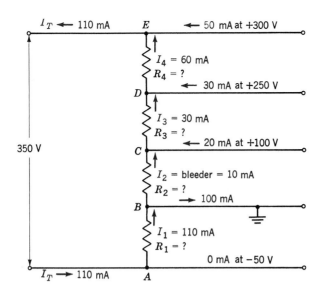

Figure 10-6   Voltage-divider circuit for Self-Test 10-11.

Total current = bleeder current + load current

At B: $I_T = I_1 = I_b$ + load current

$I_T = 10 +$ _____

$I_T = 110$ mA = _____ A

At C: $I_3 = I_2$ + current from load

= 10 + _____

$I_3 =$ _____ mA

At D: $I_4 =$ _____ + current from load

= 30 + _____

$I_4 =$ _____ mA

At E: $I_T =$ _____ + current from load

= 60 + _____

$I_T =$ _____ mA = _____ A

2. Find the sectional resistances.

For R_1: The voltage drop between A and B = 50 − 0.

$V_1 = 50$ V

$R_1 = \dfrac{50}{?}$

$R_1 =$ _____ Ω   *Ans.*

For R_2: The voltage drop between B and C = _____ V.

$R_2 = \dfrac{100}{?}$

$R_2 =$ _____ Ω   *Ans.*

For R_3: The voltage drop between _____ and _____

= 250 − _____

= _____ V

$$R_3 = \frac{?}{0.03}$$

$$R_3 = \underline{\qquad} \ \Omega \qquad Ans.$$

For R_4: The voltage drop between D and E = $\underline{\qquad}$ V.

$$R_4 = \frac{?}{0.06}$$

$$R_4 = \underline{\qquad} \ \Omega \qquad Ans.$$

Exercises

Applications

1. Find the resistances of sections R_1, R_2, and R_3 of the voltage divider shown in Fig. 10-7a.

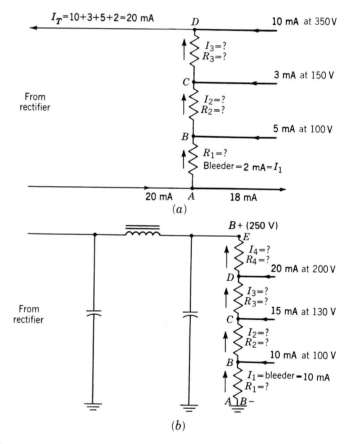

Figure 10-7

2. Find the resistances of sections R_1, R_2, R_3, and R_4 of the voltage divider shown in Fig. 10-7b. *Hint:* Find the current distribution starting from point A and working up to point E.
3. A power supply is to deliver 45 mA at 150 V, 30 mA at 100 V, and 10 mA at 50 V. Assuming a bleeder current of 5 mA, calculate the resistances of the sections of the voltage divider needed.

4. Design a voltage divider to deliver 70 mA at 350 V, 40 mA at 250 V, and 20 mA at 100 V if the power-supply transformer is rated at 155 mA and operates at 90 percent of its rated value.
5. A power supply is to deliver 60 mA at 250 V, 5 mA at 100 V, and 2 mA at 50 V. The transformer is rated at 80 mA and operates at 90 percent of its rated value. Design a voltage divider to deliver the required currents and voltages.
6. In the voltage divider shown in Fig. 10-8, the bias bleeder current is 60 mA. Find the values of R_1, R_2, R_3, and R_4.

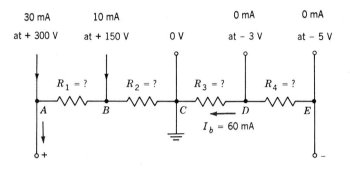

Figure 10-8

Resistance Measurement by the Voltage-Comparison Method. The Potentiometer Rule

JOB 10-6

In Fig. 10-9, the value of the unknown resistor R_x may be found by connecting it in series with a known standard resistance R_k and measuring the voltage drops across each.

$$V_x = I_x \times R_x \qquad (4\text{-}4)$$
$$V_k = I_k \times R_k \qquad (4\text{-}5)$$
$$V_T = I_T \times R_T \qquad (4\text{-}7)$$
$$I_T = I_x = I_k \qquad (4\text{-}1)$$

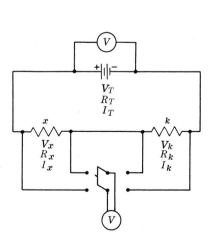

Figure 10-9 Voltage-comparison method of measuring resistance. X = unknown resistance; K = known standard resistance; R_T = total resistance.

Dividing Eq. (4-4) by Eq. (4-5) and canceling out the equal currents $I_x = I_k$,

$$\frac{V_x = \cancelto{1}{I_x} \times R_x}{V_k = \cancelto{1}{I_k} \times R_k}$$

we obtain the

$$\frac{V_x}{V_k} = \frac{R_x}{R_k}$$ Formula 10-10

Similarly, by dividing Eq. (4-4) by Eq. (4-7), we obtain the

$$\frac{V_x}{V_T} = \frac{R_x}{R_T}$$ Formula 10-11

These formulas mean that, in a series circuit, the voltage across any resistance depends on the value of the resistance—a large voltage appearing across a large resistance and a small voltage appearing across a small resistance. This is discussed in greater detail in Job 12-2.

Example 10-12

Using Fig. 10-9, find the value of the unknown resistance K if the known resistance $R_k = 1000\ \Omega$, $V_x = 100$ V, and $V_k = 25$ V.

Solution

Given: $R_k = 1000\ \Omega$ Find: $R_x = ?$
          $V_x = 100$ V
          $V_k = 25$ V

$$\frac{V_x}{V_k} = \frac{R_x}{R_k}$$ (10-10)

$$\frac{100}{25} = \frac{R_x}{1000}$$

$$25 \times R_x = 100 \times 1000$$

$$R_x = \frac{100{,}000}{25} = 4000\ \Omega \qquad Ans.$$

Example 10-13

A horizontal-hold control circuit of a television receiver is essentially that shown in Fig. 10-10. Find (*a*) the voltage drop across the 27-kΩ resistor, (*b*) the voltage drop across the 75-kΩ resistor, and (*c*) the voltage drop across the 68-kΩ resistor.

Solution

a. Find the total resistance of the circuit.

$$R_T = R_1 + R_2 + R_3$$ (4-3)

$$R_T = 27\ \text{k}\Omega + 75\ \text{k}\Omega + 68\ \text{k}\Omega = 170\ \text{k}\Omega \qquad Ans.$$

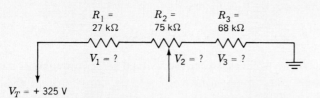

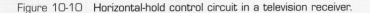

Figure 10-10 Horizontal-hold control circuit in a television receiver.

Find V_1.

$$\frac{V_1}{V_T} = \frac{R_1}{R_T} \qquad (10\text{-}11)$$

$$\frac{V_1}{325} = \frac{27 \text{ k}\Omega}{170 \text{ k}\Omega}$$

$$170 \times V_1 = 325 \times 27$$

$$V_1 = \frac{8775}{170} = 51.6 \text{ V} \qquad Ans.$$

b. Find V_2.

$$\frac{V_1}{V_2} = \frac{R_1}{R_2} \qquad (10\text{-}10)$$

$$\frac{51.6}{V_2} = \frac{27 \text{ k}\Omega}{75 \text{ k}\Omega}$$

$$27 \times V_2 = 51.6 \times 75$$

$$V_2 = \frac{3870}{27} = 143.3 \text{ V} \qquad Ans.$$

c. Find V_3.

$$\frac{V_1}{V_3} = \frac{R_1}{R_3} \qquad (10\text{-}10)$$

$$\frac{51.6}{V_3} = \frac{27 \text{ k}\Omega}{68 \text{ k}\Omega}$$

$$27 \times V_3 = 51.6 \times 68$$

$$V_3 = \frac{3508.8}{27} = 129.9 \text{ V} \qquad Ans.$$

The Potentiometer Rule

When a number of resistances are connected in series, the voltage drop across any one of them may be found by using another form of formula (10-11).

$$\frac{V_x}{V_T} = \frac{R_x}{R_T} \qquad (10\text{-}11)$$

By cross-multiplying, we get

$$V_x \times R_T = R_x \times V_T$$

Dividing both sides by R_T will give the

$$V_x = \frac{R_x}{R_T} \times V_T$$ Formula 10-12

In effect, this formula says that the voltage across any part of a series circuit is some fractional part of the total voltage. The numerator of the fraction is the resistance of the part of the circuit and the denominator is the total resistance of the circuit.

Self-Test 10-14

A voltage divider in the base circuit of a transistor is shown in Fig. 10-11. Find the voltage drop across $R_{B'}$.

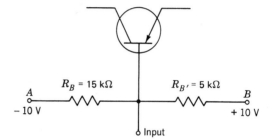

Figure 10-11 Voltage divider in the base circuit of a transistor.

Solution

1. Since R_B and $R_{B'}$ are in ————, the potentiometer rule may be used. The total voltage between the points B and A is the difference in potential between these points. Thus,

$$V_T = +10 - (-10) = \text{————— V}$$

2. Formula (10-12) may now be rewritten using the symbols given in our circuit. Thus,

$$V_x = \frac{R_x}{R_T} \times V_T \qquad (10\text{-}12)$$

will become

$$V_{RB'} = \frac{R_{B'}}{R_T} \times \text{—————}$$

3. Substituting values will give

$$V_{RB'} = \frac{5}{5 + ?} \times 20$$

$$V_{RB'} = \frac{5 \text{ k}\Omega}{20 \text{ k}\Omega} \times 20$$

$$V_{RB'} = \text{————— V} \qquad Ans.$$

Exercises

Practice

In Probs. 1 to 6, R_k and R_x are connected in series. Find the missing values in each problem.

Problem	R_k, Ω	V_k, V	V_x, V	R_x, Ω	V_T, V
1	500	10	25	?	
2	50	15.5	70.2	?	
3	10	36.4	40.3	?	
4	100	15	?	250	
5	1000	?	?	2000	100
6	15,000	?	?	20,000	120

7. In a circuit similar to that shown in Fig. 10-10, $R_1 = 22,000\ \Omega$, $R_2 = 50,000\ \Omega$, and $R_3 = 100,000\ \Omega$. Find (a) V_1, (b) V_2, and (c) V_3 if $V_T = 200$ V.
8. In the picture-control circuit shown in Fig. 10-12, find the voltage drop across R_1 and R_2.

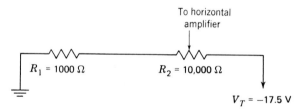

To horizontal amplifier

$R_1 = 1000\ \Omega$

$R_2 = 10,000\ \Omega$

$V_T = -17.5$ V

Figure 10-12

9. A brightness-control circuit in the Admiral TV chassis model NA1-1A is essentially that shown in Fig. 10-13. Find the voltage between ground and point A.

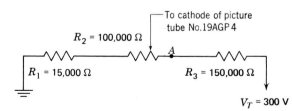

To cathode of picture tube No.19AGP4

$R_2 = 100,000\ \Omega$

$R_1 = 15,000\ \Omega$

A

$R_3 = 150,000\ \Omega$

$V_T = 300$ V

Figure 10-13   Brightness-control circuit in a television receiver.

10. In a circuit similar to that shown in Fig. 10-11, $R_B = 35\ \text{k}\Omega$, $R_{B'} = 5\ \text{k}\Omega$, and the total voltage from A to B is 30 V. Find the voltage drop across $R_{B'}$.

Job 10-6   Resistance Measurement by the Voltage-Comparison Method. The Potentiometer Rule   **365**

11. The volume control of a transistorized phono amplifier is essentially a 470-kΩ and a 5-kΩ resistor in series across 9.5 V. Find the voltage drop across the 5-kΩ resistor.
12. In a brightness-control circuit, the total resistance of the potentiometer equals 5 MΩ, and V_T equals 145 V. Find the cathode voltage when the potentiometer arm has covered (a) 1 MΩ and (b) 3.5 MΩ.

JOB 10-7

Resistance Measurement Using the Wheatstone Bridge

The Wheatstone bridge is a device for obtaining accurate measurements of resistance. A diagrammatic sketch of the circuit is shown in Fig. 10-14. The resistors R_1, R_2, R_3, and R_x are arranged in two parallel branches. R_1, R_2, and R_3 are variable *known* resistors, and R_x is the *unknown* resistance. A galvanometer G is connected across the two branches between points B and D. A battery supplies the total current I_T.

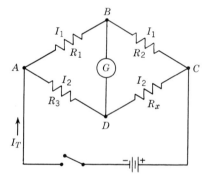

Figure 10-14 Wheatstone-bridge circuit.

Operation of the Bridge

1. When the circuit is closed, a current I_T will flow through the circuit which divides at point A, current I_1 flowing through the branch ABC (R_1 and R_2), and current I_2 flowing through the branch ADC (R_3 and R_x).
2. The variable resistors are adjusted until the galvanometer indicates no deflection. Usually R_1 and R_2 are fixed at some convenient ratio, and then R_3 is varied until the galvanometer reads zero.
3. When there is no deflection of the galvanometer, it indicates that there is no current flowing between points B and D and therefore B and D are at the same voltage level. Under these conditions, we find that

$$\text{Voltage across } AB = \text{voltage across } AD \quad \text{or} \quad I_1 \times R_1 = I_2 \times R_3$$
$$\text{Voltage across } BC = \text{voltage across } DC \quad \text{or} \quad I_1 \times R_2 = I_2 \times R_x$$

Dividing one equation by the other gives us the fundamental equation for the Wheatstone bridge.

$$\frac{I_1 \times R_1}{I_1 \times R_2} = \frac{I_2 \times R_3}{I_2 \times R_x}$$

Canceling out the currents I_1 and I_2, we obtain

$$\frac{R_1}{R_2} = \frac{R_3}{R_x}$$

Formula 10-13

Example 10-15

When a Wheatstone bridge was used, the following readings were obtained: $R_1 = 1000\ \Omega$, $R_2 = 10{,}000\ \Omega$, and $R_3 = 67.4\ \Omega$. Find the unknown resistance.

Solution

Given: $R_1 = 1\ \mathrm{k}\Omega$ Find: $R_x = ?$
  $R_2 = 10\ \mathrm{k}\Omega$
  $R_3 = 67.4\ \Omega$

$$\frac{R_1}{R_2} = \frac{R_3}{R_x} \qquad (10\text{-}13)$$

$$\frac{1\ \cancel{\mathrm{k}\Omega}}{10\ \cancel{\mathrm{k}\Omega}} = \frac{67.4}{R_x}$$

$$R_x = 67.4 \times 10 = 674\ \Omega \qquad \textit{Ans.}$$

Exercises

Practice

Find the value of R_x for each problem.

Problem	R_1, Ω	R_2, Ω	R_3, Ω	R_x, Ω
1	1000	10,000	84.3	?
2	10,000	1000	952.7	?
3	100	10	60.4	?
4	10	1	175.3	?
5	1	10	128.9	?
6	5000	10,000	823.7	?
7	100	1000	46.5	?
8	1000	100	9.16	?
9	10	1	1.6	?
10	1	10	35.79	?

(See CD-ROM for Test 10-2)

Equivalent Circuits

The Thevenin and the Norton equivalent circuits are very useful in reducing the complexity of circuits for analysis. Each of these equivalents has its own advantages, depending on the particular circuit that is being analyzed. Since both equivalents have the same resistance, the factor determining which to use is whether the terminals analyzed will be a short or an open circuit.

The Thevenin equivalent finds the voltage with the terminals being analyzed open. The Norton equivalent finds the current with the terminals shorted.

Our first example (Example 10-16) has three objectives: (1) to develop skills with the Thevenin equivalent, (2) to increase our knowledge about the output characteristics of circuits, and (3) to apply the Thevenin equivalent to a voltage divider.

The problem as phrased is for a voltage divider power supply. The general concepts and procedures would be the same whether it were an amplifier or any other type of linear circuit rather than a power supply. The Thevenin equivalent resistance that is obtained is sometimes called the *output resistance* of the power supply. This equivalent resistance is important because it defines how loads interact with the power supply.

Example 10-16

Figure 10-15 shows the circuit of a power supply that delivers two different voltages; 30 and 9 V. Each voltage has its own load which is represented by the load resistors R_L9 and R_L30. Find the Thevenin equivalent circuit at output terminals A and B of the 9-V supply.

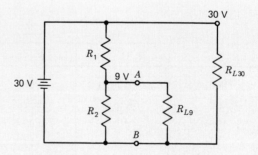

Figure 10-15 Dual voltage supply with voltage divider output.

Solution

1. First redraw the circuit to that shown in Fig. 10-16. This shows the circuit in the form in which it will be analyzed and the final form that is sought.
2. Find the Thevenin equivalent resistance. Short the 30-V battery. Notice that this removes the second load, R_L30. The resistance between terminals A and B will be the result of R_1 in parallel with R_2.

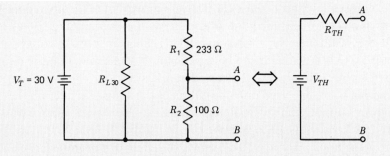

Figure 10-16   Power supply circuit with Thevenin equivalent for terminals A and B.

$$R_{TH} = \frac{R_1 \times R_2}{R_1 + R_2} = \frac{100 \times 233}{100 + 233} = 70 \ \Omega$$

3. Find the Thevenin equivalent voltage. This will be the voltage across R_2.

$$V_{TH} = V_2 = \frac{R_2}{R_1 + R_2} \times V_T \qquad (10\text{-}12)$$

$$V_{TH} = \frac{100}{233 + 100} \times 30$$

$$= \frac{100}{333} \times 30 = 9 \ \text{V}$$

The Thevenin equivalent circuit shown in Fig. 10-16 is now completely specified. The output voltage is the 9 V that is desired. It will be easy to determine the effects of loading on the output voltage. The Thevenin equivalent is used as in Fig. 10-17. The load is placed between the two terminals A and B. This forms a very simple one-loop circuit which can be analyzed for any value of load.

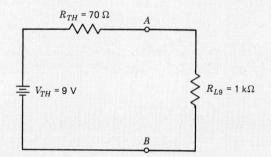

Figure 10-17   Power supply equivalent circuit with load attached.

4. Find the output voltage and the current through the load if the load value is 1 kΩ.

$$I_L = \frac{V_{TH}}{R_{TH} + R_L}$$

$$I_L = \frac{9}{70 + 1 \ \text{k}\Omega} = 8.4 \ \text{mA}$$

$$V_{RL}9 = I_L \times R_L = 8.4 \ \text{mA} \times 1 \ \text{k}\Omega = 8.4 \ \text{V}$$

This example shows how to use the Thevenin theorem to greatly reduce the complexity of a very practical circuit. It also shows that if you ignore 70 Ω relative to 1 kΩ, the result may be misleading. The change in output voltage from 9 V unloaded to 8.4 V loaded could be significant. It also shows that the output resistance of a power supply must be considered.

Example 10-16 analyzed only the effects of the voltage divider on the output of the power supply. The internal impedance of the power supply was ignored. Figure 10-18 shows the change in the circuit if a source resistance R_S for the power supply is included. Example 10-17 shows the effect of this impedance and the use of both theorems in the same problem.

Example 10-17

Find the Norton equivalent for the output of the power supply in Fig. 10-18.

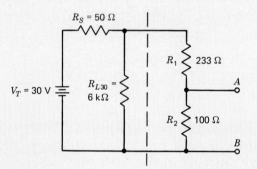

Figure 10-18   Voltage divider power supply with internal resistance R_s.

Solution

This is a complicated series-parallel network. Consider the ways in which the circuit could be simplified. The dotted line divides the circuit into two parts. If the part on the left were converted into a Thevenin equivalent with Thevenin voltage V_{TH} and Thevenin resistance R_{TH}, the entire circuit would be only one loop. This would greatly simplify the analysis and would improve the understanding of the circuit operation (see Fig. 10-19).

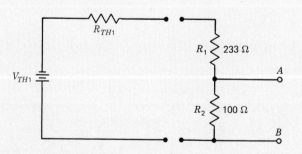

Figure 10-19   Power supply with Thevenin equivalent for part of circuit.

1. Make a Thevenin equivalent circuit for the left part of the circuit shown in Fig. 10-18. Assume that the load on the 30-V line is 6 kΩ.

$$V_{TH} = \frac{R_L}{R_L + R_S} \times V_T = \frac{6 \text{ k}\Omega}{6 \text{ k}\Omega + 50} \times 30 = 29.75 \text{ V}$$

$$R_{TH} = \frac{R_S \times R_L 30}{R_S + R_L 30} = \frac{50 \times 6 \text{ k}\Omega}{50 + 6 \text{ k}\Omega} = 49.59$$

Notice that in the above calculations the load on the 30-V line affects the values of the Thevenin equivalent circuit. This means that if the 30-V load changes, some of this change will show up in the 9-V power supply. In Example 10-16 the two supplies were independent. Now we find that the inclusion of an internal resistance in the power supply causes a connection between the two loads.

2. Connect the Thevenin equivalent to the remainder of the circuit and determine the Norton equivalent for the power supply. The resulting circuit is shown in Fig. 10-19. When the Norton current is found by shorting terminals A and B, the resulting circuit is only the series connection of V_{TH}, R_{TH}, and R_1. This is easy to analyze.

$$I_N = \frac{V_{TH}}{R_{TH} + R_1} = \frac{29.75}{49.59 + 233} = 105 \text{ mA}$$

$$R_N = \frac{R_2 \times (R_1 + R_{TH})}{R_2 + R_1 + R_{TH}} = \frac{100 \times (233 + 49.59)}{100 + 233 + 49.59} = 73.86 \text{ } \Omega$$

The final Norton equivalent circuit is shown in Fig. 10-20.

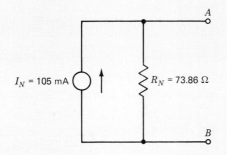

Figure 10-20 Norton equivalent for 9-V power supply.

Exercises

Applications

1. Solve for current z in Fig. 9-21. *Hint:* Make an equivalent circuit for circuit points g and f.
2. Solve Example 9-13 for current x by making a Thevenin equivalent of the parallel network between points p and b. This makes the circuit a simple one-loop problem.
3. In Fig. 9-28, solve the circuit for current z through R_5. *Hint:* Use a Norton equivalent.
4. Make a Norton equivalent of Fig. 6-9 to find the current I_3 in R_3.
5. In Fig. 6-16, find the current in R_5 using a Norton's equivalent.

DC Equivalent Circuits for Self-Biased Transistor Circuits

Solid-state diodes.   As we learned in Job 1-1, many metals have free electrons in the outer shells of their atoms. The ability of these free electrons to move about makes most metals fine conductors of electricity. On the other hand, insulators such as glass and rubber do not have many free electrons. A semiconductor material such as germanium has some free electrons whose number can be considerably increased by the addition of small amounts of arsenic atoms as an impurity. These many loosely held electrons in the outer shells of the germanium atoms cause it to be referred to as *N germanium*. This does not mean that it is negatively charged, but merely that there are free electrons present.

If gallium is added as an impurity instead of arsenic, the outer shells of the germanium atoms will be filled except for one electron. This missing electron constitutes a "hole" which the atom wants to fill to reach a stable state. This material is known as *P germanium*. Because N germanium wants to "give" electrons and P germanium wants to "accept" them, both materials act as fairly good electron conductors.

Now suppose that we press these two materials together, as shown in Fig. 10-21. The free electrons at the junction will meet with the positive holes and cancel each other's charge. After a very short time, the field produced by this zero-charged area repels both the electrons and the holes and acts as a barrier to both. Of course, no current will flow. In Fig. 10-22*a*, we have connected the minus (−) terminal of a battery to the N material, and the plus (+) terminal to the P material. The forward push of the many electrons from the battery helps the free electrons to overcome the barrier, and the combination acts as a good conductor. Connected in this way, the combination is said to be *forward-biased*, and the current flows through the load resistor.

In Fig. 10-22*b*, we have reversed the polarity, connecting the + of the battery to the N material and the − of the battery to the P material. The + battery pulls the free electrons away from the junction, and the − battery pulls the positive holes away from the junction. This has the effect of widening the barrier at the junction, and no current will flow. This is called *reverse bias*.

This combination is called a *diode*, the symbol for which is shown in Fig. 10-23. When connected as shown, current will flow only from left to right.

Transistors.   A transistor is very similar to a semiconductor diode except that the transistor has two junctions instead of one. In Fig. 10-24, the *emitter* is an N-type material which forms a junction with a P-type material called the *base*. This base

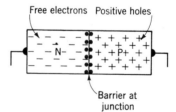

Figure 10-21   A barrier is formed at the junction of N- and P-type materials.

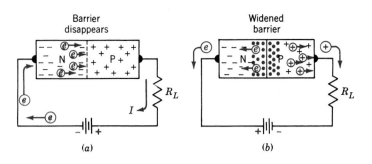

(a)   (b)

Figure 10-22 (a) The barrier disappears under forward biasing, and the junction conducts.
(b) The barrier is widened under reverse biasing, and no current flows.

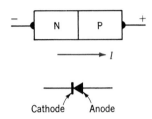

Figure 10-23 Diode symbol.

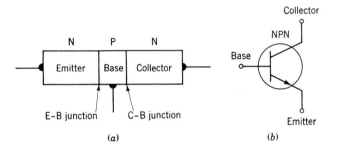

(a)   (b)

Figure 10-24 (a) NPN transistor. (b) Symbol for NPN transistor.

in turn forms another junction with the *collector*, an N-type material. Such a combination is called an *NPN-type* transistor.

If the emitter and collector are made of P-type material and the base is N-type as shown in Fig. 10-25, the transistor is called a *PNP type*. The only difference between the two types is the direction in which the current flows in the emitter. In the symbols, the electron current is *against* the arrowheads.

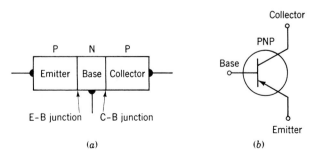

(a)   (b)

Figure 10-25 (a) PNP transistor. (b) Symbol for PNP transistor.

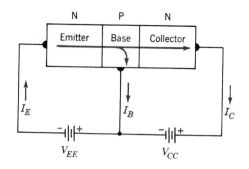

Figure 10-26 Electron flow in an operating NPN transistor.

Transistor operation.   In Fig. 10-26, when the voltage $-V_{EE}$ is applied, the electrons move out into the emitter lead and enter the emitter material. These electrons, together with the free electrons of the emitter material, are pushed toward the base region. The greater the forward bias, the greater the number of electrons that enter the base region. The number of holes in the base region is very small because it is so very thin and because it is so slightly doped. A few electrons will find holes and combine to form a small base current. The rest of the electrons (about 98 percent) will not find a hole and will be pushed across the junction by the electrons newly arriving from the emitter. Once they get past the collector-base junction, they are attracted by the large collector voltage (V_{CC}) and become part of the collector current. They return to the battery, join up with I_B, enter V_{EE}, and start around again. As you can see, the emitter current is the sum of the base current and the collector current. An increase in the forward bias will cause an increase in the emitter current and a corresponding increase in the collector current and the base current. Conversely, a decrease in the forward bias results in a decrease of both base and collector current.

Transistor current gain.   As an amplifying device, the transistor is now replacing many of the functions of the vacuum tube. In Fig. 10-27a, a small change in the input signal to the grid controls a large change in the current that flows through the load. Similarly, in Fig. 10-27b, a small change in the input signal to the base of the transistor controls a large change in the current that flows through the collector and the load. This base-to-collector current gain is defined as β (beta).

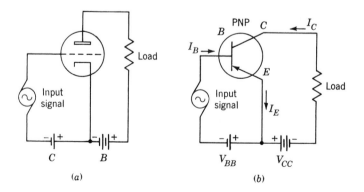

Figure 10-27 Comparison of (a) triode circuit and (b) transistor circuit. Small changes in the input signal control large currents through the load.

$$\beta = \frac{\Delta I_C}{\Delta I_B}$$ Formula 10-14

where β = current gain
ΔI_C = change in dc collector current
ΔI_B = change in dc base current

In order to solve a transistor circuit by the equivalent-resistance circuit-analysis method, we must first find the collector-to-base resistance R_{CB} for the particular transistor. For the self-biased circuit shown in Fig. 10-28a, the formula for R_{CB} is

$$R_{CB} = \frac{R_B}{\beta}$$ Formula 10-15

where R_{CB} = collector-to-base resistance
R_B = base-bias resistor
β = current gain

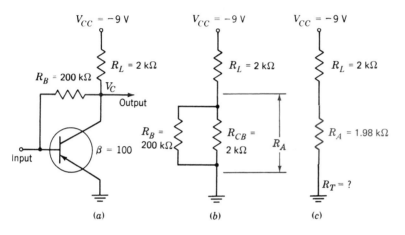

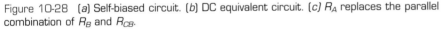

Figure 10-28   (a) Self-biased circuit. (b) DC equivalent circuit. (c) R_A replaces the parallel combination of R_B and R_{CB}.

Self-Biased Circuits

Example 10-18

Solve the circuit shown in Fig. 10-28a for (a) the load current I_{RL}, (b) the voltage across R_L, and (c) the collector voltage V_C.

Solution

1. Find R_{CB}.

$$R_{CB} = \frac{R_B}{\beta}$$ (10-15)

$$R_{CB} = \frac{200 \text{ k}\Omega}{100} = 2 \text{ k}\Omega$$

2. Find the load current I_{RL}. This is the current that will flow through the series-parallel circuit shown in Fig. 10-28b. This circuit is further simplified into that shown in Fig. 10-28c, in which R_A replaces the parallel combination of R_B and R_{CB} of Fig. 10-28b.

a. Find R_A in Fig. 10-28*b*. Since R_B and R_{CB} are in parallel,

$$R_A = \frac{R_B \times R_{CB}}{R_B + R_{CB}} \qquad (5\text{-}5)$$

$$= \frac{200 \text{ k}\Omega \times 2 \text{ k}\Omega}{200 \text{ k}\Omega + 2 \text{ k}\Omega}$$

$$= \frac{400 \times 10^6}{202 \times 10^3} = 1.98 \times 10^3$$

$$R_A = 1.98 \text{ k}\Omega$$

b. Find R_T in Fig. 10-28*c*. Since R_L and R_A are in series,

$$R_T = R_L + R_A \qquad (4\text{-}3)$$
$$= 2 + 1.98 = 3.98 \text{ k}\Omega$$

c. Find the total current in the circuit of Fig. 10-28*c*.

$$I_T = \frac{V_{CC}}{R_T} = \frac{9}{3.98 \times 10^3} = 2.26 \times 10^{-3}$$

$$I_T = 2.26 \text{ mA}$$

d. Find I_{RL}.

$$I_{RL} = I_T = 2.26 \text{ mA} \qquad Ans. \qquad (4\text{-}1)$$

3. Find the voltage across R_L.

$$V_{RL} = I_{RL} \times R_L \qquad (2\text{-}1)$$
$$= 2.26 \times 10^{-3} \times 2 \times 10^3 = 4.52 \text{ V} \qquad Ans.$$

4. Find the collector voltage V_C. As indicated in Fig. 10-28*a*, the collector voltage V_C, referred to ground, is equal to the difference between V_{CC} and V_{RL}.

$$V_C = -V_{CC} + V_{RL} = -9.0 + 4.52 = -4.48 \text{ V} \qquad Ans.$$

Other solid-state devices.   Many new solid-state devices are being used in power control. These devices can be thought of as equivalent transistor circuits. Some of the more popular devices are shown in Fig. 10-29. The symbol, the voltage-current characteristics, and the equivalent transistor circuit are shown for the silicon-controlled rectifier (SCR), diac, and triac. These are diode devices with special characteristics.

The SCR is a single-direction device; i.e., current flows in only one direction. When a voltage is applied to the gate, the SCR can be turned on like a switch. The voltage-current curve shows that when this happens the voltage across the device drops and the current through it increases. It is in the "on" state. To turn it off, the current must be reduced below the "holding current." This happens automatically in ac circuits when the voltage goes through zero.

The forward voltage at which the SCR turns on by itself is called the *forward breakover voltage* $V_{(BR)F}$. Devices are usually rated by the maximum allowable instantaneous value of repetitive peak off-state voltage that may be applied, V_{DRM}. This voltage is lower than the forward breakover voltage. It corresponds to the peak ac voltage that can be applied to a device that is not turned on.

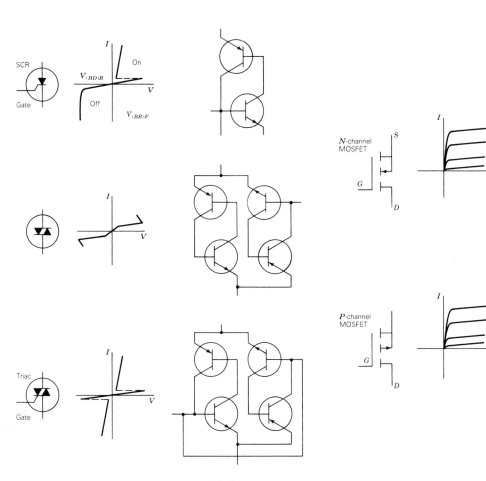

Figure 10-29 Solid-state power control devices.

The backward voltage at which the diode starts to conduct by itself is the *reverse breakdown voltage*, $V_{(BD)R}$. This condition should be avoided; therefore, the devices are usually rated by V_{RRM}, or repetitive peak reverse voltage. This is the maximum allowable instantaneous value of a repetitive reverse voltage that may be applied.

A triac is like two SCRs in opposite directions. It can be turned on for voltages in either direction as shown by its current-voltage curves.

A diac is like a triac except that it cannot be triggered on through a gate. It is like two diodes in opposite directions. They are usually used in ac circuits to produce trigger signals to other devices such as SCRs.

Another class of device that has become popular for power control is the field-effect transistor (FET). FETs made with the process known as metal-oxide-semiconductor (MOS) are known as MOSFETs. The main circuit for a MOSFET is between the source *(S)* and the drain *(D)*. The voltage from drain to source is called V_{DS}. The current in this circuit is controlled by a voltage applied between the gate *(G)* and the source. This voltage is designated V_{GS}.

When a voltage is applied between the gate and the source, the device provides an almost constant current. MOSFETs are very useful for the control of current to a device. They are frequently used in motor control circuits.

MOSFETS can be made with *P*-type or *N*-type materials in the channel from drain to source. These devices are known as *P*-channel MOSFETs and *N*-channel

MOSFETs, respectively. Notice from the voltage-current diagrams that the type of channel determines the polarity of the gate signal needed to turn on the MOSFET.

Exercises

Applications

1. In a circuit similar to that shown in Fig. 10-28a, $V_{CC} = -10$ V, $R_L = 5$ kΩ, $R_B = 100$ kΩ, and $\beta = 50$. Find (a) the load current I_{RL}, (b) the voltage across R_L, and (c) the collector voltage V_C.
2. Repeat Exercise 1 with the following values: $V_{CC} = -10$ V, $R_B = 80$ kΩ, $R_L = 2$ kΩ, and $\beta = 20$.
3. In a circuit similar to that shown in Fig. 10-29a, $V_{CC} = -13$ V, $R_B = 1$ MΩ, $R_L = 40$ kΩ, $R_E = 2$ kΩ, and $\beta = 100$. Find (a) the collector current I_C, (b) the emitter current I_E, (c) the voltage across R_L, (d) the collector voltage V_C, (e) the voltage across R_E, and (f) the collector-to-emitter voltage V_{CE}.
4. Repeat Exercise 3 with the following values: $V_{CC} = -14$ V, $R_B = 0.5$ MΩ, $R_L = 20$ kΩ, $R_E = 5$ kΩ, and $\beta = 50$.

JOB 10-10

DC Equivalent Circuts for Fixed-Bias Transistor Circuits

In the fixed-bias circuit shown in Fig. 10-30a, the load resistor R_L is included in the beta-dependent loop, and it must therefore be included in the formula for R_{CB}. For the fixed-bias circuit shown in Fig. 10-30a, the formula for R_{CB} is

$$R_{CB} = \frac{R_B}{\beta} - R_L \qquad \text{Formula 10-16}$$

where R_{CB} = collector-to-base resistance
    R_B = base-bias resistor
    R_L = collector or load resistance
    β = current gain

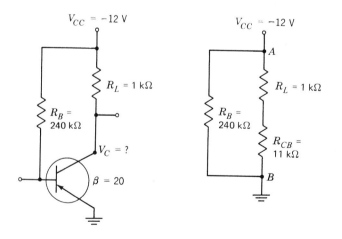

Figure 10-30 (a) Fixed-bias circuit. (b) DC equivalent circuit.

Fixed-Bias Circuits

Example 10-19

Solve the circuit shown in Fig. 10-30a for (a) the collector current I_C, (b) the voltage across R_L, and (c) the collector voltage V_C.

Solution

1. Find R_{CB}.

$$R_{CB} = \frac{R_B}{\beta} - R_L \qquad (10\text{-}16)$$

$$= \frac{240}{20} - 1$$

$$R_{CB} = 12 - 1 = 11 \text{ k}\Omega$$

2. Find I_C. The collector current is the current through the series group of R_L and R_{CB} as shown in Fig. 10-30b. Since R_B is in parallel with this series group, the voltage across this group is equal to the total voltage $V_{CC} = -12$ V.

$$I_C = \frac{V}{R_T}$$

$$I_C = \frac{12}{11 + 1} = \frac{12}{12 \times 10^3}$$

$$= 1 \times 10^{-3} = 1 \text{ mA} \qquad Ans.$$

3. Find V_{RL}.

$$V_{RL} = I_C \times R_L$$

$$V_{RL} = 0.001 \times 1000 = 1 \text{ V} \qquad Ans.$$

4. Find V_C. As shown in Fig. 10-30a, the collector voltage V_C, referred to ground, is equal to the difference between V_{CC} and V_{RL}.

$$V_C = -12 + 1 = -11 \text{ V} \qquad Ans.$$

Fixed-Bias Circuits with Emitter Resistance

Example 10-20

Solve the circuit shown in Fig. 10-31a (on page 380) for (a) the emitter current I_E, (b) the emitter voltage V_E, (c) the collector current I_C, (d) the voltage across R_L, and (e) the collector voltage V_C.

Solution

1. Find R_{CB}.

$$R_{CB} = \frac{R_B}{\beta} - R_L \qquad (10\text{-}16)$$

$$= \frac{100}{25} - 1$$

$$R_{CB} = 4 - 1 = 3 \text{ k}\Omega$$

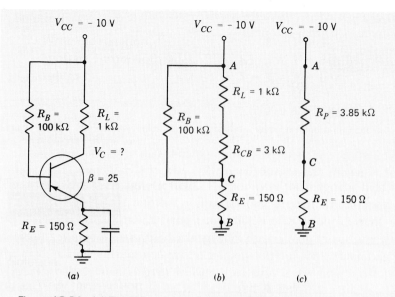

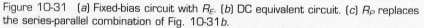

Figure 10-31 (a) Fixed-bias circuit with R_E. (b) DC equivalent circuit. (c) R_P replaces the series-parallel combination of Fig. 10-31b.

2. Find I_E, the emitter current. This is the current that flows through the series-parallel circuit shown in Fig. 10-31b and which must be simplified into the circuit shown in Fig. 10-31c. In Fig. 10-31b, the series resistance of R_{CB} and R_L is equal to 3 kΩ + 1 kΩ = 4 kΩ. This 4 kΩ is in parallel with the 100 kΩ of R_B. The total resistance of this parallel combination is

$$R_P = \frac{100 \times 4}{100 + 4} \qquad \text{(5-5)}$$

$$R_P = \frac{400}{104}$$

$$= 3.85 \text{ k}\Omega$$

Now we can find the total resistance between points A and B.

$$R_T = R_P + R_E = 3.85 + 0.15 = 4 \text{ k}\Omega$$

The emitter current I_E is equal to the total current that flows between points A and B.

$$I_E = I_T = \frac{V_{CC}}{R_T}$$

$$I_E = \frac{10}{4 \times 10^3} = 2.5 \text{ mA} \qquad Ans.$$

3. Find the emitter voltage V_E. This is the voltage, referred to ground, which is equal to the drop across R_E.

$$V_E = I_E \times R_E$$
$$V_E = 0.0025 \times 150 = 0.375 \text{ V} \qquad Ans.$$

4. In Fig. 10-31b, the collector current is the current flowing between points A and C, and is equal to the voltage between A and C divided by the resis-

tance of the series group made of R_L and R_{CB}. The voltage between A and C (V_{AC}) is equal to the difference between V_{CC} and V_E.

$$V_{AC} = V_{CC} - V_E$$
$$V_{AC} = 10 - 0.375 = 9.625 \text{ V}$$

The resistance of the series group between A and C was found to be 4 kΩ. Therefore,

$$I_C = \frac{V_{AC}}{R_{AC}} = \frac{9.625}{4000} = 2.4 \text{ mA} \qquad Ans.$$

5. Find V_{RL}. In Fig. 10-31a,

$$V_{RL} = I_C \times R_L$$
$$V_{RL} = 0.0024 \times 1000 = 2.4 \text{ V} \qquad Ans.$$

6. Find V_C.

$$V_C = -V_{CC} + V_{RL}$$
$$V_C = -10 + 2.4 = -7.6 \text{ V} \qquad Ans.$$

Exercises

Applications

1. In a circuit similar to that shown in Fig. 10-30a, $V_{CC} = -10$ V, $R_L = 1$ kΩ, $R_B = 200$ kΩ, and $\beta = 20$. Find (a) the collector current I_C, (b) the voltage across R_L, and (c) the collector voltage V_C.
2. Repeat Prob. 1 with the following values: $V_{CC} = -8$ V, $R_L = 2$ kΩ, $R_B = 120$ kΩ, and $\beta = 30$.
3. In a circuit similar to that shown in Fig. 10-31a, $V_{CC} = -8$ V, $R_L = 1$ kΩ, $R_B = 200$ kΩ, $R_E = 200$ Ω, and $\beta = 100$. Find (a) the emitter current I_E, (b) the emitter voltage V_E, (c) the collector current I_C, (d) the voltage across R_L, and (e) the collector voltage V_C.
4. Repeat Prob. 3 with the following values: $V_{CC} = -10$ V, $R_L = 2$ kΩ, $R_B = 100$ kΩ, $R_E = 140$ Ω and $\beta = 20$.

Transistor Equivalent Circuits

The electrical properties of semiconductors are very complicated. The real properties are represented in models so that reasonable analysis can be made. The most popular model is called the *H-parameter model* (Fig. 10-32 on page 382). This model is a mathematical model that relates the voltages and currents at the three terminals of the transistor.

In the model, one of the terminals is common between the input and the output. If you look at the input terminal pair, you will notice that the model is a Thevenin equivalent circuit. The model on the output pair is a Norton equivalent circuit.

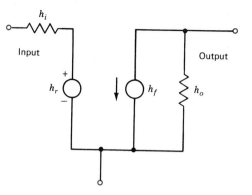

Figure 10-32   General *H*-parameter model for a transistor.

The terms for each of the components of the model are:

h_i—input impedance
h_r—reverse transfer voltage ratio
h_f—forward current transfer ratio
h_o—output conductance

Any of the three terminals of the transistor (the emitter, the base, or the collector) may be taken as the common terminal in the model. These three possible configurations are shown in Fig. 10-33. Beside each model is shown the *H*-parameter equivalent circuit. Notice that the general form of the model does not change. The only difference is that the second subscript is added to each of the parameter names to indicate which terminal is common. For example, the input impedance

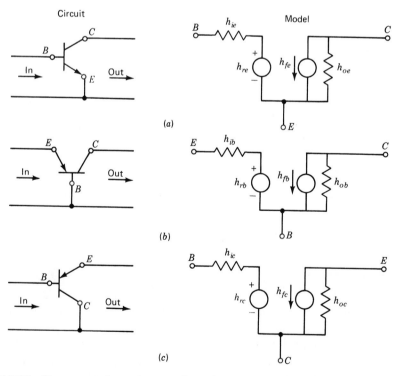

Figure 10-33   Three general transistor configurations and their associated *H*-parameter models. (*a*) common emitter; (*b*) common base; (*c*) common collector.

is labeled h_{ie} for the common emitter, h_{ib} for the common base, and h_{ic} for the common collector. The common emitter configuration is the most commonly used. To give a better understanding of the model, this configuration is further analyzed.

In the common-emitter configuration the input is between the base and the emitter. The input impedance h_{ie} is the impedance that would be measured between the base and the emitter terminals at the operating point of the transistor. This is expressed as:

$$h_{ie} = \frac{v_{be}}{i_b}$$

This is simply Ohm's law: $R = V/I$.

The reverse voltage transfer ratio h_{re} is the Thevenin voltage that is measured between the base and the emitter due to the voltage that is between the collector and the emitter. It is called "reverse" because the voltage comes from the output and is measured at the input. It is expressed as

$$h_{re} = \frac{v_{be}}{v_{ce}}$$

The forward current transfer ratio is the Norton output current due to the current that comes into the input (base). It is the ratio

$$h_{fe} = \frac{i_c}{i_b}$$

For the common emitter, the forward current transfer ratio h_{fe} is the beta of the transistor.

The last parameter of the model is the output conductance. This is also just like the conductance in Ohm's law:

$$h_{oe} = \frac{i_c}{v_{ce}}$$

This would be the reciprocal of the Norton equivalent resistance at the output. This is the only subtle difference between the Norton model and the H-parameter model—h_o is expressed as a conductance in the H-parameter model and as a resistance in the Norton model. There is no essential difference between the two.

Both upper- and lowercase subscripts are used for transistor H-parameter values. The uppercase subscripts represent the dc values of the parameter, and the lowercase represents the ac or signal parameter values.

Typical values of H parameters for a modern transistor can be seen from those for the 2N3726 silicon NPN transistor. There are two classifications for the parameters, dc and signal (ac). The dc beta for the transistor varies with collector current as shown in Table 10-1.

Table 10-1

I_C mA	H_{FE} min	H_{FE} max
0.01	80	—
0.1	120	—
1.0	135	350
50	115	—

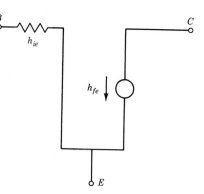

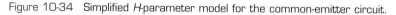

Figure 10-34   Simplified *H*-parameter model for the common-emitter circuit.

Notice that there is a value of collector current for which the H_{FE} peaks. Not all of the *H* parameters are specified for dc; H_{FE} is usually the only one of interest.

The *H* parameters for the signal frequencies are:

h_{ie}—11.5 kΩ max
h_{re}—1500 × E_6
h_{fe}—135 min, 420 max at 1 mA I_C
h_{oe}—80 × 10⁻⁶ mhos

The values of h_{re} and h_{oe} are very small. This leads to a further approximation that is often made to the *H*-parameter model for the common-emitter configuration. The reverse transfer voltage h_{re} is so small for modern transistors that it is ignored. The output conductance h_{oe} is also very small (output resistance is very high), and hence it also is neglected. This leaves the model shown in Fig. 10-34.

Use of the simplified model of Fig. 10-34 is very easy, except that it has in it a symbol that is rather unusual, h_{ie}. In the previous discussion of transistors, the concept of beta was well covered; however, when it is written as a circuit symbol as above, it is called a *dependent source*. This means that the value of the current cannot be set independently but is determined by some other value in the circuit, in this case the base current. Some caution must be taken when calculating with it. Example 10-21 will show how the *H*-parameter model is used in a circuit and for calculation.

Example 10-21

Figure 10-35 shows a very elementary common-emitter amplifier. It is designed so that there is 1 mA of direct current flowing in the collector. Find the dc values of V_{ce}, I_b, and I_e. Use a 2N3726 transistor and assume that the transistor has the minimum beta.

Solution

1. If 1 mA of current is the collector, that current must flow through R_C. The voltage drop across R_C is:

$$V_{rc} = I_C \times R_C = 1 \text{ mA} \times 10 \text{ k}\Omega$$
$$= 10 \text{ V}$$

the supply voltage is 15 V, and V_{rc} is 10 V; therefore:

$$V_{ce} = 15 - 10 = 5 \text{ V}$$

2. The current flowing in the collector is H_{fe} times the base current; therefore, to find the base current, perform the following calculation:

$$I_b = \frac{I_C}{H_{fe}}$$

$$= \frac{1 \text{ mA}}{135}$$

$$= 7.4 \text{ } \mu A$$

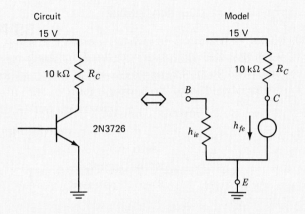

Figure 10-35 A common-emitter amplifier circuit and its associated H-parameter model.

3. The emitter current is the sum of the collector current and the base current.

$$I_e = 1 \text{ mA} + 7 \text{ } \mu A$$
$$= 1.007 \text{ mA}$$

For a modern high-gain transistor of this type, the example shows that the base current can usually be ignored with respect to the collector current. This gives the approximation that the emitter current is equal to the collector current. As shown above, this is a very good approximation.

Example 10-22

Using the same circuit as that in Example 10-21, find the V_{be} and V_{rc} for a signal of 1 μA into the base of the transistor. Notice that in this example we are interested in the ac, or signal values. The parameters used are thus the signal parameters that are expressed with lowercase symbols.

Solution

1. First find the collector current. Assume the minimum value of h_{fe}:

$$i_C = h_{fe} \times i_b = 135 \times 1 \times 10^{-6} \text{ A}$$
$$= 0.135 \text{ mA}$$

2. Now find v_{rc}.

$$v_{rc} = i_c \times R_c = 0.135 \text{ mA} \times 10 \text{ k}\Omega$$
$$= 1.35 \text{ V}$$

3. The v_{be} is caused by the base current flowing through h_{ie}. Therefore:

$$v_{be} = i_b \times h_{ie} = 1 \times 10^{-6} \text{ A} \times 11.5 \text{ k}\Omega$$
$$= 11.5 \text{ mV}$$

For this amplifier an 11.5-mV input signal is amplified to 1.35 V across the collector resistor.

Self-Test 10-23

1. The most popular model of a transistor is the ——————— model.
2. In this model the input is a ——————— equivalent and the output is a ——————— equivalent.
3. On the input the value of h_i is the input ———————. It is the ratio ———————.
4. The output parameters h_o is the output ——————— obtained from the ratio ———————.
5. For the common-emitter configuration, the forward current gain h_{ie} is also known as ———————.
6. The Thevenin voltage on the input is called ——————— and is the ratio ———————.
7. When the parameter values are given, the uppercase values represent ——————— values.
8. Parameter values given in lowercase are ——————— or ——————— values.
9. The most popular configuration of a transistor is the ———————.

Exercises

Applications

1. Assume that the H_{fe} of the transistor is 250. Use the circuit in Example 10-21 to find the base current.
2. If the base current is 5 μA and the collector-to-emitter voltage in Example 10-21 is 10 V, what is the beta of the transistor?
3. Using the circuit shown in Fig. 10-35, find the ac base to emitter voltage V_{be} if the beta of the transistor is 250 and the base signal is 1 μA.
4. Using the conditions stated in Prob. 3, find the signal voltage across R_C. (Notice that for the elementary common-emitter amplifier, the beta of the transistor does not affect the base voltage but does affect the collector circuit values.)

(See CD-ROM for Test 10-3)

Attenuators

An *attenuator pad* is a combination of resistors whose purpose is (1) to reduce the source voltage to the lower voltage required by the load but (2) to keep the original source current unchanged.

The first of these requirements may be met very easily by inserting a resistance in series with the source as explained in Example 4-24 in Job 4-7. In this example, a lamp rated at 12 V and 40 Ω is to operate from a 24-V source. Therefore, the original 24-V source voltage must be reduced to the 12 V required by the lamp. This reduction is accomplished by the insertion of a 40-Ω resistor in series with the lamp. If the lamp had been connected directly to the 24-V source, the current would have been

$$I_L = \frac{V_L}{R_L} = \frac{24}{40} = 0.6 \text{ A}$$

But, with the 40-Ω series resistor added,

$$R_T = R_1 + R_L \qquad\qquad (4\text{-}3)$$
$$R_T = 40 + 40 = 80 \ \Omega$$

and

$$I_L = \frac{V_L}{R_T} = \frac{24}{80} = 0.3 \text{ A}$$

Notice that although the voltage was reduced from 24 to 12 V, the current drawn from the source was *also reduced* from 0.6 to 0.3 A. An attenuator circuit would also reduce the voltage from 24 to 12 V but *not* change the original current. Actually, this means that the resistance that the load presents to the source would remain unchanged. Thus, if the lamp resistance were 40 Ω, an attenuator circuit would reduce the voltage but would *not* change the total resistance of the circuit, and would therefore *not* change the original current. Now how can this be done?

Consider the circuit shown in Fig. 10-36. $V_T = 24$ V, and $I_T = V_T \div R_T$ or $^{24}/_{40} = 0.6$ A. We wish to have $V_L = 12$ V, but *keep* $I_T = 0.6$ A. In order to do this, we shall insert resistors R_1 and R_2 as shown in Fig. 10-37 on page 388. The current drawn by the lamp when $V_L = 12$ V is

$$I_L = \frac{V_L}{R_L} = \frac{12}{40} = 0.3 \text{ A}$$

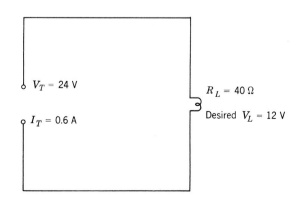

$V_T = 24$ V

$I_T = 0.6$ A

$R_L = 40 \ \Omega$

Desired $V_L = 12$ V

Figure 10-36 A 24-V source is to produce a load voltage of 12 V.

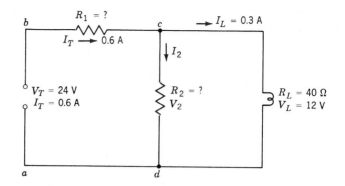

Figure 10-37   An L pad is inserted in the circuit of Fig. 10-36.

By Kirchhoff's first law,

$$I_2 = I_T - I_L$$
$$I_2 = 0.6 - 0.3 = 0.3 \text{ A}$$

Since V_2 and V_L are in parallel, $V_2 = V_L = 12$ V.
Therefore,

$$R_2 = \frac{V_2}{I_2} = \frac{12}{0.3} = 40 \ \Omega \qquad \textit{Ans.}$$

In order to find R_1, apply Kirchhoff's second law, and trace around the circuit in the direction *abcd*.

$$V_T - V_1 - V_2 = 0$$
$$24 - V_1 - 12 = 0$$
$$24 - 12 = V_1$$
$$V_1 = 12 \text{ V}$$

R_1 may now be found.

$$R_1 = \frac{V_1}{I_1} = \frac{12}{0.6} = 20 \ \Omega \qquad \textit{Ans.}$$

Check: If we are correct, the total resistance of the series-parallel combination of R_1, R_2, and R_L will exactly equal the original R_L. Since R_2 and R_L are in parallel, the effective resistance of the group equals

$$\frac{R_2 \times R_L}{R_2 + R_L} = \frac{40 \times 40}{40 + 40} = \frac{1600}{80} = 20 \ \Omega$$

This 20-Ω effective resistance is in series with $R_1 = 20 \ \Omega$. R_T therefore equals $20 + 20 = 40 \ \Omega$, which is exactly what the total resistance was *before* the insertion of the attenuator resistors. If the total resistance of the entire combination is 40 Ω, then $I_T = V_T \div R_T$, or $I_T = {}^{24}\!/_{40} = 0.6$ A. Notice that I_T is *still* 0.6 A but that V_L is now reduced to 12 V. This combination of R_1 and R_2 when inserted as shown in Fig. 10-37 is called an *L pad* owing to its resemblance to an inverted letter L.

Example 10-24

Insert an L pad in the circuit shown in Fig. 10-38 which will reduce the load voltage to 4 V while maintaining a constant resistance to the source.

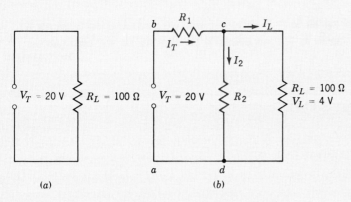

Figure 10-38 (a) Original circuit. (b) Circuit with L pad inserted.

Solution

Insert R_1 and R_2 as shown in Fig. 10-38b. The total current I_T *before* the insertion of the pad was

$$I_T = \frac{V_T}{R_T} = \frac{20}{100} = 0.2 \text{ A}$$

If the voltage at the load is to be 4 V,

$$I_L = \frac{V_L}{R_L} = \frac{4}{100} = 0.04 \text{ A}$$

Find I_2 by Kirchhoff's first law.

$$I_2 = I_T - I_L = 0.2 - 0.04 = 0.16 \text{ A}$$

Find R_2. Since R_2 and R_L are in parallel, $V_2 = V_L = 4$ V.

$$R_2 = \frac{V_2}{I_2} = \frac{4}{0.16} = 25 \text{ } \Omega \qquad \textit{Ans.}$$

Find R_1. Apply Kirchhoff's second law to find V_1 by tracing around the circuit in the direction *abcd*.

$$V_T - V_1 - V_2 = 0$$
$$20 - V_1 - 4 = 0$$
$$V_1 = 16 \text{ V}$$

Therefore,

$$R_1 = \frac{V_1}{I_1} = \frac{16}{0.2} = 80 \text{ } \Omega \qquad \textit{Ans.}$$

Check: The total resistance of R_1, R_2, and R_L connected as shown must equal R_L or 100 Ω.

$$R_T = R_1 + \frac{R_2 \times R_L}{R_2 + R_L}$$

$$= 80 + \frac{25 \times 100}{25 + 100}$$

$$= 80 + \frac{2500}{125}$$

$$R_T = 80 + 20 = 100 \qquad \textit{Check}$$

Self-Test 10-25

The resistance of a 40-V source is 50 Ω. In order to reduce the volume to a 50-Ω speaker, an L pad is inserted to reduce the voltage to 4 V. Find R_1 and R_2.

Solution

The diagram for the circuit is shown in Fig. 10-39. In Fig. 10-39a, the total current *before* the pad is inserted is

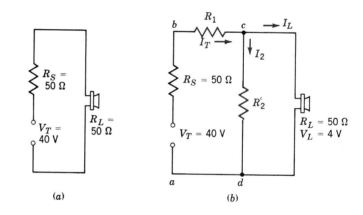

Figure 10-39 (a) Original circuit. (b) Circuit with L pad inserted.

1. $I_T = \dfrac{V_T}{R_T} = \dfrac{V_T}{R_S + ?}$

2. $I_T = \dfrac{40}{?}$

3. $I_T = $ _____ A

In Fig. 10-39b, after R_1 and R_2 are in the circuit,

4. $I_L = \dfrac{V_L}{R_L} = \dfrac{4}{50} = $ _____ A

Find I_2 by Kirchhoff's first law. At point c,

5. $I_2 = I_T - I_L = 0.4 - $ _____ $= 0.32$ A

6. Find R_2. Since $V_2 = V_L = $ _____ V,

390 Chapter 10 Applications of Series and Parallel Circuits

7. $R_2 \ = \dfrac{V_2}{I_2} = \dfrac{4}{0.32} = $ _____ Ω

Find R_1. Apply Kirchhoff's second law to find V_1 by tracing around the circuit in the direction *abcd*.

8. $V_T - V_S - V_1 - $ _____ $= 0$

9. $V_T - I_T R_S - V_1 - V_2 = 0$

10. $40 - ($ _____ $)(50) - V_1 - 4 = 0$

11. $40 - $ _____ $- V_1 - 4 = 0$

12. _____ $- V_1 = 0$

13. $V_1 = $ _____ V

Therefore,

14. $R_1 = \dfrac{V_1}{I_1} = \dfrac{V_1}{I_T} = \dfrac{16}{?}$

15. $R_1 = $ _____ Ω *Ans.*

Check: The total resistance of R_1, R_2, and R_L connected as shown must equal R_L or 50 Ω.

16. $R_T = R_1 + \dfrac{R_2 \times ?}{R_2 + ?}$

17. $R_T = $ _____ $+ \dfrac{12.5 \times 50}{12.5 + 50}$

18. $R_T = 40 + $ _____

$R_T = 50 \ \Omega$ *Check*

Exercises

Practice

Insert an L pad into each of the following circuits. Find the values of R_1 and R_2 which will provide the required reduction in voltage.

Problem	Circuit Fig. No.	V_T, V	R_S, Ω	R_L, Ω	V_L, V
1	10-38*a*	60	...	15	12
2	10-39*a*	60	60	60	15
3	10-38*a*	120	...	10	60
4	10-39*a*	120	100	100	30
5	10-39*a*	120	60	60	24
6	10-39*a*	100	20	20	5
7	10-39*a*	250	100	100	100
8	10-38*a*	6	...	10	1.5
9	10-39*a*	40	100	100	5
10	10-39*a*	6.3	20	20	2.1

T-Type Attenuators

As we have seen, the L-type attenuator maintains a constant resistance to the source even though the voltage across the load is changed. However, the position of the source and the load may not be interchanged, because the resistance presented on one side of the pad is different from the resistance on the other side. If it is desired that the pad present a constant resistance on *each* side of it, a *T type* of pad is needed. This is a symmetrical pad and offers the *same* resistance on both sides. When this type of pad is used, the source and the load *may* be interchanged, because the resistance presented by each side of the pad is constant.

In the T type, $R_1 = R_3$ and $R_S = R_L$.

Example 10-26

In Fig. 10-40a, it is desired to reduce the voltage across R_L to 20 V. Design a T-type attenuator pad to achieve this result.

Solution

In Fig. 10-40a, the total current *before* the pad is inserted is

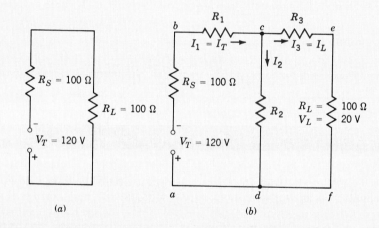

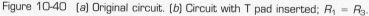

Figure 10-40 (a) Original circuit. (b) Circuit with T pad inserted; $R_1 = R_3$.

$$I_T = \frac{V_T}{R_T} = \frac{V_T}{R_S + R_L} = \frac{120}{100 + 100} = \frac{120}{200} = 0.6 \text{ A}$$

In Fig. 10-40b, after R_1, R_2, and R_3 have been inserted,

$$I_L = \frac{V_L}{R_L} = \frac{20}{100} = 0.2 \text{ A}$$

Find I_2 by Kirchhoff's first law operating at point c.

$$I_2 = I_1 - I_L = 0.6 - 0.2 = 0.4 \text{ A}$$

R_1, R_2, and R_3 may be found by applying Kirchhoff's second law to different loops of the circuit of Fig. 10-40b. Trace the circuit *abcd*, assuming the currents to be in the directions indicated on the diagram.

$$120 - 0.6(100) - 0.6R_1 - 0.4R_2 = 0$$
$$120 - 60 - 0.6R_1 - 0.4R_2 = 0$$
$$-0.6R_1 - 0.4R_2 = -60 \qquad (1)$$

Now trace the circuit *cefdc*.

$$-0.2R_3 - 20 + 0.4R_2 = 0$$
$$-0.2R_3 + 0.4R_2 = 20$$

but since $R_1 = R_3$, we can substitute R_1 for R_3 in this equation.

$$-0.2R_1 + 0.4R_2 = 20 \qquad (2)$$

Solve equations (1) and (2) simultaneously.

$$
\begin{array}{lll}
-0.6R_1 - 0.4R_2 & = -60 & (1) \\
-0.2R_1 + 0.4R_2 & = 20 & (2)
\end{array}
$$

Add:
$$-0.8R_1 \qquad\qquad = -40$$

$$R_1 = \frac{-40}{-0.8} = 50 \ \Omega \qquad Ans.$$

and
$$R_3 = 50 \ \Omega \qquad Ans.$$

Substitute this value for R_1 in equation (2).

$$-0.2R_1 + 0.4R_2 = 20$$
$$-0.2(50) + 0.4R_2 = 20$$
$$-10 + 0.4R_2 = 20$$
$$0.4R_2 = 20 + 10$$
$$0.4R_2 = 30$$

$$R_2 = \frac{30}{0.4} = 75 \ \Omega \qquad Ans.$$

Check: The total resistance of R_1, R_2, and R_L connected as shown must equal R_L, or 100 Ω.

$$\text{The resistance from } c \text{ to } f = R_3 + R_L$$
$$= 50 + 100 = 150 \ \Omega$$

This 150 Ω is in parallel with R_2. The resistance of this group is equal to

$$\frac{75 \times 150}{75 + 150} = \frac{11{,}250}{225} = 50 \ \Omega$$

This 50 Ω is in series with R_1. The total resistance is

$$R_T = 50 + R_1$$
$$R_T = 50 + 50 = 100 \ \Omega$$

which is equal to R_L. *Check*

Exercises

Applications

Insert a T pad into the following circuits similar to that shown in Fig. 10-40*a*. Find the values of R_1, R_2, and R_3 which will provide the required reduction in voltage.

Problem	V_T, V	R_S, Ω	R_L, Ω	V_L, V
1	120	60	60	30
2	12	200	200	4
3	60	60	60	15
4	100	400	400	30
5	60	10	10	5
6	200	400	400	20
7	24	100	100	1.5

(See CD-ROM for Test 10-4)

<hr>

JOB 10-13

Power and Resistor Size

In all the previous work there has been only small concern for the power that is dissipated in a resistor and the size of that resistor. The greater the power used by a device, the greater the heat produced. If this heat cannot be dissipated to the surrounding atmosphere, the resistor becomes hotter and hotter until it burns up. To overcome this difficulty, resistors are manufactured and rated according to their ability to dissipate the heat generated within them. High-power resistors are much larger than low-power resistors.

Resistors are manufactured in standard power sizes. These standards are developed by groups such as the military and the Electronics Industry Associations (EIA). These standards are necessary so that the design and manufacture of electronic hardware will be simplified.

Resistor standards are grouped in many ways. One group is according to the level of power that they can handle. This general grouping has three major categories. The first is called "general-purpose" resistors. These are resistors that can handle up to 2 W. The second group is called "medium power" resistors. This group can handle 2 to 6 W. The third group is called "high-power" resistors. This group will handle 6 to 210 W.

The general-purpose group is probably the most standard of the three groups. The most popular type of resistor in this group is the carbon composition type. The standard sizes are ⅛, ¼, ½, 1, and 2 W. Notice that each size is twice as high in heat dissipation ability as the next-lower size.

For the next two groups, one of the most popular types of resistor is the wirewound resistor. Standards for this group give the following sizes: 3, 5, 7, 10, 15, 26, 55, 113, 159, and 210 W.

It is usually possible to find a manufacturer who either produces another standard or manufactures nonstandard devices. This may be an advantage in some situations, but replacement and cost may be a problem.

Once you have determined the power in a resistor by use of Eqs. (7-1), (7-4), or (7-5), you must select the size of resistor that will handle that, or more power. This is similar to the selection of wire sizes.

Example 10-27

Find the size of resistor needed for the resistors shown in Fig. 6-20.

Solution

1. For resistor R_1, the power is 30 V $\times$ 3 A = 90 W. From the list of standard sizes, the next-*larger* size is 113 W. Thus, although the resistor generates only 90 W, we must choose a 10-Ω 113-W resistor.
2. For resistor R_2, the power is 3.75 V $\times$ 0.75 A = 2.81 W. Choose the next-larger size of standard resistor—a 5-Ω 3-W resistor.

Continuing as above, we can develop the following tabulation of resistor powers and sizes:

R, No.	R, Ω	Power, W	Size, W
1	10	90	113
2	5	2.81	3
3	10	5.63	7
4	24	5.06	7
5	8	40.5	55
6	20	180	210

Exercises

Practice

1. Find the proper resistor power size for the resistors shown in Fig. 10-37.
2. Find the proper resistor power size for the resistors shown in Fig. 10-39.
3. Find the proper resistor power size for the resistors shown in Fig. 4-8.
4. Find the proper resistor power size for the resistors shown in Fig. 4-9.
5. Find the proper resistor power size for the resistors shown in Fig. 5-21.
6. Find the proper resistor power size for the resistors shown in Fig. 6-12.

Assessment

1. A 0–5-A ammeter has a resistance of 5 Ω. Find the shunt needed to extend the range to 0–100 A.

2. A 0–1-mA milliammeter was increased to read 0–75 mA. Find the true current when the meter is reading 0.6 mA.

3. A 0–10-mV millivoltmeter with a resistance of 5 Ω is extended to read 250 V. What is the true voltage when the meter reads 7.25 mV?

4. In the voltage divider network shown in Fig. 10-41, use the currents and voltage in the figure to determine the value of each resistor, the total current, and the total voltage.

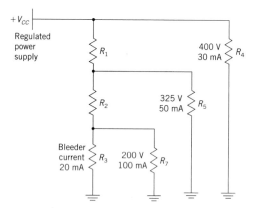

Figure 10-41

5. Determine the voltage at B in the circuit shown in Fig. 10-42.

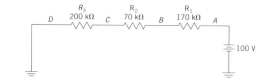

Figure 10-42

6. What is a very important use of a Wheatstone bridge?

7. What is the main advantage to finding the Thevenin values for a complex circuit?

8. In a circuit similar to that shown in Fig. 10-28, $V_{CC} = -15$ V, $R_L = 3$ kΩ, $R_B = 175$ kΩ, and $\beta = 150$. Find the load current, the voltage drop across the load, and the collector voltage.

9. In a circuit similar to that shown in Fig. 10-30a, $V_{CC} = -20$, $R_L = 1.5$ kΩ, $R_B = 100$ kΩ, and $\beta = 30$. Find I_C, E_{load}, and V_C.

10. If a transistor is designed for a current of 3 mA through R_C, what is the voltage drop across R_C? Use Example 10-21 as a model.

11. What is the purpose of an attenuator pad?

12. What is one possible hazard of using a resistor that does not have a wattage rating as large as needed?

INTERNET
ACTIVITIES

Internet Activity A

View the following site to learn about the basic design of a moving-coil galvanometer. **http://www.usd.edu/phys/Labs/4_1_2/4_1_2c.html.** Using the information provided at this site, list the basic parts of a moving-coil galvanometer.

Internet Activity B

Using the site listed above, determine how the scale of a galvanometer is commonly marked.

Chapter 10 Solutions to Self-Tests and Reviews

Self-Test 10-2

1. 100
2. 750
3. 49
4. new, old
5. 100
6. NA
7. 49
8. 7.5

Self-Test 10-9

1. 1000
2. 100
3. 10
4. NA
5. 100
6. 10

Job 10-4 Review

Ammeter Range

a. shunt

b. shunt
c. meter
d. meter
e. new
f. old

Reading Ammeters

1. N
   a. I_S
   b. I_S

Voltmeter Range

1. R_{meter}
2. new
3. old
4. less
5. sensitivity

Self-Test 10-11

1. 20, 100, 0.1, 10
   At B: 100, 0.11
   At C: 20, 30
   At D: I_3, 30, 60
   At E: I_4, 50, 110, 0.11

2. For R_1: 0.11, 455
   For R_2: 100, 0.01,
   10,000
   For R_3: C, D, 100, 150,
   150, 5000
   For R_4: 50, 50, 833

Self-Test 10-14

1. series, $+20$
2. V_T
3. 15, 5

Self-Test 10-23

1. H-parameter
2. Thevenin, Norton
3. impedance, V_{in}/I_{in}
4. conductance, I_{out}/V_{out}
5. beta
6. $h_r, V_{in}/V_{out}$
7. dc
8. ac signal

9. common emitter

Self-Test 10-25

1. R_L
2. 100
3. 0.4
4. 0.08
5. 0.08
6. 4
7. 12.5
8. V_2
9. NA
10. 0.4
11. 20
12. 16
13. 16
14. 0.4
15. 40
16. R_L, R_L
17. 40
18. 10

11

chapter

Efficiency

Learning Objectives

1. Write percents as decimals and decimals as percent.
2. Write fractions as percents.
3. Find percentage, base, or rate using the formula $P = B \times R$.
4. Use conversion factors to change from one unit of electric and mechanical power to another.
5. Use a formula to determine the efficiency of a machine.
6. Find the output and input of an electric device.
7. Determine the maximum transfer of power from a source to a load.

JOB 11-1

Checkup on Percent (Diagnostic Test)

The efficiency of electric machinery is expressed as a percent. The concept of percent is used throughout the fields of electricity and electronics to describe and compare electrical effects. Can you solve the following problems that use percent? If you have difficulty with any of these problems, turn to Job 11-2, which follows.

Exercises

Applications

1. What is the possible error in a resistor marked 1000 Ω if the indicated error (tolerance) is 10%?
2. A solenoid exerts a pull of 0.9 lb at its rated voltage. What pull is exerted at 85% of its rated voltage?
3. A 15-A fuse carried a 15% overload for 3 s. What current flowed through the fuse during this time?
4. The voltage to a relay might vary up to 10% of the rated value. Is a voltage of 200 V within the normal variation if the rated voltage is 230 V?
5. The peak voltage of an ac wave is 141% of the ac meter reading. Find the peak voltage if the meter reads 50 V.
6. What is the efficiency of a motor if it uses 20 units of energy and delivers 18 units of energy?
7. What is the efficiency of a transmission line if the power supplied is 4500 W and the power delivered is 4200 W?
8. A transformer has an efficiency of 96%. If the transformer uses 60 W, what power will it deliver?
9. The National Electrical Code specifies a maximum of 2% voltage drop in a house line. What is the minimum voltage at a load if the supply voltage is 120 V?

10. A television mechanic bought $125 worth of capacitors at a 35% discount. What did he pay for the tubes?

11. If he was also allowed 2% discount for payment within 10 days, what was his actual cost?

12. A television mechanic charged the list price of $2.25 for a capacitor. If the item cost her $1.80, what was her percent of profit?

13. The television mechanic estimated that her overhead expenses equaled 20% of her cost. If she charges the list price of $4 for an item, what is the most that she should pay for it?

14. Some storage batteries are shipped dry and filled with electrolyte by the seller. In making up a batch of electrolyte to contain 40% acid, how much acid should be used to make 32 oz of electrolyte?

Brushup on Percent

The word *percent* is used to indicate some portion of 100. A mark of 92% on an examination indicates that the student got 92 points out of a possible 100 points. This may also be written as a decimal (0.92) or as a common fraction ($\frac{92}{100}$), both of which are read as 92 *hundredths*. A percent indicated by a percent sign (%) cannot be used in a calculation until it has been changed to an equivalent decimal.

Changing percents to decimals. Since 92% means $\frac{92}{100}$, the change to a decimal is made by dividing 92 by 100. This is easily accomplished by moving the decimal point two places to the left, as shown in Job 2-6.

RULE To change a percent to a decimal, move the decimal point two places to the left and drop the percent sign.

Example 11-1

Change the following percents to decimals.

Solution

$$32\% = 32.\% = 0.32 \quad Ans.$$
$$8\% = 8.\% = 0.08 \quad Ans.$$
$$16\frac{1}{2}\% = 16.5\% = 0.165 \quad Ans.$$

Exercises

Practice

Change the following percents to decimals:

1. 38%	2. 60%	3. 6%	4. 19%	5. 4%
6. 26.4%	7. 3.6%	8. 100%	9. 125%	10. 0.5%

| 11. 16⅔% | 12. ¾% | 13. 62.5% | 14. 1% | 15. 12.5% |
| 16. 0.9% | 17. 2.25% | 18. 5½% | 19. 4¼% | 20. ½% |

Changing decimals to percents. To change a decimal to a percent is to reverse the process of changing a percent to a decimal.

RULE To change a decimal to a percent, move the decimal point two places to the right and add a percent sign.

Example 11-2

Change the following decimals to percents:

Solution

$$0.20 = 20.\% = 20\% \quad Ans.$$
$$0.06 = 6.\% = 6\% \quad Ans.$$
$$0.125 = 12.5\% \quad Ans.$$
$$1.1 = 110.\% = 110\% \quad Ans.$$
$$2 = 2.00 = 200.\% = 200\% \quad Ans.$$

Exercises

Practice

Express the following decimals as percents:

1. 0.50	2. 0.04	3. 0.2	4. 0.075	5. 1.45
6. 0.008	7. 1.00	8. 0.87	9. 0.625	10. 0.092
11. 0.222	12. 0.15	13. 0.7	14. 3	15. 0.055

Changing common fractions to percents

RULE To change a common fraction to a percent, express the fraction as a decimal; then move the decimal point two places to the right and add a percent sign.

The decimal equivalents for many of the common fractions may be found in Table 2-3.

Example 11-3

Express the following fractions as percents.

Solution

$$\frac{3}{4} = 0.75 = 75\% \qquad Ans.$$
$$\frac{1}{2} = 0.50 = 50\% \qquad Ans.$$
$$\frac{5}{8} = 0.625 = 62.5\% \qquad Ans.$$

If the fraction is not on the decimal equivalent chart, change the fraction to a decimal by dividing the numerator by the denominator, as shown in Job 2-5.

Example 11-4

Express $\frac{8}{25}$ as a percent.

Solution

$$\frac{8}{25} = 25\overline{)8.00} \quad \begin{array}{r} 0.32 = 32\% \qquad Ans. \\ \hline \end{array}$$

$$
\begin{array}{r}
0.32 = 32\% \qquad Ans. \\
25\overline{)8.00} \\
-7\,5 \\
\hline
50 \\
-50 \\
\hline
0
\end{array}
$$

Exercises

Practice

Express the following common fractions as percents:

1. $\frac{1}{4}$	2. $\frac{3}{8}$	3. $\frac{2}{5}$	4. $\frac{4}{5}$	5. $\frac{5}{8}$
6. $\frac{3}{16}$	7. $\frac{3}{10}$	8. $\frac{3}{5}$	9. $\frac{3}{6}$	10. $\frac{9}{10}$
11. $\frac{13}{20}$	12. $\frac{3}{32}$	13. $\frac{2}{3}$	14. $\frac{5}{6}$	15. $\frac{5}{11}$
16. $\frac{3}{7}$	17. $\frac{16}{48}$	18. $\frac{12}{30}$	19. $\frac{2}{9}$	20. $\frac{13}{15}$

Using percent in problems. There are three parts to every problem involving percent.

1. The *base B* is the entire amount.
2. The *rate R* is the percent of the base.
3. The *part P* is the portion of the base.

These three parts are combined in the following

$$B \times R = P$$

Formula 11-1

Finding the part

Example 11-5

An electrician earning $340 per week got a 6½% increase in pay. Find the amount of the increase.

Solution

Given: $B = \$340$ Find: $P = ?$
        $R = 6\frac{1}{2}\% = 0.065$

The rate of 6½% must be changed to a working decimal, as shown above. In the formula, $R = 0.065$.

$$P = B \times R$$
$$P = 340 \times 0.065 = \$22.10 \qquad Ans.$$

Example 11-6

The resistance of a resistor depends on the accuracy with which it is manufactured. This is shown on the resistor as a fourth band of color called the *tolerance value*. What are the upper and lower values of resistance that might be expected on a resistor marked 3000 Ω and 5% tolerance?
Note: The tolerance colors are: gold = 5%; silver = 10%; black (or no color) = 20%.

Solution

Given: $B = 3000 \ \Omega$ Find: Tolerance = $P = ?$
        $R = $ tolerance $= 5\% = 0.05$ Upper value = ?
                                               Lower value = ?

1. Find the permitted tolerance.

$$P = B \times R$$
$$P = 3000 \times 0.05 = 150 \ \Omega \qquad Ans.$$

2. Find the upper and lower values.

Upper value = $3000 + 150 = 3150 \ \Omega$ *Ans.*
Lower value = $3000 - 150 = 2850 \ \Omega$ *Ans.*

Hint for Success

Letters commonly have more than one meaning. In this situation P does not mean power but means part or percentage. This is because the most recent definition, $P = B \times R$, supersedes previous usage. There is no new definition for V, so your concept of voltage remains correct.

Self-Test 11-7

The voltage lost in a line supplying a motor is 5 percent of the generator voltage V_G of 220 V. Find (*a*) the voltage lost and (*b*) the voltage supplied to the motor load.

Solution

Given: $V_G = $ base $B = \dfrac{(1)}{(2) \ \text{(decimal)}}$ Find: $V_l = $ part $P = ?$
        $R = 5\% = $ $V_L = ?$

a.

$$220 \times \frac{B \times R}{\underset{(3)}{}} = P$$
$$= P$$
$$P = \underset{(4)}{} \text{ V}$$
$$V_I \quad \underline{\quad (5) \quad} \text{ V} \quad Ans.$$

(11-1)

b.

$$V_L = V_G - \frac{(6)}{\underset{(7)}{}}$$
$$V_L \quad \underline{\quad (7) \quad} - 11$$
$$V_L \quad \underline{\quad (8) \quad} \text{ V} \quad Ans.$$

(6-3)

Exercises

Practice

1. How much is 22 percent of 150?
2. How much is 6 percent of $500?
3. How much is 2.5 percent of $300?

Applications

4. What is the amount of sales tax on an order of $7.50 if the tax rate is 3%?
5. A TV bench worker earned $375 per week. He received an increase of 15%. Find the amount of the increase.
6. A TV mechanic used 30% of a 1000-ft roll of push-back wire. How many feet of wire did she use?
7. A 15-A fuse carried a 15% overload for 2 s. What current flowed through the fuse during this time?
8. What is the possible error in a 500-Ω resistor if it is marked with a silver band indicating only 10% accuracy?
9. The effective value of an ac wave is 70.7% of the peak value. What is the effective voltage of a wave that reaches a peak of 165 V?
10. In a brightness-control circuit, 6% of the 525 horizontal lines were blanked out. How many lines were blanked out?
11. Out of a lot of 200 transistors, 2½% were rejected as being defective. How many were rejected?
12. The voltage drop in a line supplied by a 220-V generator is 2% of the generator voltage. Find the voltage drop and the voltage supplied to the load.
13. How many watts of power are lost in a transformer rated at 250 W if 3% of the energy is lost as heat?
14. What is the interest for 1 year at 14½% on $125?
15. If you are allowed a discount of 15% on a bill amounting to $127.85, how much is the discount?
16. A zener diode rated at 10 W is used as a regulator in a power supply as shown in Fig. 11-1. The power supply delivers 20 V to the load. Find the largest

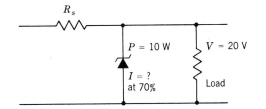

Figure 11-1   A zener diode used as a regulator in a power supply.

current that the zener should be allowed to pass if only 70% of the rated current is permitted.

Finding the rate

Example 11-8

20 is what percent of 80?

Solution

Given: $B = 80$ Find: $R = ?$
         $P = 20$

$$B \times R = P \qquad\qquad (11\text{-}1)$$
$$80 \times R = 20$$
$$R = \frac{20}{80} = \frac{1}{4} = 0.25 = 25\% \qquad Ans.$$

Example 11-9

In transmitting 50 hp by a belt system, 1.2 hp was lost due to slippage. What percent of the power was lost?

Solution

Given: $B = 50$ hp   Find: $R = ?$
         $P = 1.2$ hp

$$B \times R = P$$
$$50 \times R = 1.2$$
$$R = \frac{1.2}{50} = 0.024 = 2.4\% \qquad Ans.$$

Exercises

Practice

1. 30 is what percent of 120?
2. 30 is what percent of 80?
3. What percent of 60 is 24?
4. What percent of 85 is 12?
5. What percent of 25 is 0.5?
6. What percent of 30 is 40?
7. 150 is what percent of 100?
8. 24.5 is what percent of 70?
9. What percent of 25½ is 8½?
10. What percent of 18.5 is 3.5?

Applications

11. A discount of $2 was given on a bill of $16. Find the rate of discount.
12. If three tubes out of a lot of 90 tubes are defective, what percent is defective?

13. If 100 m of a 250-m roll of wire has been used, what percent has been used?
14. A signal generator costing $65 was sold at a loss of $25. Find the rate of loss.
15. In transmitting 32 hp by a belt system, 0.72 hp was lost due to slippage. What percent of the power was lost?
16. A team played 18 games and won 15 of them. What percent of the games played did the team win?
17. An electrician used 40 ft of BX cable from a 200-ft-long coil. What percent did he use?
18. A woman earning $380 per week received an increase of $45 per week. Find the percent of increase.
19. The voltage loss in a supply line is 5.5 V. If the generator voltage is 110 V, find the rate of loss.
20. A 15-A fuse carried 18 A for 2 s. What was the percent of overload?

Finding the base

Example 11-10

5 is 25% of what number?

Solution

Given: $P = 5$ Find: $B = ?$

$R = 25\% = 0.25$

$$B \times R = P \qquad (11\text{-}1)$$
$$B \times 0.25 = 5$$

$$B = \frac{5}{0.25} = 20 \qquad Ans.$$

Example 11-11

A mechanic is able to save $18.50 each week. This amount is equal to 12½% of his weekly wages. How much does he earn each week?

Solution

Given: P = amount saved = $18.50 Find: B = total wages = ?

R = percent saved = 12½% = 0.125

$$B \times R = P \qquad (11\text{-}1)$$
$$B \times 0.125 = 18.50$$

$$B = \frac{18.50}{0.125} = \$148 \qquad Ans.$$

Practice

1. 10 is 50% of what number?
2. 20 is 4% of what number?
3. 16 is 40% of what number?
4. 25 is 2.5% of what number?
5. 70 is 3½% of what number?

Applications

6. The voltage loss in a line is 3 V. If this is 2% of the generator voltage, what is the generator voltage?
7. A mechanic received a 15% increase in her wages amounting to $35.75. What were her wages before the increase?
8. A motor has an output of 5 hp. This amount is 85% of the power put into the motor. Calculate the power input.
9. A fuse carried 16.5 A. This was 110% of the fuse rating. What is the fuse rating?
10. What must be the generator voltage if only 98% of the generator voltage is delivered to a 117-V line?

Review of Percent Problems

RULE To change a percent to a decimal, move the decimal point _____(a)_____ places to the _____(b)_____ and drop the percent sign.

RULE To change a decimal to a percent, move the decimal point _____(c)_____ places to the _____(d)_____ and add a percent sign.

RULE To change a common fraction to a percent, express the fraction as a decimal; then move the decimal point _____(e)_____ places to the right and add a _____(f)_____ sign.

All percent problems contain three parts:

1. The base *B* is the _____ amount.
2. The rate *R* is a percent of the _____.
3. The part *P* is the portion of the _____.
4. The relationship among these parts is given by the formula

$$\underline{\qquad} \times \underline{\qquad} = \underline{\qquad}$$

11-1

Exercises

Practice

1. Change the following percents to decimals:
   - *a.* 62%
   - *b.* 3%
   - *c.* 5.6%
   - *d.* 0.8%
   - *e.* 116%
   - *f.* 4½%
   - *g.* 6¼%.

2. Change the following decimals to percents:
   - *a.* 0.4
   - *b.* 0.08
   - *c.* 0.6
   - *d.* 2.00
   - *e.* 0.625
   - *f.* 0.045
   - *g.* 5

3. Change the following fractions to percents:
   - *a.* ¾
   - *b.* ⅗
   - *c.* 3/7
   - *d.* 3/10
   - *e.* 3/13
   - *f.* ⅙
   - *g.* 24/52

4. What is 18% of 96?
5. How much is 4.5% of $2000?
6. 20 is what percent of 120?
7. What percent of 30 is 12?
8. 8 is 40% of what number?
9. How much is 3¼% of $1500?
10. 10 is 2½% of what number?
11. What percent of 38.4 is 8?
12. How much is 0.6% of 75

Applications

13. In the transmission of 6 hp by a belt system, 0.09 hp was lost owing to slippage. What percent was lost?
14. Of the 60 workers in a shop, 80% got production bonuses. How many got a bonus?
15. According to the National Electrical Code, the maximum allowable voltage drop to a power load is 3%. If the supply voltage is 120 V, find the minimum voltage at the input terminals of a motor requiring 20 A and located 200 ft from the supply.
16. A woman received 150% of her hourly wage for every hour that she worked above 40 h. If she worked 50 h in a certain week at a base rate of $6.50, what was her overtime pay? What was her total salary? How much was withheld from her salary for social security at 6.2% tax rate?
17. A man earning $285 per week received an increase of 8%. Find the amount of the increase and his new salary.
18. Emitter bias resistors usually have a wattage rating about 80% higher than the calculated wattage. What should be the wattage rating of an emitter bias resistor developing 0.5 W?
19. The cost of rewinding an armature is $74. If the electrician wishes to make a profit of 20%, at what price must the customer be billed?

(See CD-ROM for Test 11-1)

Conversion Factors for Electric and Mechanical Power

A *motor* is a device that uses electric power and converts this power into the mechanical power of a rotating shaft. The electric power supplied to a motor is measured in watts or kilowatts; the mechanical power delivered by a motor is measured in horsepower. One horsepower is equivalent to 746 W of electric power. For most calculations, it is sufficiently accurate to consider 1 hp equal to 750 W.

A *generator* is a device that uses mechanical power and converts this power into electric power. The mechanical power supplied to a generator is measured in horsepower; the electric power delivered by a generator is measured in watts or kilowatts.

Tables of conversion factors. Tables 11-1 and 11-2 are based on the following relationships.

$$1 \text{ kW} = 1000 \text{ W}$$
$$1 \text{ hp} = 750 \text{ W}$$

Table 11-1

	To Change	To	Multiply By
1	Kilowatts (kW)	Watts (W)	1000
2	Horsepower (hp)	Watts (W)	750
3	Kilowatts (kW)	Horsepower (hp)	$\frac{4}{3}$
4	Horsepower (hp)	Kilowatts (kW)	$\frac{3}{4}$

Table 11-2

	To Change	To	Divide By
5	Watts (W)	Kilowatts (kW)	1000
6	Watts (W)	Horsepower (hp)	750

Items 3 and 4 are derived as follows:

Since hp = W/750 (item 6) and W = kW × 1000 (item 1), we can substitute (kW × 1000) for watts in item 6. This gives

$$\text{hp} = \frac{\text{kW} \times 1000}{750}$$

or

$$\text{hp} = \text{kW} \times \frac{4}{3}$$

Similarly, since kW = W/1000 (item 5) and W = hp × 750 (item 2), we can substitute (hp × 750) for watts in item 5. This gives

$$\text{kW} = \frac{\text{hp} \times 750}{1000}$$

or

$$\text{kW} = \text{hp} \times \frac{3}{4}$$

Example 11-12

1. $1.5 \text{ kW} = 1.5 \times 1000 = 1500 \text{ W}$ *Ans.*
2. $\frac{1}{2} \text{ hp} = \frac{1}{2} \times 750 = 375 \text{ W}$ *Ans.*
3. $1\frac{1}{2} \text{ kW} = 1\frac{1}{2} \times \frac{4}{3} = \frac{3}{2} \times \frac{4}{3} = 2 \text{ hp}$ *Ans.*
4. $1\frac{1}{2} \text{ hp} = 1\frac{1}{2} \times \frac{3}{4} = \frac{3}{2} \times \frac{3}{4} = \frac{9}{8} = 1.125 \text{ kW}$ *Ans.*
5. $1800 \text{ W} = 1800 \div 1000 = 1.8 \text{ kW}$ *Ans.*
6. $1125 \text{ W} = 1125 \div 750 = 1.5 \text{ hp}$ *Ans.*

Exercises

Practice

Change the following units of measurement.

1. 6.5 kW to watts
2. 2 hp to watts
3. 2300 W to kilowatts
4. 2625 W to horsepower
5. 0.05 kW to watts
6. 1 kW to horsepower
7. 1 hp to kilowatts
8. $1\frac{3}{4}$ kW to horsepower
9. $2\frac{1}{4}$ hp to kilowatts
10. 7.5 kW to horsepower
11. $\frac{1}{8}$ hp to kilowatts
12. $\frac{3}{4}$ hp to watts
13. 4000 W to horsepower
14. $4\frac{1}{2}$ hp to kilowatts
15. 10 kW to horsepower
16. 8.75 kW to horsepower

Efficiency of Electric Apparatus

Most machines are designed to do a specific job. A motor takes in electric energy and delivers mechanical energy in the form of a rotating shaft. If it could change *all* the electric energy into mechanical energy, it would be said to be 100% efficient. Unfortunately, not all the electric energy put into the motor appears as mechanical energy at the shaft. Some of the energy is used in overcoming friction, and some appears as heat energy. *No energy is lost*—it merely does not appear at the shaft as *useful* energy. The amount of useful energy (the power output) is therefore always *less* than the energy received by the machine (the power input). This means that the power output is always some fractional part of the power input. This fraction, obtained by dividing the output by the input, is called the *efficiency* of the machine. It is always expressed as a percent by multiplying the fraction by 100. In general, efficiency means the ability of a device to pass on the energy it receives without loss. The efficiency of electric devices is generally large, ranging from 75 to 98%.

about electronics

We pass under them almost everyday, and we hardly notice them until we need them to find the way out. They cost the United States a billion dollars a year. You guessed it—exit signs. One sign alone takes from 44 to 350 kWh of electricity each year. Now there's a more efficient sign that could save U.S. companies 800 kW of electricity each year. The U.S. Postal Service liked the idea and replaced more than 30,000 of their old signs, thereby preventing the emission of more than 5000 tons of carbon per year. When you're buying televisions, appliances, computers, other office equipment, or home electronics, look for the ENERGY STAR® label, the symbol for energy efficiency. It means that these products provide equal or better performance while using less energy, and both you and the environment will benefit.

ENERGY STAR® is a U.S. registered mark.

$$\text{Efficiency} = \frac{\text{output}}{\text{input}} \qquad \text{Formula 11-2}$$

Hint for Success

You must always be careful with notation. In this case *P* is used to represent power and does not represent part or percent. You must use your knowledge to recognize the equation $P = I \times V$ as a power equation.

In this formula, the output and the input must *both* be expressed in the *same units of measurement*. This will ordinarily necessitate a change in the units of measurement, since for a motor,

Output is measured in horsepower.

Input is measured in watts or kilowatts.

For a generator,

Output is measured in watts or kilowatts.

Input is measured in horsepower.

CALCULATOR HINT

If you are using a programmable calculator, you cannot have more than one meaning for a symbol. You cannot use *P* for both power and percent. You will have to use different names or labels.

Example 11-13

Find the efficiency of a motor that receives 4 kW and delivers 4 hp.

Solution

Given: Output = 4 hp   Find: Percent eff = ?
               Input = 4 kW

1. Express all measurements in the same kind of units.

$$\text{Output} = 4 \text{ hp} = 4 \times 750 = 3000 \text{ W}$$
$$\text{Input} = 4 \text{ kW} = 4 \times 1000 = 4000 \text{ W}$$

2. Find the efficiency.

$$\text{Eff} = \frac{\text{output}}{\text{input}} = \frac{3000}{4000} = 0.75 = 75\% \qquad \textit{Ans.}$$

Self-Test 11-14

A generator receives 7 hp and delivers 20 A at 230 V. Find its efficiency.

Solution

1. Given: Output $\begin{cases} V = \underline{\hspace{2cm}} \text{ V} \\ I = \underline{\hspace{2cm}} \text{ A} \end{cases}$ Find: Percent eff = ?
              Input = $\underline{\hspace{2cm}}$ hp

2. Express all measurements in the same kind of unit.

Output: $P = I \times V = 20 \times 230 = 4600$ _____

Input: 7 hp = $7 \times 750 = 5250$ _____

3. Find the efficiency.

$$\text{Eff} = \frac{?}{?}$$

$$\text{Eff} = \frac{4600}{5250} = 0.876 = \underline{\hspace{2cm}} \% \qquad \textit{Ans.}$$

Self-Test 11-15

A transformer requires a current of 46 mA at 120 V. If its power output is 5 W, find the efficiency of the transformer.

Solution

Given: Output = _____(1)_____ W   Find: Plate eff = ?

$$\text{Input} = \begin{cases} V = \underline{\quad(2)\quad} \text{ V} \\ I = \underline{\quad(3)\quad} \text{ mA} \end{cases}$$

Express all measurements in the same kind of unit.

Input: $P = I \times V = \dfrac{(4)}{(5)} \times 120$

             Input $P = 5.52$ _____(5)_____

Output: $P = 5$ _____(6)_____

        Plate eff $= \dfrac{\text{output}}{(7)}$

        Plate eff $= \dfrac{5}{5.52} = 0.906 = \underline{\quad(8)\quad}$ %   *Ans.*

Exercises

Applications

1. A motor rated at 2 hp (delivers 2 hp) receives 1.8 kW of energy. Find its efficiency.
2. A motor delivers 3 hp and receives 2.4 kW. Find its efficiency.
3. A "power" transformer draws 0.5 kW from a line and delivers 480 W. Find its efficiency.
4. A filament transformer draws 60 W and supplies 50 W to the tube filaments. Find its efficiency.
5. A generator rated at 10 kW (delivers 10 kW) receives 15 hp. Find its efficiency.
6. A generator delivers 1¼ kW and receives 2 hp. What is its efficiency?
7. A transmission line receives 230 kW and delivers 210 kW. What is the efficiency of transmission?
8. A shunt motor takes 24 A at 220 V and delivers 5 hp. Find the watt output of the motor and its efficiency.
9. A 2-hp motor requires 17 A at 110 V. What percent of the input is delivered at the shaft? What percent is wasted?
10. A "power" transformer draws 1.4 A from a 117-V line. What is the efficiency of the transformer if it delivers 235 V at 0.65 A?

JOB 11-5

Finding the Output and Input of an Electric Device

The formula for the percent efficiency may be used to find the values of both the output and the input of electric devices. The percent efficiency must be expressed as a decimal by moving the decimal point two places to the left.

Finding the output

Hint for Success

The main point of this problem is efficiency. In order to solve it, you must know unit conversion. In the equation the value of ¾ is assumed to be understood as the factor that converts horsepower to kilowatts. The clue to this is that there is an equation with different units on each side. If you have trouble understanding an equation like this, use the clue to direct you to conversion factors for help.

Example 11-16

Find the kilowatt output of a generator if it receives 6 hp and operates at an efficiency of 90%.

Solution

Given: Input = 6 hp   Find: Output = ?
         Eff = 90% = 0.90

1. Write the formula. $Eff = \dfrac{output}{input}$ (11-2)

2. Substitute. $\dfrac{0.90}{1} = \dfrac{output}{6}$

3. Cross-multiply.   Output = $6 \times 0.90 = 5.4$ hp

But the output of a generator must be measured in kilowatts. Therefore,

$$5.4 \text{ hp} = 5.4 \times \frac{3}{4} = \frac{16.2}{4} = 4.05 \text{ kW} \qquad Ans.$$

Example 11-17

A motor has an efficiency of 87%. If it draws 20 A from a 220-V line, what is its horsepower output?

Solution

Given: Input I = 20 A   Find: Output = ?
         Input V = 220 V
             Eff = 87% = 0.87

1. Find the wattage input.

$$P = I \times V = 20 \times 220 = 4400 \text{ W input}$$

2. Find the wattage output.

$$Eff = \frac{output}{input} \qquad (11\text{-}2)$$

$$\frac{0.87}{1} = \frac{output}{4400}$$

$$\text{Output} = 4400 \times 0.87 = 3828 \text{ W output}$$

But the output of a motor must be measured in horsepower. Therefore,

$$3828 \text{ W} = 3828 \div 750 = 5.1 \text{ hp} \qquad Ans.$$

Finding the input

Example 11-18

How much power is needed to operate a 5-kW generator if its efficiency is 92%?

Solution

Given: Output = 5 kW   Find: Input = ?
          Eff = 92% = 0.92

$$\text{Eff} = \frac{\text{output}}{\text{input}} \qquad\qquad (11\text{-}2)$$

$$\frac{0.92}{1} = \frac{5}{\text{input}}$$

$$0.92 \times \text{input} = 5$$

$$\text{Input} = \frac{5}{0.92} = 5.43 \text{ kW}$$

But the input to a generator is measured in horsepower. Therefore,

$$5.43 \text{ kW} = 5.43 \times \frac{4}{3} = \frac{21.72}{3} = 7.24 \text{ hp} \qquad Ans.$$

CALCULATOR HINT

Many times, certain types of assumed knowledge are built into a programmable calculator. For example, if you want to convert from horsepower to kilowatts, your calculator may have a set of built-in conversion factors. Unfortunately, these factors may not be exactly what you want; you may have to convert horsepower to watts and then watts to kilowatts.

Self-Test 11-19

How much current is drawn by a 2½-hp motor if it has an efficiency of 90% and operates on 110 V?

Solution

1. Given: Output = 2½ _____   Find: Input I = ?
             Eff = 90% = (decimal)
           Input V = 110 V

2. Find the power input.

$$\text{Eff} = \frac{\text{output}}{\text{input}} \qquad\qquad (11\text{-}2)$$

$$\frac{0.90}{1} = \frac{2.5}{?}$$

$$0.90 \times \underline{\qquad\qquad} = 2.5$$

$$\text{Input} = \frac{2.5}{?}$$

$$\text{Input} = 2.78 \underline{\qquad\qquad}$$

But the input to a motor is measured in watts or _____. Therefore,

$$2.78 \text{ hp} = 2.78 \times \underline{\qquad\qquad} \text{ W}$$

or $2.78 \text{ hp} = \underline{\qquad\qquad} \text{ W} \qquad Ans.$

3. Find the current input.

$$P = I \times V \qquad (7\text{-}10)$$
$$\underline{\qquad\qquad} = I \times 110$$
$$I = \underline{\qquad\qquad} A \qquad Ans.$$

Exercises

Applications

1. Find the horsepower output of a motor drawing 3 kW and operating at an efficiency of 90%.
2. An 80% efficient "power" transformer delivers 50 W. Find its power input.
3. Find the horsepower output of a motor drawing 5 A at 110 V and operating at an efficiency of 80%.
4. A motor operates at an efficiency of 92% and draws 5.5 A from a 110-V line. Find the horsepower output.
5. A transmission line operating at an efficiency of 98% receives 20 A at 230 V. Find the power delivered.
6. How much power is delivered by a transformer operating at an efficiency of 88% if it draws 0.08 kW?
7. Find the input to a 3-hp motor if its efficiency is 85%.
8. Find the horsepower input to a generator delivering 5 kW at an efficiency of 90%.
9. A generator delivers 20 A at a voltage of 110 V. Find its output in watts. If it has an efficiency of 90%, find the horsepower input.
10. How much current is drawn by a 2-hp motor if it has an efficiency of 87% and operates at 110 V?
11. What voltage is necessary to operate a 3-hp motor if it draws 11.37 A and its efficiency is 90%?
12. Find the kilowatt output of a generator which uses 9 hp and has an operating efficiency of 92%.

JOB 11-6

Load Matching for Power Transfer

This section combines the concepts of Thevenin equivalents, power, and efficiency to give a better understanding of how power is transferred from one circuit to another. Assume that you have an amplifier, or battery, or generator, or any other power source. Make a Thevenin equivalent at the output terminals of the device. This would result in the circuit shown in Fig. 11-2. The Thevenin resistance is usually called the *output impedance* or the *source resistance*. This represents all the resistors inside the power source.

A load R_L is attached to the power output terminals. Since we have made a Thevenin equivalent of the power source, the resulting circuit is a simple one-loop circuit. The Thevenin voltage drives the output and puts current through R_{TH} and the load R_L. This current causes power losses in R_{TH} and R_L. The power lost in R_{TH} is lost as heat inside the power source. The power loss in the load is the desired out-

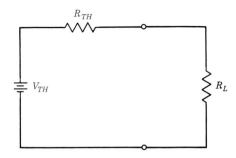

Figure 11-2 Power source Thevenin equivalent with load attached.

put power. The efficiency is the ratio of the desired output power P_L to the total power delivered by the voltage source $(P_{TH} + P_L)$.

Assume that V_{TH} is 12 V and R_{TH} is 10 Ω. Table 11-3 shows how the power delivered to the load and the efficiency change as the load resistance changes.

Table 11-3

R_L, Ω	R_{TH}, Ω	$P_{R_{TH}}$, W	P_{R_L}, W	Eff, %
6	10	5.63	3.38	38
8	10	4.44	3.56	45
10	10	3.60	3.60	50
12	10	2.98	3.57	55
14	10	2.58	3.50	58

> **Hint for Success**
>
> A very important use of mathematics is in drawing general conclusions from model analyses. This example shows how you can draw a general conclusion—that maximum power transfer occurs when the load matches the internal impedance—by analyzing a simple model of the concept.

Table 11-3 indicates that as the load increases, the efficiency of the circuit increases. This occurs because as the load resistance increases, the voltage across the load also increases. Since it is a series circuit, the current through R_L and R_{TH} is equal. Using the power equation $P = V \times I$, we find that the power output increases relative to the power lost in the source. Careful now! This does not means that the power keeps increasing!

As R_L increases, the current in the loop will decrease, thus producing the decreasing power in R_{TH} as seen in Table 11-3. Notice that the power in the load reaches a maximum (3.60 W) when R_{TH} is equal to R_L. (See Example 15-4).

If R_L is less than R_{TH}, increasing R_L increases the voltage across R_L and the power in R_L increases. When $R_L = R_{TH}$, the power in the load is equal to the power lost in the source. When R_L becomes larger than R_{TH}, the current decreases so quickly that the power in the load decreases. This experiment helps us to formulate the following rule:

RULE For maximum transfer of power from a source to a load, the source resistance must equal the load resistance.

1. The output impedance of the source is the same as the _____ impedance.
2. As the load impedance increases the power _____ increases.
3. The maximum power is delivered to the load when the load resistance _____ or _____ the output resistance.

Exercises

Applications

1. Calculate the loop current, the power in R_L and R_{TH}, and the efficiency for the circuit shown in Fig. 11-2, using the values of R_L given in the table.
2. A power supply has an open circuit voltage of 9 V and an output impedance of $2\ \Omega$. What is the maximum power that it can deliver to a load?
3. A motor runs on 5 V. At maximum speed it requires ¼ W. What is the largest source impedance that the supply could have and still run the motor at maximum speed?

JOB 11-7

Review of Conversion Factors and Efficiency

$$1\ \text{kW} = \underline{\quad(1)\quad}\ \text{W}$$
$$1\ \text{hp} = 750\ \underline{\quad(2)\quad}$$
$$\text{W} = \text{kW} \times \underline{\quad(3)\quad}$$
$$\text{W} = \text{hp} \times \underline{\quad(4)\quad}$$
$$\text{hp} = \text{kW} \times \underline{\quad(5)\quad}$$
$$\text{kW} = \text{hp} \times \underline{\quad(6)\quad}$$
$$\text{kW} = \text{W} \div \underline{\quad(7)\quad}$$
$$\text{hp} = \text{W} \div \underline{\quad(8)\quad}$$

The *efficiency* of a machine is the ratio of the power _____(9)_____ to the power _____(10)_____ and is usually expressed as a percent.

$$\text{Eff} = \frac{\text{output}}{\text{input}}$$

11-2

Units of measurement.   For a generator:

Output is in kilowatts or _____(11)_____ or voltamperes
Input is in _____(12)_____

For a motor:

Output is in _____(13)_____
Input is in kilowatts or watts or _____(14)_____

Exercises

Practice

1. Change (*a*) 500 W to kilowatts; (*b*) 500 W to horsepower; (*c*) 1.5 kW to watts; (*d*) 1.5 hp to watts; (*e*) 10 kW to horsepower; and (*f*) 10 hp to kilowatts.

Applications

2. Find the efficiency of a 1½-hp induction motor using 1.3 kW of power.
3. Find the efficiency of a generator if it delivers 20 A at 120 V and uses 3.5 hp.
4. What is the efficiency of a ¾-hp motor if it draws 5.5 A from a 110-V line?
5. Find the horsepower output of a motor drawing 5 A at 110 V if it operates at an efficiency of 85%.
6. Find the kilowatt output of a generator that uses 12 hp and operates at an efficiency of 92%.
7. Find the power needed to operate a ¾-hp motor if its efficiency is 90%.
8. How much current is drawn from a 120-V line by a 3-hp motor if it operates at an efficiency of 87%?
9. A 120-V 60-cycle generator has an internal resistance of 3.5 Ω. What is the maximum power that can be delivered to a load?
10. If the generator of Prob. 9 has an efficiency of 92%, what is the maximum power that has to be delivered to the generator? How many amperes?

(See CD-ROM for Test 11-2)

Assessment

chapter 11

1. Convert the following units.
   a. 75 hp to kilowatts
   b. 18.75 kW to horsepower
   c. 1500 hp to watts
   d. 300,000 W to horsepower
2. A 125-hp motor uses 111.9 kW on energy. At 746 W per horsepower, how efficient is this motor?

3. A single-phase 240-V motor converts 33.9091 kW of electrical energy into mechanical energy. If the motor is 88% efficient, what is the horsepower rating of the motor? (*Hint:* Use 746 W per horsepower.)
4. Very often, technicians will install impedance-matching transformers at speakers as part of an audio system. Why do they do this?

INTERNET
ACTIVITIES

Internet Activity A

Use the site given here to find the formula for amperes when horsepower is known for an ac/1-phase 115-V or 120-V circuit.
http://www.elec-toolbox.com/electform.htm

Internet Activity B

Find the efficiency of nonlinear power supplies and explain why they are being used in new electronic equipment. Use the information at this site.
http://www.ee.uts.edu.au/~venkat/pe_html/ch07s1/ch07s1p1.htm

Chapter 11 Solutions to Self-Tests and Reviews

Self-Test 11-7

1. 220
2. 0.05
3. 0.05
4. 11
5. 11
6. V_l
7. 220
8. 209

Review of percent problems

a. two
b. left
c. two
d. right
e. two
f. percent
1. total
2. base
3. base
4. **B, R, P**

Self-Test 11-14

1. 230, 20, 7
2. W, W
3. output, input, 87.6

Self-Test 11-15

1. 5
2. 120
3. 46

4. 0.046
5. W
6. W
7. input
8. 90.6

Self-Test 11-19

1. hp, 0.90
2. input, input, 0.90, hp, kilowatts, 750, 2085
3. 2085, 19

Self-Test 11-20

1. Thevenin
2. efficiency
3. equals, matches

Job 11-7 Review

1. 1000
2. W
3. 1000
4. 750
5. ⅓
6. ¾
7. 1000
8. 750
9. output
10. input
11. watts
12. horsepower
13. horsepower
14. voltamperes

Resistance of Wire

chapter

Learning Objectives

1. Write a ratio.
2. Solve proportions.
3. Use the AWG table to find the appropriate wire size.
4. Determine the resistance of wires of different materials.
5. Determine the resistance of wires of different lengths.
6. Determine the resistance of wires of different areas.
7. Determine the resistance of wires of different diameters, lengths, and material.
8. Find the length of a wire that produces a given resistance.
9. Find the diameter of wire that produces a given resistance.
10. Use the temperature coefficient of resistance to find resistance at a different temperature.

JOB 12-1

Checkup on Ratio and Proportion (Diagnostic Test)

Many complicated ideas may be expressed quite simply when written in formula form. For example, the rule "In a series circuit, the voltage is directly proportional to the resistance" may be written as $V_1/V_2 = R_1/R_2$. Or "The resistance of a wire is inversely proportional to its area" may be written as $R_1/R_2 = A_2/A_1$. The mathematical concepts of *ratio* and *proportion* enable us to compare voltages, currents, etc., and to show how they depend on each other. We shall be using these ideas in the next and succeeding jobs. Let us check up on what we know about ratio and proportion. If you have any difficulty with any of the following problems, turn to Job 12-2.

Exercises

Applications

1. What is the ratio of 12 to 30 cm?
2. Two pulleys have diameters of 20 and 5 in, respectively. What is the ratio of their diameters?
3. If the number of turns of wire on the primary of a bell-ringing transformer is 360 turns and the secondary has 40 turns, what is the turns ratio of primary to secondary?
4. The power factor (PF) of an ac circuit is described as the ratio of its resistance R to its impedance Z. Find the PF if $R = 500 \ \Omega$ and $Z = 550 \ \Omega$.
5. What is the ratio of 1 hp to 1 kW?
6. A wire 30 ft long has a resistance of 0.02 Ω. If the resistance is directly proportional to the length, find the resistance of 10 ft of this wire.

7. If 60 ft of conduit costs $9.50, how much does 150 ft cost?
8. The ratio of acid to water in the electrolyte of a storage battery is 2:3. If 30 oz of acid is used to prepare a batch of electrolyte, how much water should be mixed with it?
9. In a simple transformer, the voltage is directly proportional to the number of turns. In a simple step-down transformer, the primary has 240 turns and the secondary has 30 turns. Find the voltage of the secondary if the primary voltage is 120 V.
10. The resistance of a wire is inversely proportional to its cross-sectional area. If a wire with an area of 5200 units has a resistance of 2 Ω, find the resistance of a wire of the same material with an area of 100 units.
11. In a parallel circuit, the current is inversely proportional to the resistance. Write a formula to state this fact.
12. In a series circuit, the voltage drops are directly proportional to the resistances. What is the voltage across a 100-Ω resistance if the voltage across a 500-Ω resistance is 10 V?

Brushup on Ratio and Proportion

Ratio. A *ratio* is the comparison of two quantities by division. The two quantities must be the same *kind* of thing. That is, we can compare two weights, two lengths, or two voltages, but we cannot compare a current with a voltage, or a length with a weight.

Units of a ratio The comparison, or ratio, of 2 yd to 1 ft is not 2 to 1. This would mean that 2 yd is only twice as large as 1 ft, when actually it is six times as large. Therefore, in order to be comparable, two quantities must be measured in the *same units* of measurement. Since 2 yd and 1 ft are both lengths, we may compare them by division.

$$\frac{2 \text{ yd}}{1 \text{ ft}} = \frac{6 \cancel{\text{ft}}}{1 \cancel{\text{ft}}} = \frac{6}{1}$$

Since the units of measurement are identical, they cancel out. Thus, a ratio itself has no units of measurement. Another way to write this ratio is 6:1, in which the ratio symbol (:) replaces the fraction bar. However, it is written, the ratio is read as "6 to 1."

RULE To find the ratio of two similar quantities
1. Express them in the same units of measurement
2. Form a fraction using the first quantity as the numerator and the second quantity as the denominator.
3. Express in lowest terms.

Order of a ratio. The order in which a ratio is expressed is as important as the numbers themselves. The ratio *must* be expressed in the same order in which the comparison is made.

Example 12-1

The pinion in Fig. 12-1 has 15 teeth and is in mesh with a gear having 60 teeth. Find (*a*) the ratio of gear teeth to pinion teeth, and (*b*) the ratio of pinion teeth to gear teeth.

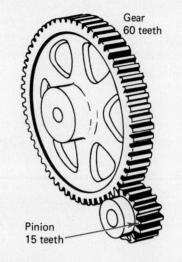

Figure 12-1 Two spur gears in mesh.

Solution

a. $\dfrac{\text{Gear}}{\text{Pinion}} = \dfrac{60}{15} = \dfrac{4}{1} = 4:1$ *Ans.*

This means that for every 4 teeth on the gear, there is 1 tooth on the pinion.

b. $\dfrac{\text{Pinion}}{\text{Gear}} = \dfrac{15}{60} = \dfrac{1}{4} = 1:4$ *Ans.*

This means that for every 1 tooth on the pinion, there are 4 teeth on the gear.

> ## Hint for Success
>
> Ratios are easiest to work with when they are in their simplest form.

Example 12-2

Find the ratio of 1 mV to 1 V.

Solution

Given: The quantities 1 mV and 1 V   Find: Ratio = ?

$\text{Ratio} = \dfrac{1\ \text{mV}}{1\ \text{V}} = \dfrac{1\ \text{mV}}{1000\ \text{mV}} = \dfrac{1}{1000}$ or 1:1000 (read as "1 to 1000")

Example 12-3

Find the ratio of 1 hp to 1 kW.

Solution

Given: The quantities 1 hp and 1 kW   Find: Ratio = ?

If it is difficult to change one unit to the other or vice versa, then change both units to a third unit. Since 1 hp = 750 W and 1 kW = 1000 W,

$$\text{Ratio} = \frac{1 \text{ hp}}{1 \text{ kW}} = \frac{750 \text{ W}}{1000 \text{ W}} = \frac{3}{4} \text{ or 3:4 (read as "3 to 4")} \quad \textit{Ans.}$$

Not all ratios are expressed as fractions. It is sometimes more convenient to express a ratio as a decimal or as a percent.

Example 12-4

The power factor (PF) of an ac circuit is the ratio of its resistance R to its impedance Z. Find the PF of a circuit if $R = 2.46 \ \Omega$ and $Z = 30 \ \Omega$.

Solution

Given: $R = 2.46 \ \Omega$ Find: PF = ?
$Z = 30 \ \Omega$

$$\text{PF} = \frac{R}{Z} = \frac{2.46}{30} = 0.082$$

Since PF is often expressed as a percent,

$$\text{PF} = 0.082 \text{ or } 8.2\% \quad \textit{Ans.}$$

CALCULATOR HINT

Remember, when you change a fraction to a decimal, divide:

2 . 4 6 ÷ 3 0 =

0.082 or 8.2%

Example 12-5

The efficiency of a motor is the ratio of the output to the input and is described as a percent. Find the efficiency of a motor whose output is 5 hp and whose input is 4 kW.

Solution

Given: Output = 5 hp   Find: Eff = ?
Input = 4 kW

$$\text{Eff} = \frac{\text{output}}{\text{input}} = \frac{5 \text{ hp}}{4 \text{ kW}}$$

$$= \frac{5 \times 750}{4 \times 1000} = \frac{3750 \text{ W}}{4000 \text{ W}} = 0.9375$$

$$\text{Eff} = 0.9375 \text{ or } 93.75\% \quad \textit{Ans.}$$

Practice

Find the ratio of the quantities in each exercise.

1. 3 in to 12 in
2. 6 ft to 2 ft
3. 18 V to 27 V
4. 105 turns to 20 turns
5. 3 ft to 18 in
6. 2 kHz to 500 Hz
7. 0.4 MΩ to 100,000 Ω
8. 2 hp to 3 kW

Applications

9. In Fig. 12-2, find (*a*) the ratio of D_2 to D_1 and (*b*) the ratio of D_1 to D_2.

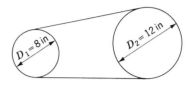

Figure 12-2

10. Find the teeth ratio of two gears if the first has 35 teeth and the second has 30 teeth.
11. In Fig. 12-3, the turns ratio of the transformer is the ratio of the number of turns on the primary coil N_p to the number of turns on the secondary coil N_s. Find the turns ratio.
12. In Fig. 12-3, find the ratio of the secondary turns N_s to the primary turns N_p.

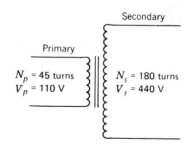

Figure 12-3 Step-up transformer.

13. In Fig. 12-3, find the ratio of the primary voltage V_p to the secondary voltage V_s.
14. If the primary voltage V_p of a transformer is 125 V and the secondary voltage V_s is 20 V, find the ratio of V_p to V_s.
15. Find the efficiency of a motor if the output is 2 hp and the input is 1.8 kW.
16. The quality Q of a coil is the ratio of its reactance X to its resistance R. Find the Q of an RF coil if its reactance is 2500 Ω and its resistance is 25 Ω.
17. Find the Q of a tuned circuit if $X = 12,000$ Ω and $R = 70$ Ω.
18. If there are 10 mm in each centimeter (cm) of metric length, find the ratio of 14 mm to 49 cm.
19. β, the current gain of a transistor, is defined as the ratio of the collector cur-

rent I_C to the base current I_B. Find the β of a transistor if $I_B = 5\ \mu A$ and $I_C = 200\ \mu A$.

20. Find the PF of an ac circuit if $R = 56.8\ \Omega$ and $Z = 65\ \Omega$.
21. The multiplying power of a meter is the ratio of the new voltage to the old voltage. Find the multiplying power of a voltmeter reading 10 V if it is extended to read 100 V.
22. Find the multiplying power of a 150-V 150,000-Ω voltmeter that has been extended to read 750 V.
23. What is the ratio of the distance on a drawing to the actual distance if the scale of the drawing is ¼ in equals 1 ft?
24. If 34 oz of acid is mixed with 51 oz of water to make the electrolyte for a storage battery, find (*a*) the ratio of acid to water and (*b*) the ratio of water to acid.
25. A generator rated at 117 V supplies 115 V to a motor some distance away. Find (*a*) the line loss in volts, (*b*) the ratio of the line loss to the rated voltage in percent, and (*c*) the ratio of the load voltage to the rated voltage in percent.

Proportion.   Consider the following statement. If 3 pencils cost 8 cents, then 6 pencils will cost 16 cents. At the given rate, the cost depends only on the number of pencils bought. The more pencils bought, the larger the cost. The fewer pencils bought, the smaller the cost. When two quantities depend on each other, the relationship between them may be stated as a *proportion*.

A proportion is a mathematical statement that two ratios are equal. In our problem, the two quantities are the number of pencils N and the cost C.

First purchase:	$N_1 = 3$ pencils	$C_1 = 8$ cents
Second purchase:	$N_2 = 6$ pencils	$C_2 = 16$ cents

The ratio of the number of pencils is

$$\frac{N_1}{N_2} = \frac{3}{6} = \frac{1}{2}$$

The ratio of the costs is

$$\frac{C_1}{C_2} = \frac{8}{16} = \frac{1}{2}$$

Since the two ratios are equal, the proportion may be written mathematically as

$$\frac{N_1}{N_2} = \frac{C_1}{C_2}$$

and is an example of a *direct proportion*.

Direct proportion.   When two quantities depend on each other so that one increases as the other increases or one decreases as the other decreases, they are said to be *directly proportional* to each other. Notice that the two ratios are compared in the *same order* so that the items in the first situation (N_1 and C_1) are in the numerator and the items in the second situation (N_2 and C_2) are in the denominator. The proportion may be written as

$$\frac{C_1}{C_2} = \frac{N_1}{N_2} \quad \text{or} \quad \frac{C_2}{C_1} = \frac{N_2}{N_1} \quad \text{or} \quad \frac{N_2}{N_1} = \frac{C_2}{C_1} \quad \text{or} \quad \frac{N_1}{N_2} = \frac{C_1}{C_2}$$

It is not important which ratio is written first. However, once the first ratio is written, the second ratio must be written *in the same order*.

RULE **To set up a direct proportion between two variables:**

1. Make a ratio of one of the variables.
2. Make a ratio of the second variable in the same order.
3. Set the two ratios equal to each other.

Example 12-6

Write as a proportion: The weight of a pipe is directly proportional to its length.

Solution

First pipe	Second pipe
Length = L_1	Length = L_2
Weight = W_1	Weight = W_2

Since the ratio of the lengths (L_1/L_2) and the ratio of the weights (W_1/W_2) are equal, the proportion may be written as

$$\frac{L_1}{L_2} = \frac{W_1}{W_2} \quad \text{or} \quad \frac{W_1}{W_2} = \frac{L_1}{L_2}$$

or, since the ratio of the lengths (L_2/L_1) and the ratio of the weights (W_2/W_1) are equal, the proportion may also be written as

$$\frac{L_2}{L_1} = \frac{W_2}{W_1} \quad \text{or} \quad \frac{W_2}{W_1} = \frac{L_2}{L_1}$$

RULE **To solve problems involving direct proportion:**

1. Set up the proportion.
2. Substitute the values.
3. Solve by cross multiplication.

Example 12-7

In the series circuit shown in Fig. 12-4, the voltage across any resistor is directly proportional to the resistance of the resistor. Find the value of V_2.

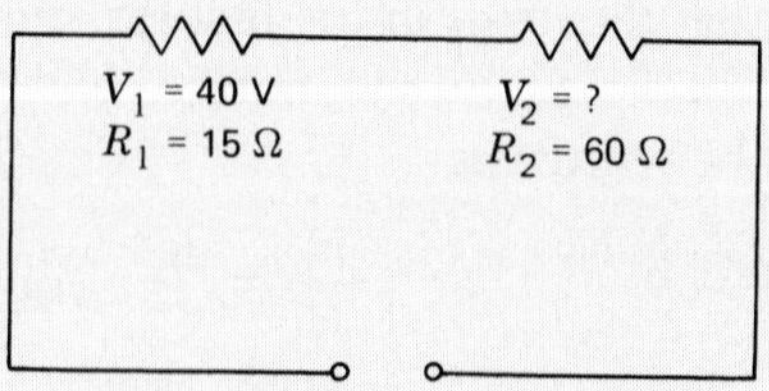

V1 = 40 V
R1 = 15 Ω
V2 = ?
R2 = 60 Ω

Figure 12-4

Solution

1. Set up the direct proportion.

$$\frac{V_1}{V_2} = \frac{R_1}{R_2}$$

2. Substitute numbers.

$$\frac{40}{V_2} = \frac{15}{60}$$

3. Cross-multiply.

$$15 \times V_2 = 40 \times 60$$

4. Solve for V_2.

$$V_2 = \frac{2400}{15} = 160 \text{ V} \qquad \textit{Ans.}$$

Self-Test 12-8

An automobile can travel 280 mi in 7 h. At the same rate of speed, how long will it take to travel 200 mi?

Solution

This is an example of a direct proportion, since the distance increases as the time ___(a)___ and the distance traveled depends on the ___(b)___ spent traveling.

Given: $D_1 = $ ___(c)___ mi   Find: $T_2 = ?$

   $T_1 = $ ___(d)___ h

   ___(e)___ $= 200$ mi

1. Set up the direct proportion.

$$\frac{D_1}{D_2} = \frac{?}{?}$$

2. Substitute numbers.

$$\frac{280}{200} = \frac{7}{?}$$

3. Cross-multiply.

$$280 \times T_2 = 200 \times \underline{\qquad}$$

4. Solve for T_2.

$$T_2 = \frac{1400}{?}$$

$$T_2 = \underline{\qquad} \text{ h} \qquad \textit{Ans.}$$

Exercises

Applications

1. An 8-ft-long steel beam weighs 2200 lb. What would be the weight of a similar beam 10 ft long?

2. If 1 gross (144) of resistors costs $3.90, how much do 36 resistors cost?
3. An airplane can travel 2040 mi in 6 h. At the same rate, how long would it take the plane to travel 1530 mi?
4. The shadows cast by vertical poles are directly proportional to the heights of the poles. If a pole 9 ft high casts a shadow 12 ft long, how high is a tree that casts a shadow 36 ft long?
5. The volume of a gas is directly proportional to the temperature. If a gas occupies 100 ft^3 at 70°F, what volume will it occupy at 50°F, assuming the pressure to be constant?
6. If 14 lb of cement is used to make 70 lb of concrete, how many pounds of concrete can be made with 30 lb of cement?
7. Using a diagram similar to Fig. 12-4, find R_1 if $V_1 = 20$ V, $V_2 = 36$ V, and $R_2 = 1800$ Ω.
8. If a 500-ft length of wire has a resistance of 30 Ω, find the resistance of an 850-ft length of the same wire. Use the fact that the resistance of a wire is directly proportional to its length.
9. In a transformer, the voltage is directly proportional to the number of turns. In a diagram similar to that shown in Fig. 12-3, find the secondary voltage V_s if $N_p = 160$ turns, $V_p = 18$ V, and $N_s = 400$ turns.
10. The Elco stereo amplifier model 3080 uses a series-circuit base-bias voltage divider similar to that shown in Fig. 12-5. Find the value of the bias voltage V_2.

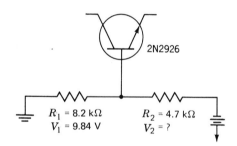

Figure 12-5 Series-circuit base-bias voltage divider.

Inverse proportion.   When two gears are meshed as shown in Fig. 12-6, the smaller the gear, the faster it turns. When gear 1 turns once, its 40 teeth must engage 40 teeth on gear 2. Therefore, gear 2 must turn *twice* in order for 40 teeth (2 × 20) to mesh with the 40 teeth on gear 1.

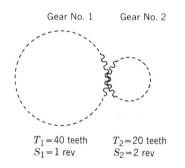

Gear No. 1   Gear No. 2

$T_1 = 40$ teeth $T_2 = 20$ teeth
$S_1 = 1$ rev $S_2 = 2$ rev

Figure 12-6 The speed is inversely proportional to the number of teeth.

When two quantities depend on each other so that one *increases* as the other *decreases* or one *decreases* as the other *increases*, they are said to be *inversely proportional* to each other. This may be written mathematically as

$$\frac{T_1}{T_2} = \frac{S_2}{S_1}$$

Notice that the two ratios are compared in the *opposite*, or *inverse*, order. The proportion will be correct regardless of which ratio is written first, *provided* the second ratio is written in the *opposite order*. Thus, the proportion may be written as

$$\frac{T_1}{T_2} = \frac{S_2}{S_1} \quad \text{or} \quad \frac{T_2}{T_1} = \frac{S_1}{S_2} \quad \text{or} \quad \frac{S_1}{S_2} = \frac{T_2}{T_1} \quad \text{or} \quad \frac{S_2}{S_1} = \frac{T_1}{T_2}$$

RULE To set up an inverse proportion between two variables:

1. Make a ratio of one variable.
2. Make a ratio of the second variable in the opposite order.
3. Set the two ratios equal to each other.

Example 12-9

Write as a proportion: "The speeds of pulleys are inversely proportional to the diameters."

Solution

The diagram for the problem is shown in Fig. 12-7.

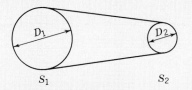

Figure 12-7

$$\frac{D_1}{D_2} = \frac{S_2}{S_1} \quad \text{or} \quad \frac{S_1}{S_2} = \frac{D_2}{D_1} \quad \text{or} \quad \frac{D_2}{D_1} = \frac{S_1}{S_2} \quad \text{or} \quad \frac{S_2}{S_1} = \frac{D_1}{D_2}$$

RULE To solve problems involving inverse proportion:

1. Set up the proportion.
2. Substitute the values.
3. Solve by cross multiplication.

Example 12-10

The resistance of a length of wire is inversely proportional to its cross-sectional area. A certain copper wire has a resistance of 32 Ω and a cross-sectional area of 20 units. Find the resistance of another wire of the same length whose area is 100 units.

Solution

Given: $R_1 = 32\ \Omega$ Find: $R_2 = ?$
            $A_1 = 20$
            $A_2 = 100$

$$\frac{R_1}{R_2} = \frac{A_2}{A_1}$$

$$\frac{32}{R_2} = \frac{100}{20}$$

$$100R_2 = 32 \times 20$$

$$R_2 = \frac{640}{100} = 6.4\ \Omega \qquad Ans.$$

Self-Test 12-11

In a parallel circuit, the currents are inversely proportional to the resistances. In the circuit shown in Fig. 12-8, find I_2.

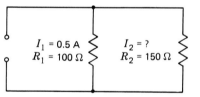

$I_1 = 0.5\ A$ $I_2 = ?$
$R_1 = 100\ \Omega$ $R_2 = 150\ \Omega$

Figure 12-8

Solution

1. The variables in this problem are the current and the ——————.
2. Set up a ratio comparing the current I_1 to I_2.

$$\frac{?}{?}$$

3. Since the currents are *inversely* proportional to the resistances, set up an *inverse* ratio of resistances.

$$\frac{?}{?}$$

4. Since the two variables are proportional, the ratios may be set —————— to each other.

$$\frac{I_1}{I_2} = \frac{?}{?}$$

5. Substitute numbers.

$$\frac{0.5}{I_2} = \frac{150}{?}$$

6. Solve for I_2.

$$150 I_2 = 0.5 \times 100$$
$$I_2 = \frac{50}{?}$$
$$I_2 = \underline{\quad\quad} A \qquad Ans.$$

Exercises

Practice

The speeds of pulleys are inversely proportional to their diameters. Using Fig. 12-7,

1. Find D_1 if $S_1 = 40$ rpm, $S_2 = 100$ rpm, and $D_2 = 2$ in.
2. Find S_1 if $D_1 = 5$ in, $D_2 = 6$ in, and $S_2 = 120$ rpm.

The resistance of a wire is inversely proportional to its cross-sectional area.

3. Find R_2 if $A_1 = 200$, $R_1 = 50$ Ω, and $A_2 = 150$.
4. Find A_2 if $A_1 = 200$, $R_1 = 50$ Ω, and $R_2 = 10$ Ω.
5. Find A_1 if $R_1 = 20$ Ω, $A_2 = 10,000$, and $R_2 = 1$ Ω.
6. Find R_1 if $A_1 = 30$, $A_2 = 600$, and $R_2 = 16$ Ω.

In a parallel circuit, the currents are inversely proportional to the resistances. Using a diagram similar to Fig. 12-8,

7. Find I_1 if $I_2 = 0.06$ A, $R_1 = 30$ Ω, and $R_2 = 40$ Ω.
8. Find R_2 if $I_1 = 2$ A, $I_2 = 4.5$ A, and $R_1 = 90$ Ω.
9. Find R_1 if $I_1 = 2.25$ A, $I_2 = 3.5$ A, and $R_2 = 0.6$ Ω.

In a transformer, the currents in the primary and secondary coils are inversely proportional to the voltages. Using a diagram similar to Fig. 12-3,

10. Find V_s if $V_p = 120$ V, $I_p = 5$ A, and $I_s = 2$ A.
11. Find I_p if $V_p = 120$ V, $V_s = 15$ V, and $I_s = 2$ A.
12. Find I_s if $V_p = 120$ V, $I_p = 4$ A, and $V_s = 6$ V

Review of Ratio and Proportion

1. A *ratio* is the _____ of two quantities of the same kind.
2. A ratio is indicated by the _____ of the quantities.
3. A ratio may be written as (1) 8 to 2, (2) 8:2, or (3) 8⁄2 and is always read as _____.
4. The two quantities of a ratio must always be measured in the _____ units of measurement.
5. A *proportion* is a mathematical statement that two ratios are _____.
6. In a *direct proportion*, the two quantities increase or decrease together. This is indicated by comparing the two ratios in the __(same/opposite)__ order. Thus, if A is directly proportional to B,

7. $$\frac{A_1}{A_2} = \frac{B_1}{B_2} \quad \text{or} \quad \frac{B_1}{B_2} = \frac{?}{?}$$

8. or $$\frac{A_2}{A_1} = \frac{?}{?}$$

9. or $$\frac{B_2}{B_1} = \frac{?}{?}$$

10. In an *inverse proportion*, as one quantity increases, the other ——————, and vice versa. This is indicated by comparing the two ratios in the ——————— order. Thus, if C is inversely proportional to D,
   (same/opposite)

11. $$\frac{C_1}{C_2} = \frac{D_2}{D_1} \quad \text{or} \quad \frac{D_2}{D_1} = \frac{?}{?}$$

12. or $$\frac{C_2}{C_1} = \frac{?}{?}$$

13. or $$\frac{D_1}{D_2} = \frac{?}{?}$$

Exercises

Practice

1. Find the ratio of (*a*) 6 to 18 in, (*b*) 1 ft to 2 yd, (*c*) 100 mA to 1 A, (*d*) 500 Hz to 0.8 kHz, and (*e*) 4 hp to 4 kW.
2. In a series circuit, the voltage across a resistance is directly proportional to the resistance. Using a circuit similar to that shown in Fig. 12-4, find R_1 if $V_1 = 90$ V, $V_2 = 48$ V, and $R_2 = 80 \ \Omega$.

In Probs. 3 to 6, A is directly proportional to B.

3. Find B_2 if $A_1 = 15$, $A_2 = 6$, and $B_1 = 90$.
4. Find B_1 if $A_1 = 35$, $A_2 = 45$, and $B_2 = 63$.
5. Find A_2 if $A_1 = 27$, $B_1 = 3$, and $B_2 = 8$.
6. Find A_1 if $A_2 = 56$, $B_1 = 4$, and $B_2 = 7$.

In a parallel circuit similar to that shown in Fig. 12-8, the current in each branch is inversely proportional to the resistance.

7. Find I_1 if $I_2 = 0.4$ A, $R_1 = 100 \ \Omega$, and $R_2 = 150 \ \Omega$.
8. Find I_2 if $I_1 = 1.8$ A, $R_1 = 225 \ \Omega$, and $R_2 = 75 \ \Omega$.

In a transformer, the currents in the primary and secondary coils are inversely proportional to the voltages. Using a diagram similar to that shown in Fig. 12-3,

9. Find V_p if $V_s = 2200$ V, $I_p = 50$ A, and $I_s = 2.5$ A.
10. Find I_p if $V_p = 24$ V, $V_s = 120$ V, and $I_s = 0.2$ A.

(See CD-ROM for Test 12-1)

JOB 12-4

The American Wire Gage Table

Wires are manufactured in standard sizes which are listed in Table 12-1 and are known as the American Wire Gage (AWG). It has been found that calculations involving these wires may be simplified by expressing their diameters and cross-sectional areas in new units of measurement.

Table 12-1 American Wire Gage Table

Resistance of bare annealed copper wire at 20°C (68°F)

AWG no.	Diameter d, mil	Area d^2, cmil	Resistance Ω/1000 ft.	AWG no.	Diameter d, mil	Area d^2, cmil	Resistance Ω/1000 ft.
0000	460.0	211,600	0.0490	19	35.89	1288	8.051
000	409.6	167,800	0.0618	20	31.96	1022	10.15
00	364.8	133,100	0.0779	21	28.46	810.1	12.80
0	324.9	105,500	0.0983	22	25.35	642.4	16.14
1	289.3	83,690	0.1239	23	22.57	509.5	20.36
2	257.6	66,360	0.1563	24	20.10	404.0	25.67
3	229.4	52,630	0.1970	25	17.90	320.4	32.37
4	204.3	41,740	0.2485	26	15.94	254.1	40.81
5	181.9	33,100	0.3133	27	14.20	201.5	51.47
6	162.0	26,250	0.3951	28	12.64	159.8	64.90
7	144.3	20,820	0.4982	29	11.26	126.7	81.83
8	128.5	16,510	0.6282	30	10.03	100.5	103.2
9	114.4	13,090	0.7921	31	8.928	79.70	130.1
10	101.9	10,380	0.9989	32	7.950	63.21	164.1
11	90.74	8234	1.260	33	7.080	50.13	206.9
12	80.81	6530	1.588	34	6.305	39.75	260.9
13	71.96	5178	2.003	35	5.615	31.52	329.0
14	64.08	4107	2.525	36	5.000	25.00	414.8
15	57.07	3257	3.184	37	4.453	19.83	523.1
16	50.82	2583	4.016	38	3.965	15.72	659.6
17	45.26	2048	5.064	39	3.531	12.47	831.8
18	40.30	1624	6.385	40	3.145	9.888	1049

Mils. The diameters of wires are expressed in terms of *mils* instead of inches.

$$1000 \text{ mil} = 1 \text{ in} \quad \text{or} \quad 1 \text{ mil} = \frac{1}{1000} \text{ in}$$

Changing Units

To change inches to mils: Multiply the number of inches by 1000.
To change mils to inches: Divide the number of mils by 1000.

> **Hint for Success**
>
> There are many wire gage standards. Only AWG is used in the National Electrical Code.

Example 12-12

A wire has a diameter of 0.162 in. Change the diameter to mils and find the AWG number.

Solution

0.162 in = 0.162 x 1000 = 162 mil   *Ans.*

In the second column of the AWG table marked "Diameter, mil," read down until you find 162 mil. Read to the left to the column marked "AWG No." to find No. 6 wire.   *Ans.*

Example 12-13

Express the diameter of No. 18 wire in mils and inches.

Solution

Read down in the first column until you find No. 18. Read to the right to the second column to find 40.30 mil.

$$40.30 \text{ mil} = 40.30 \div 1000 = 0.0403 \text{ in} \qquad \textit{Ans.}$$

Exercises

Practice

Using Table 12-1, find the missing values in the following problems.

Problem	Gage no.	Diameter, mil.	Diameter, in.
1	?	28.46	?
2	10	?	?
3	?	?	$\frac{1}{4}$
4	?	?	0.05082
5	?	?	$\frac{162}{1000}$
6	?	204.3	?
7	14	?	?
8	12	?	?
9	?	40.30	?
10	?	?	$\frac{1}{16}$

Circular mils. The circular mil (cmil) is the unit of area which is used to measure the cross-sectional area of wires. The circular-mil area is used rather than the ordinary units of area because of the ease of obtaining the circular-mil area when the diameter is given in mils.

RULE To obtain the circular-mil area of a wire, express the diameter in mils and then square the resulting number.

$$A = D_m{}^2 \qquad \qquad \text{Formula 12-1}$$

where A = area, cmil
         D_m = diameter, mil

In order to reduce the number of digits needed to write the circular-mil areas of conductors over 4/0, we use the term MCM, which means *thousands* of circular mils. Thus, 360 MCM means 360,000 cmil.

Example 12-14

Find the circular-mil area A of No. 12 wire.

Solution

Read across to the right from No. 12 in the first column to find the circular-mil area in column 3.

$$\text{Area of No. 12 wire} = 6530 \text{ cmil} \qquad Ans.$$

Example 12-15

Find the circular-mil area A of a wire with a diameter of 0.5 in.

Solution

Change inches to mils.

$$0.5 \text{ in} = 0.5 \times 1000 = 500 \text{ mil} = D_m$$
$$A = D_m^2 \qquad\qquad\qquad (12\text{-}1)$$
$$A = (500)^2 = 25 \times 10^4$$
$$A = 250 \times 10^3$$
$$A = 250 \text{ MCM} \qquad Ans.$$

Example 12-16

A cable is to be replaced by a rectangular busbar 1 in wide and ¼ in thick. Find the circular-mil area.

Solution

1. Change inches to mils.

$$1 \text{ in} = 1 \times 1000 = 1000 \text{ mil}$$
$$\tfrac{1}{4} \text{ in} = 0.25 \times 1000 = 250 \text{ mil}$$

2. Find the area.

$$A = L \times W$$
$$A = 1000 \times 250$$
$$A = 250{,}000 \text{ mil}^2$$

3. Convert square mils to circular mils. The area of any circle is given by the formula

$$0.7854 d^2 = A$$

In mils, $0.7854 d_m^2 = A_{\text{mil}^2}$

Substituting A_{cmil} for d_m^2 will give

$$0.7854 A_{\text{cmil}} = A_{\text{mil}^2}$$

or

$$A_{cmil} = \frac{A_{mil^2}}{0.7854}$$

Formula 12-2

$$A_{cmil} = \frac{250,000}{0.7854}$$

$$A_{cmil} = 318,000 \text{ cmil}$$

$$A_{cmil} = 318 \text{ MCM} \qquad Ans.$$

RULE To obtain the diameter of a wire in mils, find the square root of the circular-mil area.

$$D_m = \sqrt{\text{cmil}}$$

Formula 12-3

CALCULATOR HINT

To find the square root of a number on a graphing calculator, you must press the square root key before entering the number.

Example 12-17

Find the diameter of a wire with a cross-sectional area of 25 MCM.

Solution

$$D_m = \sqrt{\text{cmil}} \qquad (12\text{-}3)$$

$$D_m = \sqrt{25,000}$$

$$D_m = 158 \text{ mil} \qquad Ans.$$

(Review Job 7-8 on square root.)

If the diameter is wanted in inches,

$$158 \text{ mil} = \frac{158}{1000} = 0.158 \text{ in} \qquad Ans.$$

Exercises

Practice

Find the missing values in the following problems:

Problem	Diameter, in.	Diameter, mil.	Area, cmil
1	?	10	?
2	0.05	?	?
3	?	?	4096
4	?	60	?
5	0.032	?	?
6	?	?	9500
7	?	87.2	?
8	¼	?	?
9	?	?	30,000
10	0.102	?	?

11. A rectangular busbar measures 1¼ in by ¼ in. Find its circular-mil area.
12. A rectangular busbar measures ¾ in by ⅜ in. Find the diameter (in mils) of an equivalent round wire.

There is a pattern in the AWG table that is helpful to use as a "rule-of-thumb" guide for wire size selection. A wire that is 3 sizes larger than another wire has half the resistance, twice the weight, and twice the area of the other wire. A wire that is 10 sizes larger than another has one-tenth the resistance, 10 times the weight, and 10 times the area of the other wire.

Resistances of Wires of Different Materials

All materials differ in their atomic structure and therefore in their ability to resist the flow of an electric current. The measure of the ability of a specific material to resist the flow of electricity is called its *specific resistance*.

The *specific resistance K* of a material is the resistance offered by a wire of this material which is 1 ft long with a diameter of 1 mil. The specific resistance of different materials is given in Table 12-2.

Table 12-2 Specific Resistance of Materials in Ohms Per Mil-Foot at 20°C			
Material	**K**	**Material.**	**K**
Silver	9.7	Tantalum	93.3
Copper	10.4	German silver (18%)	200
Gold	14.7	Monel metal	253
Aluminum	17.0	Manganin	265
Tungsten	34.0	Magnesium	276
Brass	43.0	Constantan	295
Iron (pure)	60.0	Nichrome	600
Tin	69.0	Nickel	947

Hint for Success

When looking up specific resistances, be sure that the units are the same as those used in this book. If they are not, all equations will have to be modified for the different units.

RULE The resistance of a wire is directly proportional to the specific resistance of the material.

$$\frac{R_1}{R_2} = \frac{K_1}{K_2}$$

Formula 12-4

where R_1 and K_1 = resistance and specific resistance of a wire of one material, respectively

R_2 and K_2 = resistance and specific resistance of a wire of another material, respectively

Example 12-18

A copper wire has a resistance of 5 Ω. What is the resistance of a Nichrome wire of the same length and cross-sectional area?

Solution

Given: Copper wire, $R_1 = 5\ \Omega$ Find: R_2 of Nichrome = ?
$K_1 = 10.4$
Nichrome wire, $K_2 = 600$

$$\frac{R_1}{R_2} = \frac{K_1}{K_2}$$ (12-4)

$$\frac{5}{R_2} = \frac{10.4}{600}$$

$$10.4R_2 = 5 \times 600$$

$$10.4R_2 = 3000$$

$$R_2 = \frac{3000}{10.4} = 288\ \Omega \quad Ans.$$

Exercises

Applications

1. A copper wire has a resistance of 12 Ω. What is the resistance of an aluminum wire of the same length and diameter?
2. A tungsten wire has a resistance of 40 Ω. What is the resistance of a Nichrome wire of the same length and diameter?
3. A Nichrome wire has a resistance of 125 Ω. What is the resistance of an iron wire of the same length and diameter?
4. A copper wire has a resistance of 2.5 Ω. What is the resistance of a manganin wire of the same length and diameter?

Resistances of Wires of Different Lengths

Consider two wires of the same material and diameter but of different lengths. A length of wire may be considered to be made of a large number of small lengths all connected in series. The total resistance of the length of wire is then equal to the sum of the resistances of all the small lengths. Therefore, the longer the wire, the greater its resistance.

RULE The resistance of a wire is directly proportional to its length.

$$\frac{R_1}{R_2} = \frac{L_1}{L_2} \qquad \text{Formula 12-5}$$

where R_1 and L_1 = resistance and length of first wire, respectively
R_2 and L_2 = resistance and length of second wire, respectively

Hint for Success

Be sure that the lengths are expressed in the same units of measure.

Example 12-19

If 1000 ft of No. 16 copper wire has a resistance of 4 Ω, find the resistance of 1800 ft of the same wire.

Solution

Given: $R_1 = 4\ \Omega$ Find: $R_2 = ?$
$L_1 = 1000$ ft
$L_2 = 1800$ ft

$$\frac{R_1}{R_2} = \frac{L_1}{L_2} \qquad (12\text{-}5)$$

$$\frac{4}{R_2} = \frac{1000}{1800}$$

$$10R_2 = 4 \times 18$$

$$R_2 = \frac{72}{10} = 7.2\ \Omega \qquad Ans.$$

Self-Test 12-20

Find the resistance of 700 ft of No. 20 copper wire. What current flows through the wire when there is a voltage drop of 14.2 V across the ends of the wire?

Solution

Given: $L = \underline{\quad(1)\quad}$ ft of No. 20 copper wire   Find: $R = ?$
$V = \underline{\quad(2)\quad}$ $I = ?$

1. Find the resistance of No. 20 copper wire. From Table 12-1, No. 20 wire has a resistance of $\underline{\quad(3)\quad}$ Ω/1000 ft.

2. Find the resistance of 1 ft of this wire.
$$\frac{R}{\text{ft}} = \frac{10.15}{1000} = \underline{\quad(4)\quad} \ \Omega/\text{ft}$$

3. Find the resistance of $\underline{\quad(5)\quad}$ ft of this wire.
$$R = \frac{R}{\text{ft}} \times \text{ft} = 0.010\ 15 \times 700 = \underline{\quad(6)\quad} \ \Omega \qquad Ans.$$

4. Find the current.
$$V = IR \qquad\qquad\qquad\qquad (2\text{-}1)$$
$$14.2 = I \times \underline{\quad(7)\quad}$$
$$I = \underline{\quad(8)\quad} \ \text{A} \qquad Ans.$$

Exercises

Applications

1. If 1000 ft of No. 12 copper wire has a resistance of 1.6 Ω, find the resistance of 2300 ft of this wire.
2. Find the resistance of 800 ft of No. 16 copper wire.
3. Find the resistance of 400 ft of No. 20 copper wire.
4. What is the resistance of the primary coil of a transformer if it is wound with 600 ft of No. 20 copper wire?
5. The coil of an electromagnet is wound with 300 ft of No. 16 copper wire. What is the resistance of the coil? What current will be drawn from a 6-V battery?
6. The coil of an outboard motor is wound with 100 ft of No. 18 copper wire. What current will it draw from a 6-V battery?
7. The four field windings of a motor are connected in series. If each winding uses 50 ft of No. 32 copper wire, find the total resistance of the field windings. What current is drawn when the motor is used on a 110-V line?
8. Find the resistance of a two-wire service line 100 ft long if No. 8 copper wire is used.
9. What is the resistance of a telephone line 1 mi long if No. 18 copper wire is used? What is the voltage drop across this line if the current is 200 mA?
10. Find the resistance of 10 ft of No. 22 copper wire. What is the resistance of the same length of No. 22 Nichrome wire? If this Nichrome wire is used as the resistance element in an electric furnace, what current is drawn from a 110-V line?

JOB 12-7

Resistances of Wires of Different Areas

The flow of electricity in a wire is very similar to the flow of water in a pipe. A large-diameter pipe can carry more water than a small-diameter pipe. Similarly, a large-diameter wire can carry more current than a small-diameter wire. But the ability of a wire to carry a large current means that its resistance is small. Thus, a large-diameter wire has a small resistance whereas a small-diameter wire has a large resistance. As we have learned, when two quantities depend on each other so that one increases as the other decreases, and vice versa, the relationship is called an *inverse ratio*.

RULE

The resistance of a wire is inversely proportional to its cross-sectional area.

$$\frac{R_1}{R_2} = \frac{A_2}{A_1} \qquad \text{Formula 12-6}$$

where R_1 and A_1 = resistance and area of first wire, respectively
R_2 and A_2 = resistance and area of second wire, respectively

Example 12-21

A wire has a resistance of 20 Ω and a cross-sectional area of 320 cmil. Find the resistance of a wire of the same length but with an area of 800 cmil.

Solution

Given: $R_1 = 20\ \Omega$ Find: $R_2 = ?$
$A_1 = 320\ \text{cmil}$
$A_2 = 800\ \text{cmil}$

$$\frac{R_1}{R_2} = \frac{A_2}{A_1} \qquad (12\text{-}6)$$

$$\frac{20}{R_2} = \frac{800}{320}$$

$$800 R_2 = 320 \times 20$$

$$R_2 = \frac{6400}{800} = 8\ \Omega \qquad Ans.$$

> **Hint for Success**
>
> Be sure that the areas are expressed in the same units of measure.

Example 12-22

A certain length of No. 36 copper wire has a resistance of 200 Ω. Find the resistance of the same length of No. 30 wire.

Solution

Given: First wire: No. 36 = 200 Ω = R_1 Find $R_2 = ?$
Second wire: No. 30

1. From Table 12-1, find the circular-mil area of each wire.

A_1 of No. 36 = 25 cmil
A_2 of No. 30 = 100 cmil   (approx)

2. Set up the inverse ratio.

$$\frac{R_1}{R_2} = \frac{A_2}{A_1} \qquad (12\text{-}6)$$

$$\frac{200}{R_2} = \frac{100}{25}$$

$$100 R_2 = 200 \times 25$$

$$R_2 = \frac{5000}{100} = 50\ \Omega \qquad Ans.$$

Exercises

Applications

1. If $R_1 = 21\ \Omega$, $A_1 = 10$ MCM, and $A_2 = 4$ MCM, find R_2.
2. A wire has a resistance of 30 Ω and a cross-sectional area of 100 cmil. Find the resistance of the same length of a wire with an area of 320 cmil.
3. A certain length of No. 30 wire has a resistance of 50 Ω. Find the resistance of the same length of No. 40 wire.
4. A certain length of No. 36 wire has a resistance of 1.5 Ω. Find the resistance of the same length of No. 34 wire.
5. A certain length of No. 12 wire has a resistance of 3.2 Ω. Find the resistance of the same length of No. 18 wire.

(See CD-ROM for Test 12-2)

JOB 12-8 Resistances of Wires of Any Diameter, Any Length, or Any Material

In the last three jobs we learned that

1. The resistance is directly proportional to the specific resistance K of the material.
2. The resistance is directly proportional to the length L.
3. The resistance is inversely proportional to the area A.

These three concepts may be combined into one rule.

Hint for Success

The units of K determine the units to be used for L and A.

RULE The resistance R of a wire is equal to the specific resistance K multiplied by the length in feet L, and divided by the circular-mil area of the wire A.

$$R = \frac{K \times L}{A}$$

Formula 12-7

where
R = resistance of wire, Ω
K = specific resistance for a particular material
L = length of wire, ft
A = area of wire, cmil

Example 12-23

Find the resistance of a tungsten wire 10 ft long whose diameter is 0.005 in.

Solution

Given: Tungsten $K = 34$ Find: $R = ?$

$$L = 10 \text{ ft}$$
$$D = 0.005 \text{ in}$$

1. Change the diameter to mils.

$$0.005 \text{ in} = 0.005 \times 1000 = 5 \text{ mil}$$

2. Find the circular-mil area.

$$A = D_m{}^2 = 5^2 = 25 \text{ cmil} \qquad (12\text{-}1)$$

3. Find the resistance of the wire.

$$R = \frac{K \times L}{A} \qquad (12\text{-}7)$$

$$R = \frac{34 \times 10}{25} = \frac{340}{25} = 13.6 \ \Omega \qquad \textit{Ans.}$$

Example 12-24

An electric heater is made with a heating element of 4 ft of No. 30 Nichrome wire. (*a*) What is the resistance of the heating coil? (*b*) What current will it draw from a 120-V line? (*c*) How much power will it use?

Solution

The diagram for the problem is shown in Fig. 12-9.

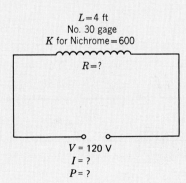

$L = 4$ ft
No. 30 gage
K for Nichrome = 600

$R = ?$

$V = 120$ V
$I = ?$
$P = ?$

Figure 12-9

a. Find the resistance of the heating coil.
  1. Find the circular-mil area. From Table 12-1,

$$\text{No. } 30 = 100.5 \text{ cmil} = 100 \text{ cmil} \quad \text{(approx)}$$

2. Find the resistance of the coil.

$$R = \frac{K \times L}{A} \qquad (12\text{-}7)$$

$$R = \frac{600 \times 4}{100} = \frac{2400}{100} = 24\ \Omega \qquad Ans.$$

b. Find the current drawn.

$$V = IR \qquad (2\text{-}1)$$
$$120 = I \times 24$$
$$I = \frac{120}{24} = 5\ A \qquad Ans.$$

c. Find the power used.

$$P = I \times V \qquad (7\text{-}1)$$
$$P = 5 \times 120 = 600\ W \qquad Ans.$$

Exercises

Practice

1. Find the resistance of a 0.1-in-diameter copper wire which is 100 ft long.
2. Find the resistance of 150 ft of No. 32 copper wire.
3. Find the resistance of 10 ft of ⅛-in-diameter silver wire.
4. What is the resistance of 20 ft of No. 20 gage manganin wire? What current will flow through the wire if the voltage is 6 V?
5. Find the resistance of 30 ft of No. 24 iron wire. What current will flow through the wire if it is used as a resistance element across 110 V? What is the power used?
6. What is the resistance of ½ ft of No. 40 Nichrome wire?
7. A winding made of 400 ft of No. 30 copper wire is placed across 6 V. Find (a) the current drawn and (b) the power used.
8. Find the resistance of 200 ft of copper annunciator wire whose diameter is 0.04 in.
9. The ½-in-diameter copper leads from a welding control are each 1 ft long. What is the voltage drop along these leads when the welding current is 2500 A?
10. What is the resistance of an aluminum conductor 162 mil in diameter and 500 ft long?

JOB 12-9 Finding the Length of Wire Needed
to Make a Certain Resistance

Example 12-25

How many feet of No. 20 gage Nichrome wire are required to make a heater coil of 10 Ω resistance?

Solution

Given: K for Nichrome = 600   Find: $L = ?$

                    Gage = No. 20

                    $R = 10\ \Omega$

1. Find the circular-mil area. From Table 12-1, for No. 20 gage,

$$A = 1022 \text{ cmil}$$

2. Find the length of the wire.

$$R = \frac{K \times L}{A} \qquad\qquad (12\text{-}7)$$

$$\frac{10}{1} = \frac{600 \times L}{1022}$$

$$600L = 1022 \times 10$$

$$L = \frac{10{,}220}{600} = 17.03 \text{ ft} \qquad Ans.$$

Example 12-26

How much resistance must be placed in series with a 50-V 50-Ω lamp in order to operate it from a 110-V line? How many feet of No. 18 German silver wire are required to make this limiting resistor?

Solution

The diagram for the circuit is shown in Fig. 12-10.

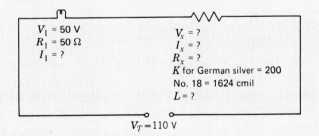

$V_1 = 50$ V
$R_1 = 50\ \Omega$
$I_1 = ?$

$V_x = ?$
$I_x = ?$
$R_x = ?$
K for German silver = 200
No. 18 = 1624 cmil
$L = ?$

$V_T = 110$ V

Figure 12-10

1. Find the series current.

$$V_1 = I_1 \times R_1 \qquad\qquad (4\text{-}4)$$

$$50 = I_1 \times 50$$

$$I_1 = \frac{50}{50} = 1 \text{ A}$$

Therefore, $I_x = I_1 = 1 \text{ A}$ (4-1)

2. Find V_x.

$$V_T = V_1 + V_x \qquad\qquad (4\text{-}2)$$

$$110 = 50 + V_x$$

$$V_x = 110 - 50 = 60 \text{ V}$$

3. Find R_x.

$$V_x = I_x \times R_x \qquad \text{(4-5)}$$
$$60 = 1 \times R_x$$
$$R_x = 60 \ \Omega \qquad Ans.$$

4. Find the length of wire needed to make a 60-Ω resistor.

$$R = \frac{K \times L}{A} \qquad \text{(12-7)}$$
$$\frac{60}{1} = \frac{200 \times L}{1624}$$
$$200L = 1624 \times 60$$
$$L = \frac{97,440}{200} = 487.2 \ \text{ft} \qquad Ans.$$

Hint for Success

The units used to measure length are determined by K.

Exercises

Applications

1. What is the resistance of an aluminum wire 200 ft long if its circular-mil is 2000?
2. What is the resistance of a 100-ft-long constantan wire if its diameter is 80 mil?
3. How many feet of 0.01-in-diameter copper wire are required to make a resistance of 10 Ω?
4. A 0.02-in-diameter Nichrome wire is used for the coils of an electric heater. If the resistance of the coils is 18 Ω, how many feet of wire are needed? What current and power are drawn when the heater is used on a 110-V line?
5. How many feet of No. 28 copper wire are needed to make an ammeter shunt of 0.2 Ω?
6. What would be the length of the wire in Exercise 5 if manganin wire were used instead of copper?
7. A subway-car heater is made of No. 24 iron wire. It is to operate on 550 V and 10 A. How many feet of wire are needed to make the heater?
8. What resistance must be inserted in series with an MZ4616 zener diode in order to operate it from two dry cells connected in series? The diode takes 2.3 V at 0.3 A. What length of No. 20 iron wire is needed to make the resistance?
9. Which of the following has the greater resistance: (*a*) 50 ft of No. 20 iron wire or (*b*) 10 ft of No. 18 Nichrome wire?
10. How many feet of No. 14 copper wire are required to make 0.8 Ω of resistance?

JOB 12-10 Finding the Diameter of Wire
Needed to Make a Certain Resistance

Example 12-27

What must be the diameter and gage number of a Nichrome heating element 40 ft long if the resistance is to be 22 Ω?

Solution

Given: K for Nichrome = 600   Find: D = ?

R = 22 Ω   Gage No. = ?

L = 40 ft

1. Find the required circular-mil area.

$$R = \frac{K \times L}{A}$$ (12-7)

$$\frac{22}{1} = \frac{600 \times 40}{A}$$

$$22A = 24{,}000$$

$$A = \frac{24{,}000}{22}$$

$$= 1091 \text{ cmil}$$

2. Find the diameter in mils.

$$D_m = \sqrt{\text{cmil}}$$ (12-2)

$$D_m = \sqrt{1091}$$

$$= 33.03 \text{ mil} \qquad Ans.$$

3. Find the gage number. This may be found by looking up the table for either the diameter of 33.03 mil or the area of 1091 cmil.

No. 19 = 35.89 mil   No. 19 = 1288 cmil

? = 33.03 mil   ? = 1091 cmil

No. 20 = 31.96 mil   No. 20 = 1022 cmil

Using either method, the closest gage number is No. 20. However, always choose the *smaller* gage number so as to select a wire of *larger* area, and therefore smaller resistance and larger current-carrying ability. Therefore, use No. 19 wire.   *Ans.*

> **Hint for Success**
>
> The problems in Jobs 12-5 through 12-10 are all based on equation 12-7. They are just different ways of using the same equation.

Example 12-28

Find the size of copper wire to be used in a two-wire system 100 ft long if the total resistance of the system is to be 1.3 Ω.

Solution

The diagram for the problem is shown in Fig. 12-11 on page 448. When the wire is made of copper, we can use Table 12-1 to simplify our calculations. If we can find the resistance per 1000 ft, we can look in the table to find the wire which has this resistance. The total length L_1 of the two-wire system equals $2 \times 100 = 200$ ft.

$$\Omega/\text{ft} = \frac{1.3}{200}$$

$$= 0.0065 \ \Omega/\text{ft}$$

$$\Omega/1000 \text{ ft} = 0.0065 \times 10^3$$

$$= 6.5 \ \Omega/1000 \text{ ft}$$

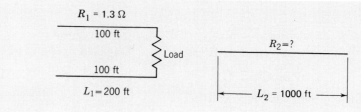

Figure 12-11

In column 4 of Table 12-1, the next *smaller* resistance per 1000 ft is 6.385 Ω. Reading to the left to column 1, we find the gage number to be No. 18.   *Ans.*

Exercises

Applications

1. What diameter of constantan wire should be used to make a coil of 10 Ω resistance if only 3 ft of wire is used?

2. What must be the diameter of an aluminum wire 2000 ft long if its resistance is to be 3 Ω?

3. What must be the diameter of an iron wire 40 ft long if the resistance is to be 50 Ω?

4. What size of copper wire should be used for a single line 200 ft long if the allowable resistance is 1 Ω?

5. What diameter of copper wire should be used for a two-wire system 500 ft long if the total resistance is not to exceed 10 Ω?

6. The heating coil of an electric heater using 4 ft of Nichrome wire is to have a resistance of 24 Ω. What gage wire should be used?

7. A rheostat is to be wound with 500 ft of German silver wire to produce a resistance of 60 Ω. What gage wire should be used?

8. What gage of annealed copper wire at 68° is required to have a maximum of 0.05 Ω for a two-wire system that is 500 ft long?

9. For a second branch of the system in Exercise 8 it is required that the resistance for a 250-ft branch circuit be no more than 0.5 Ω. What size of wire must be used?

10. Suppose that the circuit in Exercise 9 is grounded at both ends so that there is only one line in the circuit. If the wire resistance is still the same, what size of wire can now be used?

Temperature Coefficients of Resistance

This job discusses the effects of temperature on wires and wire selection. First, the properties of the wire itself and how temperature affects the resistance of the wire are covered. It will also be shown how these effects can be either constructive or destructive.

Materials whose resistance increases as the temperature increases are said to have a positive temperature coefficient (PTC). Materials whose resistance decreases as temperature increases have a negative temperature coefficient (NTC). Carbon is one substance that has a negative temperature coefficient. The carbon atoms' rhythmic vibrations tend to assist rather than obstruct the movement of the free electron.

Second, the fact that wire can be viewed as a thermal system and how this affects the selection of the appropriate wire are discussed. All of these topics are related to requirements of the National Electrical Code (NEC).

Resistance Change Can Destroy Current-Driven Devices

In general, the resistance of a conductor increases as temperature increases. Higher temperature gives the electrons the energy to go where *they* want to go; this means that it will take a higher voltage to make them go where *you* want them to go. If higher voltage is required to achieve a particular current, the resistance of the material will be higher.

The increase in resistance with temperature can be a self-destructive property. If the current through the conductor is constant, as the temperature of the conductor increases the power dissipated also increases. This happens because the I^2R power dissipated in a conductor increases as its resistance increases. The increase in power dissipated increases the temperature. This can cause the conductor to burn out and/or start a fire. The "victims" can be wire connections and connectors.

The change in resistance per unit of change in temperature is called the *temperature coefficient of resistance*. It is commonly represented by the Greek letter α (alpha). If the temperature goes from T_1 to T_2, the resistance goes from $R(T_1)$ to $R(T_2)$. The equation for the new value of resistance is:

$$\begin{aligned} R(T_2) &= R(T_1)\,[1 + \alpha_{T1}\,(T_2 - T_1)] \qquad\qquad \text{12-8} \\ &= R(T_1) + R(T_1) \times \alpha_{T1}\,(T_2 - T_1) \end{aligned}$$

The equation for the temperature coefficient, derived from the above equation, is

$$\alpha = \dfrac{\dfrac{(R_2 - R_1)}{R_1}}{T_2 - T_1}$$

$$\alpha = \dfrac{\text{Relative change resistance}}{\text{Change temperature}}$$

The coefficient α depends upon the material and the temperature. Table 12-3 shows the coefficients for different materials at a temperature of 20°C. For pure metals near room temperature the coefficient is approximately 0.004 for temperatures in degrees Centigrade and 0.0023 for degrees Fahrenheit.

Table 12-3

Temperature Coefficient of Resistance (per °C) at 20°C

Metal	α
Copper	0.0039
Aluminum	0.0040
Mild Steel	0.0045
Lead	0.0040

The National Electrical code (see Table 12-4) uses different values of α because it assumes materials commonly used in electrical conductors and a temperature of 75°C. The notes provided by the NEC at the foot of that table are given here. They show that formula 12-8 is used by the NEC with a temperature T_1 of 75°C (167°F). A warning is given that the table applies only to the particular wire materials and constructions indicated. This emphasizes the need for caution in these matters.

> These resistance values are valid *only* for the parameters as given. Using conductors having coated strands, different stranding types, and, especially, other temperatures changes the resistance.
>
> Formula for temperature change: $R_2 = R_1 [1 + \alpha(T_2 - 75)]$ where $\alpha_{cu} = 0.00323$, $\alpha_{AL} = 0.00330$.
>
> Conductors with compact and compressed stranding have about 9% and 3%, respectively, smaller bare conductor diameters than those shown. See Table 5A [NEC table] for actual compact cable dimensions.
>
> The IACS conductivities used bare copper = 100%, aluminum = 61%.
>
> Class B stranding is listed as well as solid for some sizes. Its overall diameter and area are that of its circumscribing circle.
>
> The construction information is per NEMA WC8-1988. The resistance is calculated per National Bureau of Standards Handbook 100, dated 1966, and Handbook 109, dated 1972.

Temperature is commonly expressed in degrees Centigrade. Figure 12-12 will help you visualize the relationship between Centigrade and Fahrenheit. The Centigrade scale is defined so that 0°C is the freezing point of water and 100°C is the boiling point of water. The relationship between Centigrade and Fahrenheit is a straight line. A cable temperature of 75°C won't boil water, but you would not want to hold onto it for very long!

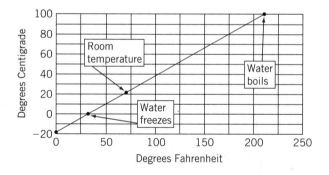

Figure 12-12

Resistance Change Can Protect Voltage-Driven Devices.   The increase in resistance due to temperature can also be used to protect a device. If the device is voltage-determined, the power dissipated is E^2/R. In this situation, as the temperature increases (causing the resistance to increase), the power dissipated is reduced. This is most commonly seen in devices such as light bulbs. Since the bulb is across a voltage supply, the voltage remains constant. When a bulb turns on, the resistance is low and there is a high surge of current. As the filament heats up, the resistance increases and limits the current. This is an example of why fuses are more likely to blow when a device is initially turned on. This is also a reason to have a slow-blow fuse. If the fuse can hold long enough to allow the temperature to limit the current, there will not be an overload.

Table 12-4 Conductor Properties (NEC Table 8)

Size AWG/ kcmil	Area, Cir. Mils	Conductors				DC Resistance at 75°C (167°F)		
		Stranding		Overall		Copper		Aluminum
		Quantity	Diam., in	Diam., in	Area, in²	Uncoated Ω/kft	Coated Ω/kft	Ω/kft
18	1,620	1	—	0.040	0.001	7.77	8.08	12.8
18	1,620	7	0.015	0.046	0.002	7.95	8.45	13.1
16	2,580	1	—	0.051	0.002	4.89	5.08	8.05
16	2,580	7	0.019	0.058	0.003	4.99	5.29	8.21
14	4,110	1	—	0.064	0.003	3.07	3.19	5.06
14	4,110	7	0.024	0.073	0.004	3.14	3.26	5.17
12	6,530	1	—	0.081	0.005	1.93	2.01	3.18
12	6,530	7	0.030	0.092	0.006	1.98	2.05	3.25
10	10,380	1	—	0.102	0.008	1.21	1.26	2.00
10	10,380	7	0.038	0.116	0.011	1.24	1.29	2.04
8	16,510	1	—	0.128	0.013	0.764	0.786	1.26
8	16,510	7	0.049	0.146	0.017	0.778	0.809	1.28
6	26,240	7	0.061	0.184	0.027	0.491	0.510	0.808
4	41,740	7	0.077	0.232	0.042	0.308	0.321	0.508
3	52,620	7	0.087	0.260	0.053	0.245	0.254	0.403
2	66,360	7	0.097	0.292	0.067	0.194	0.201	0.319
1	83,690	19	0.066	0.332	0.087	0.154	0.160	0.253
1/0	105,600	19	0.074	0.373	0.109	0.122	0.127	0.201
2/0	133,100	19	0.084	0.419	0.138	0.0967	0.101	0.159
3/0	167,800	19	0.094	0.470	0.173	0.0766	0.0797	0.126
4/0	211,600	19	0.106	0.528	0.219	0.0608	0.0626	0.100
250	—	37	0.082	0.575	0.260	0.0515	0.0535	0.0847
300	—	37	0.090	0.630	0.312	0.0429	0.0446	0.0707
350	—	37	0.097	0.681	0.364	0.0367	0.0382	0.0605
400	—	37	0.104	0.728	0.416	0.0321	0.0331	0.0529
500	—	37	0.116	0.813	0.519	0.0258	0.0265	0.0424
600	—	61	0.099	0.893	0.626	0.0214	0.0223	0.0353
700	—	61	0.107	0.964	0.730	0.0184	0.0189	0.0303
750	—	61	0.111	0.998	0.782	0.0171	0.0176	0.0282
800	—	61	0.114	1.030	0.834	0.0161	0.0166	0.0265
900	—	61	0.122	1.094	0.940	0.0143	0.0147	0.0235
1000	—	61	0.128	1.152	1.042	0.0129	0.0132	0.0212
1250	—	91	0.117	1.289	1.305	0.0103	0.0106	0.0169
1500	—	91	0.128	1.412	1.566	0.00858	0.00883	0.0141
1750	—	127	0.117	1.526	1.829	0.00735	0.00756	0.0121
2000	—	127	0.126	1.632	2.092	0.00643	0.00662	0.0106

Not only does the resistance increase with temperature but the temperature coefficient of resistance often increases also. Table 12-5 shows the temperature coefficient of resistance for tungsten. Notice that the temperature coefficient increases with temperature but not as drastically as resistance does. This is because the temperature coefficient is related to the slope of the above curve. The resistance curve is almost a straight line, curving only slightly upward. It is the slight upward curve that causes the coefficient to increase.

Table 12-5

Temperature Coefficient of Tungsten (per °C)

Temperature	Coefficient
18	0.0045
500	0.0057
1000	0.0089

Exercises

Practice

1. Describe how temperature affects the electrons in a metal.
2. Describe the temperature coefficient of resistance.
3. What is the approximate temperature coefficient for metals?
4. What is the equation for the relationship of the resistance of a wire at a temperature T_2 to the resistance at a temperature T_1?
5. Describe how temperature can cause the destruction of a current-driven device.
6. What are the defining temperatures for the Centigrade scale?
7. Under what conditions can an increasing temperature be protective to a device?

Calculating Resistance   Formula 12-8 is repeated here.

$$R(T_2) = R(T_1) [1 + \alpha_{T1} (T_2 - T_1)]$$

In most industrial applications of this formula the practitioner refers to a table. For example, the NEC provides Table 8 in Chapter 9 of the code (given in this chapter as Table 12-4). The heading for the table is shown in Fig. 12-13.

The analysis starts by selecting a conductor size (see item 1 in Fig. 12-13). The conductor size can be specified as an AWG number and/or as a cross-sectional area in cmil. These values are in the first and second columns respectively. The structure of the conductor is further specified according to the number of conductor strands and the size of the insulation (see item 2). A single-strand conductor is a solid conductor. A multistrand conductor is often referred to as a *flexible wire*.

The only remaining wire parameters are associated with the type of metal used in the conductor, aluminum or copper (see item 4). Copper cable can be either coated or uncoated. Knowing the size of the wire, its structure in number of strands, the type of metal used in the wire, and whether copper is coated or not allows you to find the value of the resistance $R(T_1)$ for the reference temperature of the table (see item 5). The resistance is given in ohms per thousand feet.

CALCULATOR HINT

Practical applications of the theory of wire size determination require extensive logic specified by NEC. Special calculators have been built to include this logic.

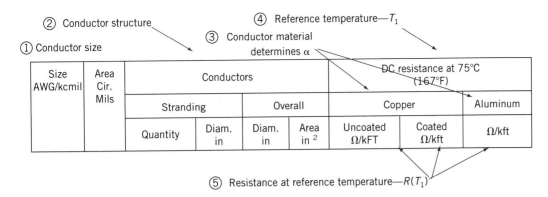

② Conductor structure
④ Reference temperature—T_1
③ Conductor material
determines α
① Conductor size

Size AWG/kcmil	Area Cir. Mils	Conductors				DC resistance at 75°C (167°F)		
		Stranding		Overall		Copper		Aluminum
		Quantity	Diam. in	Diam. in	Area in²	Uncoated Ω/kFT	Coated Ω/kft	Ω/kft

⑤ Resistance at reference temperature—$R(T_1)$

Figure 12-13

The temperature coefficient of resistance for the wire α is determined by the type of metal and is provided in the footnote: for copper α_{cu} = 0.00323, for aluminum α_{AL} = 0.00330.

The value of α is determined by the type of material (see item 3). The values to be used are specified in footnotes to the table.

> ## RULE
>
> **Formula for temperature change:** $R_2 = R_1[1 + \alpha(T_2 - 75)]$
> **where:** α_{cu} = 0.00323, α_{AL} = 0.00330.

Example 12-29

A distribution system uses a solid No. 14 uncoated copper wire. Find the resistance of a run of 350 ft of wire at a temperature of 130°F. Also, find the temperature coefficient of resistance for the wire.

Solution

Start by using Table 12-4 on page 451. The easiest answer to get is the value of α. The problem states that the wire is copper; therefore, for copper α_{cu} = 0.00323.

The table provides the resistance for a wire at 75°C. Therefore, you will have to find the resistance at that temperature, $R(75°C)$, and convert that answer to a resistance for 130°F.

The table provides a resistance in ohms per kilofoot of wire. This will have to be converted to a resistance for a 350-ft run.

Start by referring to the table for a wire size of No. 14. Two values are given: copper 7 strands and 1 strand. Since the wire is solid, select the 1 strand entry. You find out that the wire is uncoated copper, so you move to that column and read the value 3.07. This is the value of $R(75°C)$ in ohms per 1000 feet.

To find the value of $R(130°F)$, you must first convert 130°F to degrees Centigrade.

$$°C = (F - 32) \times 5/9$$
$$= 54.4$$

The required resistance is obtained from

$$R(T_2) = R(T_1) [1 + \alpha_{T1} (T_2 - T_1)]$$
$$R(130°F) = 3.07 [1 + 0.00323 (54.4 - 75)]$$
$$= 2.87 \ \Omega \text{ per 1000 feet}$$

The run for the wire is 350 ft. The wire resistance comes from a length of wire that goes 350 ft to the device and 350 ft back to the source. The total wire length is therefore 700 ft. The wire resistance is

$$R = 2.87 \times \frac{700}{1000}$$
$$= 2.01 \ \Omega$$

Wire as a Thermal System

When a wire or cable is deployed, the state of the wire depends upon the temperature of the surrounding air (ambient temperature), the temperature of the conductor, and the properties of the cover or insulation. The essential components are shown in Fig. 12-14.

The heat is generated in the wire core as described earlier. The conductor rises to temperature TC. The heat passes through the insulation and goes into the surrounding atmosphere. This is a two-phase process in which the heat first moves through the insulation and then is transported between the ambient atmosphere and the surface of the insulation. The temperature of the air surrounding the wire is called the *ambient temperature* (TA).

This general concept of heat transport is summarized in a group of tables that ease the use of the theory. An example is how the NEC addresses conductor ampacities in Section 315. NEC Tables 310-16 to 310-19 are examples of how the theory is applied. NEC Table 310-16 is given here as Table 12-6. As mentioned in the title, the table is "Based on Ambient Temperature of 30°C (86°F)." The table consists of two parts: one for copper wire and one for aluminum wire, reflecting the difference in properties of the core material, as discussed earlier.

The top portion of the table indicates the various temperature ratings that the wire and the insulation are designed to handle.

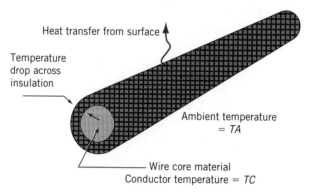

Figure 12-14

Table 12-6

Allowable Ampacities of Insulated Conductors Rated 0 through 2000 V, 60° to 90°C (140° to 194°F); Not More Than Three Current-Carrying Conductors in Raceway or Cable or Earth (Directly Buried), Based on Ambient Temperature of 30°C (86°F); NEC Table 310-16

| Size | Temperature Rating of Conductor. See Table 310-13 | | | | | | Size |
| | 60°C (140°F) | 75°C (167°F) | 90°C (194°F) | 60°C (140°F) | 75°C (167°F) | 90°C (194°F) | |
	Types	Types	Types	Types	Types	Types	
AWG, kcmil	TW*, UF*	FEPW* RH*, RHW* THHW* THW* THWN* XHHM* USE*, ZW*	TBS SA SIS FEP* FEPB*, M I RHH* RHW-2 THHN* THHW* THW-2 THWN-2 USE-2, X HH XHHW* XHHW-2 ZW-2	TW*, UF*	RH*, RHW* THHW* THWN* XHHW* USE*	TBS SA SIS THHN* THHW* THW-2 THWN-2 RHH* RHW-2 USE-2 XHH XHHW XHHW-2 ZW-2	AWG, kcmil
	Copper			**Aluminum or Copper-Clad Aluminum**			
18	—	—	14	—	—	—	—
16	—	—	18	—	—	—	—
14	20*	20*	25*	—	—	—	—
12	25*	25*	30*	20*	20*	25*	12
10	30	35*	40*	25	30*	35*	10
8	40	50	55	30	40	45	8
6	55	65	75	40	50	60	6
4	70	85	95	55	65	75	4
3	85	100	110	65	75	85	3
2	95	115	130	75	90	100	2
1	110	130	150	85	100	115	1
1/0	125	150	170	100	120	135	1/0
2/0	145	175	195	115	135	150	2/0
3/0	165	200	225	130	155	175	3/0
4/0	195	230	260	150	180	205	4/0

Table 12-6 Continued

Temperature Rating of Conductor. See Table 310-13

Size	60°C (140°F) Types	75°C (167°F) Types	90°C (194°F) Types	60°C (140°F) Types	75°C (167°F) Types	90°C (194°F) Types	Size
250	215	255	290	170	205	230	250
300	240	285	320	190	230	255	300
350	260	310	350	210	250	280	350
400	280	335	380	225	270	305	400
500	320	380	430	260	310	350	500
600	355	420	475	285	340	385	600
700	385	460	520	310	375	420	700
750	400	475	535	320	385	435	750
800	410	490	555	330	395	450	800
900	435	520	585	355	425	480	900
1000	455	545	615	375	445	500	1000
1250	495	590	665	405	485	545	1250
1500	520	625	705	435	520	585	1500
1750	545	650	735	455	545	615	1750
2000	560	665	750	470	560	630	2000

Correction Factors

Ambient Temp, °C	For ambient temperatures other than 30°C (86°F), multiply the allowable ampacities shown above by the appropriate factor shown below.						Ambient Temp, °F
21–25	1.08	1.05	1.04	1.08	1.05	1.04	70–77
26–30	1.00	1.00	1.00	1.00	1.00	1.00	78–86
31–35	0.91	0.94	0.96	0.91	0.94	0.96	87–95
36–40	0.82	0.88	0.91	0.82	0.88	0.91	96–104
41–45	0.71	0.82	0.87	0.71	0.82	0.87	105–113
46–50	0.58	0.75	0.82	0.58	0.75	0.82	114–122
51–55	0.41	0.67	0.76	0.41	0.67	0.76	123–131
56–60	—	0.58	0.71	—	0.58	0.71	132–140
61–70	—	0.33	0.58	—	0.33	0.58	141–158
71–80	—	—	0.41	—	—	0.41	159–176

*Unless otherwise specifically permitted elsewhere in this code, the overcurrent protection for conductor types marked with an asterisk (*) shall not exceed 15 A for No. 14, 20 A for No. 12, and 30 A for No. 10 copper; or 15 A for No. 12 and 25 A for No. 10 aluminum and copper-clad aluminum after any correction factors for ambient temperature and number of conductors have been applied.

Exercises

Practice

1. What are the three thermal processes associated with a wire?
2. What properties affect the ampacity of a conductor?

Applications

1. What is the resistance per 1000 ft of a solid No. 12 uncoated copper wire?
2. A No. 12 flexible wire has seven strands. What is the resistance per 1000 ft for a coated copper wire?
3. If the same wire in Prob. 2 were made of aluminum, what would be the resistance of 1000 ft?
4. What is the resistance of 500 ft of No. 3 stranded aluminum wire?
5. A two-wire distribution circuit 500 ft long operates at 75°C and can have no more than 2 Ω of resistance. What size of aluminum wire can you use?
6. If the circuit of Prob. 5 were to use uncoated copper wire, what size could you use?
7. If the wire of Prob. 6 were coated, what size could you use?
8. A 600-V RHW-2 type copper conductor is used in a circuit that has an ambient temperature of 30°C and operates at temperatures up to 90°C. What is the ampacity of a No. 8 wire in this branch?
9. If the conductor in Prob. 8 were aluminum, what would be its ampacity?
10. A 600-V cable operates in a circuit with a 60°C temperature rating. If the ambient temperature is 30°C, what size of copper wire is required to provide 30 A?

Review of Resistance of Wires

1. A *mil* is the unit used to measure the _____ of round wires. A *mil* is equal
2. to one-_____ of an inch. A *circular mil* (cmil) is the unit used to measure
3. the cross-sectional _____ of wires.

$$A = D_m{}^2 \qquad\qquad\qquad 12\text{-}1$$

$$D_m = \sqrt{\text{cmil}} \qquad\qquad\qquad 12\text{-}3$$

4. where A = area of the wire, _____
5. D_m = diameter of the wire, _____

To convert square mils to circular mils, we use the formula

6.
$$A_{\text{cmil}} = \frac{A_{\text{mil2}}}{?} \qquad\qquad\qquad 12\text{-}2$$

The specific resistance K of a wire is the resistance of a wire 1 ft long with a

7. diameter of 1 _____. The specific resistances of different materials are given in Table 12-2.

The resistance of a wire is directly proportional to its specific resistance.

8.
$$\frac{R_1}{R_2} = \frac{K_?}{K_?} \qquad\qquad\qquad 12\text{-}4$$

The resistance of a wire is directly proportional to its length.

9.
$$\frac{R_1}{R_2} = \frac{L_?}{L_?}$$

12-5

The resistance of a wire is inversely proportional to its area.

10.
$$\frac{R_1}{R_2} = \frac{A_?}{A_?}$$

12-6

The general formula for the resistance of any wire is

11.
$$R = \frac{K \times L}{?}$$

12-7

where R = resistance of the wire, Ω

              K = specific resistance of the material

12. L = length of the wire, _____

13. A = area of the wire, _____

$$R(T_2) = R(T_1) [1 + \alpha_{T1} (T_2 - T_1)]$$
$$= R(T_1) + R(T_1) \times \alpha_{T1} (T_2 - T_1)$$

12-8

where T_1 = original temperature

            T_2 = new temperature

    $R(T_1)$ = resistance at original temperature

    $R(T_2)$ = resistance at new temperature

        α = temperature coefficient of resistance

Exercises

Applications

1. Find the diameter in mils and gage number of the wires whose diameters are (a) 0.025 in, (b) 0.102 in, (c) ⅛ in, and (d) $^{128}\!/_{1000}$ in.
2. Find the area in circular mils of wires whose diameters are (a) 15 mil, (b) 40 mil, (c) 0.01 in, and (d) 0.204 in.
3. A constantan wire has a resistance of 50 Ω. What is the resistance of a manganin wire of the same length and area?
4. Find the resistance of a two-wire line 500 ft long using No. 20 copper wire. Find the voltage drop across this line if the current is 150 mA.
5. A length of No. 12 copper wire has a resistance of 3.6 Ω. What is the resistance of the same length of No. 8 wire?
6. Find the resistance of 6 in of No. 20 Nichrome wire.
7. A resistance element designed to operate at 110 V and 6 A is to be made of No. 30 Nichrome wire. Find the length of wire needed.
8. What diameter of iron wire 12 ft long is needed to make a resistance of 80 Ω?
9. What is the resistance per 1000 ft for a No. 14 stranded aluminum wire at 167°F?
10. A No. 1/0 1000-V cable operates in a 30°C ambient temperature. If the cable temperature can go to 167°F and the cable is copper-clad aluminum, what is its maximum ampacity?

Assessment
chapter 12

1. A rectangular busbar measures 1¼ by ¼ in. What is its volume if the total length is 36 in?
2. A piece of copper wire has a resistance of 24 Ω. What is the resistance of the wire if the diameter of the wire is doubled? (*Hint:* Make sure that you determine the new cross-sectional area, πr^2.)
3. In Ques. 1, if another identical piece of wire were to be connected at each end, and the new resistance measured, what would the value be?
4. You measure a roll for No. 12 copper wire and find that it has a resistance of 14 Ω. How much longer is a piece of No. 10 copper wire if they both have the same resistance?
5. What is the resistance of a copper feeder conductor that is 0.813 inches in diameter?
6. How many feet of No. 20 Nichrome wire is required for 240-V, 2200-W heating element?
7. What is the size of a copper conductor that is 500 ft in length and has a resistance of 500.963 mΩ?

INTERNET ACTIVITIES

Internet Activity A

Go to the following site and find the applications in which TFFN cable can be used. **http://www.southwire.com/sw/wirecabl/wc-catal.htm**

Internet Activity B

Use the following site to locate five suppliers for No. 8 RHW cable. **http://www.piap.ch/country/usa.html**

Chapter 12 Solutions to Self-Tests and Reviews

Self-Test 12-8

a. increases
b. time
c. 280
d. 7
e. D_2
1. T_1, T_2
2. T_2
3. 7
4. 280, 5

Self-Test 12-11

1. resistance
2. I_1, I_2

3. R_2, R_1
4. equal, R_2, R_1
5. 100
6. 150, 0.33

Job 12-3 Review

1. comparison
2. division
3. "8 to 2"
4. same
5. equal
6. same
7. A_1, A_2
8. B_2, B_1
9. A_2, A_1

10. decreases, opposite
11. C_1, C_2
12. D_1, D_2
13. C_2, C_1

Self-Test 12-20

1. 700
2. 14.2
3. 10.15
4. 0.010 15
5. 700
6. 7.1
7. 7.1
8. 2

Job 12-12 Review

1. diameters
2. thousandth
3. areas
4. cmil
5. mil
6. 0.7854
7. mil
8. 1, 2
9. 1, 2
10. 2, 1
11. A
12. ft
13. cmil

1. Find wire size to prevent excessive voltage drop.
2. Find minimum wire size to supply a given load.
3. Determine ampacity of wires.

Maximum Current-Carrying Capacities of Wire

Heat is produced whenever a current flows in a wire. The larger the current, the greater the amount of heat that is produced. This heat must be given up to the surrounding atmosphere, or the wire may get hot enough to burst into flame.

The heat produced in a wire depends not only on the current but also on the resistance to that flow of current. Thick wires, because of their low resistance, can carry more current than thin wires before they dangerously overheat. But exactly how many amperes can be safely carried by a particular wire? Can a No. 20 wire carry 30 A, or is a No. 12 or larger wire required? Many tests have been made by the National Board of Fire Underwriters to determine the largest current that may be safely carried by wires of different sizes and different insulation. The National Electrical Code sets values like these shown in Table 13-1 on page 462. This table gives the *maximum* current permitted in any size of wire insulated as indicated.

Using the Table of Allowable Current Capacities

Example 13-1

What is the largest current that can be safely carried by No. 12 rubber-covered wire?

Solution

1. Locate No. 12 wire in column 1 of Table 13-1.
2. Read across to the right to column 5 for rubber-covered wire.
3. The answer (25 A) means that No. 12 rubber-covered wire should never be permitted to carry more than 25 A.   *Ans.*

Example 13-2

What is the allowable current-carrying capacity of No. 14 varnished-cambric-insulated wire?

Table 13-1

Allowable Current-Carrying Capacities of Insulated Copper Conductors (Ampacities)

National Electrical Code 1981

(1) AWG No.	(2) Diameter, mil	(3) Cross Section, cmil	(4) Resistance, Ω/1000 ft	Allowable Carrying Capacity, A		
				(5) Rubber-Covered, Type RH	(6) Varnished Cambric, Type V	(7) Asbestos, Type AF
14	64.1	4107	2.53	20	25	30
12	80.8	6530	1.59	25	30	40
10	101.9	10,380	0.999	35	40	55
8	128.5	16,510	0.628	50	55	70
6	162.0	26,250	0.395	65	70	95
4	204.3	41,740	0.249	85	95	120
3	229.4	52,630	0.197	100	110	145
2	257.6	66,370	0.156	115	125	165
1	289.3	83,690	0.124	130	145	190
0	325	105,500	0.0983	150	165	225
00	364.8	133,100	0.0779	175	190	250
000	409.6	167,800	0.0618	200	215	285
0000	460	211,600	0.0490	230	250	340

Solution

1. Locate No. 14 wire in column 1.
2. Read across to the right to column 6 for varnished-cambric-insulated wire.
3. The answer is 25 A.

Example 13-3

What is the smallest rubber-covered wire that should be used to carry 16 A?

Solution

1. Read down column 5 for rubber-covered wire until you find a number *larger* than 16 A. This will be 20 A.
2. Read to the left until you get to column 1. No. 14 gage is the smallest wire that can be used.   *Ans.*

Example 13-4

What gage of wire should be used if the circular-mil area required is 3000 cmil?

Solution

1. Read down column 3 until you get to the number that is *just larger* than 3000 cmil. This number is 4107 cmil.
2. Read to the left until you get to column 1. No. 14 gage is the smallest wire that may be used.   *Ans.*

Example 13-5

What gage of wire should be used if the resistance required is 0.5 Ω/1000 ft?

Solution

1. Read down column 4 until you get to a number that is *just smaller* than 0.5 Ω. This number is 0.395 Ω.
2. Read to the left until you get to column 1. No. 6 gage is the smallest wire that may be used.   *Ans.*

Self-Test 13-6

Find the voltage drop along 1000 ft of No. 10 rubber-covered wire if it is carrying its maximum current.

Solution

1. Find the maximum current for No. 10 rubber-covered wire. This is found in column 5 to be _____(1)_____ A.
2. Find the resistance per 1000 ft of No. 10 rubber-covered wire. Read across to the right from No. 10 in column 1 to find _____(2)_____ Ω in column _____(3)_____.
3. Find the voltage drop.

$$V = IR \hspace{4cm} (2\text{-}1)$$
$$V = \underline{\quad(4)\quad} \times 0.999 = \underline{\quad(5)\quad} \text{ V} \hspace{1cm} Ans.$$

In summary, always choose the wire
1. That can carry _____(6) (more/less)_____ current than is required
2. That has the _____(7) (smaller/larger)_____ circular-mil area
3. That has the _____(8) (smaller/larger)_____ resistance

Exercises

Applications

1. What is the maximum current that can be safely carried by No. 8 varnished-cambric-insulated wire?
2. What is the smallest rubber-covered wire that should be used to carry 50 A?
3. What gage wire should be used if the circular-mil area required is 18,000 cmil?
4. What gage wire should be used if the required resistance is not to exceed 0.91 Ω/1000 ft?

5. Find the voltage drop along 1000 ft of No. 8 varnished-cambric-insulated wire when it is carrying its maximum current.

6. What is the smallest varnished-cambric-insulated wire that should be used to carry 62 A safely?

7. What is the resistance per foot of the smallest rubber-covered wire that should be used to carry 66 A safely?

8. What size of asbestos-covered wire should be used to carry 42 A safely?

9. What size of wire should be used if the circular-mil area required is 42,000 cmil?

10. What is the diameter in mils and inches of the smallest rubber-covered wire that can be used to carry 80 A safely?

JOB 13-2

Finding the Minimum Size of Wire to Supply a Given Load

The minimum size of wire that can be used in any installation depends on two factors: (1) the total current load and (2) the voltage drop in the wire. In this job we shall determine the size of wire needed to carry the given current. In the next job, we shall take the voltage drop into consideration.

Hint for Success

The NEC covers only wire selections for distribution of electrical power at voltages from low to high.

Procedure

1. Determine the total current load.
2. Consult Table 13-1 on page 462 to find the smallest wire to carry this load.

Example 13-7

What is the smallest size of rubber-covered wire that can be used to supply two 10-Ω heating elements operated in parallel from a 220-V line?

Solution

Given: $R_1 = 10\ \Omega$ Find: Smallest size of rubber-covered wire = ?
        $R_2 = 10\ \Omega$
        $V_T = 220$ V

1. Find the total resistance.

$$R_T = \frac{R}{N} = \frac{10}{2} = 5\ \Omega \tag{5-4}$$

2. Find the total current.

$$I_T = \frac{V_T}{R_T} = \frac{220}{5} = 44\text{ A} \tag{4-7}$$

3. Find the gage number to handle 44 A.
   a. Read down column 5 until you get to the first number *larger* than 44 A. This number is 50 A.
   b. Read to the left from 50 A until you get to column 1, where we read the gage number as No. 8.   *Ans.*

Example 13-8

What is the smallest size of slow-burning weatherproof cable that may be used to carry current to a 10-hp motor from a 100-V source?

Solution

Given: $P = 10$ hp   Find: Smallest size of slow-burning cable = ?
          $V = 100$ V

1. Find the watts of power to be delivered by the cable. Assume an input to the motor equal to 110 percent of the rated horsepower.

$$\text{Input hp} = 1.1 \times 10 = 11 \text{ hp}$$
$$\text{Watts} = \text{hp} \times 750 = 11 \times 750 = 8250 \text{ W}$$

2. Find the current drawn.

$$P = I \times V \qquad\qquad (7\text{-}1)$$
$$8250 = I \times 100$$
$$I = \frac{8250}{100} = 82.5 \text{ A}$$

3. Find the gage number to handle 82.5 A. *Note:* For any wire other than rubber-covered or varnished-cambric-insulated wire, use the data for asbestos-covered wire in column 7.
   a. Read down column 7 until you get to the first number *larger* than 82.5 A. This number is 95 A.
   b. Read to the left from 95 A until you get to column 1, where we read the gage number as No. 6.   *Ans.*

CALCULATOR HINT

Some calculators have tables of built-in units and units conversion.

Exercises

Applications

1. What is the minimum size of rubber-covered wire that can be used to supply a load of 60 lamps, each of which draws 0.91 A?
2. What is the minimum size of asbestos-covered wire that can be used to supply a load of three 5-Ω electric ovens operated in parallel from a 110-V line?
3. What is the minimum size of varnished-cambric-covered wire that can be used to supply a load of thirty 200-W lamps and twenty 100-W lamps if they are all operated in parallel from the same 110-V line?
4. A 30-hp motor is to be used on a 220-V line. If the efficiency of the motor is 90%, what is the smallest size of varnished-cambric-covered wire that can be used?
5. A 220-V generator supplies motors drawing 5, 10, and 12 kW of power. What is the minimum size of rubber-covered wire that can be used?
6. A 110-V line powers a computer consuming 50 W and a monitor consuming 300 W. What size of type RH cable is needed for a cluster of six computer stations?
7. Each computer station in Exercise 6 is given a 200-W light fixture. What size of RH wire is needed now?

8. The conduit is not large enough for the required size of RH type wire in Exercises 6 and 7. What size of type V wire can be used?
9. What is the smallest cable size of each type of wire that can be used to supply a 30-kW 220-V oven?
10. What is the minimum of size VF wire that can supply 7.44 kW at 120 V?

| JOB 13-3 | Finding the Size of Wire Needed to Prevent Excessive Voltage Drops |

A wire not only must be able to carry a current without overheating but must also be able to carry this current without too many volts being used up in the wire itself. If the voltage drop in the line is too large, the voltage available at the load will be too small for proper operation of the load. For example, a 117-V generator is used to supply current to a bank of lamps some distance away. If the supply line should use up 17 V in supplying this current, then only 100 V would be available at the lamps and they would be dim. Not only that, but the loss in the line represents wasted power and a very inefficient system. To cover this, the National Electrical Code stipulates that the voltage drop in a line must not be more than 2% of the rated voltage for lighting purposes and not more than 5% of the rated voltage for power installations.

Hint for Success

Voltage drop limits are imposed so as to ensure adequate power available at outlets.

Example 13-9

A bank of lamps draws 32 A from a 117-V source 300 ft away. Find the smallest gage of rubber-covered wire that can be safely used.

Solution

The diagram for the circuit is shown in Fig. 13-1.

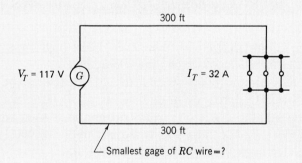

V_T = 117 V (G) I_T = 32 A

300 ft
300 ft
Smallest gage of *RC* wire = ?

Figure 13-1

1. Find the allowable voltage drop. For lighting purposes, the Code permits only a 2% drop. Therefore,

$$\text{Allowable drop} = 2\% \text{ of } 117 = 0.02 \times 117 = 2.34 \text{ V}$$

2. Find the line resistance for this drop.

$$V_l = I_l \times R_l \qquad (6\text{-}1)$$

$$2.34 = 32 \times R_l$$

$$R_l = \frac{2.34}{32} = 0.0731 \ \Omega$$

3. Find the circular-mil area of the wire that has this resistance. Since there are two wires, $L = 2 \times 300 = 600$ ft. Also, the K for copper is 10.4

$$R = \frac{K \times L}{A} \qquad (12\text{-}7)$$

$$\frac{0.0731}{1} = \frac{10.4 \times 600}{A}$$

$$0.0731 \times A = 10.4 \times 600$$

$$0.731 \ A = 6240$$

$$A = \frac{6240}{0.0731} = 85{,}360 \text{ cmil}$$

4. Find the gage number of a wire that has a circular-mil area that is *larger* than 85,360 cmil.
   a. Read down column 3 until you get to a number *just larger* than 85,360. This number is 105,500.
   b. Read across to the left to find gage No. 0 in column 1.   *Ans.*

The steps involved in the last problem can be combined as follows: The voltage drop in the line V_d is given as

$$V_d = I \times R \qquad (2\text{-}1)$$

Also,
$$R = \frac{K \times L}{A} \qquad (12\text{-}7)$$

Substituting,
$$V_d = I \times \frac{KL}{A}$$

For two wires, and substituting CM for A,

$$V_d = \frac{2KIL}{CM} \qquad \text{Formula 13-1}$$

or, solving for CM,

$$CM = \frac{2KIL}{V_d} \qquad \text{Formula 13-2}$$

where V_d = allowable voltage drop, V
K = specific resistance
L = length of the line, ft
I = line current, A
CM = area of the wire, cmil

Example 13-10

A 117-V circuit is run 300 ft to serve a load of 10 A. Using No. 12 copper wire, find (*a*) the voltage drop, and (*b*) the voltage at the load.

Solution

From Table 12-1, for No. 12 wire, $CM = 6530$.

a.
$$V_d = \frac{2KIL}{CM} \qquad (13\text{-}1)$$

$$= \frac{2(10.4)(10)(300)}{6530}$$
$$V_d = 9.6 \text{ V} \qquad Ans.$$

b.
$$V_L = V_G - V_1 \qquad (6\text{-}3)$$
$$= 117 - 9.6 = 107.4 \text{ V} \qquad Ans.$$

Example 13-11

A bank of lathes is operated by individual motors in a machine shop. The motors draw a total of 60 A at 110 V from the distributing panel box. What size of rubber-covered wire is required for the two-wire line between the panel box and the switchboard located 100 ft away if the switchboard voltage is 115 V?

Solution

The diagram for the circuit is shown in Fig. 13-2.

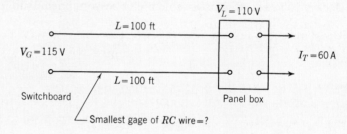

Figure 13-2

1. Find the line drop between the switchboard and the panel box.
$$V_G = V_l + V_L \qquad (6\text{-}2)$$
$$115 = V_l + 110$$
$$V_l = 115 - 110 = 5 \text{ V}$$

2. Find the circular-mil area of the wire.
$$CM = \frac{2KIL}{V_d} \qquad (13\text{-}2)$$
$$= \frac{2(10.4)(60)(100)}{5}$$
$$= 25,000 \text{ cmil}$$

3. Find the gage number of a wire that has a circular-mil area that is *larger* than 25,000 cmil.
   a. Read down column 3 until you get to a number *just larger* than 25,000 cmil. This number is 26,250.
   b. Read across to the left to find gage No. 6 in column 1. This No. 6 wire can safely carry the 60 A without overheating, since its maximum current-carrying capacity is 65 A as found in column 5 for rubber-covered wire. *Ans.*

Self-Test 13-12

A 10-hp motor is operated at an efficiency of 90% from a 232-V source 100 ft away. What is the minimum size of slow-burning insulated wire that can be used for the line supplying the motor?

Solution

The diagram for the circuit is shown in Fig. 13-3.

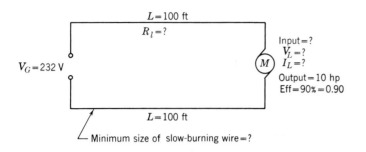

Figure 13-3

1. Find the allowable voltage drop. For power installations, the code permits a 5% drop. Therefore,

$$V_l = \text{allowable drop} = 5\% \text{ of} \underline{\hspace{2cm}}$$
$$V_l = \underline{\hspace{2cm}} \times 232 = \underline{\hspace{2cm}} \text{ V}$$

2. Find the voltage at the motor load.

$$V_L = V_G - \underline{\hspace{2cm}} \qquad (6\text{-}3)$$
$$V_L = 232 - \underline{\hspace{2cm}} = 220.4 \text{ V}$$

3. Find the input to the motor.

$$\text{Eff} = \frac{\text{output}}{\text{input}} \qquad (11\text{-}2)$$

$$\underline{\hspace{1.5cm}} = \frac{10 \times 750}{\text{input}}$$

$$0.90 \text{ input} = \underline{\hspace{2cm}}$$

$$\text{Input} = \frac{7500}{0.90} = \underline{\hspace{2cm}} \text{ W}$$

4. Find the current drawn.

$$P_L = I_L \times V_L \qquad (7\text{-}1)$$
$$8333 = I_L \times \underline{\hspace{2cm}}$$

$$I_L = \frac{8333}{220.4} = \underline{\hspace{2cm}} \text{ A}$$

5. Find the circular-mil area of the wire.

$$CM = \frac{?}{V_d}$$

$$= \frac{2(10.4)(37.8)(100)}{?}$$

$$= \underline{\hspace{2cm}} \text{ cmil}$$

6. Find the gage number of a wire that has a circular-mil area that is <u>(smaller/larger)</u> than 6778 cmil.
   a. Read down column 3 until you get to a number just <u>(smaller/larger)</u> than 6778 cmil. This number is —————.
   b. Read across to the left to find gage No. ————— in column —————. This No. 10 wire can safely carry the required 37.8 A without overheating, since its maximum current-carrying capacity is ————— A as found in column ————— for slow-burning wire.   *Ans.*

Exercises

Applications

1. What is the minimum size of rubber-covered wire that can be used to carry 40 A if the allowable voltage drop is 10 V/1000 ft of wire?
2. A 50-ft two-wire extension line is to carry 30 A with a voltage drop of only 5 V. What is the smallest size of weatherproof cable that can be used?
3. A bank of lamps draws 90 A from a 110-V line. What size of rubber-covered wire should be used if the voltage drop along 150 ft of the two-wire system must not exceed 3 V?
4. A bank of lamps draws 50 A from a 117-V source 40 ft away. Find the smallest gage rubber-covered wire that can be used for this two-wire system if the voltage drop must not exceed 2% of the voltage at the source.
5. A 20-kW motor load is 100 ft from a 230-V source. If the allowable voltage drop is 5%, what is the smallest size of varnished-cambric-covered wire that can be used?
6. A load 400 ft from a generator requires 80 A. The generator voltage is 115.6 V, and the load requires 110 V. What is the smallest size of rubber-covered wire that can be used?
7. A 10-hp motor is located 20 ft from a 225-V generator. What is the smallest size of slow-burning wire that can be used?
8. A 10-hp motor operates at an efficiency of 85% at a distance of 100 ft from a 122-V generator. What is the smallest size of varnished-cambric-covered wire that can be used?
9. A generator supplies 5.5 kW at 125 V to a motor 500 ft away. The motor delivers 6.3 hp at an efficiency of 90%. Find (*a*) the watt input to the motor, (*b*) the watts lost in the line, (*c*) the current delivered to the line, (*d*) the voltage drop in the line, and (*e*) the smallest size of rubber-covered wire that can be used.
10. Calculate the smallest size of weatherproof wire required to conduct current to one hundred 220-Ω lamps in parallel that are located 125 ft from a generator delivering 112 V. The lamps must receive 110 V.
11. A portable tree-cutting saw requires 4 A at 120 V. If the line drop in the No. 10 cable used must not exceed 3% of the motor voltage, what is the maximum length of cable that can be used? *Hint:* Use Table 13-1 to get the resistance per foot.
12. A two-wire No. 10 AWG branch circuit carries 10 A from the main distribution panel to a load 100 ft away. Find (*a*) the total resistance of the line wires,

(b) the voltage at the load when the voltage at the panel is 120 V, and (c) the maximum distance that this branch could carry a peak load of 15 A if the line drop must not exceed 3%.

Temperature and Wire Size

The basic concept of the effects of temperature on resistance of conductors was introduced in Job 12-11. This job will extend these concepts to the problems of temperature effects on ampacity and wire sizing. Data from NEC Table 310-16, given in this book as Table 12-6, will be used. In Job 12-11 it was shown that the ampacity of a conductor depends on the wire size, the wire material, the type of cover or coating, the ambient temperature, and the temperature of the cable. The situation is more complicated than was described in earlier jobs, but at last you have come to the real world.

Example 13-13

What is the ampacity of a No. 8 RH copper conductor in an ambient temperature of 30°C with a conductor operating temperature of 75°C?

Solution

Table 12-6 gives the temperature dependence of conductor ampacities.
1. Locate the RH type copper conductor in the third column from the left in the table.
2. The maximum allowable conductor temperature is 75°C.
3. Go down the first column to the No. 8 size conductor.
4. Go right to the third column, where you will see a maximum allowable ampacity of 50 A.

Note that the footnote to Table 12-6 provides additional limitations for overcurrent protection for No. 14, No. 12, and No. 10 copper and No. 12 and No. 10 aluminum- and copper-clad aluminum conductors. The numbers in the table are given because these are the starting values for calculations involving correction factors. Multiple correction factors can be applied to these numbers, however, that will not be covered in this text.

Example 13-14

What is the allowable ampacity for a No. 12 type TW copper cable in a 30°C ambient temperature and an operating temperature of 60°C?

Solution

1. Locate the TW type copper conductor in the second column from the left in the table.
2. The maximum allowable conductor temperature is 60°C.
3. Go down the first column to the No. 12 size conductor.

4. Go right to the second column, where you will see a maximum allowable ampacity of 25 A.
5. The asterisk leads to the footnote, which states that the overcurrent protection for a No. 12 copper conductor cannot exceed 20 A; therefore, the maximum allowable current is 20 A.

Exercises

Practice

For the following problems the ambient temperature is 86°F. Find the ampacity for each cable.

1. A No. 4 copper conductor operating at 194°C.
2. An aluminum No. 2 conductor operating at 75°C.
3. A No. 10 copper conductor operating at 60°C.
4. A No. 1 copper-clad aluminum conductor operating at 90°C.
5. A No. 8 aluminum conductor operating at 140°F.
6. A No. 2 copper conductor at 194°F.
7. An aluminum 2/0 conductor operating at 167°F.
8. A 000 copper cable operating at 194°F.
9. A No. 10 aluminum cable operating at 60°C.
10. A No. 6 copper-clad aluminum conductor operating at 194°F.

Example 13-15

What is the maximum allowable ampacity for the conductor in Example 13-13 if the ambient temperature goes to 38°C?

Solution

1. Find the solution as in Example 13-13.
2. Go to the "Correction Factors" section of the table.
3. Go down the first column for "Ambient Temp, °C" to the range of temperatures that includes 38°C (36–40).
4. Move to the right to the third column and read the correction factor, 0.91.
5. Multiply the answer from Example 13-13 by the correction factor $(50)(0.91) = 45.5$.

If the ambient temperature had been given in degrees Fahrenheit, you would have used the far right column to find the appropriate temperature range.

Exercises

Practice

Find the ampacities for the following cables.

1. No. 8 copper conductor operating at 90°C in an ambient temperature of 43°C.
2. A No. 1 copper cable operating at 140°F in an ambient temperature of 120°F.
3. An aluminum No. 000 cable operating at 194°F in an ambient temperature of 122°F.

4. A No. 3 copper-clad aluminum cable operating at 167°F in an ambient temperature of 87°F.
5. A No. 3 copper cable operating at 167°F in an ambient temperature of 87°F.
6. A No. 3 copper cable operating at 140°F in an ambient temperature of 87°F.
7. A No. 12 copper cable operating at 60°C in an ambient temperature of 35°C.
8. A No. 12 copper cable operating at 60°C in an ambient temperature of 44°C.
9. A No. 14 copper cable operating at 75°C in an ambient temperature of 48°F.
10. A No. 4 aluminum conductor operating at 75°C in an ambient temperature of 125°F.

Review of Size of Wiring

The size of a wire depends on (1) the total current load and (2) the voltage drop in the wires. To find the smallest wire that may be used in a given job:

1. Find the total current drawn by the load.
2. Find the voltage drop in the supply wires.
3. Find the circular-mil area of the required wire.
4. Look up Table 13-1 to find the wire that has the next-larger circular-mil area.
5. Look up the table to find the wire that can carry more current than that required by the load.
6. Choose the larger of the two wires found in steps 4 and 5.

Exercises

Applications

1. What is the maximum current that can be safely carried by No. 10 varnished-cambric-covered wire?
2. What is the smallest asbestos-covered (type A) wire that should be used to carry 56 A?
3. What gage number wire should be used if the circular-mil area required is 52,000 cmil?
4. What is the diameter in mils of the smallest rubber-covered wire that should be used to carry 22 A?
5. What size of varnished-cambric-insulated wire is needed to supply a 5-hp motor operating at an efficiency of 90% from a 220-V line?
6. What size of rubber-covered wire is needed to supply a bank of ten 200-W lamps and two ½-hp motors connected in parallel across a 110-V line?
7. Find the size of rubber-covered wire needed to supply two 8-Ω heaters in parallel across 122 V.
8. What is the smallest gage of rubber-covered wire that can be used to carry a current of 32 A if the allowable voltage drop is 6 V/1000 ft of wire?
9. A 50-A 112-V load is located 200 ft from a switchboard delivering 115 V. Find the smallest varnished-cambric-covered wire that can be safely used in this two-wire system.

10. A lamp bank draws 60 A from a 230-V source 500 ft away. Assuming a maximum voltage drop of 2%, what is the smallest rubber-covered wire that can be used.

11. A 5-hp motor operates at an efficiency of 87% at a distance of 50 ft from a 232-V source. Find the smallest size of rubber-covered wire that can be safely used.

12. The 1¼-hp motor of an industrial floor-scraping machine operates at an efficiency of 87% from a 220-V line. If the line drop in the No. 14 gage wire used must not exceed 5% of the line voltage, find the maximum radius of operation of the scraper.

(See CD-ROM for Test 13-1)

1. Two No. 10 type RH copper conductors feed a 120-V motor that draws 22 A of current. What will be the measured voltage at the motor when it is running if the motor is 250 ft from the panel? If the motor is disconnected, and the voltage is measured at the same point, what will be the measured voltage?

2. What size of conductor is needed to feed power to a 220-V single-phase, 25 hp motor located near the panel?

3. A 100-ft light pole at a baseball diamond supports four banks of lights each of which draws 5000 W of power. If the pole is located 400 ft from the nearest 120-V source, what is the minimum size of wire that can be used to limit voltage drop to the lights to ≤ 3%?

INTERNET
ACTIVITIES

Internet Activity A

Compare the ampacities of the Alpha wire products listed at the site given here with those in Table 13-1. Compare them in terms of capacity and temperature capabilities. **http://www.alphawire.com/tech_368.htm**

Internet Activity B

Use the site listed here to compare wire sizes used in Europe and in the United States. **http://www.canford.co.uk/metawg.htm**

Chapter 13 Solutions to Self-Tests and Reviews

Self-Test 13-6

1. 35
2. 0.999
3. 4
4. 35
5. 34.97
6. more
7. larger
8. smaller

Self-Test 13-12

1. 232, 0.05, 11.6
2. V_l, 11.6
3. 0.90, 7500, 8333
4. 220.4, 37.8
5. $2KIL$, 11.6, 6778
6. larger
   a. larger
   b. 10, 1, 55, 7

14

1. Use trigonometric functions to solve problems.
2. Find acute angles of a right triangle.
3. Find the sides of a right triangle

Trigonometric Functions of a Right Triangle

Trigonometry is the study of the relationships that exist among the sides and angles of a triangle. These relationships will form a basic mathematical tool used in the solution of ac electrical problems.

Angles.   An angle is a figure formed when two lines meet at a point. In Fig. 14-1, the lines *OA* and *OC* are the sides of the angle. They meet at the point *O*, which is the vertex of the angle.

Naming angles.   The angle shown in Fig. 14-1 may be named in three ways. (1) Use three capital letters, setting them down in order from one end of the angle to the other. Thus the angle may be named $\angle AOC$ or $\angle COA$. (2) Use only the capital letter at the vertex of the angle, as $\angle O$. (3) Use a small letter or number inside the angle, as $\angle 1$.

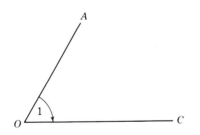

Figure 14-1   Naming an angle. Angle AOC or angle O or angle 1.

Measuring angles.   When a straight line *turns* about a point from one position to another, an angle is formed. When the minute hand of a clock turns from the 1 to the 3, the hand has turned through an angle approximately equal to $\angle AOC$. The size of an angle is measured only by the *amount of rotation of a line about a fixed point*. Suppose we place a small watch on the face of a large clock, as shown in Fig. 14-2 on page 478. When the hand of the large clock turns from 1 to 3, it forms the angle *AOC*. But in this *same time*, the hands of the small watch have formed the angle *BOD*, which is exactly equal to $\angle AOC$. Thus we see that the lengths of the sides have no effect on the size of the angle. The size of an angle depends *only* on the amount of rotation between the sides of the angle. The end of a line that makes one complete turn will describe a circle. In Fig. 14-3, a circle is divided into

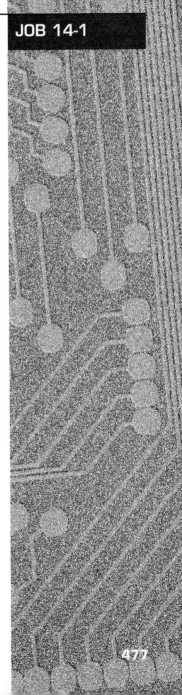

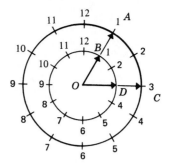

Figure 14-2   The angle formed by the rotation of the minute hand from 1 to 3 o'clock is the same on any clock face. ∠*BOD* = ∠*AOC*.

360 parts. The angle formed by a rotation through ¹⁄₃₆₀ part of the circle is called 1 *degree* (1°).

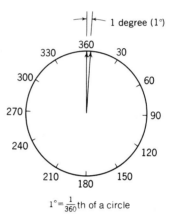

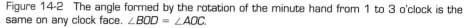

Figure 14-3   An angle is measured by the amount of rotation of a line about a fixed point.

A *right angle* is an angle formed by a rotation through one-fourth of a circle, as ∠*ABC* in Fig. 14-4. Since a complete circle contains 360°, a right angle contains ¼ × 360, or 90°.

An *acute angle* is an angle containing *less* than 90°, as ∠*AOC* in Fig. 14-1.

Triangles.   A triangle is a closed plane figure with three sides. Triangles are named by naming the three vertices of the triangle using capital letters. The letters are then given in order around the triangle. Thus in Fig. 14-4, the triangle is named △*ACB* or △*CBA* or △*BAC* or △*ABC* or △*BCA* or △*CAB*.

A *right triangle* is a triangle that contains a right angle, as △*ABC* in Fig. 14-4. The little square at ∠*B* is used to indicate a right angle.

Naming the sides of a right triangle.   In trigonometry, the sides of a right triangle are named depending on which of the *acute* angles are used. In Fig. 14-4, For angle *A:*

1. The side opposite the right angle is called the *hypotenuse* (*AC*).
2. The side opposite angle *A* is called the *opposite side* (*BC*).
3. The side of angle *A* that is *not* the hypotenuse is called the *adjacent side* (*AB*).

For angle *C:*

1. The side opposite the right angle is called the *hypotenuse* (*AC*).
2. The side opposite angle *C* is called the *opposite side* (*AB*).
3. The side of angle *C* that is *not* the hypotenuse is called the *adjacent side* (*BC*).

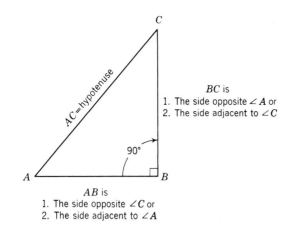

Figure 14-4 Naming the sides of a right triangle.

BC is
1. The side opposite ∠*A* or
2. The side adjacent to ∠*C*

AB is
1. The side opposite ∠*C* or
2. The side adjacent to ∠*A*

The Trigonometric Formulas

The tangent of an angle. In addition to degrees, the size of an angle may also be described in terms of the lengths of the sides of a right triangle formed from the angle. Refer to Fig. 14-5, in which each space represents one unit of length. From various points on one side of angle *A*, lines have been drawn making right angles with the other side. These lines have formed the right triangles *ABE, ACF,* and *ADG*. Notice that all these triangles contain the same angle *A*.

	In △*ABE*	In △*ACF*	In △*ADG*
Hypotenuse =	*AB* = 5	*AC* = 10	*AD* = 12.5
Side opposite ∠*A* =	*BE* = 3	*CF* = 6	*DG* = 7.5
Side adjacent ∠*A* =	*AE* = 4	*AF* = 8	*AG* = 10

In each triangle, let us divide the side *opposite* ∠*A* by the side *adjacent* to ∠*A*.

In △*ABE*:
$$\frac{\text{Side opposite} \angle A}{\text{Side adjacent} \angle A} = \frac{BE}{AE} = \frac{3}{4} = 0.75$$

In △*ACF*:
$$\frac{\text{Side opposite} \angle A}{\text{Side adjacent} \angle A} = \frac{CF}{AF} = \frac{6}{8} = 0.75$$

In △*ADG*:
$$\frac{\text{Side opposite} \angle A}{\text{Side adjacent} \angle A} = \frac{DG}{AG} = \frac{7.5}{10} = 0.75$$

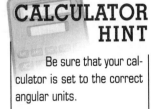

CALCULATOR HINT

Be sure that your calculator is set to the correct angular units.

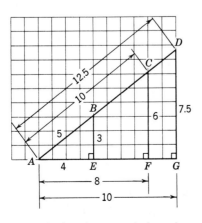

Figure 14-5 Diagram used to obtain the trigonometric formulas.

Notice that this particular division of the opposite side by the adjacent side always results in the *same* answer *regardless* of the size of the triangle. This is true because *all* the triangles contain the same angle *A*. The ratio of these sides remains constant because they all describe the same angle *A*. This constant number describes the size of angle *A* and is called the *tangent of angle A*. Therefore, if the tangent of an *unknown* angle were calculated to be 0.75, then the angle would be equal to angle *A*, or about 37°. If the angle changes, then the number for the tangent of the angle will also change. However, the tangent of every angle is a specific number that never changes. These numbers are found in Table 14-1.

RULE

The tangent of an angle = $\dfrac{\text{side opposite the angle}}{\text{side adjacent to the angle}}$

$$\tan \angle = \frac{o}{a}$$
Formula 14-1

The sine of an angle.   In each triangle shown in Fig. 14-5, divide the side *opposite* angle *A* by the side called the *hypotenuse*.

In $\triangle ABE$: $\quad \dfrac{\text{Side opposite } \angle A}{\text{Hypotenuse}} = \dfrac{BE}{AB} = \dfrac{3}{5} = 0.6$

In $\triangle ACF$: $\quad \dfrac{\text{Side opposite } \angle A}{\text{Hypotenuse}} = \dfrac{CF}{AC} = \dfrac{6}{10} = 0.6$

In $\triangle ADG$: $\quad \dfrac{\text{Side opposite } \angle A}{\text{Hypotenuse}} = \dfrac{DG}{AD} = \dfrac{7.5}{12.5} = 0.6$

Once again, the resulting quotients are fixed regardless of the size of the triangles. This number is called the *sine of angle A* and is *another* way to describe the size of the angle. If the angle changes, then the number for the sine of the angle will also change. However, the sine of every angle is a specific number which never changes. These numbers are also found in Table 14-1.

RULE

The sine of an angle = $\dfrac{\text{side opposite the angle}}{\text{hypotenuse}}$

$$\sin \angle = \frac{o}{h}$$
Formula 14-2

The cosine of an angle.   In each triangle shown in Fig. 14-5, divide the side *adjacent* to angle *A* by the *hypotenuse*.

In $\triangle ABE$: $\quad \dfrac{\text{Side adjacent } \angle A}{\text{Hypotenuse}} = \dfrac{AE}{AB} = \dfrac{4}{5} = 0.8$

In $\triangle ACF$: $\quad \dfrac{\text{Side adjacent } \angle A}{\text{Hypotenuse}} = \dfrac{AF}{AC} = \dfrac{8}{10} = 0.8$

In $\triangle ADG$: $\quad \dfrac{\text{Side adjacent } \angle A}{\text{Hypotenuse}} = \dfrac{AG}{AD} = \dfrac{10}{12.5} = 0.8$

Table 14-1

Values of the Trigonometric Functions

Angle	Sin	Cos	Tan	Angle	Sin	Cos	Tan
0°	0.0000	1.0000	0.0000	46°	0.7193	0.6947	1.0355
1°	0.0175	0.9998	0.0175	47°	0.7314	0.6820	1.0724
2°	0.0349	0.9994	0.0349	48°	0.7431	0.6691	1.1106
3°	0.0523	0.9986	0.0524	49°	0.7547	0.6561	1.1504
4°	0.0698	0.9976	0.0699	50°	0.7660	0.6428	1.1918
5°	0.0872	0.9962	0.0875	51°	0.7771	0.6293	1.2349
6°	0.1045	0.9945	0.1051	52°	0.7880	0.6157	1.2799
7°	0.1219	0.9925	0.1228	53°	0.7986	0.6018	1.3270
8°	0.1392	0.9903	0.1405	54°	0.8090	0.5878	1.3764
9°	0.1564	0.9877	0.1584	55°	0.8192	0.5736	1.4281
10°	0.1736	0.9848	0.1763	56°	0.8290	0.5592	1.4826
11°	0.1908	0.9816	0.1944	57°	0.8387	0.5446	1.5399
12°	0.2079	0.9781	0.2126	58°	0.8480	0.5299	1.6003
13°	0.2250	0.9744	0.2309	59°	0.8572	0.5150	1.6643
14°	0.2419	0.9703	0.2493	60°	0.8660	0.5000	1.7321
15°	0.2588	0.9659	0.2679	61°	0.8746	0.4848	1.8040
16°	0.2756	0.9613	0.2867	62°	0.8829	0.4695	1.8807
17°	0.2924	0.9563	0.3057	63°	0.8910	0.4540	1.9626
18°	0.3090	0.9511	0.3249	64°	0.8988	0.4384	2.0503
19°	0.3256	0.9455	0.3443	65°	0.9063	0.4226	2.1445
20°	0.3420	0.9397	0.3640	66°	0.9135	0.4067	2.2460
21°	0.3584	0.9336	0.3839	67°	0.9205	0.3907	2.3559
22°	0.3746	0.9272	0.4040	68°	0.9272	0.3746	2.4751
23°	0.3907	0.9205	0.4245	69°	0.9336	0.3584	2.6051
24°	0.4067	0.9135	0.4452	70°	0.9397	0.3420	2.7475
25°	0.4226	0.9063	0.4663	71°	0.9455	0.3256	2.9042
26°	0.4384	0.8988	0.4877	72°	0.9511	0.3090	3.0777
27°	0.4540	0.8910	0.5095	73°	0.9563	0.2924	3.2709
28°	0.4695	0.8829	0.5317	74°	0.9613	0.2756	3.4874
29°	0.4848	0.8746	0.5543	75°	0.9659	0.2588	3.7321
30°	0.5000	0.8660	0.5774	76°	0.9703	0.2419	4.0108
31°	0.5150	0.8572	0.6009	77°	0.9744	0.2250	4.3315
32°	0.5299	0.8480	0.6249	78°	0.9781	0.2079	4.7046
33°	0.5446	0.8387	0.6494	79°	0.9816	0.1908	5.1446
34°	0.5592	0.8290	0.6745	80°	0.9848	0.1736	5.6713
35°	0.5736	0.8192	0.7002	81°	0.9877	0.1564	6.3138
36°	0.5878	0.8090	0.7265	82°	0.9903	0.1392	7.1154
37°	0.6018	0.7986	0.7536	83°	0.9925	0.1219	8.1443
38°	0.6157	0.7880	0.7813	84°	0.9945	0.1045	9.5144
39°	0.6293	0.7771	0.8098	85°	0.9962	0.0872	11.4300
40°	0.6428	0.7660	0.8391	86°	0.9976	0.0698	14.3010
41°	0.6561	0.7547	0.8693	87°	0.9986	0.0523	19.0810
42°	0.6691	0.7431	0.9004	88°	0.9994	0.0349	28.6360
43°	0.6820	0.7314	0.9325	89°	0.9998	0.0175	57.2900
44°	0.6947	0.7193	0.9657	90°	1.0000	0.0000	
45°	0.7071	0.7071	1.0000				

Here, too, the resulting quotients are fixed regardless of the size of the triangles. This number is called the *cosine of angle A* and is a third way to describe the size of the angle. If the angle changes, then the number for the cosine of the angle will also change. However, the cosine of every angle is a specific number that never changes. These numbers are found in Table 14-1 on page 481.

RULE

The cosine of an angle $= \dfrac{\text{side adjacent to the angle}}{\text{hypotenuse}}$

$$\cos \angle = \frac{a}{h}$$

Formula 14-3

Example 14-1

Find the sine, cosine, and tangent of $\angle A$ in the triangle shown in Fig. 14-6.

Figure diagram: triangle with vertices A, B, C. AC = 13 in, BC = 5 in, AB = 12 in, right angle at B.

Figure 14-6

Solution

1. Name the sides using $\angle A$ as the reference angle.

$$\text{Hypotenuse} = AC = 13 \text{ in}$$
$$\text{Opposite side} = BC = 5 \text{ in}$$
$$\text{Adjacent side} = AB = 12 \text{ in}$$

2. Find the values of the three functions.

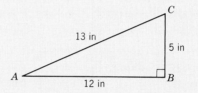

$$\sin \angle A = \frac{o}{h} \qquad \cos \angle A = \frac{a}{h} \qquad \tan \angle A = \frac{o}{a}$$

$$\sin \angle A = \frac{5}{13} \qquad \cos \angle A = \frac{12}{13} \qquad \tan \angle A = \frac{5}{12}$$

$$\sin \angle A = 0.384 \qquad \cos \angle A = 0.923 \qquad \tan \angle A = 0.417 \qquad Ans.$$

Example 14-2

Find the sine, cosine, and tangent of $\angle B$ in the triangle shown in Fig. 14-7.

Solution

1. Name the sides, using $\angle B$ as the reference angle.

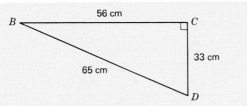

Figure 14-7

$$\text{Hypotenuse} = BD = 65 \text{ cm}$$
$$\text{Opposite side} = CD = 33 \text{ cm}$$
$$\text{Adjacent side} = BC = 56 \text{ cm}$$

2. Find the values of the three functions.

$$\sin \angle B = \frac{o}{h} \qquad \cos \angle B = \frac{a}{h} \qquad \tan \angle B = \frac{o}{a}$$

$$\sin \angle B = \frac{33}{65} \qquad \cos \angle B = \frac{56}{65} \qquad \tan \angle B = \frac{33}{56}$$

$$\sin \angle B = 0.5077 \qquad \cos \angle B = 0.8615 \qquad \tan \angle B = 0.5893 \quad \textit{Ans.}$$

Exercises

Practice

Calculate the sine, cosine, and tangent of the angles named in each triangle in Fig. 14-8.

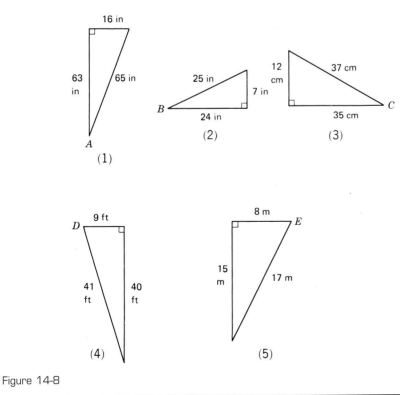

Hint for Success

Looking up trigonometric functions in a table is the equivalent of the inverse trigonometric function.

Figure 14-8

Using the Table of Trigonometric Functions

Since the values of the trigonometric functions of an angle never change, it is possible to set these values down in a table such as Table 14-1. (Functions of angles for decimal degrees may be found in the appendix.)

Example 14-3

Find the value of sin 50°.

Solution

Follow down the column marked "Angle" in Table 14-1 until you reach 50°. Read across to the right to find the number 0.7660 in the sine column. Therefore,

$$\sin 50° = 0.7660 \quad \textit{Ans.}$$

Example 14-4

Find the value of tan 30°.

Solution

Follow down the column marked "Angle" until you get to 30°. Read across to the right to find the number 0.5774 in the tangent column. Therefore,

$$\tan 30° = 0.5774 \quad \textit{Ans.}$$

Finding the Angle When the Function Is Given

Example 14-5

Find $\angle A$ if $\cos A = 0.7314$.

Solution

Follow down the column marked "Cos" until you find the number 0.7314. Read across to the left to find 43° in the column marked "Angle." Therefore,

$$\angle A = 43° \quad \textit{Ans.}$$

CALCULATOR HINT

Try using inverse trigonometric functions to make your computations easier.

Example 14-6

Find $\angle B$ if $\tan B = 1.8807$.

Solution

Follow down the column marked "Tan" until you find the number 1.8807. Read across to the left to find 62° in the column marked "Angle." Therefore,

$$\angle B = 62° \quad \textit{Ans.}$$

Practice

Find the number of degrees in each angle.

1. tan A = 0.3249
2. sin B = 0.6428
3. cos C = 0.1736
4. cos D = 0.9877
5. sin E = 0.5000
6. tan F = 1.0724
7. tan G = 0.5774
8. cos H = 0.7071
9. sin J = 0.8660
10. cos K = 0.9397

Finding the Angle When the Function Is Not in the Table

Example 14-7

Find $\angle A$ if tan A = 0.5120.

Solution

The number 0.5120 is not in the table under the column "Tan" but lies between 0.5095 (27°) and 0.5317 (28°). Choose the number closest to 0.5120.

tan 27 = 0.5095

tan A = 0.5120   difference = 0.0025

tan 28 = 0.5317   difference = 0.0197

The smaller difference indicates the closer number. Therefore,

$\angle A = 27°$ *Ans.*

Exercises

Practice

Find the number of degrees in each angle correct to the nearest degree.

1. tan A = 0.2700
2. cos B = 0.7500
3. sin C = 0.8500
4. sin D = 0.2350
5. cos E = 0.4172
6. tan F = 1.9120
7. tan G = 0.2783
8. cos H = 0.1645
9. sin J = 0.7250
10. sin K = 0.9645

JOB 14-3

Finding the Acute Angles of a Right Triangle

In order to find the value of an angle in any problem, it is only necessary to find the number for the sine *or* the cosine *or* the tangent of the angle. If we know *any one* of these values, we can determine the angle by finding the number in the appropriate column of the table.

Example 14-8

Find $\angle A$ and $\angle B$ in the triangle of Fig. 14-9.

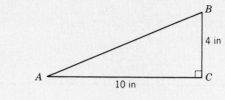

Figure 14-9

Hint for Success

Trigonometric functions are just names for the ratios of the sides of a right triangle.

CALCULATOR HINT

For calculators that use algebraic notation for problem entry, remember to use parentheses for calculated trigonometric arguments. This ensures that the argument of the function will be determined before the trigonometric function is evaluated.

Solution

1. Name the sides that have values, using $\angle A$ as the reference angle.

   4 in = the side *opposite* $\angle A$
   10 in = the side *adjacent* $\angle A$

2. We can find $\angle A$ if we can find *any* of the three functions of the angle. However, the only function that can *actually* be found is the one for which we have *known* values. In this problem, the only known values are those for the opposite and adjacent sides. The only formula that uses these particular sides is the *tangent* formula.

$$\tan A = \frac{o}{a} = \frac{4}{10} = 0.4000 \qquad (14\text{-}1)$$

3. Find the number in the tangent table closest to 0.4000.

   $\angle A = 22°$ (nearest angle)   *Ans.*

4. Since the two acute angles of any right triangle always total 90°,

$$\angle B = 90° - \angle A \qquad 14\text{-}4$$
$$\angle B = 90° - 22° = 68° \qquad Ans.$$

Example 14-9

The phase angle in an ac circuit may be represented by $\angle\theta$ (angle theta), as shown in Fig. 14-10. Find $\angle\theta$ and $\angle B$.

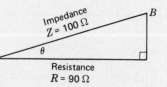

Figure 14-10 The angle theta (θ) is the phase angle in an inductive ac circuit.

Solution

1. Name the sides that have values, using $\angle \theta$ as the reference angle.

$$100 \, \Omega = \text{hypotenuse}$$
$$90 \, \Omega = \text{side adjacent } \angle \theta$$

2. Choose the trigonometric formula that uses the adjacent side and the hypotenuse.

$$\cos \theta = \frac{a}{h} = \frac{90}{100} = 0.9000 \qquad (14\text{-}3)$$

3. Find the number in the cosine table closest to 0.9000.

$$\angle \theta = 26° \text{ (nearest angle)} \qquad Ans.$$

4. Since the two acute angles of a right triangle always total 90°,

$$\angle B = 90° - \angle \theta \qquad (14\text{-}4)$$
$$\angle B = 90° - 26° = 64° \qquad Ans.$$

Self-Test 14-10

An electrician must bend a pipe to make a 3½-ft rise in a 5-ft horizontal distance. What is the angle at each bend?

Solution

The diagram describing the conditions is shown in Fig. 14-11.

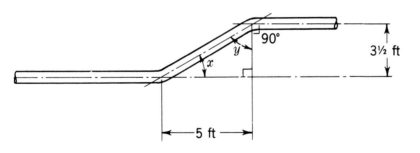

Figure 14-11

1. Name the sides that have values, using $\angle x$ as the reference angle.

Side opposite $\angle x$ = _____ ft
Side adjacent $\angle x$ = _____ ft

2. Choose the trigonometric formula that uses the opposite side and the adjacent side. The only formula that uses these two sides is the _____.

$$\tan x = \frac{?}{?} \qquad (14\text{-}1)$$

$$\tan x = \frac{3.5}{5} = \underline{\hspace{2cm}}$$

3. Find the number in the _____ table that is closest to 0.7000.

$$\angle x = \underline{\hspace{2cm}}°\quad Ans.$$

4. Since the two acute angles of a right triangle always total _____°,

$$\angle y = 90° - \underline{\hspace{2cm}}$$

(14-4)

$$\angle y = 90° - 35° = \underline{\hspace{2cm}}$$

5. The angle at the top bend $= \angle y +$ _____, or

$$\text{Angle at top bend} = 55° + 90°$$

Exercises

Practice

1. Find $\angle A$ and $\angle B$ in the right triangle shown in Fig. 14-12.

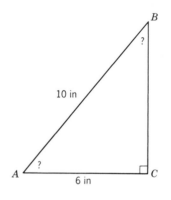

Figure 14-12

Using Fig. 14-12, find $\angle A$ and $\angle B$ if

2. $AC = 100$ ft and $BC = 70$ ft
3. $BC = 4$ in and $AB = 8$ in
4. $BC = 40\ \Omega$ and $AC = 25\ \Omega$
5. $AC = 200\ \Omega$ and $AB = 350\ \Omega$
6. $BC = 300$ W and $AB = 1000$ W
7. $AC = 600\ \Omega$ and $AB = 960\ \Omega$
8. $BC = 7.5$ V and $AC = 12.5$ V
9. $BC = 12.4$ A and $AB = 67.8$ A
10. $AC = 62.5\ \Omega$ and $BC = 100\ \Omega$

Applications

11. At what angles must a pipe be bent in order to make a 7-ft rise in a 3-ft horizontal distance?
12. At what angles must a pipe be bent in order to make a 4-ft 6-in rise in a 2-ft 3-in horizontal distance?
13. A car rises 50 ft while traveling along a road 1000 ft long. At what angle is the road inclined to the horizontal?

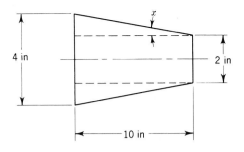

Figure 14-13 Angle *x* is equal to one-half the taper angle.

14. Find angle *x* in the taper shown in Fig. 14-13.
15. A guy wire 120 ft long reaches from the top of a pole to a point 64 ft from the foot of the pole. What angle does the wire make with the ground?

(See CD-ROM for Test 14-1)

JOB 14-4

Finding the Sides of a Right Triangle

Example 14-11

Find (*a*) side *BC* and (*b*) $\angle B$ of the right triangle shown in Fig. 14-14.

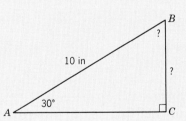

Figure 14-14

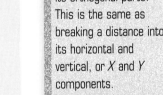

Hint for Success

Sine and cosine functions break a phasor into its orthogonal parts. This is the same as breaking a distance into its horizontal and vertical, or *X* and *Y* components.

Solution

  a. Find the side *BC*.
  1. Select the angle to be used ($\angle A = 30°$).
  2. Select the side to be found (*BC*).
  3. Select one other side whose value is known (*AB* = 10 in).
  4. Name these two sides.

$$BC = \text{side } opposite\ \angle A$$
$$10 \text{ in} = AB = hypotenuse$$

  5. Select the correct trigonometric formula. It will be the formula that uses these two sides—the *opposite* and the *hypotenuse*. Only the *sine* formula uses the *opposite* side and the *hypotenuse*.

$$\sin A = \frac{o}{h} \qquad (14\text{-}2)$$

$$\sin 30° = \frac{BC}{10}$$

From Table 14-1, sin 30° = 0.5000. Therefore,

$$\frac{0.5000}{1} = \frac{BC}{10}$$

$$BC = 0.5000 \times 10 = 5 \text{ in} \qquad Ans.$$

b. Find $\angle B$.

$$\angle B = 90° - \angle A \qquad (14\text{-}4)$$

$$\angle B = 90° - 30° = 60° \qquad Ans.$$

Example 14-12

Find side *BC* in the right triangle shown in Fig. 14-15.

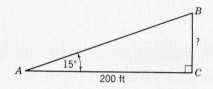

Figure 14-15

Solution

Find the side *BC*.

1. Select the angle to be used ($\angle A = 15°$).
2. Select the side to be found (*BC*).
3. Select one other side whose value is known ($AC = 200$ ft).
4. Name these two sides.

$$BC = \text{the side } opposite \ \angle A$$
$$200 \text{ ft} = AC = \text{the side } adjacent \ \angle A$$

5. Select the correct trigonometric formula. It will be the formula that uses these two sides—the *opposite* and the *adjacent*. Only the tangent formula uses the *opposite* side and the *adjacent* side.

$$\tan A = \frac{o}{a} \qquad (14\text{-}1)$$

$$\tan 15° = \frac{BC}{200}$$

From Table 14-1, tan 15° = 0.2679. Therefore,

$$\frac{0.2679}{1} = \frac{BC}{200}$$

$$BC = 0.2679 \times 200 = 53.58 \text{ ft} \qquad Ans.$$

Example 14-13

The relationship among the impedance Z, the resistance R, and the capacitive reactance X_C of an ac circuit is shown in Fig. 14-16. Find the impedance.

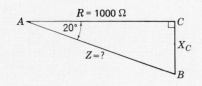

Figure 14-16

Solution

Find the impedance Z represented by side AB.

1. Select the angle to be used ($\angle A = 20°$).
2. Select the side to be found (AB).
3. Select one other side whose value is known ($AC = 1000\ \Omega$).
4. Name these two sides.

$$AB = \textit{hypotenuse}$$
$$1000\ \Omega = AC = \text{side } \textit{adjacent } \angle A$$

5. Select the correct trigonometric formula. It will be the formula that uses these two sides—the *adjacent* and the *hypotenuse*. Only the cosine formula uses the *adjacent* side and the *hypotenuse*.

$$\cos A = \frac{a}{h} \qquad\qquad (14\text{-}3)$$

$$\cos 20° = \frac{1000}{AB}$$

From Table 14-1, $\cos 20° = 0.9397$. Therefore,

$$\frac{0.9397}{1} = \frac{1000}{AB}$$
$$AB \times 0.9397 = 1000$$
$$AB = \frac{1000}{0.9397} = 1064\ \Omega$$

Impedance $Z = 1064\ \Omega$ *Ans.*

> **CALCULATOR HINT**
>
> Trigonometric operations using the tangent can be done by changing the display mode between rectangular and polar.

Self-Test 14-14

As shall be discussed in Job 18-1, the current and voltage of an ac circuit do not always appear at the same time. They are said to be "out of phase." In Fig. 14-17a, the current "leads" the voltage by a phase angle of 37°. In order to cope with this situation, the current "phasor" $I = 20$ A is resolved into its "in-phase" component I_x and its "reactive" component I_y, as shown in Fig. 14-17b. Find I_x and I_y. A further discussion of the resolution of phasors may be found in Job 19-6.

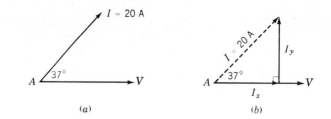

(a)   (b)

Figure 14-17 (a) The current I "leads" the voltage V by 37°. (b) I is resolved into its "in-phase" component I_x and its "reactive" component I_y.

Solution

Find I_x. In Fig. 14-17b,
1. The angle to be used = _____.
2. The side to be found = _____.
3. Another side whose value is known is I = _____.
4. Name these two sides using $\angle A = 37°$ as the reference angle.

$$I_x = \text{the side} \underline{\hspace{1.5cm}} \angle A$$
$$I = 20 = \text{the} \underline{\hspace{1.5cm}}$$

5. Select the correct trigonometric formula that uses these two particular sides. The adjacent side and the hypotenuse are connected by the _____ formula.

$$\cos \angle A = \frac{a}{h} \qquad\qquad (14\text{-}2)$$

$$\cos 37° = \frac{?}{20}$$

From Table 14-1, $\cos 37° = $ _____. Therefore,

$$\frac{0.7986}{1} = \frac{I_x}{20}$$

and
$$I_x = 0.7986 \times 20 = \underline{\hspace{1.5cm}} \text{ A} \qquad Ans.$$

Find I_y.
1. Using $\angle A = 37°$, the sides to be used are I_y and _____.
2. Using $\angle A$ as the reference angle, these sides are named as follows:

$$I_y = \text{the side} \underline{\hspace{1.5cm}} \angle A$$
$$I = 20 = \text{the} \underline{\hspace{1.5cm}}$$

3. The only formula that uses these two sides is the _____ formula.

$$\sin A = \frac{?}{h}$$

$$\sin 37° = \frac{I_y}{?}$$

From Table 14-1, $\sin 37° = $ _____. Therefore,

$$\frac{0.6018}{1} = \frac{I_y}{20}$$

and
$$I_y = 0.6018 \times 20 = \underline{\hspace{1.5cm}} \text{ A} \qquad Ans.$$

Practice

Use the triangle shown in Fig. 14-14 for the following exercises.

1. Find AC and $\angle B$ if $AB = 20$ in and $\angle A = 60°$.
2. Find BC and $\angle B$ if $AB = 26$ ft and $\angle A = 40°$.
3. Find AC and $\angle A$ if $BC = 75$ ft and $\angle B = 50°$.
4. Find AC and $\angle B$ if $AB = 400$ W and $\angle A = 28°$.
5. Find AB and $\angle B$ if $BC = 75.5$ Ω and $\angle A = 30°$.
6. Find AB and $\angle B$ if $AC = 30$ Ω and $\angle A = 25°$.
7. Find BC and $\angle A$ if $AC = 500$ ft and $\angle B = 28°$.
8. Find AC and $\angle A$ if $AB = 36.8$ in and $\angle B = 35°$.
9. Find BC and $\angle B$ if $AB = 475$ Ω and $\angle A = 15°$.
10. Find AC and $\angle A$ if $AB = 92.8$ V and $\angle B = 15°$.

Applications

11. A guy wire reaches from the top of an antenna pole to a point 12 m from the foot of the pole and makes an angle of 70° with the ground. How long is the wire? How tall is the pole?
12. Use Fig. 14-13. If the small diameter equals 3 in, the length equals 12 in, and angle x equals 4°, find the large diameter.
13. Four holes are drilled evenly spaced on the edge of a 12-cm-diameter circle. Find the distance between the centers of the holes.
14. How long is each side of the largest square bar that can be made from a piece of 6-cm-diameter round stock?
15. A ladder 10 m long leans against the side of a building and makes an angle of 65° with the ground. How high up the building does it reach?
16. In an ac circuit, the current $I = 10$ A leads the voltage by 53°. Resolve the current into its in-phase current I_x and its reactive current I_y, and calculate their values.
17. In an ac circuit, the current $I = 14$ A leads the voltage by 25°. Resolve the current into its in-phase current I_x and its reactive current I_y, and calculate their values.
18. In an impedance triangle similar to that shown in Fig. 14-16, find the reactance X_C if $R = 2000$ Ω and angle $A = 20°$.
19. Find the impedance Z in the triangle used for Prob. 18.
20. In Fig. 14-18 on page 494, find dimensions a, b, c, d, and e.

Review of Trigonometry

Formulas

1. Tangent of an angle $= \dfrac{?\ \text{side}}{?\ \text{side}}$.

$$\tan \angle = \frac{o}{a} \qquad \text{14-1}$$

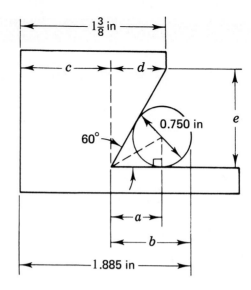

Figure 14-18

2. Sine of an angle $= \dfrac{?\quad \text{side}}{?}$.

$$\sin \angle = \dfrac{o}{h} \qquad \text{14-2}$$

3. Cosine of an angle $= \dfrac{?\quad \text{side}}{?}$.

$$\cos \angle = \dfrac{a}{h} \qquad \text{14-3}$$

4. The sum of the two acute angles of a right triangle equals _____.

$$\angle A + \angle B = 90° \qquad \text{14-4}$$

Procedure for Finding Angles in a Right Triangle

1. Name the sides which have values, using the angle to be found as the reference angle.
2. Choose the trigonometric formula that uses these sides.
3. Substitute values, and divide.
4. Find the number in the table closest to this quotient. Be certain to look in the column indicated by step 2.
5. Find the angle in the angle column corresponding to this number.
6. Use formula (14-4) to find the other acute angle.

Procedure for Finding Sides in a Right Triangle

1. Select the angle to be used.
2. Select the side to be found.
3. Select one other side whose value is known.
4. Name these two sides.
5. Select the correct trigonometric formula that uses these sides.
6. Substitute values using Table 14-1.
7. Solve the equation.

Exercises

Practice

Find the values of the following functions.

1. sin 36°   2. cos 78°   3. tan 69°   4. sin 52°   5. cos 25°

Find angle A, correct to the nearest degree.

6. sin A = 0.5878   7. cos A = 0.4226   8. tan A = 5.7500
9. cos A = 0.9800   10. sin A = 0.7200   11. tan A = 0.3113
12. tan A = 0.1340   13. cos A = 0.2868   14. sin A = 0.7240

Use the triangle shown in Fig. 14-14 for the following exercises:

15. Find $\angle A$ if BC = 30 and AC = 40.
16. Find $\angle B$ if AC = 50 and AB = 100.
17. Find $\angle B$ if BC = 25 and AB = 75.
18. Find $\angle A$ if BC = 16 and AB = 65.
19. Find $\angle B$ if AC = 22.5 and BC = 14.
20. Find $\angle A$ if AC = 4.5 and AB = 20.5.
21. Find BC if AB = 2200 W and $\angle A$ = 25°.
22. Find AC if AB = 600 W and $\angle A$ = 8°.
23. Find AB if BC = 28.6 and $\angle B$ = 42°.
24. Find AB if AC = 9.3 Ω and $\angle A$ = 34°.
25. Find BC if AC = 750 Ω and $\angle A$ = 48°.
26. Find AC if BC = 17.6 ft and $\angle B$ = 60°.
27. Find AC if AB = 4000 W and $\angle B$ = 45°.
28. Find BC if AB = 2500 W and $\angle A$ = 45°.
29. Find AB if BC = 142 V and $\angle A$ = 37°.
30. Find AC if AB = 85.8 Ω and $\angle A$ = 83°.

Applications

31. A pipe must be bent to provide a 6-ft rise in a 1-ft horizontal distance. Find the angle at each bend.
32. The foot of a ladder 30 ft long rests on the ground 12 ft from the side of a building. What angle does the ladder make with the ground?
33. In a taper similar to that shown in Fig. 14-13, the large diameter equals 0.75 in and the small diameter equals 0.47 in. Find angle x if the length equals 2.8 in.
34. How long is a ladder that reaches 8 m up the side of a building if the angle between the ladder and the ground is 65°?
35. Use Fig. 14-13. If the small diameter equals 1.25 in, the length equals 2.25 in, and angle x equals 6°, find the large diameter.
36. How many inches of wire are needed to wind a single-layer coil of 20 turns around a core whose cross section is a square with a diagonal equal to 1.414 in?
37. A rectangle measures 30 by 70 cm. Find the angle that the diagonal makes with the longer side.
38. Find the length of the diagonal in Exercise 37.
39. In an impedance triangle similar to that shown in Fig. 14-16, find the imped-ance Z if R = 150 Ω and angle A = 50°.
40. In an inductive ac circuit, the current I = 15 A leads the voltage by 30°. Resolve the current into its in-phase current I_x and its reactive current I_y, and calculate their values.

(See CD-ROM for Test 14-2)

Assessment

1. A fire service ladder is leaning against a building. If it is 7-ft from the building at the base and reaches 28 ft up the building, what is the inside angle at the base in relation to the ground?

2. A "golden triangle" is a right triangle that has a "sides ratio" of 3:4:5. For example, an angle that is 30 ft on one side and 40 ft on the other side and has a hypotenuse length of 50 ft is considered a golden triangle. What are the sine, cosine, and tangent functions for both of the inside acute angles? (*Note:* The golden triangle is widely used in construction because it is an easy and accurate way to make sure that the sides of a foundation are square (at 90° to each other).

3. A roof has a 2.5:4 rise-to-run ratio. What angle of roof is this in relation to the ground?

4. A 60-ft flag pole casts a 100-ft shadow on level ground. What is the elevation, in degrees, above the horizon of the sun?

INTERNET

ACTIVITIES

Internet Activity A

Electronic synthesizers create music by using sinusoidal functions of different frequencies. Different textures become created when these functions are added together (superimposed). At this Web site, you will see how superimposed functions make interesting music. The site is **http://www.hofstra.edu/~matscw/trig/trigex2.html.** Try out questions 33 and 35.

Internet Activity B

Strengthen and extend your understanding of trigonometry. Find the Greek source of the word *trigonometry*. List traditional and modern uses of trigonometry. **http://www.acts.tinet.ie/trigonometry_645.html**

Chapter 14 Solutions to Self-Tests and Reviews

Self-Test 14-10

1. 3½, 5
2. tangent, *o, a*, 0.7000
3. tangent, 35
4. 90, $\angle x$, 55°
5. 90°, 145

Self-Test 14-14

To find I_x

1. 37°
2. I_x
3. 20
4. adjacent, hypotenuse

5. cosine, I_x, 0.7986, 16

To find I_y

1. 20
2. opposite, hypotenuse
3. sine, *o*, 20, 0.6018, 12

Job 14-5 Review

1. opposite, adjacent
2. opposite, hypotenuse
3. adjacent, hypotenuse
4. 90°

15 chapter

Introduction to AC Electricity

Learning Objectives

1. Determine values from a graph.
2. Draw a graph from a table of data.
3. Determine how to generate an ac voltage with a magnetic field.
4. Find the instantaneous values, maximum values, and phase angles of an ac wave.
5. Determine the effective value of an ac wave.
6. Determine ac values with an oscilloscope.
7. Write vectors and phasors for ac waves.
8. Use the pythagorean theorem.
9. Determine power.
10. Find harmonics for harmonic analysis.

JOB 15-1

Graphs

In Fig. 15-1, the current that flows through the resistor depends on the voltage that is applied across the ends of the resistor. If the voltage changes, then the current will also change. Let us prepare a table of values of different voltages and the resulting currents using Ohm's law and show them in Table 15-1.

Quantities like voltages and currents whose values may change are called *variables*. An *independent variable*, like voltage, is one whose value is changed in order to observe the effect on another *dependent variable*, like the current. Tables of information like Table 15-1 are fairly easy to understand and interpret. A glance at the data tells us that the current will double if the voltage is doubled. However, as the information given in a table becomes more complicated, it becomes more difficult to understand and interpret. The relationships between the two variables

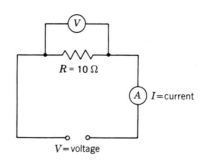

Figure 15-1   Changes in the independent variable (voltage) cause corresponding changes in the dependent variable (current).

Table 15-1

Voltage, V	Resistance, R	Current, I
0	10	0
20	10	2
40	10	4
60	10	6
80	10	8
100	10	10

may be made more evident by presenting the data in the form of a picture, or *graph.*

A graph is a picture that shows the effect of one variable on another. Graphs are used throughout the electronics industry to present information in simple form, to describe the operation of circuits, and to illustrate relationships that cannot be shown easily by data presented in tabular form.

The scale of a graph. Graphs are drawn on paper ruled with uniformly spaced horizontal and vertical lines. Every fifth line or every tenth line may be heavier than the rest in order to make it easier to read the graph. Figure 15-2, which shows the graph of the data in Table 15-1, is drawn on paper ruled 10 boxes to the inch. The

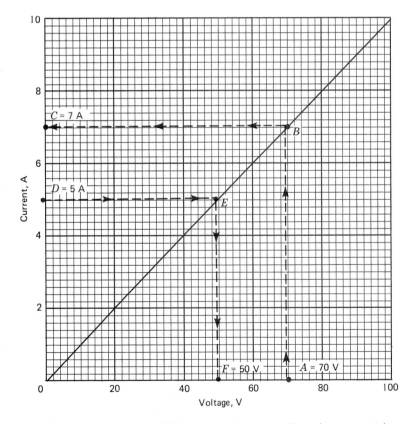

Figure 15-2 The current through a 10-Ω resistor depends on the voltage across it.

two heavy lines at right angles to each other are the *base lines*, or *reference lines*. The horizontal line, or *abscissa*, is generally used to describe the independent variable, and the vertical base line, or *ordinate*, is used for the dependent variable. The value of each box along each base line must be indicated on the graph. This is called the *scale* for that variable. The highest value of the variable, together with the space available for the graph, will determine the scale to be used.

For the voltage or abscissa: If we had 100 boxes, we could make each box equal to 1 V and indicate the scale up to 100 V. However, only 50 boxes are available. Therefore, in order to show 100 V, each box must have a value of 100 ÷ 50, or 2 V for every box. Only every tenth box is numbered in order to keep the graph neat and uncluttered.

For the current or ordinate: The vertical current scale need not be the same as the horizontal scale. Actually, since the highest value of current that must be indicated is only 10 A, we would have a graph only 5 boxes high if we used the same scale. This would cramp the graph and make it difficult to read. Since 50 boxes are available, in order to show 10 A, each box must have a value of 10 ÷ 50 , or 0.2 A for each box. Only every tenth box is numbered.

Reading graphs.   As one variable changes, the corresponding value of the other variable can be read from the graph.

Example 15-1

Using Fig. 15-2, find (*a*) the value of the current when the voltage is 70 V, and (*b*) the value of the voltage when the current is 5 A.

Solution

Notice that this information is not given in the table. Without a graph, the information could be obtained only by substitution in the formula for Ohm's law. To get these data from the graph,

a. 1. Locate 70 V on the horizontal voltage scale at *A*.
   2. Read straight up along this vertical line until the graph is reached at *B*.
   3. Read horizontally to the left along this line to reach the vertical current scale at *C*.
   4. Read the value of the current at point *C* as 7 A.   *Ans.*
b. 1. Locate 5 A on the vertical scale at point *D*.
   2. Read horizontally to the right until the graph is reached at point *E*.
   3. Follow straight down along this vertical line to reach the horizontal voltage scale at point *F*.
   4. Read the value of the voltage at point *F* as 50 V.   *Ans.*

Example 15-2

If a constant voltage is applied across the ends of different resistors, the resulting current may be calculated by Ohm's law. This information may then be shown as a graph such as Fig. 15-3, in which a constant voltage of 10 V was applied across different resistances. Using this graph, find the current when the resistance is (*a*) 5 Ω and (*b*) 7 Ω.

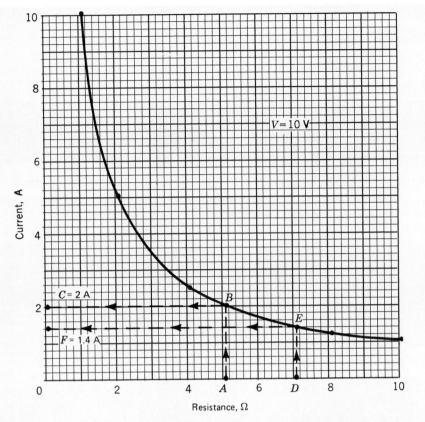

Figure 15-3   Relationship between current and resistance at a constant voltage.

Hint for Success

Remember to put the independent variable on the horizontal axis.

Solution

a. 1. Locate 5 Ω at *A* on the horizontal resistance scale.
   2. Read straight up to the graph at *B*.
   3. Read horizontally to the left to reach the vertical current scale at *C*.
   4. Read the value of the current at point *C* as 2 A.   *Ans.*
b. 1. Locate 7 Ω at *D* on the horizontal resistance scale.
   2. Read straight up to the graph at *E*.
   3. Read horizontally to the left to reach the vertical current scale at *F*.
   4. Read the value of the current at point *f* as slightly more than 1.4 A.
                                                               Ans.

Plotting Graphs

Example 15-3

An experiment was performed relating the average forward power and the average forward current for a 1N4719 diode. The data obtained are shown in the following table. Plot the graph.

Point number	1	2	3	4	5	6
Average Forward power, W	0	1	2	3	4	5
Average Forward current, A	0	0.6	1.45	2.25	3.1	4

Solution

The finished curve is shown in Fig. 15-4.

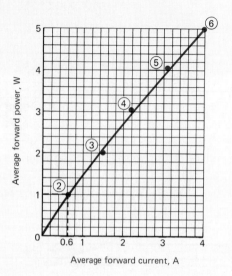

Figure 15-4 Power dissipation versus current curve for a 1N4719 diode.

1. Draw two baselines at right angles to each other.
2. Label the horizontal line as the average forward current in amperes and the vertical line as the average forward power in watts.
3. Select appropriate scales for each. Since the maximum current is 5 A, we can use 10 boxes to represent 1 A. This will require $5 \times 10 = 50$ boxes. Number every tenth box. On the power scale we can use 10 boxes to represent 1 W. This will require 40 boxes. Number every tenth box.
4. Plot the individual points. To locate point 1, put a point at the origin where the zero is located. To locate point 2, find 1 W on the vertical scale. Draw a line straight across. Now find the corresponding current (0.6 A) on the horizontal scale. Draw a line straight up. These two lines will intersect at point 1. In the same manner, plot points 3 to 6. Notice that our choice of values for the current scale makes each box represent 0.1 A, and on the vertical scale each box represents 0.1 W.
5. Draw a smooth curve through the points. If not all the points fall on this smooth curve, draw the curve so that the points are as close to the curve as possible and that there are as many points over the line as under it.

Graphs with positive and negative values. Many graphs have negative as well as positive values to be considered, and these values must be located on the graph. To do this, the two base lines are extended as shown in Fig. 15-5 to form the horizontal axis XX' and the vertical axis YY'. The two axes meet at the *origin O*. Values along the X axis measured to the *right* of YY' are *positive;* values measured to the *left* of YY' are *negative.* Values along the Y axis measured *upward* from XX' are *positive;* values *downward* from XX' are *negative.* For example, in Fig. 15-5, the points are located depending on the values of X and Y. These values are called the *coordinates* of the point. All points are located starting from the origin O. To locate point

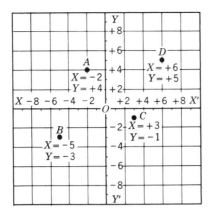

Figure 15-5 Locating points with positive and negative coordinates.

A ($X = -2$, $Y = +4$), move two units to the *left* along the X axis, since $X = -2$. Then move four units *up* along this line, since $Y = +4$. To locate point B ($X = -5$, $Y = -3$), move five units to the *left* along the X axis, since $X = -5$. Then move three units *down* along this line, since $Y = -3$. To locate point C ($X = +3$, $Y = -1$), move three units to the *right* along the X axis, since $X = +3$. Then move one unit *down* along this line, since $Y = -1$. To locate point D ($X = +6$, $Y = +5$), move six units to the *right* along the X axis, since $X = +6$. Then move five units *up* along this line, since $Y = +5$.

Example 15-4

A storage battery (6.6 V) with an internal resistance of 0.75 Ω is used to supply a variable load. Plot the curve for the power delivered to the load against the load resistance, using the following data.

Load resistance, Ω	0.005	0.01	0.02	0.03	0.04
Power delivered, W	50	67	96	117	130
Point number	1	2	3	4	5

Load resistance, Ω	0.05	0.06	0.07	0.08	0.09
Power delivered, W	140	145	147	146	145
Point number	6	7	8	9	10

Load resistance, Ω	0.10	0.11	0.12	0.13	0.14
Power delivered, W	143	140	137	135	132
Point number	11	12	13	14	15

What conclusion can you reach about the power transferred when the load resistance equals the source resistance?

Solution

The curve is shown in Fig. 15-6 on page 504. Every two boxes along the load resistance or X axis are worth 0.01 Ω. Each box along the watts or Y axis is

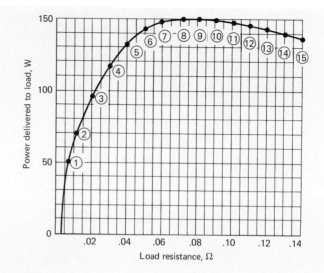

Figure 15-6   Power delivered to a load plotted against load resistance.

worth 10 W. Notice that the power delivered is a maximum when the load resistance is equal to the source resistance.

Exercises

Practice

1. Figure 15-7 is a graph showing the current taken by an incandescent lamp at various voltages. Find the current taken at the following voltages: (*a*) 30 V, (*b*) 45 V, (*c*) 50 V, (*d*) 65 V, (*e*) 80 V, (*f*) 100 V, and (*g*) 120 V.

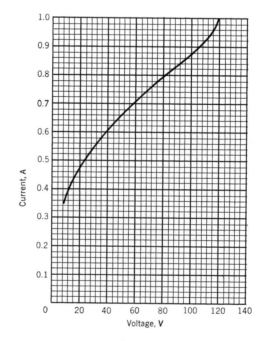

Figure 15-7   Current taken by an incandescent lamp at various voltages.

2. Figure 15-8 is a graph showing the resistance in ohms per 1000 ft of copper wires of various diameters. Referring to Table 12-1 for the diameter in mils of the gage numbers, find the resistance per 1000 ft of (*a*) No. 0, (*b*) No. 4, (*c*) No. 6, (*d*) No. 8, (*e*) No. 10, and (*f*) No. 12. Find the resistance per 1000 ft of wires whose diameters are (*g*) 150 mil, (*h*) 220 mil, (*i*) 250 mils, and (*j*) 0.3 in.

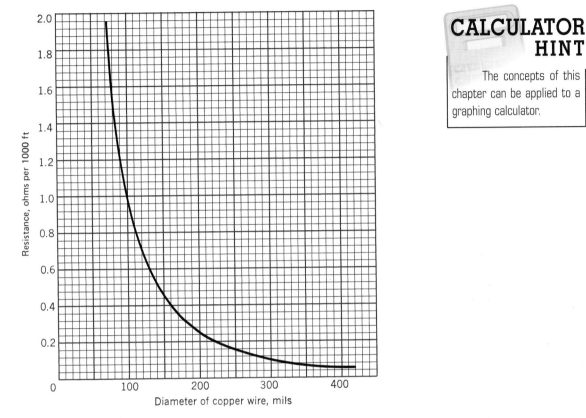

Figure 15-8 Graph of a portion of the AWG table, Table 12-1.

Plot a graph for each of the following exercises:

3. The volume of a cube when the side is changed:

Side, in	0	1	2	3	4	5
Volume, in³	0	1	8	27	64	125

4. The distance covered by a falling body for increasing units of time:

Time, s	1	2	3	4	5	6	7	8	9	10
Distance, ft	16	64	144	256	400	576	784	1024	1296	1600

5. The efficiency of an engine against its horsepower output. What is the maximum efficiency of the engine? At what horsepower output is this efficiency obtained?

Horsepower output, hp	0	10	20	30	40	50	60	70
Efficiency, %	0	30	50	64	78	80	78	64

6. The power used by a 100-Ω resistor carrying different currents:

Current, A	0.1	0.2	0.3	0.4	0.5	0.6	0.7	0.8	0.9	1.0
Power, W	1	4	9	16	25	36	49	64	81	100

7. In an inductive circuit, the current does not rise to its Ohm's law value instantaneously. The growth of the current in a certain coil is shown by the following data. Plot the curve.

Time, s	0	0.1	0.2	0.3	0.4	0.5	0.6	0.7	0.8	0.9	1.0
Current, A	0	4.2	6.6	8.0	8.9	9.3	9.6	9.8	9.9	10	10

8. The time constant RC of a capacitive circuit describes the time required to charge a capacitor. Plot the charge curve from the following data:

Time, s	0.5RC	1RC	2RC	3RC	4RC	5RC
Percentage of applied voltage	39	63	86	95	98	99

9. The inductive reactance of a 0.01-H coil at different frequencies:

Frequency, Hz	60	100	250	500	1000
Inductive reactance, Ω	3.7	6.3	15.7	31.4	62.8

10. The capacitive reactance for a 4-μF capacitor at different frequencies:

Frequency, Hz	25	50	60	120	240
Capacitive reactance, Ω	1590	795	663	331	166

11. The collector voltage V_C against the collector current I_C for a 2N265 transistor:

V_C, V	5	10	15	20	25	30
I_C, mA	14.1	7.2	4.9	4.1	3.7	3.2

12. The sine of an angle against the number of degrees in the angle:

Angle, degrees	0	30	45	60	90	120	135	150	180
Sine of the angle	0	0.5	0.7	0.86	1.0	0.86	0.7	0.5	0

Angle, degrees	210	225	240	270	300	315	330	360
Sine of the angle	−0.5	−0.7	−0.86	−1.0	−0.86	−0.7	−0.5	0

JOB 15-2

The Generation of an AC Voltage

Magnetism. The phenomenon of magnetism was discovered about 100 B.C., when it was observed that a peculiar stone had the property of attracting bits of iron to it. This natural magnet was called a *lodestone*. The lodestone was used to create other magnets artificially by stroking pieces of iron with it.

Magnetic poles. Every magnet has two points opposite each other which attract pieces of iron best. These points are called the *poles* of the magnet: the north pole

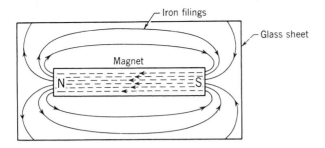

Figure 15-9   The iron filings indicate the pattern of the lines of force around a bar magnet.

and the south pole. Just as similar electric charges repel each other and opposite charges attract each other, similar magnetic poles repel each other and unlike poles attract each other.

A magnet evidently attracts a bit of iron because of some force that exists around the magnet. This force is called the *magnetic field*. Although it is invisible to the naked eye, it can be shown to exist by its effect on bits of iron. Place a sheet of glass or some other nonmagnetic material over a bar magnet as shown in Fig. 15-9. Sprinkle some iron filings over the glass, and tap it gently. The filings will fall back into a definite pattern which describes the field of force around the magnet. The field evidently seems to be made up of *lines* of force which appear to leave the magnet at the north pole, travel through the air around the magnet to the south pole, and continue through the magnet to the north pole to form a *closed loop* of force. The stronger the magnet, the greater the number of lines of force and the larger the area covered by the field.

Electromagnetism.   In 1819 Hans Christian Oersted, a Danish physicist, discovered that a field of magnetic force exists around a wire carrying an electric current. In Fig. 15-10 a wire is passed through a piece of cardboard and connected through a switch to a dry cell. With the switch open (no current flowing), sprinkle iron filings on the cardboard and tap it gently. The filings will fall back haphazardly. Now close the switch, which will permit a current to flow in the wire. Tap the cardboard again. This time, the magnetic effect of the current in the wire will cause the filings to fall back into a definite pattern of concentric circles with the wire as the center of the circles. Every section of the wire has this field of force around it in a plane perpendicular to the wire, as shown in Fig. 15-11 on page 508.

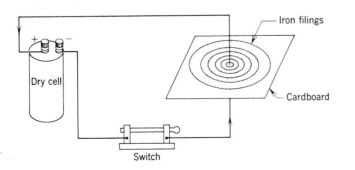

Figure 15-10   A circular pattern of magnetic force exists around a wire carrying an electric current.

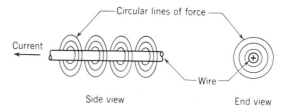

Side view   End view

Figure 15-11   The circular fields of force around a wire carrying a current are in planes that are perpendicular to the wire.

The strength of the magnetic field. The ability of the magnetic field to attract bits of iron depends on the number of lines of force present. The strength of the magnetic field around a wire carrying a current depends on the current, since it is the current that produces the field. The greater the current, the greater the strength of the field. A large current will produce many lines of force extending far from the wire, whereas a small current will produce only a few lines close to the wire as shown in Fig. 15-12.

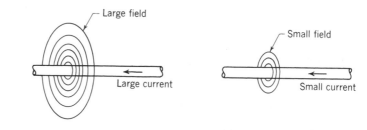

Figure 15-12   The strength of the magnetic field around a wire carrying a current depends on the amount of current.

Electromagnetic induction. Michael Faraday is credited with the discovery, in 1831, of the basic principle underlying the operation of ac machinery. Simply stated, he discovered that if a conductor "cut across" lines of magnetic force, or if lines of force "cut across" a conductor, an electromotive force (emf) or voltage would be induced across the ends of the conductor. Figure 15-13 represents a magnet with its lines of force streaming from the north to the south pole. A conductor *C*, which can be moved between the poles, is connected to a galvanometer, which is a very sensitive meter used to indicate the presence of an electromotive force. When the conductor is stationary, the galvanometer indicates zero emf. If the wire is *moved about outside* the magnetic field at position 1, the galvanometer will still indicate zero. However, if the conductor is moved to the *left* to position 2, so that it *cuts across the lines of magnetic force*, the galvanometer pointer will deflect to *A*. This indicates that an emf was induced in the conductor because lines of force were "cut." Upon reaching position 2, the galvanometer pointer will swing back to zero because no lines of force are being cut. Now suppose the conductor is moved to the *right* through the lines of force, back to position 1. During this *movement*, the pointer will deflect to *B*, indicating that an emf has again been induced in the wire, but in the *opposite direction*. If the wire is held *stationary* in the middle of the field of force at position 3, the galvanometer reads zero. If the conductor is moved up or down *parallel* to the lines of force so that *none are cut*, no emf will be induced. From experiments similar to these, Faraday deduced the following:

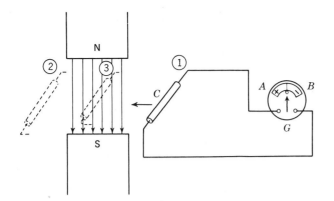

Figure 15-13 When a conductor cuts lines of force, an emf is induced in the conductor. The direction in which the conductor cuts the lines determines the direction of the induced emf.

1. When lines of force are cut by a conductor or lines of force cut a conductor, an emf is induced in the conductor.
2. There must be a relative *motion* between the conductor and the lines of force in order to induce an emf.
3. Changing the direction of the cutting will change the direction of the induced emf.

Generating an alternating emf.   Since a voltage is induced in a conductor when lines of force are cut, the amount of the induced emf depends on the number of lines cut in a unit time. In order to induce an emf of 1 V, a conductor must cut 100,000,000 lines of force per second. To obtain this great number of "cuttings," the conductor is formed into a loop and rotated on an axis at great speed as shown in Fig. 15-14. The two sides of the loop become individual conductors in series, each side of the loop cutting lines of force and inducing twice the voltage that a single conductor would induce. In commercial generators, the number of "cuttings" and the resulting emf are increased by (1) increasing the number of lines of force by using more magnets or stronger electromagnets, (2) using more conductors or loops, and (3) rotating the loops faster.

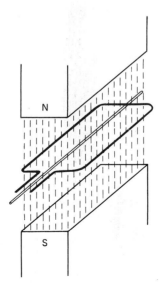

Figure 15-14 The rotating loop cuts the lines of force to produce an alternating emf.

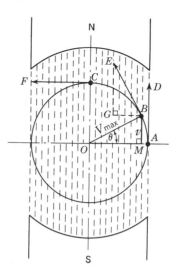

Figure 15-15   The conductor rotating at a constant speed through a uniform magnetic field.

Let us follow a single conductor as it rotates at a uniform speed through a uniformly distributed field of force. In Fig. 15-15, the lines *AD, BE,* and *CF* are all equal to *OC* and represent the direction in which the conductor is moving at points *A, B,* and *C,* respectively. At *A,* the conductor is moving parallel to the lines of force. No lines of force are cut, and zero volts are induced. At *C,* the conductor has rotated through 90° and is moving in the direction *CF.* This motion is directly across the lines of force and produces the maximum number of cuttings and the maximum voltage V_{max}. This maximum voltage may be represented by the distance *CF.* Since *OC* is equal to *CF,* the maximum voltage may be represented by the radius of the circle *OC.* At *B,* the conductor has rotated through some angle θ (theta) and is moving in the direction shown by *BE.* This motion from *B* to *E* may be considered to be made up of a motion from *B* to *G* and then from *G* to *E.* However, only one of these motions is useful in cutting lines of force. The motion *GE* is parallel to the lines of force and produces no cuttings. Therefore, the distance *BG* represents the total cuttings produced by the motion *BE. BG* therefore represents the voltage produced at the instant that the conductor is passing through point *B.* It can be shown by geometry that $BG = BM$. Thus, the vertical line drawn from the conductor at any instant perpendicular to the base represents the voltage induced in the conductor at that instant. The voltage at any instant of time evidently depends on the position of the conductor at that instant and is known as the *instantaneous* voltage *v.*

1. Only that part of the motion of a conductor directly across the field cuts lines and produces voltage.
2. The number of lines cut and the voltage produced are zero at 0° and increase to a maximum at 90°.
3. The vertical distance drawn from the conductor at any point to the horizontal base line represents the voltage produced at the instant that the conductor is passing through that point.

The complete picture of the voltage produced by a rotating conductor can now be drawn. Draw a circle, and divide the circumference into 12 parts, each 30° apart, as shown in Fig. 15-16. A horizontal baseline is drawn as shown and labeled in degrees to correspond to the positions of the conductor. The vertical distances rep-

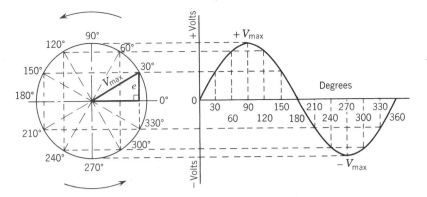

Figure 15-16 The voltage wave produced by a conductor rotating at a constant speed through a uniform magnetic field.

resent the voltages produced. As the conductor rotates, the vertical distance at each point is drawn at the corresponding point on the graph. Notice that the motion from 0 to 180° was to the *left*. The motion from 180 to 360° was to the *right* and produced a voltage in the *opposite* direction. This is indicated by *negative* voltages drawn *below* the base line. This graph gives a complete picture of all the changes in voltage during one complete rotation.

1. At 0°, the emf is 0 V.
2. At 90°, the emf is the greatest and is known as $+V_{max}$.
3. At 180°, the emf is again 0 V. The direction of the rotation and voltage changes at this point.
4. At 270°, the emf is again a maximum, but in the *opposite* direction, and is known as $-V_{max}$.
5. At 360°, the emf is 0 V.
6. The voltage passes twice through both the zero value and the maximum value during the cycle.
7. The maximum voltages are equal in value but are of opposite sign.
8. The instantaneous voltage means the voltage at any instant during an ac cycle. The symbol is v.

The sine wave. Consider the triangle *OBM* in Fig. 15-15. *OB* is a radius of the circle and is equal to the maximum voltage V_{max}. *BM* is the instantaneous voltage at any single θ, given as v. By trigonometry,

$$\sin \theta = \frac{v}{V_{max}} \qquad \text{Formula 15-1}$$

or

$$v = V_{max} \times \sin \theta \qquad \text{Formula 15-2}$$

This formula says that the value of the voltage at any instant depends on the maximum voltage and the *sine of the angle at that instant*. If the maximum voltage is 1 V,

$$v = 1 \times \sin \theta$$

or

$$v = \sin \theta \qquad \text{Formula 15-3}$$

We can now plot a graph of the changes in the voltage as a conductor rotates in a circle through a uniform field at a constant speed. Figure 15-17 on page 512 shows the graph obtained by plotting the angles against the values of the sine of

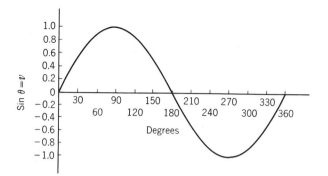

Figure 15-17   The voltage is proportional to the sine of the angle in a sine wave.

the angles. Compare this curve with the curve of Fig. 15-16. They are identical. Most commercial generators are designed to produce alternating waves of this type. They are called *sine waves* because the voltage at any instant is proportional to the sine of the angle at that instant.

Cycle, frequency.   An alternating wave may be an ac voltage or an alternating current. During one *cycle* the wave will pass through a complete series of positive and negative values. In one cycle, as shown in Fig. 15-18, an ac voltage starts at 0°, rises to a positive maximum at 90°, and falls to zero at 180°; then it *reverses its polarity*, rises to a negative maximum at 270°, and falls to zero at 360°. An alternating current starts at 0°, rises to a positive maximum at 90°, and falls to zero at 180°; then it *reverses its direction*, rises to a negative maximum at 270°, and falls to zero at 360°.

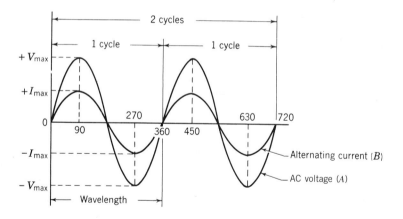

Figure 15-18   A cycle describes all the changes of voltage or current during a rotation of 360°.

Each cycle of an ac wave may be repeated over and over. The frequency f of an ac wave means the number of complete cycles that occur *in one second*. In the United States the standard frequency for light and power is 60 cycles/s (60 Hz); in Europe it is 50 cycles (50 Hz). Audio frequencies range between 30 and 15,000 Hz. Radio and television frequencies range between 15,000 and 890,000,000 Hz.

Wavelength.   The distance traveled by an ac wave during one complete cycle is called its *wavelength*. Its symbol is the Greek letter lambda (λ). This distance may be expressed in any convenient unit of length. However, since most scientific work uses the metric system, the wavelength is usually measured in terms of *meters*. One meter is equal to 39.37 in.

The speed of radio waves is the same as the speed of light, 186,000 mi/s. This speed may be expressed in the metric system in terms of meters per second. Since 1 mi = 5820 ft, and 1 m = 39.37 in,

$$
\begin{aligned}
186,000 \text{ mi} &= 186,000 \times 5280 \text{ ft} \\
&= 186,000 \times 5280 \times 12 \text{ in} \\
&= \frac{186,000 \times 5280 \times 12}{39.37} \text{ m} \\
&= 300,000,000 \text{ m}
\end{aligned}
$$

Therefore, the speed of a radio wave may be expressed as 300,000,000 m/s.

If we divide the speed in meters per second by the frequency in cycles per second, the units of time will cancel out as shown below.

$$
\frac{\text{Meters}}{\text{Second}} \div \frac{\text{cycles}}{\text{second}}
$$

Inverting and multiplying,

$$
\frac{\text{Meters}}{\frac{\text{Second}}{1}} \times \frac{\frac{1}{\text{second}}}{\text{cycles}} = \frac{\text{meters}}{\text{cycle}}
$$

The number of meters per cycle is the wavelength.

$$
\lambda = \frac{300,000,000}{f} \qquad \text{Formula 15-4}
$$

where λ = wavelength, m
f = frequency, Hz

If the frequency is expressed in kilohertz,

$$
\lambda = \frac{300,000}{f} \qquad \text{Formula 15-5}
$$

where λ = wavelength, m
f = frequency, kHz

Example 15-5

What is the wavelength of radio station WCBS, which broadcasts at a frequency of 880 kHz?

Solution

Given: Station WCBS   Find: λ = ?
f = 880 kHz

$$
\lambda = \frac{300,000}{f} \qquad (15\text{-}5)
$$

$$
\lambda = \frac{300,000}{880} = 340.9 \text{ m} \qquad \textit{Ans.}
$$

CALCULATOR HINT

Calculations can be facilitated by storing the constants in a memory location and retrieving them as needed.

Example 15-6

What is the frequency of a radio wave if its wavelength is equal to 297 m?
What New York City station broadcasts at this frequency?

Solution

Given: $\lambda = 297$ m   Find: $f = ?$
Name of New York station = ?

$$\lambda = \frac{300,000}{f} \tag{15-5}$$

$$297 = \frac{300,000}{f}$$

$$297 \times f = 300,000$$

$$f = \frac{300,000}{297} = 1010 \text{ kHz} \quad Ans.$$

The station is WINS.   *Ans.*

JOB 15-3   Instantaneous Values, Maximum Values,
and Phase Angles of an AC Wave

Since alternating voltage and current waves change in amount and direction at regular intervals, the voltage or current at any instant of time must be calculated. These values are called the *instantaneous values*. For example, in Fig. 15-15, when the conductor has rotated through some angle θ to reach point B, the instantaneous value is indicated by the vertical line BM. The radius of the circle OB indicates the maximum value of the wave. By trigonometry, we obtain

$$\sin \theta = \frac{v}{V_{max}} \tag{15-1}$$

or
$$v = V_{max} \times \sin \theta \tag{15-2}$$

If this changing voltage wave is impressed across a resistance, each instantaneous voltage v will produce its own value of instantaneous current i. Thus, as shown in Fig. 15-18, the ac voltage wave A produced the ac wave B. This current wave will have the same frequency as the voltage wave which produced it. Using this current wave, we obtain

$$\sin \theta = \frac{i}{I_{max}} \qquad \text{Formula 15-6}$$

or
$$i = I_{max} \times \sin \theta \qquad \text{Formula 15-7}$$

Example 15-7

An ac wave has a maximum value of 100 mA. Find the instantaneous current at 70°.

Solution

Given: $I_{max} = 100$ mA   Find: $i = ?$
$\theta = 70°$

$$i = I_{max} \times \sin \theta \qquad\qquad (15\text{-}7)$$
$$i = 100 \times \sin 70° = 100 \times 0.9397 = 93.97 \text{ mA} \qquad Ans.$$

Maximum values. The maximum value of an ac voltage (V_{max}) or an alternating current (I_{max}) is reached twice during each cycle. The positive maximums occur at 90°, and the negative maximums occur at 270°. By solving Eqs. (15-2) and (15-7) for V_{max} and I_{max} we obtain the following formulas:

$$V_{max} = \frac{v}{\sin \theta} \qquad\qquad \text{Formula 15-8}$$

$$I_{max} = \frac{i}{\sin \theta} \qquad\qquad \text{Formula 15-9}$$

Example 15-8

An ac voltage wave has an instantaneous value of 70 V at 30°. (*a*) Find the maximum value of the wave. (*b*) Can this wave be impressed across a capacitor whose breakdown voltage is 125 V?

Solution

Given: $v = 70$ V   Find: $V_{max} = $?
         $\theta = 30°$

a. $V_{max} = \dfrac{v}{\sin \theta} = \dfrac{70}{\sin 30°} = \dfrac{70}{0.5000} = 140$ V   *Ans.*

b. No, since 140 V is larger than the breakdown voltage of 125 V.

Hint for Success

It is conventional that positive angles are created by a counter-clockwise rotation, and negative angles by a clockwise rotation.

Phase angles. The phase angle θ or phase of an ac cycle refers to the value of the electrical angle at any point during the cycle. These angles may be found by solving formula (15-1) or (15-6).

Example 15-9

An ac voltage wave has a maximum value of 155.5 V. Find the phase angle at which the instantaneous voltage is 110 V.

Solution

Given: $V_{max} = 155.5$ V   Find: $\theta = $?
         $v = 110$ V

$$\sin \theta = \frac{v}{V_{max}} = \frac{110}{155.5} = 0.7074 \qquad (15\text{-}1)$$
$$\theta = 45° \qquad Ans.$$

CALCULATOR HINT

Remember that in this book, angle measurements are expressed in degrees.

Practice

1. What is the wavelength of an ac wave with a frequency of 30,000 Hz?
2. Find the frequency of a carrier wave with a wavelength of 600 m.
3. Find the instantaneous voltage at 50° in a wave with a maximum value of 165 V.
4. Find the maximum value of an ac wave if the instantaneous voltage is 50 V at 35°.
5. Find the phase angle at which an instantaneous voltage of 72 V appears in a wave with a maximum value of 250 V.
6. Find the instantaneous current at 85° in a wave with a maximum value of 26 A.
7. Find the maximum value of an ac wave if the instantaneous current is 9 A at 12°.
8. Find the phase angle at which an instantaneous current of 3.5 A appears in a wave with a maximum value of 20 A.
9. The maximum current of the current wave of a transmitter is 10 A. At what instant (angle) will the instantaneous current be (*a*) 5 A, (*b*) 6 A, and (*c*) 8.66 A?
10. The maximum voltage of an ac wave is 100 V. Find the instantaneous voltage at (*a*) 0°, (*b*) 15°, (*c*) 30°, (*d*) 45°, (*e*) 60°, (*f*) 75°, and (*g*) 90°. What is the average of these voltages?

JOB 15-4

Effective Value of an AC Wave

An alternating current is a current that is continually changing in amount and direction. In Fig. 15-19, the current starts at 0 A at 0 time. After 1 s, it reaches 10 A but drops down to 0 after 2 s. At this point, the current reverses its direction and increases negatively to −10 A at 3 s, after which it drops down to 0 after 4 s. The *average* current during these 4 s equals 0 + 10 + 0 − 10 + 0 divided by 5, or 0 A. Obviously, even though the average of all the currents is zero, this ac wave can be effective in running a motor or lighting a lamp. But how many amperes are actually flowing? If an ac ammeter is used to measure this current, it will read 7.07 A. This meter reading is called the *effective value* of all the changes that occur during one cycle of the wave.

> ## Hint for Success
>
> Effective value is an equal heat concept.

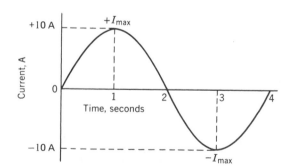

Figure 15-19 AC wave.

One of the effects of the passage of any electric current through a resistance, whether it be alternating or direct, is the production of heat. To determine the *worth*, or *effectiveness*, of an ac wave, we must compare its effect with the effect of a direct current.

RULE If an ac wave produces as much heat as 1 A of direct current, we say that the ac wave is as effective as 1 A of direct current.

Thus, as in Fig. 15-20, an ac wave may start at zero and pass through many values of current up to a maximum of 10 A, continuing to fall to zero, change direction, rise to a negative maximum of −10 A, and fall to zero again. If the *effect* of all these changes is to produce only as much heat as 7.07 A of direct current would produce, then the wave is said to have an *effective* value of only 7.07 A. The effective value of any current wave can be calculated by averaging the *heating* effects of the many individual instantaneous currents. Since $P = I^2R$, the heating effect on any resistance depends on the *square* of the current. Thus,

$$I_{dc}^2 = \text{average of } i^2$$

By taking the square root of both sides, we obtain

$$I_{dc} = \sqrt{i_{av}^2}$$

Since the effective alternating current is to have the same heating effect as the direct current,

$$I_{ac} = I_{dc} = \sqrt{i_{av}^2}$$

This equation says that the effective value of an ac wave is equal to the square root of the average (mean) of the squares of the instantaneous currents. For this reason it is sometimes called the *rms* (root mean square) value. This gives us a method for calculating the effective value of any ac wave. One-quarter of a cycle is divided into a number of equal parts, and the instantaneous current is calculated at each point. These currents are squared, and the average found. The square root of this average is equal to the effective value of the wave. The results of many calculations for waves of different maximums indicate that the effective value of a sine-wave current is always 0.707 times the maximum value of the wave.

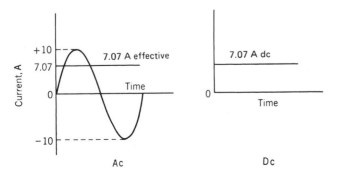

Figure 15-20   Effective values of alternating and direct currents.

$$I = 0.707 \times I_{max} \qquad \text{Formula 15-10}$$

where
$$I = \text{effective value of an ac wave}$$
$$I_{max} = \text{maximum value of an ac wave}$$

Since alternating currents are the result of ac voltages, the effective value of a voltage wave has an identical relation to the maximum voltage.

$$V = 0.707 \times V_{max} \qquad \text{Formula 15-11}$$

where
$$V = \text{effective value of an ac voltage wave}$$
$$V_{max} = \text{maximum value of an ac voltage wave}$$

Notice that the effective values are written as I and V, with no subscripts. Alternating-current meters are designed and calibrated to indicate these effective values. Unless otherwise specified, a value of an alternating current or ac voltage always means the effective value.

Formulas (15-10) and (15-11) may be transformed to obtain formulas for I_{max} and V_{max}.

$$I = 0.707 \times I_{max}$$

or
$$I_{max} = \frac{I}{0.707}$$

$$I_{max} = 1.414 \times I \qquad \text{Formula 15-12}$$

Similarly,
$$V = 0.707 \times V_{max}$$

or
$$V_{max} = \frac{V}{0.707}$$

$$V_{max} = 1.414 \times V \qquad \text{Formula 15-13}$$

Thus, an effective voltage of 120 V is the effective voltage of a wave whose maximum equals $1.414 \times 120 = 170$ V. This means that if an electric heater were used on a 120-V ac line, it would produce only as much heat as would be produced on a 120-V dc line, even though the ac line reaches the maximum of 170 V twice in each cycle.

Example 15-10

An alternating current has a maximum value of 50 A. Find (a) the effective current and (b) the instantaneous current at 10°.

Solution

Given: $I_{max} = 50$ A   Find: $I = ?$
$\theta = 10°$ $i = ?$

a.
$$I = 0.707 \times I_{max} \qquad (15\text{-}10)$$
$$I = 0.707 \times 50 = 35.35 \text{ A} \qquad Ans.$$

b.
$$i = I_{max} \times \sin \theta \qquad (15\text{-}7)$$
$$i = 50 \times \sin 10°$$
$$i = 50 \times 0.1736 = 8.68 \text{ A} \qquad Ans.$$

Example 15-11

An ac voltage wave has an effective value of 110 V. Find (*a*) the maximum value and (*b*) the instantaneous value at 40°.

Solution

Given:
$$V = 110 \text{ V} \qquad \text{Find: } V_{max} = ?$$
$$\theta = 40° \qquad\qquad v = ?$$

a.
$$V_{max} = 1.414 \times V = 1.414 \times 110 = 156 \text{ V} \qquad \textit{Ans.}$$

b.
$$v = V_{max} \times \sin \theta \qquad\qquad (15\text{-}2)$$
$$v = 156 \times \sin 40°$$
$$v = 156 \times 0.6428 = 99.6 \text{ V} \qquad \textit{Ans.}$$

Self-Test 15-12

An ac wave has an instantaneous value of 12.95 A at 15°. Find (*a*) the maximum value and (*b*) the effective value.

Solution

Given:
$$i = 12.95 \text{ A} \qquad \text{Find: } I_{max} = ?$$
$$\theta = \underline{\quad(1)\quad} \qquad\qquad I = ?$$

a.
$$I_{max} = \frac{i}{?} \qquad\qquad (15\text{-}9)$$
$$= \frac{12.95}{\sin 15°} = \frac{12.95}{?}$$
$$I_{max} = \underline{\qquad\qquad} \text{ A} \qquad \textit{Ans.}$$

b.
$$I = 0.707 \times \underline{\qquad\qquad} \qquad\qquad (15\text{-}10)$$
$$I = 0.707 \times \underline{\qquad\qquad} = 35.35 \text{ A} \qquad \textit{Ans.}$$

Exercises

Practice

Find the values indicated in each exercise.

Exercise	Maximum Value	Effective Value	Phase Angle	Instantaneous Value
1	35 A	?	30°	?
2	456 V	?	40°	?
3	?	440 V	50°	?
4	?	25 A	60°	?
5	155 V	?	30°	?
6	10 A	?	45°	?
7	?	110 V	65°	?
8	?	20 A	75°	?
9	?	?	50°	26.81 A
10	?	?	15°	120.3 V
11	100 V	?	?	43.84 V
12	20 A	?	?	5.5 A

13. A sine wave has an instantaneous value of 10 V at 30°. What is its value at 60°?

Applications

14. What must be the minimum breakdown-voltage rating of a capacitor if it is to be used on a 110-V ac line?
15. An electric stove draws 7.5 A from a 120-V dc source. (*a*) What is the maximum value of an alternating current which will produce heat at the same rate? (*b*) Find the power drawn from the ac line.
16. An underground cable is designed to operate safely at an effective voltage of 2200 V. What dc voltage will the line carry safely?
17. A 50-Ω resistor in an ac circuit is subjected to a peak voltage of 169 V. Find the effective power consumed.

(See CD-ROM for Test 15-1)

JOB 15-5

Measuring Alternating Current with an Oscilloscope

One of the most widely used electronic instruments for the measurement of ac signals is the oscilloscope. The oscilloscope makes graphs plotting voltage against time. A timing system inside the instrument forms the horizontal time axis. The signal voltage applied to the oscilloscope forms the vertical axis of the graph. Figure 15-19 is repeated here for your convenience.

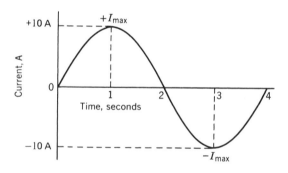

The figure shows how it possible to measure the amplitude and the frequency of the applied signal. The amplitude can be measured directly from the graph. The frequency is found by measuring the period of the ac signal. The period is the time it takes for the ac signal to repeat itself. In other words, the period is the time that it takes for the signal to make one complete cycle.

In Fig. 15-19, the period is 4 s. After 4 s the ac signal looks exactly like the part that is shown. If the frequency were twice as high as that in Fig. 15-19, the signal would repeat itself in half the time. This would be represented on the graph by dividing all of the times by 2. The signal would be zero at 0, 1, and 2 s. The maximum positive value would be at ½ s and the maximum negative value at 1½ s. Evidently, as the frequency increases the period decreases. This leads to the following rule:

RULE   The frequency equals 1 divided by the period.

$$f = \frac{1}{\text{period}} \qquad \text{Formula 15-14}$$

Example 15-13

Find the frequency of the signal shown in Fig. 15-19.

Solution

1. Measure the period of the signal from the graph. It measures 4 s.
2. Calculate the frequency.

$$f = \frac{1}{\text{period}} = \frac{1}{4} \text{ cycles per second or Hertz} \qquad \textit{Ans.}$$

In power applications, frequency is usually referred to as cycles per second or just cycles. We usually say that a house has 60-cycle power. In electronic applications, the term *Hertz* is used. The two terms are identical in meaning.

When using an oscilloscope, the picture should always be made as large as possible in order to increase the accuracy of the readings. This is accomplished by adjusting the scales on the graph electronically. The horizontal time base is calibrated in time per division. This is adjusted until one cycle takes as much of the screen as possible. The vertical scale is adjusted with the gain control. This is calibrated in volts per division. It should be adjusted until the picture is as large as possible. The adjusted picture should always appear similar to that in Fig. 15-21. The scales are not plotted on the graph as in Fig. 15-19. Instead, the scales are determined from the knob settings of the oscilloscope.

Example 15-14

The knobs of an oscilloscope are adjusted to obtain the picture shown in Fig. 15-21. If the gain is set at 5 V per division, what is the peak voltage?

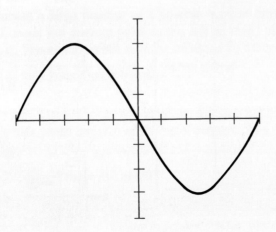

Figure 15-21   Sine wave displayed on an oscilloscope face.

Solution

1. In order to measure the peak voltage, draw a horizontal line from the peak of the picture to the scale that is in the center of the graph. Count the number of divisions to the point at which the line hits the scale. In this example there are three divisions.
2. The knob setting tells us that the gain is 5 V per division. Multiply to obtain the voltage:

$$3 \text{ divisions} \times 5 \text{ V per division} = 15 \text{ V peak} \qquad \textit{Ans.}$$

Example 15-15

The time base is set to 1 ms per division. What is the frequency?

Solution

1. Refer to Fig. 15-21 on page 521. Count the number of divisions for one cycle. In this example there are 10 divisions.
2. Get the time scale from the knob setting. It is 1 ms per division.
3. Find the period by multiplying the number of divisions per cycle by the horizontal scale:

$$10 \text{ divisions} \times 1 \text{ ms per division} = 10 \text{ ms} \qquad \textit{Ans.}$$

4. Find the frequency.

$$f = \frac{1}{\text{period}} \qquad (15\text{-}14)$$

$$f = \frac{1}{10 \text{ ms}} = \frac{1}{0.01 \text{ s}}$$

$$f = 100 \text{ Hz or cycles per second} \qquad \textit{Ans.}$$

Self-Test 15-16

1. An _____ draws the graph of an ac signal electronically.
2. The horizontal axis is the _____ axis of the graph.
3. The scale of the horizontal axis is determined by the _____ _____ generator that is inside the oscilloscope.
4. The vertical axis shows the _____ of the signal as a function of time.
5. The scale for the vertical axis is determined by the _____ setting of the oscilloscope vertical amplifier.
6. The frequency of an ac signal is found on an oscilloscope by measuring the _____ of the waveform.
7. The frequency is calculated by taking the reciprocal of _____.

Exercises

Practice

For the following exercises, use the graph in Fig. 15-21 to calculate your answers.

1. If the gain setting is 30 V per division, what is the peak voltage of the signal?
2. The time base is set at 200 μs per division. What is the frequency of the signal?
3. The gain is 50 mV per division and the time base is 500 ms per division. Find the amplitude and the frequency of the signal.
4. If the gain is 30 V per division and the time base is 1 ms per division, find the amplitude and the frequency of the signal.
5. The vertical scale is 100 V per division. What is the peak-to-peak voltage of the waveform?
6. The horizontal scale is 5 ms per division. What is the time at which the voltage reaches its lowest value?
7. The horizontal scale is 1 ms per division. What is the time at which the voltage reaches its highest positive value?
8. The signal is a 1-V peak-to-peak sine wave. If the horizontal scale is 0.5 ms per division, what is the voltage at 1.5 ms?
9. The signal is a standard 1-V peak-to-peak. What is the scale of the vertical amplifier?
10. The frequency of the sine wave is 1 kHz. What is the horizontal scale?

(See CD-ROM for Test 15-2)

Vectors and Phasors

In Job 15-3 you were introduced to the concept of phase angles for an ac wave. In this and the following jobs the concept will be expanded upon. You will learn how different ac signals can be related to each other and added together. This is done through the concept of vectors or phasors. First you will learn the difference between scalar and vector quantities. Next you will learn about the pythagorean theorem for adding two vectors that are at right angles to each other. This is the basis for adding vectors with an angle between them. In the chapters that follow, these concepts will be used to analyze specific electronic devices such as inductors, transformers, capacitors, and multiphase circuits.

As we heave learned, an ac voltage consists of a number of different instantaneous voltages, each instant of time giving rise to a different value of voltage. If this voltage is impressed across a resistor, each instantaneous voltage will cause an instantaneous current to flow at that time. The increasing and decreasing voltages and currents are shown in Fig. 15-22 on page 524. The current that flows as a result of an ac voltage will be an alternating current with the same frequency as the ac voltage. The voltage and the current both start at zero and rise to their maximum values, reaching them at the same instant. The voltage and current continue to rise and fall in step with each other throughout the entire cycle. We say that the voltage and current are "in phase" or "in step" with each other. It is difficult to draw these curves whenever we wish to indicate this or any other condition, and so we shall use a method in which the voltage and current are indicated by straight lines that are drawn to a definite length and in a definite direction.

Vectors. A *scalar quantity* is one that has size and is independent of direction. Examples of scalar quantities are length and time.

CALCULATOR HINT

Save the scale multiplier for the axes in memory registers to save time in calculations.

Hint for Success

Phasors or vectors are graphic representations of a number.

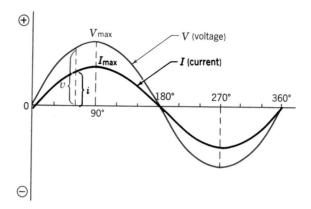

Figure 15-22 The voltage and current are in phase in a purely resistive circuit.

A *vector quantity* is one that has size *and* direction. Examples of vector quantities are force and acceleration. Vectors are used whenever we wish to show the amount and *direction* of a quantity. For example, *both* the amount and direction must be indicated to describe adequately a 10-lb force that acts straight up. If 1 in represents 10 lb, then Fig. 15-23*a* describes this force. However, if the force were acting straight *down*, it would be described as shown in Fig. 15-23*b*. We shall use vectors to show the amount and *time* at which a voltage or current is acting. Vectors that are drawn in the *same direction* will indicate that they are happening at the *same time*, or are "in phase." Vectors that are drawn in different directions will indicate that they are happening at *different times* or are "out of phase." In electricity, since different directions really represent *time* expressed as a phase relationship, an electrical vector is called a *phasor*.

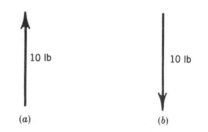

Figure 15-23 A vector indicates both amount and direction.

In a circuit containing only resistance, we have seen that the voltage and current occur at the *same time*, or are in phase. To indicate this condition by means of phasors, all that is necessary is to draw the phasors for the voltage and the current in the *same direction*. The value of each is indicated by the *length* of the phasor.

The current vector is usually drawn on a horizontal line and is used as a reference line for the other vectors (phasors) in the same diagram. Draw a line *AC* from left to right as in Fig. 15-24 to represent the current phasor. Place an arrowhead on the phasor at *C* pointing to the right. Point *A* is the "tail" of the phasor, and point *C* is the "head" of the phasor. Since the voltage occurs in phase with the current, the voltage phasor must be drawn in the *same direction* as the current phasor. Draw a line starting from the original point *A* to the right to point *V* with an arrow pointing to the right. This voltage phasor is drawn larger than the current phasor if the voltage is larger than the current. When phasors are drawn in the same

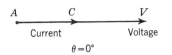

Current   Voltage

$\theta = 0°$

Figure 15-24   The voltage and current phasors are "in phase" in a circuit containing only resistance.

direction, the angle between the phasors is 0°. This angle is called the *phase angle* and is denoted by the Greek letter theta (θ).

Example 15-17

A 22-Ω electric iron is operated from a 110-V 60-Hz line. Draw a phasor diagram, and find the current and power used by the iron.

Solution

Given: $V = 110$ V   Find: $I = ?$
         $R = 22\ \Omega$ $P = ?$

1. Draw the phasor diagram. Since the iron may be assumed to be made of a purely resistive element, the phasor diagram will be the same as that shown in Fig. 15-24. It is not drawn to scale.
2. In a purely resistive circuit, Ohm's law may be used.

$$V = IR \qquad (2\text{-}1)$$
$$110 = I \times 22$$
$$I = \frac{110}{22} = 5\text{ A} \qquad Ans.$$

3. Find the power.

$$P = I \times V \qquad (7\text{-}1)$$
$$P = 5 \times 110 = 550\text{ W} \qquad Ans.$$

Exercises

Applications

Draw a phasor diagram for each exercise (not to scale).

1. What is the hot resistance of a tungsten lamp if it draws 2 A from a 120-V ac line?
2. Find the current drawn by a 50-Ω toaster from a 120-V ac line.
3. What is the voltage needed to operate a 600-W neon sign that has a resistance of 20 Ω?
4. An electric soldering iron draws 0.8 A from a 120-V 60-Hz line. What is its resistance? How much power will it consume?
5. What ac voltage is required to force 0.02 A through an 8000-Ω radio resistor? What is the power used?
6. Find the current and power drawn from a 110-V 60-Hz line by a tungsten lamp with a hot resistance of 275 Ω.
7. What is the current drawn by a 200-W incandescent lamp from a 110-V 60-Hz line? What is the hot resistance of the lamp?

8. Find the power used by a 24-Ω soldering iron which draws 5 A.
9. Find the voltage needed to operate a 500-W electric percolator if it draws 4.5 A. What is its resistance?
10. A heater rated at 10 kW operates from a 120-V line. What is the resistance of the heater?

The Pythagorean Theorem

The total voltage across an ac circuit is found by adding the voltages *even though the voltages do not appear at the same phase!* The formulas for this total voltage and all our work in ac power are applications of the *pythagorean theorem.*

In Fig. 15-25, squares are drawn on each of the three sides of the right triangle. The length of the sides of each square is equal to the length of the side of the triangle on which it is drawn. Pythagoras discovered that the *sum of the areas of the squares on the two legs of a right triangle is exactly equal to the area of the square erected on the hypotenuse.* This is true for *any right triangle.* Thus,

$$\text{Area I} + \text{area II} = \text{area III}$$
$$9 + 16 = 25$$

Since the area of a square is equal to a side times itself,

$$\text{Area I} = a \times a = a^2$$
$$\text{Area II} = b \times b = b^2$$
$$\text{Area III} = h \times h = h^2$$

The theorem may be stated as the following rule.

RULE The sum of the squares of the legs of a right triangle is equal to the square of the hypotenuse.

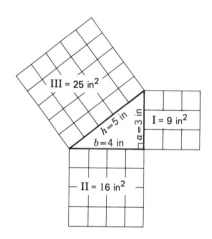

Figure 15-25 Pictorial representation of the pythagorean theorem: $a^2 + b^2 = h^2$.

Hint for Success

The pythagorean concept is a distance measure that is common in mathematics because it has the properties of distance that we feel intuitively and have taken as a property of our number system.

$$a^2 + b^2 = h^2 \qquad \text{Formula 15-15}$$

where a, b = legs of a right triangle

           h = hypotenuse of a right triangle

The formula may be solved for h by taking the square root of both sides.

$$\sqrt{a^2 + b^2} = \sqrt{h^2}$$
$$\sqrt{a^2 + b^2} = h \qquad \qquad \text{15-15a}$$

Example 15-18

Find the hypotenuse of a right triangle whose altitude is 5 in and whose base is 12 in.

Solution

The diagram is shown in Fig. 15-26.

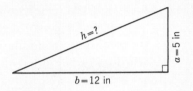

Figure 15-26

$$h = \sqrt{a^2 + b^2}$$
$$h = \sqrt{5^2 + 12^2}$$
$$h = \sqrt{25 + 144}$$
$$h = \sqrt{169} = 13 \text{ in} \qquad Ans.$$

Example 15-19

Find the altitude a of a right triangle whose hypotenuse h is 17 in and whose base b is 15 in.

Solution

Given: $h = 17$ in   Find: $a = $?

             $b = 15$ in

$$a^2 + b^2 = h^2$$
$$a^2 + 15^2 = 17^2$$
$$a^2 + 225 = 289$$
$$a^2 = 289 - 225$$
$$a^2 = 64$$
$$a = \sqrt{64} = 8 \text{ in} \qquad Ans.$$

CALCULATOR HINT

For calculators that have polar-rectangular conversion, pythagorean calculations can be made by doing the built-in conversion between those coordinate systems.

Self-Test 15-20

An electrician's toolbox measures 16 in × 12 in × 10 in. What is the length of the largest extension bit holder that can be placed in the tool box?

Solution

The toolbox is shown in Fig. 15-27.

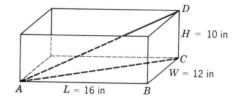

Figure 15-27   *AD* is the longest dimension in the toolbox.

1. Find the length of the diagonal *AC*. Triangle *ABC* is a right triangle with the right angle at point _____.

$$h = \sqrt{a^2 + b^2} \qquad (15\text{-}15a)$$
$$AD = \sqrt{16^2 + (\underline{\qquad})^2}$$
$$AC = \sqrt{256 + \underline{\qquad}}$$
$$AC = \sqrt{400} = \underline{\qquad} \text{ in}$$

2. Find the length of the bit holder *AD* using triangle _____, in which the right angle is at point _____.

$$h = \sqrt{a^2 + b^2} \qquad (15\text{-}15a)$$
$$AD = \sqrt{10^2 + (\underline{\qquad})^2}$$
$$AD = \sqrt{100 + 400} = \sqrt{\underline{\qquad}}$$
$$AD = \underline{\qquad} \text{ in} \qquad Ans.$$

Exercises

Practice

Find the unknown side in each of the following right triangles:

Exercise	a	b	h
1	6	8	?
2	7	24	?
3	?	63	65
4	33	?	65
5	14	22.5	?
6	17.5	6	?
7	?	20	20.5
8	6.5	?	42.5

Applications

9. A guy wire stretches from the top of a 40-m-high pole to a stake in the ground 60 m from the foot of the pole. Find the length of the wire.
10. A 10-ft ladder is placed against a wall with the foot of the ladder 6 ft from the base of the wall. How high above the ground will the ladder reach?
11. A doorway measures 3 by 7 ft. What is the diameter of the largest circular table that will pass through the doorway?
12. A conduit must be bent to provide a rise of 6 ft in a horizontal distance of 4 ft 6 in. What is the length of conduit between the bends?
13. Find the length of electrical conduit needed to include the 5-ft offset as shown in Fig. 15-28.

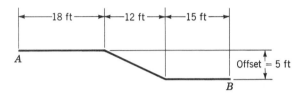

Figure 15-28   Find the length of conduit from *A* to *B*.

14. Find the distance across the corners of a square nut measuring 3¼ in on a side.
15. Find the distance *X* in Fig. 15-29.

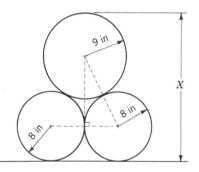

Figure 15-29

(See CD-ROM for Test 15-3)

Power

In a purely resistive circuit, the voltage and current are in phase. The power is equal to the product of *V* and *I* and is expressed in *voltamperes* (VA) or kilovoltamperes (kVA). In those ac circuits in which the voltage and current are *not* in phase, the useful or actual power is equal to the voltage multiplied by that portion of the current in phase with it. The amount of this in-phase current may be obtained from Fig. 15-30. The current in an ac circuit may be considered to be made up of two

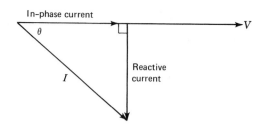

Figure 15-30   An alternating current is made up of two components.

components: a component in phase with the voltage and a component out of phase by 90°. By trigonometry, the in-phase component equals the current I multiplied by the cosine of the phase angle.

$$\textbf{In-phase current} = I \times \cos \theta \qquad \text{Formula 15-16}$$

The general formula for power in an ac circuit is

$$W = V \times \textbf{in-phase current} \qquad \text{Formula 15-17}$$

Hint for Success

Power is the rate at which energy is used.

By substitution, we obtain the

Formula $\qquad W = V \times I \times \cos \theta \qquad$ Formula 15-18

where W = effective or true power, W
V = voltage, V
I = current, A
θ = phase angle between voltage and current

The multiplier $\cos \theta$ is called the *power factor* (PF) of the circuit. W is the true watts and VI is the apparent watts, or voltamperes. Substituting PF for $\cos \theta$ in formula (15-18) gives

$$W = V \times I \times PF \qquad \text{Formula 15-19}$$

Since the PF is always less than 1 ($\cos 0° =$ a maximum of 1, decreasing to $\cos 90° = 0$), the true power will always be *less* than the apparent power VI.

Solving formula (15-19) for PF, we obtain

$$PF = \frac{W}{V \times I} \qquad \text{Formula 15-20}$$

or $\qquad PF = \dfrac{\textbf{true power}}{\textbf{apparent power}} \qquad$ Formula 15-21

The power factor may be expressed as a decimal or as a percent. For example, a PF of 0.8 may be written as 80%. In this sense, the PF describes the *portion* of the voltampere input which is actually effective in operating the device. Thus, an 80% PF means that the device uses only 80% of the voltampere input in order to operate. The higher the PF, the greater is the efficiency of the device in using the voltampere input.

Low-power-factor circuits occur when the phase angle between the current and voltage is large. This angle increases as the total reactance in the circuit increases. Practically pure resistive circuits (incandescent lamps) have high power factors of 95 to 100%. The power factors of circuits containing large induction motors (inductive reactance) range between 85 and 90%. Fractional-horsepower induction motors have power factors between 60 and 75%. Low power factors just mean wasted power, and should be increased whenever possible. We can increase a low PF caused by a lagging inductive reactance by adding an *opposite, leading* capacitive reactance, and vice versa. This will be discussed in greater detail in Chap. 18.

In a pure inductance, since the voltage leads the current by 90°, the power will be

$$W = V \times I \times \cos 90°$$
$$W = V \times I \times 0$$
$$W = 0 \text{ W}$$

Thus, the average power used by a pure inductance is zero. Actually, the inductance uses power to build up its magnetic field during one quarter of a cycle, but it delivers an equal amount of power back to the source while the field is collapsing during the second quarter of its cycle. The net result is that zero power is used by the inductance. A perfect inductance can be considered to be just like a perfect flywheel, which accumulates power during one revolution and delivers an equal amount of power back to the engine during its second revolution.

From the definition of the effective ac ampere (Job 15-4), power equals the effective current squared, multiplied by the resistance of the circuit, or

$$W = I^2 R$$

This is true for any single-phase ac circuit and is independent of the phase relation between the current and the voltage. Thus, in a pure inductance, in which $R = 0$,

$$W = I^2 R = I^2 \times 0 = 0 \text{ W} \qquad \textit{Check}$$

Exercises

Applications

Draw a phasor diagram for each exercise (not to scale).

1. Find the current sent through a 0.03-H coil by a voltage of 188.4 V at a frequency of 1 kHz.
2. Find the current and effective power drawn by a 200-mH coil which is connected to a 31.4-V source at a frequency of 1000 Hz.
3. What voltage is needed to force 0.08 A through an inductance of 0.5 H at a frequency of 100 Hz?
4. What voltage at 10 kHz is necessary to send a current of 20 mA through an inductance of 50 mH? Find (*a*) the PF, and (*b*) the effective power consumed.
5. What must be the reactance of a filter choke in order for it to pass 80 mA of current when the voltage is 240 V? What must be the inductance of the choke if the frequency is 60 Hz?
6. The coil of an electromagnet operating at 12 V 60 Hz requires 1.5 mA to operate. What is the reactance of the coil?
7. If the coil has a reactance of 8 kΩ at 60 Hz and a resistance of 1 kΩ, what is the power factor of the coil?
8. What is the current through the coil of Exercise 7 when placed in a 12-V 60-Hz circuit?
9. How much power is dissipated in the coil of Exercise 8?
10. What is the phase angle between the voltage and the current for the coil in Exercise 7?

Harmonic Analysis

In the previous jobs, it has been assumed that there is only one frequency in the voltage signal. This job assumes that the voltage is composed of the previous single frequency plus multiples of this basic frequency. The basic frequency is called the *fundamental.* The multiples of the fundamental are called *harmonics.*

Power systems in the United States have a fundamental frequency of 60 Hz. Figure 15-31 shows the relationship between the fundamental and the harmonics for a 60-Hz voltage. Part *a* shows the fundamental and the second harmonic. The second harmonic is twice the fundamental, or 120 Hz. When the fundamental goes through one cycle, the second harmonic goes through two cycles (see Fig. 15-31*a*.)

The third harmonic is 180 Hz. It is shown relative to the fundamental in Fig. 15-31*b*. The third harmonic goes through three cycles for each cycle of the fundamental. The fourth harmonic is shown in Fig. 15-31*c*.

Hint for Success

Harmonics in electricity are the same as harmonics in music.

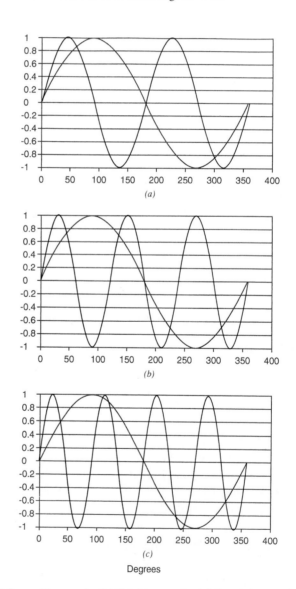

Degrees

Figure 15-31 (*a*) Second harmonic. (*b*) Third harmonic. (*c*) Fourth harmonic.

There are two main reasons why harmonics are of interest. First, it can be shown that any voltage on an electrical system can be represented as the sum of harmonics. The effects of network components, such as filters, can be understood in terms of frequency response. Frequency response is just a unified way of calculating the response of the circuit to each of the harmonics. Frequency response is most commonly referred to in hi-fi systems. This type of analysis is called working in the frequency domain—i.e., looking at the frequency aspects of a voltage rather than the time aspects. The relationship between the time and frequency domains is unique and is called *Fourier analysis*.

The second important reason for learning about harmonics is because of the effects of nonlinear components on voltages. Nonlinear devices create harmonics from a fundamental. The most common nonlinear devices are power supplies. Most of them use diodes, which are nonlinear devices. These diodes produce surges or pulses of current on the branch circuits. These pulses are full of harmonics. We will later see how the sum of harmonics can produce a pulse and noiselike voltage.

Summary

1. The fundamental is the lowest, or basic, frequency of a voltage.
2. Harmonics are multiples of the fundamental frequency.
3. Any voltage can be represented by the sum of its harmonics.
4. The sum of any set of harmonics is a waveform.
5. For a linear system the effects of a device can be determined by the device's response to the harmonic frequencies. This is called the *frequency response*.
6. Harmonics are generated by nonlinear devices.
7. The relationship between the time form of a waveform and its frequency form is unique.
8. The relationship between time and frequency representations is called Fourier analysis.

Square Waves

Example 15-21

Figure 15-32 on page 534 shows a square wave that has the values 1 and -1. It provides an interesting example of harmonic analysis. Fourier analysis shows that such a wave form is composed only of odd harmonics—i.e., 3, 5, 7, etc. The magnitude of each harmonic is $\frac{1}{n}$, where n is the number of the harmonic. The third harmonic has magnitude $\frac{1}{3}$, the magnitude of the fifth harmonic is $\frac{1}{5}$, etc.

Figure 15-32 also shows the approximation composed of the first four components. These are the fundamental and the third, fifth, and seventh harmonics. Notice how closely the sum of only four harmonics represents the original square wave. The average of the high and low values is about 0.8 instead of 1, but other than that the approximation is recognizable as a square wave. If we just adjust the gain on our amplifier, we can bring the average up to 1 V.

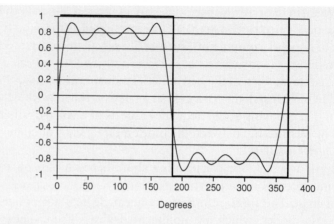

Figure 15-32

What happens if we change or remove some of the harmonics? Suppose that the fundamental is reduced by one-half. What will happen to the square wave? Figure 15-33 shows the result. The edges still go up rapidly, but the center of the plateaus droop. The high level and the low level of a square wave are like dc voltages. A square wave is created by a switch that switches between two dc sources. If you want to maintain these levels, then you must keep all the frequencies that are near dc—i.e., the fundamental and the lowest harmonics.

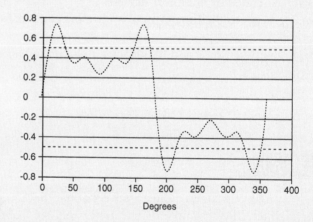

Figure 15-33

What happens if the high frequencies are removed? Figure 15-34 shows the waveform generated by the fundamental and the third harmonic. The entire fifth and seventh harmonics have been removed. The result still looks like a square wave even if only two sine waves are used. The effect of removing high frequencies is that the transitions, or edges, do not go up as fast. High frequencies give the rapid changes. Notice that the average of the high levels is higher than it was in Fig. 15-33. This is because the fundamental and low harmonics reproduce those levels.

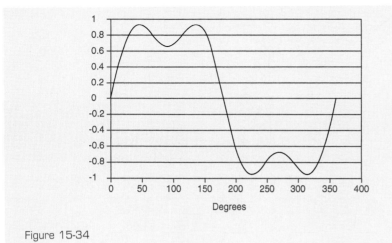

Figure 15-34

Summary

High frequencies are generated by rapid shapes in the waveform. They determine how fast the level of a waveform changes. However, low frequencies are required so that dc levels can be maintained. When pulses resemble a sine wave, only the odd harmonics are needed to reproduce the waveform.

Applications

It has usually been assumed that the concepts of harmonic analysis are required only by engineers. However, as technology becomes increasingly a part of our daily lives, engineering techniques move into our lives also. The following are only two of the basic, but common, examples of how harmonic analysis enters our modern life.

Power distribution. This example shows how pulses can arise in power systems. Many common devices used in today's equipment are nonlinear and create pulse signals on power branch circuits. This helps to explain why harmonic analysis is becoming more important in modern power system operation.

The advances in technology all seem to cause greater consumption of electrical power, along with a collection of all types of electrical and electronic equipment. For example, we use dimmer switches to control the light level. We may put a TV in every room. We are being told that our children's future will be bleak if we do not provide them with computers to enhance their education. These devices all illustrate how "noise" gets into the power distribution system.

Dimmer switches create pulses of power like those shown in Fig. 15-35 on page 536. The switch turns on for only part of a cycle and turns off when the voltage goes through zero. This produces the pulses shown in the figure. By controlling the amount of time the switch is on, we can control the brightness of the light. You may have noticed the effects of a dimmer switch on your TV picture.

about electronics

What's the "skin effect," and why is this guy still alive to talk about it? Twice a day the Museum of Science in Boston encapsulates someone in a cage to show visitors that lightning is a very short, high frequency alternating current, rather than direct current. Current generates a magnetic field. With alternating current, there is a delay in the magnetic field's response to the change in current. The "old" magnetic field pushes the current toward the outside of the cage (which is the conductor). That effect increases as the frequency increases. At very high frequencies, the whole current flows in a very narrow "skin" on the conductor.

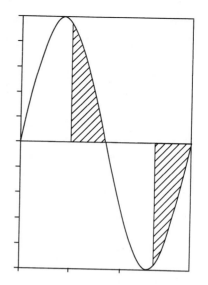

Figure 15-35

Power supplies for computers and television sets generally behave like full-wave rectifiers. Such rectifiers create pulses of current that occur near the peaks of a sine wave. The resulting current pulses look something like those in Fig. 15-36.

Thus, we see that pulses from common devices introduce harmonics that are tied to the fundamental of the ac power supply. In Job 19-5 we will see that these harmonics can be considered EMI and must be filtered out. In Job 19-6 we will see how these harmonics can contribute to neutral line current.

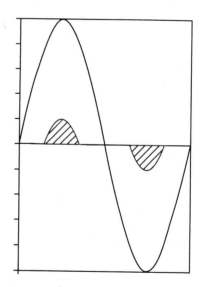

Figure 15-36

Digital signals.   Digital signals usually look something like the square wave of Fig. 15-32. Computers and other digital systems usually operate with fixed states, such as 0, 1, −1. Assume that the square pulse of Fig. 15-32 is put into a cable and that for some reason only half of the fundamental comes out. The resulting waveform would look something like Fig. 15-33. The output has a droop in the high and low levels of the signal.

Digital systems are desirable because we know what the signal should be. It should be 0, 1, or −1. We could, for example, say that any voltage above 0.5 V should be 1 and any voltage less than −0.5 should be −1. Any other voltage is 0. This solves the problem. Or does it?

The 0.5-V level is called a "fiduciary" level. This is the level at which we make a decision. For example the −1 fiduciary level is −0.5 V. The droop in the input signal crosses the fiduciary levels. This causes false signals, as shown in Fig. 15-37. The high-frequency components cause a fast rise in the signal. This causes the signal to cross the 0.5-V fiduciary level, and the output goes to a +1.

The reduced level of the fundamental causes a droop large enough for the signal to go back below the fiduciary level. This causes the output to go back to 0. As the droop goes back up, it causes the signal to again cross the fiduciary level and the output again goes to a +1. The signal then transitions back to a −1, and on the way it causes the output to go to 0 again. The same thing happens for the negative part of the square wave. The result is a set of pulses that can confuse the digital system.

The output is related to what it should be, but the relationship is very poor. It would be very easy for a digital system to misunderstand the signal. In addition, there are many more pulses than there should be, which could cause a problem if the system were counting pulses. This also increases the high-frequency noise of the system.

CALCULATOR HINT

For both speed and ease, harmonic analysis is generally done with a computer.

Application Summary

These applications showed how harmonics are generated in power distribution and indicated some effects that can come from them. The second example showed how distortions introduced in a signal can affect digital systems.

Harmonics are generated by nonlinear devices. We also learned that in power systems nonlinear devices generally produce harmonics that are related to the fundamental frequency of the power system. Also, data-transmission media can cause distortions in digital signals. These distortions can affect the logic of the detection device.

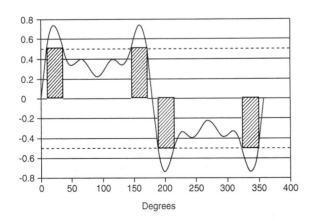

Degrees

Figure 15-37

Practice

1. Explain how an SCR dimmer switch or motor controller creates pulses of power in a branch circuit.
2. What is the frequency of the seventh harmonic of a 60-Hz line voltage?
3. What is the relationship that gives the amplitude of the harmonics for the square wave of Fig. 15-32?

True or false?

4. Any electrical waveform can be represented by a sum of harmonics.
5. The highest frequency that can have a full cycle in a waveform is called the fundamental.
6. A filter can be used to shape a signal by removing or reducing harmonics.
7. A square wave can be made from the sum of sine waves.
8. Removal of components of a signal can cause proper operation of a digital signal.
9. Nonlinear devices can create harmonics of a signal.
10. The relationship between the time representation of a signal and the frequency representation of the signal is called harmonic analysis.

1. Figure 15-38 is a graph of a series *RC* time constant function. What is the percent of total amperes measured for each of the 1-s intervals?

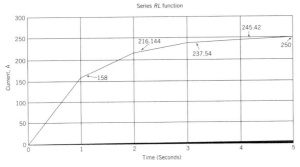

Figure 15-38

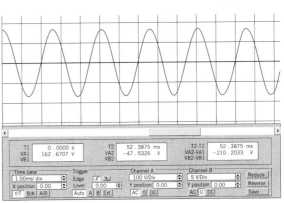

Figure 15-39

2. The maximum voltage of a 1-kHz ac sine wave is 28 V. How long does it take the waveform to go from +28 V to −28 V?

3. Why is the rms value of a sine waveform also called the effective value?

4. What is the frequency of the ac waveform shown in Fig. 15-39?

5. An AM/FM radio draws 6 A at 120 V. Draw a vector for each of these values.

6. A 28-ft extension ladder (fully extended) reaches a point on a wall 27 ft from the ground. How far from the wall is the base of the ladder?

7. A circuit has 5000 VA delivered to it, and only 3000 W of true power is consumed. Draw vectors showing the relationship between apparent power (VA), true power (watts), and reactive power (var) in this inductive circuit.

8. Draw a sine wave and show the second and third harmonic on the same fundamental waveform.

INTERNET
A C T I V I T I E S

Internet Activity A

Here is a good question to ponder: In alternating current, do hot and neutral switch? **http://www.madsci.org/posts/archives/may98/893818667.Eg.q.html**
Be sure to go to the response:
http://www.madsci.org/posts/archives/may98/893818667.Eg.r.html

Internet Activity B

Students using this book surely have never considered "why electricity is impossible to understand." Find reasons why others have problems caused by "some electrical misconception articles."
http://www.eskimo.com/~billb/miscon/miscon.html

Chapter 15 Solutions to Self-Tests and Reviews

Self-Test 15-12

1. $15°$
a. $\sin$, θ, 0.2588, 50
b. I_{max}, 50

Self-Test 15-16

1. oscilloscope
2. time
3. time-base
4. amplitude
5. gain
6. period
7. period

Self-Test 15-20

1. *B*, 12, 144, 20
2. *ACD*, *C*, 20, 500, 22.3

Inductance and Transformers

16

chapter

1. Explain how inductance is created.
2. Determine the reactance of a coil.
3. Write solutions for series resistance and inductance.
4. Determine real inductor losses and quality.
5. Find the inductance of a coil.
6. Determine the inductance of transmission lines.
7. Write equations for ideal transformers.
8. Find the current values in a transformer.
9. Write equations for impedance transformation in a transformer.
10. Write equations for a real transformer.
11. Determine the efficiency of a transformer.
12. Find the properties of real cables for power transmission.

This chapter explores the concept of inductance and inductive devices. Inductance in the ideal sense is developed first. The ideal inductor is then made more realistic by including resistance in the model. The properties of this real inductor such as phase angle and Q are then studied.

Examples of inductance help us to understand real-world effects. The inductor examples looked at relate to parallel-wire transmission lines. The model covers concepts from an extension cord to high-tension power lines.

A transformer is a more complex inductive device. This device is first developed in its ideal form. The basic properties of voltage, current, and impedance transformation by this ideal transformer are then studied. The ideal transformer is then made more real by developing a model of a transformer that consists of three real inductors.

The real-transformer model is then applied to some common test procedures that are used with transformers. These are the open-circuit or no-load test and the short-circuit test. The transformer model helps us to understand why the tests perform the way they do and what they measure.

JOB 16-1

Inductance

In this chapter no new mathematics is required beyond that already developed, but new concepts are developed. The primary new concept to understand is that of voltage and current being out of phase with each other. For the resistive devices that were previously analyzed, the voltage and the current were in phase. In an inductor the voltage leads the current by 90°. This means that when we look at power as voltage times current we get a different interpretation. Because of the angle between the voltage and the current the power in a pure inductor is not dissipated as heat as in a resistor but is stored and later returned to the circuit; i.e., *an inductor stores energy in a magnetic field.*

Another major difference in concept is the Ohm's law relationship between voltage and current. In the earlier work this relationship resulted in resistance. For the inductor the phase difference between voltage and current means that this is not a standard resistor; therefore, we call the relationship reactance. Reactance reduces the amount of current, as does a resistor.

Basic ideas

Creating Magnetic Flux

In Chap. 15 we learned that a magnetic flux field exists around a wire carrying a current. The strength of the field is proportional to the current through the wire. We also found that when the flux changes, a voltage is induced into a coil.

To see more about how inductance works, look at Fig. 16-1. A block of material with a hole in the center has wires wrapped around both ends. As a current I_1 flows through the turns of wire, a magnetic field is created in the material. The current creates a *magnetomotive force* (mmf) that acts like a magnetic voltage. This mmf causes a flux to flow in a magnetic circuit. The flux is similar to a magnetic current. The symbol for mmf is H. The symbol for flux is ϕ.

Experience tells us that the more turns of wire we put on the form, the stronger the magnet will be. This means that the potential, or mmf, must be increasing as the number of turns increases. If the number of turns of wire is N_1, we get the relationship for the mmf as

$$H = N_1 \times I_1$$

The *magnetic flux* is like a magnetic current in the material. The total flux is the sum of all the flux in the area of the coil. The symbol ϕ is used for the total flux. The density of the flux is the total flux divided by the area of the coil, A, in which the flux is concentrated. The flux density has the symbol B and is given by the equation

$$B = \frac{\phi}{A}$$

The amount of flux that is created depends upon the type of material that is used in the magnetic circuit. A block of wood does not make as good a magnet as does a block of steel. B depends upon a property of the material called the *magnetic permeability*, denoted by the symbol μ. The relationship between B and H is

$$B = \mu H$$

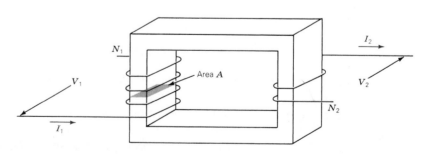

Figure 16-1   Creating magnetic fields and induced voltages.

In summary, the product of the number of turns of wire and the amount of current in the turns determines the potential, mmf, of the magnetic field. The density of flux, or current, that flows depends upon the material (μ). The total flux ϕ is the product of the flux density B and the area A of the magnetic circuit.

Inducing Voltages

In Chap. 15 it was shown that if the flux in a magnetic circuit does not change, no voltage is induced into an electric circuit. The changing of the flux can be caused either by an electrical change in the current I_1 or by a mechanical movement of a coil. A transformer stays in a fixed place, but the flux changes when an ac voltage is applied. For other devices a fixed magnetic field can be used to generate a voltage by rotating a coil as in the generator in your car. Either way the effect is the same. The coil into which the voltage is induced sees a time rate of change of the flux field.

In Fig. 16-1 the turns of wire N_2 sense the flux in the magnetic circuit. If that flux changes, a voltage is induced into these turns. If one turn of wire picks up 1 V, more turns of wire should get a larger voltage. The relationship is linear, i.e.,

$$V_2 = N_2 \times \text{the time change of } \phi$$

> **Hint for Success**
>
> To convert magnetic energy to electric energy requires that there be a time rate of change of the flux field. This can be caused by changes in the electric current or by mechanical changes in the geometry, or both.

Inductance

The induced voltage or back emf acts to oppose the change in magnetic flux. This appears very natural because if it were not true, then once we started to increase the flux the back emf would cause it to continue to increase and soon the whole world would be consumed in a single flux field. The inductance of a circuit is defined as the relationship between the induced voltage and the rate of change in current that produces it. The symbol for inductance is L and the basic unit is the henry (H). By definition, if the current changes at the rate of 1 ampere (A) per second and a voltage of 1 V is induced into the circuit, the inductance is 1 H.

From Fig. 16-1 we can see that there are different types of inductance. If we remove the second set of turns of wire in Fig. 16-1, we have a single electric circuit and a single magnetic circuit. The magnetic flux links only the coils of wire N_1. This device is called an inductor or a choke coil. The inductance that we would measure is called the *self-inductance* because it causes an effect on the driving current I_1 itself.

Now put back the second coil of wire N_2. I_1 can now cause a voltage to be induced into the second coil. The inductance that relates I_1 to V_2 is called the *mutual inductance*.

The structure shown in Fig. 16-1 is a transformer. A transformer is a complex device composed of two self-inductors, one at either end, and a mutual inductor. We shall study transformers in greater detail later.

JOB 16-2

Review of Inductance

1. The field characteristic that causes a magnetic flux is the mmf. The symbol for mmf is _____.
2. The factors that affect the magnitude of H are the number of __TURNS__ in the coil and the magnitude of the __CURRENT__

3. The flux density is represented by the symbol _____. It is the total flux ϕ divided by the _____ of the magnetic circuit.
4. The flux density B depends on the material used. The relationship is $B = \mu \times H$, where μ is the magnetic _____.
5. The inductance relates time rate of change of the _____ in the coil to the induced _____.
6. When a coil induces a voltage into itself, the inductance is called _____.
7. When a coil induces a voltage into another coil the inductance is called _____.
8. The unit of inductance is the _____ which is defined as the inductance that will induce _____ volt when the current changes at the rate of _____ ampere per second.
9. The induced voltage _____ the change in current.
10. Two ways to make the flux change are to use an _____ voltage or to _____ the coil in the flux field.

JOB 16-3

Reactance of a Coil

An ac voltage impressed across a coil will produce an alternating current of the same frequency as the voltage. The changing current will produce changing flux in the coil. The changing flux will induce an emf in the coil. This emf is a *back emf* which acts to oppose the original voltage. This opposition, called the inductive reactance, will reduce the current below that which would flow if there were no back emf. This reactance also prevents the current from appearing at the same phase as the voltage. The current will be pushed back in time. We say that the current "lags" behind the voltage that produces it. In a perfect coil—one which has only inductance and zero resistance—the current will lag behind the voltage by ¼ cycle. We say that the current lags the voltage by 90° as shown in Fig. 16-2.

Phasor diagram. Since the voltage and current are out of phase by 90°, the phasors must be drawn in two different directions 90° apart. In addition, we must show which phasor is the "leading" phasor and which is the "lagging" phasor. The

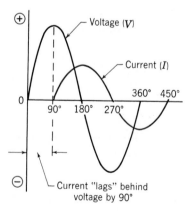

Figure 16-2   The voltage and current are *out of phase* in an ac circuit containing only inductance.

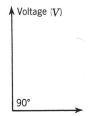

Figure 16-3 The voltage phasor leads the current phasor by 90° in a purely inductive circuit.

current is used as the reference line upon which to draw the phasor diagram. Draw a line with an arrow pointing to the right as shown in Fig. 16-3. This will represent the current I in the circuit. Since the voltage leads the current by 90°, we draw the voltage pointing upward.

Ohm's law shows the relationship between voltage and current, $V = I \times R$, for a resistance. In an inductance we saw that there is a relationship between time rate of change of current and voltage. The difference between an inductor and a resistor is that for a resistance the voltage and the current are in phase while for an inductance the voltage and the current are 90° out of phase with each other. Because of this phase relationship no power is dissipated in a pure inductor. Energy is stored in the magnetic field and then returned to the circuit.

When Ohm's law is applied to inductance, the relationship between voltage and current is called *reactance* instead of resistance. The inductor is said to *react* to the applied current. The symbol for reactance is X. This different symbol helps us remember that there is a relationship between voltage and current but that the energy is only stored and not dissipated.

There are two important factors that affect the amount of reactance. In the discussion on inductance we saw that the materials used, the number of turns in the coil, and the physical dimensions of the inductor relate the induced voltage to the current that causes the voltage. These factors define the amount of inductance. The larger the inductance the larger the reactance. The henry is the definition of the inductance, L.

The second factor is the rate of change of the current in the coil. In an ac circuit the higher the frequency the faster the flux changes. This means that the reactance increases as the frequency increases. The factor that relates the reactance to the frequency is $2\pi = 6.28$.

Putting all this together, we arrive at the equation for the reactance.

$$X_L = 6.28fL \qquad \text{Formula 16-1}$$

where X_L = inductive reactance, Ω
         f = frequency, Hz
         L = inductance, H
    6.28 = constant of proportionality

In electricity and electronics we often come across the factor $2\pi f$. A special symbol is used for this factor so that we do not have to write so much. The symbol is ω.

Another thing that we want to do is remind ourselves that the voltage leads the current by 90°. To do this, we use another symbol, j. When we use these new symbols, the formula for the reactance becomes

$$X_L = j\omega \times L \qquad (16\text{-}1a)$$

This form of the formula is easy to remember and carries more information.

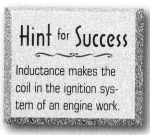

Hint for Success

Inductance makes the coil in the ignition system of an engine work.

Example 16-1

A 20-mH coil is in a tank circuit operating at a frequency of 1500 kHz. Find its inductive reactance.

Solution

Given: $L = 20$ mH   Find: $X_L = ?$
           $f = 1500$ kHz

1. Change units of measurement.

$$20 \text{ mH} = 20 \times 10^{-3} \text{ H}$$
$$1500 \text{ kHz} = 1500 \times 10^3 \text{ Hz}$$

2. Find the inductive reactance.

$$X_L = 6.28fL \qquad (16\text{-}1)$$
$$X_L = 6.28 \times 1500 \times 10^3 \times 20 \times 10^{-3}$$
$$X_L = 6.28 \times 1500 \times 20$$
$$X_L = 188{,}400 \ \Omega \qquad Ans.$$

Example 16-2

The primary coil of a power transformer has an inductance of 150 mH. (*a*) Find its inductive reactance at a frequency of 60 Hz. (*b*) What current will it draw from a 117-V line?

Solution

Given: $L = 150$ mH   Find: $X_L = ?$
           $f = 60$ Hz $I_L = ?$
           $V = 117$ V

a. $X_L = 6.28fL \qquad (16\text{-}1)$
                    $X_L = 6.28 \times 60 \times 150 \times 10^{-3}$
                    $X_L = 56.5 \ \Omega \qquad Ans.$

b. Since the only resistance in the circuit is the inductive reactance, the formula for Ohm's law, $V = IR$, may be rewritten as the

$$\boldsymbol{V_L = I_L \times X_L} \qquad \text{Formula 16-2}$$

$$117 = I_L \times 56.5$$

$$I_L = \frac{117}{56.5} = 2.07 \text{ A} \qquad Ans.$$

Example 16-3

What must be the inductance of a coil in order that it have a reactance of 942 Ω at a frequency of 60 kHz?

Solution

Given: $X_L = 942\ \Omega$ Find: $L = ?$
         $f = 60$ kHz $= 60 \times 10^3$ Hz

$$X_L = 6.28fL \qquad\qquad (16\text{-}1)$$
$$942 = 6.28 \times 60 \times 10^3 \times L$$
$$942 = 376.8 \times 10^3 \times L$$
$$L = \frac{942}{376.8 \times 10^3}$$
$$L = 2.5 \times 10^{-3}\ \text{H}$$
$$L = 2.5\ \text{mH} \qquad Ans.$$

CALCULATOR HINT

Using exponential notation or engineering notation makes data entry and readout easier.

Self-Test 16-4

A tuning coil in a radio transmitter has an inductance of 300 μH. At what frequency will it offer a reactance of 3768 Ω?

Solution

Given: $L = 300\ \mu$H   Find: $f = ?$
      $X_L = \underline{\quad (1) \quad}\ \Omega$

$$X_L = 6.28fL \qquad\qquad (16\text{-}1)$$
$$3768 = 6.28 \times f \times \underline{\quad (2) \quad}$$
$$3768 = \underline{\quad (3) \quad} \times 10^{-6} \times f$$
$$f = \frac{3768}{1884 \times 10^{-6}}$$
$$f = 2 \times \underline{\quad (4) \quad}\ \text{Hz}$$
$$f = \underline{\quad (5) \quad}\ \text{MHz} \qquad Ans.$$

Exercises

Applications

1. A 0.7-H coil is in series with a 5-kΩ resistor in a transistorized bass-boost circuit. Find the inductive reactance of the coil at (a) 100 Hz and (b) 2 kHz.
2. A loudspeaker coil of 2 H inductance is operating at a frequency of 1 kHz. Find (a) its inductive reactance and (b) the current flowing if the voltage across the coil is 40 V.
3. A 20-H Stancor C1515 choke in the filter circuit of a power supply operates at a frequency of 60 Hz. Find (a) its inductive reactance and (b) the current flowing if the voltage across the coil is 150 V.
4. An RF choke coil with an inductance of 5.5 mH operates at a frequency of 1200 kHz. Find (a) its inductive reactance and (b) the current flowing if the voltage across the coil is 41.5 V.

5. A Miller 7825 line filter choke used in a noise-control circuit of a flasher sign has an inductance of 0.6 mH. Find its inductive reactance at a frequency of 10 kHz.

6. The primary coil of an antenna transformer has an inductance of 50 μH. At what frequency will the reactance equal 314 Ω?

7. What must be the inductance of a coil in order that it have a reactance of 1884 Ω at 60 Hz?

8. An antenna circuit has an inductance of 150 μH. What is its reactance to a 1000-kHz signal?

9. An RF coil in an FM receiver has an inductance of 100 μH. What is its reactance at 100 MHz?

10. What is the reactance of the 10-mH coil in the high-pass filter shown in Fig. 16-4 to (a) a 3000-Hz AF current and (b) a 600-kHz RF current?

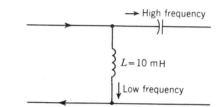

Figure 16-4 Simple high-pass filter.

11. A 1-mH coil in the primary of an IF transformer is resonant to 456 kHz. Find its inductive reactance at this frequency.

12. A 0.5-mH coil in the oscillator circuit of a continuous-wave transmitter operates at 30 MHz. Find its reactance at this frequency.

13. What must be the inductance of a coil in order that it have a reactance of 10,000 Ω at 600 kHz?

14. A transmitter tuning coil must have a reactance of 95.6 Ω at 3.8 MHz. What must be the inductance of the coil?

15. A choke coil of negligible resistance is to limit the current through it to 25 mA when 40 V are impressed across it at 1000 kHz. Find its inductance.

(See CD-ROM for Test 16-1)

JOB 16-4

Series Resistance and Inductance

Figure 16-5 shows a 100-Ω resistor connected in series with an inductance whose reactance is 100 Ω at a frequency of 60 Hz. A series current of 0.85 A produces a voltage drop of 85 V across both the resistor and the inductance. In a series circuit, the total voltage is ordinarily found by adding the voltages across all the parts of the circuit. This rule was used in dc circuits, but can we use it for ac circuits? Since the current in a series circuit remains unchanged throughout the circuit, we may draw the waveforms for the currents and voltages across each part of the circuit on the same drawing. Figure 16-6 shows the relationship of each voltage to the unchanging current. The voltage across the resistor V_R is *in phase* with the current I;

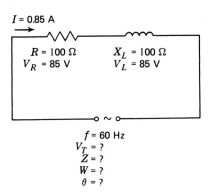

Figure 16-5 An ac series circuit containing resistance and inductance.

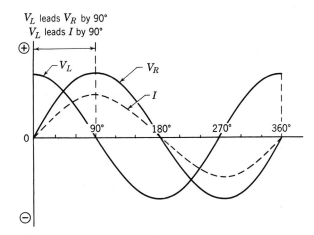

Figure 16-6 Voltages and currents in a series ac circuit containing resistance and inductance.

that is, V_R and I reach their maximum and minimum values *at the same time.* The voltage across the inductance V_L *leads* the current I by 90°; that is, V_L reaches its maximum 90° *before I* reaches its maximum. Also, V_L *leads* V_R by 90°; that is, V_L reaches its maximum 90° *before* V_R reaches its maximum.

To obtain the total voltage, we must add the voltages across all parts of the circuit. But how are we going to add voltages that do not happen at the same time? The only way to do this is to add the *instantaneous* voltages that *do* occur at the same time. Thus, in Fig. 16-7 on page 550, the instantaneous values of v_R and v_L are added for different instants of time.

At 0°, $v_R = 0$ and v_L is a maximum. Therefore,

$$V_T = v_R + v_L$$
$$V_T = 0 + v_L$$
$$V_T = v_L \qquad \text{(point } a\text{)}$$

At 45°, $v_R = v_L$ and

$$V_T = v_R + v_L$$
or $$V_T = \text{twice the value of either} \qquad \text{(point } b\text{)}$$

At 90°, $v_L = 0$ and v_R is a maximum. Therefore,

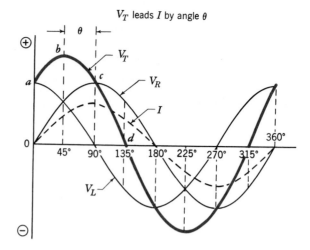

Figure 16-7 The total voltage in a series ac circuit is obtained by adding the instantaneous voltages, instant by instant.

$$V_T = v_R + v_L$$
$$V_T = v_R + 0$$
$$V_T = v_R \qquad \text{(point } c\text{)}$$

At 135°, $v_R = v_L$, but they are of opposite polarity. Therefore,

$$V_T = v_R + (-v_L)$$
$$V_T = 0 \qquad \text{(point } d\text{)}$$

By continuing in this manner, adding the voltages instant by instant, the waveform for the total voltage is obtained as shown in Fig. 16-7. Notice that the maximum value of V_T (at 45°) is *still leading* the current maximum. In this particular problem, since R and X_L are equal, the angle of lead is equal to 45°. For other values of R and X_L, angle θ will change.

Apparently, then, although we *do* add the voltages to get the total voltage, the addition is not just a simple arithmetic addition. We can see this more clearly if we draw the voltages and currents as phasors on the same unchanging current base as shown in Fig. 16-8a. This phasor diagram shows exactly the same relationships that were shown in Fig. 16-6 in wave form. Since the current in a series circuit is constant, the current is used as the reference line upon which to draw the phasor diagram. The voltage across the resistor V_R is still in phase with the current I, and the voltage across the inductance V_L still leads the current I by 90°.

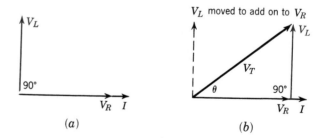

Figure 16-8 (a) Phasor diagram for a series ac circuit containing resistance and inductance. (b) V_T represents the phasor sum of V_R and V_L.

Addition of Phasors

RULE   Phasors are added by placing the tail of each phasor on the head of the preceding phasor and drawing it in its original direction and length.

RULE   The sum of the phasors is the distance from the origin of the phasors to the head of the final phasor.

In Fig. 16-8b, the total voltage V_T is obtained by adding the phasor V_L to the phasor V_R. Place the tail of V_L on the head of V_R and draw it in its original direction and length. The distance from the origin of the phasors to the head of the final phasor is the *sum* of the phasors. In this instance, the phasor V_T represents the sum of the phasors V_R and V_L. The phase angle θ is the angle by which the total voltage leads the current. In this problem, the angle is 45° and is the same angle shown on the waveform diagram of Fig. 16-7.

By applying the pythagorean theorem to Fig. 16-8b, we obtain the

$$V_T{}^2 = V_R{}^2 + V_L{}^2 \qquad\qquad \text{Formula 16-3}$$

Total impedance.   To find the impedance of the series ac circuit, we must add R and X_L vectorially as was done with the voltages. This is shown in Fig. 16-9. The voltages in Fig. 16-9a are replaced by their Ohm's law values in Fig. 16-9b. Now, by dropping out the common factor of the current I, we obtain the *impedance triangle* of Fig. 16-9c.

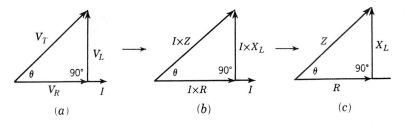

(a)   (b)   (c)

Figure 16-9 Z represents the phasor sum of R and X_L.

Applying the pythagorean theorem to Fig. 16-9c, we obtain the

$$Z^2 = R^2 + X_L{}^2 \qquad\qquad \text{Formula 16-4}$$

and

$$Z = \sqrt{R^2 + X_L{}^2} \qquad\qquad \text{Formula 16-5}$$

And, by trigonometry,

$$\cos\theta = \frac{a}{h}$$

or

$$\cos\theta = \frac{R}{Z} \qquad\qquad \text{Formula 16-6}$$

or $$\text{PF} = \frac{R}{Z}$$ Formula 16-7

The effective power is still given by

$$W = V \times I \times \cos \theta \qquad (15\text{-}18)$$
or
$$W = V \times I \times \text{PF} \qquad (15\text{-}19)$$
or
$$W = I^2R \qquad (7\text{-}4)$$

Example 16-5

In Fig. 16-5, find (*a*) the total voltage, (*b*) the impedance, (*c*) the phase angle, (*d*) the power factor, and (*e*) the power.

Solution

a.
$$V_T^2 = V_R^2 + V_L^2 = 85^2 + 85^2 \qquad (16\text{-}3)$$
$$= 7225 + 7225$$
$$= 14{,}450$$
$$V_T = \sqrt{14{,}450} = 120 \text{ V} \qquad Ans.$$

b.
$$Z = \sqrt{R^2 + X_L^2} \qquad (16\text{-}5)$$
$$Z = \sqrt{100^2 + 100^2} = \sqrt{10^4 + 10^4} = \sqrt{2 \times 10^4}$$
$$Z = 1.41 \times 10^2 = 141 \ \Omega \qquad Ans.$$

c.
$$\cos \theta = \frac{R}{Z} = \frac{100}{141} = 0.709 \qquad (16\text{-}6)$$
$$\theta = 45° \qquad Ans.$$

d.
$$\text{PF} = \frac{R}{Z} = 0.709 = 70.9\% \ lagging^* \qquad (16\text{-}7)$$

e.
$$W = V \times I \times \text{PF} \qquad (15\text{-}19)$$
$$= 120 \times 0.85 \times 0.709 = 72.3 \text{ W} \qquad Ans.$$

f.
$$W = I^2R = (0.85)^2 \times 100 = 72.3 \text{ W} \qquad Check$$

In this series circuit, the total current of 0.85 A *lags* the total voltage of 120 V by 45°.

Example 16-6

An inductance of 0.17 H and a resistance of 50 Ω are connected in series across a 110-V 60-Hz line. Find (*a*) the inductive reactance, (*b*) the impedance, (*c*) the total current, (*d*) the voltage drop across the resistor and the coil, (*e*) the phase angle, (*f*) the power factor, and (*g*) the power used.

---

*In this book, a leading or lagging PF will always refer to a *current* which will lead or lag the voltage.

Solution

Given:
$$L = 0.17 \text{ H}$$
$$R = 50 \ \Omega$$
$$V_T = 110 \text{ V}$$
$$f = 60 \text{ Hz}$$

Find:
$$X_L = ?$$
$$Z = ?$$
$$I_T = ?$$
$$V_R = ?$$
$$V_L = ?$$
$$\theta = ?$$
$$PF = ?$$
$$W = ?$$

CALCULATOR HINT

Series R_L calculations are pythagorean calculations.

a. $\qquad X_L = 6.28fL = 6.28 \times 60 \times 0.17 = 64 \ \Omega \qquad$ *Ans.*

b. $\qquad Z = \sqrt{R^2 + X_L^2} = \sqrt{50^2 + 64^2} \qquad$ (16-5)
$$= \sqrt{2500 + 4096}$$
$$Z = \sqrt{6596} = 81 \ \Omega \qquad \textit{Ans.}$$

c. Since the total opposition is the impedance Z, formula (4-7) becomes

$$\boldsymbol{V_T = I_T \times Z}$$
$$110 = I_T \times 81$$
$$I_T = \frac{110}{81} = 1.36 \text{ A} \qquad \textit{Ans.}$$

d. Since

$$I_T = I_R = I_L = 1.36 \text{ A} \qquad (4\text{-}1)$$

$$V_R = I_R \times R_R \qquad\qquad V_L = I_L \times X_L$$
$$V_R = 1.36 \times 50 = 68 \text{ V} \qquad V_L = 1.36 \times 64 = 87 \text{ V}$$

e. $\qquad \cos \theta = \dfrac{R}{Z} = \dfrac{50}{81} = 0.6173 \qquad$ (16-3)
$$\theta = 52° \qquad \textit{Ans.}$$

f. $\qquad PF = \dfrac{R}{Z} = 0.6173 = 61.7\% \textit{ lagging} \qquad \textit{Ans.} \qquad$ (16-7)

g. $W = V \times I \times PF = 110 \times 1.36 \times 0.617 = 92.5 \text{ W} \qquad$ *Ans.* (15-19)

h. $\qquad W = I^2R = (1.36)^2 \times 50 = 92.5 \text{ W} \qquad$ *Check*

In this series circuit, the total current of 1.36 A lags the total voltage of 110 V by 52°.

Self-Test 16-7

Figure 16-10 may be used to represent the plate circuit of an amplifier with transformer coupling. Find the value of V_p when the frequency of the applied voltage is (*a*) 100 Hz and (*b*) 10 Hz.

Solution

a. At 100 hz,

$$X_L = 6.28fL \qquad (16\text{-}1)$$
$$X_L = 6.28 \times 100 \times 12.75 = \underline{\quad(1)\quad} \ \Omega$$

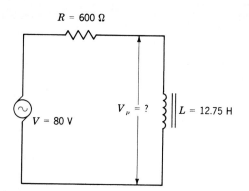

Figure 16-10 Representation of the plate circuit of a transformer-coupled amplifier.

The factor that determines whether we should include the 600-Ω resistance in the calculation for Z is called the ____(2)____ of the circuit. (See page 558.)

$$Q = \frac{X_L}{R} = \frac{8000}{600} = \underline{\quad(3)\quad}$$

Since Q is greater than 5, we ____(4)(should/should not)____ include the resistance in the calculation for Z. Therefore, $Z = \underline{\quad(5)\quad}\ \Omega$.
Find the current.

$$V = IZ$$
$$80 = I \times 8000$$
$$I = \underline{\quad(6)\quad}\ A$$

Find V_p.

$$V_p = I \times X_L$$
$$= 0.01 \times 8000 = \underline{\quad(7)\quad}\ V \qquad Ans.$$

b. At 10 Hz,

$$X_L = 6.28 \times \underline{\quad(8)\quad} \times 12.75 = \underline{\quad(9)\quad}\ \Omega$$

The Q of the circuit is now $800/\underline{\quad(10)\quad}$, or

$$Q = \underline{\quad(11)\quad}$$

Since $Q = 1.33$, the resistance ____(12)(should/should not)____ be included in the calculation for Z.

$$Z = \sqrt{R^2 + X_L^2} = \sqrt{(600)^2 + (800)^2}$$
$$= \sqrt{36 \times 10^4 + \underline{\quad(13)\quad} \times 10^4}$$
$$= \sqrt{(36 + 64) \times 10^4}$$
$$Z = \sqrt{100 \times 10^4} = \underline{\quad(14)\quad}\ \Omega$$

Find the current.

$$V = IZ$$
$$80 = I \times 1000$$
$$I = \underline{\quad(15)\quad}\ A$$

Find V_p.

$$V_p = I \times X_L$$
$$= 0.08 \times \underline{\quad(16)\quad} = \underline{\quad(17)\quad}\ V \qquad Ans.$$

Exercises

Applications

1. A resistance of 5 Ω is in series with a coil whose inductive reactance is 12 Ω. If the total voltage is 104 V, find (a) the impedance, (b) the total current, (c) the voltage drop across each part, (d) the phase angle, (e) the power factor, and (f) the power.

2. A 112-V 60-Hz ac voltage is applied across a series circuit of a 50-Ω resistor and a 100-Ω inductive reactance. Find (a) the impedance, (b) the total current, (c) the voltage drop across each part, (d) the phase angle, (e) the power factor, and (f) the power.

3. A 66-V 220-Hz ac voltage is applied across a series circuit of a 20-Ω resistor and a 0.05-H coil. Find the total current and the phase angle.

4. A tuning coil has an inductance of 48 μH and a resistance of 20 Ω. Find its impedance to a frequency of 100 kHz.

5. A fluorescent-lamp ballast has an inductance of 0.4 H and a resistance of 80 Ω. If the supply frequency is 60 Hz, find (a) the reactance of the ballast, (b) the impedance of the ballast, and (c) the voltage across the ballast when 0.5 A flows through it.

6. A transformer-coupled amplifier contains a 25,000-Ω resistance and a 20-H coil. At what frequency will the voltage across the coil equal that across the resistance? What current will flow if the impressed voltage is 35 V?

7. Part of the oscillator circuit for a continuous-wave transmitter is shown in Fig. 16-11. Find the current from point A to point B if the voltage drop is 70 V.

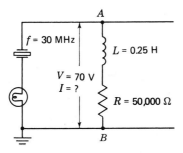

Figure 16-11   Portion of the oscillator circuit for a continuous-wave transmitter.

8. A 20-Ω resistor is in series with a 0.03-H dimmer coil. If the 230-V 60-Hz ac voltage is applied to the circuit, find (a) the current and (b) the power used.

9. A lightning-protector circuit contains a 63.7-mH coil in series with a 7-Ω resistor. What current will flow when it is tested with a 110-V 60-Hz ac voltage?

10. The coil of a telephone relay has a resistance of 500 Ω and an inductance of 0.32 H. When operated at a frequency of 500 Hz, find (a) the reactance of the coil, (b) the impedance of the coil, and (c) the voltage that must be impressed across the coil in order to operate the relay at its rated current of 5 mA.

11. The output voltage of an audio oscillator is 37.7 V at 3000 Hz. It is applied to a series circuit of 200 Ω resistance and 200 mH inductance. Find the impedance of the circuit and the current that flows.

12. A filter choke coil is connected in series with a 400-Ω resistor. When the voltage across the circuit is 120 V, the current is 0.12 A. Find the inductance of the coil if the frequency is 60 Hz. *Hint:* See Job 16-7.

13. To measure the inductance of an audio choke, a 2000-Ω resistor is connected in series with the choke. A 110-V 60-Hz voltage is impressed across the circuit, and the current is measured at 10 mA. Find the inductance of the coil.

14. A 40-V emf at 1000 Hz is impressed across a loudspeaker of 5000 Ω resistance and 1.5 H inductance. Find the current and power drawn.

15. Find the inductive reactance of a single-phase motor if the line voltage is 220 V, the line current is 20 A, and the resistance of the motor coils is 8 Ω. What is the angle of lag?

(See CD-ROM for Test 16-2)

Real Inductor Losses and Quality

In a "pure" coil, the opposition to the flow of an alternating current is its reactance. Actually, of course, every coil is made of wire which has some resistance. If this resistance is small in comparison with the reactance, it may be neglected, and the total opposition to the flow of current through the coil is equal to its inductive reactance. If the ohmic resistance of the coil is large, it must be added on to the reactance of the coil to obtain the total opposing effect. This total opposition is called the *impedance Z* of the coil. The addition of the ohms of resistance and the ohms of inductive reactance is *not* accomplished by simple addition but by *phasor (vector) addition.*

RULE The impedance of a coil is the phasor sum of the resistance and the reactance.

$$Z^2 = R^2 + X_L^2 \qquad \text{Formula 16-8}$$

or

$$Z = \sqrt{R^2 + X_L^2} \qquad \text{16-9}$$

where Z = impedance, Ω
R = dc resistance, Ω
X_L = inductive reactance, Ω

The comparative values of the inductive reactance and the resistance are described by a value called the Q, or *quality*, of the coil. This figure of merit expresses the ability of the coil to act as a reactor.

RULE The *Q* of a coil is the ratio of its inductive reactance to its effective resistance.

$$Q = \frac{X_L}{R} \qquad \text{Formula 16-10}$$

If the Q of a coil is greater than 5, then the resistance may be neglected in the calculation of the impedance. If the Q is smaller than 5, then the resistance must be added to the reactance by formula (16-9) to obtain the impedance. Also, if the ratio R/X_L is larger than 5, the reactance may be neglected and the impedance is equal to the resistance.

Example 16-8

A coil has a resistance of 5 Ω and an inductive reactance of 12 Ω at a certain frequency. Find (a) the Q of the coil and (b) the impedance of the coil.

Solution

Given: $R = 5\ \Omega$ Find: $Q = ?$
             $X_L = 12\ \Omega$ $Z = ?$

a. $Q = \dfrac{X_L}{R} = \dfrac{12}{5} = 2.4$ *Ans.*   (16-10)

b. Since Q is less than 5, the resistance *must* be included in the calculation of the impedance.

$$Z = \sqrt{R^2 + X_L^2} \qquad (16\text{-}9)$$
$$= \sqrt{5^2 + 12^2}$$
$$Z = \sqrt{25 + 144} = \sqrt{169} = 13\ \Omega \qquad Ans.$$

Example 16-9

A coil has a resistance of 10 Ω and an inductive reactance of 70 Ω at a certain frequency. Find (a) the Q of the coil and (b) the impedance of the coil. (c) If the resistance is neglected, what is the percent of error?

Solution

Given: $R = 10\ \Omega$ Find: $Q = ?$
             $X_L = 70\ \Omega$ $Z = ?$
                              Percent of error $= ?$

a. $Q = \dfrac{X_L}{R} = \dfrac{70}{10} = 7$ *Ans.*   (16-10)

b. Since Q is larger than 5, the resistance may be neglected and the impedance is equal to the inductive reactance.

$$Z = 70\ \Omega \qquad Ans.$$

c. If we include the resistance in the calculation of the impedance,

$$Z = \sqrt{R^2 + X_L^2} \qquad (16\text{-}9)$$
$$Z = \sqrt{10^2 + 70^2} = \sqrt{100 + 4900} = \sqrt{5000}$$
$$Z = 70.7\ \Omega$$

CALCULATOR HINT

A calculator that computes algebraically makes reactive circuit computations easier.

The error introduced by *not* including R is equal to $70.7 - 70 = 0.7\ \Omega$. The percent of error is

$$\frac{0.7}{70.7} \times 100 \text{ equals } 0.99 \text{ or } 1\%$$

This error is well within the normal human error incurred in merely taking measurements and is therefore unimportant.

Example 16-10

The field coils of a loudspeaker have a resistance of 6000 Ω and an inductance of 1.592 H. Find (*a*) the inductive reactance at 800 Hz, (*b*) the Q of the coils, (*c*) the impedance of the coils, and (*d*) the current flowing if the voltage across the coils is 40 V.

Solution

Given: $R = 6000\ \Omega$ Find: $X_L = ?$
$L = 1.592$ H $Q = ?$
$f = 800$ Hz $Z = ?$
$V = 40$ V $I = ?$

a. $X_L = 6.28fL = 6.28 \times 800 \times 1.592 = 8000\ \Omega$ *Ans.*

b. $Q = \dfrac{X_L}{R} = \dfrac{8000}{6000} = 1.33$ *Ans.*   (16-10)

c. Since Q is less than 5, the resistance *must* be included in the calculation of the impedance.

$$Z = \sqrt{R^2 + X_L^2} \tag{16-9}$$
$$Z = \sqrt{6000^2 + 8000^2}$$
$$Z = \sqrt{(6 \times 10^3)^2 + (8 \times 10^3)^2}$$
$$Z = \sqrt{36 \times 10^6 + 64 \times 10^6}$$
$$Z = \sqrt{(36 + 64) \times 10^6}$$
$$Z = \sqrt{100 \times 10^6}$$
$$Z = 10 \times 10^3 = 10^4 = 10{,}000\ \Omega \quad \textit{Ans.}$$

d. Since the total opposition is the impedance, the formula for Ohm's law may be rewritten as the

$$V = IZ \qquad \text{Formula 16-11}$$
$$40 = I \times 10{,}000$$
$$I = \frac{40}{10{,}000} = 0.004 \text{ A} \qquad \textit{Ans.}$$

Exercises

Practice

1. Find the Q of a coil if $R = 30\ \Omega$ and $X_L = 120\ \Omega$.
2. Find the Q of a coil if $X_L = 6000\ \Omega$ and $R = 1000\ \Omega$.

3. Find the Q of a coil at 100 Hz if $R = 200\ \Omega$ and $L = 10$ H.
4. Find the impedance of a coil if its resistance is 12 Ω and its reactance is 35 Ω.
5. Find (*a*) the Q and (*b*) the impedance of a coil if its resistance is 100 Ω and its reactance is 1000 Ω.

Application

6. An antenna circuit has an inductance of 30 μH and a resistance of 20 Ω. Find the impedance to a 500-kHz signal.
7. A 40-V emf at a frequency of 1 kHz is impressed across a loudspeaker of 5000 Ω resistance and 1.5 H inductance. Find (*a*) the inductive reactance, (*b*) the impedance, and (*c*) the current.
8. A 120-V 60-Hz line is connected across a 10-H choke coil whose resistance is 400 Ω. Find (*a*) the inductive reactance, (*b*) the Q of the coil, (*c*) the impedance, and (*d*) the current.
9. The primary of an AF transformer has a resistance of 100 Ω and an inductance of 50 mH. Find (*a*) the inductive reactance at 1 kHz and (*b*) the impedance.
10. A 3000-Ω resistor has an inductance of 10 mH. Find its impedance at (*a*) 500 Hz, (*b*) 5 kHz, (*c*) 500 kHz, and (*d*) 1500 kHz.
11. The primary of an IF coupling transformer has an inductance of 5 mH and a resistance of 100 Ω. If the voltage across the primary is 10 V at 456 kHz, what is the current flowing in the primary?
12. An AF amplifier circuit uses an audio choke of 100 mH inductance and 3000-Ω resistance. Find the impedance to (*a*) 500 Hz, (*b*) 1000 Hz, and (*c*) 5000 Hz.

(See CD-ROM for Test 16-3)

about electronics

Pictured is an iron-core transformer with primary and secondary voltage ratings. The unit is an important application of mutual inductance, even though the primary and secondary are not physically connected to each other. Power in the primary is coupled into the secondary by the magnetic field linking the two windings. It works like this: The primary winding inductance is connected to a voltage source that produces alternating current. The secondary winding inductance is connected across the load resistance. Thus power is transferred from the primary (where the generator is connected) to the secondary. There, the induced secondary voltage can produce current in the load resistance that is connected across the secondary wiring inductance.

JOB 16-6

Measuring the Inductance of a Coil

The inductance of a coil may be calculated by the use of several formulas involving specific dimensions of the coil. However, these dimensions are not always

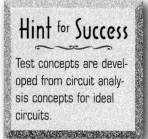
easily obtained, and other methods are substituted. One method uses a standard inductance and a circuit similar to the Wheatstone bridge. Another method obtains the resonant frequency of the combination of the coil with a known capacity, and the inductance is calculated from the formula for the resonant frequency given in Job 18-4. In a third method, called the *impedance* method, an ac voltage is impressed across the coil, and voltage, frequency, and current are measured. The resistance of the coil is obtained by use of an ohmmeter. A coil of the type that can be measured this way is called a *high Q coil*. A high-quality coil has low resistance.

Example 16-11

What is the inductance of a coil that draws 30 mA from a 120-V 60-Hz ac source? The resistance of the coil is 400 Ω.

Solution

Given: $I = 30$ mA $= 0.03$ A   Find: $L = ?$
                     $V = 120$ V
                     $f = 60$ Hz
                     $R = 400$ Ω

1. Find the impedance.

$$V = IZ \tag{16-11}$$
$$120 = 0.03 \times Z$$
$$Z = \frac{120}{0.03} = 4000 \ \Omega$$

2. Compare the R and the Z. When the resistance is small in comparison with the impedance, it may be neglected completely, making $X_L = Z$. The resistance is small when Z/R is more than 5.

$$\frac{Z}{R} = \frac{4000}{400} = 10$$

Therefore, since Z/R is larger than 5, the resistance is small when compared with the impedance and $X_L = Z$. If Z/R had been less than 5, the reactance X_L would have been found by applying the formula $Z^2 = R^2 + X_L^2$ of Eq. (16-8).

3. Find the inductance.

$$X_L = 6.28 \times 60 \times L \tag{16-1}$$
$$4000 = 376.8 \times L$$
$$L = \frac{4000}{376.8} = 10.6 \text{ H} \qquad Ans.$$

Self-Test 16-12

What is the inductance of a coil that draws 0.4 A from a 120-V 60-Hz ac source? The resistance of the coil is 100 Ω.

Solution

Given: $I = 0.4$ A   Find: $L = ?$
$V = 120$ V
$f = 60$ Hz
$R = 100\ \Omega$

1. Find the impedance.

$$V = IZ \qquad\qquad (16\text{-}11)$$
$$120 = 0.4 \times Z$$
$$Z = \frac{120}{0.4} = 300\ \Omega$$

2. Compare Z and R.

$$\frac{Z}{R} = \frac{300}{100} = 3$$

R __(1)(must/need not)__ be included in the calculation to find XL.

3. Find X_L.

$$R^2 + X_L^2 = Z^2 \qquad\qquad (16\text{-}8)$$
$$X_L^2 = Z^2 - \underline{\quad(2)\quad}$$
$$X_L^2 = \sqrt{Z^2 - R^2}$$
$$= \sqrt{(300)^2 - (100)^2}$$
$$= \sqrt{\underline{\quad(3)\quad} \times 10^4 - \underline{\quad(4)\quad} \times 10^4}$$
$$= \sqrt{\underline{\quad(5)\quad} \times 10^4}$$
$$= \underline{\quad(6)\quad} \times 10^2 = \underline{\quad(7)\quad}\ \Omega$$

4. Find the inductance.

$$X_L = 6.28 \times 60 \times L \qquad\qquad (16\text{-}1)$$
$$\underline{\quad(8)\quad} = 376.8 \times L$$
$$L = \frac{283}{376.8} = \underline{\quad(9)\quad}\ \text{H} \qquad Ans.$$

> ### CALCULATOR HINT
>
> Notice that most of the basic mathematics of circuit analysis is essentially the same. The difference is how the computations are linked together in a problem. This idea forms the concept of "object" programming, in which each basic type of computation is an object.

Exercises

Practice

1. What is the inductance of a coil whose resistance is 200 Ω if it draws 0.1 A from a 120-V 60-Hz line?
2. What is the inductance of a coil whose resistance is 500 Ω if it draws 20 mA from a 110-V 60-Hz line?
3. What is the inductance of a coil whose resistance is 300 Ω if it draws 10 mA from a 50-V 1000-Hz ac source?
4. What is the inductance of a coil whose resistance is 200 Ω if it draws 100 mA from a 50-V 1-kHz source?
5. What is the inductance of a coil whose resistance is 50 Ω if it draws 0.55 A from a 110-V 60-Hz line?
6. What is the inductance of a coil whose resistance is 100 Ω if it draws 0.4 A from a 120-V 100-Hz source?

Review of Coils and Inductance

1. When an ac voltage is impressed across a coil,
   a. The resulting current is an ————— current.
   b. This changing current produces changing fields of force which —————
      the wires of the coil.
   c. These cuttings induce a ————— emf in the coil.
2. The *inductance* of a coil is a measure of its ability to produce a ————— emf
   when the current through it is changing. The symbol for inductance is L.
3. Unit of inductance. A coil has an inductance of 1 ————— if a current changing at the rate of 1 A/s can induce a back emf of 1 V in the coil.
4. The reactance of a coil is the opposition of the coil to the passage of a
   ————— current.

$$X_L = 6.28fL \qquad\qquad 16\text{-}1$$

5. The Q of a coil is the comparison of its inductive reactance with its effective
   resistance.

$$Q = \frac{X_L}{R} \qquad\qquad 16\text{-}10$$

6. The impedance A of a coil is the ————— sum of its resistance and its reactance. If Q is ————— than 5, the resistance may be neglected and $Z = X_L$. If
   Q is ————— than 5, the resistance must be added to the reactance to find Z.

$$Z^2 = R^2 + X_L^2 \qquad\qquad 16\text{-}8$$

or
$$Z = \sqrt{R^2 + X_L^2} \qquad\qquad 16\text{-}9$$

7. To measure the inductance of a coil, the ac voltage is impressed across the coil,
   and the voltage, frequency, and current are measured. The resistance of the coil
   is measured with an —————. The inductance is calculated as follows:
   a. Find the impedance.

$$V = IZ \qquad\qquad 16\text{-}11$$

   b. Find the reactance.

$$R^2 + X_L^2 = Z^2 \qquad\qquad 16\text{-}8$$

   c. Find the inductance.

$$X_L = 6.28fL \qquad\qquad 16\text{-}1$$

   d. If the ratio of Z to R is ————— than 5, the resistance may be neglected and
      $X_L = Z$.

Exercises

Practice

1. Find the inductive reactance of a 0.2-H choke coil at (*a*) 100 Hz, (*b*) 1000 Hz,
   (*c*) 10 kHz, and (*d*) 100 kHz.
2. A 5-V 100-kHz ac voltage is impressed across an RF choke whose inductance
   is 10 mH. Find (*a*) the reactance and (*b*) the current that flows.

3. Find the inductive reactance of a 50-μH coil at a frequency of 10 MHz.
4. What must be the inductance of a coil in order that it have a reactance of 8000 Ω at 800 Hz?
5. Find the Q of a coil if $R = 80\ \Omega$ and $X_L = 4800\ \Omega$.
6. Find the impedance of a coil if its resistance is 39 Ω and its reactance is 80 Ω.
7. Find the impedance of a coil to a frequency of 60 Hz if its resistance is 30 Ω and its inductance is 0.2 mH.
8. A coil has an inductance of 50 μH and a resistance of 5 Ω. Find (*a*) the reactance to a 500-kHz frequency, (*b*) the impedance, and (*c*) the current flowing if the voltage is 3 V.
9. A coil of 100 Ω resistance draws 100 mA from a 25-V 60-Hz source. Find its inductance.
10. A coil having a Q of 50 draws 10 mA when connected to a 15-V 1-kHz power supply. Find its inductance.
11. Use the formula $V_{av} = L\,(I/t)$, where V_{av} is the average voltage induced in a circuit of L henries by a *change* of current of I amperes in t seconds. If a current changing at the rate of 200 mA/s in a coil induces a counter emf of 50 mV, what is the inductance of the coil?
12. When the current changes from 0.6 A to 0.9 A in 1.5 s, the average voltage induced in a coil is 0.3 V. Find the inductance of the coil.

(See CD-ROM for Test 16-4)

Transmission Lines

In this job we shall apply the concepts of induction to distribution systems composed of parallel lines. This is commonly used in power cords, in high-voltage transmission lines, and in some signal transmission concepts.

Inductance of a Single Wire

Start the analysis by looking at the inductance of a single wire carrying current as in Fig. 16-12. In this case a current is flowing through the wire in such a manner that the current density is uniform throughout the wire. The current in the center of the wire creates flux that links to the outside of the wire. In this way there is a self-inductance. An interesting result is that the self-inductance is independent of the size of the wire. There is a uniform inductance per unit length of the wire, L_1. This inductance per unit length is

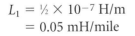

$$L_1 = \tfrac{1}{2} \times 10^{-7}\ \text{H/m}$$
$$= 0.05\ \text{mH/mile}$$

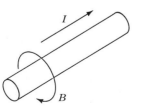

Figure 16-12   Magnetic field and current flow in a single wire.

Example 16-13

A No. 12 wire is used to carry 6.5 A of 60-Hz power to a shed 80 m from the distribution point. What is the inductive reactance of the line?

Solution

The inductance is independent of the wire size or the current carried; therefore, the only values required to calculate the reactance are the frequency and the length of the line.

1. First write the equation for inductance.

$$X_L = 6.28fL$$

2. Find the total inductance. To find the total inductance multiply the inductance per unit length times the length of the line.

$$L = 0.5 \times 10^{-7} \text{ H/m} \times 80 \text{ m}$$
$$= 4 \times 10^{-6} \text{ H}$$

3. Find the reactance.

$$X_L = 6.28 \times 60 \text{ Hz} \times 0.5 \times 10^{-7} \text{ H}$$
$$= 1.88 \times 10^{-5} = \Omega$$

Inductance of Parallel Wires

A two-wire distribution system is composed of two of the above wires as shown in Fig. 16-13. One wire carries the current and the other carries the return current. The two wires are another form of the transformer type of system that we saw in Fig. 16-1. There is an inductance for each line and a mutual inductance. The mutual inductance creates a new characteristic to the problem. The inductance is now dependent upon the geometry of the problem, i.e., the size of the wires and the separation distance of the wires.

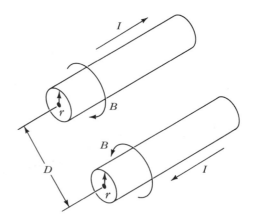

Figure 16-13   Current flow and magnetic fields for a parallel-wire transmission line.

Assume that the two wires are the same size, each with a radius r, and that the separation distance of the two wires is D. The total inductance of the transmission system is the sum of the three parts: the self-inductance of wire one, L_1; the self-inductance of wire two, L_2; and the mutual inductance of the two wires, L_{1-2}. The formula for this total is

$$L = L_1 + L_2 + L_{1-2}$$
$$= 1.482 \times \log \frac{D}{0.7788r} \text{ mH/mile}$$
$$L = 4 \times 10^{-7} \log \frac{D}{0.7788r} \text{ H/m}$$

Example 16-14

A cable bus is used to run two No. 00 cables to supply 75 A the length of a factory floor. The spacing of the cables is 1¾ in and the length of the run is 800 ft. Find the inductance, the reactance, the voltage drop, and the Q of the cable.

Solution

1. Write the equations and formulas that are to be used.

$$L_c = 1.482 \times \log \frac{D}{0.7788r} \text{ mH/mile}$$
$$1 \text{ mile} = 5280 \text{ ft}$$
$$L = L_c \text{ times the length of the cable}$$
$$X_L = 2\pi \times F \times L$$
$$R = R_{\text{line1}} \times R_{\text{line2}}$$
$$Z = \sqrt{R^2 + X_L^2}$$
$$Q = \frac{X_L}{R}$$

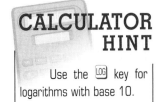

CALCULATOR HINT

Use the [LOG] key for logarithms with base 10.

2. Write what is known about the problem. For a 00 gage cable the diameter is 365 mils; therefore, the radius of the wire is 0.1825 in.

$$800 \text{ ft} = 800/5280 = 0.1515 \text{ mile}$$

3. Calculate the inductance per unit length.

$$L_c = 1.482\log \frac{D}{0.7788r} \text{ mH/mile}$$
$$= 1.482\log \frac{1.75}{0.7788 \times 0.1825}$$
$$= 1.62 \text{ mH/mile}$$

4. Calculate the total inductance of the line.

$$L = 1.62 \text{ mH/mile} \times 0.1515 \text{ mile} = 0.245 \text{ mH}$$

5. Calculate the total resistance. A 00 gage cable has a resistance of 0.0779 Ω per 1000 ft. The run is 800 ft but there is one cable going down and one coming back. The total length of cable is therefore 1600 ft.

$$R = 0.0779 \times 1600/1000$$
$$= 0.125 \text{ Ω}$$

6. Find the inductive reactance of the line.

$$X_L = 2\pi \times F \times L$$
$$= 6.28 \times 60 \times 0.245/1000$$
$$= 0.0923 \text{ Ω}$$

7. Find the impedance of the line due to the resistance and the inductance.

$$Z = \sqrt{R^2 + X_L^2}$$
$$Z = \sqrt{0.125^2 + 0.923^2}$$
$$Z = 0.155 \ \Omega$$

8. Find the voltage drop for the line. Use Ohm's law and the total impedance to find total voltage drop.

$$V = 75 \ A \times 0.155 \ \Omega$$
$$= 11.64 \ V$$

9. Find the Q of the cable.

$$Q = \frac{X_L}{R}$$
$$= \frac{0.092}{0.125}$$
$$= 0.741$$

We see that the quality or Q of the circuit is very low. This is the way we would want it. The distribution system is not supposed to be an inductor.

The voltage drop is larger than would normally be expected. The resistive component would predict a drop of 9.375 V. There is an additional 2.265-V drop due to the inductance of the line. As we shall see later, a cable also has a capacitance which counteracts the inductive reactance. This will reduce the reactive voltage drop. The important implication of the inductive voltage drop is that it indicates the amount of energy that is stored in the magnetic field around the cable. This energy is the source of electromagnetic interference (EMI) that becomes increasingly important when electronic control and computer communications are used in factory environments.

Exercises

Applications

1. A type A No. 8 wire is used to supply 35 A to a water pump 1300 ft from the electric meter. If the wire spacing is 0.5 in, what is the inductance in mH for the line?
2. A 600-ft No. 12 line with a spacing of 0.35 in supplies a motor. What is the reactance of the line?
3. A 5-mile high-voltage power line is constructed using No. 2 wire. (*a*) If the spacing of the lines is 28 in, what is the inductance of the line? (*b*) If the line carries 14 A, what is the voltage drop due to the inductance and resistance of the line?

JOB 16-9

Introduction to Ideal Transformers

It is cheaper to transmit electric energy at high voltages than at low voltages because of the smaller loss of power in the line at high voltages. For this reason, the

220 V that is delivered by an ac generator may be stepped up to 2200 or even 220,000 V for transmission over long distances. At its destination, the voltage is stepped down to 240 V for industrial users and to 120 V for ordinary home and power users. The changes in the voltage continue. Elsewhere, the 120-V supply is reduced to 20 V to operate a toy train or to 12 V to operate a doorbell. All these changes in voltage are made by an extremely efficient electric device called a *transformer*.

Hint for Success

Ideal transformers are useful for analysis of general transformer concepts.

Basic construction.   As shown in Fig. 16-14, a transformer consists of (1) the *primary coil* which *receives* energy from an ac source, (2) the *secondary coil* which *delivers* energy to an ac load, and (3) a *core* on which the two coils are wound. The core is generally made of some highly magnetic material, although cardboard, ceramics, and other nonmagnetic materials are used for the cores of some radio and television transformers.

Principle of operation.   An alternating current will flow when an ac voltage is applied to the primary coil of a transformer. This current produces a field of force which changes as the current changes. The changing magnetic field is carried by the magnetic core to the secondary coil, where it cuts cross the turns of that coil. These "cuttings" induce a voltage in the secondary coil. In this way, an ac voltage in one coil is transferred to another coil, even though there is no electrical connection between them. The number of lines of force available in the primary is determined by the primary voltage and the number of turns on the primary—each turn producing a given number of lines. Now, if there are *many turns on the secondary*, each line of force will cut *many turns* of wire and *induce a high voltage*. If the *secondary contains only a few turns*, there will be few cuttings and a *low induced voltage*. The secondary voltage, then, depends on the number of secondary turns as compared with the number of primary turns. If the secondary has twice as many turns as the primary, the secondary voltage will be twice as large as the primary voltage. If the secondary has half as many turns as the primary, the secondary voltage will be one-half as large as the primary voltage. This is stated as the following rule:

RULE   The voltage on the coils of a transformer is directly proportional to the number of turns on the coils.

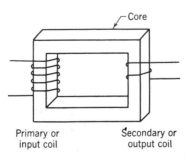

Figure 16-14 Basic transformer construction.

$$\frac{V_p}{V_s} = \frac{N_p}{N_s} = a$$

Formula 16-12

where V_p = voltage on primary coil
V_s = voltage on secondary coil
N_p = number of turns on primary coil
N_s = number of turns on secondary coil

The ratio V_p/V_s is called the *voltage ratio* (VR). The ratio N_p/Ns is called the *turns ratio* (TR). By substituting these terms in formula (16-7), we obtain an equivalent statement.

$$\mathbf{VR = TR}$$

Formula 16-13

Nomenclature. A voltage ratio of 1:3 (read as "1 to 3") means that for each volt on the primary, there are 3 V on the secondary. This is called a *step-up* transformer. A step-up transformer *receives a low voltage* on the primary and *delivers a high voltage* from the secondary. A voltage ratio of 3:1 (read as 3 to 1) means that for 3 V on the primary, there is only 1 V on the secondary. This is called a *step-down* transformer. A step-down transformer *receives a high voltage* on the primary and *delivers a low voltage* from the secondary.

The autotransformer. Instead of the two-winding, four-terminal standard transformer, the autotransformer (Fig. 16-15) consists of just a *single* winding. This winding may be tapped at any point along its length to provide a set of three terminals (*A*, *B*, *C*). The winding *AB* is the primary, and the entire winding *AC* is the secondary. Even though the windings have an electrical connection at *B*, the principle of operation remains the same. When an ac voltage is applied to the primary winding *AB*, the lines of force around *AB* link the turns between *A* and *C*, inducing a higher or lower voltage according to the formula *VR = TR*.

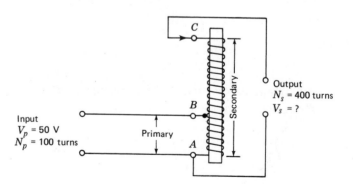

Figure 16-15   An autotransformer.

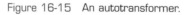

Example 16-15

A bell transformer reduces the primary voltage of 120 V to the 18 V delivered by the secondary. If there are 180 turns on the primary and 27 turns on the secondary, find (*a*) the voltage ratio and (*b*) the turns ratio.

Solution

The diagram for the problem is shown in Fig. 16-16.

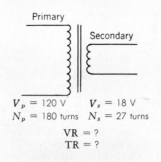

$V_p = 120$ V $V_s = 18$ V
$N_p = 180$ turns $N_s = 27$ turns
VR = ?
TR = ?

Figure 16-16

a. $\text{VR} = \dfrac{V_p}{V_s} = \dfrac{120}{18} = \dfrac{20}{3}$ (reads as "20 to 3")   *Ans.*

b. $\text{TR} = \dfrac{N_p}{N_s} = \dfrac{180}{27} = \dfrac{20}{3}$ (read as "20 to 3")   *Ans.*

Note: The ratios are always expressed in fractional form, even if the fraction can be reduced to a whole number.

Example 16-16

Find the voltage delivered by the secondary of the autotransformer shown in Fig. 16-15.

Solution

Given: $V_p = 50$ V   Find: $V_s = ?$
         $N_p = 100$ turns
         $N_s = 400$ turns

$$\frac{V_p}{V_s} = \frac{N_p}{N_s}$$ (16-12)

$$\frac{50}{V_s} = \frac{1\cancel{0}\cancel{0}}{4\cancel{0}\cancel{0}}$$

$$V_s = 200 \text{ V} \qquad \textit{Ans.}$$

CALCULATOR HINT

Use a programmable calculator to save the equations and make these computations easier to enter and to check.

Example 16-17

A "power" transformer has 99 turns on the primary and 315 turns on the secondary. What voltage will it deliver if the primary voltage is 110 V?

Solution

Given: $N_p = 99$ turns   Find: $V_s = ?$
                $N_s = 315$ turns
                $V_p = 110$ V

$$\frac{V_p}{V_s} = \frac{N_p}{N_s} \qquad (16\text{-}12)$$

$$\frac{110}{V_s} = \frac{99}{315}$$

$$99 V_s = 315 \times 110$$

$$V_s = \frac{315 \times 110}{99} = 350 \text{ V} \qquad Ans.$$

Self-Test 16-18

A Stancor P-6011 transformer used in a TV set has a voltage ratio of 11:35. If the primary has 242 turns, how many turns must be wound on the secondary?

Solution

Given: VR $= 11{:}35$ Find: $N_s = ?$
            $N_p = 242$

1. $\text{VR} = \dfrac{V_p}{?}$

and since $\dfrac{V_p}{V_s} = \dfrac{N_p}{N_s}$ (16-12)

2. $\overline{} = \dfrac{N_p}{N_s}$

                $\dfrac{11}{35} = \dfrac{242}{N_s}$

3. $N_s = \dfrac{242 \times 35}{11} = \underline{}$ turns   *Ans.*

Exercises

Applications

1. A "power" transformer has 85 turns on the primary and 255 turns on the secondary. What voltage will it deliver if the primary is connected to a 120-V source?

2. A filament transformer reduces the 110 V on the primary to 10 V on the secondary. Find (*a*) the voltage ratio and (*b*) the turns ratio.

3. The Stancor P-6293 universal-type power transformer steps down the voltage from 120 to 2.5 V. Find (*a*) the voltage ratio and (*b*) the turns ratio.

4. A Stancor A-4773 transformer whose turns ratio is 1:3 is used as a plate-to-grid coupling transformer in an amplifier circuit. What is the secondary voltage if the primary voltage is 15 V?

5. A 24:1 welding transformer has 25 turns on the secondary. How many turns are there on the primary?

6. A UTC LS-185 plate transformer steps up the voltage from 100 to 2500 V. If there are 50 turns on the primary, how many turns are on the secondary?

7. Find the voltage at the spark plugs if a 6-V alternator is connected to a coil with a primary winding of 50 turns and a secondary winding of 50,000 turns.

8. A coil with a primary winding of 100 turns must deliver 4800 V. If the primary is connected to a 6-V source, find the number of turns on the secondary.

9. The output from a 2N1097 transistor is to be matched to a 13.9-Ω voice coil by a 12:1 matching transformer. If the primary voltage is 18 V, find the voltage across the voice coil.

10. A transformer whose primary is connected to a 120-V source delivers 10 V. If the number of turns on the secondary is 20 turns, find the number of turns on the primary. How many extra turns must be added to the secondary if it must deliver 35 V?

11. A toy train transformer is connected to a 120-V 60-Hz source. The secondary has 60 turns and delivers 24 V. How many turns are on the primary?

12. The secondary coil of a transformer has 100 turns, and the secondary voltage is 5 V. If the turns ratio is 22:1, find (a) the voltage ratio, (b) the primary voltage, and (c) the primary turns.

13. A step-down transformer is wound with 3750 turns on the primary and 60 turns on the secondary. What is the delivered voltage if the high-voltage side is 15,000 V?

14. The 117-V primary of a transformer has 250 turns. Two secondaries are to be provided to deliver (a) 12.6 V and (b) 35 V. How many turns are needed on each secondary?

15. A power transformer with 100 turns on the primary is to be connected to a 120-V source of supply. Separate secondary windings are to deliver (a) 2.5 V, (b) 6.3 V, and (c) 600 V. Find the number of turns on each secondary.

16. A transformer bank is used to transform 2000 kVA (kilovoltamperes) from 14,000 to 4000 V. Find (a) the turns ratio of the transformer, and (b) the primary current.

Current in a Transformer

In the modern transformer, the power delivered to the primary is transferred to the secondary with practically no loss. For all practical purposes, the power input to the primary is equal to the power output of the secondary, and the transformer is assumed to operate at an efficiency of 100 percent. Thus,

$$\textbf{Power input} = \textbf{power output} \qquad \text{Formula 16-14}$$

Since

$$\textbf{Power input} = V_p \times I_p \qquad \text{Formula 16-15}$$

and

$$\textbf{Power output} = V_s \times I_s \qquad \text{Formula 16-16}$$

$$V_p \times I_p = V_s \times I_s \qquad \text{Formula 16-17}$$

Hint for Success

Transformer current values are important because of heating effects in the transformer.

By dividing both sides of the equation by $V_s \times I_p$ and canceling out identical terms,

$$\frac{V_p \times \overset{1}{\cancel{I_p}}}{V_s \times \cancel{I_p}} = \frac{\overset{1}{\cancel{V_s}} \times I_s}{\cancel{V_s} \times I_p}$$

we obtain the

$$\frac{V_p}{V_s} = \frac{I_s}{I_p}$$

Formula 16-18

This formula indicates that the current ratio in a transformer is *inversely proportional* to the voltage ratio. If the *voltage* ratio *increases*, the *current* ratio will *decrease*. If the *voltage* ratio *decreases*, the *current* ratio will *increase*.

In addition, since

$$\frac{V_p}{V_s} = \frac{N_p}{N_s}$$

we may substitute N_p/N_s for V_p/V_s in Eq. (16-13). This gives the

$$\frac{N_p}{N_s} = \frac{I_s}{I_p}$$

Formula 16-19

Example 16-19

The Stancor model P6293 universal-type power transformer delivers 36 W of power to a rectifier circuit. If the primary voltage is 120 V, how much current is drawn by the transformer?

Solution

Given: Power output $= 36$ W   Find: $I_p = ?$
$V_p = 120$ V

Power input $=$ power output   (16-14)
$V_p \times I_p =$ power output
$120 \times I_p = 36$

$$I_p = \frac{36}{120} = 0.3 \text{ A} \qquad Ans.$$

Example 16-20

The primary of a transformer is connected to a 110-V 60-Hz line. The secondary delivers 250 V at 0.1 A. Find (*a*) the current in the primary and (*b*) the power input to the primary.

Solution

The diagram for the problem is shown in Fig. 16-17.

a.
$$\frac{V_p}{V_s} = \frac{I_s}{I_p} \qquad (16\text{-}18)$$

$$\frac{110}{250} = \frac{0.1}{I_p}$$

$$110 \times I_p = 250 \times 0.1$$

$$I_p = \frac{25}{110} = 0.227 \text{ A} \qquad Ans.$$

b. Power input $= V_p \times I_p = 110 \times 0.227 = 24.97$ W   *Ans.*

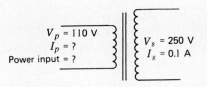

$V_p = 110$ V
$I_p = ?$
Power input $= ?$

$V_s = 250$ V
$I_s = 0.1$ A

Figure 16-17

Example 16-21

A bell transformer with 300 turns on the primary and 45 turns on the secondary draws 0.3 A from the 110 V line. Find (*a*) the current delivered by the secondary, (*b*) the voltage delivered by the secondary, and (*c*) the power delivered by the secondary.

Solution

Given: $N_p = 300$ turns   Find: $I_s = ?$
          $N_s = 45$ turns $V_s = ?$

$$I_p = 0.3 \text{ A} \qquad \text{Power output} = ?$$
$$V_p = 110 \text{ V}$$

a.
$$\frac{N_p}{N_s} = \frac{I_s}{I_p} \qquad (16\text{-}14)$$

$$\frac{300}{45} = \frac{I_s}{0.3}$$

$$45 \times I_s = 300 \times 0.3$$

$$I_s = \frac{90}{45} = 2 \text{ A} \qquad Ans.$$

b.
$$\frac{V_p}{V_s} = \frac{N_p}{N_s} \qquad (16\text{-}7)$$

$$\frac{110}{V_s} = \frac{300}{45}$$

$$300 \times V_s = 110 \times 45$$

$$V_s = \frac{4950}{300} = 16.5 \text{ V} \qquad Ans.$$

c.   Power output $= V_s \times I_s = 16.5 \times 2 = 33$ W   *Ans.*

Self-Test 16-22

A 120:24-VR transformer draws 1.5 A. Find the secondary current.

Solution

Given: VR = 120:24   Find: I_s = ?

I_p = 1.5 A

1. $$\frac{V_p}{V_s} = \frac{I_?}{I_?}$$ (16-18)

2. But $$\frac{V_p}{V_s} = VR = \frac{?}{?}$$

Therefore, we can substitute this ratio in formula (16-18).

3. $$\frac{120}{24} = \frac{I_s}{?}$$

4. $$24 \times I_s = 120 \times \underline{\qquad}$$

5. $$I_s = \frac{180}{24} = \underline{\qquad} A \qquad Ans.$$

Exercises

Applications

1. A bell transformer draws 20 W from a line. What is the secondary current if the secondary voltage is 10 V?
2. A Thermador model 5A6086 power transformer delivers 22.5 W of power. If the primary voltage is 112.5 V, how much current is drawn by the primary?
3. A 120-V 60-Hz line supplies power to a Stancor model P-6297 universal-type transformer. If the secondary delivers 480 V at 0.04 A, find (*a*) the primary current and (*b*) the power input.
4. A bell transformer with 240 turns on the primary and 30 turns on the secondary draws 0.25 A from a 120-V line. Find (*a*) the secondary current, (*b*) the secondary voltage, and (*c*) the secondary power.
5. A transformer is wound with 2200 turns on the primary and 150 turns on the secondary. (*a*) If it delivers 2 A, what is the primary current? (*b*) If the primary voltage is 110 V, what is the secondary voltage?
6. A 230:110-VR step-down transformer in a stage-lighting circuit draws 10 A from the line. Find the current delivered.
7. A filament transformer delivers 1.5 A at 6.3 V. If V_p is 110 V, find (*a*) I_p and (*b*) the power input.
8. A transformer with 2400 turns on the primary and 480 turns on the secondary draws 8.5 A from a 230-V line. Find (*a*) I_s, (*b*) V_s, and (*c*) the power output.
9. A transformer has 120 turns on the primary and 1500 turns on the secondary. (*a*) If it delivers 0.4 A, what is the primary current? (*b*) If the primary voltage is 120 V, what is the secondary voltage?
10. A 9:2 step-down transformer draws 1.8 A. Find the I_s.
11. A substation transformer reduces the voltage from the transmission-line voltage of 150,000 to 4400 V. If the transmission line carries 15 A, what current will the transformer deliver?
12. A step-down transformer with a turns ratio of 50,000:250 has its primary connected to a 27,000-V transmission line. If the secondary is connected to a 7.5-Ω load, find (*a*) the secondary voltage, (*b*) the secondary current, (*c*) the primary current, and (*d*) the power output.

Impedance Changing Using Transformers

The maximum transfer of energy from one circuit to another will occur when the impedances of the two circuits are equal, or *matched*. If the two circuits have unequal impedances, a transformer may be used as an impedance-changing device. In Fig. 16-18 a generator on the primary side of a transformer is coupled to a load Z_s on the secondary side of the transformer. In the circuit V_p and I_p are the voltage and current looking into the ideal transformer. At the output the voltage V_s is put across the load and the resulting current in the load is I_s. Multiplying Eqs. (16-18) and (16-19) for an ideal transformer gives

$$\frac{N_p}{N_s} \times \frac{N_p}{N_s} = \frac{V_p}{V_s} \times \frac{I_s}{I_p}$$

or

$$\left(\frac{N_p}{N_s}\right)^2 = \frac{V_p}{I_p} \times \frac{I_s}{V_s}$$

But $V_p/I_p = Z_{eq}$ and $I_s/V_s = 1/Z_s$; therefore,

$$\left(\frac{N_p}{N_s}\right)^2 = \frac{Z_{eq}}{Z_s} \qquad\qquad 16\text{-}20$$

This equation shows that the square of the turns ratio relates the load impedance, Z_s, to an impedance that would be seen at the input to the transformer. Figure 16-18b shows the equivalent circuit that would be seen by the generator on the primary side of the transformer. We say that the impedance Z_{eq} is the load impedance *referred* to the input.

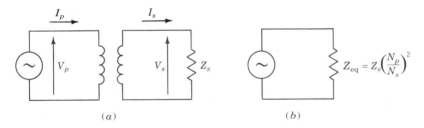

Figure 16-18   (a) Load Z_s attached to secondary of ideal transformer. (b) Equivalent load seen by the power source.

Example 16-23

Find the turns ratio that will transform a 10-Ω load into a 400-Ω referred impedance on the primary.

Solution

First write the equation that will be used.

$$\left(\frac{N_p}{N_s}\right)^2 = \frac{Z_{eq}}{Z_s}$$

Next indicate the values that are known: $Z_p = 4000\ \Omega$, $Z_s = 10\ \Omega$, and substitute into the equation.

$$\left(\frac{N_p}{N_s}\right)^2 = \frac{4000}{10}$$

Now solve for the turns ratio.

$$\frac{N_p}{N_s} = \sqrt{400} = 20$$

Example 16-24

What load does a 2400-V generator see if it drives a 25-kVA 2400:240-V 60-Hz distribution transformer?

Solution

First find the transformer turns ratio using Eq. (16-12).

$$\frac{V_p}{V_s} = \frac{2400}{240} = 10 = \frac{N_p}{N_s}$$

We know that the output voltage of the transformer is 240 V and that the transformer is designed to deliver 25 kVA; therefore, we can use Eq. (7-5) to find the load impedance.

$$Z_s = \frac{V^2}{P} = \frac{240^2}{2500} = 23.0 \ \Omega$$

The generator is attached to the primary side of the transformer. The generator "looks" into terminals 1-2 of the transformer. The impedance that it sees is the load referred to the primary. By Eq. (16-20)

$$Z_{se} = Z_s \times \left(\frac{N_p}{N_s}\right)^2 = 2.3 \times 100 = 230 \ \Omega$$

CALCULATOR HINT

The square operator is a post operator; it is entered after the number that it operates on.

Example 16-25

A 1:20 step-up transformer is used to match a microphone with a grid-circuit impedance of 40,000 Ω. Find the impedance of the microphone.

Solution

Given: $N_p/N_s = \frac{1}{20}$ Find: $Z_p = ?$
  $Z_s = 40,000 \ \Omega$

$$\left(\frac{N_p}{N_s}\right)^2 = \frac{Z_p}{Z_s}$$

$$\left(\frac{1}{20}\right)^2 = \frac{Z_p}{40,000}$$

$$\frac{1}{400} = \frac{Z_p}{40,000}$$

$$Z_p = \frac{40,000}{400} = 100 \ \Omega \qquad Ans.$$

Exercises

Applications

1. Find the turns ratio of a transformer used to match a 1600-Ω load to a 4-Ω load.

2. Find the turns ratio of a transformer used to match a 60-Ω load to a 540-Ω line.
3. The impedance of the output circuit of a power stage is 8000 Ω. What is the turns ratio of a transformer used to transfer the power to a 500-Ω line supplying a public address system?
4. A 2N265 transistor works into a load impedance of 9000 Ω. Find the turns ratio of the transformer needed to feed into a 10-Ω voice coil.
5. Find the turns ratio of a microphone transformer required to couple a 20-Ω microphone to a 500-Ω line.
6. Find the turns ratio of a microphone transformer required to couple a 20-Ω microphone to a grid circuit of 50,000 Ω impedance.
7. A 55:1 output transformer is used to match an output tube to a 4-Ω coil. Find the impedance of the output circuit.
8. Two 2N406 output transistors work in push-pull into a load impedance of 14,000 Ω. Find the required turns ratio of a transformer to match the output to an 8-Ω voice coil.
9. Find the turns ratio of a Stancor model A-8101 transformer which is used to match a 500-Ω line to an 8-Ω voice coil.
10. Find the turns ratio of the transformer needed to match a load of 4500 Ω to two 9-Ω speakers in parallel.
11. What would be the turns ratio in Exercise 10 if there were three 9-Ω speakers in parallel?
12. Find the turns ratio of the transformer used to match a 4200-Ω impedance to a 500-Ω line supplying a distant auditorium loudspeaker.
13. Find the turns ratio of the transformer needed to match a 50-Ω Amperite model PGL dynamic microphone to a 500-Ω line.
14. The secondary load of a step-down transformer with a turns ratio of 6 to 1 is 800 Ω. Find the impedance of the primary.

Review of Ideal Transformers

1. A transformer transmits energy from one circuit to another by means of electromagnetic induction between two coils.
2. The _____ coil is connected to the source of supply. The *secondary* coil is connected to the _____.
3. A step-up transformer _____ the voltage and decreases the current. A step-down transformer decreases the voltage and _____ the current.
4. The turns ratio of a transformer is the comparison (by division) of the number of turns on the _____ with the number of turns on the _____.

$$TR = \frac{N_p}{N_s}$$

The voltage ratio of a transformer is the comparison (by division) of the voltage on the primary with the voltage on the secondary.

$$VR = \frac{V_p}{V_s}$$

5. In a 100% efficient transformer

a. $$\text{Power input} = \text{power output} \qquad \text{16-14}$$

b. $$\text{Power input} = V_p \times I_p \qquad \text{16-15}$$

c. $$\text{Power output} = V_s \times I_s \qquad \text{16-16}$$

d. The voltage is directly proportional to the number of turns.

$$\frac{V_p}{V_s} = \frac{?}{?} \qquad \text{16-12}$$

e. The voltage is inversely proportional to the current.

$$\frac{V_p}{V_s} = \frac{?}{?} \qquad \text{16-18}$$

f. The number of turns is inversely proportional to the current.

$$\frac{N_p}{N_s} = \frac{?}{?} \qquad \text{16-19}$$

6. For an ideal transformer the load impedance is referred to the primary by the square of the _____.

Exercises

Applications

1. A bell transformer reduces the voltage from 120 to 15 V. If there are 22 turns on the secondary, find (a) the number of turns on the primary and (b) the turns ratio.

2. Find the voltage at the spark plugs if a 12-V alternator is connected to a coil with 80 turns on the primary and 40,000 turns on the secondary.

3. A UTC LS-185 plate transformer has a turns ratio of 1:25. If there are 70 turns on the primary, how many turns are on the secondary?

4. If the turns ratio of a transformer is 20:3, find the primary voltage if the secondary voltage is 24 V.

5. The 110-V primary of a transformer has 500 turns. Two secondaries are to be provided to deliver (a) 22 V and (b) 5 V. How many turns are needed for each secondary?

6. A power transformer delivers 50 W of power. If the primary voltage is 110 V, find the primary current.

7. A transformer connected to a 120-V 60-Hz line delivers 750 V at 200 mA. Find (a) the primary current and (b) the power drawn by the primary.

8. A 5:1 transformer draws 0.5 A from a 120-V line. Find (a) the secondary current, (b) the secondary voltage, and (c) the power output.

9. A transformer with 1500 turns on the primary and 375 turns on the secondary draws 3.5 A from a 115-V line. Find (a) the secondary current, (b) the secondary voltage, and (c) the power output.

10. A transformer draws 250 mA from a 120-V line and delivers 80 mA at 350 V. Find its efficiency.

11. A transformer drawing 150 W from the line operates at an efficiency of 90% and delivers 50 V. Find (a) the power delivered and (b) the secondary current.

12. A 2N321 amplifier feeds into the 500-Ω primary of an audible automobile signal minder. Find the turns ratio of the transformer needed to match it with a 3.2-Ω speaker.

13. A 1:30 step-up transformer is used to match a 50-Ω microphone to a grid circuit. Find the impedance of the grid circuit.
14. A step-down transformer with a turns ratio of 45,000:150 has its primary connected to a 72,000-V transmission line. If the secondary is connected to a 20-Ω load, find (*a*) the secondary voltage, (*b*) the secondary current, (*c*) the primary current, and (*d*) the power output.

(See CD-ROM for Test 16-5)

Real Transformers

In Job 16-4 it was shown that a real inductor has some loss in it. This loss is represented by a resistor in series with the ideal inductor. In this job we shall expand the inductor concept to represent the losses that occur in a real transformer. The mechanical model in Job 16-1 was described as a transformer. It had three major parts: an input winding, an output winding, and a mutual inductive circuit. As you would expect when you make the transformer model more realistic, each of these components has electrical losses. These losses are due to the resistance of the windings and energy that is lost in the core material.

> **Hint for Success**
>
> This model of a real transformer is only a first step toward reality, but it is still very useful.

A better electrical model than the ideal transformer model will help us understand how real transformers operate. It is also useful for understanding how some transformer tests operate. When we develop a transformer model, we can assume that we are working on either the primary side or the secondary side of the transformer. This is the same as asking which winding you are going to attach your instruments to. The development will start by looking at a transformer from the primary side. When this is completed, we will show the model as seen from the secondary. It should be emphasized that for these models the primary is where you put the voltage in and the secondary is where you take it out. Sometimes in testing it is desirable to turn the transformer around and use the secondary as the input. In this case the model would assume that the secondary is the primary.

The Basic Model

The model of an ideal transformer of the type we studied above is shown in Fig. 16-19 on page 580. It has N_1 turns on the primary and N_2 turns on the secondary. The corresponding induced voltages across the windings are E_1 and E_2. Assume that you have your instruments connected to the primary side of the transformer. As you look in through the primary terminals you see effects caused by the primary and the secondary of the transformer. The effects from the secondary are said to be *referred* to the primary. They are called referred because you see them after they have been transformed by the turns ratio of the ideal transformer.

The primary is an inductor that has losses in its windings and in its self-induced flux linkages. We represent the losses by a resistor R_1 and the primary inductance by the inductor L_1. This gives the circuit shown in Fig. 16-19b. The model shows that when a current I_1 flows in the primary a part of the voltage is lost in the primary inductor before it can be applied to the ideal transformer.

The mutual flux that links the primary and secondary windings is caused by currents flowing in the primary and the back emf of the secondary circuits. Figure 16-19c shows a node where Kirchhoff's current law can be applied to determine the result of these two currents. The resultant current causes the mutual flux.

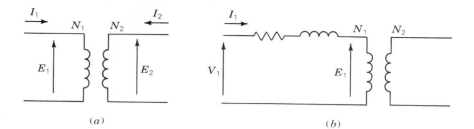

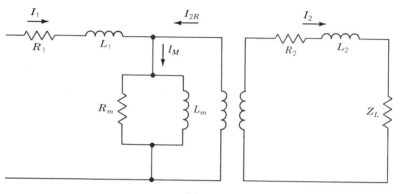

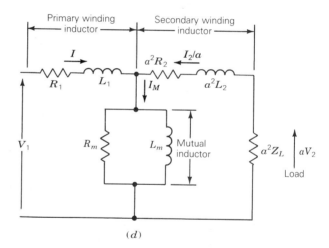

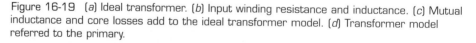

Figure 16-19 (a) Ideal transformer. (b) Input winding resistance and inductance. (c) Mutual inductance and core losses add to the ideal transformer model. (d) Transformer model referred to the primary.

   The current I_1 is the current that is coming in from the primary inductor. The current I_{2R} represents the secondary current that opposes the input and is referred to the primary. These two currents come together and create the current I_m which is the mutual current that creates the mutual flux linkages. A part of this mutual current flows through the mutual inductance L_m of the transformer. The rest of the current represents losses in the mutual coupling that are represented by a resistor R_m.

   In Job 16-10 we saw how impedances are changed by the turns ratio of an ideal transformer. Apply this concept to bring the secondary inductor and load into the primary circuit. When this is done, we have the model of a transformer referred to

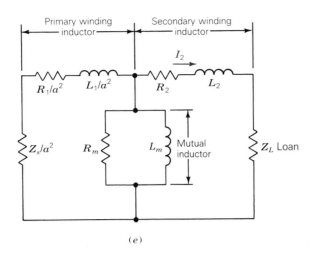

Figure 16-19 *(Continued)* (e) Transformer model referred to the secondary.

the primary as shown in Fig. 16-19*d*. The load impedance Z_L is also transformed by the ideal transformer.

In the same manner we can transform the impedances in the primary circuit to the secondary and have the model in Fig. 16-19*e* of a transformer referred to the secondary. In this model the input of the source circuit Z_s is transformed by the ideal transformer also.

Testing Transformers

In this job you will learn how to combine mathematics, test instruments, and circuit models to achieve practical results with transformers. Circuit models help you think about, understand, and reason about real-life situations. The model of the transformer is a "picture" that relates all of the things that we know about transformers that are important to what we are doing. A model can change as the intended use of the model changes.

The purpose of the models shown in this section is to assist in field testing of transformers. When a device that we are testing is bad, there are major changes in the values of some of the components. A part shorts out or burns out causing a large change in some value that we are measuring. Because the changes in component values are so large our model values do not have to be very accurate in order for us to find the problem. We learn to make assumptions about component values that ease the calculations. This will be done in the examples.

Two common tests can be made easily on transformers. They are called the open-circuit or no-load tests, and short-circuit tests. In an open-circuit test a voltage is applied to one winding and the other is left open; i.e., no load is attached to it. In a short-circuit test the second winding is shorted. Both of these tests are easily performed and provide useful information about the transformer under test. To understand the results of these tests, we will use the basic model of a transformer which we have developed.

In a real transformer there are two primary ways in which power is lost, in the core of the transformer and in the windings. The winding losses are usually called "copper" losses because the windings are from copper wire. The short-circuit tests primarily measure copper losses. The open-circuit test primarily checks for core losses. The model helps us understand why this is so.

Open-Circuit, No-Load Test

The setup for an open-circuit test is shown in Fig. 16-20a. A transformer is driven from a power source and the input current, input power, and input voltage are measured. The secondary is left open; i.e., no load is applied. The test voltage is usually applied to the winding that has a voltage rating equal to the source voltage that is available. If the test unit is a step-up transformer, care must be taken to protect personnel from the higher output voltage.

The equivalent-circuit model for the open-circuit test is shown in Fig. 16-20b. The model is referred to the primary because that is where the instruments are attached.

Other tests can be conducted at the same time that the open-circuit test is conducted. For example, instead of just leaving the secondary open a voltmeter can be connected to it. This allows a measurement of the turns ratio by comparing the output and input voltages.

A second additional measurement of the test unit may be made to determine the dc resistance of the input winding. This is done with a standard ohmmeter. Since the test is at dc there are no inductive effects in the measurements. The impedance of the mutual inductance will be zero at dc. This shorts out the core-loss resistance of the mutual inductance. This means that the ohmmeter will measure only the resistance of the input windings. For direct current the only resistance that can be seen is R_1.

With the secondary open no current will flow in it. This means that the current, I_1, into the transformer will be considerably less than normal. In most cases this current will probably be less than 10 percent of the rated value. The voltages will be at rated values but the current values will be much less than normal. This

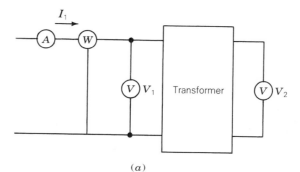

(a)

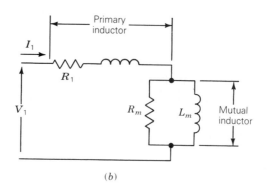

(b)

Figure 16-20 (a) Instrument setup for open-circuit test. (b) Model for open-circuit test.

reduced current will drastically reduce the copper losses. There will be no copper loss in the secondary because there is no current flowing. The copper loss in the primary varies as the square of the current. Since the input current is only about 10% (0.1) of normal, the power loss in the input winding will be only about 1% (0.01) of normal. This means that for the open-circuit tests the copper losses can usually be ignored.

In the following examples we will look at various situations.

Example 16-26

Negligible primary winding resistance. A 120-V, 60-Hz step-down transformer is measured in a no-load test. The results of the test are: $P_{in} = 75$ W, $I_1 = 1.5$ A, and the output voltage is $V_2 = 10$ V. Find the core-loss resistance, the reactance of the mutual inductance, and the turns ratio for the transformer.

Since the current is so much less than normal in the no-load test we can ignore the losses in the primary winding and the voltage drop across it. When this is done the only power loss is the core loss in R_m. The power meter which takes into consideration the phase difference of the voltage and current is therefore measuring the power that is lost in the core, P_c. This gives the relationship

$$P_c = \text{power loss in core}$$
$$= \text{power input with secondary open}$$

The value of the core-loss resistor is found from the relationship for the power in a resistor:

$$R_m = \frac{V_p^2}{P_{in}}$$

If the output voltage is measured also, it is possible to estimate the turns ratio of the transformer. If the voltage drop in the primary inductor is negligible and there is no current in the secondary, the input voltage and the output voltages are related through the ideal transformer.

$$\frac{V_p}{V_s} = \frac{N_p}{N_s} = \alpha$$

Solution

Since the input windings are ignored, the model shows that the only power loss is in the core-loss resistor. Using the power relationship the resistance is

$$R_m = \frac{(120)^2}{75}$$
$$= 192 \ \Omega$$

To find the reactance of the mutual inductance first find the current in the inductor L_m. Remember that the current in the inductor is 90° out of phase with the current in the resistor. The sum of the current in the resistor and the current in the inductor equals the input current. Because of the phase difference the current addition is by vectors. First find the current in the core-loss resistor.

CALCULATOR HINT

It may be useful to store α and α square as constants.

$$I_{R_m} = \frac{V_1}{R_m}$$

$$= \frac{120}{192} = 0.625 \text{ A}$$

Now use vectors to find the current in the mutual inductance.

$$I_c = \sqrt{I_1^2 - I_{R_m}^2}$$

$$= \sqrt{1.5^2 - 0.625^2}$$

$$= 1.364 \text{ A}$$

The reactance of the mutual inductance is

$$X_m = \frac{V_1}{I_c}$$

$$= \frac{120}{1.364} = 88.003 \ \Omega$$

The turns ratio is obtained from the measurements of the input and the output voltages.

$$\alpha = \frac{N_p}{N_s} = \frac{V_1}{V_2}$$

$$= \frac{120}{10} = 12$$

Example 16-27

We have said that the effects of the input inductor can be ignored. This is true only under the conditions of the no-load test where the currents in the transformer are very low. When the transformer is operating at normal values, the current in the primary causes a noticeable change in the circuit values. In this example we shall look at the effect of the primary winding resistance on the open-circuit test.

First measure the resistance of the input windings with an ohmmeter. Since you now know the value of R_1, you can calculate how much power is lost in the input winding. The total power into the transformer is the power in the input winding and the power lost in the core; therefore,

$$P_c = P_{in} - I_1^2 \times R_1$$

Assume the same transformer as the above example except now include a primary winding resistance of 0.4 Ω. Find the core-loss resistance.

Solution

Use the equation above to find power in the core resistance.

$$P_c = P_{in} - I_1^2 \times R_1$$
$$P_c = 75 - 1.5^2 \times 0.4$$
$$= 74.1 \ \Omega$$

The difference between the inclusion of the winding resistance and the exclusion is only 0.1 Ω. This is hardly enough to be concerned about.

Short-Circuit Test

The instruments used in a short-circuit test are similar to those used in the open-circuit test. The difference is that the secondary is shorted instead of being left open. It is important to notice that shorting out the secondary can be very dangerous. This would cause a very large current to flow unless the input voltage is reduced. The input voltage is generally reduced to about 3 to 15% of the normal voltage so that the secondary current will be at about the rated value. In the open-circuit test the voltages were kept at normal levels. In the short-circuit test the currents are kept at the normal level.

The test circuit is shown in Fig. 16-21a. The instruments on the primary side are arranged as in the open-circuit test. The voltmeter on the secondary of the open-circuit test can be replaced with an ammeter. The measurements taken are the input power, the input voltage, the input current, and the output current.

The core-loss resistance and inductance can be ignored because the voltage levels are only about 10% of normal. The powers in these elements vary as the square of voltage across them. This means that the power will be only about 1% of normal. The input and output currents will be about normal; therefore, the input and output windings and their losses are at normal values.

The model for the transformer referred to the primary with the core mechanisms ignored is shown in Fig. 16-21b. It can be seen that the primary and secondary windings are difficult to separate in this model. It is therefore usually assumed that the values of the winding and the referred winding are equal. This is in general a very reasonable assumption because of the impedance transformation properties of an ideal transformer.

The short-circuit test is used primarily to measure copper losses. It also indicates the voltage drop in the transformer caused by the input and output windings.

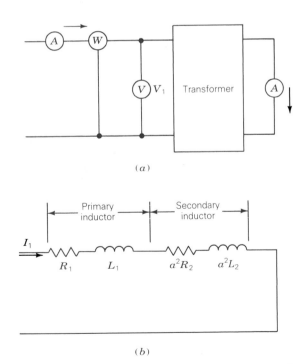

Figure 16-21 (a) Instrument setup for short-circuit test. (b) Model for short-circuit test.

Example 16-28

A 25-kVA, 440/220-V 60-Hz transformer is subjected to a short-circuit test. The results are $P_{in} = 1030$ W, $V_1 = 42$ V, $I_1 = 57$ A. Find the equivalent core-loss resistance for the transformer and the equivalent winding inductance referred to the primary.

Solution

The power loss in the short-circuit test is assumed to be due to the copper losses only. For this test the power dissipated is 1030 W. The equivalent copper-loss resistance is

$$R_{eq} = \frac{P_{in}}{I_1^2} = \frac{1030}{57^2}$$
$$= 0.317 \ \Omega$$

Next find the equivalent inductance of the windings. To do this, first find the total input impedance.

$$Z_{eq} = \frac{V_1}{I_1} = \frac{42}{57}$$
$$= 0.737 \ \Omega$$

The equivalent winding reactance is then determined by vectors.

$$X_{eq} = \sqrt{Z_{eq}^2 - R_{eq}^2} = \sqrt{0.737^2 = 0.317^2}$$
$$= 0.665 \ \Omega$$

Exercises

Applications

1. A transformer is tested under no-load conditions. When the transformer is powered from a 60-Hz 110-V source, the power meter indicates that the transformer is consuming 10 W. What is the equivalent core resistance?
2. The input current to the transformer of Exercise 1 is 0.75 A. What is the current that causes the mutual flux?
3. Again using the transformer of Exercises 1 and 2, what is the mutual inductance for the transformer?
4. A step-down transformer is used to reduce 120 V in to 10 V out. Under open-circuit test conditions the input power is 6.45 W and the input current is 2.1 A. Find the equivalent core-loss resistance, the mutual inductance, and the turns ratio.
5. A 440-V transformer is used to supply a 110-V service. In a short-circuit test the power input is 680 W, the input voltage is 42 V, and the input current is 57 A. Find the equivalent copper-loss resistance referred to the input.
6. For the test in Exercise 5 what is the equivalent winding reactance referred to the input?

Efficiency of a Transformer

In Job 11-4 we learned that the efficiency of an electric machine is equal to the ratio of the power output to the power input. This ratio is expressed as a percent by multiplying it by 100. In general, the efficiency of any device is the ratio of its output to its input and describes the effectiveness of the device in utilizing the energy supplied to it. Thus, a transformer which delivers *all* the power put into it would have an efficiency of 100%. In the last job, we assumed that transformers have an efficiency of 100% and deliver all the energy that they receive. Actually, because of copper and core losses, the efficiency of even the best transformer is less than 100%.

$$\text{Eff} = \frac{\text{power output}}{\text{power input}} \qquad \text{Formula (11-2)}$$

> **Hint for Success**
>
> Efficiency is reduced only by resistance. The loss may not be wire resistance but a loss in core material. In the model these losses are represented as resistors.

Example 16-29

A plate transformer draws 30 W from a 117-V line and delivers 300 V at 90 mA. Find its percent efficiency.

Solution

Given: Power input = 30 W   Find: Percent eff = ?
$$V_s = 300 \text{ V}$$
$$I_s = 90 \text{ mA} = 0.09 \text{ A}$$

$$\text{Eff} = \frac{\text{power output}}{\text{power input}} \qquad (11\text{-}2)$$

$$\text{Eff} = \frac{300 \times 0.09}{30} = 0.9 = 90\% \qquad Ans.$$

Example 16-30

A transformer whose efficiency is 80% draws its power from a 120-V line. If it delivers 192 W, find (*a*) the power input and (*b*) the primary current.

Solution

Given: Eff = 80% = 0.80   Find: Power input = ?
$$V_p = 120 \text{ V} \qquad I_p = ?$$
Power output = 192 W

a.
$$\text{Eff} = \frac{\text{power output}}{\text{power input}} \qquad (11\text{-}2)$$

$$0.80 = \frac{192}{\text{power input}}$$

$$\text{Power input} = \frac{192}{0.80} = 240 \text{ W} \qquad Ans.$$

b.
$$\text{Power input} = V_p \times I_p \qquad (16\text{-}15)$$
$$240 = 120 \times I_p$$
$$I_p = \frac{240}{120} = 2 \text{ A} \qquad Ans.$$

In the above efficiency examples, measurements of input and output power are made under actual conditions. These measurements are accurate but they do not provide insight into what is happening inside the transformer or what will be the efficiency under different conditions. In the development below, the model of a real transformer is used to answer these questions. This helps you to know whether the transformer is in good working order and to know what effect varying load conditions will have on the efficiency of the system.

An ideal transformer can be made more real by using either the model of Fig. 16-19*d* or that of Fig. 16-19*e*. The "copper" losses for the transformer can be calculated from the short-circuit test and the core losses from the open-circuit test. These values are then substituted into the transformer model. The example below shows how to use both the open- and the short-circuit tests to measure a transformer's internal parameters. It is then shown how to combine these results to calculate the efficiency for any load condition.

Example 16-31

A 25-kVA 440/220-V 60-Hz transformer is tested with the following results:

Type of Test	P_{in}	V_{in}	I_{in}
Primary open circuit	710	220	9.6
Secondary short circuit	20	4	14

What is the efficiency of the transformer when it delivers 113 A to the 220-V load? Ignore the effects of the transformer inductance.

Solution

We want to create a model of the transformer and the load referred to the primary. This model will allow us to calculate the power lost in the transformer and the power dissipated in the load. To do this, first use the short-circuit test to find the equivalent winding-loss resistance.

$$R_{\text{eq}} = \frac{P_{\text{in}}}{I_{\text{in}}} = \frac{20}{14^2} = 0.102 \ \Omega$$

Because the short-circuit test was done with the secondary shorted, the values calculated are referred to the primary, which is what we want. There is no need to transform these values.

We must now find the distribution of resistance between the primary and secondary windings. This is done by using the assumption that the primary and secondary losses are equal so that

$$R_1 = \frac{R_{\text{eq}}}{2} = a^2 \times R_2 = 0.051 \ \Omega$$

This completes the determination of the winding losses. The next step is to determine what the core losses are. This is done with the open-circuit data. Notice that the open-circuit test was done with the transformer backward; i.e., the 220-V side was used as the input. This means that the results calculated will be relative to the secondary. They will later have to be referred to the primary for our final model. From the data we get the core-loss resistance as seen from the secondary as follows:

$$R_{m2} = \frac{V_{in}^2}{P_{in}} = \frac{220^2}{710} = 68.2 \ \Omega$$

Use the turns ratio, a, to refer this value to the input. First find the turns ratio from the specifications for the transformer. The transformer is designed to reduce 440 V to 220 V. This means that the turns ratio is 220/440 = ½ and

$$R_m = R_{m2} \times \alpha^2 = 17.0 \ \Omega$$

The load that we want to use is 113 A at 220 V. This represents a resistance of 220/113 = 1.95 Ω. When this value is also referred to the primary side the value is

$$R_{L1} = 1.95 \times a^2 = 0.487 \ \Omega$$

These values can be put into the model shown in Fig. 16-22a. The core-loss resistor in this model is so much larger than the other resistors that we will ignore it for now. This makes the computation easier but it overestimates the efficiency because it ignores the power lost in the core. For the purpose of this example this is not a serious error. When this is done the model becomes that of Fig. 16-22b.

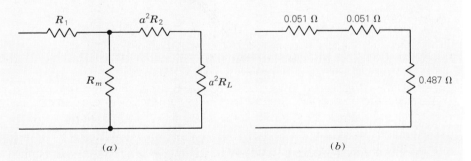

Figure 16-22 (a) Model for Example 16-31. (b) Calculated model values.

From the model we see that current coming into the transformer goes through the two winding losses and then the load. This sounds reasonable. Power loss is I^2R; therefore, the loss in R_1 and a^2R_2 represents the loss in the transformer. The loss in the load is the power out of the transformer. The current through each of these resistors is the same; therefore, it cancels out of the calculation. The efficiency of the transformer is calculated using the relationship: Power in = power out + loss in transformer.

$$Eff = \frac{power\ out}{power\ in} = \frac{power\ out}{power\ out + power\ loss}$$

$$= \frac{R_{L1}}{R_{L1} + R_1 + \alpha^2 \times R_2} = \frac{0.487}{0.487 + 0.051 + 0.051}$$

$$= 0.827 = 82.7\%$$

Exercises

Applications

1. What is the efficiency of a transformer if it draws 800 W and delivers 700 W?
2. A toy transformer draws 150 W from a 110-V line and delivers 24 V at 5 A. Find its efficiency.
3. A transformer draws 1.5 A at 120 V and delivers 7 A at 24 V. Find (*a*) the power input, (*b*) the power output, and (*c*) the efficiency.
4. A power transformer draws 96 W and delivers 420 V at 200 mA. Find (*a*) the efficiency and (*b*) the primary current if the primary voltage is 120 V.
5. In Fig. 16-23, an impedance-matching transformer couples an output transistor delivering 2.1 W to a voice coil which receives 1.60 W. Find its efficiency.

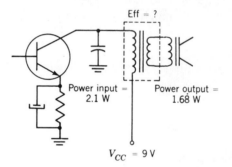

Figure 16-23   Finding the efficiency of an impedance-matching transformer.

6. A transformer that draws 1000 W from a 230-V line operates at an efficiency of 92% and delivers 50 V. Find (*a*) the watts delivered and (*b*) the secondary current.
7. A transformer that delivers 10,000 W at an efficiency of 96% draws its power from a 2000-V line. Find (*a*) the power input and (*b*) the primary current.
8. A transformer delivers 750 V at 120 mA at an efficiency of 90 percent. If the primary current is 0.8 A, find (*a*) the power input and (*b*) the primary voltage.
9. A transformer delivers 660 V at 98 mA at an efficiency of 84 percent. If the primary current is 875 mA, find (*a*) the power input and (*b*) the primary voltage.
10. The three secondary coils of a power-supply transformer deliver 100 mA at 350 V, 2 A at 2.5 V, and 1.2 A at 12.6 V. What is the efficiency of the transformer if it draws 60 W from the 117-V line?
11. A 60-Hz transformer is used to drop 220 V to 110 V. The transformer is tested with the following tests and results. For a no-load test with the input on the 110-V side the power in is 355 W, the voltage in is 110 V, and the current in is 6.7 A. The short-circuit test is run from the primary side of the transformer. For this test the power in is 50 W, the input voltage is 1.5 V, and the input cur-

rent is 35 A. Find the core-loss resistance and the winding resistance for the transformer model referred to the primary.

12. For the transformer in Exercise 11 find the efficiency of the transformer for the following loads: (*a*) 40 A, (*b*) 60 A, (*c*) 80 A. Why does the efficiency drop as the current increases?

Real Cables and Power Distribution

Task 16-1.1   Task Background

This task is based on cable tests and measurements made by Westinghouse.[*] You should try to remember the concepts and the quantities in this chapter. They relate to your real-world work.

Earlier jobs discussed cable resistance and cable magnetic induction. They are brought together in this job to show that actual cables behave just as you have been shown. There are four important lessons in this task. First, the theory is real. Second, the cable properties that dominate depend upon the size of the cable. Third, a very important outcome of mathematics is not the number but the idea. Fourth, the interpretation of data depends on the scale of values that is applied to it.

In the earlier jobs the emphasis was on calculating a value or values. An important idea in mathematics that is often overlooked is that a numerical value is not always the most important. Often the important lesson of mathematics is the characteristic represented in the mathematical relationships and the theory. It is these characteristics that give us guidance as we do our jobs. We develop rules of thumb. We develop insight about how to solve problems or about what is happening. We develop experience.

One way to present data is to create a table of numbers organized in a particular manner. In this chapter most of the data will be ordered by cable size. All the needed information and concepts are in the table; however, they are hard to see in that format. The overall concept can be seen better in a graph. Neat and detailed graphs are not necessary for the lessons we are trying to teach. A quick sketch is often sufficient.

Two theoretical concepts of cables will be confirmed. The first is that cable resistance varies inversely with the cross-sectional area of the cable. The second is that inductance is independent of the size of the cable. Guidelines will be developed to show that for cables larger than a number 1, the cable inductance may be important. The ability of models to predict real-life situations depends on what is included in the model. Models are valid only if they accurately estimate the real problem.

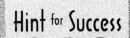

Hint for Success

Mathematics has value only to the extent that it describes real world phenomena. The better the description, the more complicated the model.

The Westinghouse Data

Part of the Westinghouse data is shown in Table 16-1 on page 592. The data are for average characteristics of products from several manufacturers of copper and aluminum conductors and cable. Thus, the table is representative of real applications and is useful for understanding general concepts in the workplace. If you are

[*]*Westinghouse Consulting Application Guide*, 10th Edition, 1991–1992, Westinghouse Electric Corporation, July 1991.

Table 16-1

Properties of Cables in Conduit

a.

	Copper			
	Magnetic		**Nonmagnetic**	
AWG	**R**	**X**	**R**	**X**
14	0.3130	0.00780	0.3130	0.00624
12	0.1968	0.00730	0.1968	0.00584
10	0.1230	0.00705	0.1230	0.00564
8	0.0789	0.00691	0.0789	0.00553
6	0.0490	0.00640	0.0490	0.00512
4	0.0318	0.00591	0.0318	0.00473
2	0.0203	0.00548	0.0203	0.00438
1	0.0162	0.00533	0.0162	0.00426
1/0	0.0130	0.00519	0.0129	0.00415
2/0	0.0104	0.00511	0.0103	0.00409
3/0	0.00843	0.00502	0.00803	0.00402
4/0	0.00696	0.00489	0.00666	0.00391

b.

	Aluminum			
	Magnetic		**Nonmagnetic**	
AWG	**R**	**X**	**R**	**X**
14	NA	NA	NA	NA
12	NA	NA	NA	NA
10	NA	NA	NA	NA
8	NA	NA	NA	NA
6	0.0833	0.00509	0.0833	0.00407
4	0.053	0.00490	0.0530	0.00392
2	0.0335	0.00457	0.0335	0.00366
1	0.0267	0.00440	0.0267	0.00352
1/0	0.0212	0.00410	0.0212	0.00328
2/0	0.017	0.00396	0.0170	0.00317
3/0	0.0138	0.00386	0.01380	0.00309
4/0	0.01103	0.00381	0.01097	0.00305

working with a specific cable and you have information for that cable type, then it may be desirable to use that data. Otherwise, Table 16-1 is sufficient.

The table is for both copper and aluminum conductors in a conduit like that shown in Fig. 16-24. The table includes cable size, shown by the cable number in the first column, and data for both resistance R and reactance X.

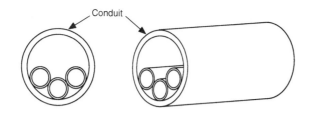

Figure 16-24

Cable Resistance versus Area

Example 16-32

Jobs 12-8 and 13-3 discussed the subject of line drop as a function of the cable size. They showed that the cable resistance varies inversely with the cross-sectional area of the cable. Use Eq. (12-7),

$$R = \frac{K \times L}{A} \qquad (12\text{-}7)$$

to match the Westinghouse data to the theory.

Solution

To do this, create Table 16-2 in the following manner. Table 16-1 gives wire size, which can be converted to cross-sectional area by using column 1 of Table 13-1 and the cross section in cmil in column 3. Table 16-2 has the wire size, area, and the resistance per 100 ft from the Westinghouse data.

Table 16-2

Resistance for Copper Cable in Magnetic Conduit

AWG	Area	R
14	4107	0.3130
12	6530	0.1968
10	10380	0.1230
8	16510	0.0789
6	26250	0.0490
4	41740	0.0318
2	66360	0.0203
1	83690	0.0162
1/0	105500	0.0130
2/0	133100	0.0104
3/0	167800	0.00843
4/0	211600	0.00696

The data from this table are plotted in Fig. 16-25 on page 594. In this figure the cable resistance in ohms per 100 ft is plotted against the area of the cable. The resistance decreases as the inverse of the area, as given by Eq. (12-7).

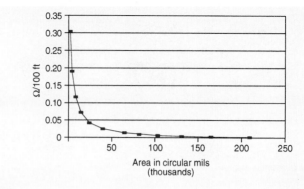

Figure 16-25

Cable Reactance versus Area

Job 16-7 stated that the inductance is independent of the cable size. To check this, Table 16-3 was created the same way Table 16-2 was created. It appears that the data in the table show that the reactance decreases as the cable gets larger. This is not what the earlier theory stated.

Table 16-3

Reactance for Copper Cable in Magnetic Conduit

AWG	Area	X
14	4107	0.00780
12	6530	0.00730
10	10380	0.00705
8	16510	0.00691
6	26250	0.00640
4	41740	0.00591
2	66360	0.00548
1	83690	0.00533
1/0	105500	0.00519
2/0	133100	0.00511
3/0	167800	0.00502
4/0	211600	0.00489

The complication is the scale on which we are looking at the reactance data. If you look very closely at the data, you see one thing. If you step back and see it in proportion to your problem, you see something else. To clarify the situation, data in the table are plotted in Fig. 16-26. Also given is the resistance plot of Fig. 16-25. The two lower curves use the Y axis on the left. The upper X curve uses the Y axis on the right. The right axis represents the raw data. The left axis represents the scale on which we will apply the data.

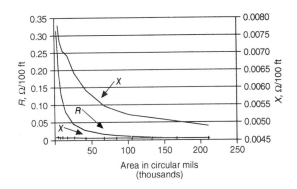

Figure 16-26

The upper X curve shows that the reactance does appear to decrease as the cable gets larger. The amount of this variation, however, is very little in comparison to the total cable impedance. This can be seen by looking at the two lower curves. These curves both use the same scale on the left. In this case, the resistance is decreasing as the cable gets larger; however, the lower X curve looks like a straight line. Using this scale, the reactance does not appear to change as the cable size changes.

The important point of this example is that you must always keep things in perspective. The problem is to find the correct perspective. The left scale is correct because the situations to which we want to apply the theory usually relate to impedance. The cable impedance is composed of both R and X; therefore, the left scale, which applies to both R and X, is the correct scale to use.

Summary

When working with cable impedance, you can assume that the inductance of a cable does not change with cable area.

Specific Resistance of Cable Materials

Example 16-33

This example will illustrate that the theory in Job 13-3 is correct. In that theory the cable resistance depends upon a property of the material called the specific resistance K. This constant should not depend upon the size of the cable. It should depend only on the material used to make the cable.

We need a formula that can be used to determine the specific resistance of the material for the data in the table. The table represents "typical" values for all cables tested. The value of K determined in this manner is also a "typical" value.

To calculate K, first rearrange Eq. (12-7) as follows.

$$R = \frac{K \times L}{A} \qquad (12\text{-}7)$$

$$K = \frac{R}{L} \times A \qquad \text{Formula 6-21}$$

The last equation can be used to determine a value of K for each cable size. There are many ways for a typical value of K to be obtained. We could pick some wire size, calculate K, and assume that is the typical value. For real experimental data, calculating K for different wire sizes commonly yields different values. We must then decide which value is the best or how to determine a "typical" value that is appropriate.

In this example there are two questions. First, is the specific resistance independent of cable size? Second, is there a constant for each type of cable material?

Solution

To answer these questions, the preceding formula was used to create Table 16-4.

Table 16-4

Calculated Specific Resistance

AWG	Aluminum	Copper
14		1285.491
12		1285.104
10		1276.74
8		1302.639
6	2186.625	1286.25
4	2212.22	1327.332
2	2223.06	1347.108
1	2234.523	1355.778
1/0	2236.6	1360.95
2/0	2262.7	1370.93
3/0	2315.64	1347.434
4/0	2333.948	1409.256

Figure 16-27 is a plot of the calculated values of K for both aluminum and copper cables. You can see that there is some variation in specific resistance, but on the scale in which we are interested, both curves can be assumed constant.

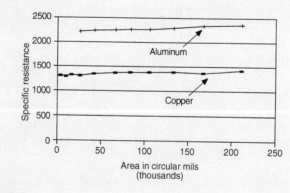

Figure 16-27

We also know that aluminum is a poorer conductor of electricity than copper. The specific resistance of aluminum cable is about 50% higher than copper. The values in Table 12-2 give

$$\frac{K_{Al}}{K_{Cu}} = \frac{17.0}{10.4} = 1.63$$

The units of measure for Table 16-4 are different. We can look at the table of data and assume a value of 2200 for aluminum and 1340 for copper. These values give

$$\frac{K_{Al}}{K_{Cu}} = \frac{2200}{1340} = 1.64$$

We can conclude from Fig. 16-27 that the theory does fit actual cables.

Cable Impedance

Example 16-34

In considering both the resistive and reactive components of cable imped-ance, is the reactive component important?

Solution

The data for copper wire in magnetic conduit from Table 16-1 has been repeated in columns 2 and 3 of Table 16-5. The fourth column is the calcu-lated total impedance, Z. The data from the table are plotted in Fig. 16-28.

Table 16-5

Cable Impedance for Copper Cable in Magnetic Conduit

AWG	R	X	Z	Error
14	0.3130	0.00780	0.313097	0.03%
12	0.1968	0.00730	0.196935	0.07%
10	0.1230	0.00705	0.123202	0.16%
8	0.0789	0.00691	0.079202	0.38%
6	0.0490	0.00640	0.049416	0.85%
4	0.0318	0.00591	0.032345	1.71%
2	0.0203	0.00548	0.021027	3.58%
1	0.0162	0.00533	0.017054	5.27%
1/0	0.0130	0.00519	0.013998	7.67%
2/0	0.0104	0.00511	0.011588	11.42%
3/0	0.00843	0.00502	0.009811	16.39%
4/0	0.00696	0.00489	0.008506	22.21%

CALCULATOR HINT

Large amounts of information are often best expressed by using graphics to show relationships.

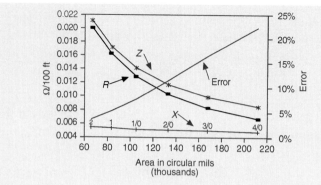

Figure 16-28

It is common to think of the resistive component of a cable as the only important part of the impedance. The last column of the table shows the error when using this assumption. The error occurs because you should be using Z instead of R. The size of the error for a number 4/0 cable is

$$\text{Error} = Z - R$$
$$= 0.008506 - 0.00696$$
$$= 0.001546 \qquad Ans.$$

The error as a percent of the number you are using is

$$\text{Percent error} = \frac{\text{error}}{R}$$
$$= \frac{0.001546}{0.00696}$$
$$= 0.2221$$
$$= 22.21\% \qquad Ans.$$

For cables smaller than number 1, the error is probably too small to be concerned with. As the cable gets larger, the error can be appreciable. Thus, reactance is probably more important in feeder cables than in branch circuits.

Effect of Conduit on Cable Reactance

Figures 15-10 to 15-12 in Job 15-2 show that electrical flow in a wire causes magnetic fields around it. Those figures applied to Fig. 16-24 explain how the magnetic fields that surround the cable in the conduit will interact with the conduit. Job 16-1 discussed how these fields create self-inductance and mutual inductance. Job 16-1 also showed that the magnetic permeability of the material in the magnetic circuit affects the inductance of the electrical circuit. The conduit acts as an inductive material that, like a transformer, can affect the inductance measured in the cable.

Table 16-1 has data for copper cable in both magnetic and nonmagnetic conduits. This example will determine if the theory that the material affects inductance is correct. The data in Table 16-6 were used to create Fig. 16-29. Job 16-1 indicated that a magnetic material in the conduit should cause a higher inductance and hence a greater reactance. Figure 16-29 shows that this is true for cables of all sizes.

Table 16-6

Reactance of Copper Cable in Conduits

AWG	Magnetic	Nonmagnetic
14	0.00780	0.00624
12	0.00730	0.00584
10	0.00705	0.00564
8	0.00691	0.00553
6	0.00640	0.00512
4	0.00591	0.00473
2	0.00548	0.00438
1	0.00533	0.00426
1/0	0.00519	0.00415
2/0	0.00511	0.00409
3/0	0.00502	0.00402
4/0	0.00489	0.00391

The effects of cables do not end at the outside of the cable. Nature causes a "leakage" of fields. This leakage couples into the surrounding conduit. This process is one way electromagnetic interference can arise in industrial locations.

Task 16-1.2   Applications

Power Distribution

The preceding task confirmed that mathematical concepts apply to real cables. The question now is what these concepts mean and when should they be used. This task will look at the basic problems of delivering power. The NEC does not specify voltage losses in cabling. It does make recommendations about the amount of voltage loss to allow. It suggests that there be a maximum of 5% loss from the service entrance to the connected load. This loss is to be divided between the feeders and the branch circuits. If each loss is essentially equal, then there should be only a 2 to 3% loss in each part.

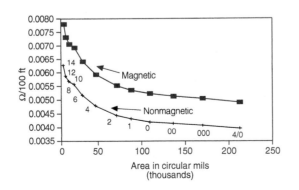

Figure 16-29

Section 220-2 of the NEC states standard voltages to be used for calculations that are made relative to section 220. It states that the voltages are 120, 120/240, 208Y/120, 240, 480Y/277, 480, and 600 V.

A Problem: 115-V, 7½-HP Motor

A branch circuit is to be wired to power a 7½-hp motor with a full load current of 80 A. The distance from the panel board to the motor is 150 ft.

One approach for finding the proper cable size was given in Job 13-2. Assume we want to use type RH cable. The cable size can then be found from Table 13-1, going down column 5 until we come to AWG No. 4. This cable can handle 85 A and, therefore, is large enough to handle the motor.

Job 13-3 went one step further. It took into consideration the voltage drop. In that job it was assumed that only the resistive component of the cable drop was important. The cable length is 150 ft, but the circuit length is 300 ft because we have one wire going to the load and one coming back. Both these wires have a voltage drop. For 300 ft of No. 4 wire with a current of 80.0 A, the total voltage drop is 7.8 V. This uses the resistance value of Table 16-1, $R = 0.0318\ \Omega$ per 100 ft, or $0.0954\ \Omega$ for 300 ft. At this point we have a cable that meets code for the current and we know the cable voltage drop.

However, in this problem we take the restrictions further and look at the recommendations of the NEC for voltage drop in feeder and distribution circuits. Then when we increase the size of the cable we will look at the effect of reactance on the cable.

Figure 16-28 shows that the error caused by ignoring the reactance of a No. 4 cable is probably less than 2% (1.7%, to be exact). This implies that we can use the simple calculation that considers only the resistance.

The NEC recommendation is that there be no more than 5 percent loss of voltage in the feeder and branch circuits. For the number 4 cable the percent loss is

$$\text{Voltage loss} = \frac{7.6}{115} = 6.6\%$$

This loss is the branch circuit loss only. It does not include the feeder circuit. It would be preferable to have only about 2.5% loss, so that the other 2.5% could be lost in the feeder circuits. This means that we have to increase the size of the distribution cable so that the motor will get sufficient power to operate properly.

Let's jump up to a 2/0 cable.

$$V_{\text{drop}} = \frac{2 \times 150}{100} \times 0.0104 \times 80$$
$$= 2.5\ \text{V}$$

A 2.5-V drop gives

$$\text{Voltage loss} = \frac{2.5}{115} = 2.2\%$$

This appears to be much better than the 2.5% goal that we had. We may be able to use a smaller cable, but looking at Fig. 16-28, we see that there is about 10% error caused by ignoring the reactance. It is advisable to check the cable including this factor.

For a 2/0 cable, the impedance per 100 ft is

$$Z = \sqrt{0.0104^2 + 0.00511^2}$$
$$= 0.011588$$

This gives a voltage drop of

$$V_{\text{drop}} = \frac{2 \times 150}{100} \times 0.011588 \times 80$$
$$= 2.8 \text{ V}$$

which is a 2.4% voltage loss in the branch circuit. This percentage loss is acceptable, and it is much closer to our goal than when we ignored the reactance of the cable. It should reduce our desire to use a smaller cable.

Summary

This section showed a more complex, but complete, process for selecting cable for a particular application. It covered Job 12-4, which asks only whether the cable can carry the desired current. It then applied the NEC recommendations and Job 13-3 to look at the voltage drop in the branch circuit. It then went further and used Job 16-7 and this job to look at the effects of inductance.

This is an extreme case of a very large motor for a single-phase, 115-V circuit. The length is longer than in most home wiring situations, but the problem does start to define the limits of house wiring and the start of industrial applications. We have to provide sufficient voltage to operate a motor properly. This requirement called for a larger cable. When the cable size was increased, the reactance of the cable in the conduit was an important factor.

Task 16-1.3 Related Jobs

This job has brought together ideas from many previous jobs you have done. These related jobs are listed next. For each previous job, there is a brief description of how it relates to this job. If you have had problems with anything in this job, you may want to review these sections.

Job 12-4: AWG table. In Job 12-4 you learned about standard wire sizes, cross-sectional area, and resistance per unit length.

Job 12-5: Specific resistance. Job 12-5 introduced the idea that the material of which the conductor is made is the fundamental determinant of the resistance of the conductor. You should be particularly interested in the K values for copper and aluminum.

Job 12-6: Resistance and lengths. Job 12-6 is really the basis for using the resistance-per-unit-length values.

Job 12-7: Resistance and areas. Job 12-7 can be used to relate wire sizes. To do this, recall from Job 12-4 that wire size number and area are uniquely related.

Job 12-8: $R = (K \times L)/A$. This job is a combination of all the preceding jobs. It ties all these relationships into one formula.

Job 12-9: Wire length for resistance. This job shows another way of looking at the relationship between length and resistance.

Job 12-10: Diameter of wire. Job 12-10 is really a study of the effects of wire size on resistance.

Job 12-11: Temperature coefficients of resistance. Job 12-11 shows how resistance changes with temperature.

Job 13-1: Maximum current. In Job 13-1 NEC limits on the current-carrying capability of wire are introduced.

Job 13-2: Minimum-size wire. Job 13-2 evaluates the smallest wire gage that will carry a given current load.

Job 13-3: Voltage drop. Job 13-3 determines a wire size that will produce a given voltage drop. This job assumes the cable has resistance only.

Job 13-4: Temperature and wire size: Job 13-4 shows how wire ampacity varies with temperature.

Job 15-2: Field around a wire. The introduction to Job 15-2 indicates the fields that surround a wire. This leads to the concepts of cable inductance and interaction of cable and conduit.

Job 16-1: Inductance. The basic concepts of inductance are introduced in Job 16-1.

Job 16-7: Transmission lines. Job 16-7 develops the concept of inductance per unit length and the fact that inductance is independent of the size of the conductor.

Exercises

Practice

1. Using Eq. (16-21) and the data in Table 16-1, calculate the specific resistance for copper using the data for the following cable sizes: 4, 2, 1, 0.
2. Using Eq. (16-21) and the data in Table 16-1, calculate the specific resistance for aluminum using the data for the following cable sizes: 4, 2, 1, 0.
3. Repeat the analysis of the problem in Task 16-1.2 for the 7½-hp motor using aluminum wire, Table 16-1*b*.

1. A choke provides 30 Ω of opposition to current flow in a 120-V 60-Hz circuit. What is the inductance of the coil?

2. Two 25-Ω series resistors are in series with a 30-mH and 49.577-mH coil in a 200 V 60-Hz circuit. Determine (a) total resistance, (b) total inductance, (c) total inductive reactance, (d) total impedance, (e) total current flow in the circuit, (e) voltage drop across each component, (f) phase angle, (g) power factor, (h) true power, (i) reactive power, and (j) apparent power.

3. A 6.366-mH coil is operating at 2.5 kHz. If the Q value of this coil is 100, what is the internal resistance of the coil?

4. What is the inductance of a coil if it draws 235 μA at 5 V when operating at a frequency of 10 kHz?

5. Do the effects of transmission inductance and mutual inductance play a greater role in low-frequency or high-frequency circuits? Explain your answer.

6. A step-down transformer delivers 120 V and has a turns ratio of 4:1. What is the primary voltage, and how many turns are there in the primary?

7. A 480- to 120-V step-down transformer is connected to a 25-Ω load. What is the current in the primary windings?

8. An 8-Ω speaker load is connected to an amplifier that has 8 kΩ of internal impedance. If an impedance-matching transformer is used, what must be the turns ratio to achieve maximum transfer of power from the source to the load?

9. When performing an open circuit test of a 240-V 60-Hz transformer, you determine that it is consuming 18 W of power. What is the equivalent core resistance of the transformer?

10. A 96% efficient transformer supplies 250 VA when fully loaded. What is the primary power being taken from the input line?

INTERNET
A C T I V I T I E S

Internet Activity A

When the winding resistances of medium and large power transformers are measured, three main problems must be overcome. What are they?
http://www.avointl.com/products/transformer/xtra/guides/problem.html

Internet Activity B

Use a professional calculator to specify wire and cable. Identify the elements covered in the summary of this calculator. http://www.aes-soft.com/

Chapter 16 Solutions to Self-Tests and Reviews

Job 16-2 Review

1. H
2. turns, current
3. B, area
4. permeability
5. current, voltage
6. self-inductance
7. mutual inductance
8. henry, 1, 1
9. opposes
10. ac, move

Self-Test 16-4

1. 3768
2. 300×10^{-6}
3. 1884
4. 10^6
5. 2

Self-Test 16-7

1. 8000
2. Q

3. 13.3
4. should not
5. 8000
6. 0.01
7. 80
8. 10
9. 800
10. 600
11. 1.33
12. should
13. 64
14. 1000
15. 0.08
16. 800
17. 64

Self-Test 16-12

1. must
2. R^2
3. 9
4. 1
5. 8

6. 2.83
7. 283
8. 283
9. 0.75

Job 16-7 Review

1. a. alternating
   b. cut
   c. back
2. back
3. H
4. changing
6. phasor, larger, smaller
7. ohmmeter
   d. larger

Self-Test 16-18

1. V_s
2. VR
3. 770

Self-Test 16-22

1. I_s, I_p
2. $\dfrac{120}{24}$
3. 1.5
4. 1.5
5. 7.5

Job 16-12 Review

2. primary, load
3. increases, increases
4. primary, secondary
5. d. N_p, N_s
   e. I_s, I_p
   f. I_s, I_p
6. turns ratio

Capacitance

1. Explain how capacitance is created.
2. Find the capacitance of parallel capacitors.
3. Find the capacitance of series capacitors.
4. Determine the reactance of a capacitor.
5. Write the equations for current and voltage in a capacitor.
6. Determine capacitance.
7. Write equations for a real capacitor.

JOB 17-1

Introduction to Capacitance

What is a capacitor?　A capacitor, or "condenser," is formed whenever two pieces of metal are separated by a thin layer of insulating material. To form a capacitor of any appreciable value, however, the area of the metal pieces must be quite large and the thickness of the insulating material, or *dielectric*, must be quite small.

What does a capacitor do?　When we wish to store up electricity for a little while, we "charge" the capacitor. When we "discharge" a capacitor, we draw the electrons from it to operate some device. The two plates of the capacitor shown in Fig. 17-1 are electrically neutral, since there are as many protons as electrons on each plate. The capacitor has no "charge."

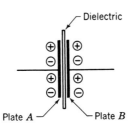

Figure 17-1　Electrically neutral capacitor.

Now let us connect a battery across the plates as shown in Fig. 17-2a on page 606. When the switch is closed (Fig. 17-2b), the positive side of the battery pulls electrons from plate A of the capacitor and deposits them on plate B. Electrons will continue to flow from A to B until the number of electrons on plate B exert a force equal to the electromotive force of the battery. The capacitor is now charged. It will remain in this condition even if the battery is removed as shown in Fig. 17-3a on page 606. This is a condition of extreme unbalance, and the electrons on plate B will attempt to return to plate A if they can. The resistance of the dielectric and the surrounding air prevents this from happening, although some electrons do

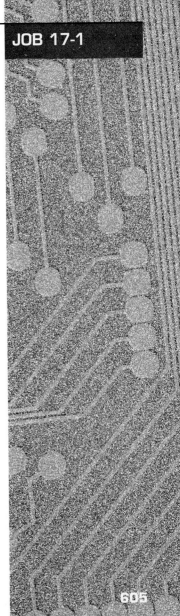

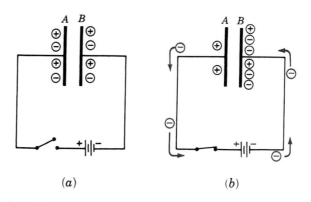

(a)   (b)

Figure 17-2   A charged capacitor has an excess of electrons on one plate. (a) neutral capacitor; (b) charged capacitor.

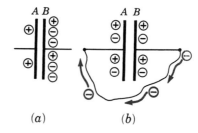

(a)   (b)

Figure 17-3   (a) A charged capacitor has more electrons on one plate than on the other. (b)
Discharging a capacitor.

Hint for Success

Capacitance developed
from the concepts
learned from experi-
ments with Leyden jars,
in which electrical
charge was put into a
bottle.

manage to "leak" off plate *B* and return to plate *A*. However, if a conductor is
placed across the plates as in Fig. 17-3*b*, then the electrons find an easy path back
to plate *A* and they will return in a rush, thus "discharging" the capacitor. There are
many uses for capacitors such as in tuning circuits, filter circuits, coupling circuits,
bypasses for alternating currents of high frequency, and blocking devices in audio
circuits. In each application, the capacitor operates by storing up electrons and discharging them at the proper time. Capacitors cannot be used in dc circuits, since
the dielectric of the capacitor acts to present an open circuit. Current will flow in
an ac circuit containing a capacitor as shown in Fig. 17-4. During the positive half
of the ac cycle, the electrons travel through the lamp and pile up on plate *A* of the
capacitor. During this time, electrons are drawn off plate *B* of the capacitor by the
ac source. During the negative half of the cycle, the direction of the electron flow
is reversed. The capacitor discharges through the lamp; the source pulls the electrons from plate *A* and piles them up on plate *B*. This action continues with each
reversal of the alternating current. There seems to be a continuous flow of
electrons through the capacitor which lights the lamp, but it is actually a flow of
electrons *around* the capacitor which operates the lamp.

Meaning of capacitance.   The measure of the ability of a capacitor to hold electrons is called its *capacitance*. Since an electron is so small, and since there are so
many of them, the *coulomb* is used as the measure of electrical quantity. In Job
1-1 we learned that a coulomb is equal to approximately 6 billion electrons. The
capacitance of a capacitor has been defined as the number of coulombs of electricity that may be held on its plates by a pressure of 1 V. If 1 coulomb is held on
the plates by a pressure of 1 V, then the capacitance is called 1 farad (F). From this,
we can get a formula to find the capacitance of a capacitor.

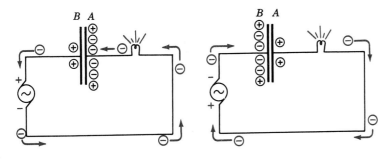

Figure 17-4 Electron flow around a capacitor in an ac circuit.

$$C = \frac{Q}{V}$$

Formula 17-1

where
- C = capacitance, F
- Q = number of electrons on plates, coulombs
- V = voltage across the plates, V

It is very inconvenient to discuss capacitances in terms of farads, since a farad is such a large unit of measurement. Always change units of capacitance into microfarads (μF) or picofarads (pF). See Job 3-7 for methods of changing units of measurement.

CALCULATOR HINT

Remember to use the divide key to calculate capacitance.

JOB 17-2

Capacitors in Parallel

In the last job we learned that capacitance depends on the number of coulombs that can be held on the plates by a pressure of 1 V. A capacitor that can hold 3 coulombs will have three times the capacitance of another that can hold only 1 coulomb if the same voltage is applied to both. What is it about a capacitor that enables one to hold more electrons than another? Obviously, the electrons must be held somewhere, and they are usually distributed on the surface of the capacitor plates. If the area of the plates is large, then many electrons can be placed on the large area, but if the plates are small in area, then only a few electrons can be held there and the capacitance will be small. The capacitance of a capacitor depends on this plate area and also on the thickness of the dielectric. The larger the plate area, the larger the capacitance. The *thinner* the dielectric, the *larger* the capacitance.

Hint for Success

Small capacitors in parallel with large capacitors are usually called *trimmer capacitors*.

What is the effect of placing capacitors in parallel? In Fig. 17-5 on page 608, C_1 and C_2 are connected in parallel. Since plate A and plate B are connected, the effect is the same as if we had one big plate with an area equal to the sum of plates A and B. Similarly, plates C and D on the other side are connected together to form one large plate equal to the sum of C and D. Since capacitance increases as we increase the plate area, we get the capacitance of the combination by adding the capacitances of the individual capacitors.

$$C_T = C_1 + C_2 + C_3 + \text{etc.}$$

Formula 17-2

where
- C_T = total capacitance
- C_1, C_2, C_3, etc. = capacitances of the individual capacitors

All capacitances must be measured in the same units.

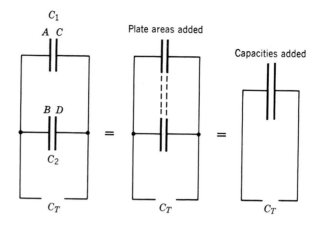

Figure 17-5   When capacitors are in parallel, the plate areas are added and the total capacitance C_T equals the sum of the individual capacitances.

Working voltage.   There is a limit to the voltage that can be applied across any capacitor. If too large a voltage is applied, it will overcome the resistance of the dielectric and a current will be forced through it from one plate to the other, sometimes burning a hole in the dielectric. In this event, a short circuit exists and the capacitor must be discarded. This applies to mica or waxed-paper capacitors. If the dielectric is air, the "short" disappears as soon as the voltage is removed. The maximum voltage that may be applied to a capacitor is known as the *working voltage* and must never be exceeded.

Example 17-1

A 0.000 35-μF tuning capacitor is in parallel with a trimmer capacitor of 0.000 075 μF. What is the total capacity?

Solution

The diagram for the circuit is shown in Fig. 17-6.

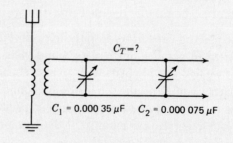

$C_T = ?$

$C_1 = 0.000\ 35\ \mu F$ $C_2 = 0.000\ 075\ \mu F$

Figure 17-6

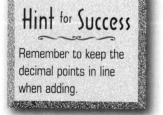

Hint for Success

Remember to keep the decimal points in line when adding.

$$C_T = C_1 + C_2 \qquad (17\text{-}2)$$
$$C_T = 0.000\ 35 + 0.000\ 075$$
$$C_T = 0.000\ 425\ \mu F \qquad Ans.$$

Example 17-2

What are the total capacitance and working voltage of a 0.0005-μF 50-V capacitor, a 0.015-μF 100-V capacitor, and a 0.000 25-μF 100-V capacitor when they are connected in parallel?

Solution

The total capacitance is the sum of the capacities:

$$
\begin{array}{r}
0.000\ 5 \\
0.015 \\
\underline{0.000\ 25} \\
C_T = \overline{0.015\ 75}\ \mu F \qquad Ans.
\end{array}
$$

Just as a chain is only as strong as its weakest link the working voltage of a group of parallel capacitors is only as great as the *smallest* working voltage. Therefore, the working voltage of the combination is only 50 V.

Exercises

Applications

1. A capacitor in a tuning circuit has a capacitance of 0.000 32 μF. When the stage is aligned, the trimmer capacitor in parallel with it is adjusted to a capacitance of 0.000 053 μF. What is the total capacitance of the combination?
2. A mechanic has the following capacitors available: 0.0003-μF 75-V; 0.000 25-μF 50-V; 0.0002-μF 50-V; 0.000 15-μF 75-V; and 0.000 05-μF 75-V. Which of these should be arranged in parallel to form a combination with a capacitance of 0.0005 μF and 75 V working voltage?
3. What is the total capacitance in parallel of the following capacitors: 35 pF, 0.005 μF, and 0.000 03 μF?
4. A capacitor of 0.0003-μF capacitance is connected in parallel with a 0.000 005-F capacitor. What is the total capacitance?
5. What is the total capacitance of a 25-pF, 0.000 06-μF, and a 0.000 000 03-F capacitor?
6. A capacitor of 0.003 μF is in parallel with another which holds a charge of 0.000 004 C at 200 V. What is the total capacitance?
7. What amount of capacitance must be added in parallel with a 0.000 55-μF capacitor in order to get a total capacitance of 0.0007 μF?
8. What is the total capacitance in parallel of two capacitors if they are charged at 120 V with 0.000 06 and 0.000 172 8 C, respectively?

Capacitors in Series

It has been found that the thicker the dielectric, or the greater the distance between the plates, the smaller the capacitance. If we were to combine capacitances so that the effective distance between the plates were increased, then the resulting capacitance would necessarily be less than before.

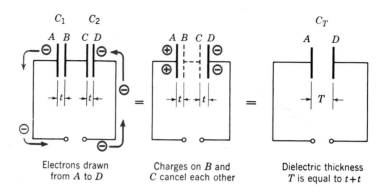

Electrons drawn | Charges on B and | Dielectric thickness
from A to D | C cancel each other | T is equal to $t+t$

Figure 17-7   Capacitors in series increase the total dielectric thickness and *decrease* the total capacitance.

What is the effect of placing capacitors in series?   In Fig. 17-7, capacitors C_1 and C_2 are connected in series. When a voltage is impressed on this combination, electrons are drawn from plate A and deposited on plate D. The charges on plates B and C are equal and opposite and therefore neutralize each other. In this event, they may be considered to be eliminated and the combination replaced by a single capacitor whose dielectric thickness is equal to the sum of the dielectric thicknesses of the original capacitors. Thus, the effect of a series combination is to increase the dielectric thickness and therefore to decrease the capacitance.

$$\frac{1}{C_T} = \frac{1}{C_1} + \frac{1}{C_2} + \frac{1}{C_3} + \text{etc.} \qquad \text{Formula 17-3}$$

All capacitances must be measured in the same units.

Notice the similarity between this formula and the formula for resistances in parallel given in Job 5-8 as formula (5-3). The similarity is continued through the formula for the total resistance in parallel of a number of equal resistors given in the same job as formula (5-4). For equal capacitors in *series*, we have

$$C_T = \frac{C}{N} \qquad \text{Formula 17-4}$$

where C_T = total capacitance
C = capacitance of one of the equal capacitors
N = number of equal capacitors

Working voltage.   Within limits, the total voltage that may be applied across a group of capacitors in series is equal to the sum of the working voltages of the individual capacitors.

> ### Hint for Success
>
> For series capacitors, the susceptances add.

Example 17-3

A 4-, a 5-, and a 10-μF capacitor are connected in series. Find the total capacitance.

Solution

Review Job 5-4, Addition of Fractions.

$$\frac{1}{C_T} = \frac{1}{C_1} + \frac{1}{C_2} + \frac{1}{C_3} \qquad (17\text{-}3)$$

$$\frac{1}{C_T} = \frac{1}{4} + \frac{1}{5} + \frac{1}{10}$$

$$\frac{1}{C_T} = \frac{11}{20}$$

$$11 \times C_T = 20$$

$$C_T = \frac{20}{11} = 1.818 \ \mu F \qquad Ans.$$

Example 17-4

A voltage-doubler circuit is shown in Fig. 17-8. Only C_1 is charged during the first half of the cycle. When the polarity reverses during the second part of the cycle, C_2 is charged to *twice* the peak voltage of the line because it is in series aiding with the line and C_1. What are the total capacitance and working voltage of the capacitor combination if C_1 and C_2 are both 100-μF 200-V capacitors?

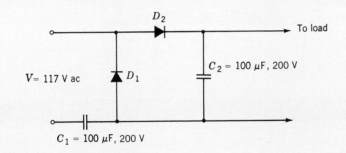

Figure 17-8   Voltage-doubler circuit.

Solution

$$C_T = \frac{C}{N} = \frac{100}{2} = 50 \ \mu F \qquad Ans.$$

$$\text{Working voltage} = 200 + 200 = 400 \text{ V} \qquad Ans.$$

Example 17-5

The frequency that beats against the incoming frequency in the superheterodyne receiver is produced by an oscillator circuit. The Colpitts oscillator shown in Fig. 17-9 on page 612 controls the frequency of oscillation by means of the variable inductance. What is the total capacity of the series combination of C_1 and C_2 if $C_1 = 0.0005 \ \mu F$ and $C_2 = 0.000 \ 25 \ \mu F$?

Solution

When only two capacitors are in series, we can use a formula very similar to the formula for finding the total resistance of two resistors in parallel.

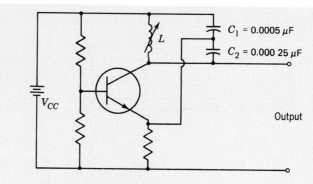

Figure 17-9 Colpitts oscillator.

$$C_T = \frac{C_1 \times C_2}{C_1 + C_2} \qquad \text{Formula 17-5}$$

In this formula, all the measurements must be in the same units. Since it is easier to use whole numbers than decimals in the formula, all measurements are changed to picofarads.

$$C_1 = 0.0005 \ \mu F = 0.0005 \times 10^6 = 500 \text{ pF}$$
$$C_2 = 0.000\ 25 \ \mu F = 0.000\ 25 \times 10^6 = 250 \text{ pF}$$
$$C_T = \frac{500 \times 250}{500 + 250} = \frac{125,000}{750} = 167 \text{ pF} \qquad Ans.$$

CALCULATOR HINT

Use the inverse function to simplify the computations.

Exercises

Practice

1. Capacitors of 3, 4, and 6 μF are connected in series. What is their total capacitance?
2. What are the total capacitance and working voltage of a voltage doubler similar to that shown in Fig. 17-8 if the circuit uses two 40-μF 175-V capacitors?
3. Find the total capacitance of the series capacitors in a Colpitts oscillator similar to that shown in Fig. 17-9 if $C_1 = 0.0004 \ \mu F$ and $C_2 = 0.000\ 02 \ \mu F$.
4. Find the total capacitance of the series capacitors in a Colpitts oscillator if (a) $C_1 = 0.01 \ \mu F$ and $C_2 = 0.001 \ \mu F$, (b) $C_1 = 0.0015 \ \mu F$ and $C_2 = 800$ pF.

Applications

5. What is the range of total capacitances available in an oscillator circuit which uses a tuning capacitor with a 35- to 350-pF range in series with a padder capacitor set at 300 pF?
6. In some vibrator power supplies operating from a storage battery, a pair of "buffer" capacitors are placed across the secondary of the transformer to reduce the voltage peaks. What are the total capacitance and working voltage of a pair of 0.0075-μF 800-V buffer capacitors in series?
7. Find the total capacitance of a 6-, a 10-, and a 15-μF capacitor in series.
8. What is the total capacitance of a 0.000 000 02-F, a 0.04-μF, and a 60,000-pF capacitor in series?
9. A series combination of a 4- and a 12-μF capacitor is connected in parallel with a 5-μF capacitor. Find the total capacitance of the combination.

10. A 30-pF capacitor is connected in series with a parallel combination of a 30- and a 60-pF capacitor. Find the total capacitance of the combination.

(See CD-ROM for Test 17-1)

Reactance of a Capacitor

We learned in Job 17-1 that as a capacitor is charged, electrons are drawn from one plate and deposited on the other. As more and more electrons accumulate on the second plate, they begin to act as an opposing voltage which attempts to stop the flow of electrons just as a resistor would do. This opposing effect is called the *reactance* of the capacitor and is measured in ohms.

What Factors Determine the Reactance?

1. The size of the capacitor is one factor. The larger the capacitor, the greater the number of electrons that may be accumulated on its plates. However, because the plate area is large, the electrons do not accumulate in one spot but spread out over the entire area of the plate and do not impede the flow of new electrons onto the plate. Therefore, a large capacitor offers a small reactance. If the capacitance were small, as in a capacitor with a small plate area, the electrons could not spread out and would attempt to stop the flow of electrons coming to the plate. Therefore, a small capacitor offers a large reactance. The reactance is therefore *inversely* proportional to the capacitance.

2. If an ac voltage is impressed across the capacitor, electrons are accumulated first on one plate and then on the other. If the frequency of the changes in polarity is low, the time available to accumulate electrons will be large. This means that a large number of electrons will be able to accumulate, which will result in a large opposing effect, or a large reactance. If the frequency is high, the time available to accumulate electrons will be small. This means that there will be only a few electrons on the plates, which will result in only a small opposing effect, or a small reactance. The reactance is therefore *inversely* proportional to the frequency.

3. A special constant of proportionality is necessary to change the current-reducing effect of the electron accumulation into ohms of reactance. This number is 2π. The formula for capacitive reactance is

$$X_C = \frac{1}{2\pi f C}$$

with C measured in farads. If the capacitance is measured in microfarads,

$$X_C = \frac{1}{2 \times 3.14\, f \times (C/1{,}000{,}000)} = \frac{1{,}000{,}000}{6.28 \times f \times C}$$

and since

$$\frac{1{,}000{,}000}{6.28} \approx 159{,}000 \text{ (approx)}$$

$$X_C = \frac{159{,}000}{f \times C} \qquad \text{Formula 17-6}$$

where X_C = capacitive reactance, Ω
   f = frequency, Hz
   C = capacitance, μF

> ### Hint for Success
>
> A capacitor operates using electric fields.

Note: The larger the capacitance, the smaller the reactance. The larger the frequency, the smaller the reactance.

Example 17-6

The antenna circuit of a two-transistor AM receiver is shown in Fig. 17-10. Find the capacitive reactance of the tuning capacitor to a frequency of 500 kHz when it is set at 300 pF.

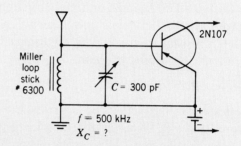

Figure 17-10 Find the capacitive reactance of the tuning capacitor.

Solution

1. Change the measurements to the required units.

$$f = 500 \text{ kHz} = 500 \times 10^3 \text{ Hz}$$
$$C = 300 \text{ pF} = 300 \times 10^{-6} \text{ } \mu\text{F}$$

2. Find the capacitive reactance.

$$X_C = \frac{159,000}{f \times C} \quad\quad (17\text{-}6)$$

$$X_C = \frac{159 \times 10^3}{500 \times 10^5 \times 300 \times 10^{-6}}$$

$$X_C = \frac{159}{15 \times 10^4 \times 10^{-6}}$$

$$X_C = 10.6 \times 10^2 = 1060 \text{ } \Omega \quad\quad Ans.$$

Example 17-7

A capacitor is formed whenever two metal pieces are separated by a dielectric. A capacitor formed by two wires of a circuit or two turns of a coil produces a *distributed capacitance*. This is extremely undesirable, because even a very small distributed capacitance can transfer energy from one circuit to another at radio frequencies, since the reactance at radio frequencies is very small. This unwanted transfer of energy represents wasted power. What is the current lost through a distributed capacitance of 100 pF formed by two parallel wires, one of which carries a 1000-kHz current? The difference in potential between the wires is 1 V.

Solution

The diagram of the circuit is shown in Fig. 17-11.

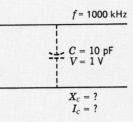

$$f = 1000 \text{ kHz}$$

$$C = 10 \text{ pF}$$
$$V = 1 \text{ V}$$

$$X_c = ?$$
$$I_c = ?$$

Figure 17-11

1. Change the measurements to the required units.

$$1000 \text{ kHz} = 10^3 \times 10^3 = 10^6 \text{ Hz}$$
$$10 \text{ pF} = 10 \times 10^{-6} = 10^{-5} \text{ } \mu\text{F}$$

2. Find the capacitive reactance.

$$X_C = \frac{159{,}000}{f \times C} = \frac{159{,}000}{10^6 \times 10^{-5}} = \frac{159{,}000}{10} = 15{,}900 \text{ } \Omega$$

3. Find the current. Since the only opposition to the flow of current around a capacitor is its reactance, the formula for Ohm's law, $V = IR$, may be rewritten as

$$V_C = I_C \times X_C \qquad \text{Formula 17-7}$$
$$1 = I_C \times 15{,}900$$

$$I_C = \frac{1}{15{,}900} = 0.000 \ 062 \text{ A} = 0.062 \text{ mA} \qquad Ans.$$

Example 17-8

Find the size of filter capacitor necessary to provide a capacitive reactance of 159 Ω at a frequency of 60 Hz.

Solution

The diagram for the circuit is shown in Fig. 17-12.

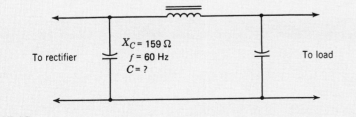

To rectifier

$$X_C = 159 \text{ } \Omega$$
$$f = 60 \text{ Hz}$$
$$C = ?$$

To load

Figure 17-12

$$X_C = \frac{159{,}000}{f \times C} \qquad (17\text{-}6)$$

$$\frac{159}{1} = \frac{159{,}000}{60 \times C}$$

$$159 \times 60 \times C = 159{,}000$$

$$C = \frac{159{,}000}{159 \times 60} = \frac{100}{6} = 16.7 \ \mu F \qquad Ans.$$

Use the commercially available 16-μF 500-V capacitor.

Self-Test 17-9

A capacitance of 30 pF draws 20 mA when connected across a 106-V source. Find the frequency of the ac voltage.

Solution

1. Given: $C = 30$ pF   Find: $f = ?$
   $I = \underline{\hspace{2cm}}$
   $V = 106$ V

2. Change the measurements to the required units.

$$30 \text{ pF} = 30 \times \underline{\hspace{2cm}} \ \mu F$$
$$20 \text{ mA} = 20 \times 10^{-3} = \underline{\hspace{2cm}} \text{ A}$$

3. Find the capacitive reactance.

$$V_C = I_C \times X_C \qquad (17\text{-}7)$$
$$106 = 0.02 \times X_C$$
$$X_C = \frac{106}{0.02} = \underline{\hspace{2cm}} \ \Omega$$

4. Find the frequency.

$$X_C = \frac{159{,}000}{f \times C} \qquad (17\text{-}6)$$
$$5300 = \frac{159 \times 10^3}{f \times 30 \times 10^{-6}}$$
$$f = \frac{159 \times 10^3}{30 \times 10^{-6} \times 53 \times 10^2}$$
$$f = \frac{159 \times 10^3}{159 \times ?}$$
$$f = 1 \times 10^3 \times \underline{\hspace{2cm}} = 1 \times 10^?$$
$$f = 1 \underline{\hspace{2cm}} \qquad Ans.$$

Exercises

Practice

1. What is the reactance of a 0.0003-μF capacitor at (*a*) 30 kHz, (*b*) 100 kHz, and (*c*) 800 kHz?

2. What is the reactance of an oscillator capacitor of 0.0004 μF to a frequency of 456 kHz?

Applications

3. A 10-μF capacitor in the emitter circuit of a 2N322 transistor used in a 12-V audio amplifier produces a voltage drop of 3 V at 1 kHz. Find the current passed by the capacitor.

4. What is the reactance of a 20-μF coupling capacitor to an audio frequency of 1 kHz? What current will flow if the voltage across the capacitor is 2.38 V as shown in Fig. 17-13?

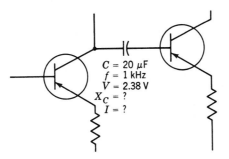

$$C = 20 \ \mu F$$
$$f = 1 \text{ kHz}$$
$$V = 2.38 \text{ V}$$
$$X_C = ?$$
$$I = ?$$

Figure 17-13

5. A 0.000 35-μF tuning capacitor is in parallel with a 0.000 05-μF trimmer capacitor. Find the total capacitance of the combination and its reactance to a frequency of 100 kHz.

6. A capacitor draws 4 A when connected across a 120-V 60-Hz line. What will be the current drawn if both the capacitance and frequency are doubled?

7. What should be the capacitance of a thyratron control circuit if the reactance must be 1590 Ω at 60 Hz?

8. A capacitor in a telephone circuit has a capacitance of 2 μF. What current flows through it when an emf of 15 V at 795 Hz is impressed across it?

9. A pocket radio uses a 10-μF coupling capacitor in the base circuit of a 2N35 NPN transistor. Find its reactance to a frequency of (a) 1 kHz, (b) 5 kHz, and (c) 20 kHz.

10. The emitter resistor for a 2N1265 transistor is bypassed with a 5-μF capacitor. What is the reactance of this capacitor to a 1.5-kHz frequency? If the voltage across the capacitor is 10.6 V, what current will flow?

11. Find the capacitive reactance between two wires if the stray capacitance between them is 8 pF and one wire carries a radio frequency of 1500 kHz.

12. A 2N218 transistor acting as an FM sound detector has its emitter resistor bypassed with a 4-μF capacitor. What is the reactance of the capacitor to the 500-Hz AF current?

13. A capacitance of 1.1 μF draws 0.05 A when connected across a 120-V line. Find the frequency of the ac voltage.

14. An antenna tuning capacitor has a capacitance of 250 pF. What current flows through it when 41.8 V are impressed across it at 7.6 MHz?

15. A 2N190 PNP transistor used as an audio amplifier has its emitter stabilizing resistor bypassed with a 50-μF capacitor. Find the reactance of the capacitor to a frequency of (a) 100 Hz, and (b) 5 kHz.

16. In the Admiral transistorized television receiver model NA1-2B, the first sound IF transistor 2SC460 has its emitter resistor bypassed with a 0.01-μF capacitor. What is its reactance to a frequency of 21.25 MHz?

17. Find the bypass capacitor required for a 2N109 audio output transistor if it is to have a reactance of 795 Ω at 10 kHz.

18. A "leading" current of 5 A is to be obtained from a 220-V 60-Hz line by means of a capacitor. Find the required capacitance.

19. A static capacitor capable of passing 40 A at 240 V and 60 Hz is added to a line to correct the power factor. What must be its capacitance?

JOB 17-5

Current and Voltage in a Capacitor

When an alternating voltage is impressed across a capacitor, the capacitor will be alternately charged and discharged. While it is charging, the flow of electrons to the plate of the capacitor is largest at the instant the charge is begun. This is so because there are no electrons already on the plate to exert an opposing force. As more and more electrons accumulate on the plate of the capacitor, they exert a greater and greater force which tends to stop the flow of electrons to the plate. When the capacitor is fully charged to the voltage of the source, the flow of the current falls to zero, since the back pressure is equal to the pressure of the charging source. Notice that the maximum current occurs when the voltage is zero, and a zero current flows when the voltage is a maximum. If the capacitor is continually charged and discharged by a source of alternating voltage, the relation between current and voltage will be that shown in Fig. 17-14. This indicates that in a capacitive circuit the current *leads* the voltage or the voltage *lags* behind the current. This condition does not exist until the circuit is in operation for a few seconds, as it is obviously impossible for the current to start at any value other than zero. However, once the circuit is in operation, the phase relations are adjusted so that the current will *lead* the voltage by ¼ cycle of 90 electrical degrees. As with an inductance, this applies only to a "perfect" capacitor—one in which there is no resistance due to the resistance of the capacitor plates or leads.

The current that flows as a result of an ac voltage across a capacitor will be an alternating current with the same frequency as the ac voltage. The current remains forever 90 electrical degrees *ahead* of the voltage. The current *leads* the voltage or

> **Hint for Success**
>
> The capacitor is the complement of the inductor. Nature requires that, for energy flow to exist, there must be both electric and magnetic fields.

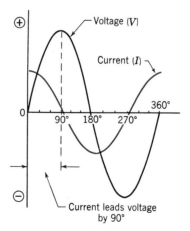

Figure 17-14 The voltage and current are *out of phase* in an ac circuit containing only capacitance.

the voltage *lags* behind the current by 90°. The phase angle θ in a capacitive circuit is $-90°$.

Phasor diagram. To express this leading current by phasors, the phasors must be drawn 90° apart. The current is used as the reference line upon which to draw the phasor diagram. Draw a line with an arrow pointing to the right as shown in Fig. 17-15. This will represent the current I in the circuit. Since the voltage *lags* the current by 90°, we shall be forced to draw the voltage phasor in such a way so as to be 90° *after* the current. Since the current phasor already points to 3 o'clock, the voltage phasor must point to 6 o'clock in order to *lag* the current vector by 90°.

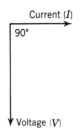

Figure 17-15   The voltage phasor lags the current phasor by 90° in a purely capacitive circuit.

The amount of current in a "pure" capacitance is found by Ohm's law. However, since a pure capacitance contains zero resistance, R is replaced by X_C. Ohm's law for a pure capacitance will then be

$$V_C = I_C \times X_C \qquad (17\text{-}7)$$

Power. Since the voltage and current in a pure capacitive circuit are 90° out of phase, the power used is equal to zero. This fact is obtained by substituting in formula (15-18) for ac power.

$$W = V \times I \times \cos \theta \qquad (15\text{-}18)$$
$$W = V \times I \times \cos 90°$$
$$W = V \times I \times 0 = 0 \text{ W}$$

Example 17-10

A 10-μF coupling capacitor in a transistorized record player passes 300 mA at a frequency of 0.4 kHz. Find (*a*) the voltage drop across the capacitor, (*b*) the power factor, and (*c*) the effective power consumed.

Solution

Given: $C = 10 \ \mu F$ *Find: V = ?*

$I = 300 \text{ mA} = 0.3 \text{ A}$
$f = 0.4 \text{ kHz} = 400 \text{ Hz}$

a. Find the reactance of the capacitor.

$$X_C = \frac{159,000}{f \times C} = \frac{159,000}{400 \times 10} = \frac{159}{4} = 40 \ \Omega \text{ (approx)} \qquad (17\text{-}6)$$

Find the voltage drop.

CALCULATOR HINT

Many computations can be facilitated with a change between polar and rectangular coordinates.

$$V_C = I_C \times X_C \qquad\qquad (17\text{-}7)$$
$$V_C = 0.3 \times 40 = 12\text{ V} \qquad Ans.$$

b.
$$\text{PF} = \cos\theta = \cos 90° = 0 \qquad Ans.$$

c. Find the effective power.
$$W = V \times I \times \cos\theta \qquad\qquad (15\text{-}18)$$
$$W = 12 \times 0.3 \times \cos 90°$$
$$W = 12 \times 0.3 \times 0 = 0\text{ W} \qquad Ans.$$

d.
$$W = I^2R = (0.3)^2 \times 0 = 0\text{ W} \qquad Check$$

Exercises

Practice

1. A voltage of 9 V at a frequency of 10 kHz is impressed across a 4-μF capacitor. Find (a) the current, (b) the PF, and (c) the effective power used.
2. What current will flow through a 0.000 015-F capacitor if the voltage across it is 10.6 V at a frequency of 100 Hz?

Applications

3. An absorption-type wave trap in a television receiver is tuned to 27.25 MHz. What is the reactance of the 47-pF capacitor in the trap?
4. What is the reactance of a 0.06-μF coupling capacitor to an audio frequency of 2 kHz? What current will flow if the voltage across the capacitor is 6 V?
5. What is the reactance of a 0.02-μF coupling capacitor to a frequency of 200 kHz? What current will flow if the voltage across the capacitor is 4 V?
6. A 4-μF bypass capacitor passes 200 mA at a frequency of 1 kHz. Find (a) the voltage drop across the capacitor and (b) the effective power consumed.
7. The potential difference between two wires having a distributed capacity of 20 pF is 2 V. Find the flow of current between the wires if one of them carries a 1000-kHz current.
8. Find the voltage across a 10-μF filter capacitor if it passes 1 A at a frequency of 60 Hz.
9. What must be the reactance of a capacitor in order for it to pass 1 A of current when the voltage is 100 V? What is the capacitance of the capacitor if the frequency is 1000 kHz?
10. The plates of a tuning capacitor are set to provide a capacitance of 200 pF. If the capacitor passes 350 mA at 3 MHz, what is the voltage drop across it?

JOB 17-6

Measurement of Capacity

Voltmeter-ammeter method. The circuit used to measure the capacitance of a capacitor is shown in Fig. 17-16. The voltmeter measures the voltage across the capacitor, and the ammeter measures the current in the circuit. Since the resistance of the capacitor is so very small when compared with its reactance, we can neglect it in this situation.

Example 17-11

Find the capacitance of the capacitor shown in Fig. 17-16.

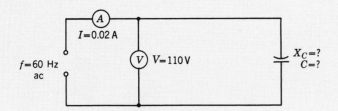

Figure 17-16 Circuit for measuring capacitance by the voltmeter-ammeter method.

Solution

1. Find the capacitive reactance.

$$V_C = I_C \times X_C \qquad (17\text{-}7)$$
$$110 = 0.02 \times X_C$$
$$X_C = \frac{110}{0.02} = 5500 \ \Omega \qquad Ans.$$

2. Find the capacitance.

$$X_C = \frac{159,000}{f \times C} \qquad (17\text{-}6)$$
$$\frac{5500}{1} = \frac{159,000}{60 \times C}$$
$$5500 \times 60 \times C = 159,000$$
$$C = \frac{159,000}{330,000} = 0.482 \ \mu F \qquad Ans.$$

Exercises

Practice

Find the capacitance of each capacitor if the measurements obtained by the voltmeter-ammeter method are given below.

Exercise	V	I	f
1	110 V	0.11 A	60 Hz
2	110 V	0.5 A	60 Hz
3	18 V	20 mA	60 Hz
4	50 V	10 mA	25 Hz
5	318 mV	20 mA	1 kHz

A Real Capacitor

Assume that a capacitor is charged to 12 V by connecting it to a battery. When the capacitor is removed from the battery, the voltage will initially remain at 12 V. With

Capacitors are meant to block a steady dc voltage while passing ac signals. Different types of capacitors are manufactured for specific values of C. Capacitors are named according to the dielectric. Common types are air, ceramic, mica, paper, film, and electrolytic capacitors. Some types are illustrated here. Capacitors used in electronic circuits are small and economical, but are a common source of trouble because they can have either an open at the conductors or a short circuit through the dielectric. Even though a capacitor is an insulator, it can be checked with an ohmmeter.

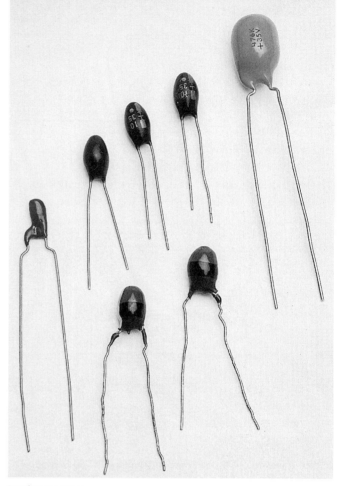

Tantalum capacitors.

Chip capacitors.

Disk ceramic capacitor.

Paper capacitor.

Air-dielectric variable capacitor.
Length is 2 in.

Typical capacitor with multiple sections.

Mica capacitor.

Film capacitor.

time the voltage will slowly decrease. The rate at which the voltage decreases depends on the quality of the capacitor.

The voltage decreases because the charge that is built up on the capacitor is slowly dissipated. The structure of a capacitor was shown in Job 17-1. The material between the plates of the capacitor in Fig. 17-2 determines its properties. The resistance of that material allows the charge on one plate to move to the other plate. This is a "leakage" resistance in the capacitor material. This can be modeled as an ideal capacitor and a resistor in parallel with it. This is shown in Fig. 17-18 on page 624. This leakage causes a change from ideal in the operation of the capacitor. We shall now look at these effects.

Figure 17-18 shows a 20-Ω resistance and a 15-Ω capacitive reactance in parallel across a 60-V 60-Hz ac source. The total current drawn by the circuit can be found by adding the currents in each branch. However, if they are not in phase, they must be added vectorially. The phase relations in the circuit are shown in Fig. 17-17a. The constant voltage is used as the reference line upon which to draw the phasor diagram. The current in the resistance I_R is in phase with the voltage, and the current in the capacitor I_C *leads* the voltage by 90°.

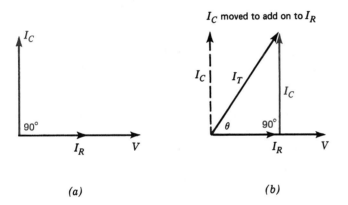

Figure 17-17   (a) Phasor diagram for a parallel ac circuit containing resistance and capacitance. (b) I_T represents the phasor sum of I_R and I_C.

Addition of the phasors.   In Fig. 17-17b, the total current I_T is obtained by adding the phase I_C to the phasor I_R. Place the tail of I_C on the head of I_R, and draw it in its original direction and length. The distance from the origin of the phasors to the head of the final phasor is the sum of the phasors. In this instance, the phasor I_T represents the sum of the phasors I_R and I_C. The angle θ is the angle by which the total current *leads* the total voltage. By applying the pythagorean theorem to Fig. 17-17b, we obtain

$$I_T{}^2 = I_R{}^2 + I_C{}^2 \qquad \text{Formula 17-8}$$

By Ohm's law,
$$V_T = I_T \times Z$$

By trigonometry,
$$\cos \theta = \frac{a}{h}$$

$$\cos \theta = \frac{I_R}{I_T}$$

or
$$PF = \frac{I_R}{I_T}$$

The power is still given by

$$W = V \times I \times \cos \theta$$

or

$$W = V \times I \times \text{PF}$$

or

$$W = I^2 \times R$$

Example 17-12

A 20-Ω resistor and a capacitor of 15 Ω capacitive reactance at 60 Hz are connected in parallel across a 60-V 60-Hz ac source. Find (*a*) the total current, (*b*) the impedance, (*c*) the phase angle, (*d*) the power factor, and (*e*) the power drawn by the circuit.

Solution

The diagram for the circuit is shown in Fig. 17-18.

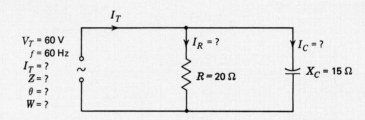

Figure 17-18   AC parallel circuit containing resistance and capacitance.

a. Find the branch currents. Since

$$V_T = V_R = V_C = 60 \text{ V} \qquad \text{(5-1)}$$

$$
\begin{array}{ll}
V_R = I_R \times R & V_C = I_C \times X_C \\
60 = I_R \times 20 & 60 = I_C \times 15 \\
I_R = \dfrac{60}{20} = 3 \text{ A} \qquad & I_C = \dfrac{60}{15} = 4 \text{ A}
\end{array}
$$

Find the total current.

$$I_T^2 = I_R^2 + I_C^2 = 3^2 + 4^2 \qquad \text{(17-8)}$$
$$= 9 + 16$$
$$I_T^2 = 25$$
$$I_T = \sqrt{25} = 5 \text{ A} \qquad Ans.$$

b.
$$V_T = I_T \times Z$$
$$60 = 5 \times Z$$
$$Z = \frac{60}{5} = 12 \ \Omega \qquad Ans.$$

c.
$$\cos \theta = \frac{I_R}{I_T} = \frac{3}{5} = 0.6000$$
$$\theta = 53° \qquad Ans.$$

d.
$$\text{PF} = \frac{I_R}{I_T} = 0.6000 = 60\% \text{ leading} \qquad Ans.$$

e. $$W = V \times I \times \cos \theta = 60 \times 5 \times \cos 53°$$
$$= 60 \times 5 \times 0.6 = 180 \text{ W} \quad \textit{Ans.}$$
$$W = I_R{}^2 \times R = (3)^2 \times 20 = 180 \text{ W} \quad \textit{Check}$$

In the parallel circuit, the total current of 5 A leads the total voltage of 60 V by 53°.

Example 17-13

The purpose of the "low-pass" circuit shown in Fig. 17-19 is not only to permit low frequencies to pass on to the load but to prevent the passage of high frequencies. Find the effectiveness of the circuit by calculating the percent of the total current in the resistor for (1) a 1-kHz audio frequency and (2) a 1000-kHz radio frequency.

Solution

The diagram for the circuit is shown in Fig. 17-19.

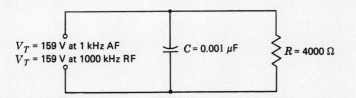

V_T = 159 V at 1 kHz AF
V_T = 159 V at 1000 kHz RF $C = 0.001\ \mu F$ $R = 4000\ \Omega$

Figure 17-19   Low-pass filter.

1. For the 1-kHz audio frequency,

a. $$X_C = \frac{159{,}000}{f \times C} = \frac{159{,}000}{10^3 \times 1 \times 10^{-3}} \qquad (17\text{-}6)$$

$$X_C = \frac{159{,}000}{1} = 159{,}000\ \Omega \qquad \textit{Ans.}$$

Find the branch currents. Since

$$V_T = V_R = V_C = 159 \text{ V} \qquad (5\text{-}1)$$

$$V_R = I_R \times R \qquad\qquad V_C = I_C \times X_C$$
$$159 = I_R \times 4000 \qquad\qquad 159 = I_C \times 159{,}000$$

$$I_R = \frac{159}{4000} = 0.04 \text{ A} \qquad I_C = \frac{159}{159{,}000} = 0.001 \text{ A} \qquad \textit{Ans.}$$

b. Find the total current.

$$I_T{}^2 = I_R{}^2 + I_C{}^2 = 0.04^2 + 0.001^2 \qquad (17\text{-}8)$$
$$= 0.0016 + 0.000\,001$$
$$I_T{}^2 = 0.001\,601$$
$$I_T = \sqrt{0.001\,601} = 0.04 \text{ A} \qquad \textit{Ans.}$$

c. Find the percent of the total current in the resistor.

$$\text{Percent} = \frac{I_R}{I_T} \times 100 = \frac{0.04}{0.04} \times 100 = 100\% \qquad \textit{Ans.}$$

That is, practically 100% of the 1-kHz audio frequency passes through the resistor.

2. For the 1000-kHz radio frequency: Since 1000 kHz is 1000 times as large as the audio frequency of 1 kHz, the X_C at 1000 kHz will be equal to 1/1000 of the X_C at 1 kHz. Therefore,

a.
$$X_C = \frac{159{,}000}{1000} = 159 \ \Omega \qquad \textit{Ans.}$$

Find the branch currents. Since

$$V_T = V_R = V_C = 159 \text{ V} \qquad (5\text{-}1)$$

$$\begin{array}{ll} V_R = I_R \times \text{R} & V_C = I_C \times X_C \\ 159 = I_R \times 4000 & 159 = I_C \times 159 \\ I_R = \dfrac{159}{4000} = 0.04 \text{ A} \qquad & I_C = \dfrac{159}{159} = 1 \text{ A} \qquad \textit{Ans.} \end{array}$$

b. Find the total current.

$$I_T^2 = I_R^2 + I_C^2 = 0.04^2 + 1^2 \qquad (17\text{-}8)$$
$$= 0.0016 + 1$$
$$I_T^2 = 1.0016$$
$$I_T = \sqrt{1.0016} = 1.001 \text{ A} \qquad \textit{Ans.}$$

c. Find the percent of the total current in the resistor.

$$\text{Percent} = \frac{I_R}{I_T} \times 100 = \frac{0.04}{1.001} \times 100 = 4\% \qquad \textit{Ans.}$$

That is, only 4% of the 1000-kHz current passes through the resistor. The circuit is evidently a good low-pass circuit. Practically all the low 1-kHz current goes through the resistor, but very little of the high 1000-kHz current gets through. The majority of the 1000-kHz RF current finds an easy path through the low reactance of the capacitor at this high frequency.

Exercises

Practice

1. An 8-Ω resistor and a 15-Ω capacitive reactance are in parallel across a 120-V 60-Hz ac line. Find (a) the total current, (b) the impedance, (c) the phase angle, (d) the power factor, and (e) the power drawn by the circuit.
2. A 26.5-Ω resistor and a 3-μF capacitor are in parallel across a 106-V 1-kHz ac source. Find (a) the total current, (b) the impedance, (c) the phase angle, (d) the power factor, and (e) the power drawn.
3. Repeat Exercise 1 for a 120-Ω resistor and a 60-Ω capacitive reactance.

Applications

4. An 8000-Ω emitter resistor in parallel with a 4-μF capacitor in a 500-Hz circuit is shown in Fig. 17-20. If the voltage drop across the combination is 8 V, find (a) the total current and (b) the impedance of the combination.

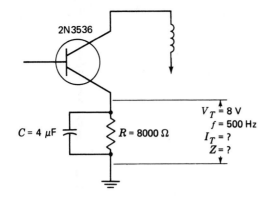

Figure 17-20

5. A 40-Ω resistor is bypassed with a 5-μF capacitor. Find the total current flowing if the voltage across the capacitor is 10.6 V at 1.5 kHz.
6. In Fig. 17-21a, find the percent of the total AF current that passes through the resistor. Find the percent of the total RF current that passes through the resistor. On the basis of these answers, may the circuit be classified as a low-pass circuit?

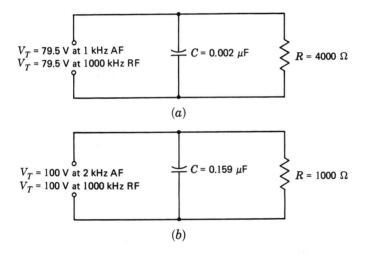

(a)

(b)

Figure 17-21

7. Repeat Exercise 6 for the circuit shown in Fig. 17-21b.
8. In a grid-leak circuit, $C = 250$ pF and $R = 1$ MΩ. If a 5-kHz AF signal causes a voltage drop of 0.6 V, find (a) the total current and (b) the impedance of the combination.
9. The grid-to-cathode capacity of 10 pF that is formed in the grid circuit of an amplifier is in parallel with the 1-MΩ grid resistor. If the signal voltage is 10 V at 318 kHz, find the current that will flow.

(See CD-ROM for Test 17-2)

Review of Capacitance

A capacitor is made of two metallic plates separated by an _____(1)_____ material, or dielectric.

A capacitor is used to store up an electric _____(2)_____.

The capacitance of a capacitor is a measure of its ability to store up _____(3)_____.

The capacitance _____(4) (increases/decreases)_____ if the *plate area increases* or the *dielectric thickness* _____(5)_____.

The working voltage is the _____(6)_____ voltage that may be placed across a capacitor before it breaks down. The working voltage in parallel is the working voltage of the _____(7) (strongest/weakest)_____ capacitor. The working voltage in series is the _____(8)_____ of the working voltages of the series capacitors.

A *farad* is the capacitance of a capacitor which can hold 1 coulomb of electricity on its plates under a pressure of 1 _____(9)_____.

The *reactance* of a capacitor is the opposition offered by the capacitor to the passage of an _____(10)_____ current. The reactance decreases as the frequency or capacitance _____(11) (increases/decreases)_____.

Distributed, or "stray," capacitance is the capacity formed when two wires run close together or when any two metal parts are separated by a thin _____(12)_____ material.

Formulas

Capacitance:
$$C = \frac{Q}{V}$$
17-1

where
C = capacitance, F
Q = charge, coulombs
V = voltage, V

Capacitors in parallel:
$$C_T = C_1 + C_2 + C_3$$
17-2

where C_T = total capacitance
C_1, C_2, C_3 = individual capacitances, all measured in the same units

Capacitors in series:
$$\frac{1}{C_T} = \frac{1}{C_1} + \frac{1}{C_2} + \frac{1}{C_3}$$
17-3

where C_T = total capacitance
C_1, C_2, C_3 = individual capacitances, all measured in the same units

Equal capacitors in series:
$$C_T = \frac{C}{N}$$
17-4

where C_T = total capacitance
C = capacitance of one of the equal capacitors
N = number of equal capacitors

Two capacitors in series:
$$C_T = \frac{C_1 \times C_2}{C_1 + C_2}$$
17-5

where C_T = total capacitance
C_1, C_2 = individual capacitances, all measured in the same units

Reactance of a capacitor:
$$X_C = \frac{159,000}{f \times C}$$
17-6

where X_C = reactance, Ω
        f = frequency, Hz
        C = capacitance, μF

Exercises

Practice

1. Find the total capacitance in parallel of a 0.0035-μF, a 0.000 000 04-F, and a 6200-pF capacitor.
2. What amount of capacitance must be added in parallel to a 0.000 35-μF capacitor to obtain a total capacitance of 0.001 15 μF?
3. Find the total capacitance in series of a 3-, a 6-, and an 8-μF capacitor.
4. What are the total capacitance and working voltage of two 25-μF 180-V capacitors used in a voltage multiplier?
5. What is the total capacitance of the capacitors in a Colpitts oscillator similar to Fig. 17-9 if C_1 = 0.0004 μF and C_2 = 0.000 35 μF?
6. An oscillator circuit contains a 0.000 35- and a 0.000 25-μF capacitor in series. Find the total capacitance.
7. Part of the first stage of a hi-fi preamplifier is shown in Fig. 17-22. Find the capacitive reactance of C_E at (a) 30 Hz, and (b) 15,000 Hz.

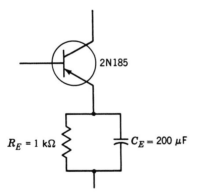

2N185

R_E = 1 kΩ C_E = 200 μF

Figure 17-22

8. Find the reactance of a 0.0003-μF tuning capacitor to a frequency of 1350 kHz.
9. Find the impedance of a capacitor if its reactance is 40 Ω and its resistance is 20 Ω.
10. Find the impedance of a capacitive circuit to a 2-kHz audio frequency if the resistance is 2000 Ω and the capacitance is 0.02 μF.
11. What cathode bypass capacitor is needed to provide a reactance of 1000 Ω at an audio frequency of 500 Hz?
12. In finding the capacitance of a capacitor by the voltmeter-ammeter method, the current was 0.004 A and the voltage was 110 V at 60 Hz. Find the capacitance.

(See CD-ROM for Test 17-3)

Assessment

1. A 100-V capacitor has a Q of 0.05. What is the capacitance of this component?
2. Three capacitors are connected in parallel. The total capacitance is 8.5 μF. If two of the capacitors are 2.2 μF, what is the value of the third capacitor?
3. You have inserted a 30-μF, a 40-μF, and a 50-μF capacitor onto a PC test breadboard. If these capacitors are connected in series, what is the total capacitance that you would measure?
4. A capacitor has 7.5788 Ω of reactance when operating at a frequency of 60 Hz. What is the capacitance of the component?
5. What amount of current will flow when a 500 μF capacitor is connected to a 350-V 100-Hz circuit?
6. If a 120-V 60-Hz source is connected to a capacitor and 1500 μA of current is flowing, what is the capacitance of the circuit?
7. A 2.653-μF capacitor is connected in parallel with a 20-Ω resistor. If the applied voltage is 120 V 1500 Hz, determine the following: (a) X_C, (b) Z, (c) I_R, (d) I_{XC}, (e) I_T, (f) $\angle\theta$, (g) power factor, (h) true power, (i) reactive power, and (j) apparent power.

INTERNET
ACTIVITIES

Internet Activity A

Capacitors store energy and therefore can be dangerous. This is particularly true in dc circuits. The following sites give some insight to this problem. How are capacitors used to supplement a car battery for audio power?
http://www.circuitcity.com/carster/tech5.html
What are some conditions under which a capacitor on a circuit board can explode, and what are the results?
http://www.ase.org.uk/safety3.html
http://www.eskimo.com/~billb/amateur/capexpt.html

Internet Activity B

What is a displacement current?
http://www.antique-radio.org/terms/capacity.html
http://www.ece.utexas.edu/~tsnyder/theory.html

Chapter 17 Solutions to Self-Tests and Reviews

Self-Test 17-9

1. 20 mA
2. 10^{-6}, 0.02
3. 5300
4. 10^{-3}, 10^3, 6, MHz

Job 17-8 Review

1. insulating
2. charge
3. electrons
4. increases
5. decreases
6. largest
7. weakest
8. sum
9. V
10. alternating
11. increases
12. insulating

18 Series AC Circuits

Learning Objectives

1. Find solutions for resistance and capacitance in series.
2. Find solutions for resistance, inductance, and capacitance in series.
3. Determine series resonance.
4. Find the inductance or capacitance required for series resonance.

JOB 18-1

Simple Series AC Circuits

The general rules for solving dc series circuits are also applicable to the solution of ac series circuits. These rules are:

1. The current in each part is equal to the current in every other part and equal to the total current.
2. The total voltage is equal to the sum of the voltages across all parts of the circuit.

As we have seen in earlier chapters, there are some differences in how we analyze circuits that contain reactive components, inductors, or capacitors. These elements induce phase differences that must be included in the calculations. This affects the way we do the calculations, but the basic rules above are still applied.

In the following jobs we shall look at the characteristics of some fundamental series ac circuits using reactive components. In Job 16-3 we saw a series combination of a resistor and an inductor as a real inductor. This was the first of the series ac circuits to be analyzed.

The following circuits do not occur in a natural way as the series resistor and inductor. These circuits are the basis for many more complex circuits and devices. An understanding of their properties will be useful in understanding these more complex devices.

JOB 18-2

Resistance and Capacitance in Series

Figure 18-1 shows a 5-Ω resistor connected in series with a capacitor with a reactance of 12 Ω at a frequency of 60 Hz. A series current of 1 A produces a voltage drop of 5 V across the resistor and 12 V across the capacitor.

The total voltage across the circuit can be found by adding the voltage drops across each part. But, just as in the last job, since the voltages are *not* in phase, they must be added vectorially. The phase relations in a capacitive circuit are shown in Fig. 18-2a. The constant series current is used as the reference line. The voltage across the resistor V_R is in phase with the current I; the voltage across the capacitor V_C *lags* the current I by 90°; V_C *lags* V_R by 90°.

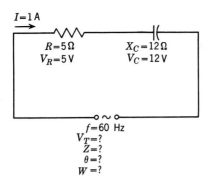

Figure 18-1 AC series circuit containing resistance and capacitance.

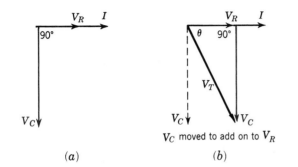

Figure 18-2 (*a*) Phasor diagram for a series ac circuit containing resistance and inductance. (*b*) V_T represents the phasor sum of V_R and V_C.

Addition of the phasors. In Fig. 18-2*b*, the total voltage V_T is obtained by adding the phasor V_C to the phasor V_R. Place the tail of V_C on the head of V_R, and draw it in its original direction and length. The distance from the origin of phasors to the head of the final phasor is the sum of the phasors. In this instance, the phasor V_T represents the sum of the phasors V_R and V_C. The phase angle θ is the angle by which the total voltage *lags* behind the current. By applying the pythagorean theorem to Fig. 18-2*b*, we obtain the

$$V_T^2 = V_R^2 + V_C^2 \qquad \text{Formula 18-1}$$

Total impedance. To find the impedance of the series ac circuit, we must add R and X_C vectorially as was done with the voltages. This is shown in Fig. 18-3 on page 634. The voltages in Fig. 18-3a are replaced by their Ohm's law values in Fig. 18-3b. Now, by dropping out the common factor of the current I, we obtain the *impedance triangle* of Fig. 18-3c.

Applying the pythagorean theorem to Fig. 18-3c, we obtain

$$Z^2 = R^2 + X_C^2 \qquad \text{Formula 18-2}$$

$$Z = \sqrt{R^2 + X_C^2} \qquad \text{Formula 18-3}$$

And, by trigonometry,

$$\cos \theta = \frac{a}{h}$$

or

$$\cos \theta = \frac{R}{Z}$$

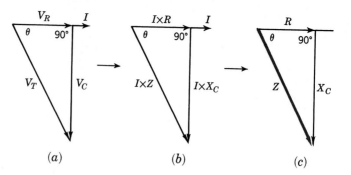

Figure 18-3 Z represents the phasor sum of R and X_C.

The power is still given by

$$W = V \times I \times \cos \theta \qquad (15\text{-}18)$$

or $\qquad W = V \times I \times \text{PF} \qquad (15\text{-}19)$

or $\qquad W = I^2 R \qquad (7\text{-}4)$

Example 18-1

In Fig. 18-1, find (*a*) the total voltage, (*b*) the impedance, (*c*) the phase angle, (*d*) the power factor, and (*e*) the power.

Solution

CALCULATOR HINT

The phase angle for a capacitor is negative and should be entered as a negative number.

a.
$$V_T{}^2 = V_R{}^2 + V_C{}^2 = 5^2 + 12^2 \qquad (18\text{-}1)$$
$$= 25 + 144$$
$$V_T{}^2 = 169$$
$$V_T = \sqrt{169} = 13 \text{ V} \qquad Ans.$$

b.
$$Z = \sqrt{R^2 + X_C{}^2} \qquad (18\text{-}3)$$
$$= \sqrt{5^2 + 12^2}$$
$$= \sqrt{25 + 144}$$
$$Z = \sqrt{169} = 13 \ \Omega \qquad Ans.$$

c.
$$\cos \theta = \frac{R}{Z} = \frac{5}{13} = 0.3846$$
$$\theta = 67° \text{ (approx)} \qquad Ans.$$

d.
$$\text{PF} = \frac{R}{Z} = 0.3846 = 38.5\% \ leading \qquad Ans.$$

e.
$$W = V \times I \times \text{PF} \qquad (15\text{-}19)$$
$$W = 13 \times 1 \times 0.385 = 5 \text{ W} \qquad Ans.$$

f.
$$W = I^2 R = (1)^2 \times 5 = 5 \text{ W} \qquad Check$$

In this series circuit, the total current of 1 A *leads* the total voltage of 13 V by 67°.

Example 18-2

A capacitance of 4 μF and a resistance of 30 Ω are connected in series across a 100-V 1-kHz ac source. Find (a) the capacitive reactance, (b) the impedance, (c) the total current, (d) the voltage drop across the resistor and the capacitor, (e) the phase angle, (f) the power factor, and (g) the power.

Solution

Given: $\quad C = 4\ \mu$F $\qquad\qquad$ Find: $X_C = ?$
$\qquad\qquad R = 30\ \Omega \qquad\qquad\qquad\qquad Z = ?$
$\qquad\qquad V_T = 100\ \text{V} \qquad\qquad\qquad I_T = ?$
$\qquad\qquad f = 1\ \text{kHz} = 1000\ \text{Hz} \qquad V_R = ?$
$\qquad\qquad\qquad\qquad\qquad\qquad\qquad\qquad V_C = ?$
$\qquad\qquad\qquad\qquad\qquad\qquad\qquad\qquad \theta = ?$
$\qquad\qquad\qquad\qquad\qquad\qquad\qquad\qquad \text{PF} = ?$
$\qquad\qquad\qquad\qquad\qquad\qquad\qquad\qquad W = ?$

a. $\quad X_C = \dfrac{159{,}000}{f \times C} = \dfrac{159{,}000}{1000 \times 4} = \dfrac{159}{4} = 40\ \Omega$ (approx)

b. $\quad Z = \sqrt{R^2 + X_C^2}$ $\qquad\qquad\qquad\qquad$ (18-3)

$\qquad\quad = \sqrt{30^2 + 40^2}$

$\qquad\quad = \sqrt{900 + 1600} = \sqrt{2500} = 50\ \Omega \qquad$ *Ans.*

c. $\qquad\qquad\qquad V_T = I_T \times Z$

$\qquad\qquad\qquad 100 = I_T \times 50$

$\qquad\qquad\qquad\qquad I_T = \dfrac{100}{50} = 2\ \text{A} \qquad$ *Ans.*

d. Since $\qquad\qquad\qquad I_T = I_R = I_C = 2\ \text{A}$ $\qquad\qquad$ (4-1)

$\quad V_R = I_R \times R_R \qquad\qquad\qquad V_C = I_C \times X_C$

$\quad V_R = 2 \times 30 = 60\ \text{V} \qquad$ *Ans.* $\qquad V_C = 2 \times 40 = 80\ \text{V} \qquad$ *Ans.*

e. $\qquad\qquad \cos\theta = \dfrac{R}{Z} = \dfrac{30}{50} = 0.6000$

$\qquad\qquad\qquad \theta = 53°$ (approx) $\qquad$ *Ans.*

f. $\qquad\qquad \text{PF} = \dfrac{R}{Z} = 0.600 = 60\%$ *leading* $\qquad$ *Ans.*

g. $\quad W = V \times I \times \text{PF} = 100 \times 2 \times 0.6 = 120\ \text{W} \qquad$ *Ans.* $\qquad$ (15-9)

h. $\qquad\qquad W = I^2 R = (2)^2 \times 30 = 120\ \text{W} \qquad$ *Check*

In this series circuit, the total current of 2 A leads the total voltage of 100 V by 53°.

Exercises

Practice

1. A 119-V 60-Hz ac voltage is applied across a series circuit of an 8-Ω resistor and a capacitor whose reactance is 15 Ω. Find (a) the impedance, (b) the total current, (c) the voltage drop across each part, (d) the phase angle, (e) the power factor, and (f) the power.

2. A 134-V 60-Hz ac voltage is applied across a series circuit of a 30-Ω resistor and a 60-Ω capacitive reactance. Find (*a*) the impedance, (*b*) the total current, (*c*) the voltage drop across each part, (*d*) the phase angle, (*e*) the power factor, and (*f*) the power.

3. A 113-V 100-Hz ac voltage is applied across a series circuit of a 100-Ω resistor and a 15.9-μF capacitor. Find (*a*) the impedance, (*b*) the total current, (*c*) the voltage drop across each part, (*d*) the phase angle, (*e*) the power factor, and (*f*) the power.

4. Find the current and angle of lead for a series circuit of a 10-Ω resistor and an 8-μF capacitor if the applied voltage is 110 V at 60 Hz.

5. In the volume-control circuit shown in Fig. 18-4, find the impedance of the control unit to frequencies of (*a*) 1000 Hz and (*b*) 10 kHz.

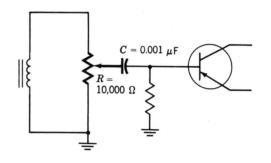

Figure 18-4

Applications

6. In the resistance-coupled stage shown in Fig. 18-5, the voltage drop between points *A* and *B* is 1.414 V. If the frequency of the current between these points is 10 kHz, find the voltage drop across the 1000-Ω resistor.

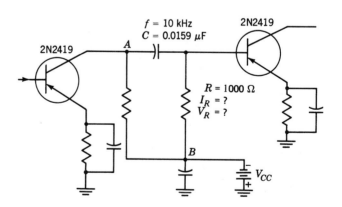

Figure 18-5   Resistance-coupled amplifier stage.

7. When the value of an unknown capacitor was calculated by the impedance method, a 300-Ω resistor was placed in series with the capacitor. A 110-V, 60-Hz ac voltage caused a current of 0.22 A to flow. Find (*a*) the impedance of the circuit, (*b*) the reactance of the capacitor, and (*c*) the capacitance of the capacitor.

8. A 10,000-Ω resistor and a capacitor are placed in series across a 60-Hz line. If the voltage across the resistor is 50 V, and across the capacitor 100 V, find (*a*) the current in the resistor, (*b*) the current in the capacitor, (*c*) the reactance of the capacitor by Ohm's law, and (*d*) the capacitance of the capacitor.

9. A 120-V source is connected to a resistance of 50 Ω in series with a capacitance of 10 μF. What frequency will permit a current of 0.8 A to flow?

10. A circuit consisting of a 40-μF capacitance in series with a rheostat is connected across a 138-V 60-Hz line. What must be the value of the resistance in order to permit a current of 2 A to flow?

Resistance, Inductance, and Capacitance in Series

Figure 18-6 shows a 16-Ω resistor, an inductive reactance of 80 Ω, and a capacitive reactance of 50 Ω connected in series across a frequency of 60 Hz. A series current of 0.5 A produces a voltage drop of 8 V across the resistor, 40 V across the inductance, and 25 V across the capacitance.

The total voltage across the entire circuit may be found by adding the voltage drops across each part. However, since the voltages are *not* in phase, they must be added vectorially. The phase relations in the circuit are shown in Fig. 18-7*a* on page 638. The constant series current is used as the reference line. The voltage across the resistor V_R is in phase with the current *I*. The voltage across the inductance V_L *leads* the current *I* by 90°. The voltage across the the capacitance V_C *lags* the current *I* by 90°. Since V_L and V_C are exactly 180° out of phase and act in exactly *opposite* directions, the voltage V_C is denoted by a minus sign.

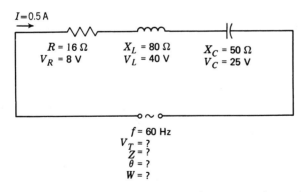

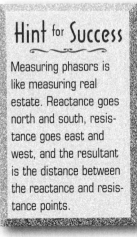

Hint for Success

Measuring phasors is like measuring real estate. Reactance goes north and south, resistance goes east and west, and the resultant is the distance between the reactance and resistance points.

Figure 18-6   AC series circuit containing resistance, inductance, and capacitance.

Addition of phasors.   When there are three phasors, it is best to add only two at a time. To add the phasor V_C to V_L, place the tail of V_C on the head of V_L and draw it in its original direction and length, which will be straight *down* as shown in Fig. 18-7*b*. Since these phasors act in opposite directions, their *sum* is actually the *difference* between the phasors as indicated by V_d. After this addition, the phasor diagram looks like Fig. 18-8*a* on page 638. Notice that the effect of the capacitor has disappeared. The 40 V of coil voltage have exactly balanced the 25 V of capacitive voltage and have left an excess of 15 V of coil voltage. This 15 V of coil voltage must now be added vectorially to the 8 V of resistance voltage. In Fig. 18-8*b*, the total

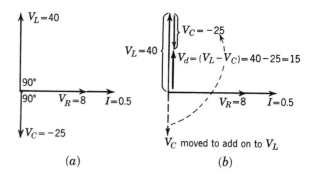

Figure 18-7 (a) Phasor diagram for a series ac circuit containing resistance, inductance, and capacitance. (b) V_d represents the phasor sum of V_L and V_C.

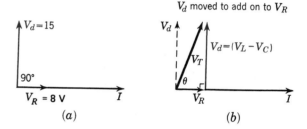

Figure 18-8 (a) V_d represents the result of adding V_C to V_L. (b) V_T represents the phasor sum of $V_R + V_L + V_C$.

voltage V_T is obtained by adding the phasor V_d to the phasor V_R. Place the tail of V_d on the head of V_R, and draw it in the proper direction. If V_C is larger than V_L, the phasor V_d will have a *downward* direction. The distance from the origin of the phasors to the head of the final phasor is the sum of the phasors. In this instance, the phasor V_T represents the sum of the phasors $V_R + V_L + V_C$. The phase angle θ is the angle by which the total voltage will lead the current. If V_C were larger than V_L, the total voltage would *lag* behind the current by this angle. By applying the pythagorean theorem to Fig. 18-8b, we obtain the

$$V_T{}^2 = V_R{}^2 + (V_L - V_C)^2 \qquad \text{Formula 18-4}$$

Total impedance.   To find the impedance of the series ac circuit, we must add R, X_L, and X_C vectorially as was done with the voltages. This is shown in Fig. 18-9, resulting in the impedance triangle of Fig. 18-9c.

Applying the pythagorean theorem to Fig. 18-9c, we obtain

$$Z^2 = R^2 + (X_L - X_C)^2 \qquad \text{Formula 18-5}$$

$$Z = \sqrt{R^2 + (X_L - X_C)^2} \qquad \text{Formula 18-6}$$

And, by trigonometry,

$$\cos \theta = \frac{a}{h}$$

or $$\cos \theta = \frac{R}{Z}$$

The power is still given by

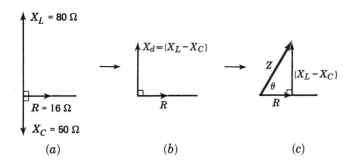

Figure 18-9 Z represents the phasor sum of $R + X_L + X_C$.

$$W = V \times I \times \cos \theta$$
or $$W = V \times I \times \text{PF}$$
or $$W = I^2 R \qquad (7\text{-}4)$$

Example 18-3

In Fig. 18-6, find (a) the total voltage, (b) the impedance, (c) the phase angle, (d) the power factor, and (e) the power.

Solution

a.
$$V_T{}^2 = V_R{}^2 + (V_L - V_C)^2 \qquad (18\text{-}4)$$
$$V_T{}^2 = 8^2 + (40 - 25)^2$$
$$V_T{}^2 = 8^2 + 15^2 = 64 + 225 = 289$$
$$V_T = \sqrt{289} = 17 \text{ V} \qquad Ans.$$

b.
$$Z = \sqrt{R^2 + (X_L - X_C)^2} \qquad (18\text{-}6)$$
$$Z = \sqrt{16^2 + (80 - 50)^2}$$
$$Z = \sqrt{16^2 + 30^2} = \sqrt{256 + 900} = \sqrt{1156}$$
$$Z = 34 \ \Omega \qquad Ans.$$

c.
$$\cos \theta = \frac{R}{Z} = \frac{16}{34} = 0.4706$$
$$\theta = 62° \text{ (approx)} \qquad Ans.$$

d.
$$\text{PF} = \frac{R}{Z} = 0.4706 = 47.1\% \text{ } lagging \qquad Ans.$$

e.
$$W = V \times I \times \text{PF} = 17 \times 0.5 \times 0.471 = 4 \text{ W} \qquad Ans.$$

f.
$$W = I^2 R = (0.5)^2 \times 16 = 4 \text{ W} \qquad Check$$

In this series circuit, the total current of 0.5 A *lags* the total voltage of 17 V by 62°.

Example 18-4

An 18-Ω resistor, a 4-μF capacitor, and a 2.5-mH inductor are connected in series across a 60-V 1-kHz ac source. Find (a) the capacitive reactance, (*b*) the inductive reactance, (*c*) the impedance, (*d*) the total current, (*e*) the voltage drop across each part, (*f*) the phase angle, (*g*) the power factor, and (*h*) the power.

Solution

CALCULATOR HINT

When using a programmable calculator, save all the known variables such as *R*, *C*, *L*, *V*, and *F*. Then create equations to solve for unknowns such as *XC*, *XL*, *I*, and *W*.

Given:
$$R = 18 \ \Omega$$
$$C = 4 \ \mu F$$
$$L = 2.5 \ \text{mH} = 0.0025 \ \text{H}$$
$$V_T = 60 \ \text{V}$$
$$f = 1 \ \text{kHz} = 1000 \ \text{Hz}$$

Find:
$$X_C = ?$$
$$X_L = ?$$
$$Z = ?$$
$$I_T = ?$$
$$V_R = ?$$
$$V_L = ?$$
$$V_C = ?$$
$$\theta = ?$$
$$\text{PF} = ?$$
$$W = ?$$

a.
$$X_C = \frac{159,000}{f \times C} = \frac{159,000}{1000 \times 4} = \frac{159}{4} = 40 \ \Omega \quad Ans.$$

b.
$$X_L = 6.28fL = 6.28 \times 1000 \times 0.0025 \quad (16\text{-}1)$$
$$= 6.28 \ \text{x} \ 2.5 = 16 \ \Omega \quad Ans.$$

c.
$$Z = \sqrt{R^2 + (X_C - X_L)^2} \quad (18\text{-}6)$$

(Notice that X_C is written first because X_C is larger than X_L.)
$$Z = \sqrt{18^2 + (40 - 16)^2} = \sqrt{18^2 + 24^2}$$
$$= \sqrt{324 + 576}$$
$$Z = \sqrt{900} = 30 \ \Omega \quad Ans.$$

d.
$$V_T = I_T \times Z$$
$$60 = I_T \times 30$$
$$I_T = \frac{60}{30} = 2 \ \text{A} \quad Ans.$$

e. Since
$$I_T = I_R = I_L = I_C = 2 \ \text{A} \quad (4\text{-}1)$$

$$V_R = I_R \times R_R \qquad\qquad V_C = I_C \times X_C$$
$$V_R = 2 \times 18 = 36 \ \text{V} \quad Ans. \qquad V_C = 2 \times 40 = 80 \ \text{V} \quad Ans.$$

$$V_L = I_L \times X_L$$
$$V_L = 2 \times 16 = 32 \ \text{V} \quad Ans.$$

f.
$$\cos \theta = \frac{R}{Z} = \frac{18}{30} = 0.6000$$
$$\theta = 53° \ (\text{approx}) \quad Ans.$$

g.
$$\text{PF} = \frac{R}{Z} = 0.6000 = 60\% \ leading \quad Ans.$$

h.
$$W = V \times I \times \text{PF} = 60 \times 2 \times 0.6 = 72 \ \text{W} \quad Ans.$$

i.
$$W = I^2R = (2)^2 \times 18 = 72 \ \text{W} \quad Check$$

In this series circuit, since the capacitive reactance is larger than the inductive reactance, the total current of 2 A *leads* the total voltage of 60 V by 53°.

Electricity enables people to enjoy healthier and longer lives, but more than half of the available commercial energy is used by only a quarter of the world's population. In some developing nations, people without electricity lack basic services. Consequently, where there is no access to health services, housing, clean water, and sewage treatment, life expectancy is low and infant mortality is high.

Impedances in Series

The total impedance of a number of devices in series must include *all* the resistances and *all* the reactances. Since all the resistances are in phase with each other,

$$R_T = R_1 + R_2 + R_3 \qquad\qquad (4\text{-}3)$$

Since all the inductive reactances are in phase with each other,

$$X_{LT} = X_{L_1} + X_{L_2} + X_{L_3} \qquad\qquad \text{Formula 18-7}$$

Since all the capacitive reactances are in phase with each other,

$$X_{CT} = X_{C_1} + X_{C_2} + X_{C_3} \qquad\qquad \text{Formula 18-8}$$

Example 18-5

In Fig. 18-10, find (*a*) the impedance, (*b*) the current, (*c*) the phase angle, (*d*) the PF, and (*e*) the power.

$$R_1 = 17\ \Omega \qquad R_2 = 2\ \Omega \qquad R_2 = 5\ \Omega \qquad x_C = 60\ \Omega$$
$$x_{L1} = 10\ \Omega \qquad x_{L2} = 18\ \Omega$$

$$V_T = 80\ \text{V}$$
$$f = 60\ \text{Hz}$$
$$Z = ?$$
$$I_T = ?$$
$$\theta = ?$$
$$PF = ?$$
$$W = ?$$

Figure 18-10

Solution

a.
$$R_T = R_1 + R_2 + R_3 = 17 + 2 + 5 = 24 \ \Omega \qquad (4\text{-}3)$$
$$X_{LT} = X_{L_1} + X_{L_2} = 10 + 18 = 28 \ \Omega \qquad (18\text{-}7)$$
$$Z = \sqrt{R^2 + (X_C - X_L)^2} \qquad (18\text{-}6)$$
$$= \sqrt{24^2 + (60 - 28)^2}$$
$$= \sqrt{24^2 + 32^2} = \sqrt{576 + 1024}$$
$$= \sqrt{1600}$$
$$Z = 40 \ \Omega \qquad Ans.$$

b.
$$V_T = I_T \times Z$$
$$80 = I_T \times 40$$
$$I_T = \frac{80}{40} = 2 \ A \qquad Ans.$$

c.
$$\cos \theta = \frac{R_T}{Z} = \frac{24}{40} = 0.6000$$
$$\theta = 53° \ (\text{approx}) \qquad Ans.$$

d.
$$PF = \frac{R_T}{Z} = 0.6000 = 60\% \ leading \qquad Ans.$$

e.
$$W = V \times I \times PF = 80 \times 2 \times 0.6 = 96 \ W \qquad Ans.$$

f.
$$W = I^2 R_T = (2)^2 \times 24 = 96 \ W \qquad Check$$

In this series circuit, the total current of 2 A *leads* the total voltage of 80 V by 53°.

Example 18-6

A rectifier delivers 200 V at 120 Hz to a filter circuit consisting of a 30-H filter choke coil and a 20-μF capacitor connected as shown in Fig. 18-11. How much 120-Hz voltage appears across the capacitor which feeds into the plate supply? Has this filter succeeded in removing the ac component from the plate supply?

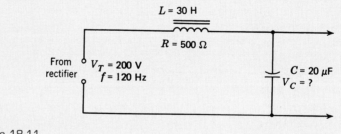

Figure 18-11

Solution

The circuit diagram is shown in Fig. 18-11.

1. Find the reactance of the coil.

$$X_L = 6.28fL \qquad (16\text{-}1)$$

$$X_L = 6.28 \times 120 \times 30$$
$$X_L = 22{,}600 \ \Omega \qquad Ans.$$

2. Find the reactance of the capacitor.

$$X_C = \frac{159{,}000}{f \times C} = \frac{159{,}000}{120 \times 20} = \frac{1590}{24} = 66 \ \Omega \qquad Ans. \qquad (17\text{-}6)$$

3. Find the total reactance effect X_d so that we may compare it with R. If X_d/R is greater than 5, we shall be able to neglect R and greatly simplify our calculations.

$$X_d = X_L - X_C$$
$$X_d = 22{,}600 - 66 = 22{,}534 \ \Omega \qquad Ans.$$

4. Find the Q of the circuit.

$$Q = \frac{X_d}{R} = \frac{22{,}534}{500} = 45+ \qquad Ans.$$

Therefore, since Q is greater than 5, we shall neglect the resistance of 500 Ω, and

$$Z = X_d = 22{,}534 \ \Omega \qquad Ans.$$

5. Find the series current.

$$V_T = I_T \times Z$$
$$200 = I_T \times 22{,}534$$
$$I_T = \frac{200}{22{,}534} = 0.0089 \ \text{A} \qquad Ans.$$

6. Find the voltage across the capacitor.

$$V_C = I_C \times X_C \qquad (17\text{-}7)$$
$$V_C = 0.0089 \times 66 = 0.59 \ \text{V} \qquad Ans.$$

Therefore this is a satisfactory filter, since only 0.59 V out of the total of 200 V of 120-Hz alternating current can get through to the plate supply.

Self-Test 18-7

The output voltage of an audio oscillator is 50 V at 3 kHz. It is applied to a series circuit of a 300-Ω resistor, a coil of 531-mH inductance and 75-Ω resistance, a 125-Ω resistor, and a 2650-pF capacitor. Find (a) the impedance, (b) the total current, and (c) the phase angle.

Solution

Given: $R_1 = 300 \ \Omega$ Find: $Z = ?$

Coil $\begin{cases} L = 531 \ \text{mH} \\ R_2 = 75 \ \Omega \end{cases}$ $I_T = ?$

           $R_3 = 125 \ \Omega$ $\theta = ?$

           $C = 2650 \ \text{pF}$

           $V_T = 50 \ \text{V}$

           $f = 3 \ \text{kHz}$

a. In order to find the impedance, we must know the values of R_T, X_L, and __(1)__.

$$R_T = R_1 + R_2 + R_3 = 300 + 75 + \underline{\quad(2)\quad} \qquad (4\text{-}3)$$
$$R_T = \underline{\quad(3)\quad} \ \Omega$$
$$X_L = 6.28fL = 6.28 \times 3 \times 10^3 \times 531 \times \underline{\quad(4)\quad} \qquad (16\text{-}1)$$
$$X_L = \underline{\quad(5)\quad} \ \Omega \qquad Ans.$$
$$X_C = \frac{159{,}000}{f \times C} = \frac{159 \times 10^3}{3 \times 10^3 \times 2650 \times (6)} \qquad (17\text{-}6)$$
$$= \frac{53 \times 10^6}{2650}$$
$$X_C = 0.02 \times 10^6 = \underline{\quad(7)\quad} \ \Omega$$
$$X_d = X_C - X_L = 20{,}000 - 10{,}000 = \underline{\quad(8)\quad} \ \Omega$$

The Q of the circuit $= X_d/R = 10{,}000/500 = \underline{\quad(9)\quad}$. Therefore, $R_T = 500\ \Omega$ __(10)(may/may not)__ be neglected.

$$Z = X_d = \underline{\quad(11)\quad} \ \Omega \qquad Ans.$$

b.
$$V_T = I_T \times Z$$
$$I_T = \frac{(12)}{10{,}000} = 0.005\ \text{A} \qquad Ans.$$

c.
$$\cos\theta = \frac{R}{Z} = \frac{(13)}{10{,}000} \qquad (18\text{-}8)$$
$$\cos\theta = \underline{\quad(14)\quad}$$
$$\theta = \underline{\quad(15)\quad} \qquad Ans.$$

In this series circuit, since the capacitive reactance is larger than the inductive reactance, the total voltage of 50 V *lags* behind the total current of 0.005 A by 87°.

Exercises

Practice

1. A 16-Ω resistor, an 83-Ω inductive reactance, and a 20-Ω capacitive reactance are in series. A 130-V 60-Hz emf is impressed on the circuit. Find (*a*) the impedance, (*b*) the series current, (*c*) the voltage drops across all the parts, (*d*) the phase angle, (*e*) the power factor, and (*f*) the power.
2. A coil of 2.07-mH inductance, a 0.3-μF capacitor, and a 36-Ω resistor are connected in series across a 127.5-V 10-kHz ac source. Find (*a*) the impedance, (*b*) the total current, (*c*) the phase angle, (*d*) the power factor, and (*e*) the power.
3. A 125-V 100-Hz power supply is connected across a 4000-Ω resistor, a 0.5-μF capacitor, and a 10-H coil connected in series. Find (*a*) the individual reactances, (*b*) the impedance, (*c*) the total current, (*d*) the phase angle, (*e*) the power factor, and (*f*) the power.

Applications

4. The antenna circuit of a radio receiver consists of a 0.2-mH inductance and a 0.0001-μF capacitance. If the resistance of the antenna is small enough to be

considered to be zero, what is the impedance of the antenna to a 1200-kHz signal? If this frequency induces a voltage of 100 μV in the antenna, what current will flow?

5. In a circuit similar to Fig. 18-11, $V_T = 250$ V, $f = 120$ Hz, $L = 25$ H, $R = 400$ Ω, and $C = 25$ μF. What amount of the 120-Hz voltage will appear across the capacitor?

6. A wave trap to eliminate a 13-kHz frequency is made of a 30-mH inductance of 40 Ω resistance and a 0.005-μF capacitor in series. What is the impedance of the circuit?

7. A 10.8-V 100-kHz emf is applied across a series circuit of a 6-Ω resistance, a 0.5-mH coil, and a 5000-pF capacitance. What is the total current?

8. A 300-Ω 100-μH resistor is in series with a capacitance of 2 μF. Find the impedance of the circuit at (*a*) 500 Hz, (*b*) 5 kHz, and (*c*) 500 kHz.

9. A 5-H coil and a 1.67-μF capacitor are in series with an adjustable resistor. What must be the value of the resistance in order to draw 0.3 A from a 120-V 60-Hz line?

10. A 2.8-mH inductance and a 9-μF capacitance are in series with a 50-Ω resistor. At what frequency will the inductive reactance equal the capacitive reactance? What is the impedance of the circuit? What current will be drawn at this frequency from a 10-V source?

11. A 4-Ω resistor and an unknown inductance are connected across a 110-V 60-Hz ac line. Find the value of the inductance if the current flowing is 20 A.

12. A resistance and a capacitor in series take 120 W at a power factor of 54.5 percent (leading) from a 110-V 60-Hz line. Find (*a*) the current, (*b*) the impedance, (*c*) the resistance, and (*d*) the value of the capacitance.

(See CD-ROM for Test 18-1)

Series Resonance

Example 18-8

A 30-H coil, a 250-Ω resistor, and a variable capacitor are connected in series across a 110-V 60-Hz line. When the capacitor is adjusted to 0.2344 μF, find the impedance of the circuit.

Solution

Given: $L = 30$ H   Find: $Z = ?$
              $R = 250$ Ω
              $C = 0.2344$ μF
              $V = 110$ V
              $f = 60$ Hz

1. Find the inductive reactance.

$$X_L = 6.28fL = 6.28 \times 60 \times 30 = 11,305 \ \Omega \qquad \textit{Ans.}$$

Hint for **Success**

Energy is always flowing back and forth between an inductor and a capacitor. At resonance the inductive energy and the capacitive energies are equal.

2. Find the capacitive reactance.

$$X_C = \frac{159,000}{f \times C} = \frac{159,000}{60 \times 0.2344} = \frac{15,900}{1.4064} = 11,305 \ \Omega \qquad Ans.$$

3. Find the impedance.

$$Z = \sqrt{R^2 + (X_C - X_L)^2} = \sqrt{250^2 + (11,305 - 11,305)^2}$$
$$= \sqrt{250^2 + 0^2}$$
$$= \sqrt{250^2}$$
$$Z = 250 \ \Omega \qquad Ans.$$

Notice that for these particular values of L and C, the inductive reactance and the capacitive reactance are exactly equal. Since their actions are directly opposed to each other, the total effect of both is equal to zero and the impedance of the circuit is equal to just the resistance of the circuit. This condition is called *resonance*. Series resonance is the condition of *smallest* circuit resistance. At resonance, since the reactance effect is zero, the *largest* amount of current will flow.

Any change in the value of either L or C would give a *different* value of X_L or X_C, whose sum would no longer be zero. Under these conditions, since the reactance effect is *larger* than zero, the current that flows is *smaller* than the flow at resonance. In addition, since the values of the reactances depend on the frequency, any change in the frequency results in *different* values of reactances whose sum again would *not* be zero.

Apparently, for any combination of L and C in series, there is only *one* frequency for which X_L can equal X_C. This frequency is called the *resonant frequency*.

At this frequency, a large current will flow in the series circuit, since the reactance is zero at the resonant frequency. At any other frequency, since the sum of X_L and X_C is *not* equal to zero, the impedance will be *larger* and the current will be *smaller*.

Why is the resonant frequency important?   The antenna of a radio receives signals from many stations, each at a different frequency. How shall we separate one of these signals from all the rest?

We can separate one frequency from the rest if we can find an L and C combination which is *resonant* to that same frequency which we are trying to separate. Only this frequency will encounter a *low impedance* and therefore will produce a *large* current. All other frequencies will encounter large impedances, and the currents at these frequencies will be practically zero. The tuner of a receiver is a series circuit of an inductance and a capacitance which can be made resonant to different frequencies by changing the values of the capacitance. Thus it will select and pass on to the amplifying system only one frequency at a time.

There are many other uses for the series resonant circuit. In the superheterodyne receiver, the oscillator must deliver a definite frequency to the mixer tube. Values of L and C are chosen to make the combination resonant to that particular frequency. Bandpass filters and acceptance circuits are other common applications of the series resonant circuit.

Calculating the resonant frequency.   At resonance, the inductive reactance is equal to the capacitive reactance. Write the equation.

$$X_L = X_C \qquad \text{Formula 18-9}$$

Substitute reactances.

$$\frac{6.28fL}{1} = \frac{159,000}{f \times C}$$

Cross-multiply.

$$f^2 \times 6.28 \times L \times C = 159,000$$

Solve for f.

$$f^2 = \frac{159,000}{6.28 \times L \times C}$$

Divide.

$$f^2 = \frac{25,318}{L \times C}$$

Take the square root of both sides.

$$f = \sqrt{\frac{25,318}{L \times C}}$$

$$f = \frac{159}{\sqrt{L \times C}} \qquad \text{Formula 18-10}$$

where f = frequency, Hz f = frequency, kHz
         L = inductance, H   or L = inductance, μH
         C = capacitance, μF C = capacitance, μF

If we use the units of measurement commonly used in radio, electronics, and television work, the formula retains the same form *but* both the frequency and the inductance are expressed in different units as shown above.

Example 18-9

Calculate the resonant frequency of a tuning circuit if the inductance is 300 μH and the capacitor is set at a capacity of 300 pF.

Solution

The diagram of the circuit is shown in Fig. 18-12. If kilohertz are desired, the 300 pF must be changed to microfarads.

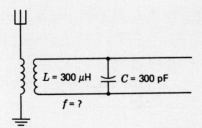

$L = 300\ \mu H$ $C = 300$ pF

$f = ?$

Figure 18-12

$$300\text{ pF} = 300 \times 10^{-6}\ \mu\text{F}$$

$$f = \frac{159}{\sqrt{L \times C}} \qquad (18\text{-}10)$$

> ### CALCULATOR HINT
>
> The value 6.28 is an approximation for 2π. Using 2π provides greater accuracy and is easier to enter in a calculator.

$$= \frac{159}{\sqrt{300 \times 300 \times 10^{-6}}} = \frac{159}{\sqrt{9 \times 10^{-2}}}$$

$$f = \frac{159}{0.3} = 530 \text{ kHz} \qquad Ans.$$

Exercises

Practice

1. A transmitting antenna has a capacitance of 200 pF, a resistance of 50 Ω, and an inductance of 200 μH. Find (*a*) the resonant frequency and (*b*) the impedance of the antenna.
2. Find the resonant frequency of a tuning circuit similar to that shown in Fig. 18-12 if $L = 250$ μH and $C = 40$ pF.
3. A series circuit consists of a 12-Ω resistance, an inductance of 0.04 H, and a capacitance of 0.16 μF. Find (*a*) the resonant frequency, (*b*) the reactance of the inductance, (*c*) the reactance of the capacitance, (*d*) the impedance, (*e*) the current at resonance if the impressed voltage is 6 V, and (*f*) the voltage drop across the capacitance.
4. What is the resonant frequency of a series circuit if the inductance is 270 μH and the capacitance is 0.003 μF?

Applications

5. What is the resonant frequency of the Hartley-type oscillator shown in Fig. 18-13 if the coil has an inductance of 40 μH and the capacitance is set at 160 pF?

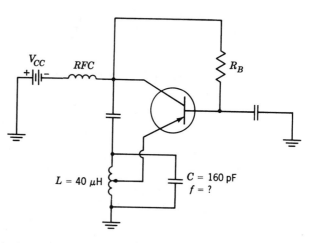

Figure 18-13 Hartley oscillator.

6. Find the resonant frequency of the series resonant section of the bandpass filter shown in Fig. 18-14.
7. Find the resonant frequency of the series resonant section of the wave trap or band-elimination filter shown in Fig. 18-15.

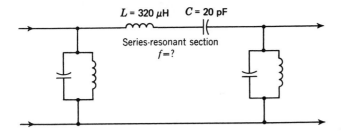

Figure 18-14 Bandpass filter.

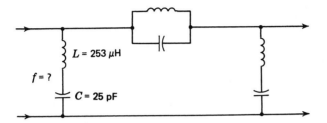

Figure 18-15 Band-elimination filter.

8. A 3-mH coil and a 40-pF capacitor are connected as shown in Fig. 18-16 to form the secondary side of an IF transformer. What is its resonant frequency? Explain why the secondary is a series tuned circuit while the primary is a parallel circuit.

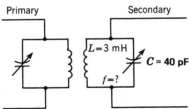

Figure 18-16 The secondary of the IF transformer is a series resonant circuit.

9. Find the resonant frequency of an antenna circuit if the inductance is 50 μH and the capacitance is 0.0002 μF.
10. Find the resonant frequency of a series circuit containing a 300-μH coil and a 365-pF capacitor.

Finding the Inductance or Capacitance Needed
to Make a Series Resonant Circuit

JOB 18-5

Formula (18-10) may be transformed to find formulas which may be used to find the inductance or capacitance needed to form a series resonant circuit at a given frequency.

$$L = \frac{25,300}{f^2} \times C$$ Formula 18-11

$$C = \frac{25{,}300}{f^2 \times L}$$

Formula 18-12

where
$$L = \text{inductance, } \mu\text{H}$$
$$C = \text{capacitance, } \mu\text{F}$$
$$f = \text{frequency, kHz}$$

Example 18-10

What value of inductance must be placed in series with a 253-pF tuning capacitor in order to provide resonance for a 500-kHz signal?

Solution

Given: $C = 253 \text{ pF} = 253 \times 10^{-6} \; \mu\text{F}$ Find: $L = ?$
$f = 500 \text{ kHz}$

$$L = \frac{25{,}300}{f^2 \times C} = \frac{25{,}300}{(500)^2 \times 253 \times 10^{-6}} \qquad (18\text{-}11)$$

$$= \frac{25{,}300}{25 \times 10^4 \times 253 \times 10^{-6}}$$

$$= \frac{100 \times 10^2}{25}$$

$$L = 400 \; \mu\text{H} \qquad Ans.$$

Example 18-11

An inductance of 40 μH is in series with a capacitor in a Hartley-type oscillator circuit. Find the value of the capacitance needed to produce resonance to a frequency of 5000 kHz.

Solution

Given: $L = 40 \; \mu\text{H}$ Find: $C = ?$
$f = 5000 \text{ kHz}$

$$C = \frac{25{,}300}{f^2 \times L} = \frac{25{,}300}{(5000)^2 \times 40} \qquad (18\text{-}12)$$

$$= \frac{25{,}300}{25 \times 10^6 \times 40}$$

$$= \frac{25{,}300}{10^9}$$

$$= 25{,}300 \times 10^{-9} \; \mu\text{F}$$

$$= 25{,}300 \times 10^{-9} \times 10^6 \text{ pF}$$

$$C = 25.3 \text{ pF} \qquad Ans.$$

Self-Test 18-12

A series circuit has a resistance of 30 Ω, an inductance of 0.382 H, and a capacitance of 0.2 μF. Find (*a*) the impedance of the circuit to a frequency of 500 Hz, (*b*) the capacitance that must be added in parallel with the 0.2-μF capacitor to produce resonance at this frequency, and (*c*) the impedance of the circuit at resonance.

Solution

Given: $R = 30\ \Omega$ Find: Z at 500 Hz
$\qquad\qquad L = 0.382$ H C to be added
$\qquad\qquad C = 0.2\ \mu$F Z at resonance
$\qquad\qquad f = 500$ Hz

a. In order to find Z, we must know X_L and ____(1)____.

$$X_L = 6.28fL = 6.28 \times 500 \times 0.382$$
$$X_L = \underline{\quad(2)\quad}\ \Omega$$
$$X_C = \frac{159{,}000}{f \times C} = \frac{159{,}000}{500 \times 0.2} = \underline{\quad(3)\quad}\ \Omega$$
$$X_d = X_C - X_L = 1590 - 1200$$
$$X_d = \underline{\quad(4)\quad}\ \Omega$$
$$Q = \frac{X_d}{R} = \frac{390}{30} = 13$$

and we ____(5)(should/should not)____ include R in our calculations for Z.

$$Z = X_d = \underline{\quad(6)\quad}\ \Omega \qquad Ans.$$

b. Find the C that will produce resonance at 500 Hz.

$$C = \frac{25{,}300}{f^2 \times L} \qquad\qquad (18\text{-}12)$$

Using the units to obtain microfarads,

$$C = \frac{25{,}300}{(0.5)^2 \times 0.382 \times 10^6}$$
$$= \frac{25{,}300}{(7) \times 10^6}$$
$$= \frac{25.3 \times 10^3}{95.5 \times (8)}$$
$$C = \underline{\quad(9)\quad}\ \mu\text{F} \qquad Ans.$$

Since 0.265 μF is needed for resonance, and we have only 0.2 μF, we must add $0.265 - 0.2 = $ ____(10)____ μF.   *Ans.*

c. At resonance, X_d will equal ____(11)____ Ω, and the only resistance in the circuit will be the original resistance of ____(12)____ Ω. Therefore, $Z = $ ____(13)____ Ω.   *Ans.*

CALCULATOR HINT

Complicated problems require values and equations, many of which are stored in a calculator. You need to learn ways of organizing and controlling this information.

Exercises

Applications

1. What value of inductance must be connected in series with a 0.0003-μF capacitor in order that the circuit be resonant to a frequency of 1000 kHz?

2. What value of capacitance must be connected in series with a 50-μH coil in order that the circuit be resonant to a frequency of 2000 kHz?
3. What value of inductance will produce resonance to 50 Hz if it is placed in series with a 20-μF capacitor?
4. What value of capacitance must be used in series with a 30-μH inductance in order to produce an oscillator frequency of 6000 kHz?
5. What value of capacitance must be added in series with a solenoid of 0.2-H inductance in order to be resonant to 60 Hz?
6. What capacity is necessary in series with a 100-μH coil to produce a wave trap for a 1200-kHz signal?
7. What is the inductance of the secondary winding of an RF transformer if it is in series with a 0.000 35-μF capacitor and is resonant to a frequency of 1000 kHz?
8. What is the capacity of an antenna circuit whose inductance is 50 μH if it is resonant to 1500 kHz?
9. A 0.000 04-μF capacitance is in series with the secondary of an RF transformer. What must be the inductance of the coil if the secondary is to be resonant to 500 kHz?
10. What must be the minimum and maximum values of the capacitor needed to produce resonance with a 240-μH coil to frequencies between 500 and 1500 kHz?

JOB 18-6

Review of Series AC Circuits

The voltages and current in a series ac circuit are not usually in phase with each other. Alternating-current voltages may not be added arithmetically but only by means of ___(1)___ addition.

The ___(2)___ is used as the reference line upon which to draw the phasor diagram.

A phasor is a straight line drawn with a definite length and in a definite direction. In ac electricity, the direction of the phasor indicates the ___(3)___ at which the voltage or current occurs in relation to another voltage or current.

Phasors drawn in the ___(4)___ direction are in phase.

Phasors drawn in ___(5)___ directions are out of phase.

Phasors are added by placing the tail of one phasor on to the ___(6)___ of another and drawing the phasor with its original length and direction.

The sum of the phasors is the phasor drawn from the ___(7)___ of the phasors to the ___(8)___ of the last phasor.

In a purely resistive circuit:

The voltage is ___(9)(in/out of)___ phase with the current.

$$V_T = V_1 + V_2 + V_3 \qquad \text{4-2}$$

$$V_T = I_T \times R_T \qquad \text{4-7}$$

$$R_T = R_1 + R_2 + R_3 \qquad \text{4-3}$$

$$P = V \times I \qquad \text{6-1}$$

In a purely inductive circuit:

The voltage ___(10)(leads/lags)___ the current by 90°.

$$V_T = V_1 + V_2 + V_3 \qquad\qquad 4\text{-}2$$

$$V_L = I_L \times \underline{\quad(11)\quad} \qquad\qquad 16\text{-}2$$

$$X_{L_T} = X_{L_1} + X_{L_2} + X_{L_3} \qquad\qquad 18\text{-}7$$

$$W = V \times I \times \underline{\quad(12)\quad}$$

$$W = V \times I \times \underline{\quad(13)\quad}$$

In a purely capacitive circuit:

The voltage ___(14)(leads/lags)___ the current by 90°.

$$V_T = V_1 + V_2 + V_3 \qquad\qquad 4\text{-}2$$

$$V_C = I_C \times \underline{\quad(15)\quad} \qquad\qquad 17\text{-}7$$

$$X_{C_T} = X_{C_1} + X_{C_2} + X_{C_3} \qquad\qquad 18\text{-}8$$

$$W = V \times I \times \cos\theta$$

$$W = \underline{\quad(16)\quad} \times \underline{\quad(17)\quad} \times PF$$

In an ac series circuit of resistance and inductance:

The total voltage ___(18)(leads/lags)___ the current by some angle θ.

$$V_T{}^2 = V_R{}^2 + \underline{\quad(19)\quad} \qquad\qquad 16\text{-}3$$

$$Z = \underline{\quad(20)\quad} \qquad\qquad 16\text{-}5$$

$$V_T = I_T \times \underline{\quad(21)\quad}$$

$$\cos\theta = \frac{R}{(22)} \qquad\qquad 16\text{-}6$$

$$PF = \frac{(23)}{Z} \qquad\qquad 16\text{-}7$$

$$W = V \times I \times \underline{\quad(24)\quad}$$

In an ac series circuit of resistance and capacitance:

The total voltage ___(25)(leads/lags)___ the current by some angle θ.

$$V_T{}^2 = V_R{}^2 + V_C{}^2 \qquad\qquad 18\text{-}1$$

$$Z = \underline{\quad(26)\quad} \qquad\qquad 18\text{-}3$$

$$V_T = I_T \times \underline{\quad(27)\quad}$$

$$\cos\theta = \frac{(28)}{Z}$$

$$PF = \underline{\quad(29)\quad}$$

$$W = V \times I \times \underline{\quad(30)\quad}$$

In an ac series circuit of resistance, inductance, and capacitance:

The total voltage will lead or lag the current, depending on the values of X_L and X_C. If X_L is larger than X_C, the voltage will ___(31)(lead/lag)___ the current. If X_L

is smaller than X_C, the voltage will ___(32)(lead/lag)___ behind the current. The angle of lead or lag is given by the angle θ.

$$V_T{}^2 = V_R{}^2 + (V_L - V_C)^2 \qquad\qquad \text{18-4}$$

$$Z = \sqrt{R^2 + \underline{\quad(33)\quad}} \qquad\qquad \text{18-6}$$

$$V_T = I_T \times \underline{\quad(34)\quad}$$

$$\cos \theta = \frac{R}{Z}$$

$$PF = \frac{R}{Z}$$

$$W = V \times I \times \cos \theta$$

Without regard for the PF of a circuit, the effective power is the power drawn by the *resistive* elements of the circuit, the formula is

$$W = \underline{\quad(35)\quad} \qquad\qquad \text{7-4}$$

Series resonance is the condition at which the inductive reactance is exactly equal to the capacitive reactance. For any given combination of coil and capacitor, there is only one frequency at which this situation can occur. This frequency is called the *resonant frequency.*

$$f = \frac{159}{\sqrt{L \times C}} \qquad\qquad \text{18-10}$$

where
f = frequency, measured in kHz
L = inductance, measured in ___(36)___
C = capacitance, measured in ___(37)___

The inductance or capacitance needed to make a circuit resonant to a given frequency is given by the formulas

$$L = \frac{25{,}300}{f^2 \times (38)} \qquad\qquad \text{18-11}$$

$$C = \frac{25{,}300}{(39) \times L} \qquad\qquad \text{18-12}$$

where all units are measured in the units given for formula (18-10).

Exercises

Practice

1. What is the voltage drop across a 3000-Ω resistor carrying 60 mA of current at a frequency of 60 Hz?
2. Find the current and power drawn by a 20-mH coil from a 125.6-V 10-kHz source.
3. A filter choke passes 60 mA of current when it is connected across a 120-V 60 Hz line. What is its inductance?
4. Find the voltage drop across a 0.05-μF capacitor if it passes 50 mA at a frequency of 1 kHz.
5. A 50-V emf at 1-kHz frequency is impressed across a 1000-Ω resistor in series with a 1-H coil. Find (*a*) the impedance, (*b*) the total current, (*c*) the voltage

drop across each part, (d) the phase angle, (e) the power factor, and (f) the power.

6. A 120-V 1-kHz ac voltage is applied across a series circuit of a 200-Ω resistor and a 1.6-μF capacitor. Find (a) the impedance, (b) the total current, (c) the voltage drop across each part, (d) the phase angle, (e) the power factor, and (f) the power.

7. An antenna circuit consists of a 10-Ω resistance, a 0.5-mH inductance, and a 50-pF capacitance. Find its impedance to a frequency of (a) 1000 kHz and (b) 500 kHz.

8. A 1000-Ω 100-μH coil is in series with a capacitance of 5 μF. Find the impedance at (a) 500 Hz, (b) 5 kHz, and (c) 500 kHz.

9. Find the frequency of operation of the Colpitts oscillator shown in Fig. 18-17. Note that C_1 and C_2 are in series.

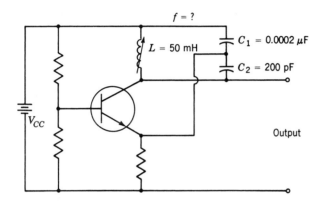

Figure 18-17 Colpitts oscillator.

10. What capacitance is needed in series with a 0.5-mH coil in order to produce resonance to 10 kHz?

(See CD-ROM for Test 18-2)

Assessment

1. A 100-Ω resistor is placed in series with a 53.052-μF capacitor and connected to a 100-Vac 60-Hz power supply. Determine the following: (a) X_C, (b) Z, (c) I, (d) E_R, (e) E_{XC}, (f) $\angle\theta$, (g) power factor, (h) true power, (i) reactive power, and (j) apparent power.

2. A 20-Ω resistor, 66.315-μF capacitor, and 79.5774-mH inductor are placed in series with a 200-Vac 60-Hz source. Determine the following: (a) X_C, (b) Z, (c) I, (d) E_R, (e) E_{XC}, (f) E_{XL}, (g) power factor, (h) true power, (i) reactive power, (j) apparent power, and (k) X_L.

3. Determine the resonant frequency if the inductor is 3.183 mH and the capacitor is 0.3183 μF.

4. What value of inductance must be connected with a 1500-pF capacitor for it to be resonant at 1 MHz?

INTERNET
ACTIVITIES

Internet Activity A

Describe the voltage-current relationships for inductors and capacitors and how they can be expressed with complex numbers.
http://www.st-and.ac.uk/~www_pa/Scots_Guide/info/signals/complex/cmplx.html

Internet Activity B

Describe a method for measuring inductance.
http://et.nmsu.edu/~etti/fall96/electronics/induct/induct.html

Chapter 18 Solutions to Self-Tests and Reviews

Self-Test 18-7

1. X_C
2. 125
3. 500
4. 10^{-3}
5. 10,000
6. 10^{-6}
7. 20,000
8. 10,000
9. 20
10. may
11. 10,000
12. 50
13. 500
14. 0.0500
15. 87°

Self-Test 18-12

1. X_C
2. 1200

---

3. 1590
4. 390
5. should not
6. 390
7. 0.0955
8. 10^3
9. 0.265
10. 0.065
11. zero
12. 30
13. 30

Job 18-6 Review

1. phasor
2. current
3. time
4. same
5. different
6. head
7. origin

---

8. head
9. in
10. leads
11. X_L
12. $\cos \theta$
13. PF
14. lags
15. X_C
16. V
17. I
18. leads
19. V_L^2
20. $\sqrt{R^2 + X_L^2}$
21. Z
22. Z
23. R
24. PF
25. lags
26. $\sqrt{R^2 + X_C^2}$
27. Z

---

28. R
29. $\dfrac{R}{Z}$
30. PF
31. lead
32. lag
33. $(X_L - X_C^2)$
34. Z
35. I^2R
36. μH
37. μF
38. C
39. f^2

Parallel AC Circuits

chapter

Learning Objectives

1. Determine the values of resistance and inductance in parallel.
2. Write the equations for parallel resistance, inductance, and capacitance.
3. Determine the resolution of phasors.
4. Use ac circuit analysis to understand power filters.
5. Find equations for parallel-series ac circuits.
6. Find equations for series-parallel ac circuits.
7. Determine the equations for parallel resonance.

JOB 19-1

Simple Parallel AC Circuits

The general rules for solving dc parallel circuits are also applicable to the solution of ac parallel circuits.

1. The voltages across all branches are equal to each other and to the total voltage.

$$V_T = V_1 = V_2 = V_3 \qquad (5\text{-}1)$$

2. The total current is equal to the sum of all the branch currents.

$$I_T = I_1 + I_2 + I_3 \qquad (5\text{-}2)$$

Parallel circuits containing only resistance. We can add the branch currents as indicated by formula (5-2) only if the branch currents are in phase. If they are out of phase, they may be added *only* by phasor addition. Let us draw the phasor diagrams for each branch of the circuit shown in Fig. 19-3 for Example 19-1 to discover the phase relationships in this type of circuit.

Phasor diagrams. The *voltage* is used as the reference line upon which to draw the phasors. In Fig. 19-1a, the current I_1 is drawn in the same direction as the voltage because the current in the purely resistive iron is in phase with the voltage. In Fig. 19-1b, the smaller current through the lamp I_2 is also drawn in the same direction as the voltage because the current through a purely resistive lamp is in phase with the voltage. Now let us draw both sets of phasors on the same voltage base as shown in Fig. 19-2a. This diagram indicates that the current in one resistance is in phase with the current in the other resistance, since they are both drawn in the

(a)   (b)

Figure 19-1 Currents in purely resistive parallel branches are in phase with the voltage.

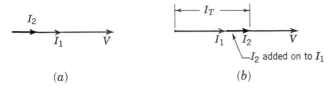

(a)   (b)

Figure 19-2 (a) Resistive branch currents are in phase with each other. (b) I_T represents the phasor sum of I_1 and I_2.

same direction. To find the total current, it is necessary only to add the two current phasors. This is done, as with any phasor quantities, by adding the tail of phasor I_2 on to the head of phasor I_1 as shown in Fig. 19-2b. The total current is then the distance from the origin of the phasors to the head of the last phasor. Since the two currents are in phase, the total current may be found by the direct arithmetical addition of the currents, using formula (5-2).

Example 19-1

A 20-Ω electric iron and a 100-Ω lamp are connected in parallel across a 120-V 60-Hz ac line. Find (a) the total current, (b) the total resistance, (c) the total power drawn by the circuit, and (d) the PF.

Solution

The diagram for the circuit is shown in Fig. 19-3.

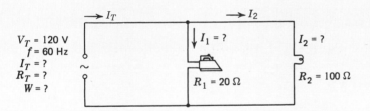

Figure 19-3

a. Find the branch currents. Since

$$V_T = V_1 = V_2 = V_3 = 120 \text{ V} \qquad (5\text{-}1)$$

$$V_1 = I_1 \times R_1 \qquad\qquad V_2 = I_2 \times R_2$$
$$120 = I_1 \times 20 \qquad\qquad 120 = I_2 \times 100$$
$$I_1 = \frac{120}{20} = 6 \text{ A} \qquad I_2 = \frac{120}{100} = 1.2 \text{ A}$$

Find the total current.

$$I_T = I_1 + I_2 = 6 + 1.2 = 7.2 \text{ A} \qquad Ans. \qquad (5\text{-}2)$$

b. $$V_T = I_T \times R_T \qquad\qquad (4\text{-}7)$$
$$120 = 7.2 \times R_T$$
$$R_T = \frac{120}{7.2} = 16.7 \ \Omega \qquad Ans.$$

c. In a purely resistive set of branch circuits, the total current is in phase with the total voltage. The phase angle is therefore equal to 0°.

$$W = V \times I \times \cos \theta = 120 \times 7.2 \times \cos 0°$$

$$W = 120 \times 7.2 \times 1 = 864 \ W \qquad Ans.$$

d. $$PF = \cos \theta = \cos 0° = 1 \qquad Ans.$$

Parallel circuits containing only inductance. This type of circuit is illustrated in Fig. 19-6 for Example 19-2.

Phasor diagrams. The *voltage* is used as the reference line upon which to draw the phasors. In Fig. 19-4a, the current I_{L_1} is drawn lagging the voltage by 90°. In Fig. 19-4b, the current I_{L_2} is also drawn lagging the voltage by 90°, since the current in *any* inductance lags the voltage by 90°. Now let us draw both sets of phasors on the same voltage base as shown in Fig. 19-5a. This diagram indicates that the current in one coil is in phase with the current in the second coil, since they are both drawn in the same direction. To find the total current, it is necessary only to add the two current phasors. This is done, as with any phasor quantities, by adding the tail of phasor I_{L_2} on the head of phasor I_{L_1} as shown in Fig. 19-5b. The total current is then the distance from the origin of the phasors to the head of the last phasor. Since the two currents are in phase, the total current may be found by the direct arithmetical addition of the currents, using formula (5-2). The difference between this circuit and the purely resistive circuit lies in the fact that the total current *lags* behind the total voltage by 90°. The phase angle $\theta = 90°$.

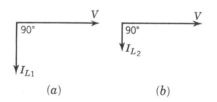

(a)   (b)

Figure 19-4 Currents in purely inductive parallel branches lag the voltage by 90°.

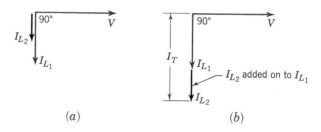

(a)   (b)

Figure 19-5 (a) Inductive branch currents are in phase with each other. (b) I_T represents the phasor sum of I_{L_1} and I_{L_2}.

Example 19-2

Two coils of 20 and 30 Ω reactance, respectively, are connected in parallel across a 120-V 60-Hz ac line. Find (*a*) the total current, (*b*) the impedance, (*c*) the power factor, and (*d*) the power drawn by the circuit.

Solution

The diagram for the circuit is shown in Fig. 19-6.

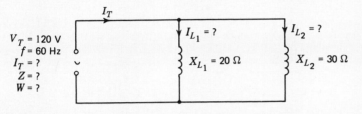

$V_T = 120$ V
$f = 60$ Hz
$I_T = ?$
$Z = ?$
$W = ?$

$I_{L_1} = ?$ $X_{L_1} = 20 \ \Omega$

$I_{L_2} = ?$ $X_{L_2} = 30 \ \Omega$

Figure 19-6

a. Find the branch currents. Since

$$V_T = V_1 = V_2 = 120 \text{ V} \qquad (5\text{-}1)$$

$$V_1 = I_1 \times X_1 \qquad\qquad V_2 = I_2 \times X_2$$
$$120 = I_1 \times 20 \qquad\qquad 120 = I_2 \times 30$$
$$I_1 = \frac{120}{20} = 6 \text{ A} \qquad\qquad I_2 = \frac{120}{30} = 4 \text{ A}$$

Find the total current.

$$I_T = I_1 + I_2 = 6 + 4 = 10 \text{ A} \qquad \textit{Ans.} \qquad (5\text{-}2)$$

b.
$$V_T = I_T \times Z$$
$$120 = 10 \times Z$$
$$Z = \frac{120}{10} = 12 \ \Omega \qquad \textit{Ans.}$$

c. In a purely inductive set of branch circuits, the total current lags behind the voltage by 90°. The phase angle is therefore equal to 90°.

$$\text{PF} = \cos \theta = \cos 90° = 0 \qquad \textit{Ans.}$$

d.
$$W = V \times I \times \cos \theta = 120 \times 10 \times \cos 90°$$
$$W = 120 \times 10 \times 0 = 0 \text{ W} \qquad \textit{Ans.}$$

$$W = I^2R = (10)^2 \times 0 = 0 \text{ W} \qquad \textit{Check}$$

> **Hint for Success**
>
> The difference between series and parallel circuits is that in a series circuit the voltage divides and in a parallel circuit the current divides.

Parallel circuits containing only capacitance.   This type of circuit is illustrated in Fig. 19-9 for Example 19-3.

Phasor diagrams.   The *voltage* is used as the reference line upon which to draw the phasors. In Fig. 19-7*a*, the current I_{C_1} is drawn *leading* the voltage by 90°. In Fig. 19-7*b*, the current I_{C_2} is also drawn *leading* the voltage by 90°, since the current in *any* capacitance leads the voltage by 90°. Now let us draw both sets of phasors on the same voltage base, as shown in Fig. 19-8*a*. This diagram indicates that

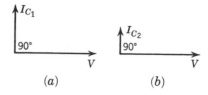

Figure 19-7 Currents in purely capacitive parallel branches lead the voltage by 90°.

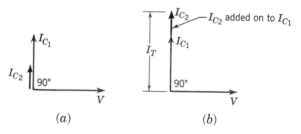

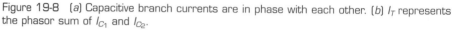

Figure 19-8 (a) Capacitive branch currents are in phase with each other. (b) I_T represents the phasor sum of I_{C_1} and I_{C_2}.

the current in one capacitor is in phase with the current in the second capacitor, since they are both drawn in the same direction. To find the total current, it is necessary only to add the two current phasors. This is done, as with any phasor quantities, by adding the tail of the phasor I_{C_2} on to the head of phase I_{C_1} as shown in Fig. 19-8b. The total current is then the distance from the origin of the phasors to the head of the last phasor. Since the two currents are in phase, the total current may be found by the direct arithmetical addition of the currents, using formula (4-2). The difference between this and the other two circuits lies in the fact that the total current *leads* the total voltage by 90°. The phase angle $\theta = 90°$.

Example 19-3

Two capacitors of 30 and 40 Ω reactance, respectively, are connected across a 120-V 60-Hz ac line. Find (a) the total current, (b) the impedance, (c) the power factor, and (d) the power drawn by the circuit.

Solution

The diagram for the circuit is shown in Fig. 19-9.

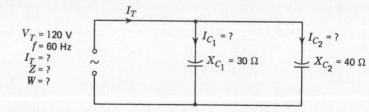

Figure 19-9

a. Find the branch currents. Since

$$V_T = V_1 = V_2 = 120 \text{ V} \qquad (5\text{-}1)$$

$$V_1 = I_1 \times X_{C_1} \qquad V_2 = I_2 \times X_{C_2}$$
$$120 = I_1 \times 30 \qquad 120 = I_2 \times 40$$
$$I_1 = \frac{120}{30} = 4 \text{ A} \qquad I_2 = \frac{120}{40} = 3 \text{ A}$$

Find the total current.

$$I_T = I_1 + I_2 = 4 + 3 = 7 \text{ A} \qquad Ans. \qquad (5\text{-}2)$$

b.
$$V_T = I_T \times Z$$
$$120 = 7 \times Z$$
$$Z = \frac{120}{7} = 17.1 \ \Omega \qquad Ans.$$

c. In a purely capacitive set of branch circuits, the total current leads the voltage by 90°. The phase angle is therefore equal to 90°.

$$PF = \cos \theta = \cos 90° = 0 \qquad Ans.$$

d.
$$W = V \times I \times \cos \theta = 120 \times 7 \times \cos 90°$$
$$W = 120 \times 7 \times 0 = 0 \text{ W} \qquad Ans.$$
$$W = I^2 R = (7)^2 \times 0 = 0 \text{ W} \qquad Check$$

Self-Test 19-4

In a parallel ac circuit,

1. The voltage across any branch is equal to the total voltage.
2. The current in a resistor is ___(in/out of)___ phase with the voltage.
3. The current in an inductance ___(leads/lags)___ the voltage by _____°.
4. The current in a capacitance ___(leads/lags)___ the voltage by _____°.
5. Currents that are in phase with each other are added ___(vectorially/ arithmetically)___ using formula (5-2).
6. Ohm's law, $V_T = I_T \times Z$, may be used for total values.
7. The power depends on the angle of lead or lag as shown by $W = V \times I \times$ _____.

Exercises

Applications

1. A 10-Ω electric heater and a 50-Ω incandescent lamp are placed in parallel across a 120-V 60-Hz ac line. Find (a) the total current, (b) the total resistance and (c) the power drawn.
2. Two toy train solenoids used in semaphore signals have inductive reactances of 24 and 48 Ω, respectively. They are connected in parallel across the 12-V winding of the power transformer. Find (a) the total current, (b) the impedance, (c) the power factor, and (d) the power drawn.
3. Two capacitors of 100 and 200 Ω capacitive reactance, respectively, are connected in parallel across a 100-V 60-Hz ac line. Find (a) the total current, (b) the impedance, (c) the power factor, and (d) the power drawn.

4. A 40-Ω soldering iron and a 100-Ω incandescent lamp are connected in parallel across a 110-V 60-Hz ac line. Find (a) the total current, (b) the total resistance, (c) the power factor, and (d) the power drawn.
5. Find the total current, impedance, and power used by the circuit shown in Fig. 19-10.

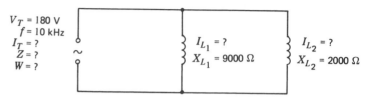

$V_T = 180$ V
$f = 10$ kHz
$I_T = ?$
$Z = ?$
$W = ?$

$I_{L_1} = ?$
$X_{L_1} = 9000\ \Omega$

$I_{L_2} = ?$
$X_{L_2} = 2000\ \Omega$

Figure 19-10

6. A mechanic replaced a leaky coupling capacitor with a parallel combination of two capacitors as shown in Fig. 19-11. If the voltage drop across the capacitors is 0.5 V at a frequency of 1 kHz, find the current in each capacitor, the total current, and the impedance of the combination.

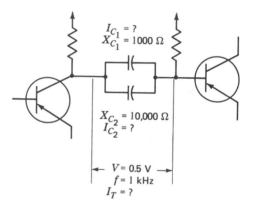

$I_{C_1} = ?$
$X_{C_1} = 1000\ \Omega$

$X_{C_2} = 10,000\ \Omega$
$I_{C_2} = ?$

$V = 0.5$ V
$f = 1$ kHz
$I_T = ?$

Figure 19-11

7. Two pure inductances of 2 and 5 H inductance are connected in parallel across a 120-V 60-Hz line. Find (a) the current in each inductance, (b) the total current, (c) the impedance, (d) the power factor, and (e) the power drawn.
8. Two pure capacitances of 0.159 and 0.04 μF are connected in parallel across a 50-V 1-kHz line. Find (a) the current in each capacitance, (b) the total current, (c) the impedance, (d) the power factor, and (e) the power drawn.

JOB 19-2

Resistance and Inductance in Parallel

Figure 19-13 shows a 24-Ω resistance and a 30-Ω inductive reactance in parallel across a 12-V 60-Hz ac source. The total current drawn by the circuit can be found by adding the currents in each branch. However, if they are not in phase, they must be added vectorially. The phase relations in the circuit are shown in Fig. 19-12a.

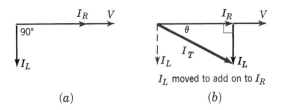

(a)

(b)

Figure 19-12 (*a*) Phasor diagram for a parallel ac circuit containing resistance and inductance. (*b*) I_T represents the phasor sum of I_R and I_L.

The constant voltage is used as the reference line upon which to draw the phasor diagram. The current in the resistance is in phase with the voltage, and the current in the inductance lags the voltage by 90°.

Addition of the phasors. In Fig. 19-12*b*, the total current I_T is obtained by adding the phasor I_L to the phasor I_R. Place the tail of I_L on the head of I_R, and draw it in its original direction and length. The distance from the origin of phasors to the head of the final phasor is the sum of the phasors. In this instance, the phasor I_T represents the sum of the phasors I_R and I_L. The phase angle θ is the angle by which the total current *lags* behind the total voltage. By applying the pythagorean theorem to Fig. 19-12*b*, we obtain

$$I_T{}^2 = I_R{}^2 + I_L{}^2 \qquad \text{Formula 19-1}$$

By Ohm's law, $\qquad\qquad V_T = I_T \times Z$

Note that the impedance Z is *not* found by the phasor addition of R and X. The formula $Z^2 = R^2 + X^2$ applies *only* to series ac circuits. In parallel circuits, the *currents* are added vectorially as shown by formula (19-1) above. The impedance Z is found by applying formula (2-1).

By trigonometry,

$$\cos \theta = \frac{a}{h}$$

$$\cos \theta = \frac{I_R}{I_T} \qquad \text{Formula 19-2}$$

or $\qquad\qquad \text{PF} = \dfrac{I_R}{I_T} \qquad \text{Formula 19-3}$

The power is still given by

$$W = V \times I \times \cos \theta$$
or $\qquad\qquad W = V \times I \times \text{PF}$
or $\qquad\qquad W = I_R{}^2 \times R \qquad\qquad (7\text{-}4)$

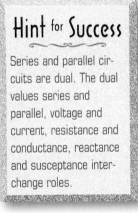

Hint for Success

Series and parallel circuits are dual. The dual values series and parallel, voltage and current, resistance and conductance, reactance and susceptance interchange roles.

Example 19-5

A toy electric train semaphore is made of a 24-Ω lamp in parallel with a solenoid coil of 30 Ω inductive reactance. If it operates from the 12-V winding of the 60-Hz power transformer, find (*a*) the total current, (*b*) the impedance, (*c*) the phase angle, (*d*) the power factor, and (*e*) the power drawn.

Solution

The diagram of the circuit is shown in Fig. 19-13.

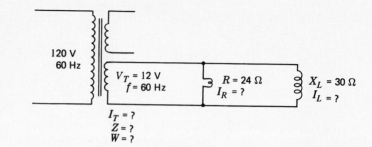

Figure 19-13 AC parallel circuit containing resistance and inductance.

a. Find the branch currents. Since

$$V_T = V_R = V_L = 12 \text{ V} \tag{5-1}$$

$$V_R = I_R \times R \qquad\qquad V_L = I_L \times X_L$$

$$12 = I_R \times 24 \qquad\qquad 12 = I_L \times 30$$

$$I_R = \frac{12}{24} = 0.5 \text{ A} \qquad\qquad I_L = \frac{12}{30} = 0.4 \text{ A}$$

Find the total current.

$$I_T{}^2 = I_R{}^2 + I_L{}^2 = 0.5^2 + 0.4^2 \tag{19-1}$$

$$= 0.25 + 0.16$$

$$I_T{}^2 = 0.41$$

$$I_T = \sqrt{0.41} = 0.64 \text{ A} \qquad Ans.$$

b.

$$V_T = I_T \times Z$$

$$12 = 0.64 \times Z$$

$$Z = \frac{12}{0.64} = 18.75 \ \Omega \qquad Ans.$$

c.

$$\cos \theta = \frac{I_R}{I_T} = \frac{0.5}{0.64} = 0.7812 \tag{19-2}$$

$$\theta = 39° \qquad Ans.$$

d.

$$\text{PF} = \frac{I_R}{I_T} = 0.7812 = 78.1\% \ lagging \qquad Ans. \tag{19-3}$$

e.

$$W = V \times I \times \cos \theta = 12 \times 0.64 \times \cos 39°$$

$$= 12 \times 0.64 \times 0.781 = 6 \text{ W} \qquad Ans.$$

$$W = I_R{}^2 \times R = (0.5)^2 \times 24 = 6 \text{ W} \qquad Check$$

In the parallel circuit, the total current of 0.64 A *lags* the total voltage of 12 V by 39°.

Example 19-6

The purpose of the "high-pass" circuit shown in Fig. 19-14 is not only to permit high frequencies to pass on to the load but to prevent the passage of low frequencies. Find the effectiveness of the circuit by calculating (*a*) the branch currents, (b) the total current, and (c) the percent of the total current in the resistor for (1) a 1-kHz audio frequency and (2) a 1000-kHz radio frequency.

Solution

The diagram for the circuit is shown in Fig. 19-14.

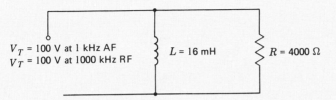

$V_T = 100$ V at 1 kHz AF
$V_T = 100$ V at 1000 kHz RF

$L = 16$ mH

$R = 4000\ \Omega$

Figure 19-14   High-pass filter.

1. For the 1-kHz audio frequency,

a.
$$X_L = 6.28\,fL = 6.28 \times 10^3 \times 16 \times 10^{-3} \qquad (16\text{-}1)$$
$$X_L = 6.28 \times 16 = 100\ \Omega \qquad Ans.$$

Find the branch currents. Since

$$V_T = V_R = V_L = 100\ \text{V} \qquad (5\text{-}1)$$

$$V_R = I_R \times R \qquad\qquad V_L = I_L \times X_L$$
$$100 = I_R \times 4000 \qquad\qquad 100 = I_L \times 100$$
$$I_R = \frac{100}{4000} = 0.025\ \text{A} \qquad I_L = \frac{100}{100} = 1\ \text{A} \qquad Ans.$$

b. Find the total current.

$$I_T{}^2 = I_R{}^2 + I_L{}^2 = 0.025^2 + 1^2 \qquad (19\text{-}1)$$
$$= 0.000\,625 + 1$$
$$I_T{}^2 = 1.000\,625$$
$$I_T = \sqrt{1.000\,625} = 1\ \text{A} \qquad Ans.$$

c. Find the percent of the total current passing through the resistor.

$$\text{Percent} = \frac{I_R}{I_T} \times 100 = \frac{0.025}{1} \times 100 = 2.5\% \qquad Ans.$$

That is, 2.5% of the 1-kHz audio frequency passes through the resistor.

2. For the 1000-kHz radio frequency: Since 1000 kHz is 1000 times as large as the audio frequency of 1 kHz, the X_L at 1000 kHz will be equal to 1000 times the X_L at 1 kHz. Therefore,

$$X_L = 1000 \times 100 = 100{,}000\ \Omega$$

Find the branch currents. Since

a.
$$V_T = V_R = V_L = 100\ \text{V} \qquad (5\text{-}1)$$

$$V_R = I_R \times R \qquad\qquad V_L = I_L \times X_L$$
$$100 = I_R \times 4000 \qquad\qquad 100 = I_L \times 100{,}000$$
$$I_R = \frac{100}{4000} = 0.025\ \text{A} \qquad I_L = \frac{100}{100\ \text{k}\Omega} = 0.001\ \text{A} \qquad Ans.$$

b. Find the total current.

$$I_T{}^2 = I_R{}^2 + I_L{}^2 = 0.025^2 + 0.001^2 \qquad\qquad (19\text{-}1)$$
$$= (25 \times 10^{-3})^2 + (1 \times 10^{-3})^2$$
$$= 625 \times 10^{-6} + 1 \times 10^{-6}$$
$$I_T{}^2 = 625 \times 10^{-6}$$
$$I_T = \sqrt{626 \times 10^{-6}} = 25 \times 10^{-3}$$
$$I_T = 0.025 \text{ A} \qquad Ans.$$

c. Find the percent of the total current passing through the resistor.

$$\text{Percent} = \frac{I_R}{I_T} \times 100 = \frac{0.025}{0.025} \times 100 = 100\% \qquad Ans.$$

That is, practically 100% of the 1000-kHz radio frequency passes through the resistor.

The circuit is evidently a good high-pass circuit, since it passes practically 100% of the high radio frequency and only 2.5% of the low audio frequency. The majority of the low audio frequency finds an easy path through the coil (I_L for the 1-kHz audio frequency equals the total current of 1 A).

Exercises

Applications

1. A 24-Ω resistor and a 10-Ω inductive reactance are in parallel across a 120-V, 60-Hz ac line. Find (*a*) the total current, (*b*) the impedance, (*c*) the phase angle, (*d*) the power factor, and (*e*) the power drawn.
2. Repeat Exercise 1 for a 1000-Ω resistor and a 100-Ω inductive reactance.
3. A 50-Ω resistor and a 0.2-H coil are in parallel across a 100-V, 100-Hz ac line. Find (*a*) the total current, (b) the impedance, (*c*) the phase angle, (*d*) the power factor, and (*e*) the power drawn.
4. In Fig. 19-15*a*, find the percent of the total AF current that passes through the resistor. Find the percent of the total RF current that passes through the resistor. On the basis of these answers, may the circuit be classified as a high-pass circuit?

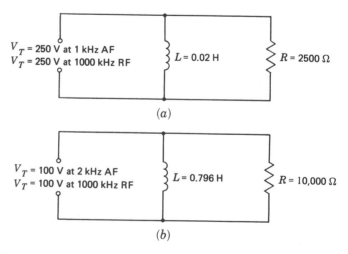

(a)

(b)

Figure 19-15

5. Repeat Exercise 4 for the circuit shown in Fig. 19-15b.
(See CD-ROM for Test 19-1)

Resistance, Inductance, and Capacitance in Parallel. The Equivalent Series Circuit

Figure 19-18 shows a 30-Ω resistor, a 40-Ω inductive reactance, and a 60-Ω capacitive reactance connected in parallel across a 120-V 60-Hz ac line. $I_R = 4$ A, $I_L = 3$ A, and $I_C = 2$ A. The total current drawn by the circuit may be found by adding the currents in each branch. However, since the currents are not in phase, they must be added vectorially. The phase relations in the circuit are shown in Fig. 19-16a. The voltage is used as the reference line upon which to draw the phasor diagram. The current in the resistor I_R is in phase with the voltage. The current in the capacitor I_C *leads* the voltage by 90°. The current in the inductance I_L *lags* the voltage by 90°. Since I_L and I_C are exactly 180° out of phase and act in exactly *opposite* directions, the current I_C is denoted by a minus sign.

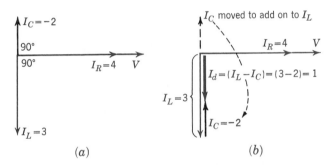

(a) (b)

Figure 19-16 (a) Phasor diagram for a parallel ac circuit containing resistance, inductance, and capacitance. (b) I_d represents the phasor sum of I_L and I_C.

Addition of phasors. When there are three phasors, it is best to add only two at a time. To add the phasor I_C to I_L, place the tail of I_C on the head of I_L and draw it in its original length and direction, which will be straight *up* as shown in Fig. 19-16b. Since these phasors are acting in *opposite* directions, their *sum* is actually the *difference* between the phasors as indicated by I_d. That is, the addition of I_C to I_L is really found by $I_d = I_L - I_C$. After this addition, the phasor diagram looks like Fig. 19-17a. Notice that the effect of the capacitor current has disappeared. The 3 A of coil current has exactly balanced the 2 A of capacitor current and has left an excess of 1 A of coil current. This 1 A of coil current must now be added to the 4 A of resistance current. In Fig. 19-17b, the total current I_T is obtained by adding the phasor I_d to the phasor I_R. Place the tail of I_d on the head of I_R, and draw it in the proper direction. If I_C were larger than I_L, the phasor I_d would have an *upward* direction. The distance from the origin of the phasors to the head of the last phasor is the sum of the phasors. In this instance, the phasor I_T represents the sum of the phasors $I_R + I_L + I_C$. The angle θ is the angle by which the total current *lags* the voltage. If I_C were larger than I_L, the total current would *lead* the voltage by this angle. By applying the pythagorean theorem to Fig. 19-17b, we obtain:

$$I_T{}^2 = I_R{}^2 + (I_L - I_C)^2 \qquad \text{Formula 19-4}$$

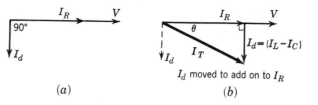

(a)   (b)

Figure 19-17 (a) I_d represents the result of adding I_C to I_L. (b) I_T represents the phasor sum of $I_R + I_C + I_L$.

By Ohm's law, $$V_T = I_T \times Z$$

By trigonometry, $$\cos \theta = \frac{a}{h}$$

$$\cos \theta = \frac{I_R}{I_T} \qquad (19\text{-}2)$$

or $$\text{PF} = \frac{I_R}{I_T} \qquad (19\text{-}3)$$

The power is still given by

$$W = V \times I \times \cos \theta$$
or $$W = V \times I \times \text{PF}$$
or $$W = I_R^2 \times R$$

Example 19-7

A 30-Ω resistor, a 40-Ω inductive reactance, and a 60-Ω capacitive reactance are connected in parallel across a 120-V 60-Hz ac line. Find (a) the total current, (b) the impedance, (c) the phase angle, (d) the power factor, and (e) the power drawn by the circuit.

Solution

The diagram for the circuit is shown in Fig. 19-18.

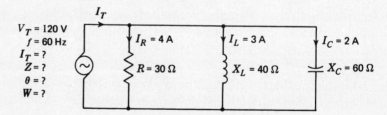

Figure 19-18 AC parallel circuit containing resistance, inductance, and capacitance.

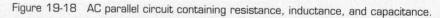

a. Find the branch currents. Since

$$V_T = V_1 = V_2 = 120 \text{ V}$$

$$V_R = I_R \times R \qquad\qquad V_L = I_L \times X_L \qquad\qquad V_C = I_C \times X_C$$
$$120 = I_R \times 30 \qquad\qquad 120 = I_L \times 40 \qquad\qquad 120 = I_C \times 60$$
$$I_R = \frac{120}{30} = 4 \text{ A} \qquad\qquad I_L = \frac{120}{40} = 3 \text{ A} \qquad\qquad I_C = \frac{120}{60} = 2 \text{ A} \qquad Ans.$$

Find the total current.

$$I_T{}^2 = I_R{}^2 + (I_L - I_C)^2 \qquad\qquad (19\text{-}4)$$
$$= 4^2 + (3 - 2)^2 = 4^2 + 1^2$$
$$= 16 + 1$$
$$I_T{}^2 = 17$$
$$I_T = \sqrt{17} = 4.12 \text{ A} \qquad Ans.$$

b.
$$V_T = I_T \times Z$$
$$120 = 4.12 \times Z$$
$$Z = \frac{120}{4.12} = 29.1 \ \Omega \qquad Ans.$$

c.
$$\cos \theta = \frac{I_R}{I_T} = \frac{4}{4.12} = 0.9708 \qquad (19\text{-}2)$$
$$\theta = 14° \qquad Ans.$$

d.
$$\text{PF} = \frac{I_R}{I_T} = 0.9708 = 97.1\% \ lagging \qquad Ans. \qquad (19\text{-}3)$$

e.
$$W = V \times I \times \cos \theta = 120 \times 4.12 \times \cos 14°$$
$$= 120 \times 4.12 \times 0.971$$
$$= 480 \text{ W} \qquad Ans.$$
$$= I_R{}^2 \times R$$
$$= (4)^2 \times 30$$
$$W = 480 \text{ W} \qquad Check$$

In the parallel circuit, the total current of 4.12 A lags the total voltage by 14°.

The Equivalent Series Circuit

Example 19-8

In the circuit shown in Fig. 19-19, find (a) the impedance, (b) the phase angle, and (c) the equivalent series circuit.

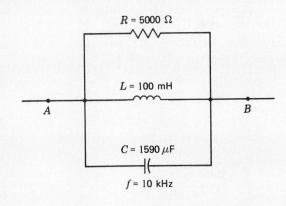

R = 5000 Ω

L = 100 mH

A   B

C = 1590 μF

f = 10 kHz

Figure 19-19

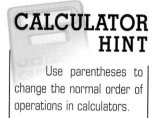

CALCULATOR HINT

Use parentheses to change the normal order of operations in calculators.

Solution

a. Find the reactances.

$$X_L = 6.28fL = 6.28 \times 10 \times 10^3 \times 100 \times 10^{-3}$$
$$X_L = 6280 \ \Omega$$

$$X_C = \frac{159{,}000}{f \times C} = \frac{159{,}000}{10 \times 10^3 \times 1590 \times 10^{-6}} = 10{,}000 \ \Omega$$

Find the branch currents. If the voltage is unknown, it may be assumed to be any value. For ease in calculation, it should be equal to, or greater than, the largest impedance. Assume $V_{AB} = 10{,}000$ V. Since

$$V_{AB} = V_R = V_L = V_C$$

$$V_R = I_R \times R \qquad\qquad V_L = I_L \times X_L$$
$$10{,}000 = I_R \times 5000 \qquad\qquad 10{,}000 = I_L \times 6280$$
$$I_R = 2 \text{ A} \qquad\qquad\qquad I_L = 1.59 \text{ A}$$

$$V_C = I_C \times X_C$$
$$10{,}000 = I_C \times 10{,}000$$
$$I_C = 1 \text{ A}$$

Find the total current.

$$I_T{}^2 = I_R{}^2 + (I_L - I_C)^2 \qquad\qquad (19\text{-}4)$$
$$I_T{}^2 = 2^2 + (1.59 - 1)^2$$
$$= 4 + 0.348$$
$$I_T{}^2 = 4.348$$
$$I_T = \sqrt{4.348} = 2.08 \text{ A}$$

Find the impedance.

$$V_T = I_T \times Z$$
$$10{,}000 = 2.08 \times Z$$
$$Z = 4800 \ \Omega \qquad Ans.$$

b. Find the phase angle.

$$\cos \theta = \frac{I_R}{I_T} = \frac{2}{2.08} \qquad\qquad (19\text{-}2)$$
$$\cos \theta = 0.9615$$
$$\theta = 16° \text{ lagging} \qquad Ans.$$

c. Find the equivalent series circuit.

The impedance from A to B is 4800 Ω with the current lagging the voltage by 16°. If the circuit between A and B had been a *series circuit*, then the 4800 Ω of impedance would have been formed from an impedance triangle as shown in Fig. 19-20b, containing resistance and inductance only because of the *lagging* current of 16°.

1. Find the equivalent series resistance. In Fig. 19-20b,

$$\cos 16° = \frac{R}{4800} \qquad\qquad (14\text{-}3)$$
$$R = 4800 \times \cos 16°$$
$$R = 4800 \times 0.9615 = 4615 \ \Omega \qquad Ans.$$

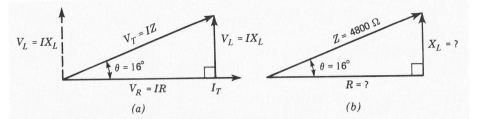

Figure 19-20 (a) I_T lags V_T by 16° in the series circuit of R and L. (b) R and X_L in series are equivalent to the parallel impedance of 4800 Ω.

2. Find the equivalent inductance. In Fig. 19-20b,

$$\sin 16° = \frac{X_L}{4800}$$
$$X_L = 4800 \times \sin 16°$$
$$X_L = 4800 \times 0.2756 = 1323 \ \Omega$$

Also, $\quad\quad X_L = 6.28fL$ $\hspace{3cm}$ (16-1)
$$1323 = 6.28 \times 10 \times 10^3 \times L$$
$$L = 0.021 \ H = 21 \ mH \quad\quad Ans.$$

Therefore, the equivalent series circuit consists of a resistance of 4615 Ω in series with an inductance of 21 mH. This circuit would produce the same load as the original parallel circuit.

Exercises

Applications

1. Find the series equivalent circuit for the circuit shown in Fig. 19-13 for Example 19-4.
2. Find the series equivalent circuit for the circuit shown in Fig. 19-18 for Example 19-6.
3. A 24-Ω resistor, a 6-Ω inductive reactance, and a 15-Ω capacitive reactance are in parallel across a 120-V 60-Hz line. Find (a) the total current, (b) the impedance, (c) the phase angle, (d) the power factor, (e) the power drawn by the circuit, and (f) the series equivalent circuit.
4. Repeat Exercise 3 for a 30-Ω resistor, a 60-Ω inductive reactance, and a 40-Ω capacitive reactance across the same line in parallel.
5. A 50-Ω resistor, a 0.02-H coil, and a 3-μF capacitor are connected in parallel across a 100-V 1-kHz ac source. Find (a) the reactance of the coil and capacitor, (b) the current drawn by each branch, (c) the total current, (d) the impedance, (e) the phase angle, (f) the power factor, (g) the power drawn by the circuit, and (h) the series equivalent circuit.
6. Repeat Exercise 5 for a 2200-Ω resistor, a 20-H coil, and a 0.8-μF capacitor in parallel across a 220-V 60-Hz ac line.
7. In a circuit similar to that shown in Fig. 19-19, $R = 1590 \ \Omega$, $L = 0.16 \ H$, and $C = 0.1 \ \mu$F. Find the series equivalent circuit.

(See CD-ROM for Test 19-2)

Resolution of Phasors

Up to this point we have been considering only circuits that contain only "pure" inductances and capacitances. Actually, such pure components do not exist. There is always some resistance in every coil or capacitor. This resistance must be taken into account whenever the resistance is an appreciable value as compared with the reactance of the component. In addition, most motor loads may be considered to be a series combination of resistance and inductance or resistance and capacitance. For example, an *induction motor* may be considered to be a series combination of resistance and inductance in which the current *lags* behind the impressed voltage. A *synchronous motor* may be considered to be a series combination of resistance and capacitance in which the current *leads* the impressed voltage. The amount of lead or lag depends on the relative amounts of resistance in series with the inductance or capacitance.

These leading and lagging currents will not be out of phase by exactly 90° but may be out of phase by *any* angle. For example, the current in branch *A* of a parallel circuit may *lead* the total voltage by 30° and the current in branch *B* may *lag* the total voltage by 50°. The phasor diagram for this condition is shown in Fig. 19-21.

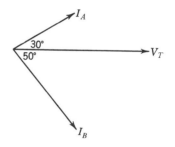

Figure 19-21   Leading and lagging branch currents in a parallel circuit.

Hint for Success

The mathematics of phasors and the mathematics of coordinate transformation are the same because they have a common mathematical basis.

Hint for Success

Note that $V^2 = V_x^2 + V_y^2$ refers to *vectors*, not *voltage*.

CALCULATOR HINT

The resolution of a phasor can be done using the rectangular-polar transformation built into most calculators.

The total current will still be the *phasor* sum of I_A and I_B. However, the pythagorean theorem may not be used because the angle between the phasors is no longer 90°. The phasor addition may be accomplished if we first resolve each phasor into its component parts which *are* 90° out of phase. Consider the phasor V in Fig. 19-22.

By the pythagorean theorem, V is the phasor sum of V_x and V_y,

$$V^2 = V_x^2 + V_y^2$$
$$V^2 = 3^2 + 4^2 = 9 + 16 = 25$$
$$V = \sqrt{25} = 5$$

When the process is reversed, a phasor V may be *resolved* into its two components whose phasor sum will be equal to the original phasor. These components are always at right angles to each other.

V_x is called the *horizontal* or x component of V.
V_y is called the *vertical* or y component of V.

The value of each component depends on the value of the total phasor and on the angle θ between the phasor and the X axis. The relationships between the components and the phasor are determined by the basic definitions of the sine and cosine of angle θ. For example, in Fig. 19-22,

$$\sin \theta = \frac{V_y}{V} \quad \text{and} \quad \cos \theta = \frac{V_x}{V}$$

Cross-multiplying each equation yields the

$$V_y = V \sin \theta \qquad \qquad \text{Formula 19-5}$$

$$V_x = V \cos \theta \qquad \qquad \text{Formula 19-6}$$

The rectangular components V_x and V_y of the phasor V in Fig. 19-22 can now be found.

$$V_y = V \sin \theta \qquad \qquad V_x = V \cos \theta$$
$$V_y = 5 \times \sin 37° \qquad \qquad V_x = 5 \times \cos 37°$$
$$V_y = 5 \times 0.6018 = 3 \qquad V_x = 5 \times 0.7986 = 4$$

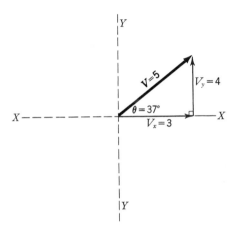

Figure 19-22   A phasor (or vector) V may be resolved into its horizontal component V_x and its vertical component V_y.

Example 19-9

A 10-lb force acts up and to the right at an angle of 30° with the horizontal. Find its horizontal and vertical components.

Solution

The diagram for the problem is shown in Fig. 19-23.

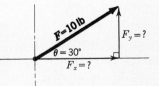

Figure 19-23

$$F_y = F \sin \theta \qquad (19\text{-}5)$$
$$F_y = 10 \sin 30°$$
$$F_y = 10(0.5000) = 5 \text{ lb} \qquad Ans.$$

$$F_x = F \cos \theta \qquad (19\text{-}6)$$
$$F_x = 10 \cos 30°$$
$$F_x = 10 (0.8660) = 8.66 \text{ lb} \qquad Ans.$$

Example 19-10

A sled is being pulled with a force of 50 lb that is exerted through a rope held at an angle of 20° with the ground as shown in Fig. 19-24. How much of this force is useful in moving the sled horizontally?

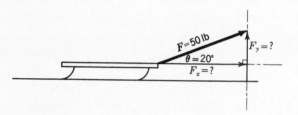

Figure 19-24

Solution

Only the horizontal component F_x is useful in moving the sled horizontally.

$$F_x = F \cos \theta \qquad (19\text{-}6)$$
$$F_x = 50 \cos 20°$$
$$F_x = 50(0.9397) = 47 \text{ lb} \qquad Ans.$$

Example 19-11

A window pole is used to pull down a window. If the force exerted through the pole is 30 lb at an angle of 72° with the horizontal, find the useful vertical component.

Solution

The diagram for the problem is shown in Fig. 19-25.

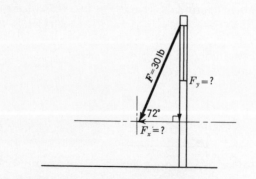

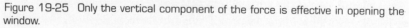

Figure 19-25 Only the vertical component of the force is effective in opening the window.

$$F_y = F \sin \theta \qquad (19\text{-}5)$$
$$F_y = 30 \sin 72°$$
$$F_y = 30(0.9511) = -28.5 \text{ lb} \qquad Ans.$$

Note: The minus sign in the answer is used to indicate that the force is acting *downward*. The direction in which a component acts is indicated by either a plus (+) or a minus (−) sign. These signs are the same as those used to locate points on a graph.

V_x components acting to the *right* are +.
V_x components acting to the *left* are −.
V_y components acting *upward* are +.
V_y components acting *downward* are −.

Example 19-12

A current of 20 A in one branch of an ac circuit leads the total voltage by 40°. Find its "in-phase" current and its "reactive" current.

Solution

The phasor diagram is shown in Fig. 19-26a.

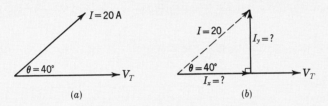

Figure 19-26 (a) Phasor diagram for a leading current. (b) The current *I* is resolved into its components I_x and I_y.

1. Resolve the current into its components I_x and I_y as shown in Fig. 19-26b. I_x is the in-phase current, since it acts in the same direction as the total voltage V_T. I_y is the reactive current, since it leads the total voltage V_T by 90°.
2. Find each component.

$$I_x = I \cos \theta \qquad (19\text{-}6) \qquad\qquad I_y = I \sin \theta \qquad (19\text{-}5)$$
$$I_x = 20 \cos 40° \qquad\qquad\qquad\quad I_y = 20 \sin 40°$$
$$I_x = 20(0.766) = 15.32 \text{ A} \qquad\quad I_y = 20(0.6428) = 12.86 \text{ A} \qquad Ans.$$

Example 19-13

A current of 10 A supplying an induction motor lags the voltage by 53°. Find the in-phase and reactive currents.

Solution

The phasor diagram is shown in Fig. 19-27a on page 678.

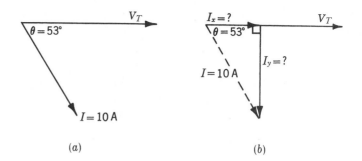

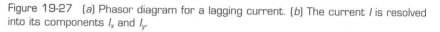

Figure 19-27 (a) Phasor diagram for a lagging current. (b) The current *I* is resolved into its components I_x and I_y.

1. Resolve the current into its in-phase component I_x and its reactive component I_y as shown in Fig. 19-27*b*.
2. Find each component.

$I_x = I \cos \theta$ (19-6) $I_y = I \sin \theta$ (19-5)

$I_x = 10 \cos 53°$ $I_y = 10 \sin 53°$

$I_x = 10(0.6018) = 6$ A $I_y = 10(0.7986) = -8$ A   *Ans.*

Exercises

Practice

Find the in-phase current I_x and the reactive current I_y for each of the following currents.

Exercise	I, A	θ
1	20	30° leading
2	10	35° leading
3	20	60° lagging
4	10	55° lagging
5	26	42° leading
6	1.8	23° leading
7	0.1	50° lagging
8	4.5	19° lagging
9	14	25° leading
10	31	28° lagging

JOB 19-5

Power Filters and Ground Loops

This job has two objectives. One is to introduce the concept of "clean" power and to show how to use filters to get clean power. The second is to introduce concepts of circuit analysis that emphasize ideas and qualitative analysis rather than numbers. As technical persons we more often deal with ideas because the costs of computation are too high. These costs are not always in dollars, but rather in time.

Alternating current power in the United States implies a nice, clean 60-Hz sine wave. In reality, there is a lot of "noise" on power lines. Noise is usually caused by motors and electronic equipment connected to the power lines. It can also be induced magnetically or capacitively into the lines from surrounding equipment and lines. Sometimes RF sources such as cellular phones can also induce signals into the lines. Other sources of noise are fluorescent lights, high-speed power-switching elements such as SCRs, thyristors, solenoids, or relays, etc. This noise is called *electromagnetic interference*, or EMI. It is sometimes called *radio frequency interference*, or RFI. The concepts are the same, but RFI includes higher frequencies than EMI. There is some overlap, but in general EMI is lower in frequency than RFI.

In the United States, because of the "radio" nature of EMI, it is regulated by the FCC (Federal Communications Commission). The European regulation agency is VDE (Verband Deutscher Elektrotechniker). Other regulators are UL, CSA, TUV, and IEC.

There are two types of EMI in the FCC regulations. One is radiated EMI. This type comes out of the device and goes through the air and could couple into the power lines. The second type is conducted EMI. This type is noise that is in the power lines. This job is concerned with conducted EMI. The FCC limits are shown in Table 19-1.

> **Hint for Success**
>
> Filters are a practical aspect of theory that is becoming more important as power quality becomes more important.

Table 19-1

FCC EMI Limits

Frequency (MHz)	Max RF Line Voltage (dB Above 1 μV)	
	Class A	Class B
0.45–1.6	60	48
1.6 –30	69.5	48

EMI that gets into electronic equipment through the power lines can cause the equipment to malfunction. Television sets and computers have EMI notices saying that the attached equipment complies with FCC Part A or FCC Part B. The manufacturers of industrial computers and digital control systems may specify EMI requirements for the input power beyond the FCC requirements.

The severity of the problem depends on the purpose for which the equipment is used and where it is used. FCC Class A systems include computing devices marketed for use in commercial, industrial, or business environments. Class B systems include equipment marketed for use at home as well as in the office.

One way to meet EMI requirements is to place a filter in the power line at the input to the electronic equipment. Filter effects are frequency-dependent effects. Basic ac power is at 60 Hz. The noise is at much higher frequencies, up to megahertz. These higher frequencies are closer to the operating frequencies of computer circuits and can interfere with them because the noise looks like data to the computer. The noise signals can also interfere with video signals, such as in television. Very low levels of noise can come in through the power supply and create lines and patterns on video displays. The eye is very sensitive and can see these very faint patterns.

Electrical circuits that selectively pass or eliminate voltages of different frequencies are called filters. There are three common types of frequency characteristics: low-pass, high-pass, and band-pass filters. These characteristics are shown in Fig. 19-28. A low-pass filter will pass low frequencies and block high frequencies. A high-pass filter will pass high frequencies and block low frequencies. If you put a low-pass and a high-pass filter together, then the low frequencies are blocked, some frequencies in the middle get through, and the high frequencies are again blocked. Only frequencies in a "band" in the middle would get through, so this is called a *band-pass filter*.

For power distribution we want the power at 60 Hz to get through to the load. We do not want the noise at high frequencies to get through. Thus, we want a low-pass filter. There are many types of low-pass filters. Two of the most basic filter configurations are called T and π. These filters are shown in Fig. 19-29a.

It is apparent that many calculations are required to evaluate the frequency response of either a T or a π circuit. Values for the entire circuit would have to be calculated at many frequencies; the results would then have to be plotted. For circuits of this complexity, you should use special mathematics and/or circuit analysis programs. The materials for this job were analyzed using PSpice. The results showed that mathematically describing these circuits is far too complicated for them to be presented properly in a text of this type. To do such calculations without these special tools is so time-consuming that it is useless in the workplace.

If we think of mathematics only as numbers and computation, then expensive mathematical computations have little value in the workplace. This is especially true for complex subjects such as *RLC* circuits, for which the computation costs are very high. What we need is a way to carry classroom study to field applications with mental "pictures" of the mathematics and the concepts it represents.

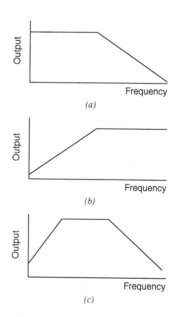

Figure 19-28   (*a*) Low-pass filter. (*b*) High-pass filter. (*c*) Band-pass filter.

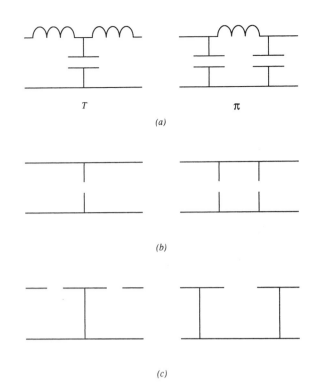

T   π

(a)

(b)

(c)

Figure 19-29 (a) Low-pass filters. (b) Low-frequency model. (c) High-frequency model.

The engineering or scientific mathematics may be complicated, but technological understanding can come from the analysis of approximate circuits. The secret of this concept is that as the frequency gets higher, the impedance of an inductor gets larger and larger and eventually looks like an open circuit compared to the rest of the circuit. Similarly for a capacitor, as the frequency gets higher, the impedance gets lower, and the capacitor starts to look like a short circuit.

Figure 19-29 shows how to apply this concept to the T and π circuits. The elements in a T circuit are in the shape of a T and the elements of a π circuit form a π. The names for the circuits come from the shape of the circuits.

At low frequencies with the inductors shorted out and the capacitors open, the T and π circuits become those in Fig. 19-29b. The circuits look like two wires connecting the input to the output. Thus, at low frequencies there is no loss.

At high frequencies, the capacitors are short circuits and the inductors are open circuits. Figure 19-29c shows these equivalent high-frequency circuits. You can see that the lower line is still continuous and connects to the output. This is good because grounding is still in effect. We have not created a safety problem. The high line, however, is now broken by the open inductors. In addition, the capacitors short out the two power lines. This combination of open and short circuits means that no voltages will get from the input to the output. Hence both the T and π circuits of Fig. 19-29 are low-pass filters. The frequency characteristics of the circuits look like Fig. 19-28a.

Notice how we developed these ideas by using mathematical concepts and simple mental calculations. This type of analysis can also be used to check more complicated calculations or to evaluate product information.

A new term, *insertion loss*, is used to specify EMI filters. This is a new concept to the power distribution field. It was used to evaluate filter performance in radio and audio systems and was carried over to other electronics areas. Now we see it in devices intended to be installed in power feeder circuits.

Figure 19-30 shows the concepts used in insertion loss. Figure 19-30*a* is the basic circuit without any device inserted into it. The voltage source creates a current I flowing through the load R_L. In Fig. 19-30*b*, the device is put into the circuit. The current through the load is now I_2. The insertion loss is defined as the ratio of I_2 to I. Insertion loss is usually expressed in decibels.

$$\text{Insertion loss} = \frac{I_2}{I}$$

$$\text{Insertion loss (dB)} = -20 \log \frac{I_2}{I}$$

The minus sign is used in the dB definition so that the loss will be expressed as a positive number rather than a negative dB value.

The following two examples give further details on using these concepts. At the technology level, the objective is not to design, research, or even analyze complex *RLC* circuits. Your responsibility is to use the devices properly, test if the systems are working properly, and—if they are not working properly—determine why they are not. This is a subtle but very important difference in perspective. The designer has to be concerned with only one thing: what the circuit should do. When the technician is called in, there is only one thing that does not have to be considered: what the circuit should do. If it were doing what it should be doing, you would not be there.

Part of the information that the technology person should be given is how the circuit is supposed to perform. Only the designer knows this, and it should be provided to you. Sometimes you are not given this information. The information provided is often designed to sell the product and/or protect the interests of the manufacturer. The important question is, Can you operate properly with the information that is provided? The answer is often no, which becomes evident in a

CALCULATOR HINT

When using log calculations that have computed arguments, be sure to include the computation in parentheses, for example, log(v/i).

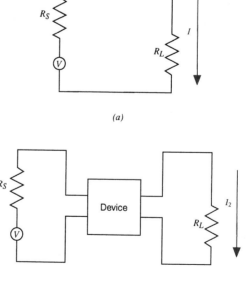

(a)

(b)

Figure 19-30 (a) Basic circuit. (b) Circuit with device inserted.

round-about way when the manufacturer complains that people are misusing the equipment or using the wrong devices. The real problem is that the information provided is not adequate for users to deploy the devices properly.

A Real EMI Filter

Example 19-14

Figure 19-31 shows the diagram given by the manufacturer for a commercial π-type EMI filter. The specifications for one model rate the device at 30.0 A, 250 V ac, and 50/60 Hz, with a leakage current of 0.45 mA. The test voltages are line to line, 1.5 kV dc, and line to ground, 2.2 kV dc.

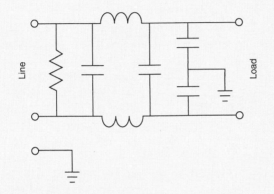

Figure 19-31   Circuit for commercial EMI filter.

The insertion loss for an EMI filter can be measured either between the two lines or between a line and ground. This is similar to 120/240 single-phase types of measurements. Measurements made between a line and ground (*L/G*) are called *common mode*. This terminology comes from electronic amplifier noise concepts. It is assumed that noise is generated between a line and ground just as the power is generated between a line and ground. The two voltages have a common origin, which leads to the term common mode.

Measurements between two lines (*L/L*) are called *differential* measurements. If it is assumed that the noise is generated between a line and ground, then if you measure the noise voltage between two lines you would be measuring the difference between the two independent common mode noises. This leads to the term differential and hence differential insertion loss.

The catalog provides the information in Table 19-2 about the insertion loss of the filter. The data have been plotted in Fig. 19-32. It can be seen either by Fig. 19-32 or by looking at the diagram of the circuit in Fig. 19-31 that at high frequencies the output of the filter goes to zero. There is nothing in the circuit model to say that the output would level off or even rise. The insertion loss data provided by the manufacturer show a rapid decline in the output and then a leveling off of the output noise. The decline is in accord with the model of the filter. The leveling off is not. What this says is that the majority of the insertion loss data provided by the company has nothing to do with

Table 19-2

EMI Insertion Loss Data

	Insertion Loss					
	0.05 MHz	**0.15 MHz**	**0.50 MHz**	**1.5 MHz**	**5.0 MHz**	**20 MHz**
L/L	8 dB	12 dB	45 dB	65 dB	65 dB	60 dB
L/G	10 dB	20 dB	32 dB	40 dB	45 dB	50 dB

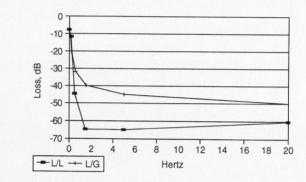

Figure 19-32 EMI filter insertion loss.

the circuit model provided. It is provided only as an indication that their device meets the FCC requirements in Table 19-1. But does it? The data are for insertion loss and the standards are in decibels above 1 μV. One is a relative measurement. The other is an absolute voltage level.

Summary

We first learned the terms *differential* and *common mode* as applied to power filters. Differential measurements are those made between the two high-voltage lines. Common mode measurements are made between one hot line and ground.

We then showed how mathematical concepts relating to inductors and capacitors can be used to simplify filter system analysis. From these simplified circuits, we were able to draw generalized performance curves without the need for lots of computation.

Even though the performance results are obtained with little effort, they can be used to evaluate company data sheets. Our example showed that there is very little relationship between the provided performance "data" and the circuit we are buying. With the lack of additional information, it is difficult to maintain, test, or evaluate a device.

An EMI Test System

Example 19-15

This example is a circuit that has been used by the FCC to test conducted emissions from test devices. Conducted emissions are noises that can come from inside the device by connecting through the power supply to the feeder lines. At frequencies above 15 MHz, the noise could be coupled to the test device's power lines from other cables that are part of the device.

Figure 19-33*a* shows how the power feeds into the test box and then to the device being tested. There are two test points on the test box. One is used to measure voltages between the neutral line and ground. The other test point measures the voltage between the hot line and ground. The metal box is grounded for safety and to eliminate noise that may be induced into the box itself. Remember that a lot of the voltages being tested are very small, but they can be lethal to electronic systems.

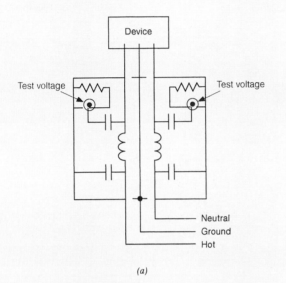

(a)

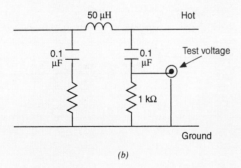

(b)

Figure 19-33 (a) FCC test system. (b) Circuit values.

Both sides of the test circuit are the same. One of them is redrawn in Fig. 19-33*b*. This circuit looks almost like a π filter. For the power lines, the circuit behaves as a π filter. The circuit behavior is somewhat different for the test voltage. Figure 19-34 shows the simplified circuits at various frequency ranges. The definitions of these frequency ranges depend on the design of the circuit. Figure 19-35 shows the results of a PSpice simulation of the test circuit; it is graphed using Quattro Pro. This information helps you recognize the three important frequency ranges for which the circuit was designed.

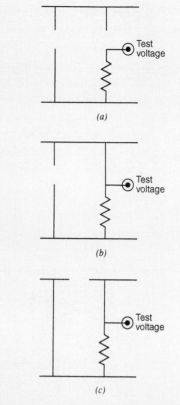

Figure 19-34　(a) Low frequencies. (b) Mid-frequencies. (c) High frequencies.

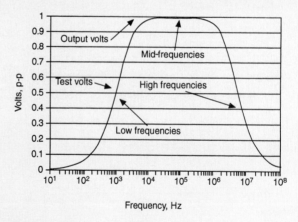

Figure 19-35　EMI test stand.

The reactance of inductors and capacitors changes as the frequency changes in a circuit. At low frequencies the reactance of an inductor is very low. It is much lower than the impedance of the rest of the circuit. Compared to the rest of the circuit, it looks like a short circuit.

As the frequency gets higher the impedance gets higher until the impedance of the inductor is much larger than the impedance of the rest of the circuit. At this point the inductor looks like an open circuit.

A capacitor works in a similar way except that at low frequencies it looks like an open circuit while at high frequencies it looks like a short circuit. This is the opposite of an inductor.

In some circuits we also have to look at combinations of reactive components but we will not consider that here. We will think only of inductors and capacitors as individual components.

The frequency at which an inductor or capacitor changes from being a short circuit to an open circuit is called the *break frequency* or corner frequency for that component. It is common practice to put each break at a distinct frequency as in Fig. 19-35. How many breaks are there in the range of frequencies we are interested in and where are they?

The ideas presented here simplify complex filters so that we can more easily analyze and understand how they operate. The results are approximate, easy to calculate, but useful. We will think of each reactive element as either an open circuit or a short circuit. They are switches that are either on or off. If you think of them as switches on the wall, how many combinations of on and off conditions can you have?

In our filter there are three reactive elements or switches. The first switch can be on or off; 2 states. The second switch is independent of the first and it can be on or off; 2 states. The third switch can also have 2 states. The total number of possible states is

$$2 \times 2 \times 2 = 2^3 = 8$$

states.

The frequency response curve in Fig. 19-35 does not show eight different frequency ranges. It only shows three; low frequencies, mid frequencies, and high frequencies. How do you know that there are only three significant ranges and how do you know where they are? What about the other 5 states? Even a simple three reactive element circuit can have tremendous complexity.

It is not your responsibility to know what the circuit designer had in mind. You should be told. You are not designing the circuit. You are only using and/or testing it. The manufacturer should give you enough information that you will be able to apply the analysis ideas shown here and determine if the circuit is working properly.

For low frequencies—i.e., below about 10 kHz—the input voltage goes directly to the output with no loss. This allows the power to be supplied to the device under test. At the same time the capacitor isolates the test voltage from the input. This action also isolates the test voltage from the power signal, which is very large compared to the noise. This approximate circuit is shown in Fig. 19-34a.

In the midfrequencies, between about 10 kHz and 10 MHz, the output and the test voltages are connected directly to the input device, as shown in Fig. 19-34b. This is the primary test range of the system. Notice that this frequency range corresponds to the frequencies restricted for conducted EMI by the FCC.

At high frequencies, Fig. 19-34c shows that the test voltage is still tied directly to the output voltage, but both of them are isolated from the input.

Summary

This example shows how to simplify complex *RLC* circuits in order to understand their operation as a function of frequency. The inductors and capacitors are treated as switches. The circuit operation is divided into frequency ranges. In each range there is a combination of *LC* "switch" settings. The resulting approximate circuit is then analyzed. This technique simplifies the analysis because it removes the inductors and capacitors and leaves only resistive circuits.

How did I know the correct set of *LC* switch settings? Simple! The designer did not tell me so I had to use PSpice to analyze the circuit. Then I knew what the designer had in mind. That is why the designer should tell you. Otherwise you have to guess!

Why didn't I use other combinations of *LC* switches? I did vary the circuit values and the model and found that indeed they could be used. They produced results that were peculiar—like those you might be called in to fix.

Exercises

Practice

1. What are EMI and RFI abbreviations for?
2. What are some causes of EMI and RFI?
3. What are the two types of EMI recognized by the FCC?
4. What are the FCC classes for EMI?
5. At high frequencies what do *L* and *C* look like?
6. At low frequencies what do *L* and *C* look like?
7. What is insertion loss?
8. What is common mode?
9. What is differential mode?
10. What is break frequency?

Hint for Success

When circuits get more complicated, it is useful to learn more advanced computation techniques and to use specialized computer software.

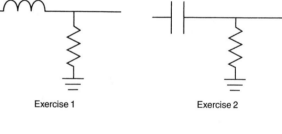

Exercise 1   Exercise 2

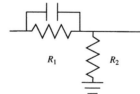

R_1 R_2 R_1 R_2

Exercise 3   Exercise 4

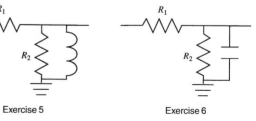

R_1 R_1

R_2 R_2

Exercise 5   Exercise 6

Figure 19-36   Circuits for Exercises 1 to 6.

Applications

11. Sketch the frequency response of each circuit in Fig. 19-36. Show the ratio of output voltage to input voltage for each circuit versus frequency. Tell whether each is a high-pass or a low-pass filter.

Parallel-Series AC Circuits

The current in any branch of an operating ac circuit is never exactly 90° out of phase with the voltage. There is always some resistance in series with the capacitance or inductance that reduces the phase angle from 90° to almost any angle. These "out-of-phase" currents must be added by phasor addition to get the total current.

Addition of out-of-phase currents.   Let us try to find the total current in a parallel circuit if the current in branch A leads the voltage by 30° and the current in branch B leads the voltage by 60°.

Solution.   The phasor diagram for the conditions stated is shown in Fig. 19-37a.
   The total current is obtained by adding current I_B to current I_A *vectorially*. Place the tail of I_B on to the head of I_A and draw it in its original direction and length as shown in Fig. 19-37b. The distance from the origin of the phasors to the head of the final phasor is the sum of the phasors, or I_T.

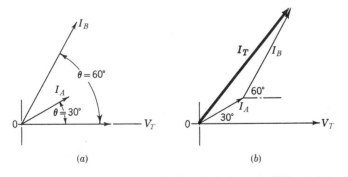

Figure 19-37   (a) Phasor diagram showing I_A leading the voltage by 30° and I_B leading the voltage by 60°. (b) I_T represents the phasor sum of I_A and I_B.

Figure 19-37b may be redrawn as shown in Fig. 19-38 on page 690 to show the x and y components of each current.

I_{A_x} and I_{B_x} are the in-phase components of I_A and I_B.
I_{A_y} and I_{B_y} are the reactive components of I_A and I_B.
I_{T_x} is the algebraic sum of all the x components.
I_{T_y} is the algebraic sum of all the y components.

$$I_{T_x} = I_{A_x} + I_{B_x} \qquad\qquad \text{Formula 19-7}$$

$$I_{T_y} = I_{A_y} + I_{B_y} \qquad\qquad \text{Formula 19-8}$$

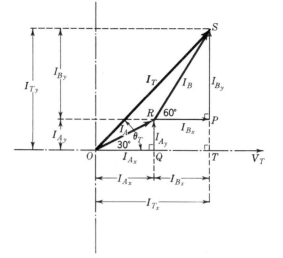

Figure 19-38 I_A and I_B have been resolved into their in-phase and reactive components. Their algebraic sums I_{T_x} and I_{T_y} are the components of the total current I_T.

By applying the pythagorean theorem to triangle OTS in Fig. 19-38, we obtain

$$I_T{}^2 = (I_{T_x})^2 + (I_{T_y})^2 \qquad \text{Formula 19-9}$$

and by trigonometry,

$$\cos \theta_T = \frac{I_{T_x}}{I_T} \qquad \text{19-10}$$

or

$$\text{PF} = \frac{I_{T_x}}{I_T} \qquad \text{19-11}$$

In actual practice, the resolution triangles ORQ and RSP are drawn on the same constant-voltage base and added as shown in the following example. This method eliminates the problem of complicated phasor diagrams resulting from combinations of leading and lagging phasors.

Example 19-16

In a parallel circuit, a current of 10 A in branch A leads the total voltage by 30°. A current of 20 A in branch B leads the total voltage by 37°. Find (*a*) the total current and (*b*) the angle by which the total current leads the total voltage.

Solution

a. 1. Draw the phasor diagram for the branch currents on the same voltage base as shown in Fig. 19-39*a*.
   2. Resolve the current in each branch into its components as shown in Fig. 19-39*b*. Calculate the components.

For branch A:

$$I_{A_x} = I_A \cos \theta \qquad (19\text{-}6)$$
$$= 10 \cos 30° = 10(0.866) = 8.66 \text{ A}$$
$$I_{A_y} = I_A \sin \theta \qquad (19\text{-}5)$$
$$= 10 \sin 30° = 10(0.5000) = 5 \text{ A}$$

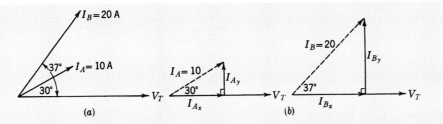
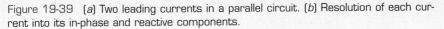

Figure 19-39 (a) Two leading currents in a parallel circuit. (b) Resolution of each current into its in-phase and reactive components.

For branch B:

$$I_{B_x} = I_B \cos \theta \qquad (19\text{-}6)$$
$$= 20 \cos 37° = 20(0.7986) = 16 \text{ A}$$
$$I_{B_y} = I_B \sin \theta \qquad (19\text{-}5)$$
$$= 20 \sin 37° = 20(0.6018) = 12 \text{ A}$$

3. Draw all the components on the same voltage base as shown in Fig. 19-40a. The total in-phase current I_{T_x} and the total reactive current I_{T_y} may now be found.

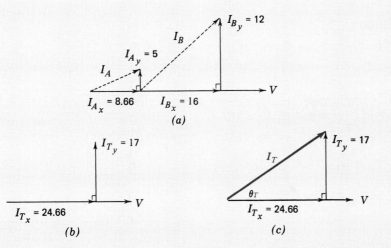

Figure 19-40 (a) The components of all the branch currents drawn on the same voltage base. (b) $I_{T_x} = I_{A_x} + I_{B_x}$ and $I_{T_y} = I_{A_y} + I_{B_y}$. (c) I_T represents the phasor sum of I_{T_x} and I_{T_y}.

$$I_{T_x} = I_{A_x} + I_{B_x} \qquad (19\text{-}7)$$
$$= 8.66 + 16 = 24.66 \text{ A}$$
$$I_{T_y} = I_{A_y} + I_{B_y} \qquad (19\text{-}8)$$
$$= 5 + 12 = 17 \text{ A}$$

4. Draw these phasors as shown in Fig. 19-40b.
5. Draw the phasor diagram for the total current by adding I_{T_x} and I_{T_y} vectorially as shown in Fig. 19-40c.

Find I_T.

$$(I_T)^2 = (I_{T_x})^2 + (I_{T_y})^2 \tag{19-9}$$
$$= (24.66)^2 + (17)^2$$
$$= 608 + 289 = 897$$
$$I_T = \sqrt{897} = 30 \text{ A } \textit{leading} \qquad \textit{Ans.}$$

b. Find the phase angle.

$$\cos \theta_T = \frac{I_{T_x}}{I_T} \tag{19-10}$$

$$\cos \theta_T = \frac{24.66}{30} = 0.822$$

$$\theta_T = 35° \textit{ leading} \qquad \textit{Ans.}$$

Example 19-17

An induction motor of 5 Ω impedance draws a current lagging by 26°. It is in parallel with a synchronous motor of 12 Ω impedance which draws a current leading by 37°. If the applied voltage is 120 V at 60 Hz, find (*a*) the current drawn by each motor, (*b*) the total current drawn, (*c*) the impedance, (*d*) the phase angle, (*e*) the power factor, and (*f*) the power drawn by the circuit.

Solution

The diagram for the circuit is shown in Fig. 19-41.

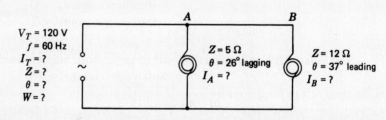

Figure 19-41

a. Find the current in each branch. Since

$$V_T = V_A = V_B = 120 \text{ V} \tag{5-1}$$

$$V_A = I_A \times Z_A \qquad\qquad V_B = I_B \times Z_B$$
$$120 = I_A \times 5 \qquad\qquad 120 = I_B \times 12$$

$$I_A = \frac{120}{5} = 24 \text{ A } \textit{lagging} \qquad I_B = \frac{120}{12} = 10 \text{ A } \textit{leading} \qquad \textit{Ans.}$$

b. The total current is found by adding the currents in the two branches. However, they are out of phase and must be added vectorially.
   1. Draw the phasor diagram for the branch currents on the same voltage base as shown in Fig. 19-42*a*.
   2. Resolve the current in each branch into its components as shown in Fig. 19-42*b*. Calculate the components.

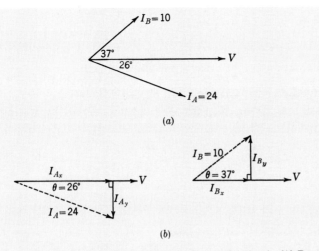

(a)

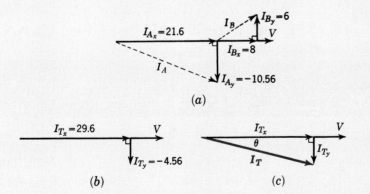

(b)

Figure 19-42 (a) Leading and lagging currents in a parallel circuit. (b) Resolution of each current into its in-phase and reactive components.

For branch A: $I_{A_x} = I_A \cos \theta$ (19-6)
                              $= 24 \cos 26° = 24(0.9) = 21.6$ A
                        $I_{A_y} = I_A \sin \theta$ (19-5)
                              $= 24 \sin 26° = 24(0.44) = -10.56$ A

The minus sign in I_{A_y} indicates a lagging reactive current.

For branch B: $I_{B_x} = I_B \cos \theta$ (19-6)
                              $= 10 \cos 37° = 10(0.8) = 8$ A
                        $I_{B_y} = I_B \sin \theta$ (19-5)
                              $= 10 \sin 37° = 10(0.6) = 6$ A

3. Draw all the components on the same voltage base as shown in Fig. 19-43a. The total in-phase current I_{T_x} and the total reactive current I_{T_y} may now be found.

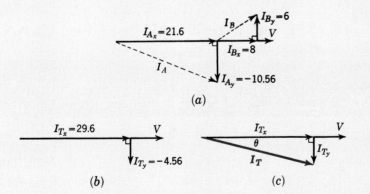

(a)

(b)   (c)

Figure 19-43 (a) The components of all the branch currents drawn on the same voltage base. (b) $I_{T_x} = I_{A_x} + I_{B_x}$ and $I_{T_y} = I_{A_y} + I_{B_y}$. (c) I_T represents the phasor sum of I_{T_x} and I_{T_y}.

$$I_{T_x} = I_{A_x} + I_{B_x}$$ (19-7)
$$= 21.6 + 8 = 29.6 \text{ A}$$

$$I_{T_y} = I_{A_y} + I_{B_y} \tag{19-8}$$
$$= -10.56 + 6 = -4.56 \text{ A}$$

The minus sign indicates a lagging reactive current.

4. Draw the phasors as shown in Fig. 19-43b.
5. Draw the phasor diagram for the total current by adding I_{T_x} and I_{T_y} vectorially as shown in Fig. 19-43c. Find I_T.

$$(I_T)^2 = (I_{T_x})^2 + (I_{T_y})^2 \tag{19-10}$$
$$(I_T)^2 = (29.6)^2 + (-4.56)^2$$
$$= 876.2 + 20.8 = 897$$
$$I_T = \sqrt{897} = 30 \text{ A } lagging \qquad Ans.$$

c. Find the impedance of the circuit.

$$V_T = I_T \times Z$$
$$120 = 30 \times Z$$
$$Z = \frac{120}{30} = 4 \ \Omega \qquad Ans.$$

d. Find the phase angle.

$$\cos \theta_T = \frac{I_{T_x}}{I_T} \tag{19-10}$$

$$\cos \theta_T = \frac{29.6}{30} = 0.986$$
$$\theta_T = 10° \ lagging \qquad Ans.$$

e. $$\text{PF} = \cos \theta_T = 98.6\% \qquad Ans.$$

f. Find the power consumed.

$$W = V \times I \times \cos \theta = 120 \times 30 \times \cos 10°$$
$$W = 120 \times 30 \times 0.986 = 3546 \text{ W} \qquad Ans.$$

In this parallel circuit, the total current of 30 A *lags* the total voltage of 120 V by 10°.

From this last example we can determine the procedure to follow in solving parallel-series ac circuits:

1. Find (*a*) the reactance, (*b*) the impedance, (*c*) the current, and (*d*) the phase angle for each branch of the parallel circuit.
2. Draw the phasor diagram for the branch currents on the same voltage base.
3. Resolve the current in each branch into its components.

$$I_y = I \times \sin \theta \tag{19-5}$$
$$I_x = I \times \cos \theta \tag{19-6}$$

4. Find the total in-phase current I_{T_x} and the total reactive current I_{T_y}.

$$I_{T_x} = I_{A_x} + I_{B_x} \tag{19-7}$$
$$I_{T_y} = I_{A_y} + I_{B_y} \tag{19-8}$$

5. Find the total current.

$$I_T^2 = (I_{T_x})^2 + (I_{T_y})^2 \tag{19-9}$$

6. Find the impedance.

$$V_T = I_T \times Z$$

7. Find the phase angle.

$$\cos \theta_T = \frac{I_{T_x}}{I_T} \qquad (19\text{-}10)$$

8. Find the power factor.

$$PF = \frac{I_{T_x}}{I_T} \qquad (19\text{-}11)$$

9. Find the total power.

$$W = V \times I \times \cos \theta$$

Example 19-18

An induction motor of 6 Ω resistance and 8 Ω inductive reactance is in parallel with a synchronous motor of 8 Ω resistance and 15 Ω capacitive reactance and a third parallel branch of 15 Ω resistance. Find (a) the total current drawn from a 150-V 60-Hz source, (b) the total impedance, (c) the phase angle, (d) the power factor, and (e) the power drawn by the circuit.

Solution

The diagram for the circuit is shown in Fig. 19-44.

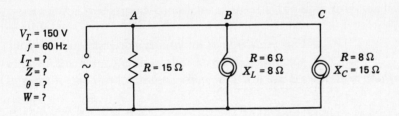

Figure 19-44

a. 1. Find the series impedance, current, and phase angle for each branch.

For branch A:

$$Z = R = 15 \ \Omega$$

$$I_A = \frac{V}{R} = \frac{150}{15} = 10 \text{ A} \qquad (2\text{-}1)$$

$$\cos \theta = \frac{R}{Z} = \frac{15}{15} = 1$$

Therefore,

$$\theta = 0°$$

The current is *in phase* with the voltage.

For branch B:

$$Z = \sqrt{R^2 + X_L{}^2} = \sqrt{6^2 + 8^2} = \sqrt{36 + 64} = \sqrt{100} = 10 \ \Omega$$

$$I_B = \frac{V}{Z} = \frac{150}{10} = 15 \text{ A}$$

$$\cos \theta = \frac{R}{Z} = \frac{6}{10} = 0.6000$$

Therefore, $\theta = 53°$

The current *lags* the voltage by 53°.

For branch *C*:

$$Z = \sqrt{R^2 + X_C^2} = \sqrt{8^2 + 15^2} = \sqrt{64 + 225} = \sqrt{289} = 17 \ \Omega$$

$$I_C = \frac{V}{Z} = \frac{150}{17} = 8.8 \text{ A}$$

$$\cos \theta = \frac{R}{Z} = \frac{8}{17} = 0.4706$$

Therefore, $\theta = 62°$

The current *leads* the voltage by 62°.

2. Draw the phasor diagram for the branch currents on the same voltage base as shown in Fig. 19-45*a*.

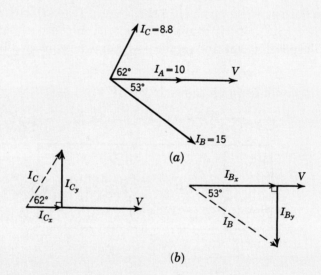

(a)

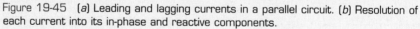

(b)

Figure 19-45 (*a*) Leading and lagging currents in a parallel circuit. (*b*) Resolution of each current into its in-phase and reactive components.

3. Resolve the current in each branch into its components as shown in Fig. 19-45*b*. Calculate the components.

For branch *A*:

$$I_{A_x} = I_A \times \cos \theta = 10 \times \cos 0° = 10 \times 1 = 10 \text{ A} \qquad (19\text{-}6)$$
$$I_{A_y} = I_A \times \sin \theta = 10 \times \sin 0° = 10 \times 0 = 0 \text{ A} \qquad (19\text{-}5)$$

For branch *B*:

$$I_{B_x} = I_B \times \cos \theta = 15 \times \cos 53° = 15 \times 0.6 = 9 \text{ A} \qquad (19\text{-}6)$$
$$I_{B_y} = I_B \times \sin \theta = 15 \times \sin 53° = 15 \times 0.8 = -12 \text{ A} \qquad (19\text{-}5)$$

Note: The minus sign in I_{B_y} indicates a lagging reactive component.

For branch C:

$$I_{C_x} = I_C \times \cos\theta = 8.8 \times \cos 62° = 8.8 \times 0.470 = 4.14 \text{ A} \quad (19\text{-}6)$$
$$I_{C_y} = I_C \times \sin\theta = 8.8 \times \sin 62° = 8.8 \times .88 = 7.74 \text{ A} \quad (19\text{-}5)$$

4. Draw all the components on the same voltage base as shown in Fig. 19-46a. The total in-phase current I_{T_x} and the total reactive current I_{T_y} may now be found.

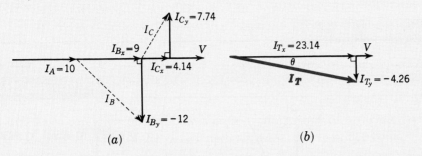

(a)   (b)

Figure 19-46 (a) The components of all the branch currents drawn on the same voltage base. (b) I_T represents the phasor sum of I_{T_x} and I_{T_y}.

$$I_{T_x} = I_{A_x} + I_{B_x} + I_{C_x} \quad (19\text{-}7)$$
$$= 10 + 9 + 4.14 = 23.14 \text{ A}$$
$$I_{T_y} = I_{A_y} + I_{B_y} + I_{C_y} \quad (19\text{-}8)$$
$$= 0 + (-12) + 7.74 = -4.26 \text{ A}$$

Note: The minus sign in I_{T_y} indicates a lagging reactive component.

5. Draw the phasor diagram for the total current by adding I_{T_x} and I_{T_y} vectorially as shown in Fig. 19-46b. Notice that I_{T_y} is drawn downward because I_{T_y} is negative. Find I_T.

$$I_T{}^2 = (I_{T_x})^2 + (I_{T_y})^2 \quad (19\text{-}9)$$
$$= (23.14)^2 + (-4.26)^2$$
$$= 535.5 + 18.1 = 553.6$$
$$I_T = \sqrt{533.6} = 23.5 \text{ A } lagging \qquad Ans.$$

b. Find the impedance.

$$V_T = I_T \times Z$$
$$150 = 23.5 \times Z$$
$$Z = \frac{150}{23.5} = 6.38 \ \Omega \qquad Ans.$$

c. Find the phase angle.

$$\cos\theta_T = \frac{I_{T_x}}{I_T} \quad (19\text{-}10)$$
$$\cos\theta_T = \frac{23.14}{23.5} = 0.9847$$

Therefore, $\qquad \theta_T = 10° \ lagging \qquad Ans.$

d. $\qquad\qquad\quad \text{PF} = \dfrac{I_{T_x}}{I_T} = 0.9847 = 98.5\% \ lagging \qquad Ans.$

e. Find the power drawn by the circuit.

$$W = V \times I \times \cos\theta = 150 \times 23.5 \times \cos 10° \quad (18\text{-}3)$$
$$W = 150 \times 23.5 \times 0.985 = 3472 \ W \qquad Ans.$$

Check:

$$W_A = I_A{}^2 \times R_A = (10)^2 \times 15 = 1500 \text{ W}$$
$$W_B = I_B{}^2 \times R_B = (15)^2 \times 6 = 1350 \text{ W}$$
$$W_C = I_C{}^2 \times R_C = (8.8)^2 \times 8 = \underline{620 \text{ W}}$$
$$W_T = W_A + W_B + W_C = \overline{3470 \text{ W}} \quad \quad \textit{Check}$$

The total current of 23.5 A *lags* the total voltage of 150 V by 10°.

Example 19-19

For the circuit shown in Fig. 19-47, find (*a*) the total current, (*b*) the total impedance, (*c*) the phase angle, (*d*) the power factor, and (*e*) the power drawn by the circuit.

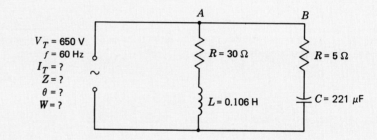

Figure 19-47

Solution

a. 1. Find the reactance, series impedance, current, and phase angle for each branch.

For branch *A:*

$$X_L = 6.28fL = 6.28 \times 60 \times 0.106 = 40 \ \Omega \quad \quad (16\text{-}1)$$
$$Z = \sqrt{R^2 + X_L{}^2} = \sqrt{30^2 + 40^2}$$
$$= \sqrt{900 + 1600}$$
$$Z = \sqrt{2500} = 50 \ \Omega$$
$$I_A = \frac{V}{Z} = \frac{650}{50} = 13 \text{ A}$$
$$\cos \theta = \frac{R}{Z} = \frac{30}{50} = 0.6000$$

Therefore,
$$\theta = 53°$$

The current of 13 A lags the total voltage by 53°.

For branch *B:*

$$X_C = \frac{159,000}{f \times C} = \frac{159,000}{60 \times 221} = 12 \ \Omega \quad \quad (17\text{-}6)$$
$$Z = \sqrt{R^2 + X_C{}^2} = \sqrt{5^2 + 12^2} \quad \quad (18\text{-}3)$$
$$= \sqrt{25 + 144}$$
$$Z = \sqrt{169} = 13 \ \Omega$$

$$I_B = \frac{V}{Z} = \frac{650}{13} = 50 \text{ A}$$

$$\cos \theta = \frac{R}{Z} = \frac{5}{13} = 0.3846$$

Therefore,
$$\theta = 67°$$

The current of 50 A leads the total voltage by 67°.

2. Draw the phasor diagram for the branch currents on the same voltage base as shown in Fig. 19-48*a*.

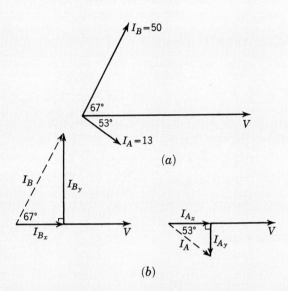

(a)

(b)

Figure 19-48 (*a*) Leading and lagging currents in a parallel circuit. (*b*) Resolution of each current into its in-phase and reactive components.

3. Resolve the current in each branch into its components as shown in Fig. 19-48*b*. Calculate the components.

For branch *A*:

$$I_{A_x} = I_A \times \cos \theta = 13 \times \cos 53° = 13 \times 0.6 = 7.8 \text{ A} \qquad (19\text{-}6)$$
$$I_{A_y} = I_A \times \sin \theta = 13 \times \sin 53° = 13 \times 0.8 = -10.4 \text{ A} \qquad (19\text{-}5)$$

For branch *B*:

$$I_{B_x} = I_B \times \cos \theta = 50 \times \cos 67° = 50 \times 0.385 = 19.25 \text{ A} \qquad (19\text{-}6)$$
$$I_{B_y} = I_B \times \sin \theta = 50 \times \sin 67° = 50 \times 0.92 = 46 \text{ A} \qquad (19\text{-}5)$$

4. Draw all the components on the same voltage base as shown in Fig. 19-49*a* on page 700. The total in-phase current I_{T_x} and the total reactive current I_{T_y} may now be found.

$$I_{T_x} = I_{A_x} + I_{B_x} = 7.8 + 19.25 = 27.05 \text{ A} \qquad (19\text{-}7)$$
$$I_{T_y} = I_{A_y} + I_{B_y} = -10.4 + 46 = 35.6 \text{ A} \qquad (19\text{-}8)$$

5. Draw the phasor diagram for the total current by adding I_{T_x} and I_{T_y} vectorially as shown in Fig. 19-49*b*. Notice that I_{T_y} is drawn *upward* because I_{T_y} is positive. Find I_T.

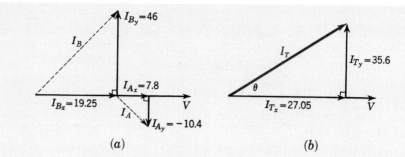

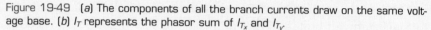

Figure 19-49 (a) The components of all the branch currents draw on the same voltage base. (b) I_T represents the phasor sum of I_{T_x} and I_{T_y}.

$$I_T^2 = (I_{T_x})^2 + (I_{T_y})^2 \qquad (19\text{-}9)$$
$$= (27.05)^2 + (35.6)^2$$
$$= 732 + 1267 = 1996$$
$$I_T = \sqrt{1999} = 44.7 \text{ A } leading \qquad Ans.$$

b. Find the impedance.

$$V_T = I_T \times Z$$
$$650 = 44.7 \times Z$$
$$Z = \frac{650}{44.7} = 14.6 \ \Omega \qquad Ans.$$

c. Find the phase angle.

$$\cos \theta_T = \frac{I_{T_x}}{I_T} \qquad (19\text{-}10)$$
$$\cos \theta_T = \frac{27.05}{44.7} = 0.605$$

Therefore, $\qquad \theta_T = 53° \ leading \qquad Ans.$

d. $\qquad \text{PF} = \dfrac{I_{T_x}}{I_T} = 0.605 = 60.5\% \ leading$

e. Find the power drawn by the circuit.

$$W = V \times I \times \cos \theta = 650 \times 44.7 \times \cos 53°$$
$$W = 650 \times 44.7 \times 0.605 = 17,540 \text{ W} \qquad Ans.$$

Check:
$$W_A = I_A^2 \times R_A = (13)^2 \times 30 = 5,070 \text{ W}$$
$$W_B = I_B^2 \times R_B = (50)^2 \times 5 = 12,500 \text{ W}$$
$$W_T = W_A + W_B = \overline{17,570 \text{ W}} \qquad Check$$

The small error is due to the approximate value of $\theta = 53°$ obtained from the trigonometric tables.

The total current of 44.6 A *leads* the total voltage of 650 V by 53°.

Exercises

Find (*a*) the total current, (*b*) the impedance, (*c*) the phase angle, and (*d*) the power drawn by each circuit shown in Fig. 19-50.

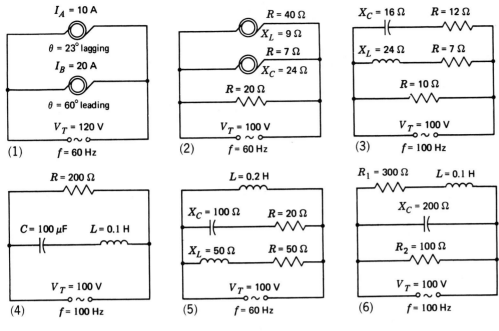

Figure 19-50

(See CD-ROM for Test 19-3)

Series-Parallel AC Circuits

Example 19-20

Solve the circuit shown in Fig. 19-51 on page 702 for (*a*) the equivalent series impedance, (*b*) the total current, (*c*) the phase angle, (*d*) the power factor, and (*e*) the power drawn by the circuit.

Solution

1. The parallel branches *A*, *B*, and *C* are solved in the same way in which we solved the branches in Example 19-18.
   a. Find the reactance, series impedance, current, and phase angle for each branch. Since the voltage across the section from *D* to *E* is unknown, a voltage may be assumed.
   b. Draw the phasor diagram for the branch currents on the same voltage base.
   c. Resolve the current in each branch into its in-phase and reactive components.
   d. Draw all the components on the same voltage base. Find the total in-phase current I_{T_x} and the total reactive current I_{T_y}.
   e. Find the total current for the parallel branches by adding I_{T_x} and I_{T_y} vectorially.

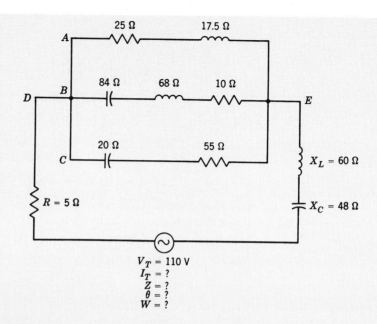

Figure 19-51 Series-parallel ac circuit.

$V_T = 110$ V
$I_T = ?$
$Z = ?$
$\theta = ?$
$W = ?$

f. Find the parallel impedance.
g. Find the phase angle for the parallel circuit.
2. The impedance found in step *f* above must now be resolved into its equivalent series resistance and reactance.
3. These are now combined with the other series resistances and reactances to get the total series impedance.
4. The phase angle, total current, and power are found as in any series circuit.

The solution is as follows:

1. a. Find the series impedance, current, and phase angle for each branch. Assume 100 V across DE.

For branch *A*:

$$Z = \sqrt{R^2 + X_L^2} = \sqrt{(25)^2 + (17.5)^2}$$
$$Z = \sqrt{625 + 306} = 30.5 \ \Omega$$
$$I_A = \frac{V}{Z} = \frac{100}{30.5} = 3.28 \text{ A}$$
$$\cos \theta = \frac{R}{Z} = \frac{25}{30.5} = 0.8197$$
$$\theta = 35° \ lagging$$

For branch *B*:

$$Z = \sqrt{R^2 + (X_C - X_L)^2} = \sqrt{(10)^2 + (84 - 68)^2}$$
$$= \sqrt{(10)^2 + (16)^2}$$
$$Z = \sqrt{100 + 256} = 18.9 \ \Omega$$
$$I_B = \frac{V}{Z} = \frac{100}{18.9} = 5.30 \text{ A}$$

$$\cos\theta = \frac{R}{Z} = \frac{10}{18.9} = 0.5291$$
$$\theta = 58° \text{ leading}$$

For branch C:

$$Z = \sqrt{R^2 + X_C^2} = \sqrt{(55)^2 + (20)^2}$$
$$Z = \sqrt{3025 + 400} = 58.5 \text{ }\Omega$$
$$I_C = \frac{V}{Z} = \frac{100}{58.5} = 1.71 \text{ A}$$
$$\cos\theta = \frac{R}{Z} = \frac{55}{58.5} = 0.9400$$
$$\theta = 20° \text{ leading}$$

b. Draw the phasor diagram for the branch currents on the same voltage base as shown in Fig. 19-52a.

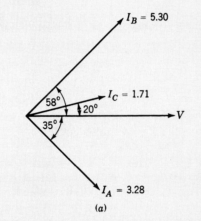

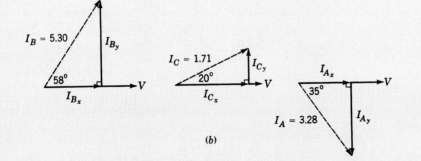

Figure 19-52 (a) Leading and lagging currents in a parallel circuit. (b) Resolution of each current into its in-phase and reactive components.

c. Resolve the current in each branch into its components as shown in Fig. 19-52b. Calculate the components.

For branch A:

$$I_{A_x} = I_A \times \cos\theta = 3.28 \times \cos 35° = 3.28 \times 0.819 = 2.68 \text{ A}$$
$$I_{A_y} = I_A \times \sin\theta = 3.28 \times \sin 35° = 3.28 \times 0.574 = -1.88 \text{ A}$$

For branch *B*:

$$I_{B_x} = I_B \times \cos \theta = 5.30 \times \cos 58° = 5.30 \times 0.53 = 2.82 \text{ A}$$
$$I_{B_y} = I_B \times \sin \theta = 5.30 \times \sin 58° = 5.30 \times 0.848 = 4.49 \text{ A}$$

For branch *C*:

$$I_{C_x} = I_C \times \cos \theta = 1.71 \times \cos 20° = 1.71 \times 0.94 = 1.61 \text{ A}$$
$$I_{C_y} = I_C \times \sin \theta = 1.71 \times \sin 20° = 1.71 \times 0.342 = 0.59 \text{ A}$$

    d. Draw all the components on the same voltage base as shown in Fig. 19-53*a*. The total in-phase current I_{T_x} and the total reactive current I_{T_y} may now be found.

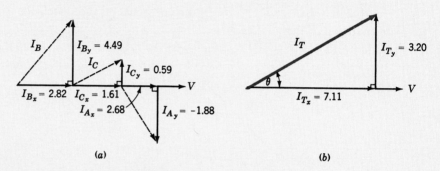

Figure 19-53 (*a*) The components of all the branch currents drawn on the same voltage base. (*b*) I_T represents the phasor sum of I_{T_x} and I_{T_y}.

$$I_{T_x} = I_{A_x} + I_{B_x} + I_{C_x} = 2.68 + 2.82 + 1.61 = 7.11 \text{ A}$$
$$I_{T_y} = I_{B_y} + I_{C_y} - I_{A_y} = 4.49 + 0.59 - 1.88 = 3.20 \text{ A}$$

    e. Draw the phasor diagram for the total current by adding I_{T_x} and I_{T_y} vectorially as shown in Fig. 19-53*b*. Notice that I_{T_y} is drawn *upward* because the total of all the *y* components is positive. Find I_T.

$$I_T{}^2 = (I_{T_x})^2 + (I_{T_y})^2$$
$$= (7.11)^2 + (3.20)^2$$
$$= 50.6 + 10.2$$
$$= 60.8$$
$$I_T = \sqrt{60.8} = 7.8 \text{ A} = I_{DE}$$

    f. Find the parallel impedance.

$$Z_{DE} = \frac{V_{DE}}{I_{DE}} = \frac{100}{7.8} = 12.8 \text{ } \Omega$$

    g. Find the phase angle for the parallel circuit between *D* and *E*.

$$\cos \theta = \frac{I_{T_x}}{I_T} = \frac{7.11}{7.8} = 0.9115$$
$$\theta = 24° \text{ } leading$$

The total parallel impedance = 12.8 Ω at 24° *leading*.

2. The equivalent series impedance is made up of the resistive and reactive components into which this parallel impedance must be resolved.

$$Z_x = R = Z \times \cos \theta$$
$$R = 12.8 \times \cos 24° = 12.8 \times 0.914$$
$$R = 11.7 \ \Omega \qquad Ans.$$
$$Z_y = X_C = Z \times \sin \theta \quad (X_C \text{ because of the } leading \text{ current})$$
$$X_C = 12.8 \times \sin 24° = 12.8 \times 0.407$$
$$X_C = 5.26 \ \Omega \qquad Ans.$$

3. Find the total series impedance. The components of Z_{DE} are now combined with the other parts of the series circuit as shown in Fig. 19-54.

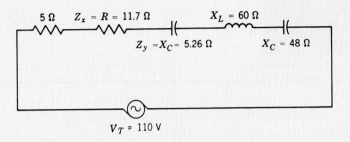

$V_T = 110$ V

Figure 19-54 The components of Z_{DE} are parts of the equivalent series circuit.

$$R_T = 5 + 11.7 = 16.7 \ \Omega$$
$$X_C = 5.26 + 48 = 53.26 \ \Omega$$
$$X_L = 60 \ \Omega$$
$$Z = \sqrt{R^2 + (X_L - X_C)^2}$$
$$= \sqrt{(16.7)^2 + (60.0 - 53.26)^2}$$
$$= \sqrt{(16.7)^2 + (6.74)^2}$$
$$= \sqrt{279 + 45.5} = \sqrt{324.5}$$
$$Z = 18 \ \Omega \qquad Ans.$$

4. Find the total series current.
$$I_T = \frac{V_T}{Z} = \frac{110}{18} = 6.11 \text{ A} \qquad Ans.$$

5. Find the phase angle.
$$\cos \theta = \frac{R}{Z} = \frac{16.7}{18} = 0.926$$
$$\theta = 22° \ lagging$$

6. Find the power factor.
$$\text{PF} = \frac{R}{Z} = 0.926 = 92.6\% \ lagging$$

7. Find the power.
$$W = V \times I \times \cos \theta = 110 \times 6.11 \times \cos 22°$$
$$= 110 \times 6.11 \times 0.926 = 622 \text{ W} \qquad Ans.$$
Check: $\quad W = I_R^2 \times R = (6.11)^2 \times 16.7 = 623 \text{ W} \qquad Check$

The total series current of 6.11 A lags the total voltage of 110 V by 22°.

Practice

Solve each of the circuits shown in Fig. 19-55 for (a) the equivalent series impedance, (b) the total current, (c) the phase angle, (d) the power factor, and (e) the
power drawn by the circuit.

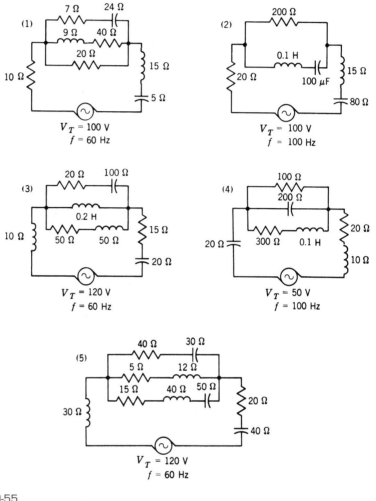

Figure 19-55

Parallel Resonance

In a series circuit, it is possible to adjust the values of a coil and a capacitor so that
their reactances will be equal for a definite frequency. The frequency at which the
inductive and capacitive reactances are equal is the resonant frequency. Since the
actions of the reactances are directly opposed to each other, the total reactance is
zero and the impedance of the circuit becomes just the resistance of the circuit.
Under these conditions, the current in the circuit will be a maximum.

Now let us arrange the coil and capacitor in *parallel* as shown in Fig. 19-56.

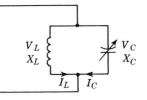

Figure 19-56 Parallel-resonant circuit.

At the resonant frequency,

$$V_L = V_C \qquad\qquad (5\text{-}1)$$
$$X_L = X_C \qquad\qquad (18\text{-}9)$$

By dividing Eq. (5-1) by Eq. (18-9), we obtain

$$\frac{V_L}{X_L} = \frac{V_C}{X_C}$$

or $\qquad\qquad I_L = I_C \qquad\qquad$ 19-12

Since the currents are exactly opposed to each other, the total current is equal to $I_L - I_C$ or practically zero. At the resonant frequency, then, since the total current is very nearly equal to zero, the impedance of the circuit to currents at that frequency will be very large. At any other frequency, since X_L is not equal to X_C, the currents will no longer be equal. The sum of the unequal currents may be quite large, which indicates a low impedance. To summarize, a parallel resonant circuit will offer a very large impedance to currents at the resonant frequency and a low impedance to currents at all other frequencies.

Uses. Just as a series resonant circuit is able to *accept* currents at the resonant frequency and reject all others, a parallel resonant circuit is able to *reject* currents at the resonant frequency and accept all others. This makes it possible to reject or "trap" a wave of a definite frequency in antenna and filter circuits. It is also a convenient method for obtaining the high impedance required in the primary of coupling transformers.

Resonant frequency. The formula for the resonant frequency of a parallel circuit is the same as that for a series circuit.

$$F = \frac{159}{\sqrt{L \times C}} \qquad\qquad 18\text{-}10$$

where $\qquad f =$ frequency, kHz
$\qquad\qquad L =$ inductance, μH
$\qquad\qquad C =$ capacitance, μF

Example 19-21

A 200-μH coil and a 50-pF capacitor are connected in parallel to form a "wave trap" in an antenna. What is the resonant frequency that the circuit will reject?

Solution

Given: $\qquad L = 200\ \mu$H $\qquad\qquad$ Find: $f = ?$
$\qquad\qquad C = 50\ \text{pF} = 50 \times 10^{-6}\ \mu$F

$$f = \frac{159}{\sqrt{L \times C}} \qquad (18\text{-}10)$$

$$= \frac{159}{\sqrt{200 \times 50 \times 10^{-6}}}$$

$$= \frac{159}{\sqrt{10^4 \times 10^{-6}}} = \frac{159}{\sqrt{10^{-2}}} = \frac{159}{10^{-1}}$$

$$f = 159 \times 10 = 1590 \text{ kHz} \qquad Ans.$$

Finding the Inductance or Capacitance Needed to Produce Resonance

Example 19-22

A 0.1-mH coil and a variable capacitor are connected in parallel to form the primary of an IF transformer as shown in Fig. 19-57. If the circuit is to be resonant to 456 kHz, what must be the value of the capacitor?

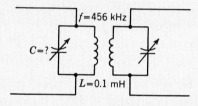

Figure 19-57

Solution

1. Change 0.1 mH to 100 μH.
2. Find the required capacitance.

$$C = \frac{25,300}{f^2 \times L} \qquad (18\text{-}12)$$

$$= \frac{25,300}{(456)^2 \times 100}$$

$$C = \frac{25,300}{207,900 \times 100} = 0.0012 \ \mu\text{F} \qquad Ans.$$

Exercises

Practice

1. Find the resonant frequency of a wave trap using a 45-pF capacitor and a 20-μH inductance.
2. Find the resonant frequency of a band-stop filter made of a 160-μH coil and a 40-pF capacitor in parallel.
3. The inductance of a parallel resonant circuit used as a wave trap in an antenna circuit is 100 μH. What must be the value of the parallel capacitance in order to reject an 800-kHz wave?

4. The capacitor of a high-impedance primary of a transformer tuned to 460 kHz is 10 pF. What is the value of the inductance?

Applications

5. The tank circuit of an oscillator contains a coil of 320 μH. What is the value of the capacitance at the resonant frequency of 1000 kHz?
6. The tank circuit of an impedance-coupled AF amplifier circuit uses an inductance of 10 H. Find the value of the capacitance necessary to produce resonance at (a) 500 Hz and (b) 1 kHz.
7. What is the inductance of the coil in a 23.4-MHz trap of a video IF amplifier which uses a capacitor of 50 pF?
8. An IF coil in a superheterodyne receiver resonates at a frequency of 455 kHz. Find the inductance of the coil if the capacitor is valued at 50 pF.
9. A wave trap in the plate circuit of an IF stage is to resonate at 27.25 MHz. Find the capacity needed to produce resonance with a coil valued at 0.85 μH.

about electronics

Piezoelectric ceramic snow skis absorb vibrations automatically to give the skier a smoother ride. When the skis flex, the ceramic material sends that data to control circuitry, mechanically dampening the vibrations of the skis.

In skis, unwanted vibrations cause the edges to lift off the snow, causing the skier to experience reduced control and responsiveness. The ACX piezo control module adds damping to the ski, thus reducing vibration and putting the ski back on the snow. The skier can ski at faster speeds, and over a wider variety of conditions, and the skis provide a smoother ride, more responsive turning, and greater stability.

JOB 19-9

Review of Parallel AC Circuits

In a purely resistive circuit:

$$V_T = V_1 = V_2 = V_3 \qquad\qquad 5\text{-}1$$

$$V_T = I_T \times R_T \qquad\qquad 4\text{-}7$$

$$I_T = I_1 + I_2 + I_3 \qquad\qquad 5\text{-}2$$

$$W = V \times I \times \underline{\quad(1)\quad} \qquad\qquad 15\text{-}18$$

The total current is $\underline{\quad(2)(\text{in/out of})\quad}$ phase with the total voltage. In a purely inductive circuit:

$$V_T = V_1 = V_2 = V_3 \qquad\qquad 5\text{-}1$$

$$V_T = I_T \times \underline{\quad(3)\quad}$$

$$I_T = I_1 + I_2 + I_3 \qquad\qquad 5\text{-}2$$

$$W = V \times I \times \cos\theta \qquad\qquad 15\text{-}18$$

The total current ___(4)(leads/lags)___ the total voltage by 90°. In a purely capacitive circuit:

$$V_T = V_1 = V_2 = V_3 \qquad\qquad 5\text{-}1$$

$$V_T = I_T \times Z$$

$$I_T = I_1 + I_2 + I_3 \qquad\qquad 5\text{-}2$$

$$W = V \times I \times \cos\theta \qquad\qquad 15\text{-}18$$

The total current ___(5)(leads/lags)___ the total voltage by 90°. In an ac parallel circuit of resistance and inductance:

$$I_T{}^2 = I_R{}^2 + \underline{\quad(6)\quad} \qquad\qquad 19\text{-}1$$

$$V_T = I_T \times Z$$

$$\cos\theta = \frac{I_R}{(7)} \qquad\qquad 19\text{-}2$$

$$PF = \frac{(8)}{I_T} \qquad\qquad 19\text{-}3$$

$$W = V \times I \times \underline{\quad(9)\quad} \qquad\qquad 15\text{-}18$$

or

$$W = V \times I \times \underline{\quad(10)\quad} \qquad\qquad 15\text{-}19$$

The total current ___(11)(leads/lags)___ the total voltage by angle θ. In an ac parallel circuit of resistance and capacitance:

$$I_T{}^2 = I_R{}^2 + \underline{\quad(12)\quad} \qquad\qquad 19\text{-}4$$

$$V_T = I_T \times \underline{\quad(13)\quad}$$

$$\cos\theta = \frac{(14)}{I_T} \qquad\qquad 19\text{-}2$$

$$PF = \frac{I_R}{(15)} \qquad\qquad 19\text{-}3$$

$$W = V \times I \times \cos\theta \qquad\qquad 15\text{-}18$$

or

$$W = \underline{\quad(16)\quad} \qquad\qquad 15\text{-}19$$

The total current ___(17)(leads/lags)___ the total voltage by angle θ. In an ac parallel circuit of resistance, inductance, and capacitance:

$$I_T{}^2 = I_R{}^2 + (I_L - \underline{\quad(18)\quad})^2 \qquad\qquad 19\text{-}4$$

$$V_T = I_T \times Z$$

$$\cos\theta = \frac{I_R}{I_T} \qquad\qquad 19\text{-}2$$

$$PF = \frac{(19)}{(20)} \qquad\qquad 19\text{-}3$$

$$W = V \times I \times \cos\theta \qquad\qquad 15\text{-}18$$

$$W = V \times I \times PF \qquad\qquad 15\text{-}19$$

The total current will lead or lag the total voltage, depending on the values of I_L and I_C. If I_L is larger than I_C, the current will __(21)(lead/lag)__ the voltage. If I_L is smaller than I_C, the current will __(22)(lead/lag)__ the voltage. The angle of lead or lag is given by the angle θ.

Parallel-Series AC Circuits

1. Find the reactance, impedance, current, and phase angle for each branch of the parallel circuit.
2. Draw the phasor diagram for the branch currents on the same voltage base.
3. Resolve the current in each branch into its components.

$$I_y = I \times \underline{\quad (23) \quad} \qquad\qquad 19\text{-}5$$

$$I_x = I \times \underline{\quad (24) \quad} \qquad\qquad 19\text{-}6$$

(y components of lagging currents are negative.)

4. Find the total in-phase and total reactive currents.

$$I_{T_x} = I_{A_x} + I_{B_x} \qquad\qquad 19\text{-}7$$

$$I_{T_y} = I_{A_y} + \underline{\quad (25) \quad} \qquad\qquad 19\text{-}8$$

5. Find the total current.

$$I_T{}^2 = (I_{T_x})^2 + \underline{\quad (26) \quad} \qquad\qquad 19\text{-}9$$

6. Find the impedance.

$$V_T = I_T \times Z$$

7. Find the phase angle.

$$\cos \theta_T = \frac{(27)}{I_T} \qquad\qquad 19\text{-}10$$

8. Find the power factor.

$$PF = \frac{I_{T_x}}{(28)} \qquad\qquad 19\text{-}11$$

9. Find the total power.

$$W = V \times I \times \cos \theta \qquad\qquad 15\text{-}18$$

or

$$W = V \times I \times PF \qquad\qquad 15\text{-}19$$

Series-Parallel AC Circuits

1. For the parallel branches: Repeat steps 1 through 7 for parallel-series circuits.
2. Resolve the parallel impedance found in step 6 above into its equivalent series resistance and reactance.
3. Combine these with the other resistances and reactances to get the total series impedance.
4. The total current, phase angle, and power are found as in any series circuit.

Parallel resonance.   A parallel resonant circuit will offer a very large impedance to currents at the resonant frequency and a low impedance to currents at all other frequencies.

$$f = \frac{159}{\sqrt{L \times C}} \qquad \text{18-10}$$

where L = inductance, measured in ___(29)___
C = capacitance, measured in ___(30)___
f = frequency, measured in ___(31)___

Finding the Inductance or Capacitance Needed to Produce a Resonant Circuit

$$L = \frac{25{,}300}{f^2 \times (32)} \qquad \text{18-11}$$

$$C = \frac{25{,}300}{f^2 \times L} \qquad \text{18-12}$$

where L, C, and f are measured in the same units as called for in formula (18-23).

Exercises

Applications

1. Two capacitors of 500 and 750 Ω reactance, respectively, are connected in parallel across a 25-V 25-Hz ac source. Find (a) the total current, (b) the impedance, and (c) the power drawn by the circuit.

2. A 0.01-H coil and a 5000-Ω resistor are connected in parallel to form a filter circuit. Find the percent of the total current passing through the resistor for (a) a 1-kHz AF frequency and (b) a 1000-kHz RF frequency. (c) Is this filter a high-pass or a low-pass filter?

3. A 200-Ω resistor and a 0.1-H coil are connected in parallel across a 100-V 1-kHz ac source. Find (a) the current in each branch, (b) the total current, (c) the impedance of the circuit, (d) the phase angle, (e) the power factor, and (f) the power drawn by the circuit.

4. A 500-Ω resistor in an emitter circuit is bypassed with a 5-μF capacitor. If a 1-kHz frequency causes a voltage drop of 10 V across the resistor, find (a) the current in each branch, (b) the total current, and (c) the impedance of the combination.

5. A 3000-Ω resistor, a 1200-Ω inductive reactance, and an 800-Ω capacitive reactance are connected in parallel across a 240-V line. Find (a) the total current, (b) the impedance, (c) the phase angle, (d) the power factor, and (e) the power drawn by the circuit.

6. A 0.5-μF capacitor, a 1-H coil, and a 2000-Ω resistor are connected in parallel across a 220-V 60-Hz ac line. Find (a) the total current, (b) the impedance, (c) the phase angle, (d) the power factor, and (e) the power drawn by the circuit.

7. In a circuit similar to that used for Exercise 1 in Fig. 19-50, I_A = 8 A with θ = 30° lagging, I_B = 15 A with θ = 60° leading, and V_T = 120 V at 60 Hz. Find (a) the total current, (b) the impedance, (c) the phase angle, (d) the power factor, and (e) the power drawn by the circuit.

8. In a circuit similar to that used for Exercise 6 in Fig. 19-50, R_1 = 100 Ω, L = 0.2 H, X_C = 500 Ω, R_2 = 200 Ω, V_T = 100 V, and f = 100 Hz. Find (a) the total current, (b) the impedance, (c) the phase angle, (d) the power factor, and (e) the power drawn by the circuit.

9. A 0.001-μF capacitor and a coil are connected in parallel to form the primary of an IF transformer similar to that shown in Fig. 19-57. What must be the

inductance of the coil in order for the circuit to be resonant to a frequency of 460 kHz?

10. The collector circuit of the 2SC563 mixer transistor in the Emerson model 984 television receiver contains a 1-μH coil in parallel with a capacitance and resonant to 21.25 MHz. What is the capacitance?

11. Solve the circuit shown in Fig. 19-58 for (a) the equivalent series impedance, (b) the total current, (c) the phase angle, and (d) the power drawn by the circuit.

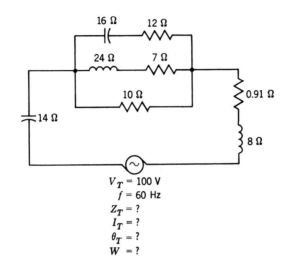

$V_T = 100$ V
$f = 60$ Hz
$Z_T = ?$
$I_T = ?$
$\theta_T = ?$
$W = ?$

Figure 19-58

12. A capacitor, a 24-Ω resistor, and an inductive reactance of 20 Ω are connected in parallel across a 120-V 60-Hz supply. The total line current is read as 6.4 A lagging. Find (a) the total reactive current, (b) the capacitor current, and (c) the total power factor.

(See CD-ROM for Test 19-4)

1. Two parallel capacitors, each 450 μF, are connected in parallel with three parallel resistors of 30 Ω, 40 Ω, and 50 Ω. What is the total current in the circuit if the frequency drops to 13.8524 Hz?

2. A 120-V ac 60-Hz supply is connected to a parallel circuit containing a 25-Ω resistor and a 92.84-mH inductor. What is the total VA supplied to the circuit and the true power that is consumed?

3. A 100-Ω resistor, 4.974-F capacitor, and a 25.8627-mH inductor are connected in parallel and connected to a 240-V 400-Hz motor generator set. Determine whether the circuit is more inductive or more capacitive, and determine the total current flow in the circuit.

4. The two currents (I_X, I_Y) found for each of the 10 exercises in Job 19-4 are called the *horizontal* and *vertical* components of the total current. Can you think of any other situation in which the ability to determine the horizontal and/or vertical component might be useful?

5. Name three types of electronic equipment that you feel might be affected by EMI.

6. In diagrams 1 through 6 of Fig. 19-50 determine the following: (a) Z, (b) I_T, (c) $\angle\theta$, (d) PF, (e) true power, (f) reactive power, and (g) apparent power.

7. Draw each of the circuits in Fig. 19-55 as either a series *RC* or a series *RL* circuit showing their respective equivalent components for simplicity.

8. Using the formulas $X_L = 2\pi f L$ and $X_C = 1/(2\pi f C)$, derive your own formula for determining resonant frequency.

INTERNET
ACTIVITIES

Internet Activity A

What causes transients? **http://www.calpoly.edu/~abratton/what.htm**

Internet Activity B

Switching power supplies are common in modern electrical systems. What types of problems can they cause? Go to:
http://www.en.polyu.edu.hk/~mw/pcbwww/pcb.html
and select "Sources and victims."

Chapter 19 Solutions to Self-Tests and Reviews

Self-Test 19-4

2. in
3. lags, 90
4. leads, 90
5. arithmetically
7. $\cos \theta$

Job 19-9 Review

1. $\cos \theta$
2. in

3. Z
4. lags
5. leads
6. $I_L{}^2$
7. I_T
8. I_R
9. $\cos \theta$
10. PF
11. lags
12. $I_C{}^2$

13. Z
14. I_R
15. I_T
16. $V \times I \times PF$
17. leads
18. I_C
19. I_R
20. I_T
21. lag
22. lead

23. $\sin \theta$
24. $\cos \theta$
25. I_{B_y}
26. $(I_{T_y})^2$
27. I_{T_x}
28. I_T
29. μH
30. μF
31. kHz
32. C

Alternating-Current Power

1. Find the power and power factor in ac circuits.
2. Determine the total power drawn by reactive loads.
3. Determine the power drawn by resistive, inductive, and capacitive loads.
4. Find the capacitance needed for power factor correction.
5. Determine neutral power line current by using circuit analysis.

JOB 20-1

Power and Power Factor

The power in any electric circuit is obtained by multiplying the voltage by the current flowing at that time. In a dc circuit, the unchanging voltage V is multiplied by the unchanging current I to give the power P. The formula for this was given in Job 7-1 as $P = V \times I$. In an ac circuit, the voltage and current are constantly changing. The power at any instant of time is obtained by multiplying the instantaneous voltage by the instantaneous current.

$$P_i = v \times i \qquad \text{Formula 20-1}$$

Power in a resistive circuit. In a purely resistive circuit, each instantaneous current occurs at the same time as the instantaneous voltage which produced it. Since the worth of all the instantaneous values is the effective value, the power in a resistive circuit is found by multiplying the effective voltage by the effective current.

$$P = V \times I \qquad (7\text{-}1)$$

Power in a reactive circuit. In an inductive circuit, the current will lag behind the voltage by some angle θ as shown in Fig. 20-1a. The power in this circuit will *not* be equal to the product of the voltage and the current, since they do not act at the same time. The actual power is equal to the voltage multiplied by *only that portion of the line current which is in phase with the voltage.* In Fig. 20-1b, OE represents the line voltage. OA represents the total line current as measured by an ammeter. This total line current may be resolved into its two component parts: OB, a component in phase with the voltage (the effective current), and BA, a component 90° out of phase with the voltage (the reactive current). By multiplying each of the current phasors of Fig. 20-1b by the line voltage, we can obtain a phasor diagram for the power in an *inductive* circuit as shown in Fig. 20-1c.

The phasor diagram of Fig. 20-1c indicates that the apparent power VA is made of two component parts—the effective power W and the reactive power (var). The apparent power may be likened to the power delivered to a flywheel. The *portion* of the apparent power that is delivered to the shaft to operate some device is similar to the effective power. The *portion* of the apparent power which is delivered by the flywheel *back* to the engine to keep it running is similar to the reactive power.

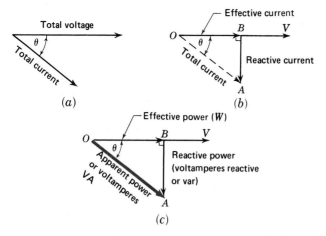

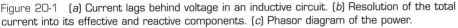

Figure 20-1 (a) Current lags behind voltage in an inductive circuit. (b) Resolution of the total current into its effective and reactive components. (c) Phasor diagram of the power.

The reactive power does no work itself. In an inductive electric circuit it represents the power stored in the magnetic field (similar to the flywheel) and then returned to the line as the field collapses. This power merely moves back and forth between the coil and the line. In Fig. 20-1c, *OA* represents the *apparent power* as measured by a voltmeter and an ammeter. It is measured in voltamperes, *not* watts. *OB* represents the *effective*, or *true, power* and is measured by a wattmeter in watts. This is the power that does the work. *AB* represents the reactive, or *var*, power that does *no* work. It is measured in voltamperes reactive (var).

$$VA = V \times I \qquad \text{20-2}$$

Formulas

$$\cos \theta = \frac{W}{VA} \qquad \text{20-3}$$

By cross multiplication,

$$W = VA \times \cos \theta \qquad \text{20-4}$$

Solving for *VA*,

$$VA = \frac{W}{\cos \theta} \qquad \text{20-5}$$

Also, since

$$\sin \theta = \frac{\text{var power}}{VA}$$

$$\textbf{var power} = \textbf{VA} \times \sin \theta \qquad \text{20-6}$$

where   *VA* = apparent power, VA
      V = voltage, V
      I = current, A
     W = true power, W
      θ = phase angle of the circuit, degrees
var power = voltamperes reactive, var

Power factor.   The ratio of the effective power as read by a wattmeter *W* to the apparent power *VA* is called the *power factor* (PF).

$$PF = \frac{W}{VA}$$ Formula 20-7

By comparing formulas (20-3) and (20-7), we can see that the power factor is equal to the cosine of angle θ.

$$PF = \cos\theta$$ Formula 20-8

By substituting PF for $\cos\theta$ in formulas (20-4) and (20-5), we obtain

$$W = VA \times PF$$ 20-9

Formulas

$$VA = \frac{W}{PF}$$ 20-10

The power factor may be expressed as a decimal or as a percent. For example, a PF of 0.8 may be written as 80%. In this sense, the PF describes the *portion* of the voltampere input which is actually effective in operating the device. Thus, an 80% PF means that the device uses only 80% of the voltampere input in order to operate, and 20% is wasted as reactive power. This is uneconomical. Generator capacity, copper size, fusing, etc., must be provided for the apparent power although part of it does not work. Low power factors may be improved by the methods discussed in Job 20-4. Circuits with

1. Resistance only will have a high PF with I in phase with V
2. High inductive reactance will have a low PF and a lagging current
3. High capacitive reactance will have a low PF and a leading current
4. Equal amounts of inductive and capacitive reactance will have a high PF with I in phase with V

Example 20-1

Find the power factor of a washing machine motor if it draws 5 A and 440 W from a 110-V 60-Hz line.

Solution

Given: $I = 5$ A   Find: PF = ?
$V = 110$ V
$W = 440$ W
$f = 60$ Hz

$$PF = \frac{W}{VA} = \frac{440}{110 \times 5} = \frac{440}{550} = 0.8 = 80\% \quad Ans. \quad (20\text{-}7)$$

Example 20-2

A single-phase capacitor-type motor operates at a power factor of 70% leading. The line voltage is 110 V, and the line current is 10 A. Find (*a*) the apparent power and (*b*) the true power taken in watts.

Solution

Given: PF = 70% = 0.70   Find: VA = ?
$I = 10$ A W = ?
$V = 110$ V

a.	$VA = V \times I = 10 \times 110 = 1100 \text{ VA}$	*Ans.*	(20-2)
b.	$W = VA \times \text{PF} = 1100 \times 0.70 = 770 \text{ W}$	*Ans.*	(20-9)

CALCULATOR HINT

Because power is a physical and measurable complex number, power calculations are best done on calculators that can do complex arithmetic.

Example 20-3

A 5-kVA single-phase ac generator operates at a power factor of 90%. Find (*a*) the full-load power supplied and (*b*) the full-load current if the terminal voltage is 120 V.

Solution

Given: $VA = 5000 \text{ VA}$ Find: $W = ?$
         $\text{PF} = 90\% = 0.9$ $I = ?$
         $V = 120 \text{ V}$

a.
$$W = VA \times \text{PF} \qquad (20\text{-}9)$$
$$= 5000 \times 0.9 = 4500 \text{ W} \qquad Ans.$$

b.
$$P = I \times V \qquad (7\text{-}1)$$
$$4500 = I \times 120$$
$$I = \frac{4500}{120} = 37.5 \text{ A} \qquad Ans.$$

Example 20-4

An impedance coil of 3 Ω resistance and 4 Ω inductive reactance is connected across a 24-V 60-Hz ac source. Find (*a*) the PF, (*b*) the current drawn, and (*c*) the effective power consumed by the coil.

Solution

Given: $R = 3 \text{ }\Omega$ Find: $\text{PF} = ?$
         $X_L = 4 \text{ }\Omega$ $I = ?$
         $V = 24 \text{ V}$ $W = ?$
         $f = 60 \text{ Hz}$

a. Find the impedance.
$$Z = \sqrt{R^2 + X_L^2} = \sqrt{3^2 + 4^2} = \sqrt{9 + 16} = \sqrt{25} = 5 \text{ }\Omega$$

Find the PF.
$$\text{PF} = \frac{R}{Z} = \frac{3}{5} = 0.60 = 60\% \qquad Ans. \qquad (16\text{-}7)$$

b. Find the current drawn.
$$V_T = I_T \times Z$$
$$24 = I_T \times 5$$
$$I_T = \frac{24}{5} = 4.8 \text{ A} \qquad Ans.$$

c. Find the effective power.
$$W = V \times I \times \text{PF} \qquad (15\text{-}19)$$
$$= 24 \times 4.8 \times 0.6 = 69 \text{ W} \qquad Ans.$$

Exercises

Applications

1. Find the power factor of a refrigerator motor if it draws 288 W and 3 A from a 120-V 60-Hz line.

2. A single-phase circuit has unity power factor. Find the true power in watts if the applied voltage is 120 V and the line current is 20 A.

3. Find the effective power used by a capacitor-type jigsaw motor operating at a power factor of 75% if it draws 4 A at 120 V.

4. The lights and motors in a shop draw 16 kW of power. The power factor of the entire load is 80%. Find the voltamperes of power delivered to the shop.

5. A circuit consumes 4 kW. If the line current is 40 A and the voltage is 110 V, find (*a*) the apparent power in voltamperes, and (*b*) the power factor of the circuit.

6. A capacitance of 5 Ω resistance and 12 Ω reactance is connected across a 117-V 60-Hz ac line. Find the power factor and the effective power.

7. A motor operating at 90% PF draws 270 W from a 120-V line. Find the current drawn.

8. Find the PF of a motor in an air-conditioning unit if it draws 500 W and 5 A from a 120-V line.

9. A motor operates on a 220-V line at a power factor of 0.8 lagging. If it consumes 8 kW, find the motor line current.

10. A certain load draws 90 kW and 100 kVA from an ac supply. At what power factor does it operate?

11. An industrial load draws 15 A from a 230-V line at a PF of 85 percent. Find the voltamperes of apparent power and the effective power taken from the line.

12. A 40-V emf at 1 kHz is impressed across a loudspeaker of 5000 Ω resistance and 1.5 H inductance. Find (*a*) the impedance, (*b*) the PF, (*c*) the current drawn, and (*d*) the effective power drawn by the speaker.

13. A 10-hp motor operates at an efficiency of 80% and a PF of 90%. Find the voltamperes of apparent power delivered to the motor. *Hint:* Find the watt input to the motor by the efficiency formula (11-2).

14. An inductive load operating at a phase angle of 53° draws 1200 W from a 120-V line. Find the current taken.

15. Find the current drawn from a 230-V line by a motor if it uses 6 kW of power at a PF of 0.65.

16. A 120-V 2-hp motor operates at an efficiency of 80%. Find the power input to the motor. If the current drawn by the motor is 20 A, find the PF.

17. A motor draws 600 W and 1000 VA from a 120-V 60-Hz ac line. Find the reactive voltamperes of the motor.

18. A single-phase alternator delivers 80 kW at 80% power factor. Find the kilovoltampere output of the alternator.

19. A wattmeter connected in a 120-V inductive circuit reads 720 W. Find the line current if the power factor of the circuit is 60%.

20. A single-phase motor draws 25 A when connected to a 220-V ac supply. If the power factor of the motor is 80% lagging, find (*a*) the voltamperes supplied, (*b*) the power input, and (*c*) the reactive voltamperes.

Farmers trying to raise crops in regions without bees often rent bees to carry out pollination—which is expensive. A new, "bee-less" pollination method allows farmers to apply a pollen mixture to the plants. This mixture is electrically charged to help pull the pollen to the stigma. Charged grains "stick" to plants five times better than uncharged pollen.

Total Power Drawn by Combinations of Reactive Loads

As we learned in Job 7-2, the total power in a circuit may be found by adding the power taken by the individual parts. If the power drawn by one branch is not in phase with the power drawn by another branch, however, the addition of the power must be made by *phasor addition*. This will be done, as in Job 19-4, by resolving the power into its components and adding the components.

Resolving the apparent power into its components. In Fig. 20-1*c*,

$$\sin \theta = \frac{\text{var power}}{VA} \quad \text{and} \quad \cos \theta = \frac{W}{VA}$$

Cross-multiplying each equation yields the

	$\text{\textbf{var power} = \textbf{VA} \times \textbf{sin } \theta}$	20-6
Formulas	$\textbf{W} = \textbf{VA} \times \textbf{cos } \theta$	20-4
or	$\textbf{W} = \textbf{VA} \times \textbf{PF}$	20-9

Also, using the tangent function, we obtain

$$\tan \theta = \frac{\text{\textbf{var power}}}{\textbf{W}} \qquad \text{Formula 20-11}$$

Applying the pythagorean theorem, we obtain

$$(\textbf{VA})^2 = \textbf{W}^2 + (\textbf{var } \textbf{P})^2 \qquad \text{Formula 20-12}$$

> **Hint for Success**
>
> Notice how often you have problems that use the pythagorean theorem. Review these problems and note that each use represents an underlying complex number. If you believe in the pythagorean theorem, then you believe that imaginary numbers exist and are a real phenomenon.

Combinations of devices of different power factors

Example 20-5

Find (*a*) the total effective power, (*b*) the total apparent power, (*c*) the total PF, and (*d*) the total current of the combination shown in Fig. 20-2.

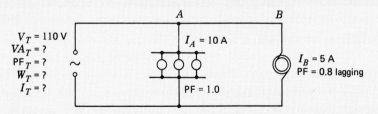

Figure 20-2

Solution

When the power in one branch is not in phase with the power in another branch, the apparent powers must be resolved into their components and the components added as shown below.

a. Find the total effective power.

 1. Find the apparent power *VA*, the effective power *W*, the phase angle θ, and the var power (var *P*) for each load.

For branch *A*: $VA_A = IA \times V_A = 110 \times 10 = 1100 \text{ VA}$ (20-2)
 $W_A = VA_A \times PF_A = 1100 \times 1 = 1100 \text{ W}$ (20-9)

Since $\text{PF}_A = \cos \theta = 1.0 \qquad \theta = 0°$ (20-8)

$$\text{var } P_A = VA_A \times \sin \theta \tag{20-6}$$
$$= 1100 \times \sin 0°$$
$$= 1100 \times 0$$
$$\text{var } P_A = 0 \text{ var}$$

For branch B: $VA_B = I_B \times V_B = 5 \times 110 = 550 \text{ VA}$ (20-2)
                    $W_B = VA_B \times \text{PF}_B = 550 \times 0.8 = 440 \text{ W}$ (20-9)

Since $\text{PF}_B = \cos \theta = 0.8 \qquad \theta = 37° \text{ lagging}$ (20-8)

$$\text{var } P_B = VA_B \times \sin \theta \tag{20-6}$$
$$= 550 \times \sin 37°$$
$$= 550 \times 0.6$$
$$\text{var } P_B = -330 \text{ var}$$

2. Draw the phasor diagram for the apparent powers on the same voltage base as shown in Fig. 20-3a.
3. Resolve the power in each branch into its components as shown in Fig. 20-3b.

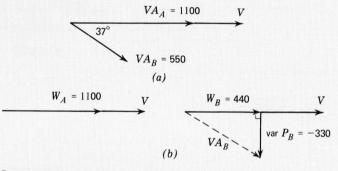

(a)

(b)

Figure 20-3   (a) VA_A and VA_B are not in phase, since they act at different power factors. (b) Resolution of each apparent power into its effective and reactive components.

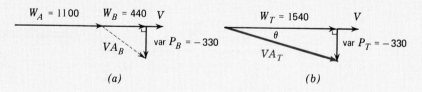

(a)   (b)

Figure 20-4   (a) The components of all the branch powers drawn on the same voltage base. (b) VA_T represents the phasor sum of $W_T +$ var P_T.

4. Draw all the components on the same voltage base as shown in Fig. 20-4a. The total effective power W_T and the total reactive power var P_T may now be found by

$$\boldsymbol{W_T = W_A + W_B} \qquad\qquad \text{Formula 20-13}$$

$$\textbf{var } \boldsymbol{P_T = } \textbf{var } \boldsymbol{P_A + } \textbf{var } \boldsymbol{P_B} \qquad\qquad \text{Formula 20-14}$$

$$W_T = W_A + W_B = 1100 + 440 = 1540W \qquad (20\text{-}13)$$
$$\text{var } P_T = \text{var } P_A + \text{var } P_B = 0 + (-330) = -330 \text{ var} \qquad (20\text{-}14)$$

b. Find the total apparent power. Draw the phasor diagram for the total power by adding W_T and var P_T vectorially as shown in Fig. 20-4b. Notice that var P_T is drawn *downward* because var P_T is negative.

Applying the pythagorean theorem, we obtain

$$VA_T{}^2 = W_T{}^2 + \text{var } P_T{}^2 \qquad \text{20-15}$$

Formulas

$$PF_T = \frac{W_T}{VA_T} \qquad \text{20-16}$$

Find VA_T.

$$VA_T{}^2 = W_T{}^2 + \text{var } P_T{}^2 \qquad (20\text{-}15)$$
$$= (1540)^2 + (-330)^2$$
$$VA_T{}^2 = 237 \times 10^4 + 11 \times 10^4$$
$$= 248 \times 10^4$$
$$VA_T = \sqrt{248 \times 10^4} = 1575 \text{ VA} \qquad Ans.$$

c. Find the total PF.

$$PF_T = \frac{W_T}{VA_T} = \frac{1540}{1575} = 0.977 = 97.7\% \text{ } lagging \qquad Ans. \text{ (20-16)}$$

d. Find the total current.

$$VA_T = V_T \times I_T \qquad (20\text{-}2)$$
$$1575 = 110 \times I_T$$
$$I_T = \frac{1575}{110} = 14.3 \text{ A} \qquad Ans.$$

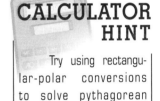

CALCULATOR HINT

Try using rectangular-polar conversions to solve pythagorean problems.

Alternative Solution

The phase angle and the apparent power may be found by applying the tangent formula to the right triangle shown in Fig. 20-4b. This solution is simpler, since it does not involve the labor of squaring numbers and finding the square root of the sum.

$$\tan \theta_T = \frac{\text{var } P_T}{W_T} \qquad \text{20-17}$$

Formulas

$$VA_T = \frac{W_T}{PF_T} \qquad \text{20-18}$$

a. Find the phase angle in Fig. 20-4b.

$$\tan \theta_T = \frac{\text{var } P_T}{W_T} = \frac{330}{1540} = 0.214 \quad \text{therefore} \quad \theta = 12° \quad (20\text{-}17)$$

$$PF = \cos \theta = \cos 12° = 0.978 = 97.8\% \text{ } lagging \qquad Ans. \text{ (20-8)}$$

b. Find the total apparent power.

$$VA_T = \frac{W_T}{PF_T} = \frac{1540}{\cos 12°} = \frac{1540}{0.978} = 1574 \text{ VA} \qquad Ans. \text{ (20-18)}$$

Example 20-6

Find (*a*) the total effective power, (*b*) the total power factor, and (*c*) the total apparent power for the circuit shown in Fig. 20-5.

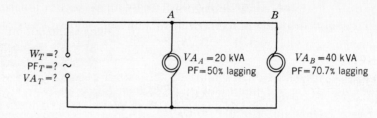

$$W_T = ?$$
$$PF_T = ?$$
$$VA_T = ?$$

$VA_A = 20$ kVA
PF = 50% lagging

$VA_B = 40$ kVA
PF = 70.7% lagging

Figure 20-5

Solution

a. 1. Find the effective power and the var power for each load.

For load *A*:

$$PF = \cos \theta = 0.500 \quad \text{therefore} \quad \theta = 60° \quad (20\text{-}8)$$
$$W_A = VA_A \times PF_A = 20 \times 0.5 = 10 \text{ kW} \quad (20\text{-}9)$$
$$\text{var } P_A = VA_A \times \sin \theta = 20 \times \sin 60° = 20 \times 0.866 = -17.32 \text{ kvar}$$

For load *B*:

$$PF = \cos \theta = 0.707 \quad \text{therefore} \quad \theta = 45° \quad (20\text{-}8)$$
$$W_B = VA_B \times PF_B = 40 \times 0.707 = 28.28 \text{ kW} \quad (20\text{-}9)$$
$$\text{var } P_B = VA_B \times \sin \theta = 40 \times \sin 45° = 40 \times 0.707 = -28.28 \text{ kvar}$$

2. Draw the phasor diagram for the apparent powers on the same voltage base as shown in Fig. 20-6*a*.

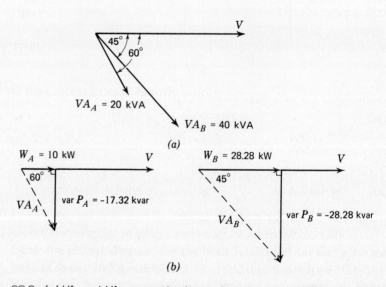

Figure 20-6 (*a*) VA_A and VA_B are not in phase, since they act at different power factors. (*b*) Resolution of each apparent power into its effective and reactive components.

3. Resolve the power in each branch into its components as shown in Fig. 20-6b.
4. Draw all the components on the same voltage base as shown in Fig. 20-7a.

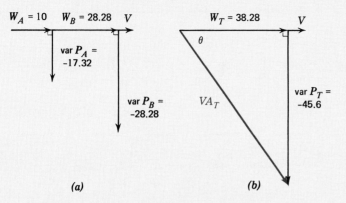

Figure 20-7 (a) The components of all the branch powers drawn on the same voltage base. (b) VA_T represents the phasor sum of $W_T +$ var P_T.

The total effective power W_T and the total reactive power var P_T may now be found.

$$W_T = W_A + W_B = 10 + 28.28 = 38.28 \text{ kW} \qquad Ans. \qquad (20\text{-}13)$$
$$\text{var } P_T = \text{var } P_A + \text{var } P_B = (-17.32) + (-28.28) = -45.6 \text{ kvar}$$

b. Find the total power factor by adding W_T and var P_T vectorially as shown in Fig. 20-7b.

$$\tan \theta_T = \frac{\text{var } P_T}{W_T} = \frac{-45.6}{38.28} = -1.191 \qquad \text{therefore} \qquad \theta = 50° \quad (20\text{-}17)$$
$$\text{PF} = \cos \theta = \cos 50° = 0.643 = 64.3\% \text{ lagging} \qquad Ans. \qquad (20\text{-}8)$$

c. Find the total apparent power.

$$VA_T = \frac{W_T}{\text{PF}_T} = \frac{38.28}{\cos 50°} = \frac{38.28}{0.643} = 59.5 \text{ kVA} \qquad Ans. \qquad (20\text{-}18)$$

Example 20-7

An inductive load taking 10 A and 2000 W from a 220-V line is in parallel with a motor taking 1400 W at a PF of 50% lagging. Find (a) the total effective power, (b) the total PF, (c) the total apparent power, and (d) the total current drawn by the circuit.

Solution

The diagram for the circuit is shown in Fig. 20-8 on page 726.

a. 1. Find the apparent power, the effective power, the phase angle, and the var power for each load.

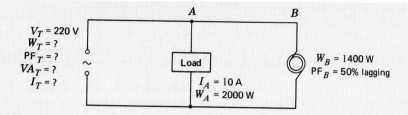

Figure 20-8

For load A:

$$VA_A = I_A \times V_A = 10 \times 220 = 2200 \text{ VA} \tag{20-2}$$
$$W_A = 2000 \text{ W}$$
$$\cos \theta = \frac{W}{VA} = \frac{2000}{2200} = 0.909 \quad \text{therefore} \quad \theta = 25° \text{ lagging} \tag{20-3}$$
$$\text{var } P_A = VA_A \times \sin \theta = 2200 \times \sin 25° = 2200 \times 0.423 = -931 \text{ var}$$

For load B:

$$W_B = 1400 \text{ W}$$
$$VA_B = \frac{W}{PF} = \frac{1400}{0.5} = 2800 \text{ VA} \tag{20-10}$$
$$PF_B = \cos \theta = 0.500 \quad \text{so} \quad \theta = 60° \tag{20-8}$$
$$\text{var } P_B = VA_B \times \sin \theta \tag{20-6}$$
$$= 2800 \times \sin 60° = 2800 \times 0.866$$
$$\text{var } P_B = -2425 \text{ var}$$

2. Find the total effective and reactive power.

$$W_T = W_A + W_B = 2000 + 1400 = 3400 \text{ W} \quad \textit{Ans.}$$
$$\text{var } P_T = \text{var } P_A + \text{var } P_B = (-931) + (-2425) = -3356 \text{ var}$$

b. Find the phase angle and the PF.

$$\tan \theta_T = \frac{\text{var } P_T}{W_T} = \frac{3356}{3400} = 0.987 \quad \text{therefore} \quad \theta = 45° \tag{20-17}$$
$$PF = \cos \theta = \cos 45° = 0.707 = 70.7\% \text{ lagging} \quad \textit{Ans.} \tag{20-8}$$

c. Find the total apparent power.

$$VA_T = \frac{W_T}{PF_T} = \frac{3400}{0.707} = 4809 \text{ VA} \quad \textit{Ans.} \tag{20-18}$$

d. Find the total current.

$$I_T = \frac{VA_T}{V_T} = \frac{4809}{220} = 21.8 \text{ A} \quad \textit{Ans.} \tag{20-2}$$

Summary

Total power drawn by combinations of reactive loads

1. Find the apparent power for each load using any of the following formulas.

$$VA = V \times I \tag{20-2}$$
or
$$VA = \frac{W}{PF} \tag{20-10}$$
or
$$VA = \frac{W}{\cos \theta} \tag{20-5}$$

2. Find the effective and reactive power for each load.

$$W = VA \times PF \qquad (20\text{-}9)$$

or
$$W = VA \times \cos\theta \qquad (20\text{-}4)$$

and
$$\text{var } P = VA \times \sin\theta \qquad (20\text{-}6)$$

Note: If the PF or the phase angle is not given, it may be found by either of the following formulas:

$$PF = \frac{W}{VA} \qquad (20\text{-}7)$$

$$\cos\theta = \frac{W}{VA} \qquad (20\text{-}3)$$

3. Find the total effective power.

$$W_T = W_A + W_B \qquad (20\text{-}13)$$

4. Find the total reactive power.

$$\text{var } P_T = \text{var } P_A + \text{var } P_B \qquad (20\text{-}14)$$

5. Find the phase angle θ.

$$\tan\theta_T = \frac{\text{var } P_T}{W_T} \qquad (20\text{-}17)$$

6. Find the total PF.

$$PF = \cos\theta \qquad (20\text{-}8)$$

7. Find the total apparent power.

$$VA_T = \frac{W_T}{PF_T} \qquad (20\text{-}18)$$

8. Find the total current drawn, using any of the following formulas.

$$VA_T = V_T \times I_T \qquad (20\text{-}2)$$

$$\boldsymbol{W_T = V_T \times I_T \times \cos\theta_T} \qquad \text{Formula 20-19}$$

$$\boldsymbol{W_T = V_T \times I_T \times PF_T} \qquad \text{Formula 20-20}$$

Exercises

Applications

1. A refrigerator motor drawing 6 A at 80% PF lagging is in parallel with a washing machine motor drawing 8 A at 80% PF lagging from a 110-V line. Find (*a*) the total effective power, (*b*) the total PF, (*c*) the total apparent power, and (*d*) the total current drawn.
2. Motor *A* draws 10 A and 800 W from a 110-V 60-Hz ac line. Motor *B* draws 6 A and 480 W from the same line in parallel. Find (*a*) the PF of each motor, (*b*) the total effective power, (*c*) the total PF, (*d*) the total apparent power, and (*e*) the total current drawn.
3. A purely resistive lamp load (PF = 1) drawing 8 A from a 110-V 60-Hz line is in parallel with an induction motor taking 10 A at 70% PF. Find (*a*) the total effective power, (*b*) the total PF, (*c*) the total apparent power, and (*d*) the total current drawn.

4. A lamp bank (PF = 1) drawing 1200 W from a 110-V 60-Hz line is in parallel with an induction motor taking 8 A at a power factor of 90%. Find (*a*) the total effective power, (*b*) the total PF, (*c*) the total apparent power, and (*d*) the total current.

5. Find (*a*) the total effective power, (*b*) the total PF, and (*c*) the total apparent power for the circuit shown in Fig. 20-9.

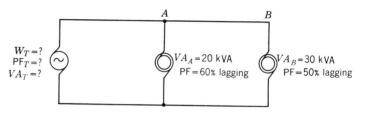

Figure 20-9

6. Repeat Exercise 4 for a 2-kW lamp load in parallel with a motor operating at a PF of 60% lagging and drawing 3 kW from a 110-V 60-Hz line.

7. Repeat Exercise 4 for the following circuit: A motor drawing 5 kW at 80% PF lagging is in parallel with a second motor drawing 8 kW at 70% PF lagging from a 220-V line.

8. Repeat Exercise 4 for the circuit shown in Fig. 20-10.

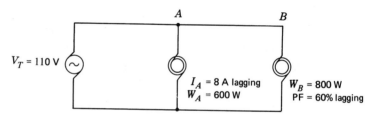

Figure 20-10

9. A 2-kW lamp load, a 70-kVA motor operating at a PF of 60% lagging, and a 40-kVA motor operating at a PF of 70% lagging are connected in parallel. Find (*a*) the total effective power, (*b*) the total PF, and (*c*) the total apparent power.

10. Repeat Exercise 4 for the circuit shown in Fig. 20-11.

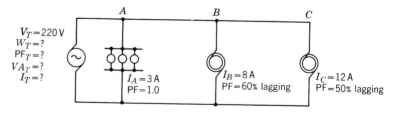

Figure 20-11

(See CD-ROM for Test 20-1)

Power in capacitive loads. When the current is out of phase with the voltage, only that portion of the current which is in phase with the voltage is useful in producing usable power.

In an inductive circuit, the current lags behind the voltage, but in a capacitive circuit, the current leads the voltage. In such a circuit, the phasor diagrams will be very similar to those of an inductive circuit, except that now there will be a *leading* current and a *leading* power. All the formulas developed for the inductive circuits in the last job will also apply to a capacitive circuit. The only difference is that the reactive power will *lead* in a capacitive circuit.

Example 20-8

A lamp bank drawing 1 kW of power is in parallel with a synchronous motor drawing 2 kW at a leading PF of 80% from a 220-V 60-Hz ac line. Find (*a*) the total effective power, (*b*) the total PF, (*c*) the total apparent power, and (*d*) the total current drawn.

Solution

The diagram of the circuit is shown in Fig. 20-12.

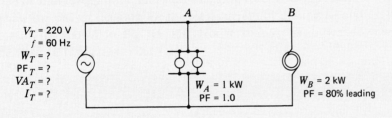

Figure 20-12

a. 1. Find the apparent power, the effective power, the phase angle, and the var power for each load.

For load *A*:

$$W_A = 1 \text{ kW} = 1000 \text{ W} \quad \text{(given)}$$

$$VA_A = \frac{W_A}{\text{PF}} = \frac{1000}{1} = 1000 \text{ VA} \tag{20-10}$$

$$\text{PF} = \cos\theta = 1.0 \quad \text{therefore} \quad \theta = 0° \tag{20-8}$$

$$\text{var } P_A = VA_A \times \sin\theta = 1000 \times \sin 0° = 1000 \times 0 = 0 \text{ var} \tag{20-6}$$

For load *B*:

$$W_B = 2 \text{ kW} = 2000 \text{ W} \quad \text{(given)}$$

$$VA_B = \frac{W_A}{\text{PF}} = \frac{2000}{0.8} = 2500 \text{ VA} \tag{20-10}$$

$$\text{PF} = \cos\theta = 0.8 \quad \text{therefore} \quad \theta = 37° \text{ leading} \tag{20-8}$$

$$\text{var } P_B = VA_B \times \sin\theta = 2500 \times \sin 37° = 2500 \times 0.6 = 1500 \text{ var}$$

> **Hint for Success**
>
> Notice that the manual computation algorithms used to calculate phase angles lose the sign information. The angle is always positive, and you have to supply the information as to whether the angle is leading or lagging.

2. Find the total effective and reactive power.

$$W_T = W_A + W_B = 1000 + 2000 = 3000 \text{ W}$$
$$\text{var } P_T = \text{var } P_A + \text{var } P_B = 0 + 1500 = 1500 \text{ var} \qquad (20\text{-}14)$$

b. Find the phase angle and the PF.

$$\tan \theta_T = \frac{\text{var } P_T}{W_T} = \frac{1500}{3000} = 0.5 \qquad \text{therefore} \qquad \theta = 27° \qquad (20\text{-}17)$$
$$\text{PF}_T = \cos \theta = \cos 27° = 89.1\% \text{ } leading \qquad (20\text{-}8)$$

c. Find the total apparent power.

$$VA_T = \frac{W_T}{\text{PF}_T} = \frac{3000}{0.891} = 3367 \text{ VA} \qquad Ans. \qquad (20\text{-}18)$$

d. Find the total current drawn.

$$I_T = \frac{VA_T}{V_T} = \frac{3367}{220} = 15.3 \text{ A} \qquad Ans. \qquad (20\text{-}2)$$

Example 20-9

A 10-kVA induction motor operating at 85% lagging PF and a 5-kVA synchronous motor operating at 68.2% leading PF are connected in parallel across a 220-V 60-Hz ac line. Find (a) the total effective power, (b) the total PF, (c) the total apparent power, and (d) the total current drawn.

Solution

The diagram for the circuit is shown in Fig. 20-13.

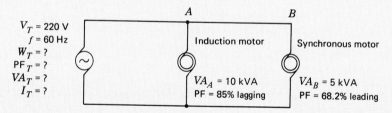

Figure 20-13

a. 1. Find the apparent power, the effective power, the phase angle, and the var power for each load.

For the induction motor A:

$$VA_A = 10 \text{ kVA} \qquad \text{(given)}$$
$$W_A = VA_A \times \text{PF}_A = 10 \times 0.85 = 8.5 \text{ kW} \qquad (20\text{-}9)$$
$$\text{PF}_A = \cos \theta = 0.85 \qquad \text{therefore} \qquad \theta = 32° \qquad (20\text{-}8)$$
$$\text{var } P_A = VA_A \times \sin \theta = 10 \times \sin 32° = 10 \times 0.53 \qquad (20\text{-}6)$$
$$\text{var } P_A = -5.3 \text{ kvar } lagging$$

For the synchronous motor B:

$$VA_B = 5 \text{ kVA} \qquad \text{(given)}$$

$$W_B = VA_B \times PF_B = 5 \times 0.682 = 3.41 \text{ kW} \qquad (20\text{-}9)$$
$$PF_A = \cos \theta = 0.682 \qquad \text{therefore} \qquad \theta = 47° \qquad (20\text{-}8)$$
$$\text{var } P_B = VA_B \times \sin \theta = 5 \times \sin 47° = 5 \times 0.73 = 3.65 \text{ kvar } \textit{leading}$$

2. Draw the phasor diagram for the apparent powers on the same voltage base as shown in Fig. 20-14a.

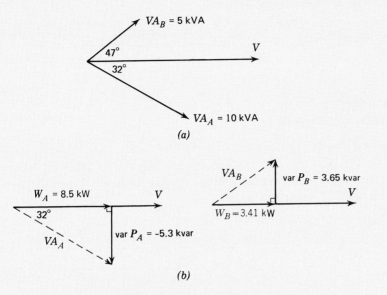

(a)

(b)

Figure 20-14 (a) Leading and lagging apparent powers. (b) Resolution of each power into its effective and reactive components.

3. Resolve the power in each branch into its components as shown in Fig. 20-14b.

4. Draw all the components on the same voltage base as shown in Fig. 20-15a. The total effective power W_T and the total reactive power var P_T may now be found.

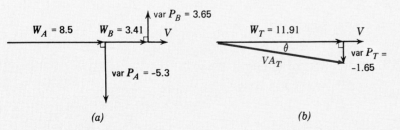

(a)   (b)

Figure 20-15 (a) The components of all the branch powers drawn on the same voltage base. (b) VA_T represents the phasor sum of $W_T +$ var P_T.

$$W_T = W_A + W_B = 8.5 + 3.41 = 11.91 \text{ kW} \qquad \textit{Ans.} \qquad (20\text{-}13)$$
$$\text{var } P_T = \text{var } P_A + \text{var } P_B = (-5.3) + 3.65 = -1.65 \text{ kvar } \textit{lagging}$$

b. Find the total PF by adding W_T and var P_T vectorially as shown in Fig. 20-15b.

CALCULATOR HINT

If problems are set up using complex arithmetic, the mathematics retains the phase angle sign information and there is less chance of error.

$$\tan \theta_T = \frac{\text{var } P_T}{W_T} = \frac{-1.65}{11.91} = -0.1385 \quad \text{so} \quad \theta = 8° \quad (20\text{-}17)$$
$$\text{PF} = \cos \theta = \cos 8° = 0.99 = 99\% \text{ lagging} \quad (20\text{-}8)$$

c. Find the total apparent power.

$$VA_T = \frac{W_T}{PF_T} = \frac{11.91}{0.99} = 12.03 \text{ kVA} = 12{,}030 \text{ VA} \quad \text{Ans.}$$

d. Find the total current drawn.

$$I_T = \frac{VA_T}{V_T} = \frac{12{,}030}{220} = 54.7 \text{ A } \text{lagging} \quad \text{Ans.} \quad (20\text{-}2)$$

Example 20-10

A synchronous motor drawing 10 A at 60% leading PF from a 110-V line is in parallel with an induction motor drawing 1 kW at 80% PF lagging. Find (a) the total effective power, (b) the total PF, (c) the total apparent power, and (d) the total current drawn.

Solution

The diagram for the circuit is shown in Fig. 20-16.

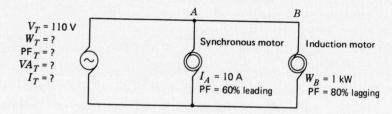

Figure 20-16

a. 1. Find the apparent power, the effective power, the phase angle, and the var power for each load.

For the synchronous motor A:

$$VA_A = I_A \times V_A = 10 \times 110 = 1100 \text{ VA} \quad (20\text{-}2)$$
$$W_A = VA_A \times PF_A = 1100 \times 0.6 = 660 \text{ W} \quad (20\text{-}9)$$
$$PF_A = \cos \theta = 0.600 \quad \text{therefore} \quad \theta = 53° \quad (20\text{-}8)$$
$$\text{var } P_A = VA_A \times \sin \theta = 1100 \times \sin 53° = 1100 \times 0.8 \quad (20\text{-}6)$$
$$\text{var } P_A = 880 \text{ var } \textit{leading}$$

For the induction motor B:

$$W_B = 1 \text{ kW} = 1000 \text{ W} \quad \text{(given)}$$
$$VA_B = \frac{W_B}{PF_B} = \frac{1000}{0.8} = 1250 \text{ VA} \quad (20\text{-}10)$$
$$PF_B = \cos \theta = 0.800 \quad \text{therefore} \quad \theta = 37° \quad (20\text{-}8)$$
$$\text{var } P_B = VA_B \times \sin \theta = 1250 \times \sin 37° = 1250 \times 0.6 \quad (20\text{-}6)$$
$$\text{var } P_B = -750 \text{ var } \textit{lagging}$$

2. Find the total effective and reactive power.

$$W_T = W_A + W_B = 660 + 1000 = 1660 \text{ W} \qquad Ans.$$
$$\text{var } P_T = \text{var } P_A + \text{var } P_B = (+880) + (-750) = 130 \text{ var } leading$$

b. Find the phase angle and the PF.

$$\tan \theta_T = \frac{\text{var } P_T}{W_T} = \frac{130}{1660} = 0.078 \qquad so \qquad \theta = 4° \qquad (20\text{-}17)$$
$$\text{PF}_T = \cos \theta = \cos 4° = 0.998 = 99.8\% \; leading \qquad Ans.$$

c. Find the total apparent power.

$$VA_T = \frac{W_T}{\text{PF}_T} = \frac{1660}{0.998} = 1663 \text{ VA} \qquad Ans. \qquad (20\text{-}18)$$

d. Find the total current drawn.

$$I_T = \frac{VA_T}{V_T} = \frac{1663}{110} = 15.1 \text{ A } leading \qquad Ans. \qquad (20\text{-}2)$$

Exercises

Applications

1. An induction motor drawing 400 VA at 80% lagging PF is in parallel with a synchronous motor drawing 700 VA at 90% leading PF. Find (a) the total effective power, (b) the total PF, and (c) the total apparent power.
2. Repeat Exercise 1 for a circuit containing an induction motor that draws 100 kVA at a PF of 85% in parallel with a capacitive load that draws 80 kVA at a PF of 70%.
3. A capacitive load drawing 20 A at 70% PF from a 120-V 60-Hz line is in parallel with an induction motor drawing 4 kW at 70% PF. Find (a) the total effective power, (b) the total PF, (c) the total apparent power, and (d) the total current drawn.
4. Repeat Exercise 3 for a circuit containing a 60% PF induction motor drawing 1 kW in parallel with a synchronous motor rated at 120 V, 15 A, and 70% PF.
5. Find the total PF of a parallel combination of a 5-kW lamp load, a 10-kW inductive load (PF = 80%), and an 8-kW capacitive load (PF = 90%).
6. Find (a) the total effective power, (b) the total PF, (c) the total apparent power, and (d) the total current drawn by the circuit shown in Fig. 20-17.

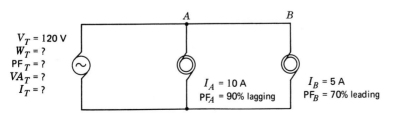

$V_T = 120 \text{ V}$
$W_T = ?$
$PF_T = ?$
$VA_T = ?$
$I_T = ?$

A B

$I_A = 10 \text{ A}$
$PF_A = 90\%$ lagging

$I_B = 5 \text{ A}$
$PF_B = 70\%$ leading

Figure 20-17

7. Repeat Exercise 6 for the circuit shown in Fig. 20-18 on page 734.

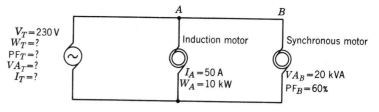

Figure 20-18

8. An induction motor draws 20 A at a PF of 80% from a 120-V line. What is the total PF of the circuit if (*a*) a lamp load drawing 15 A is connected in parallel and (*b*) a capacitive load drawing 5 A at a PF of 60% leading is connected in parallel?

9. Repeat Exercise 6 for a circuit containing an induction motor taking 4 A at 80% PF lagging, a synchronous motor taking 8 A at 50% PF leading, and a lamp load taking 6 A at 100% PF, all connected in parallel across a 120-V line.

10. Repeat Exercise 6 for a circuit containing a 100-W inductive load at a PF of 80%, a 600-W capacitive load at a PF of 60%, and an induction motor drawing 1 kW at a PF of 90%, all connected in parallel across a 120-V line.

(See CD-ROM for Test 20-2)

JOB 20-4

Power-Factor Correction

Consider the three circuits shown in Fig. 20-19. Each circuit draws the same effective power, but at decreasing PFs.

For the circuit of Fig. 20-19*a:*

$$W = V \times I \times PF \qquad (18\text{-}4)$$
$$1100 = 110 \times I \times 1$$
$$I = \frac{1100}{110} = 10 \text{ A}$$

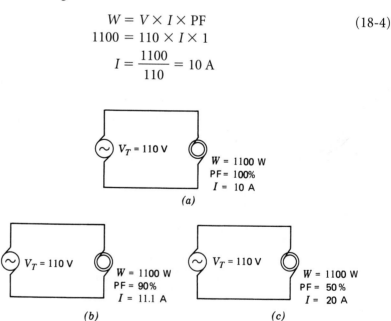

Figure 20-19 In order to produce a constant effective power, the current must increase as the power factor decreases.

For the circuit of Fig. 20-19b:

$$W = V \times I \times \text{PF} \qquad\qquad (18\text{-}4)$$
$$1100 = 110 \times I \times 0.9$$
$$1100 = 99 \times I$$
$$I = \frac{1100}{99} = 11.1 \text{ A}$$

For the circuit of Fig. 20-19c:

$$W = V \times I \times \text{PF} \qquad\qquad (18\text{-}4)$$
$$1100 = 110 \times I \times 0.5$$
$$1100 = 55 \times I$$
$$I = \frac{1100}{55} = 20 \text{ A}$$

By an investigation of the amount of current drawn in each of the circuits, we can see that as the PF decreases, more and more current must be supplied in order to produce the same effective power. Now, regardless of the current that is drawn to provide this 1100 W of effective power, the consumer pays for only 1100 W. Therefore, if the power company is forced to send 20 instead of 10 A to provide 1100 W of power, it must provide heavier wires to carry the larger current. This is expensive. In addition, since heating losses depend on the *square* of the current, these losses will be higher at low PF than at high PF. For these reasons, the power company demands an extra premium payment if the PF falls below a certain value for a particular installation.

A low PF is generally due to the large consumption of power by underloaded induction motors that take a *lagging* current. This will produce a large, wasteful, *lagging* reactive power. In order to correct this low power factor, an equal amount of *leading* reactive power should be connected into the circuit. To do this, synchronous motors, which operate at unity, or leading power factors are placed in parallel with the inductive load. The leading reactive kilovars supplied by the synchronous motor compensate for the lagging reactive kilovars and result in an improvement of the total power factor. Any improvement in the PF will release supply capacity, increase efficiency, and generally improve the operating characteristics of the system.

Capacitors, which take a leading current, may also be used for this purpose. They are rated in kilovars instead of microfarads. Standard ratings are either 15 or 25 kvar. Although it is desirable to raise the power factor to 100%, any standard capacitor that raises the power factor from 90 to 95% is considered acceptable.

> **Hint for Success**
>
> Power factor correction is becoming a legal and economic issue.

Example 20-11

An induction motor draws 1300 W and 1500 VA from an ac line. (*a*) What is its power factor? (*b*) What is the value of the reactive voltamperes, or vars, drawn by the motor? (*c*) How many vars of leading reactive power are needed to raise the power factor to unity?

Solution

a.
$$\text{PF} = \frac{W}{VA} = \frac{1300}{1500} = 0.8667 = 86.7\% \qquad Ans. \qquad (20\text{-}7)$$

b. 1. $$\text{Since } \cos \theta = \text{PF} \qquad (20\text{-}8)$$
$$\cos \theta = 0.8667$$
$$\theta = 30°$$

2. $$\text{var power} = VA \times \sin \theta \qquad (20\text{-}6)$$
$$= 1500 \times \sin 30°$$
$$= 1500 \times 0.5 = 750 \text{ var} \qquad Ans.$$

c. The leading var power should equal the lagging var power, or 750 var. *Ans.*

Example 20-12

An induction motor takes 15 kVA at 220 V and 80% lagging PF. What must be the PF of a 10-kVA synchronous motor connected in parallel in order to raise the total PF to 100%, or unity?

Solution

Given: $VA = 15$ kVA   Find: To get a PF = 1,
$PF = 80\%$ *lagging* $\Big\}$ induction motor   PF of synchro-
$V = 220$ V   nous motor = ?
$VA = 10$ kVA synchronous motor

1. Find the var P of the induction motor.

Since

$$PF = \cos \theta = 0.8 \qquad \theta = 37° \qquad (20\text{-}8)$$
$$\text{var } P = VA \times \sin \theta \qquad (20\text{-}6)$$
$$\text{var } P = 15 \times \sin 37° = 15 \times 0.6 = -9 \text{ kvar } lagging$$

The phasor diagram for the induction motor is shown in Fig. 20-20a.

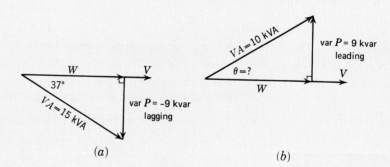

(a)   (b)

Figure 20-20   (a) Phasor diagram for the induction motor. (b) Phasor diagram for the synchronous motor.

2. To adjust the PF to unity means that cos $\theta = 1.0$. At unity PF, if cos $\theta = 1.0$, θ will equal 0°. When the phase angle is 0°,

$$\text{var } P = VA \times \sin \theta \qquad (20\text{-}6)$$
$$\text{var } P = VA \times \sin 0°$$
$$\text{var } P = VA \times 0 = 0 \text{ var}$$

Thus, to adjust a circuit to unity PF, all that is required is to make the total reactive power equal to 0 var. Since we already have 9 kvar *lagging* in the circuit, we must bring this down to zero by adding 9 kvar *leading*. This *leading* reactive power must come from the synchronous motor. The phasor diagram for the synchronous motor must be as shown in Fig. 20-20b. We can find the phase angle for the synchronous motor from this diagram.

3.

$$\sin \theta = \frac{\text{var } P}{VA} \qquad \text{20-21}$$

Formulas

$$\sin \theta = \frac{\text{var } P \text{ of induction motor}}{VA \text{ of synchronous motor}} \qquad \text{20-22}$$

$$\sin \theta = \frac{9}{10} = 0.9000$$

$$\theta = 64°$$

4. Find the PF of the synchronous motor.

$$PF = \cos \theta = \cos 64° \qquad (20\text{-}8)$$
$$= 0.438$$
$$PF = 43.8\% \qquad Ans.$$

Example 20-13

A 220-V 50-A induction motor draws 10 kW of power. An 8-kVA synchronous motor is placed in parallel with it in order to adjust the PF to unity. What must be the PF of the synchronous motor?

Solution

Given: $\left.\begin{array}{l} V = 220 \text{ V} \\ I = 50 \text{ A} \\ W = 10 \text{ kW} \end{array}\right\}$ induction motor   Find: To get a PF = 1.0, PF of synchronous motor = ?

$VA = 8$ kVA   synchronous motor

1. Find the var P of the induction motor.

$$VA = I \times V = 50 \times 220 = 11{,}000 \text{ VA} = 11 \text{ kVA}$$

Since

$$\cos \theta = \frac{W}{VA} \qquad (20\text{-}3)$$

$$\cos \theta = \frac{10}{11} = 0.909$$

$$\theta = 25°$$

$$\text{var } P = VA \times \sin \theta = 11 \times \sin 25° \qquad (20\text{-}6)$$

$$\text{var } P = 11 \times 0.423 = -4.65 \text{ kvar } lagging$$

2. Draw the phasor diagram for the induction motor as shown in Fig. 20-21a.

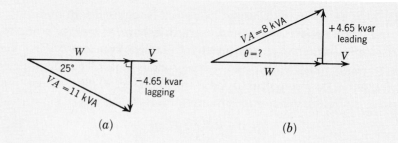

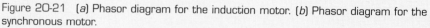

Figure 20-21 (a) Phasor diagram for the induction motor. (b) Phasor diagram for the synchronous motor.

To adjust the PF of the circuit to unity, the phasor diagram for the synchronous motor must indicate a *leading* power as shown in Fig. 20-21*b*.

3. Find the phase angle for the synchronous motor.

$$\sin \theta = \frac{\text{var } P}{VA} = \frac{4.65}{8} = 0.581$$

$$\theta = 36°$$

4. Find the PF of the synchronous motor.

$$PF = \cos \theta = \cos 36° = 0.809 = 80.9\% \qquad Ans. \qquad (20\text{-}8)$$

Note: If only the effective power W of the synchronous motor is given, find angle θ for the synchronous motor by using the

$$\tan \theta = \frac{\textbf{lagging var } P}{\textbf{W of synchronous motor}} \qquad \text{Formula 20-23}$$

Then $$PF = \cos \theta \qquad\qquad\qquad (20\text{-}8)$$

Example 20-14

An induction motor takes 64 kW at 80% PF from an ac line. Find (*a*) the apparent power, (*b*) the phase angle, (*c*) the lagging var power, (*d*) the number of standard 15- or 25-kvar capacitors needed to improve the PF without producing a leading PF, and (*e*) the new PF.

Solution

Given: $W = 64$ kW   Find: $VA = ?$
        PF = 80% *lagging*   Phase angle = ?
                                           Lagging var power = ?
                                                    $C = ?$
                                              New PF = ?

a. Find the apparent power of the induction motor.

$$VA = \frac{W}{PF} = \frac{64 \text{ kW}}{0.8} = 80 \text{ kVA} \qquad Ans. \qquad (20\text{-}10)$$

b. $$PF = \cos \theta = 0.800 \qquad\qquad (20\text{-}8)$$
$$\theta = 37° \qquad Ans.$$

c. Find the var P of the induction motor.
$$\text{var } P = VA \times \sin \theta = 80 \times \sin 37° \qquad (20\text{-}6)$$
$$= 80 \times 0.6 = -48 \text{ kvar} \qquad Ans.$$

d. Two 25-kvar capacitors would produce a leading PF, and so we use one 15-kvar and one 25-kvar capacitor to balance the lagging 48 kvar.

e. This will still leave 8 kvar of reactive power in the circuit ($48 - 40 = 8$).
$$\sin \theta = \frac{\text{var } P}{VA} = \frac{8}{80} = 0.1000 \qquad (20\text{-}21)$$
$$\theta = 6°$$
$$\text{New PF} = \cos \theta = \cos 6° = 0.9945 \qquad (20\text{-}8)$$
$$\text{PF} = 99.5\% \qquad Ans.$$

Power-Factor Correction with Capacitors

Figure 20-22*a* shows the phasor diagram for a circuit with a given power factor (θ_1), and Fig. 20-22*b* shows the phasor diagram for the circuit with an improved power factor (θ_2). In Fig. 20-22*a*,

$$\tan \theta_1 = \frac{\text{kvar}_1}{\text{kW}}$$

or
$$\text{kvar}_1 = \text{kW} \times \tan \theta_1$$

In Fig. 20-22*b*,
$$\tan \theta_2 = \frac{\text{kvar}_2}{\text{kW}}$$

or
$$\text{kvar}_2 = \text{kW} \times \tan \theta_2$$

The *difference* between these kvar values will be the capacitor kvar (C kvar) needed to improve the power factor from the first situation (θ_1) to the improved second situation (θ_2).

$$C \text{ kvar} = \text{kvar}_1 - \text{kvar}_2$$
$$= \text{kW} \times \tan \theta_1 - \text{kW} \times \tan \theta_2$$

$$\mathbf{C \text{ kvar} = kW (\tan \theta_1 - \tan \theta_2)} \qquad 20\text{-}24$$

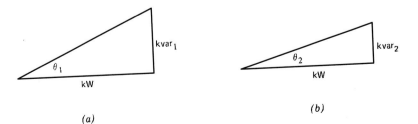

(a)   *(b)*

Figure 20-22   (*a*) Phasor diagram for the original circuit. (*b*) Phasor diagram for the circuit with improved PF.

Example 20-15

A plant has a load of 300 kW with an average PF of 80%. How much capacitor kvar is required to raise the power factor 90%? How many standard 25-C kvar capacitors would be needed?

Solution

Given: $\quad\quad W = 300 \text{ kW}$ $\quad\quad\quad$ Find: C kilovars
$\quad\quad\quad\quad\quad$ PF = 80% $\quad\quad\quad\quad\quad\quad\quad$ needed.
$\quad\quad\quad$ Required PF = 90%

1. Find the phase angles.

$$\cos\theta_1 = \text{PF} = 80\% = 0.8000 \quad\quad\quad (20\text{-}8)$$
$$\theta_1 = 37°$$
$$\cos\theta_2 = \text{PF} = 90\% = 0.9000 \quad\quad\quad (20\text{-}8)$$
$$\theta_2 = 26°$$

2. Find the C kilovars needed.

$$C\,\text{kvar} = \text{kW}(\tan\theta_1 - \tan\theta_2) \quad\quad (20\text{-}24)$$
$$= 300(\tan 37° - \tan 26°)$$
$$= 300(0.7536 - 0.4877)$$
$$= 300(0.2659) = 79.77 \text{ kvar}$$

Use three standard 25-C kvar capacitors. $\quad$ *Ans.*

Example 20-16

Find the system capacity that has been released by the improvement in the power factor of Example 20-15.

Solution

1. Find the original VA.

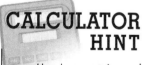
$$VA = \frac{W}{\text{PF}} = \frac{300}{0.8} = 375 \text{ kVA} \quad\quad (20\text{-}10)$$

2. Find the improved VA.

$$VA = \frac{W}{\text{PF}} = \frac{300}{0.9} = 333 \text{ kVA}$$

3. Find the released capacity. At 80% PF, the circuit required 375 kVA. At 90% PF, the circuit requires only 333 kVA.

$$\text{Released capacity} = 375 - 333 = 42 \text{ kVA at } 80\% \quad\quad \textit{Ans.}$$

Example 20-17

Examples 20-15 and 20-16 may both be solved quickly (although not so accurately) by the use of the graphs shown in Fig. 20-23.

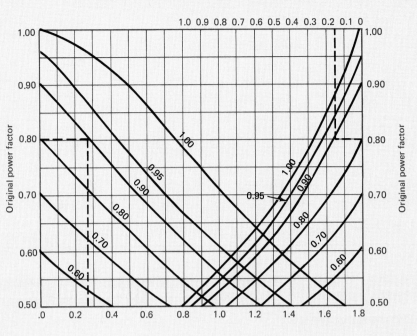

System capacity released—kVA per kilowatt load

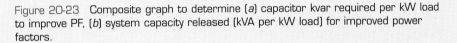

Figure 20-23 Composite graph to determine (a) capacitor kvar required per kW load to improve PF, (b) system capacity released (kVA per kW load) for improved power factors.

Find the C kilovars for Example 20-15.

1. Locate the original power factor 0.80 at the *left* of the graph.
2. Read horizontally to the curve marked 0.90 (the new PF).
3. Read down to find 0.27 C kvar per kilowatt load.
4. Multiply this 0.27 by the 300 kW to obtain 81 C kvar.   *Ans.*

Use the nearest standard value of 80 C kvar.

Find the system capacity released for Example 20-15.

1. Locate the original power factor 0.80 at the *right* side of the graph.
2. Read horizontally to the left to the curve marked 0.90 (the new PF).
3. Read *up* to find 0.15 kVA released per kilowatt load.
4. Multiply this 0.15 by the 300 kW to obtain 45 kVA released.

This means that an additional 45 kVA at 80% PF load may be added to the existing circuits as a result of the improved PF.

Self-Test 20-18

A circuit operates with a load of 10 kW at a power factor of 85%. How much capacitor kilovar is required to raise the power factor to unity? Find the amount of released capacity.

Solution

1. Find the phase angles.

$$\cos \theta_1 = PF = 85\% = \underline{\quad(1)\quad}$$ $\quad$ (20-8)

$$\theta_1 = \underline{\quad(2)\quad}°$$

$$\cos \theta_2 = PF = \underline{\quad(3)\quad}$$

$$\theta_2 = \underline{\quad(4)\quad}°$$

2. Find the C kilovars needed.

$$C\,kvar = \underline{\quad(5)\quad} \quad (\tan \theta_1 = \tan \theta_2)$$

$$= 10(\tan 32° - \tan 0°)$$

$$= 10(\underline{\quad(6)\quad} - \underline{\quad(7)\quad})$$

$$= 10(\underline{\quad(8)\quad}) = \underline{\quad(9)\quad} \; C\,kvar \quad Ans.$$

3. Find the amount of released capacity. Find the original VA.

$$VA = \frac{(10)}{(11)} \qquad (20\text{-}10)$$

$$= \frac{10}{0.85} = \underline{\quad(12)\quad} \; kVA$$

Find the improved VA.

$$VA = \frac{10}{1} = 10 \; kVA$$

Find the released capacity.

$$11.8 - 10 = \underline{\quad(13)\quad} \; kVA \quad Ans.$$

Self-Test 20-19

Check Self-Test 20-18 using the graph of Fig. 20-23.

Solution

Find the C kilovars needed.

1. Locate the original power factor 0.85 at the ___(1)(left/right)___ side of the graph.
2. Read horizontally to reach the curve marked ___(2)___.
3. Read ___(3)(up/down)___ to read ___(4)___ C kvar per kilowatt load.
4. Multiply this 0.63 by ___(5)___ kW to get ___(6)___ C kvar.

Find the amount of increased capacity.

1. Locate the original power factor 0.85 at the ___(7)(left/right)___ side of the graph.
2. Read horizontally to the left to reach the curve marked ___(8)___.
3. Read ___(9)(up/down)___ to read ___(10)___ released kVA per kilowatt load.
4. Multiply this 0.18 by ___(11)___ to get ___(12)___ kVA.

Example 20-20

When operating at full load, an induction motor draws 800 W and 4 A from a 220-V 60-Hz line. Find (*a*) the PF of the motor, (*b*) the lagging reactive power, (*c*) the apparent power drawn by a capacitor in order to raise the PF to unity, (*d*) the current drawn by this capacitor, (*e*) the reactance of the capacitor, and (*f*) the capacitance of this capacitor.

Solution

Given: $W = 800$ W   Find: PF = ?
         $I = 4$ A   var P = ?
         $V = 220$ V   VA of capacitor = ?
         $f = 60$ Hz I_C = ?
                                        X_C = ?
                                           C = ?

a. Find the PF of the motor.

1. $VA = V \times I = 220 \times 4 = 880$ VA   (20-2)

2. $\cos \theta = \dfrac{W}{VA} = \dfrac{800}{880} = 0.909$ *Ans.*   (20-3)

                 $\theta = 25°$

b. Find the lagging reactive power.

$$\text{var } P = VA \times \sin \theta = 880 \times \sin 25° \qquad (20\text{-}6)$$
$$\text{var } P = 880 \times 0.423 = -372 \text{ var } \textit{lagging} \qquad \textit{Ans.}$$

c. In order to raise the PF to unity, the capacitor must provide 372 var of reactive power leading. Since the apparent power in a capacitor is equal to the var *P*, 372 VA of capacitor power will exactly balance the lagging var *P*.

d. Find the current drawn by this capacitor.

$$VA = I \times V \qquad (20\text{-}2)$$
$$372 = I \times 220$$
$$I = \frac{372}{220} = 1.69 \text{ A} \qquad \textit{Ans.}$$

e. Find the reactance of the capacitor.

$$V_C = I_C \times X_C \qquad (17\text{-}7)$$
$$220 = 1.69 \times X_C$$
$$X_C = \frac{220}{1.69} = 130 \ \Omega \qquad \textit{Ans.}$$

f. Find the capacitance of the capacitor.

$$X_C = \frac{159{,}000}{f \times C} \qquad (17\text{-}6)$$
$$131.2 = \frac{159{,}000}{60 \times C}$$
$$131.2 \times 60 \times C = 159{,}000$$
$$C = \frac{159{,}000}{7872} = 20.2 \ \mu\text{F} \qquad \textit{Ans.}$$

> **Hint for Success**
>
> This job shows how changes in technology can affect accepted practice in a field. In this case, actions formerly permitted under code are now regarded as dangerous.

Applications

1. A 440-V line delivers 15 kVA to a load at 75% PF lagging. To what PF should a 10-kVA synchronous motor be adjusted in order to raise the PF to unity when connected in parallel?

2. A 220-V line delivers 10 kVA to a load at 80% PF lagging. What must be the PF of an 8-kW synchronous motor in parallel in order to raise the PF to unity?

3. A 220-V 20-A induction motor draws 3 kW of power. A 4-kVA synchronous motor is placed in parallel to adjust the PF to unity. What must be the PF of the synchronous motor?

4. A bank of motors draws 20 kW at 75% PF lagging from a 440-V 60-Hz line. What must be the capacity of a static capacitor connected across the motor terminals if it is to raise the total PF to 1.0?

5. A motor draws 1500 W and 7.5 A from a 220-V 60-Hz line. What must be the capacity of a capacitor in parallel that will raise the total PF to unity?

6. A 4-hp motor operates at an efficiency of 85% and a PF of 80% lagging when connected across a 120-V 60-Hz line. Find the capacity of the capacitor needed to raise the total PF to 100%. *Hint:* Find the kilowatt input to the motor, and proceed as before.

7. An inductive load draws 5 kW at 60% PF from a 220-V 60-Hz line. Find the kilovoltampere rating of the capacitor needed to raise the total PF to 100%.

8. *a.* A 30-kW motor operates at 80% PF lagging. In parallel with it is a 50-kW motor that operates at 90% PF lagging. Find (1) the total effective power, (2) the total PF, and (3) the total apparent power.

   b. Find the PF adjustment that must be made on a 20-kW synchronous motor in parallel with the motors in (*a*) in order to raise the PF of the circuit to unity.

9. An industrial plant has a load of 200 kW with an average power factor of 70%. How much capacitor kilovar is required to raise the power factor to 90%? How many standard 25-*C* kvar capacitors would be needed? Find the system capacity that would be released by this improvement in the power factor. Check the results using the graph shown in Fig. 20-23.

10. A circuit takes 150 kVA at an 80% lagging power factor. (*a*) Find the *C* kilovars needed to raise the power factor to 95%. (*b*) Find the system capacity released. (*c*) What values are obtained for (*a*) and (*b*) using the graph in Fig. 20-23?

11. One load at a factory takes 50 kVA at 50% power factor lagging, and another load connected to the same line takes 100 kVA at 86.6% power factor lagging. Find the total (*a*) effective power, (*b*) reactive power, (*c*) power factor, (*d*) apparent power, and (*e*) number and size of standard *C* kilovar capacitors needed to raise the power factor to 90%.

12. A group of induction motors takes 100 kVA at 84% power factor lagging. To improve the power factor, a synchronous motor taking 60 kVA at 70.7% power factor leading was connected in the same line. Find the total (*a*) effective power, (*b*) reactive power, (*c*) new power factor, and (*d*) apparent power.

Review of AC Power

The apparent power *VA* is the product of the voltage and the current used by a circuit as measured by ac meters.

The effective power *W* is the actual power used by the circuit as measured by a wattmeter.

The power factor (PF) of a circuit is the ratio of the effective power to the apparent power. It may be expressed as a decimal or as a percent. The PF is also equal to the cosine of the phase angle of the circuit.

$$\textbf{PF} = \textbf{cos}\ \theta \qquad\qquad 20\text{-}8$$

The relations among *VA*, *W*, and PF are given by

$$\textbf{PF} = \frac{W}{VA} \qquad 20\text{-}7 \qquad \textbf{or} \qquad \textbf{cos}\ \theta = \frac{W}{VA} \qquad 20\text{-}3$$

$$\textbf{W} = \textbf{VA} \times \textbf{PF} \qquad 20\text{-}9 \qquad \textbf{or} \qquad \textbf{W} = \textbf{VA} \times \textbf{cos}\ \theta \qquad 20\text{-}4$$

$$\textbf{VA} = \frac{W}{PF} \qquad 20\text{-}10 \qquad \textbf{or} \qquad \textbf{VA} = \frac{W}{\cos\ \theta} \qquad 20\text{-}5$$

The apparent power may be resolved into its components—the effective power *W* and the reactive power *var*—by the following formulas:

$$\textbf{W} = \textbf{VA} \times \textbf{cos}\ \theta \qquad\qquad 20\text{-}4$$

$$\textbf{var}\ P = \textbf{VA} \times \textbf{sin}\ \theta \qquad\qquad 20\text{-}6$$

The procedure for solving problems involving resistive, inductive, and capacitive loads is as follows:

1. For each load, find

    a. The apparent power, using the formula

$$\textbf{VA} = \textbf{V} \times \textbf{I} \qquad\qquad 20\text{-}2$$

or

$$\textbf{VA} = \frac{W}{PF} \qquad\qquad 20\text{-}10$$

or

$$\textbf{VA} = \frac{W}{\cos\ \theta} \qquad\qquad 20\text{-}5$$

or

$$(\textbf{VA})^2 = \textbf{W}^2 + (\textbf{var}\ P)^2 \qquad\qquad 20\text{-}12$$

   b. The effective power, using the formula

$$\textbf{W} = \textbf{VA} \times \textbf{PF} \qquad\qquad 20\text{-}9$$

or

$$\textbf{W} = \textbf{VA} \times \textbf{cos}\ \theta \qquad\qquad 20\text{-}4$$

   c. The phase angle, using the formula

$$\textbf{cos}\ \theta = \frac{W}{VA} \qquad\qquad 20\text{-}3$$

or

$$\textbf{cos}\ \theta = \textbf{PF} \qquad\qquad 20\text{-}8$$

or

$$\textbf{tan}\ \theta = \frac{\textbf{var}\ P}{W} \qquad\qquad 20\text{-}11$$

d. The var power.

$$\text{var } P = VA \times \sin \theta \qquad\qquad \text{20-6}$$

2. Find the total effective power.

$$W_T = W_A + W_B \qquad\qquad \text{20-13}$$

3. Find the total reactive power.

$$\text{var } P_T = \text{var } P_A + \text{var } P_B \qquad\qquad \text{20-14}$$

4. Find the phase angle for the entire circuit by adding W_T and var P_T vectorially. In the triangle thus formed,

$$\tan \theta_T = \frac{\text{var } P_T}{W_T} \qquad\qquad \text{20-17}$$

5. Find the total PF.

$$PF_T = \cos \theta_T \qquad\qquad \text{20-8}$$

or

$$PF_T = \frac{W_T}{VA_T} \qquad\qquad \text{20-16}$$

6. Find the total apparent power.

$$VA_T = \frac{W_T}{PF_T} \qquad\qquad \text{20-18}$$

or

$$VA_T{}^2 = W_T{}^2 + \text{var } P_T{}^2 \qquad\qquad \text{20-15}$$

7. Find the total current drawn, using the formula

$$VA_T = I_T \times V_T \qquad\qquad \text{20-2}$$

or

$$W_T = V_T \times I_T \times \cos \theta_T \qquad\qquad \text{20-19}$$

or

$$W_T = V_T \times I_T \times PF_T \qquad\qquad \text{20-20}$$

Power-Factor Correction

1. By synchronous motors:
   a. Find the var P of the induction motor given.
   b. For correction to unity PF, the leading var P of the synchronous motor must equal the lagging var P of the induction motor. Therefore, for the synchronous motor,

$$\sin \theta = \frac{\textbf{var } P \textbf{ of induction motor}}{VA \textbf{ of synchronous motor}} \qquad\qquad \text{20-22}$$

or

$$\tan \theta = \frac{\textbf{lagging var } P}{W \textbf{ of synchronous motor}} \qquad\qquad \text{20-23}$$

   from either of which the value of angle θ may be obtained.
   c. The PF of the synchronous motor is

$$PF = \cos \theta \qquad\qquad \text{20-8}$$

2. By capacitors rated in C kilovars:
   a. (1) Find the kilovar P of the induction motor load.
   　　(2) Balance this reactive power using standard 15- or 25-C kvar capacitors.

It is not necessary to achieve 100% PF; 90 to 95% is considered good practice.

b. Use the formula $C\,\text{kvar} = \text{kW}(\tan\theta_1 - \tan\theta_2)$

3. By capacitors rated in microfarads:

a. Find the var P of the induction motor given.

b. Set this var P equal to the apparent power VA drawn by the capacitor, since the apparent power of a capacitor is equal to its var P.

c. Solve for I in the formula

$$VA = I \times V \qquad\qquad\qquad 20\text{-}2$$

d. Find the reactance of the capacitor.

$$V_C = I_C \times X_C \qquad\qquad\qquad 17\text{-}7$$

e. Find the capacitance at the given frequency.

$$X_C = \frac{159{,}000}{f \times C} \qquad\qquad\qquad 17\text{-}6$$

System Capacity Released by Improved Power Factor

1. Find the original voltamperes by the formula

$$VA = \frac{W}{PF_1} \qquad\qquad\qquad 20\text{-}10$$

2. Find the new voltamperes with the improved PF.

$$VA = \frac{W}{PF_2} \qquad\qquad\qquad 20\text{-}10$$

3. Subtract. This will be the number of voltamperes newly available at the original PF.

Exercises

Applications

1. A coil of 7 Ω resistance and 24 Ω reactance is connected across a 120-V 60-Hz line. Find the PF and the current drawn.

2. A 2-hp motor operates at a PF of 80% and an efficiency of 90%. Find (a) the watt input to the motor and (b) the apparent power delivered to the motor.

3. A motor drawing 10 A at 90% PF lagging is in parallel with another motor drawing 8 A at 90% PF lagging from a 120-V line. Find (a) the total effective power, (b) the total PF, (c) the total apparent power, and (d) the total current drawn.

4. A purely resistive lamp load drawing 20 A from a 220-V 60-Hz line is in parallel with a capacitive motor taking 8 A at 34.2% PF. Find (a) W_T, (b) PF_T, (c) VA_T, and (d) I_T.

5. A 1-kW lamp load, a 20-kVA induction motor operating at a PF of 70%, and a 15-kVA motor operating at a PF of 50% lagging are connected in parallel. Find (a) W_T, (b) PF_T, and (c) VA_T.

6. An induction motor drawing 250 VA at 70% PF is in parallel with a synchronous motor drawing 400 VA at 60% PF leading. Find (a) the total effective power, (b) the total PF, and (c) the total apparent power.

7. An induction motor drawing 10 A at 80% PF is in parallel with a capacitive load drawing 1 kW at 90% PF from a 117-V 60-Hz line. Find (a) the total

effective power, (*b*) the total PF, (*c*) the total apparent power, and (*d*) the total current drawn.

8. A lamp bank drawing 10 A, a synchronous motor drawing 8 A at 75% PF leading, and an induction motor drawing 5 A at 80% PF are all connected in parallel across a 230-V 60-Hz line. Find (*a*) the total apparent power and (*b*) the total current drawn.

9. A 120-V 20-A induction motor draws 2 kW of power. A 1.5-kVA synchronous motor is placed in parallel with it to adjust the PF to unity. What is the PF of the synchronous motor?

10. An induction motor draws 1 kW from a 110-V 60-Hz line at a PF of 70%. Find (*a*) the apparent power, (*b*) the lagging var power, and (*c*) the capacitance needed in parallel to raise the total PF to 1.0.

11. An induction motor draws 1200 W and 1500 VA from an ac line. (*a*) What is its power factor? (*b*) What is the value of the reactive voltamperes drawn by the motor? (*c*) How many vars of leading reactive power are needed to raise the power factor to unity?

12. An induction motor takes 20 kVA at 220 V and 80% PF. What must be the PF of a 15-kVA synchronous motor connected in parallel in order to raise the total PF to 100%?

13. An inductive load takes 70 kW at 85% PF from an ac line. Find (*a*) the apparent power, (*b*) the phase angle, (*c*) the lagging var power, (*d*) the number of standard 15- or 25-kvar capacitors needed to improve the PF without producing a leading PF, and (*e*) the new power factor.

14. A plant has a load of 250 kW with an average PF of 85%. How much capacitor kilovar is required to raise the power factor to 90%? Find the system capacity that has been released.

15. A circuit operates with a load of 30 kW at a power factor of 80%. How much capacitor kilovar is required to raise the power factor to unity? Find the system capacity that has been released.

(See CD-ROM for Test 20-3)

JOB 20-6

Three-Wire, Single-Phase 120/240 Circuits

This job further expands some ideas that have been presented earlier. In Job 4-4 the concept of cancellation of neutral line currents was introduced for dc circuits. In Example 6-5 the same configuration was again reviewed as a series-parallel circuit. Job 15-9 introduced harmonics. In this job the concepts of these three earlier jobs will be used to analyze the effects of harmonics on the neutral line current.

This job extends these ideas to three-wire, single-phase 120/240 circuits. It will show that under "balanced" conditions, there is no current in the neutral line. The problem arises when nonlinear loads generate harmonics. The line may be balanced for the fundamental, or 60-Hz, power, but it is not balanced for the harmonics. These harmonics cause current in the neutral line and hence can cause this current to exceed its design value. Section 220-22 of NEC allows reduction of the ampacity of the neutral line to as low as 70% of the ampacity of the hot lines. This reduction may not be taken for circuits feeding discharge lighting and/or computers because these types of loads generate large harmonics.

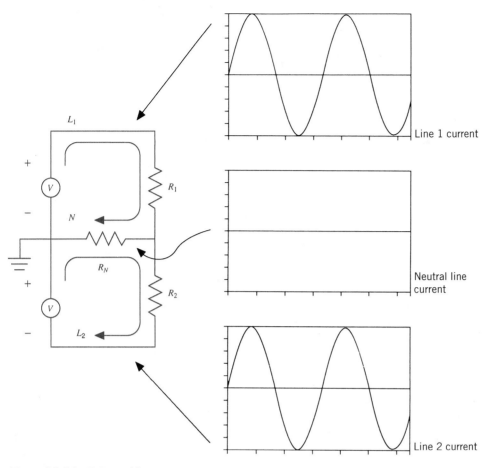

Figure 20-24 Balanced line currents.

Figure 20-24 is a model of the circuit that we are discussing. It appears from the model that there are two independent voltage sources. Actually these voltage sources represent the power that comes from the power company. This power is generally derived from transformers. The two voltages are, therefore, tied together and are exactly in phase, as shown by the top and bottom voltages in the diagram.

The two voltage sources are providing power to loads R_1 and R_2, which are assumed equal. Each of the sources causes currents to flow in each loop as shown in the figure. Because all loads are exactly equal and the voltages are exactly equal and tied together in phase, the current in each loop is exactly equal at all times.

The neutral line has a resistance R_N. The voltage across R_N is zero because the loop currents are balanced. Just as in the dc case in Job 4-4, these currents are in opposite directions in the neutral line and cancel out.

In summary, if the line voltages are exactly equal and exactly in phase, and the current in each line is exactly equal, there is no current in the neutral line. All these "exactlys" tell you that this is an ideal case—but very useful in a practical sense.

In Fig. 20-25 it is assumed that the load in circuit 1 at the top is nonlinear. It still has the same resistance, R_1, but now the nonlinear components generate second and third harmonics. This nonlinear current does not have a corresponding value in circuit 2 at the bottom. The nonlinear current must be returned to the power source through the neutral line. The resistance of the neutral line causes the voltage shown in the center graph in the figure. This voltage does not have any

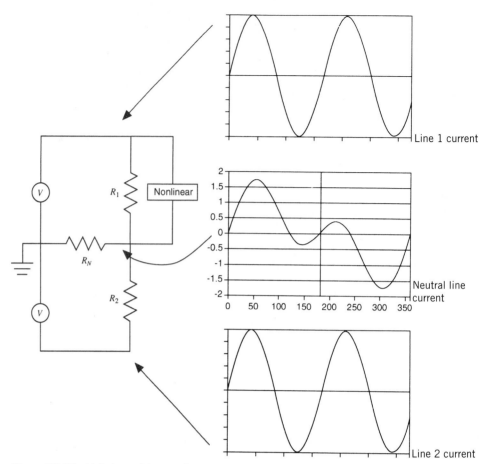

Figure 20-25 Unbalanced harmonics.

fundamental in it. The fundamental is canceled because of the balance of R_1 and R_2.

The voltage on the neutral line looks like noise. This is why it is sometimes said that you have a noisy ground line. This ground noise can affect devices plugged into both circuits. These devices are not actually "grounded" because of the resistance of the neutral line. This noise voltage then feeds into a device through the power cord.

You may ask why the noise does not appear on the line voltages in the top and bottom graphs. The reason it is not in our model is because we have assumed that line L_1 and L_2 do not have any resistance. In the real lines, there would be some resistance, and this would cause some noise to appear.

In Fig. 20-26 there is a duplicate of the nonlinear device in the second circuit. If it is possible to control the two nonlinear devices perfectly, then the currents will be exactly synchronized and will again cancel, as shown. The problem is that this synchronization is very difficult to achieve.

The fundamental is controlled by the power company. The fundamental power is all coming from the same power company; therefore, the fundamentals are better synchronized and balanced. The nonlinear devices are more independent. The size and phase of the harmonics are usually related strongly to the load on the non-linear device. These loads are generally not synchronized and are more independent. This lack of control of harmonics is why the NEC does not allow reduction of the ampacity of the neutral line for loads like fluorescent lights and computers.

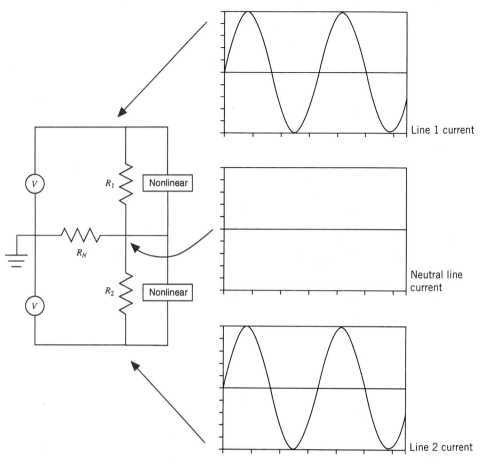

Figure 20-26 Balanced harmonics.

Summary

This job has extended the understanding of current in the neutral line to the ac 120/240 power systems. It then showed that harmonics generated by nonlinear devices create noise on the ground. These effects are related to the NEC regulations for sizing the neutral line.

Section 220-22 of the NEC has added a note for the 1993 edition* that recognizes the problems of harmonics on neutral lines. The note relates to three-phase systems, in which the problem is more difficult than in single-phase system. As we have seen, however, the problem still exists on a single-phase system. The mathematics to cover the three-phase, four-wire power system is slightly different, but the ideas and the net effect are the same as those covered here.

Harmonics are becoming a larger problem for two reasons. First, from the power distribution point of view, there is an increase in neutral line currents; second, from the digital systems perspective, noise affects their performance.

---

*"A 3-phase, 4-wire power system used to supply power to computer systems or other similar electronic loads may necessitate that the power system design allow for the possibility of high harmonic neutral currents."

1. A 240-V, single-phase, 20-hp motor is drawing 100 A of line current. What is the power factor of this motor?

2. A bank of eight 500-W lights, a 120-V 15-hp 67% PF motor, and a 120-V 25-hp 78% PF motor are connected to a 120-V load center. Determine the apparent power, true power, and total current being delivered to the load center.

3. A 240-V single-phase, 15% leading, synchronous motor drawing 1.5-kVA, along with a 240-V, 67% lagging induction motor, drawing 2.5 kVA and a 240-V, 4-kW resistance heating element are the connected load. What are the total kilovoltamperes, power factor, and line current being delivered?

4. A 230-V single-phase 10-hp motor draws 60 A. Determine (*a*) the uncorrected power factor, and (*b*) the size of capacitor required to bring the power factor up to 92%.

INTERNET
A C T I V I T I E S

Internet Activity A

Can power factor cost you money? What are the factors that affect your electric utility bill?
http://www.angelfire.com/ma/CenCap/pfFAQ.html

Internet Activity B

How can a power factor problem be caused? Are there laws restricting power factor effects?
http://members.aol.com/schock565/pfc.html

Chapter 20 Solutions to Self-Tests and Reviews

Self-Test 20-18

1. 0.85
2. 32
3. 1.000
4. 0
5. kW
6. 0.6249
7. 0.000
8. 0.6249
9. 6.25
10. *W*
11. PF
12. 11.8
13. 1.8

Self-Test 20-19

1. left
2. 1.00
3. down
4. 0.63
5. 10
6. 6.3
7. right
8. 1.00
9. up
10. 0.18
11. 10
12. 1.8

21

c h a p t e r

Three-Phase Systems

Learning Objectives

1. Write the equations for the wye connection.
2. Write the equations for the delta connection.

JOB 21-1

The Wye Connection

A three-phase circuit is simply a combination of three single-phase circuits. This combination is desirable for many reasons. In general, three-phase equipment is more economical than single-phase equipment. For any particular physical size, the horsepower ratings of three-phase motors and the kilovoltampere ratings of three-phase generators are larger than those of single-phase motors and generators. Three-phase motors are more efficient than single-phase types because three-phase power is more uniform, and does not pulsate as does single-phase power. Also, the distribution of three-phase power may be accomplished with only three-fourths of the line copper required by a single-phase system.

Generation of three-phase voltages. In Fig. 21-1a, three sets of coils are mounted 120° apart on an axis that is free to turn. The coils, with their six leads, are presented in Fig. 21-1b. The coil end at which the voltage is into the wire is called the *start* end (1, 3, 5), and the end at which the voltage is out of the wire is called the *finish* end (2, 4, 6). At the instant shown, coil X is traveling parallel to the lines of force, and the emf induced in it is zero. As the coils rotate, the voltage induced in coil X will be that shown as V_X in Fig. 21-1c. After a rotation of 120°, the emf in coil Y will be zero and starting a new cycle, shown as V_Y in Fig. 21-1c. After another 120°, coil Z will reach this neutral position, and will begin the cycle shown as V_Z in Fig. 21-1c. If we place these three waves on the same electrical time base, we get the three-phase system shown in Fig. 21-2b. Actually, this figure represents three separate waves, each 120° apart. Each of these phases may be used to operate any single-phase device. The vector diagram for these three voltages is shown in Fig. 21-2c.

The three coils, with their six leads, may be connected inside the alternator, which results in the more practical three final leads. This interconnection of the single-phase windings of generators, motors, transformers, etc., can be done by either of two standard methods—the *wye* connection or the *delta* connection. Most generators are wye-connected, but loads may be either wye-connected or delta-connected.

The wye connection. In this connection, the start ends (1, 3, 5) are connected inside the machine as shown in Fig. 21-3a, and the finish ends (2, 4, 6) are brought out as the terminals. Any loads would be connected to these terminals. The connection is shown in diagram form in Fig. 21-3b. Since this connection looks like a

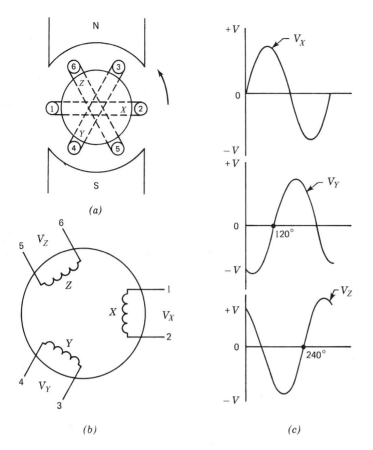

(a)

(b)   (c)

Figure 21-1 (a) Three sets of coils mounted 120° apart. (b) Diagrammatic representation of the coils. (c) Waveforms produced by the individual coils.

star, it is often called the *star* connection. The voltage between any two terminals may be described as V_{XY}, V_{XZ}, or V_{YZ}.

Line voltages in wye. Let us find the voltage between terminals X and Y in Fig. 21-3b. In this balanced system, $V_X = V_Y$ or simply the phase voltage V_p. The line

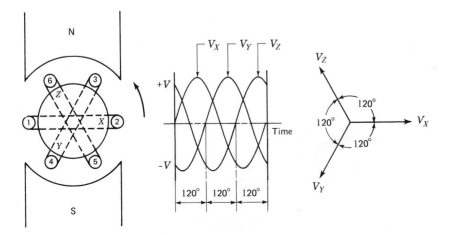

Figure 21-2 (a) Three separate coils mounted 120° apart. (b) Waveforms produced by these coils. (c) Vector diagram of the voltages.

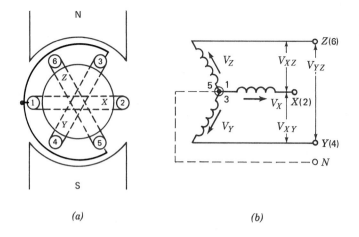

(a)   *(b)*

Figure 21-3   *(a)* In the wye connection, the start ends (1, 3, 5) are connected inside the machine; the finish ends (2, 4, 6) form the terminals. *(b)* Diagram form of the wye or *star* connection. A common lead *N* may or may not be brought out from the connection at point 0.

voltage V_{XY} is *not* simply twice the phase voltage, because the phase voltages are 120° out of phase and must be added vectorially. Let us trace through the circuit from terminal X to terminal Y as shown in Fig. 21-4a. According to Kirchhoff's laws, if the direction of the path traced through a voltage source is the *same* as that of the arrow, the sign of the voltage is *plus*; if the direction of the trace is *opposite* to the arrow, the sign of the voltage is *minus*. Since our trace through V_X indicates an opposing, or negative voltage, the vector diagram must be redrawn as shown in Fig. 21-4b.

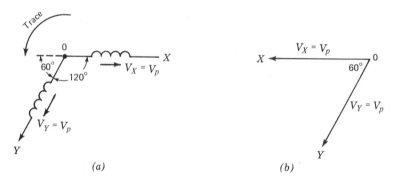

(a)   *(b)*

Figure 21-4   *(a)* The trace from X to Y indicates that the vector OX (V_x) opposes the vector OY (V_y). *(b)* True vector diagram with OX (V_x) reversed to show a negative voltage.

Add vector V_Y to vector V_X by placing the tail of V_Y on the head of V_X and drawing it in its original size and direction as shown in Fig. 21-5a. The total voltage (V_{line} or V_L) will be the resultant vector OY. Since the phase voltages V_X and V_Y are equal, each vector is described as simply V_p. Resolve the vector XY into its components as shown in Fig. 21-5b.

$$XC = V_p \cos 60° = 0.5V_p$$
$$CY = V_p \sin 60° = 0.866V_p$$

Adding the in-phase components will give the vector diagram shown in Fig. 21-5c. ($OC = OX + CX = 1V_p + 0.5V_p = 1.5V_p$.)

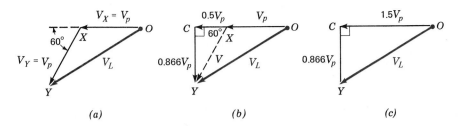

Figure 21-5 (a) *OY* is the vector sum of *OX* and *XY*. (b) Resolution of vector *XY* into its components. (c) Vector triangle resulting from the addition of the in-phase components; V_L is the vector sum of the phase voltages.

In $\Delta \, OCY$,
$$V_L{}^2 = (1.5\,V_p)^2 + (0.866\,V_p)^2$$
$$V_L = \sqrt{2.25\,V_p{}^2 + 0.75\,V_p{}^2}$$
$$= \sqrt{3\,V_p{}^2}$$
$$V_L = \sqrt{3}\,V_p$$

This relationship is true for any pair of line-to-line voltages.

$$\boldsymbol{V_L = \sqrt{3}\,V_p} \qquad\qquad \text{Formula 21-1}$$

where V_L = line voltage
V_p = phase voltage

The complete vector diagram for the three voltages and the three line voltages is shown in Fig. 21-6. Note that (1) the line voltages are equal and out of phase with each other by 120°. (2) Each phase voltage is in phase with its own phase current. (3) The phase current (which is the line current) is 30° out of phase with the line voltage at unity power factor.

Line currents in wye. The current in any line wire *X*, *Y*, or *Z* flows from the neutral point *O* through the generator coils to the lines. This means that the current I_L in any line wire is equal to the current in the phase (I_p) to which it is connected.

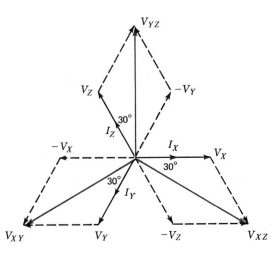

Figure 21-6 Vector diagram showing the relations between the phase and line voltages at 100% PF.

$$I_L = I_p \qquad \text{Formula 21-2}$$

Three-phase power. The power developed in each phase W_p of a wye-connected three-phase circuit is the same as the power in a single-phase circuit.

$$W_p = V_p I_p \cos \theta \qquad \text{Formula 21-3}$$

where θ is the angle between the phase current and the phase voltage. The power developed in all three phases is

$$W_T = 3 W_p = 3 V_p I_p \cos \theta \qquad \text{Formula 21-4}$$

Since it is more practical to measure line voltages and line currents than phase values, let us convert phase values to line values.

In a wye connection,

$$I_p = I_L \quad \text{and} \quad V_p = \frac{V_L}{\sqrt{3}}$$

Substituting these values in formula (21-4) gives

$$W_T = 3 \frac{V_L}{\sqrt{3}} \times I_L \times \cos \theta$$

or

$$W_T = \sqrt{3} V_L I_L \cos \theta \qquad \text{Formula 21-5}$$

Also, since $VA = W/\cos \theta$, $\qquad\qquad\qquad$ (20-5)

$$VA = \frac{\sqrt{3} V_L I_L \cancel{\cos \theta} \; \dfrac{1}{}}{\cancel{\cos \theta} \; \dfrac{}{1}}$$

CALCULATOR HINT

Notice that the types of calculations done are essentially the same as those for single-phase calculations, with the exception of an occasional $\sqrt{3}$ appropriately placed.

or

$$VA = \sqrt{3} V_L I_L \qquad \text{Formula 21-6}$$

Expressed in kilounits,

$$kW = \frac{\sqrt{3} V_L I_L \cos \theta}{1000} \qquad \text{Formula 21-7}$$

$$kVA = \frac{\sqrt{3} V_L I_L}{1000} \qquad \text{Formula 21-8}$$

The system power factor of a balanced three-phase, wye-connected system, which is also the coil power factor, is still given by the formula PF $= \cos \theta = W/VA$, where θ is the angle between the *phase* current and the *phase* voltage. This may be restated as the

$$PF = \frac{W_T}{\sqrt{3} V_L I_L} \qquad \text{Formula 21-9}$$

Note: A system is *balanced* when each of its three elements is equal to the others.

Example 21-1

If a three-phase, wye-connected generator delivers 120 V per phase, what is the line voltage between any two phases?

Solution

$$V_L = \sqrt{3}V_p \qquad\qquad (21\text{-}1)$$
$$V_L = 1.732 \times 120 = 208 \text{ V} \qquad Ans.$$

Example 21-2

A three-phase, wye-connected ac generator has a terminal voltage of 480 V and delivers a full-load current of 300 A per terminal at a power factor of 80%. Find (*a*) the full-load current per phase, (*b*) the phase voltage, (*c*) the kilovoltampere rating, and (*d*) the full-load power in kilowatts.

Solution

a. In a wye-connected system, $I_p = I_L$: $\qquad\qquad (21\text{-}2)$

$$I_p = 300 \text{ A} \qquad Ans.$$

b.
$$V_p = \frac{V_L}{\sqrt{3}} = \frac{480}{1.732} = 277 \text{ V} \qquad Ans. \qquad (21\text{-}1)$$

c.
$$kVA = \frac{\sqrt{3}V_L I_L}{1000} \qquad\qquad (21\text{-}8)$$

$$= \frac{1.732 \times 480 \times 300}{1000} = 249.4 \text{ kVA} \qquad Ans.$$

d.
$$kW = \frac{\sqrt{3}V_L I_L \cos\theta}{1000} \qquad\qquad (21\text{-}7)$$

$$= \frac{1.732 \times 480 \times 300 \times 0.8}{1000}$$

$$= 199.5 \text{ kW} \qquad Ans.$$

Alternative method

$$kW = kVA \times PF \qquad\qquad (20\text{-}9)$$
$$= 249.4 \times 0.8 = 199.5 \text{ kW} \qquad Ans.$$

Example 21-3

A balanced load of three 10-Ω resistors (PF = 100%) is connected across a 480-V three-phase line as shown in Fig. 21-7*a* on page 760. With the neutral wire omitted, the circuit is seen to be wye-connected, as shown in Fig. 21-7*b*. Find (*a*) the phase voltage for each resistor, (*b*) the current in each resistor, (*c*) the line current, (*d*) the power taken by each resistor, (*e*) the total three-phase power, and (*f*) the total kilovoltampere input to the load.

Solution

a. The phase voltage across each resistor is
$$V_p = \frac{V_L}{\sqrt{3}} = \frac{480}{1.732} = 277 \text{ V} \qquad Ans. \qquad (21\text{-}1)$$

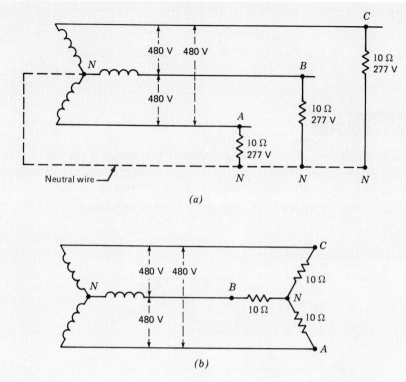

Figure 21-7

b. The current in each resistor with no reactance present (PF = 100%) is

$$I_p = \frac{V_p}{Z_p} = \frac{V_p}{R_p} = \frac{277}{10} = 27.7 \text{ A} \qquad Ans.$$

c. In a wye-connected circuit, the line and phase currents are equal.

$$I_L = I_p = 27.7 \text{ A} \qquad Ans. \tag{21-2}$$

d. The power drawn by each resistor in each phase is

$$W_p = V_p I_p \cos \theta \tag{21-3}$$
$$= 277 \text{ x } 27.7 \times 1 = 7673 \text{ W} \qquad Ans.$$

e. The total power may be found in three different ways.

$$\begin{array}{ll}
\text{W}_T = 3W_p & \text{W}_T = 3V_p I_p \cos \theta \\
\quad = 3 \times 7673 & \quad = 3 \times 277 \times 27.7 \times 1 \\
\quad = 23{,}020 \text{ W} & \quad = 23{,}020 \text{ W}
\end{array}$$

$$\begin{aligned}
\text{W}_T &= \sqrt{3} V_L I_L \cos \theta \\
&= 1.732 \times 480 \times 27.7 \\
&= 23{,}030 \text{ W} \qquad Ans.
\end{aligned}$$

f. The kilovoltampere input is

$$\frac{\sqrt{3} V_L I_L}{1000} = \frac{1.732 \times 480 \times 27.7}{1000} = 23.03 \text{ kVA} \qquad Ans.$$

The total power $W_T = 23.03$ kW is equal to the kilovoltamperes because the power factor is 100%. The phase angle is 0°.

Example 21-4

Each phase of a balanced wye-connected three-phase load consists of a resistor of 12 Ω in series with a pure inductive reactance of 5 Ω. The load is supplied with a 480-V 60-Hz three-phase line. Find (*a*) the impedance per phase, (*b*) the power factor, (*c*) the current in each branch of the load, (*d*) the line current, (*e*) the total power, and (*f*) the kilovoltamperes input. (See Fig. 21-8).

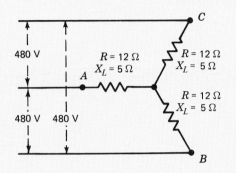

Figure 21-8

Solution

a.
$$Z_p = \sqrt{R^2 + X_L{}^2} \qquad (16\text{-}4)$$
$$= \sqrt{12^2 + 5^2} = \sqrt{169} = 13 \ \Omega \qquad Ans.$$

b.
$$\text{PF} = \cos \theta = \frac{R_p}{Z_p} = \frac{12}{13} = 0.923 = 92.3\% \qquad Ans.$$

c. Find the phase voltage.
$$V_p = \frac{V_L}{\sqrt{3}} = \frac{480}{1.732} = 277 \ \text{V} \qquad (21\text{-}1)$$

Find the phase current.
$$I_p = \frac{V_p}{Z_p} = \frac{277}{13} = 21.3 \ \text{A} \qquad Ans.$$

d.
$$I_L = I_p = 21.3 \ \text{A} \qquad Ans. \qquad (21\text{-}2)$$

e.
$$W_T = \sqrt{3} V_L I_L \cos \theta \qquad (21\text{-}5)$$
$$= 1.732 \times 480 \times 21.3 \times 0.923$$
$$= 16,344 \ \text{W} = 16.34 \ \text{kW} \qquad Ans.$$

f. $\quad \text{kVA} = \dfrac{\sqrt{3} V_L I_L}{1000} = \dfrac{1.732 \times 480 \times 21.3}{1000} = 17.71 \ \text{kVA} \qquad Ans.$

$$(21\text{-}8)$$

Example 21-5

A 220-V three-phase induction motor draws a full-load current of 35 A per terminal when wye-connected. Find (*a*) the full-load kilovoltamperes input and (*b*) the kilowatt input if the full-load power factor is 80%.

Solution

$$a. \quad kVA = \frac{\sqrt{3}V_L I_L}{1000} = \frac{1.732 \times 220 \times 35}{1000} = 13.34 \text{ kVA} \qquad Ans.$$

$$(21\text{-}8)$$

$$b. \quad kW = kVA \times PF \qquad\qquad (20\text{-}9)$$
$$= 13.34 \times 0.8 = 10.67 \text{ kW} \qquad Ans.$$

$$or \quad kW = \frac{\sqrt{3}V_L I_L \cos\theta}{1000} \qquad\qquad (21\text{-}7)$$

$$= \frac{1.732 \times 220 \times 35 \times 0.8}{1000} = 10.67 \text{ kW} \qquad Ans.$$

Exercises

Applications

1. The phase voltage of a wye-connected generator is 1200 V. What is the line voltage?

2. If the terminal voltage of a wye-connected generator is 4160 V, what is the voltage generated in each of the phases?

3. A wye-connected generator has a phase current of 40 A. Find the current in the line wires.

4. A three-phase motor takes a current of 30 A per terminal. If the line voltage is 440 V and the power factor is 80%, find the power used by the motor.

5. A wye-connected generator delivers a phase current of 20 A at a phase voltage of 265 V and a power factor of 85%. Find (a) the generator terminal voltage, (b) the power developed per phase, and (c) the total three-phase power.

6. A three-phase load draws 50 kW from a 440-V line. If the line current is 80 A, find (a) the kilovoltamperes, and (b) the power factor of the load.

7. A three-phase, 208-V 6-pole 60-Hz induction motor draws 25 A and 7.93 kW from a line while driving a 10-hp load. Find (a) the total kilovoltamperes, and (b) the power factor of the motor.

8. A certain three-phase, wye-connected alternator is rated at 1000 kVA and 2300 V at unity power factor. Find (a) the current rating per terminal, (b) the rated phase current, and (c) the rated coil voltage.

9. A three-phase, 60-Hz 6600-V wye-connected alternator delivers 4000 kW and 425 A at 6600 V. Find (a) the kilovoltamperes delivered, (b) the power factor, and (c) the coil voltage.

10. A three-phase, 60-Hz wye-connected synchronous motor draws 150 A and 500 kW from a 2300-V three-phase supply. Find (a) the kilovoltamperes, (b) the power factor, and (c) the phase voltage.

11. Each phase winding of a three-phase, wye-connected alternator is rated at 6930 V and 361 A. If the alternator is designed to operate at full load with a lagging power factor of 80%, find (a) the line voltage, (b) the line current, (c) the kilovoltampere rating at full load, and (d) the full-load power in kilowatts.

12. A three-phase, wye-connected alternator is rated at 12,500 kVA, 10,000 V, and 60 Hz. Find (a) the full-load kilowatt output of the generator at a power factor of 80% lagging, (b) the full-load line current of the generator, and (c) the voltage rating of each of the three windings.

13. Three resistive loads of 10 Ω each are connected to a three-phase 120-V 60-Hz supply. If the loads are connected in wye, find (*a*) the current in each resistor, (*b*) the current in the line, and (*c*) the power used by the system.

14. A wye-connected load of three 10-Ω resistors is connected to a 220-V three-phase supply. At 100% power factor, find (*a*) the voltage applied across each resistor, (*b*) the current in each phase, (*c*) the line current, and (d) the total power used.

15. Each coil of a three-phase 60-Hz wye-connected alternator generates 277 V. The alternator supplies a load consisting of three wye-connected impedances, each with 12-Ω resistance and 5-Ω inductive reactance. Find (*a*) the line voltage, (*b*) the impedance of each resistor, (*c*) the current through each load impedance, (d) the line current, (*e*) the power factor, and (*f*) the kilowatt load being supplied.

16. A three-phase, wye-connected load takes 80 kW of power at a power factor of 80% lagging from a three-phase 440-V 60-Hz line. Find (*a*) the phase voltage, (*b*) the total kilovoltamperes, (*c*) the line current, (d) the phase current, and (*e*) the reactive kilovoltamperes.

17. Each coil of the armature of a three-phase wye-connected alternator is rated at 50 kVA, 2400 V, 60 Hz. If the power factor of the system is 85%, find (*a*) the line voltage, (*b*) the full-load kilowatt output of the alternator, (c) the full-load line current, and (d) the full-load coil current.

18. A three-phase 60-Hz wye-connected turboalternator is supplying 14,000 kW at 13,200 V and 80% power factor to a load. Find (*a*) the kilovoltampere output, (*b*) the line current, (c) the coil current, and (d) the coil voltage.

19. Three loads, each with a series resistance of 86.6 Ω and an inductive reactance of 50 Ω, are wye-connected to a 240-V three-phase power supply. Find (*a*) the impedance of each phase, (*b*) the voltage of each phase, (c) the line current, (d) the power factor, (*e*) the three-phase power, and (*f*) the voltamperes.

20. Each phase of a balanced wye-connected, three-phase load consists of a resistor of 24 Ω in series with a pure inductive reactance of 10 Ω. If the load is supplied from a 440-V 60-Hz three-phase line, find (*a*) the impedance per phase, (*b*) the power factor, (*c*) the current in each branch of the load, (d) the line current, (*e*) the total power, and (*f*) the kilovoltampere input.

The Delta Connection

The coils of a three-phase alternator may also be connected as shown in Fig. 21-9*a*. The finish end of coil *X* (point 2) is connected to the start end of coil *Y* (point 3); then the finish end of coil *Y* (point 4) is connected to the start end of coil *Z* (point 5); finally the finish end of coil *Z* (point 6) is connected to the start end of coil *X* (point 1).

The junction of points 2 and 3 is brought out as line 1.
The junction of points 4 and 5 is brought out as line 2.
The junction of points 1 and 6 is brought out as line 3.

This is shown in diagram form in Fig. 21-9*b*. Since the configuration looks like the Greek letter delta (Δ), this connection is known as the *delta* connection. In this figure, notice that each pair of line leads is connected directly across a coil winding.

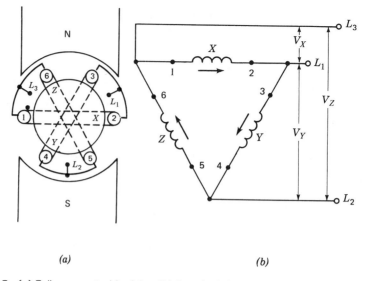

(a)   (b)

Figure 21-9   (a) Coils connected in delta. (b) Standard delta configuration.

For example, the voltage generated in phase Y is also the voltage between lines 1 and 2. Thus,

$$V_L = V_p \qquad \text{Formula 21-10}$$

Currents in delta-connected generators.   The current in any line wire is the vector sum of the currents in the two coils to which it is attached. That is, for example, I_{XY} is equal to the vector sum of I_X and I_Y in Fig. 21-10a. Let us assume that the coils are loaded so that equal currents flow in the three coils, and that these currents are in phase with their respective voltages as shown in Fig. 21-10b.

According to Kirchhoff's law, the total current leaving a junction must equal the total current entering the junction. At point A in Fig. 21-10a,

$$I_Y + I_{XY} = I_X$$
or
$$I_{XY} = I_X - I_Y$$

This means that the vector total will be obtained by *reversing* the direction of I_Y and adding *this* vector to I_X as shown in Fig. 21-10c. To add the vectors, place the tail of I_Y on the head of I_X and draw it in its original size and direction as shown

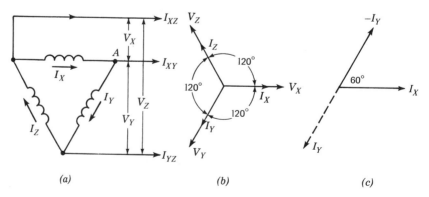

(a)   (b)   (c)

Figure 21-10   (a) Coils in delta. (b) Vector diagram of the phase currents. (c) Vector diagram for the addition of opposing currents.

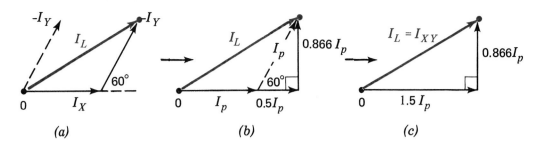

Figure 21-11   (a) Vector addition of I_X and $-I_Y$. (b) Resolution of I_Y into its components. (c) I_L is the vector sum of the phase currents.

in Fig. 21-11a. The total (I_L) will be the vector from O to the tip of I_Y. Resolve the vector I_Y into its components as shown in Fig. 21-11b. Since $I_X = I_Y =$ the phase current I_p, we may drop the individual subscripts and use just the letter I_p to describe the current vectors.

$$\text{Horizontal component} = I_p \cos 60° = 0.5I_p$$
$$\text{Vertical component} = I_p \sin 60° = 0.866I_p$$

Adding the in-phase components will give the vector diagram shown in Fig. 21-11c.

$$I_L^2 = (1.5I_p)^2 + (0.866I_p)^2$$
$$I_L = \sqrt{2.25I_p^2 + 0.75I_p^2} = \sqrt{3I_p^2}$$
$$I_L = \sqrt{3}I_p$$

A similar situation exists for the currents in the other line wires. Thus, for a balanced load,

$$I_L = \sqrt{3}I_p \qquad\qquad \text{Formula 21-11}$$

where $I_L =$ line current
  $I_p =$ phase current

Power in delta.   The power developed in each phase W_p of a delta-connected, three-phase circuit is the same as the power in a single-phase circuit.

$$W_p = V_pI_p \cos \theta \qquad\qquad (21\text{-}3)$$

where θ is the angle between the phase current and the phase voltage. The power developed in all three phases is

$$W_T = 3W_p = 3V_pI_p \cos \theta \qquad\qquad (21\text{-}4)$$

Since it is more practical to measure line voltages and line currents than phase values, let us convert phase values to line values. In a delta connection,

$$V_p = V_L \quad \text{and} \quad I_p = \frac{I_L}{\sqrt{3}}$$

Substituting these values in formula (21-4) gives

$$W_T = 3 V_L \times \frac{I_L}{\sqrt{3}} \cos \theta$$

or

$$\boldsymbol{W_T = \sqrt{3}V_LI_L \cos \theta} \qquad\qquad \text{Formula 21-5}$$

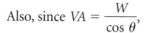
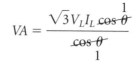
Also, since $VA = \dfrac{W}{\cos\theta}$, (20-5)

$$VA = \dfrac{\sqrt{3}\,V_L I_L \cancel{\cos\theta}\,\overset{1}{}}{\underset{1}{\cancel{\cos\theta}}}$$

or

$$\boldsymbol{VA = \sqrt{3}\,V_L I_L} \qquad\qquad \text{Formula 21-6}$$

Expressed in kilo units,

$$\mathbf{kW} = \dfrac{\sqrt{3}\,V_L I_L \cos\theta}{1000} \qquad\qquad 21\text{-}7$$

$$\mathbf{kVA} = \dfrac{\sqrt{3}\,V_L I_L}{1000} \qquad\qquad 21\text{-}8$$

$$\mathbf{PF} = \dfrac{W_T}{\sqrt{3}\,V_L I_L} \qquad\qquad 21\text{-}9$$

It should be noted that these formulas for power are identical for the wye and delta systems. This should be true, because the relations in a three-phase line are the same whether the power originates in a wye-connected or a delta-connected generator.

Example 21-6

If the current through each phase of a delta-connected generator is 30 A, find the current in the line wires.

Solution

$$I_L = \sqrt{3}\,I_p \qquad\qquad (21\text{-}11)$$
$$I_L = 1.732 \times 30 = 52 \text{ A} \qquad Ans.$$

Example 21-7

Each phase of a three-phase, delta-connected generator supplies a full-load current of 50 A at a voltage of 220 V and at a power factor of 80% lagging. Find (a) the line voltage, (b) the line current, (c) the three-phase kilovoltamperes, and (d) the three-phase power in kilowatts.

Solution

a. In a delta-connected system,

$$V_L = V_P \qquad\qquad (21\text{-}10)$$
$$V_L = 220 \text{ V} \qquad Ans.$$

b. $\qquad I_L = \sqrt{3}\,I_p = 1.732 \times 50 = 86.6 \text{ A} \qquad Ans. \qquad (21\text{-}11)$

c. $\quad \mathbf{kVA} = \dfrac{\sqrt{3}\,V_L I_L}{1000} = \dfrac{1.732 \times 220 \times 86.6}{1000} = 33 \text{ kVA} \qquad Ans. \quad (21\text{-}8)$

d. $\qquad \mathbf{kW} = \mathbf{kVA} \times \mathbf{PF} = 33 \times 0.8 = 26.4 \text{ kW} \qquad Ans. \qquad (20\text{-}9)$

Example 21-8

A balanced load of three 10-Ω resistors (PF = 1) is connected across a 480 V three-phase line as shown in Fig. 21-12a. This is a delta connection, as shown in Fig. 21-12b. Find (a) the phase voltage across each resistor, (b) the current in each resistor, (c) the line current, (d) the three-phase power in kilowatts, and (e) the total kilovoltamperes input.

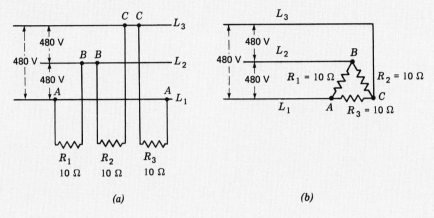

Figure 21-12

Hint for Success

Figure 21-12 is an excellent example of a practical use of topology. Both diagrams represent a delta connection, and each can be converted to the other. Each has an advantage in what it emphasizes and demonstrates.

Solution

a.
$$V_p = V_L = 480 \text{ V} \qquad Ans. \qquad (21\text{-}10)$$

b. The current in each resistor with zero reactance (PF = 1) is

$$I_p = \frac{V_p}{Z_p} = \frac{V_p}{R_p} = \frac{480}{10} = 48 \text{ A} \qquad Ans.$$

c. In a delta-connected circuit, the line current is

$$I_L = \sqrt{3}I_p = 1.732 \times 48 = 83.1 \text{ A} \qquad Ans. \qquad (21\text{-}11)$$

d. $\text{kW} = \dfrac{\sqrt{3}V_L I_L \cos\theta}{1000} = \dfrac{1.732 \times 480 \times 83.1 \times 1}{1000} = 69.1 \text{ kW} \quad (21\text{-}7)$

e. $\text{kVA input} = \dfrac{\sqrt{3}V_L I_L}{1000} = \dfrac{1.732 \times 480 \times 83.1}{1000} = 69.1 \text{ kVA} \quad (21\text{-}8)$

or $\qquad \text{kVA} = \dfrac{\text{kW}}{\text{PF}} = \dfrac{69.1}{1} = 69.1 \text{ kVA} \qquad Ans. \qquad (20\text{-}10)$

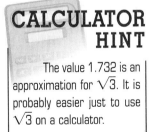

CALCULATOR HINT

The value 1.732 is an approximation for $\sqrt{3}$. It is probably easier just to use $\sqrt{3}$ on a calculator.

Example 21-9

Each phase of a balanced delta-connected load consists of a resistor of 56 Ω in series with a pure inductive reactance of 33 Ω. The load is supplied with a 220-V 60-Hz three-phase line. Find (a) the impedance per phase, (b) the power factor, (c) the phase voltage, (d) the phase current, (e) the line current, (f) the total power consumed, and (g) the kilovoltampere input.

Solution

a. $Z_p = \sqrt{R^2 + X_L^2}$
    $= \sqrt{56^2 + 33^2} = \sqrt{3136 + 1089} = \sqrt{4225} = 65\ \Omega$ *Ans.*

b. $\text{PF} = \cos\theta = \dfrac{R_p}{Z_p} = \dfrac{56}{65} = 0.862$ *Ans.*

c. $V_p = V_L = 220\ \text{V}$ *Ans.*   (21-10)

d. $I_p = \dfrac{V_p}{Z_p} = \dfrac{220}{65} = 3.38\ \text{A}$ *Ans.*

e. $I_L = \sqrt{3}\,I_p = 1.732 \times 3.38 = 5.85\ \text{A}$ *Ans.*   (21-11)

f. $\text{kW} = \dfrac{\sqrt{3}\,V_L I_L \cos\theta}{1000} = \dfrac{1.732 \times 220 \times 5.85 \times 0.862}{1000}$ (21-7)

    $= 1.92\ \text{kW}$ *Ans.*

g. $\text{kVA} = \dfrac{\text{kW}}{\text{PF}} = \dfrac{1.92}{0.862} = 2.23\ \text{kVA}$ *Ans.*   (20-10)

Example 21-10

A 220-V delta-connected, three-phase induction motor draws a full-load current of 35 A per terminal. Find (*a*) the phase voltage, (*b*) the phase current, (*c*) the full-load kilovoltampere input, and (d) the kilowatt input if the full-load power factor of the motor is 75%.

Solution

a. $V_p = V_L = 220\ \text{V}$ *Ans.*   (21-10)

b. $I_p = \dfrac{I_L}{\sqrt{3}} = \dfrac{35}{1.732} = 20.2\ \text{A}$ *Ans.*   (21-11)

c. $\text{kVA} = \dfrac{\sqrt{3}\,V_L I_L}{1000} = \dfrac{1.732 \times 220 \times 35}{1000} = 13.3\ \text{kVA}$ *Ans.*   (21-8)

d. $\text{kW} = \text{kVA} \times \text{PF} = 13.3 \times 0.75 = 9.98\ \text{kW}$ *Ans.*   (20-9)

Example 21-11

A three-phase induction motor draws 35 kW from a 220-V main. If the line current is 100 A, find (*a*) the power factor of the motor and (*b*) the reactive kilovoltampere.

Solution

a. $\text{PF} = \dfrac{\text{W}}{\text{VA}} = \text{kW} \times \dfrac{1000}{\sqrt{3}\,V_L I_L}$ (20-7)

    $= \dfrac{35 \times 1000}{1.732 \times 220 \times 100} = 0.919$ *Ans.*

b.

$$\text{PF} = \cos\theta = 0.919$$
$$\theta = 24°$$

$$\tan\theta = \frac{\text{var}}{\text{W}} \qquad (20\text{-}11)$$

or $\qquad \text{var} = \text{W} \tan\theta$

or $\qquad \text{kvar} = \text{kW} \tan\theta = 35 \times \tan 24°$
$$= 35 \times 0.4452 = 15.6 \text{ kvar} \qquad Ans.$$

Alternative method

$$\text{kVA} = \frac{\sqrt{3}V_L I_L}{1000} = \frac{1.732 \times 220 \times 100}{1000} = 38.1 \text{ kVA} \qquad (21\text{-}8)$$

$$\text{kvar} = \text{kVA} \times \sin\theta$$
$$= 38.1 \times \sin 24° = 38.1 \times 0.4067 = 15.5 \text{ kvar} \qquad Ans.$$

Note: The slight difference in the values is due to the use of the approximate phase angle of 24°. Greater accuracy would be obtained by the use of decimal trigonometric tables.

Exercises

Applications

1. The phase current in a delta-connected generator is 60 A. Find the line current.
2. If the line current in a delta-connected generator is 50 A, find the coil current.
3. If the voltage per phase of a delta-connected generator is 440 V, find the line voltage.
4. Three equal resistive heating elements in delta are supplied by a three-phase 220-V line. If 8.66 A flows through each line wire, find the resistance of each element.
5. Six 24-Ω resistors are connected in two groups: three in wye, and three in delta. If the two groups in parallel are supplied by a three-phase 240-V line, find (*a*) the current in each wye-connected resistor, and (*b*) the current in each delta-connected resistor.
6. Each phase of a three-phase, delta-connected generator supplies a full-load current of 20 A at a voltage of 120 V and at a power factor of 80% lagging. Find (*a*) the line voltage, (*b*) the line current, (*c*) the total kilovoltamperes, and (*d*) the total kilowatts.
7. A heating load consists of three 24-Ω resistive heating elements connected in delta. If the load is supplied by a 240-V three-phase line, find (*a*) the voltage across each heating element, (*b*) the current in each heater, (*c*) the line current, (*d*) the total power in kilowatts, and (*e*) the total kilovoltampere input.
8. Each of the three parts of a balanced delta-connected load consists of a resistance of 16 Ω in series with a pure inductive reactance of 12 Ω. The load is supplied from a 220-V three-phase line. Find (*a*) the impedance per phase, (*b*) the power factor, (*c*) the phase voltage, (*d*) the phase current, (*e*) the line current, (*f*) the total kilowatts, and (*g*) the total kilovoltamperes.
9. A three-phase induction motor draws 50 kW from a 440-V line. If the line current is 80 A, find (*a*) the power factor of the motor, (*b*) the phase angle, (*c*) the kilovoltampere input, and (*d*) the reactive kilovars.

10. A delta-connected load draws 38.4 kW from a three-phase supply. If the line current is 69.3 A and the power factor is 80%, find (a) the line voltage, (b) the voltage across each phase, (c) the current per phase, (d) the impedance per phase, and (e) the resistance and reactance of each phase.

11. Each coil of a three-phase generator has a current-carrying capacity of 50 A and a voltage rating of 480 V. Find the kilovoltampere rating of the alternator when it is (a) connected in delta and (b) connected in wye.

12. A three-phase, delta-connected alternator is rated at 75 kVA, 240 V, and 60 Hz. Find (a) the full-load line current, (b) the full-load line voltage, and (c) the kilovoltamperes, current, and voltage of one phase of the alternator.

JOB 21-3

Review of Three-Phase Connections

For either connection, as shown in Fig. 21-13,

$$Z \text{ per phase} = \frac{V_p}{I_p}$$

$$VA \text{ per phase} = V_p I_p$$

$$\text{Three-phase } VA = VA_T = 3V_p I_p$$

$$= \sqrt{3} V_L I_L$$

$$\text{Three-phase kVA} = \frac{\sqrt{3} V_L I_L}{1000}$$

$$\text{Power factor} = \cos \theta$$

$$\theta = \text{angle between } V_p \text{ and } I_p$$

$$\cos \theta = \frac{R_p}{Z_p}$$

$$\text{Power per phase} = W = V_p I_p \cos \theta$$

$$\text{Three-phase power} = W_T = 3V_p I_p \cos \theta$$

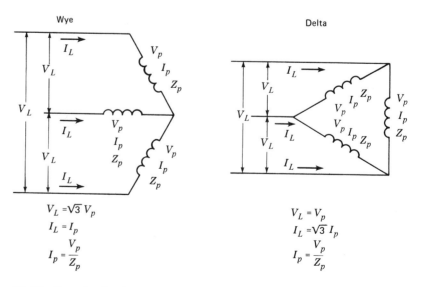

Figure 21-13 Formulas for the wye and delta connections.

$$W_T = \sqrt{3}V_L I_L \cos\theta$$

$$\text{Three-phase kW} = \frac{\sqrt{3}V_L I_L \cos\theta}{1000}$$

$$\text{PF} = \frac{W_T}{VA_T}$$

$$= \frac{W_T}{\sqrt{3}V_L I_L}$$

$$\text{kW} = \text{kVA} \times \text{PF}$$

$$\text{kvar} = \text{kVA} \times \sin\theta$$

$$\text{kvar} = \text{kW} \times \tan\theta$$

Example 21-12

A 60-kVA wye-connected alternator rated at 120 V per coil is connected to a balanced load consisting of three 3-Ω resistors connected in delta as shown in Fig. 21-14. Find (*a*) the current per phase in the load, (*b*) the current per coil in the alternator due to the load, and (*c*) the kilovoltamperes delivered by the alternator to the load.

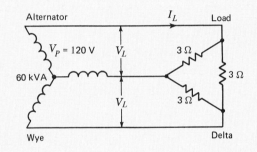

Figure 21-14

Solution

a. Find the line voltage of the alternator. In wye,

$$V_L = \sqrt{3}V_p = 1.732 \times 120 = 208 \text{ V} \qquad (21\text{-}1)$$

Find the phase voltage in the load. In delta,

$$V_p = V_L = 208 \text{ V} \qquad (21\text{-}10)$$

Find the phase current in the load.

$$I_p = \frac{V_p}{Z_p} = \frac{V_p}{R_p} = \frac{208}{3} = 69.3 \text{ A} \qquad Ans.$$

b. Find the line current to the load. In delta,

$$I_L = \sqrt{3}I_p \qquad (21\text{-}11)$$
$$= 1.732 \times 69.3 = 120 \text{ A}$$

Find the current in the alternator coil to create this line current. In wye,

$$I_p = I_L = 120 \text{ A} \qquad Ans. \qquad (21\text{-}2)$$

$$c. \qquad kVA = \frac{\sqrt{3}V_L I_L}{1000} \qquad\qquad (21\text{-}8)$$

$$= \frac{1.732 \times 208 \times 120}{1000} = 43.2 \text{ kVA} \qquad Ans.$$

Example 21-13

Each phase winding of a wye-connected generator is rated at 20 kVA and 120 V. Find (a) the line voltage, (b) the line current at full load, (c) the kilowatt output at full load and 80% power factor, and (d) the reactive kilovoltamperes at full load.

Solution

$a. \qquad\qquad V_L = \sqrt{3}V_p = 1.732 \times 120 = 208 \text{ V} \qquad Ans. \qquad (21\text{-}1)$

$b. \qquad\qquad VA = V_p I_p$
$$20{,}000 = 120 \times I_p$$

$$I_p = \frac{20{,}000}{120} = 166.7 \text{ A}$$

In wye, $\qquad\qquad I_L = I_p = 166.7 \text{ A} \qquad Ans. \qquad (21\text{-}2)$

$c. \qquad\qquad kW = \frac{\sqrt{3}V_L I_L \cos\theta}{1000} \qquad\qquad (21\text{-}7)$

$$= \frac{1.732 \times 208 \times 166.7 \times 0.8}{1000}$$

$$kW = 48 \text{ kW} \qquad Ans.$$

d. If the PF = 0.8, then

$$\cos\theta = 0.8$$
and $\qquad\qquad \theta = 37°$

$$\tan\theta = \frac{kvar}{kW}$$
or $\qquad kvar = kW \times \tan\theta$
$$= 48 \times \tan 37° = 48 \times 0.7536 = 36.2 \text{ kvar} \qquad Ans.$$

Check: If each coil is rated at 20 kVA, then the total kilovoltamperes equal $3 \times 20 = 60$ kVA.

$$kvar = kVA \times \sin\theta = 60 \times \sin 37°$$
$$= 60 \times 0.602 = 36.1 \text{ kvar} \qquad Check$$

Example 21-14

A three-phase 220-V feeder supplies two motors as shown in Fig. 21-15. (a) Find the kilovoltamperes, kilowatts, and lagging var power for the induction motor. (b) Find the kilovoltamperes, kilowatts, and leading var power for the

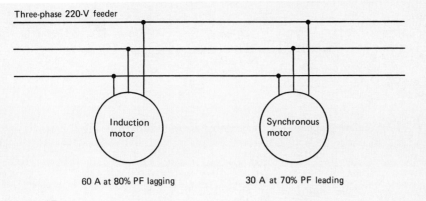

Three-phase 220-V feeder

Induction motor

Synchronous motor

60 A at 80% PF lagging

30 A at 70% PF leading

Figure 21-15

synchronous motor. (*c*) Find the total load supplied to the two motors. (*d*) Find the total var power drawn by the two motors. (*e*) Find the total power factor for the two loads. (*f*) Find the line current drawn by the two motors.

Solution

a. For the induction motor:

$$\text{kVA} = \frac{\sqrt{3}\,V_L I_L}{1000} = \frac{1.732 \times 220 \times 60}{1000} = 22.86 \text{ kVA} \qquad Ans.$$

$$\text{kW} = \frac{\sqrt{3}\,V_L I_L \cos\theta}{1000} = \frac{1.732 \times 220 \times 60 \times 0.8}{1000} = 18.29 \text{ kW} \qquad Ans.$$

If the PF = 80%, then

$$\cos\theta = 0.8$$

and $$\theta = 37°$$

$$\text{kvar} = \text{kVA} \times \sin\theta = 22.86 \times \sin 37°$$
$$= 22.86 \times 0.6 = 13.72 \text{ kvar } \textit{lagging} \qquad Ans.$$

b. For the synchronous motor:

$$\text{kVA} = \frac{\sqrt{3}\,V_L I_L}{1000} = \frac{1.732 \times 220 \times 30}{1000} = 11.43 \text{ kVA} \qquad Ans.$$

$$\text{kW} = \frac{\sqrt{3}\,V_L I_L \cos\theta}{1000} = \frac{1.732 \times 220 \times 30 \times 0.7}{1000} = 8.00 \text{ kW} \qquad Ans.$$

If the PF = 70%, then

$$\cos\theta = 0.7$$

and $$\theta = 45°$$

$$\text{kvar} = \text{kVA} \times \sin\theta = 11.43 \times \sin 45°$$
$$= 11.43 \times 0.7 = 8.00 \text{ kvar } \textit{leading}$$

c. Since the kilowatts drawn by the two motors are in phase,

$$\text{Total kW} = W_1 + W_2 = 18.29 + 8.00 = 26.29 \text{ kW} \qquad Ans.$$

d. Since the reactive powers are 180° out of phase and oppose each other,

$$\text{Total kvar} = \text{kvar}_1 - \text{kvar}_2$$
$$= 13.72 - 8.00 = 5.72 \text{ kvar } \textit{lagging} \qquad Ans.$$

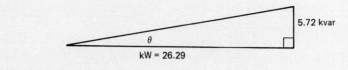

5.72 kvar

θ

kW = 26.29

Figure 21-16

e. The vector diagram for the combination is shown in Fig. 21-16.

$$\tan \theta = \frac{\text{kvar}}{\text{kW}}$$

$$\tan \theta = \frac{5.72}{26.29} = 0.2176$$

$$\theta = 12°$$

$$\text{PF} = \cos \theta = \cos 12° = 0.978 \qquad \textit{Ans.}$$

f.

$$W_T = \sqrt{3} V_L I_L \cos \theta$$

$$26{,}290 = 1.732 \times 220 \times I_L \times 0.978$$

$$I_L = 70.5 \text{ A} \qquad \textit{Ans.}$$

Exercises

Applications

1. Three coils are connected in delta across a 240-V three-phase supply. If the line current is 30 A, find (*a*) the coil current, (*b*) the coil voltage, and (*c*) the total kilovoltamperes.

2. A 120-V three-phase feeder delivers power to twenty-four 120-V 60-W lamps. If this load is balanced in three groups of eight lamps each, find the current through each line wire.

3. A 220-V three-phase feeder delivers 90 A at 80% power factor to a distribution panel. Find the kilowatts delivered.

4. The coils of a three-phase 60-Hz generator are wye-connected. Each coil is rated at 4800 VA and 120 V. Find (*a*) the line voltage, (*b*) the coil current, (*c*) the full-load line current, and (*d*) the full-load kilovoltampere rating of the generator.

5. Each phase of a three-phase alternator is rated at 42 kVA and 120 V. It is wye-connected and supplying power at its rated output to a load at 80% power factor. Find (*a*) the line current, (*b*) the line voltage, and (*c*) the total power.

6. If the alternator of Exercise 5 is now reconnected delta, and each phase delivers full load, find (*a*) the line current, (*b*) the line voltage, and (*c*) the total power.

7. If each phase of a three-phase 440-V line carries a load of 40 kW, find the total power delivered.

8. A three-phase, wye-connected generator is rated at 650 kVA, 1300 V, and 60 Hz. Find (*a*) the kilowatt output at a power factor of 80%, (*b*) the full-load line current, (*c*) the full-load current rating of each coil, and (*d*) the voltage across each coil.

9. A three-phase, delta-connected generator is rated at 650 kVA, 1300 V, and 60 Hz. Find (*a*) the kilowatt output at a power factor of 80%, (*b*) the full-load coil current, (*c*) the full-load line current, and (*d*) the voltage rating of each coil.

10. Each phase of a balanced delta-connected load consists of a resistor of 24 Ω in series with a pure inductive reactance of 7 Ω. The load is supplied from a 220-V 60-Hz three-phase line. Find (a) the impedance per phase, (b) the power factor, (c) the phase voltage, (d) the phase current, (e) the line current, (f) the total power, and (g) the kilovoltampere input.

11. A three-phase, delta-connected alternator is rated at 15 kVA and 208 V. It is supplying a balanced wye-connected load which has an impedance of 20 Ω per phase and a power factor of 80% lagging. Find (a) the line current for the given load, (b) the total power delivered to the load in kilowatts, (c) the current in one coil of the alternator due to the load, and (d) the resistance and the inductive reactance of the load per phase.

12. A delta-connected motor with an efficiency of 87% is operated at the rated output of 20 hp (1 hp = 746 W) on a three-phase line. If its phase power factor is 85% lagging and the phase voltage is 208 V, find (a) the phase current of the motor, (b) the line current, and (c) the line voltage.

13. An inductive load of 55 kVA at 74% power factor lagging is being operated from a 480-V three-phase line. In order to raise the power factor, a synchronous motor drawing 16 kW at 80% leading power factor is added in parallel. (a) Find the kilowatts and kilovars for the induction motor. (b) Find the kilovoltamperes and kilovars for the synchronous motor. (c) Find the total kilowatts and kilovars drawn by the combination of motors. (d) Find the power factor of the combination. (e) Find the total line current drawn.

14. A three-phase, three-wire 1200-V feeder supplies a plant which has a 250-kVA load operating at 75% lagging power factor. A 1200-V 100-kVA synchronous motor operating at a power factor of 80% leading is added in parallel to raise the power factor. Find (a) the kilowatts and kilovars of the inductive load, (b) the kilowatts and kilovars of the synchronous motor, (c) the total kilowatts and total kilovars of the combination, and (d) the new power factor.

(See CD-ROM for Test 21-1)

Assessment

1. Some common three-phase wye-connected distribution voltages are 4160, 8320, 12,470, and 13,200 V. Determine the line-to-neutral voltage for each of these distribution systems.

2. A 500-kVA, 69-kV-3θ delta primary to 12,470 V/7200 V wye-connected secondary is installed. If it is loaded to 80% of its full load value, determine the kilovoltamperes available and the full load current of any phase.

INTERNET
ACTIVITIES

Internet Activity A

There are certain facts that are common to both wye and delta. What are they?
http://www.engr.usask.ca/~wjr126/TechFile/3Phase.html

Internet Activity B

What is three-phase power? Should I use it? Can I get it in my house?
http://www.faqs.org/faqs/electrical-wiring/part2/section-9.html

Three-Phase Transformer Connections

chapter

1. Determine the wye and delta combinations of transformers for three-phase power connections.
2. Write the equations for balanced three-phase loads.
3. Write the equations for open delta transformers.

Overview of Three-Phase Transformer Connections

Chapter 21 examined the two most common methods of providing three-phase power, the wye and the delta connections. This chapter will show how to use transformers to change voltage levels and possibly change from a wye to a delta system, or vice versa.

In general, three single-phase transformers are required to transform a three-phase system, one transformer for each phase. Figure 22-1 shows the four basic combinations of primary and secondary configurations, in which the primary can be either a wye or a delta and the secondary can be either a wye or a delta.

The figure is also a summary of the phase and line voltages and currents for each of the combinations. The summary assumes ideal transformers and balanced phase loads. The concept of balanced loads is an ideal. Section 220-4(d) of the NEC requires that loads be balanced between branch-circuit hot legs and neutral. This is not the same as balancing phases, but it does express the concept. This is a theoretical and ideal design objective and is the model summarized in Fig. 22-1.

The fundamental measurements in three-phase systems are line values and phase values. Line values are values that are measurable in the conductors. We can determine line values by putting a voltmeter between two lines or measuring the current in any single line. We do not care what is in the load. All we are concerned with is the lines. Line values are of interest to persons responsible for the power distribution system.

For transformer systems, there are two sets of lines, one on the input and one on the output. If your responsibility is to provide the power to the transformer, all you have to do is get the power to the input side. From your point of view the important line values are the input voltage, V, and input current, I. In the summary in Fig. 22-1, all values are given in terms of V and I. Thus the summary is from the power supplier's point of view.

If you are responsible for the transformer, then you are interested in the phase values. These are the values that are measured in each of the single-phase transformers. They are the voltages across a winding or the current through the winding. The winding could be either the primary or the secondary.

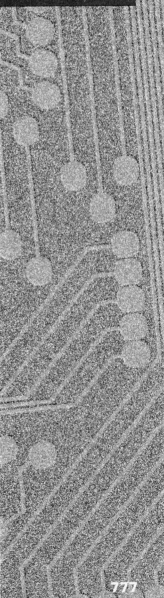

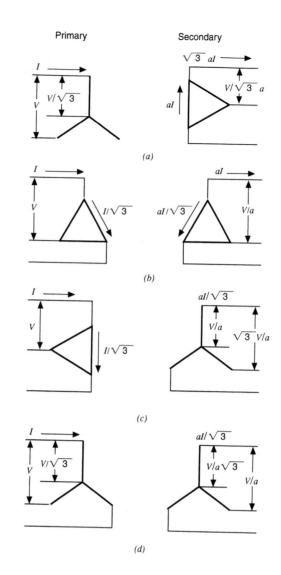

Figure 22-1

Summary of Line and Phase Values

As with all mathematics, when you put together more than one idea you may have a conflict in notation. In the earlier formulas the subscript p was used for both primary and phase. Here p will be used for phase and pr, for primary. The subscript L will be used for line values, and S will be used for secondary.

The values in Fig. 22-1 come from the following set of equations. The equation numbers following some of the equations are the original equation numbers in this book.

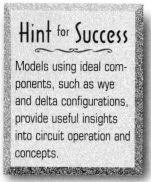

$$\text{Delta:} \quad I_L = \sqrt{3}I_p \qquad (21\text{-}11)$$
$$V_p = V_L$$
$$\text{Wye:} \quad V_L = \sqrt{3}V_P \qquad (21\text{-}1)$$
$$I_P = I_L$$

$$\text{Ideal transformer:} \quad a = \frac{N_{\text{pr}}}{N_S} = \frac{V_{\text{pr}}}{V_S} \qquad (16\text{-}12)$$

$$a = \frac{N_{\text{pr}}}{N_S} = \frac{I_S}{I_{\text{pr}}} \qquad (16\text{-}19)$$

It is easy to remember these equations. All you have to remember is $\sqrt{3}$. For a delta configuration, you know that the line current will be greater than the phase current because the line current is going to divide into two phases. For a wye you know that the line voltage is greater than the phase voltage because the line voltage is across two phases. The number you multiply by is $\sqrt{3}$.

How do you decide which of the four configurations to use? A consideration used in the selection of connection type is grounding. The wye connection has a natural ground point at the center of the wye or star. This ground is often used for the higher-voltage side of the transformer. Thus, in a step-down application you may see a wye-delta connection, whereas in a step-up it may be a delta-wye connection. The wye-wye connection is seldom used. Some common grounding techniques for three-phase systems are shown in Fig. 22-2.

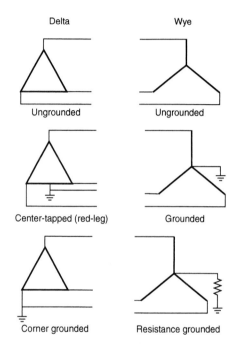

Figure 22-2 Commonly used grounding methods.

Another consideration could be reliability of power delivery. A delta connection allows one transformer to be removed from service while still providing service on all three phases. When a leg of the delta is removed, the result is called an *open delta*, or *V*, connection. This idea will be reviewed later in this chapter.

Problem Solving

The mathematics of this chapter, based on Fig. 22-1, teaches how to analyze and solve three-phase transformer circuits easily. Which of the many analysis techniques that can be used depends upon the questions you want to answer. Technical people usually measure a line or phase value and want to understand the meaning of the measurement. For these questions, the technique used here is sufficient because for balanced phases all phases are the same.

In the earlier chapters the mathematics was straightforward. There was usually only one way to solve a problem, and there was only one correct answer. In this chapter the complexity of the problems means that there is more than one way to solve them. You will learn that the answers you get will depend upon the order in which you solve for the values. The differences are caused by the number of places of accuracy in the calculations. You can argue that all that is required is to carry more places of accuracy, which has some truth but is contrary to common practice.

The traditional concept of mathematics is formulas, numbers, and proofs. In this chapter ideas are just as important. Electrical symbols represent ideas, just as the symbols of mathematics represent ideas. Mathematics is really only logic. In this chapter the biggest part of getting the answer is the thinking that goes into the development of the problem. If you do this part well, then the answers will come more easily.

Solutions Using Fig. 22-1

The first set of problems will be for the delta-delta connections of Fig. 22-1b. The assumptions of balanced loads and ideal transformers are idealistic, but they represent real-life situations and provide insight and understanding to real problems. The problems will show how to get answers quickly. Long, complicated, formal mathematical approaches are replaced by experience and understanding. It is more important that you learn how to visualize and think about a problem than learn a specific formula.

Solving these problems is like putting a puzzle together. *First* list all of the facts that you know about the problem. This is like opening the puzzle box and dumping out all the pieces. You want to take a broad look at the problem, so you turn each piece right side up and see what it looks like. These facts, or puzzle pieces, are called the "givens."

Next, list the relationships that can be used to solve the problem. This is like putting the tools you may use for the job in your toolbox. The tools for the first examples in this chapter are taken from the equations for the delta, wye, and ideal transformers. Of course, this is electronics, so you will always have $E = IR$ and other such formulas in your bag.

The *third* step is to list the values that you want to solve for. These values are called the "wants." This step is like reading your work orders. It gives you direction. Every now and then, you get so involved in the problem you forget why you are there. You have to go back and read these orders so you can get back on a productive track.

Another advantage that every master has is a "bag of tricks." We have already used this bag but just did not tell you about it. What do you think we used from this bag?*

You have set the problem up and you are ready to start the assembly. Put all the easy pieces together first. They are like the corners of the puzzle. Are there any tools that will take a given and directly give you a want? If so, use these right away. The results can then be put under the givens because they are now known values. They can be used to solve for other wants.

While placing pieces, we may see how to introduce other concepts and formulas such as the power factor, power calculations, and Ohm's law. At some point

---

*Two tricks were used. The first is the balanced-load assumption. As we shall see later, this greatly reduces the amount of work that we have to do. The second is the idea of an ideal transformer. This also reduces the amount of work. We looked at real transformers in Job 16-13. We did this because we wanted to test the transformer, and we wanted to know more about its operation. In these problems we are assuming that the transformer is all right and that it is well designed for its purpose. With these assumptions we can think of it as an ideal transformer.

there may be no direct solutions left. You may have to solve for some intermediate values not requested. You then move up the ladder to solve for another want. Remember, every time you solve for a value, it can be placed under the givens and used to solve for other values.

We will now apply these ideas to the following problems.

Summary and Review

1. What are the steps in defining a problem?
   a. List all the given facts.
   b. List the tools to use for the solution.
   c. List the values for which you want to solve.
2. What are the steps needed to solve the problem?
   a. Solve the easy parts first and use these as givens for other solutions.
   b. Look for tools that take givens to wants.
   c. Look for intermediate values that can provide steps to other solutions.

Balanced Three-Phase Loads

Your goal is to do your job and use mathematics as a tool, not to change jobs and become a mathematician. To solve problems like these, first look for simplifications that will allow you to put more time on your job and less on calculations. Balanced loading is one such assumption. If the unbalance is minor, the solution to the real problem will be close to the solution for a balanced system. This may be an acceptable answer for your purposes.

We can see symmetry in the delta and the wye transformer configurations. Each phase looks like the other phase. Balanced loading completes this symmetry. It is not necessary to solve all three phases in balanced circuits because the solution for one phase is the solution for each phase. The only difference might be in the phase of the voltage or current, but the questions that we ask are not about phase. Our questions are about the size of the voltage or the size of the current. This kind of reason and understanding is the basis of a master's trick.

The Delta-Delta Connection

A wiring diagram for a delta-delta connection is shown in Fig. 22-3a on page 782. The diagram in this form does not make the delta-delta configuration obvious. The three transformers are labeled T_1, T_2, and T_3. The three-phase power is brought into the primaries with one line at each of the junctions between the transformer primaries. This places each primary across a pair of lines. Transformer T_1 is across lines A and B, transformer T_2 is across lines A and C, and transformer T_3 is across lines B and C. The primaries are connected in a loop, triangle, or delta.

The power out is taken from the junctions between each of the secondaries. This places one secondary on each of the phases of output power. T_1 is across lines A' and B', T_2 is across lines A' and C', and T_3 is across lines B' and C'.

The delta-delta configuration can be noted by the fact that both the primaries and the secondaries of the transformers form loops. One way to notice the delta connections is that each pair of lines has a winding across it. This type of connection is called a *closed-delta* connection because it contains three transformers and they complete the entire delta. We shall later see an *open-delta* connection. If the diagram is redrawn as in Fig. 22-3b, the delta configurations are more obvious.

Job 22-2   Balanced Three-Phase Loads   **781**

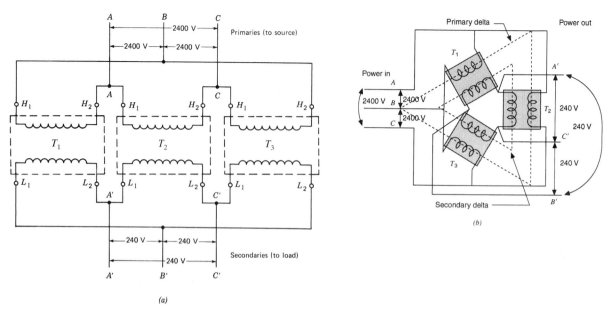

Figure 22-3 Connections for delta-delta. (*a*) Delta-delta connection for a three-phase 10:1 transformer. (*b*) Circuit in delta form.

Delta-Delta Transformer Calculations

Example 22-1

The primary of a delta-delta transformer is rated at 40 kVA and 2400 V. The transformer is designed to step the voltage down in a ratio of 10:1. The load on each phase is 200 A with a power factor of 80%. Find the voltage across each primary coil, the voltage across each secondary coil, the secondary line voltage, the kVA load on the transformer bank, the kilowatt load on the transformer bank, the current in each primary winding, the current in each secondary winding, the current in each primary line, the kVA load on each transformer, and the kVA input to the transformer.

Givens. First draw the model in Fig. 22-1*b* and fill in all the information that is given in the problem. The problem states that each of the transformers is rated at 2400 V. This is meant to imply that the voltage on the input lines is 2400 V. We learn to work with this misuse of language and logic. Just because a transformer is rated at 2400 V does not mean that it actually is operating at that value. We will assume that you were smart enough to check this out and found that the input line voltage was 2400 V.

$$V_{L\,\mathrm{pr}} = 2400 \text{ V}$$

The voltage ratio is 10:1; therefore, we know that for the ideal transformer

$$\alpha = 10$$

The current in the secondary lines is 200 A with a power factor of 80%. From this knowledge,

$$I_{L_S} = 200 \text{ A}$$

From Eqs. (20-3) and (20-8),

$$PF = \cos \theta = \frac{W}{VA} = 80\% = 0.80$$

We are also told that the transformer is rated at 40 kVA.

Tools. The problem statement says that the system has balanced loads and delta connections. This means that we can use the following tools for the delta and the transformers.

$$\text{Delta:} \quad I_L = \sqrt{3}I_p \tag{21-11}$$
$$V_p = V_L$$

$$\text{Ideal transformer:} \quad \alpha = \frac{N_{pr}}{N_S} = \frac{V_{pr}}{V_S} \tag{16-12}$$

$$\alpha = \frac{I_S}{I_{pr}} \tag{16-19}$$

All these tools are already included in Fig. 22-1b. In addition we know that the basic formulas such as Ohm's law, $E = IR$, and the power formulas, $P = V \times I \times \cos(\theta)$, etc., can be used.

Wants. The wants are listed in the sentence that starts with "Find."

a. *The voltage across each primary coil,* which is the phase voltage on the input delta
b. *The voltage across each secondary coil,* which is the phase voltage on the output delta
c. *The secondary line voltage,* which is the voltage measure across any two lines of the balanced output
d. *The kVA load on the transformer bank*
e. *The kilowatt load on the transformer bank*
f. *The current in each primary winding,* which is the phase current in any input delta winding
g. *The current in each secondary winding,* which is the phase current in any output delta winding
h. *The current in each primary line,* which is the line current for each of the input lines
i. *The kVA load on each transformer*
j. *The kVA input to the transformer*

Notice we have to translate most of the statements. We often use different terminology in different contexts. In (*a*) we are asked to find "the voltage across each primary." This is a perfectly valid way of requesting the information. We also know that the circuit we are working on is a delta configuration. In a delta, this voltage is referred to as the phase voltage.

The problem stated that the load in each phase is 200 A. It is assumed that the power factor is the same in each phase, so that this means there is a balanced load. If this were not true, then we might think that the question is asking for three sets of answers. One difficulty with this kind of question is that the person asking may have indeed wanted you to give three answers or at least to state that the one answer applies to all three phases. Here is a situa-

> **Hint for Success**
>
> The more complex a problem is, the more necessary it is to have an orderly plan of solution and to document the process.

tion in which we are not sure whether the desired response is (1) one voltage, (2) three voltages, or (3) one voltage and the statement that this one voltage applies to all three phases because the system is balanced. The last answer is the best because it indicates more knowledge.

Solution

Figure 22-4 reproduces Fig. 22-1b for the delta-delta connection. The givens are inserted in the appropriate places in Fig. 22-4a. First we are given that the input line voltage is $V = 2400$ V. Next we were told that the voltage ratio is 10. From this we know that $a = 10$. We substitute this in the appropriate places on the secondary side. The last given is the output line current and power factor.

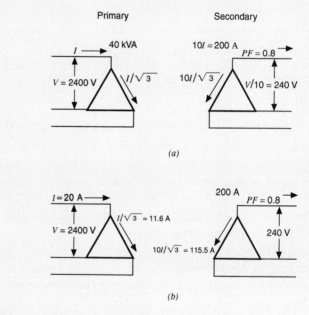

Figure 22-4

When these puzzle values are placed on the drawing, we see immediately that want (a) is just a test to see if we understand that line voltage and phase voltage are the same thing for delta connections. A simple division, 2400/10 = 240, solves for both want (b) and want (c).

If we solve for all of the values in Fig. 22-4, then we will have a complete understanding of the circuit. All other values can be derived from there. This is important to know because questions such as those about power are not indicated on the diagram.

A quick look at the diagram shows that it is very easy to get the phase current for the secondary. We can use the relationship between line current and phase current for the delta connection. We know that aI is the line current of 200 A.

$$I_p = \frac{I_L}{\sqrt{3}} = \frac{200}{\sqrt{3}} = 115.5 \text{ A} \qquad Ans.$$

This is want (*f*). Because we now know the current in the secondary winding, we can find the current in the primary winding by using the ideal transformer relationship.

$$I_{pr} = \frac{I_p}{\alpha} = \frac{115.5}{10} = 11.6 \text{ A} \qquad Ans.$$

The last calculated value to find is the input line current. We could use the relationship between line and phase currents in a delta connection.

$$\frac{I}{\sqrt{3}} = 11.6$$
$$I = 11.6 \times \sqrt{3} = 20.09$$
$$= 20 \text{ A} \qquad Ans.$$

It may be easier to notice that the output line current is just *aI*. This gives a very simple calculation for want (*h*):

$$I_{Lpr} = \frac{I_{L_S}}{\alpha} = \frac{200}{10} = 20 \qquad Ans.$$

One solution gives an answer of 20 A and the other gives 20.09 A. In this case the difference in the answers is very small. Sometimes the differences can be larger. The difference is caused by the sequence used to find the answer. It is desirable to use the solution technique that requires the fewest intermediate calculations. The most direct solution reduces rounding and approximation effects. Not only does the most direct calculation usually give the answer most quickly, but it is usually the most accurate.

The circuit is now completely solved. These values are placed on the diagram, as shown in Fig. 22-4*b*. All that remains is to solve for the other values that are derived from these fundamental circuit values. The output line powers are easily seen from the diagram.

$$kVA = \frac{V_L \times I_L}{1000} = \frac{240 \text{ V} \times 200 \text{ A}}{1000} = 48 \text{ kVA} \qquad Ans.$$

This can be converted to watts load by using the power factor.

$$kW = kVA \times PF = 48 \text{ kVA} \times 0.8 = 38.4 \text{ kW} \qquad Ans.$$

These are the answers for (*d*) and (*e*). To find the kVA for each transformer, multiply the voltage across each phase by the phase current. For an ideal transformer, it does not make any difference whether we look at the input or the output side of a phase. In our case, there is a difference because of rounding. For the output side we get

$$kVA_{per\ transformer} = \frac{V_p \times I_p}{1000} = \frac{240 \text{ V} \times 115.5 \text{ A}}{1000} = 27.7 \text{ kVA} \qquad Ans.$$

If we used the primary side, the answer would be

$$kVA = \frac{2400 \times 11.6}{1000} = 27.84 \text{ kVA}$$

This does not mean that there is 0.14-kVA loss in the transformer. It just means that when we calculated the primary phase current, we rounded 11.55 A to 11.6 A. Because of rounding and different possible sequences of calculation, the answers may not all agree exactly. One way to reduce the problem is to carry more places of accuracy in the calculations than are required. This is often not a practical solution.

CALCULATOR HINT

Rounding answers to intermediate calculations in which the computations are broken into steps in order to facilitate manual computation introduces errors. It is better to save the intermediate results in calculator memory.

The last variable to solve for is the power input.

$$\text{kVA}_{\text{input}} = \frac{\sqrt{3}\,V_L \times I_L}{1000} = \frac{1.732 \times 2400 \times 20}{1000} = 83.1 \text{ kVA} \qquad \textit{Ans.}$$

Figure 22-1 shows that the input line power and the output line power are equal.

$$\text{kVA}_{\text{in}} = V \times I = aI \times \frac{V}{a} = \text{kVA}_{\text{out}}$$

Again, this is because we have assumed a lossless transformer. At this point you may ask why the 40-kVA value was given. Sometimes there is information that is not necessary. You must learn to disregard this information.

Summary and Review

Figure 22-1b provides a basis for solving for line and phase values for a delta-delta transformer problem. For many problems, there is more than one way to solve. The most direct method provides the fastest and more precise answer. Many questions require translation of the language into the proper terminology for the given situation. It is necessary that people learn how to ask questions and to answer them so that the workers' responsibilities are clear and precise.

The steps in solving a problem are as follows:

1. List all given values and facts.
2. List all tools that can be used to solve the problem.
3. List all values that are wanted.
4. Solve the problem in the following way:
   a. Find all simple answers.
   b. Find all answers that come directly from the application of tools to givens.
   c. Find intermediate values necessary to solve for desired values.

Balanced Load Voltages and Currents

Example 22-2

Find the relationships between the phase voltages, phase currents, and line currents for a balanced load on a delta system. The secondary of a delta-delta transformer and its delta load are shown in Fig. 22-5.

Solution

For most practical situations the method of the previous example is sufficient. This example shows how to set up complex problems in a more rigorous way. It develops some results relative to phase interactions that are not possible with the previous method. It is also included because it serves as background for some later tasks.

You may want to read this example now and return to it later. It is suggested that you at least try to understand the results of this example even if you are not interested at this time in developing the skills to get the results on your own.

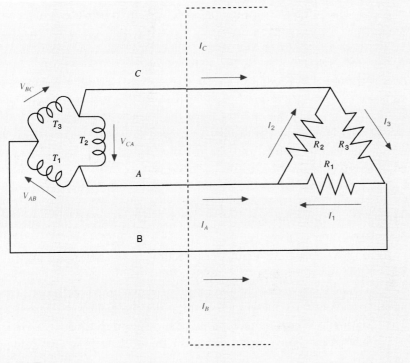

Figure 22-5

It is necessary to be orderly in this method. The most important step is probably to draw a diagram such as Fig. 22-5. On this figure you place all your definitions of the problem. In Example 22-1, all the definitions were given to you. You just had to fill in the givens and solve for the wanted values. This example asks you to solve the circuit. It is up to you to make all the definitions yourself.

Another major difference between this example and the previous example is that we are now going to look at each individual phase. We are not going to use symmetry and summary types of relationships as we did in Example 22-1. The cost in work is high, but the information and knowledge gained are also very high. If all you want is the information that you got in Example 22-1, then use that technique. If that is not enough, then try something more complex.

Step 1: Define variables. The first step is to define the variables that will be used, including names as well as directions and possibly other parameters. Arrows are used to indicate direction.

When assigning names, it is very helpful to follow a scheme or pattern. Some variables come from commonly used naming schemes. These commonly accepted names aid in understanding and should be used whenever possible.

One common naming scheme is to name the lines in a three-phase system A, B, and C. This is done in Fig. 22-5. To keep this pattern, the line currents are named I_A, I_B, and I_C. We have assigned all positive directions as going out from the transformer. Directions are arbitrary. As long as we are consistent, then if the arrow goes in the wrong direction, the calculated value will be negative. The purpose of assigning directions is to provide guidance in the development of the equations.

The next definitions are for the voltages across the windings of the secondaries of the transformers. Because the windings in a delta configuration are connected between lines, we will use the line names to identify the winding voltages. Voltage V_{CA} is the voltage for the winding connected between lines A and C. In a similar manner we define the voltages V_{BC} and V_{AB}.

The winding voltage definitions must indicate positive and negative directions. The arrows point from the negative end to the positive end of the winding. Here again the direction is arbitrary.

The variables to be identified are the currents in the phase loads. Each load is connected between a pair of lines, as are the transformer windings. Therefore, we number the load currents in accord with the transformer that determines their voltage, so the load currents are I_1, I_2, and I_3.

Again, it is necessary to define the direction in which the currents will flow. We will use the convention that current flows from positive to negative. The positive and negative ends of the loads are determined by the positive and negative ends of the transformer winding that the load is connected across. The resulting load current directions are shown in Fig. 22-5.

We have used Ohm's law to establish these definitions. In addition, we have introduced traditions about which way current flows. This has been done so that we will get positive answers.

Step 2: Define independent variables.   Independent variables are those variables that can have any desired value and are independent of all other variables in the problem. In this problem transformer windings can be taken as independent. We can adjust each phase voltage by controlling the voltages on the input to the transformer. The voltages have amplitude and phase.

The phase is determined because the circuit is a three-phase current, so these winding voltages are separated by 120°. The amplitude of all three phases will be the same. This amplitude with be V, which could be 120 V, 208 V, 220 V, or any other value that we choose. We therefore define the winding voltages as

$$V_{AB} = V \times \sin(\theta)$$

$$V_{BC} = V \times \sin(\theta + 120°)$$

$$V_{CA} = V \times \sin(\theta + 240°)$$

Step 3: Define dependent variables.   Dependent variables are determined by other values. For example, if you apply a voltage across a resistor, then the current is determined by the voltage and the size of the resistor. You could vary the voltage or change the resistor. They are assumed to be independent. The current is dependent. It is determined by the voltage and the resistor.

There are two sets of dependent variables, the phase currents in the loads and the line currents. For this problem we will assume that the loads are pure resistances. For what we want to show in this example, this is no restriction. As long as the loads are balanced, the results will be the same.

The current in any leg of the load will be determined by the voltage across that leg. We know from the delta relationship that each phase of the transformer determines the voltage for the corresponding load phase. This gives the following equations for the load currents.

$$I_1 = \frac{V_{AB}}{R_1}$$

$$I_2 = \frac{V_{CA}}{R_2}$$

$$I_3 = \frac{V_{BC}}{R_3}$$

The last set of dependent variables includes the line currents. These can be obtained by summing the currents at each node of the loads. This gives the following set of equations:

$$I_A = I_2 - I_1$$
$$I_B = I_1 - I_3$$
$$I_C = I_3 - I_2$$

These equations answer the problem. The problem was not specific as to what form the answer should have. The equations are a precise answer, but they do not make the relationships obvious to most people. Next we will look further into the answers through graphs of the functions.

Winding voltages. The winding voltages are independent variables. They were defined in step 2 above. Figure 22-6 shows these voltages. It has been assumed that the line voltage is 110 V, which gives a peak voltage of 155 V. Line voltage V_{AB} is a sine wave. Voltage V_{BC} looks the same, except that it is shifted to the right 120°. Voltage V_{CA} is shifted 240°. These are the voltages that are applied to the loads.

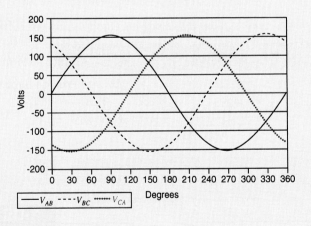

Figure 22-6

Line currents. The winding voltages cause currents to flow in the load as defined in step 3. It is assumed that the load resistances are 1 Ω each. This gives load currents that look like the voltage curves in Fig. 22-6, except they are in amperes.

The load currents create the line currents shown in Fig. 22-7 according to step 3. They are sine waves, but the shifts are different from the phase voltages and phase currents. Line current I_C is shifted 90° and line current I_A is

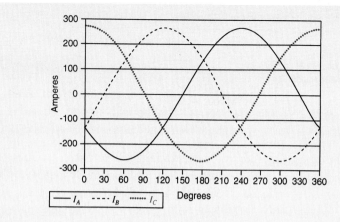

Figure 22-7

shifted 210°, a difference of 120°. Line current I_B is shifted 330°, another 120° from I_A. The peaks of each of the currents are equal.

The balanced load of the system causes the line currents to be equal in magnitude and separated by 120°. This makes it possible to use the model of the previous example if you are interested only in the amount of current in a single phase and you are not interested in the relationships between phases.

A dotted line has been drawn in Fig. 22-5 that cuts through the three line currents and separates the load from the transformer source. The current in these three lines coming from the transformer must go back to it. This means that at any given instant the sum of the currents must be zero.

Figure 22-7 shows that there is always at least one positive current and at least one negative current. It shows how the current flows to the load and back on an ac basis.

Table 22-1

Instantaneous Line Currents

Angle,°	I_A	I_B	I_C
0	135	135	−269
30	0	233	−233
60	−135	269	−135
90	−233	233	0
120	−269	135	135
150	−233	0	233
180	−135	−135	269
210	0	−233	233
240	135	−269	135
270	233	−233	0
300	269	−135	−135
330	233	0	−233
360	135	135	−269

Table 22-1 gives the instantaneous line currents for times indicated as phase angles. There may be a slight rounding error, but if you sum I_A, I_B, and I_C, the result is zero.

If you measured the peak current in each of the lines, you would get 269 A. If you measured the RMS current, it would be 190 A. All three lines would read the same. This is the information that is provided by Example 22-1, showing that Example 22-1 relates to meter measurements. The techniques of Example 22-1 are not sufficient to show that current flowing out of the power source equals the current flowing back. There is no way to add and subtract the meter readings to show this fact.

This is an excellent example of how important it is to understand how and what your meters are measuring relative to what you want to know. When your instruments measure peak, RMS, average, or other time-related values, you cannot expect the results to satisfy electronic concepts that are valid on an instantaneous basis.

This example also shows how important it is to select the model according to what it is you want to show.

Line current composition. Figure 22-7 showed the three line currents. They looked like the corresponding voltage sine waves, but shifted. Figure 22-8 shows how two sine waves can combine to form a third sine wave. The figure shows how line current I_A is formed from load currents I_2 and I_1. Table 22-2 is a table of values taken from the graph. You can see from the table that

$$I_A = I_2 - I_1$$

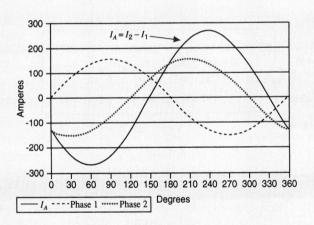

Figure 22-8

If you used the solution from Example 22-1, the phase currents would be 110 A. The equations for I_A would then give

$$I_A = I_2 - I_1$$
$$= 120 - 120$$
$$= 0$$

This is entirely the wrong answer. You may want to think that if you subtract one current from another, the difference will be smaller. This is not necessarily true for ac. If the phase difference is large enough, as in this case,

Table 22-2

Instantaneous Line and Phase Currents

Angle,°	I_A	I_1	I_2
0	135	0	135
30	0	78	78
60	−135	135	0
90	−233	156	−78
120	−269	135	−135
150	−233	78	−156
180	−135	0	−135
210	0	−78	−78
240	135	−135	0
270	233	−156	78
300	269	−135	135
330	233	−78	156
360	135	0	135

the difference is actually larger than either of the two sine waves. I_A is larger than either of the phase currents.

Summary and Review

Be orderly in the formulation of the problem. Also, draw a diagram to help you see the relationships of the variables.

To define the problem,

1. Define all variables of the problem.
2. Determine which variables are independent and assign them values.
3. Determine the dependent variables by writing equations relating them to the independent variables and other dependent variables.

Use graphs to make the relationships more obvious. Remember that the line currents are sine waves, but they are shifted differently from the phase voltage and current. At any instant in time the sum of the line currents is zero. In a balanced system, the solutions for each phase look the same, but they are shifted 120°.

If you are not interested in phase relationships or instantaneous effects, then the models in Fig. 22-1 are sufficient.

In reviewing the delta-delta connection, note that transformers in a delta connection have a winding across each phase. A closed-delta configuration has three transformers in a loop or delta. For a balanced system the solutions for each of the phases look the same.

The Wye-Wye Connection

The wye-wye connection has three single-phase transformers with their primaries and secondaries connected in a wye. A wye is recognized by the fact that there is a common connection between all three transformers. The lines are connected across two transformers.

The wye-wye transformer connection is summarized in Fig. 22-1d. In the wye connection, the line current and the phase current are equal. The line voltage splits with the factor $\sqrt{3}$ for the phase voltage. This is similar to the delta connection, where the phase voltage and the line voltage are equal and the line current splits with the factor $\sqrt{3}$ between phases.

The ideal wye-wye connection is reproduced in Fig. 22-9. The input line voltage is 4160 V and the transformer has a step-down ratio of 20:1. Find the phase voltage on the primary and the secondary phase and line voltages.

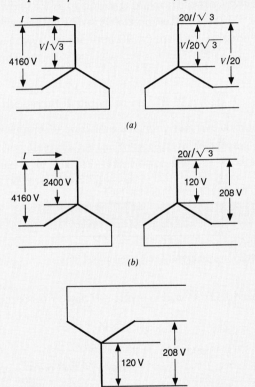

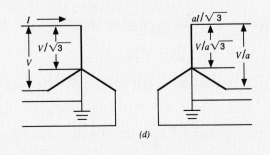

Figure 22-9

The problem does not ask for any relationships between phases or instantaneous relationships. We can, therefore, use the method of solution of Fig. 22-1. The given values are shown in Figure 22-9a. Figure 22-9b shows the calculated values.

The input phase voltage is easily calculated from the input line voltage.

$$V_p = \frac{4160}{\sqrt{3}} = 2401.77 \text{ V} \qquad Ans.$$

Traditionally, we call this 2400 V, not 2402 V. The error is less than 0.1%, which is irrelevant in practical terms. As we will see, the standard is 2400, because when we divide 2400 by 20, we get 120, which is another standard.

The second desired answer is the secondary phase voltage. The diagram shows the secondary phase voltage is the input-phase voltage divided by the voltage ratio. Thus, by using the ideal transformer (and the standard answer from the last equation),

$$V_{PS} = \frac{2400}{20} = 120 \text{ V} \qquad Ans.$$

The secondary line voltage can be obtained from the wye relationship between phase and line voltages for balanced loads.

$$V_{LS} = V_{PS} \sqrt{3} = 120 \times \sqrt{3} = 207.8 \text{ V} \qquad Ans.$$

This is rounded up to 208 V. Looking at the diagram, we see that another way to solve for the output line voltage is to get it directly from the input line voltage:

$$V_{LS} = \frac{V}{\alpha} = \frac{4160}{20} = 280 \text{ V}$$

Notice again that the more direct solution is the more accurate.

In Figure 22-9c the diagram of the secondary windings has been flipped vertically to show why section 220-2 of the NEC has a standard set of voltages for calculating called 208Y/120. For a wye (Y) connection, an output line voltage of 208 V corresponds to a phase voltage of 120 V, which leads to the notation 208Y/120.

The connection shown in Fig. 22-9b is a three-wire, three-phase system. The only voltages available are 208 V in three different phases. A natural common point is the junction of the three legs of the star. (Notice all the names we give this connection: Y, wye, star.) If this point is grounded and a fourth line is taken from the ground, then we have a four-wire, three-phase system. This system is shown in Fig. 22-9d. Both 208 V and 120 V are now available. By taking the power between any two lines, you get 208 V. If you take the power between any single line and ground, then you get 120 V.

Summary and Review

Industry standards sometimes determine what the "correct" answer to a problem should be. The wye configurations produce a line voltage of 208 V and a phase voltage of 120 V, which gives the standard 208Y/120.

Three-Phase Power from Two Transformers

Earlier we said that in general it requires three single-phase transformers for a three-phase system. In this job, we show how it is possible to transform three phases with only two single-phase transformers.

The Open-Delta Connection

The open-delta configuration transforms three-phase power in a delta configuration using only two transformers. This configuration is created by removing one of the transformers in a closed-delta configuration. All lines on the input and the output remain as before. Only the transformer is removed.

Figure 22-10 shows both the original delta and the same delta with transformer 2 removed. The two remaining transformers form a V. For this reason the configuration is sometimes called a V connection.

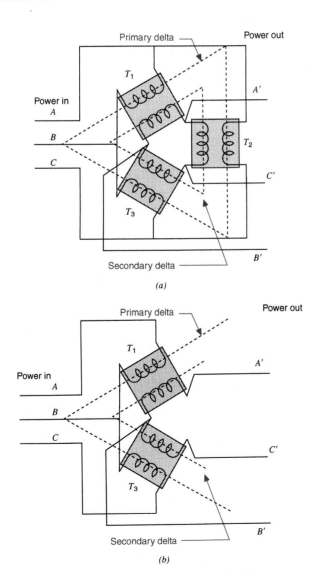

Figure 22-10

One advantage of the delta-delta power distribution system is that any single transformer in the delta configuration can be removed without interrupting the distribution of power. In Fig. 22-10, transformer 2 was removed, but any of the three transformers could have been taken out. This allows for maintenance without power interruption.

Another use of the open-delta configuration is where there is an anticipated growth in demand. The open-delta system can be installed; later, as there is a need for additional power, the third transformer can be installed without any other modification to the distribution system.

Secondary voltages in an open delta.   A three-phase source creates three separate voltages, each separated by 120°, as shown in Fig. 22-11a. The voltage vectors have been moved in Fig. 22-11b to show the delta relationship of voltages. In this form the concept of a delta is obvious.

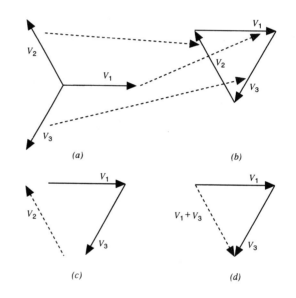

Figure 22-11

In Fig. 22-11c the voltage for transformer 2 has been removed. This voltage can be re-created by adding together the voltages for phase 1 and phase 3. This sum is shown in Fig. 22-11d. An interesting observation is that the re-created voltage is opposite in phase from that of the original transformer. The average user will never be able to tell the difference. If it is noticeable, then all that is necessary is to turn the plug around in the wall socket.

Figure 22-12 shows the line voltages for the open delta. This figure corresponds to Fig. 22-6. V_{AB} is the same as V_1 and V_{BC} is V_3. The reconstructed voltage $V_1 + V_3$ corresponds to V_{CA}. Notice that $V_1 + V_3$ is V_{CA} flipped 180°.

Summary diagram for open and closed deltas.   In this section we will create a summary diagram for the open-delta configuration that will correspond to those in Fig. 22-1.

Figure 22-13 shows an open-delta system with balanced loads attached. The load currents I_1, I_2, and I_3 in the balanced loads are equal.

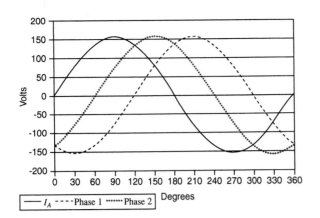

Figure 22-12

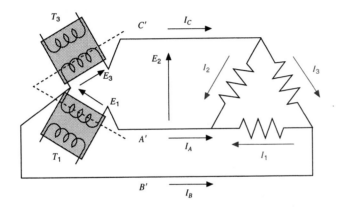

Figure 22-13

Line current I_C is the phase current for transformer T_3. Equation (21-11) gives the line current for any phase current I_P for a balanced delta system. The current in any leg of the delta load is

$$I_L = \sqrt{3}I_P \qquad\qquad (21\text{-}11)$$

The I_P in this equation is the current in a load phase and not the current in a transformer phase and I_L is the line current not the load current. Using the terminology of this section, the equation is

$$\text{Line current} = \sqrt{3}I_L$$

The preceding values are placed in Fig. 22-14a. The voltage and phase currents on the primary side can be obtained from the ideal transformer relationships.

$$V_{\text{pr}} = \alpha V$$

$$I_{\text{pr}} = \frac{I_S}{\alpha} = \frac{\sqrt{3} \times I_L}{\alpha}$$

Again we have used the fact that for an open delta, the phase current is the line current.

CALCULATOR HINT

Linking solutions directly in the calculator, preferably by using a programmable calculator, eliminates the need to use rounded intermediate calculation. This improves accuracy.

The summary in Fig. 22-14a is in terms of the load currents and voltages. This is contrary to Fig. 22-1, which is in terms of the input voltage and current. Figure 22-14a can be converted to an input summary by the following substitutions.

$$I = \frac{\sqrt{3} \times I_L}{\alpha}$$

$$V = \frac{V_S}{\alpha}$$

With these substitutions we get the summary in Fig. 22-14b, which corresponds to Fig. 22-1. Notice the similarity to Fig. 22-1b.

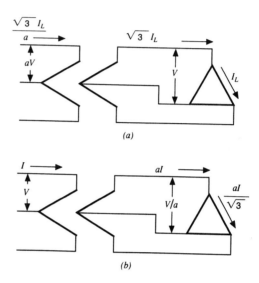

Figure 22-14

We now have two summary diagrams for the open delta. Figure 22-14a is of load voltage and current and Fig. 22-14b is in terms of branch circuit voltage and current.

Power in Delta and Open Delta

A transformer is rated by the current capacity of its windings. A natural assumption would be that an open-delta connection would be able to handle two-thirds of the power capability of a delta. Actually it will only be able to handle 58 percent, rather than 67 percent, of a delta system.

To understand why the power capability of an open delta is lower than expected, look at Fig. 22-1. In Fig. 22-15a an open-delta system delivers power to a balanced delta load. In Fig. 22-15b a full-delta transformer delivers the power to another balanced load. In both cases the line current I_C in terms of the load current is the same. The difference is that the line current for the open-delta circuit is the current in the transformer winding. In the delta in Fig. 22-15b, the line current divides in the transformers, thereby reducing the current in each transformer. The power limitation is determined by the amount of current that the transformer winding can handle.

$$I_L = \sqrt{3}I_p \qquad\qquad (21\text{-}11)$$

$$I_T = \sqrt{3}I_{LO}$$

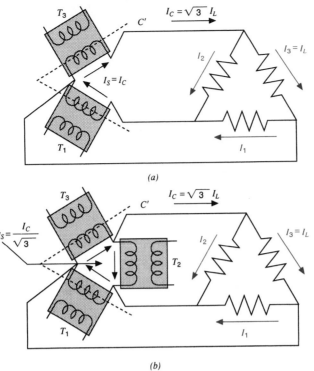

(a)

(b)

Figure 22-15

where I_T is the current capacity of the transformer winding and I_{LO} is the current in each leg of the load on the open delta system.

For the full delta, the winding current is

$$I_T = \frac{I_C}{\sqrt{3}} = \frac{\sqrt{3}I_{LF}}{\sqrt{3}} = I_{LF}$$

Combining these two equations gives

$$I_{LF} = \sqrt{3}I_{LO}$$

Since we are assuming that the two circuits work at the same voltage, the ratio of the power in the two circuits is

$$\frac{P_{LO}}{P_{LF}} = \frac{I_{LO}}{I_{LF}} = \frac{1}{\sqrt{3}}$$

$$= 0.577 = 58\%$$

Exercises

Applications

1. A transformer is required to change the incoming 4160-V line to a 480-V line voltage. What transformer ratios are required if the transformer is connected in (a) wye-delta, (b) delta-delta, (c) delta-wye, (d) wye-wye?
2. The transformer of Exercise 1 is required to deliver 60-kW balanced load. What is the power requirement for the primary and secondary windings for (a) wye-delta, (b) delta-delta. (c) delta-wye, (d) wye-wye.

3. For the transformer of Exercise 1 what is the primary line current to deliver 60-kW to the load if the transformer is connected in (a) wye-delta, (b) delta-delta, (c) delta-wye, (d) wye-wye?
4. What are the secondary line and phase currents for the transformer of Exercise 1 when it delivers 60-kW to each phase load for (a) wye-delta, (b) delta-delta, (c) delta-wye, (d) wye-wye?
5. What are the primary line and phase currents for the transformer of Exercise 1 when it delivers 60-kW to each phase load for (a) wye-delta, (b) delta-delta, (c) delta-wye, (d) wye-wye?

(See CD-ROM for Test 22-1.)

1. A 4160-V/2400-V wye-connected distribution system supplies a 208-V/120-V three-phrase, four-wire system. If the connected load is 2500 kVA at 85% PF, determine the total connected kilowatts in both the primary and the secondary phases.

2. In question 2, determine the line current in both the primary and the secondary phases.

INTERNET
ACTIVITIES

Internet Activity A

Find four ways to connect wye and delta transformers.
http://bbs.elogica.com.br/mactronic/tr.html

Internet Activity B

The NEC is administered by the NFPA. Surf the site to see NEC-related materials. **http://www.nfpa.org/**

Mathematics for Logic Control

Learning Objectives

1. Explain the difference between switching circuits and digital logic circuits.
2. State the basics of binary mathematics.
3. Explain how series-connected switches represent the AND logic function.
4. Identify the mathematical symbol for the AND operation.
5. State the rules associated with the AND logic function.
6. Complete a truth table for the AND logic function.
7. Explain how parallel-connected switches perform the OR logic function.
8. Identify the mathematical symbol for the OR operation.
9. Complete a truth table for the OR logic function.
10. State the rules associated with the OR logic function.
11. Explain the EXCLUSIVE OR logic function.
12. Define the INVERT logic function.
13. Connect switches to combine the AND and OR logic functions.
14. Complete truth tables for logic circuits which combine the AND, OR, and INVERTER logic functions.
15. Explain how relays are used in logic circuits.
16. Draw the logic symbols for the AND, OR, EXCLUSIVE OR, and INVERT logic functions.

JOB 23-1

Introduction

The very rapidly expanding field of microelectronics has caused major changes in many fields of electricity and electronics. Sophisticated control systems have been added to such entities as elevators, energy systems, and home lighting and power control, not to mention the rapid growth in the area of robotics and other manufacturing and industrial controls. Almost everyone has been affected by the new forms of binary signals and controls. The intent of this chapter is to introduce the basic concepts of the mathematics of logic that are used in these modern control systems.

The field of logic and binary arithmetic can best be learned by starting with very basic concepts and applying from that point. At first you may feel that this is a drastically new form of mathematics, but you will find out that it is actually very similar to all the mathematics that has preceded this chapter. There are rules, just like algebra, and if these rules are followed, you have fewer chances to make errors.

What we are learning is a means to analyze logical problems. This is not like the previous mathematics in which numbers represent the size of components or the value of a voltage.

In modern controls we find two common types of components used to build control circuits. One type is composed of switches and relays, while the other contains modern solid-state electronics. The former could be called "switching circuits" and the latter, "digital logic circuits."

In switching circuits the primary emphasis is on the continuity or "connectedness" of the circuit. The analysis of these circuits more closely resembles the thinking of an electrician who wires a system. The electrician follows the lines on a wiring diagram to determine where it comes from and where it goes. If the line is continuous from a switch to a motor, the electrician knows that the switch controls the motor.

The second class of circuits, digital logic, is usually voltage-level-dependent. This means that the primary interest is on various control signals and what their voltage is. There is less physical resemblance between these circuits and the "connected" switching circuits. Both types of circuit are used in modern electronics, sometimes in the same system. We shall study the fundamentals of both and see how they differ and how they are the same.

Binary Concepts

In all the previous mathematics that has been studied in this book, we have become accustomed to the use of voltage, current, and resistance in problems. Each of these quantities has been used in equations and could have different values depending on the problem. These quantities are called *variables*. Resistors, capacitors, and inductors come in many sizes. This means that there must be many numbers so that we can tell just what size they are. Because we place no limits on their size, we need a number system that can handle very large and very small values.

The fundamental characteristic of the modern mathematics is that there are only two states for each variable that we use. This goes to the other extreme of complexity. Instead of having many numbers for a variable in this new type of mathematics, we require only two numbers. This is called *binary mathematics*.

We can select any two binary conditions. Some common ones that are used are yes-no, true-false, connected-not connected, closed-open, high-voltage-low-voltage, up-down, and head-tails. Rather than develop a yes-no or an up-down system of mathematics, we use numbers to represent each of these conditions; thus false = 0 and true = 1. We can then develop a single type of binary mathematics and apply it to any of the above sets of conditions. All we have to do is substitute the descriptive terms we want for the numbers 0 and 1.

> **Hint for Success**
>
> All number systems have a base or radix, which specifies how many digits the number system uses. For binary numbers the base is 2, with 0 and 1 as the only two digits. In the decimal system, the base is 10, so there are 10 digits, which are 0, 1, 2, 3, 4, 5, 6, 7, 8, and 9.

CALCULATOR HINT

Many modern calculators have the capability to add, subtract, multiply, and divide binary numbers. Furthermore, they can convert a binary number to its decimal equivalent value or vice versa.

Self-Test 23-1

1. In the field of industrial controls two classes of logic circuit are _____ and _____.
2. Switching circuits emphasize _____ or _____.
3. Digital circuits are usually dependent on _____.
4. _____ systems have only two states or values.

Simple Switches

Consider a simple switch and see how it can be analyzed as a binary device. Figure 23-1 shows the circuit that we want to analyze. A battery is connected through a

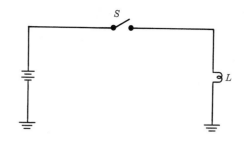

Figure 23-1 The basic switch-lamp circuit.

switch to a light. We can consider this problem in various ways. We could say the light is on or off, or the battery is turned on or turned off, or the circuit is open or closed, or the lamp is connected or not connected. These are all binary types of description of the system operation that we use in our common language or thought. They in no way confuse us. We shall now use these common concepts in a binary form of mathematical analysis.

In this problem there are two variables, the switch and the lamp. Call the variable associated with the switch S and that with the lamp L. We will say that if the switch is closed, then $S = 1$, and if the lamp is on, then $L = 1$. We thus have:

$$S = \begin{cases} 0\text{—switch open} \\ 1\text{—switch closed} \end{cases}$$

$$L = \begin{cases} 0\text{—lamp off} \\ 1\text{—lamp on} \end{cases}$$

RULE In a switching circuit an open switch is represented by a 0 and a closed switch is represented by a 1.

In the example above the output of the circuit is the lamp. If the lamp is on, there is current flowing through it and the variable is given the value 1.

RULE In a switching circuit an output that has current flowing through it has the logic value 1. If no current flows through it, it has the logic value 0.

We can see from the circuit that if $S = 1$, then $L = 1$. In words, this means that if the switch is closed ($S = 1$), the lamp is on ($L = 1$). We also see that if $S = 0$, then $L = 0$. In terms of our variables in this problem, the relationship can be stated mathematically as $S = L$.

Care must be taken in how the relationships are defined. The relationships $S = L$ is not true if we invert either one of the definitions for L or S. For example, if we decide that $S = 1$ means that the switch is open rather than closed, S is *not* equal to *L* as above. There are some cases in which people have chosen to use this different definition of *S*. Always be sure that the definitions are the same as those that are being used here.

The AND Function

The first of the basic logic function is the AND function. The word "and" is used in our common language, and we have no problem with it. There is no change at all when we go to the arithmetic of logic. Figure 23-2 shows a circuit with two single-pole-single-throw switches in series. Anyone with a basic understanding of circuits knows that both S_1 *and* S_2 must be closed so that the lamp will light. This is all that there is to the AND function. It is a mathematical tool that helps us to describe and work with the concept of AND.

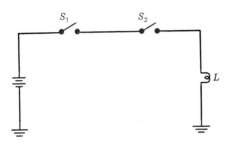

Figure 23-2 $L = S_1$ AND S_2.

Using the above conventions for the values of *S* and *L*, we can now tabulate the values as shown in Table 23-1.

Table 23-1

The AND Function

S_1	S_2	L
0	0	0
0	1	0
1	0	0
1	1	1

Table 23-1 merely states that the light will be on ($L = 1$) only if both switches are closed (S_1 AND $S_2 = 1$). The light will be off ($L = 0$) if either or both switches are open ($S_1 = 0$ or $S_2 = 0$).

The mathematical symbol for the AND function is a dot ($\cdot$). The equation relating L to S_1 and S_2 is written as:

$$L = S_1 \cdot S_2$$

This dot symbol is sometimes used to represent multiplication in arithmetic. Multiplication helps us to remember the above AND table (Table 23-1); the value of *L* is the product of the variables S_1 and S_2. This is just like the rules of algebra that we are accustomed to; 1 times 1 is 1, 1 times 0 is 0.

Now that we have an introduction to the AND function, we shall study some of its properties. It is obvious from the circuit diagram that it does not make any difference whether S_1 comes before S_2 or after S_2. In either position the operation of the switches and the lamp are exactly the same. This is also true for the mathematical equation:

$$L = S_1 \cdot S_2 = S_2 \cdot S_1$$

 RULE The AND function is independent of the order in which the variables are listed.

Another important property of the AND function can be seen from Fig. 23-2. If switch S_1 is open, it is irrelevant whether switch S_2 is open or closed, the lamp will not light. In other words, if $S_1 = 0$, the AND of S_1 and S_2 is 0. As an equation, this would be written as:

$$S \cdot 0 = 0$$

this gives the following rule:

 RULE The AND function of "0" and any variable is 0.

Another important rule for the AND function can also be seen from Fig. 23-2. If switch S_2 is closed, the lamp depends only on S_1. If S_1 is closed, the lamp is on. If S_1 is open, the lamp is off; that is, $L = S_1$. This can be written as the equation:

$$S_1 \cdot 1 = S_1$$

This gives the following rule:

 RULE The AND function of any variable and a 1 is only that variable.

Another similar property of the AND function can be seen from Fig. 23-3. Assume that S_1 is a double-pole single-throw switch and that the second pole S_1 is

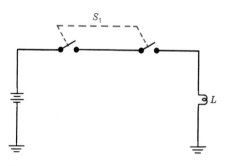

Figure 23-3 $L = S_1$ AND S_1.

switch S_2 as in Fig. 23-2. This would give the logic function $S_1 \cdot S_1$. Obviously, the second section of the switch serves no real purpose as far as the logical operation is concerned. The equation for the lamp is $L = S_1$. The equation for this is:

$$S_1 \cdot S_1 = S_1$$

We therefore have another rule for the AND function.

 RULE   The AND function of any variable and itself is only that variable.

Self-Test 23-2

1. When analyzing a switching circuit, the convention we use is that the logic value for a closed switch is _____ and an open switch is _____.

2. When an output is active or on, we say that it has a logic value _____.

3. When an output is inactive or off, we say that the logic value is _____.

4. For a logical AND function we (may/may not) take the variables in any order that we please.

5. The AND function is (dependent/independent) of the order of the variables.

6. 0 AND any variable is _____.

7. 1 AND any variable is _____.

Exercises

Practice

1. For the following exercises (a, b, and c), what is the value of L for the given values of S_1 and S_2 if $L = S_1 \cdot S_2$?
   a. $S_1 = 0$ AND $S_2 = 1$.
   b. $S_2 = 1$ AND $S_1 = 1$.
   c. $S_2 = 0$ AND $S_1 = 0$.

2. Give a circuit diagram for the logic function $S_1 \cdot S_2 \cdot S_3 = 1$.

3. What is the equation for the circuit in Exercise 2 if S_3 is permanently closed?

4. What is the equation for the circuit in Exercise 2 if S_3 is permanently open (broken)?

Truth Tables

Table 23-1 described the AND function exactly. It listed every possible combination of the input or independent variable (S_1 and S_2). Independent variables are those variables such as switches that we can set to any value as we please or as the system pleases. The dependent variables are those whose values are determined by other variables in the system. For our problem, the dependent variable is L. Tables of the form shown above are often used to describe any logic function. For this reason they are given the special name "truth table." Note the following facts about a truth table:

1. It has every possible combination of the input variables.
2. The combinations of input variables are listed in an ordered manner on the left of the table.
3. The output or independent variables are listed on the right adjacent to the combination of input variables that cause the output.

How many input combinations are there? Remember that each input can have either of two values (0 or 1). This means that S_1 can have two values and that for each of these values, S_2 can have two values; therefore, there are two times two or four possible input combinations. This is shown in Table 23-2.

Table 23-2

Combinations of Two Variables

S_1	S_2
0	0
	1
1	0
	1

Suppose that we had a third variable, S_3. For each of the two possible values for S_3, the other two variables could have any of their four possible combinations as shown in Table 23-3. The number of combinations is again doubled.

Table 23-3

Combinations of Three Variables

S_3	S_2	S_1
	0	0
0	0	1
	1	0
	1	1
	0	0
1	0	1
	1	0
	1	1

The general rule is that each additional independent variable doubles the number of combinations; in other words, the total number of input combinations is 2^N, or two times two N times. Table 23-4 shows the eight combinations for three variables.

Table 23-4

Combinations of Three Variables

No.	S_1	S_2	S_3
0	0	0	0
1	0	0	1
2	0	1	0
3	0	1	1
4	1	0	0
5	1	0	1
6	1	1	0
7	1	1	1

This leads to an important rule for the number of different states that a logic circuit can have.

RULE   The number of different combinations of input variables is 2^N, where N is the number of input variables.

The practicality of a truth table soon comes to an abrupt halt. For 3 variables there are 8 combinations, for four variables there are 16, for five variables 32, and so on. These numbers may be all right for computers, but people start to tire with 5, 6, or more variables. This still leaves us with a lot of useful applications and at least a very practical concept for problem solution.

What is the order in which the values of the independent variables are listed? It is important to have some orderly manner to list the different combinations so that none will be overlooked and so that it is easy to find any particular combination that is sought. The following is the accepted order in which to list the independent variables.

Start with the rightmost variable and give it the values 0 and then 1 while all those variables to the left are given the value 0. You then give the next variable to the left the value 1 and then repeat the combinations for those to the right. This process continues to the left until all variables have been used.

For those of you who are familiar with the above number scheme, it is called the *binary number system*. The combination of S_1, S_2, and S_3 corresponds to the decimal number listed in the number (leftmost) column.

The digit on the right is called the least significant digit (LSD). Like the rightmost number on the gasoline pump, it is the one that changes value the most rapidly. The digit on the left is the most significant digit (MSD). It changes the least rapidly.

RULE   On a truth table the input variables are arranged in the binary number sequence.

Self-Test 23-3

1. Variables that can be set to any desired value are called _____.
2. Variables that are determined by other variables are called _____.
3. The total number of input combinations for N input variables is _____.
4. The order in which the variables are listed in a truth table is called the _____ _____ system.
5. The abbreviation MSD means _____ _____ _____.
6. The abbreviation LSD means _____ _____ _____.
7. The (MSD/LSD) changes the most rapidly.
8. The (MSD/LSD) changes the least rapidly.

Exercises

Practice

1. How many possible combinations are there for three single-pole switches?
2. List the combinations for two switches S_1 and S_2.
3. If a system has 16 combinations of inputs, how many switches does it have?
4. A circuit has a limit switch, an on/off switch, and an over temperature switch. How many combinations of control switches can it have?

(See CD-ROM Test 23-1.)

JOB 23-4

The OR Functions

We are now ready to look at another logic function called the OR function. Sometimes this function is called the INCLUSIVE OR function. The reasons for this will become apparent later. Here again, mathematics is nothing more than an orderly expression of concepts that are familiar to us. Refer to the circuit in Fig. 23-4. If S_1 is on the front door and S_2 is on the back door, the light will go on if either S_1 or S_2 is closed. In other words, the light will go on if *either* the front door OR the back door is opened.

The truth table for the OR function is given in Table 23-5.

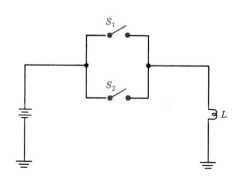

Figure 23-4 $L = S_1$ OR S_2.

Table 23-5

Truth Table for the OR Function

S_1	S_2	L
0	0	0
0	1	1
1	0	1
1	1	1

The mathematical symbol for the OR function is $+$. The rules for the OR function are similar to the rules of addition, and the $+$ signs helps us to remember this. We have $0 + 0 = 0$ and $1 + 0 = 1$. The only unusual value is the entry $1 + 1 = 1$. In this case we resort to the fact that this is a mathematical function of logic and not numbers. A 1 means that in some manner the value desired is present. Logically speaking, in Table 23-5, the 1 under S_1 means that we have S_1, or that S_1 is closed. Similarly, the 1 under S_2 means that S_2 is closed; therefore, the 1 under L should mean that we have the lamp (the lamp is on). According to this reasoning, $1 + 1 = 1$.

Another property of the OR function is that, as with the AND function, it does not make any difference which order we put the variables in. As can be seen from Fig. 23-4, interchanging S_1 and S_2, gives the same circuit performance. This means that we can write:

$$L = S_1 + S_2 = S_2 + S_1$$

This also is the same as algebra.

 RULE The OR function is independent of the order in which the variables are taken.

A very important property of the OR function can be seen from Fig. 23-4. If switch S_1 is closed, the lamp will be on no matter what position S_2 is in; that is, L is independent of S_2. This shows the desired property of the OR function; namely:

$$S_2 + 1 = 1$$

This gives the following rule:

 **RULE** Any variable ORed with a 1 is a 1.

Another property of the OR function is that if switch S_2 is open, the lamp will depend on S_1 only. This means that:

$$L = S_1 + 0 = S_1$$

Another rule of the OR function is:

RULE The OR function of any variable and a 0 is that variable only.

We have now reached a point at which the precision and clarity of mathematics help to remove the confusion of everyday language. No one had any problem with the above example. There is a different use of the word "or." Suppose that there are two substitute players that can be put into a game. The manager wants Kline or Gross to go into the game depending on what the other team does. We know by experience and common sense that Kline and Gross will not both go into the game. Only one—not both. We used the same word "or," but we had a slightly different meaning. This meaning is called the EXCLUSIVE OR. It says that one alternative, but not both alternatives, can be used. The truth table for the EXCLUSIVE OR is shown in Table 23-6.

Table 23-6

Truth Table for the EXCLUSIVE OR

S_1	S_2	L
0	0	0
0	1	1
1	0	1
1	1	0

Notice that this is exactly like the OR function except for the last line. In the last line, when both S_1 and S_2 are true, L is not true.

It should be expected that the symbol for the EXCLUSIVE OR function should be similar to that for the OR functions. The EXCLUSIVE OR symbol is $\oplus$.

It is now apparent why the OR function is sometimes called the INCLUSIVE OR as opposed to the EXCLUSIVE OR. The INCLUSIVE OR includes the situation in which you have both alternatives. The EXCLUSIVE OR function excludes that case.

The INVERT Function

The last basic function is the INVERT function. It is sometimes called the complement or NOT function. This is really the easiest of all of the functions. It merely inverts the variable; that is, it changes the value from a 1 to a 0 or from a 0 to a 1. If S_1 is true, then the inverter makes it false and vice versa. This is a single-variable function, and its truth table looks like that shown in Table 23-7.

The mathematical symbol for inversion is a bar over the variable as $\overline{S_1}$. It is read "S_1 bar" or "NOT S" It is also written S_1'. This is read "S_1 prime."

A warning is necessary about the notation of inverted quantities. If you have the term $(S_1 \cdot S_2)'$, it means that you first take S_1 AND S_2 and then take the complement of the result of that AND function. For example, if $S_1 = 1$ and $S_2 = 1$, then $S_1 S_2 = 1$. This gives $(1 \cdot 1)' = 1' = 0$.

Table 23-7

Truth Table for the INVERT Function

S_1	L
0	1
1	0

The parentheses must be used because $(S_1 S_2)'$ is not equal to $S_1 S_2'$. If you let $S_1 = 0$ and $S_2 = 1$, then:

$$(S_1 \cdot S_2)' = (0 \cdot 1)' = 0' = 1$$

while

$$S_1 \cdot S_2' = 0 \cdot 1' = 0 \cdot 0 = 0$$

These two values are not the same.

Note also that it is not equal to $S_1' S_2'$. You can use truth tables to show that $(S_1 \cdot S_2)' = S_1' + S_2'$.

Self-Test 23-4

1. The two types of OR function are —————— and ——————.
2. The OR functions are (dependent/independent) of the order of the variables.
3. Any variable OR 1 is ——————.
4. Any variable OR 0 is ——————.
5. The OR function truth table is like addition except that $1 + 1 = $ ——————.
6. The EXCLUSIVE OR function is like the OR function except that $1 + 1 = $

Exercises

Applications

1. How many input combinations are there for a truth table with four independent variables?
2. Give the truth table for $S_1 S_2 + S_3$.
3. Construct a circuit for the equation in Exercise 2.
4. Construct a truth table for the function $(S_1 S_2)' + S_3$.
5. Show that the truth table for $(S_1 S_2)'$ is the same as the truth table for the function $S_1' + S_2'$.

Summary of the Elementary Functions

The following functions have been introduced: AND, OR, EXCLUSIVE OR, and INVERT. The truth tables for these functions are given in Table 23-8 on page 814.

Table 23-8

Truth Tables for the Elementary Functions

AND			INVERT		OR			EXCLUSIVE OR		
S_1	S_2	L	S_1	L	S_1	S_2	L	S_1	S_2	L
0	0	0	0	1	0	0	0	0	0	0
0	1	0	1	0	0	1	1	0	1	1
1	0	0			1	0	1	1	0	1
1	1	1			1	1	1	1	1	0

Rules of the AND function are as follows:

 RULE The AND function is independent of the order in which the variables are listed.

 RULE The AND function of 0 and any variable is 0.

RULE The AND function of any variable and a 1 is only that variable.

RULE The AND function of any variable and itself is only that variable.

Rules of the OR function are as follows:

 RULE The OR function is independent of the order in which the variables are taken.

RULE Any variable ORed with a 1 is a 1.

RULE The OR function of any variable and a 0 is only that variable.

Facts about a truth table are as follows:

1. It has every possible combination of the input variables.
2. The combinations of input variables are listed in an ordered manner on the left of the table.
3. The output or independent variables are listed on the right, adjacent to the combination of input variables that cause the output.

RULE The number of different combinations of input variables is 2^N, where N is the number of input variables.

RULE On a truth table the input variables are arranged in the binary number sequence.

Systems of Elementary Functions

This section starts to build complex networks out of the elementary logic building blocks that have been learned in the earlier sections. It will show how logic equations for switching circuits can be obtained from the "paths" of the circuit. This means that you can take a diagram of a control circuit and trace all the paths with your finger and find the logic equation for the circuit.

The first step toward understanding the operation of a complex system is to break it into segments or blocks of elementary functions. Refer to Fig. 23-5. This circuit is the combination of the AND and the OR circuits that we analyzed previously. The two switches that are in parallel compose the function $S_2 + S_3$. The result of this composition is then in series with S_1, or in other words it is ANDed with S_1. This gives the equation for L as:

$$L = S_1(S_2 + S_3)$$

In this equation the dot that represents AND was not used. By convention, it is not used when there is no confusion as to the meaning. This is just the same as the multiplication sign in arithmetic. It would not be wrong to write $S_1 \times (S_2 + S_3)$, but it is common to write $S_1(S_2 + S_3)$.

Hint for Success

When analyzing logic functions containing the AND and OR logic functions, be sure to group the variables so that the logic function is properly represented.

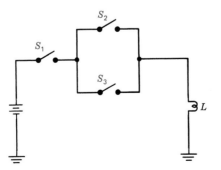

Figure 23-5 $L\ S_1(S_2 + S_3)$.

The equation can also be expanded in the same way that it could be in algebra, where we could write $S_1 \times (S_2 + S_3) = S_1 \times S_2 + S_1 \times S_3$. We can similarly write the equation for L as:

$$L = S_1 S_2 + S_1 S_3$$

Again, the dot symbol may or may not be used, depending only on whether there would be any confusion about the meaning.

When the conventions that have been chosen are used, there is a physical significance to the above equation for L. If you put your finger on the battery, the first term for L represents a path that you can follow with your finger that goes from the battery through S_1 and S_2 and then through the lamp and back to the battery. In other words, the term $S_1 S_2$ represents a path from the battery through the network and the output and back again to the battery.

In a similar manner, the term $S_1 S_3$ is a second path through the network and the load and back to the battery. Again, you can put your finger or a pencil on the battery, go through the network of switches, through the output (the lamp) and back to the battery.

In other words, we say that the lamp is on if S_1 is closed and S_2 is closed (path 1) OR if S_1 is closed and S_3 is closed (path 2). A path is a continuous line that goes from the power source through the logic elements and the output and back to the power source without intersecting itself. A path forms no loops.

When the equation for L is expressed as above, it is in the form of the sum of paths (SOP). This could also mean the sum of products, as each term of the equation for L is a set of products and the overall equation is the sum of these products. There is a slightly different definition of SOP in logic network theory; however, it is consistent with what we are saying.

Once the logic equation is obtained, it is easy to construct a truth table for the above circuit. In this case there are three input variables that can be set independently. These are S_1, S_2, and S_3. The output is the lamp L. In the truth table given in Table 23-9, the mathematical symbols have been included to emphasize the close resemblance between logic functions and conventional algebra.

Table 23-9

Truth Table for $S_1 (S_2 + S_3)$

S_1	$\times$	$(S_2$	$+$	$S_3)$	$=$	L
0	$\times$	(0	$+$	0)		0
0	$\times$	(0	$+$	1)		0
0	$\times$	(1	$+$	0)		0
0	$\times$	(1	$+$	1)		0
1	$\times$	(0	$+$	0)		0
1	$\times$	(0	$+$	1)		1
1	$\times$	(1	$+$	0)		1
1	$\times$	(1	$+$	1)		1

The only unusual entry is the last one. In conventional algebra the answer would be 2. There is no 2 in the binary system. All the other results are exactly the same

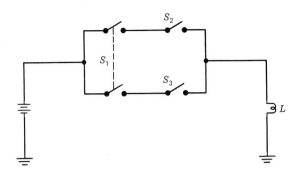

Figure 23-6 $L = S_1 S_2 + S_1 S_3$.

as in conventional algebra. It must be remembered that for the OR function $1 + 1$ is 1. When this is used, the conventional algebra can be used again.

It was shown that just as in algebra, there are alternate ways to write the equation. These alternative equations represent different sets of paths. Each equation also has a different physical representation.

Figure 23-6 shows the physical representation of the equation $L = S_1 S_2 + S_1 S_3$. If you make a truth table for this equation, you will find that it is exactly the same as the truth table for $L = S_1(S_2 + S_3)$. Check the operation of the system. Logically it is exactly the same. The difference is only in S_1. In the form of the equation for Fig. 23-6, S_1 is a double-pole single-throw switch.

The result above can be very important. It shows that the fundamental intent of logic analysis is the solution of a logic problem, not a physical one. Solving a physical problem in the mathematics of logic can help you to realize new and different configurations of hardware to achieve the same result. This can be done by changing the form of the equation.

Sometimes with complicated circuits you may find that two circuits give the same logic equation. This means that they are identical in operation and differ only in form.

For another example, analyze the circuit shown in Fig. 23-7. The circuit can be broken into two sections. The one on the left is the function $S_1 + S_2$, and the one on the right is $S_3 + S_4$. The overall circuit is the AND function of these two sections. This gives the logic equation $L = (S_1 + S_2)(S_3 + S_4)$. This says that the lamp will light if S_1 or S_2 is closed AND S_3 or S_4 is closed.

A word of caution is necessary here about the use of parentheses in these logic equations. Many times, people forget to use them and write the above equation as:

$$L = S_1 + S_2 S_3 + S_4$$

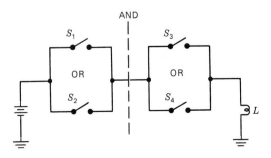

Figure 23-7 $L = (S_1 + S_2)(S_3 + S_4)$.

The problem with this is that there can be confusion about the terms in the equation. The above equation really has only three terms; S_1, $S_2 S_3$, and S_4. This is not the correct equation for L. The correct equation has four terms in it. When properly expanded the equation for L is:

$$L = (S_1 + S_2)(S_3 + S_4)$$
$$= S_1(S_3 + S_4) + S_2(S_3 + S_4)$$
$$= S_1 S_3 + S_1 S_4 + S_2 S_3 + S_2 S_4$$

The expansion of the above equation is done in exactly the same manner that algebra removes parentheses. Each term in parentheses on the left is multiplied by each term in parentheses on the right. In this equation S_1 is multiplied by S_3 and then S_4, the terms in the right-hand parentheses. The next term in the left-hand parentheses, S_2, is then multiplied by the terms in the right-hand parentheses (Table 23-10).

Table 23-10

Truth Table for $L = (S_1 + S_2)(S_3 + S_4)$

S_1	S_2	S_3	S_4	L	Path
0	0	0	0	0	
0	0	0	1	0	
0	0	1	0	0	
0	0	1	1	0	
0	1	0	0	0	
0	1	0	1	1	$S_2 S_4$
0	1	1	0	1	$S_2 S_3$
0	1	1	1	1	$S_2 S_3 + S_2 S_4$
1	0	0	0	0	
1	0	0	1	1	$S_1 S_4$
1	0	1	0	1	$S_1 S_3$
1	0	1	1	1	$S_1 S_3 + S_1 S_4$
1	1	0	0	0	
1	1	0	1	1	$S_1 S_4 + S_2 S_4$
1	1	1	0	1	$S_1 S_3 + S_2 S_3$
1	1	1	1	1	$S_1 S_3 + S_1 S_4$
					$+ S_2 S_3 + S_2 S_4$

Again, it is obvious that each of the terms in the equation corresponds to a path in the network. The lamp would be lighted if any one of the terms in the equation was a 1. It would also light if any two of the paths were connected or if all four were. The truth table (Table 23-10) shows that the 1s in the L column correspond to these situations.

An interesting point is that you cannot have a combination of three paths, only one, two, or four paths. Look at the circuit, and see whether you can reason why this is true.

Exercises

Applications

1. Give a circuit for and list the paths of the equation

$$L = S_1 S_2 + S_1 S_3$$

2. Repeat Exercise 1 for the equation

$$L = S_1(S_2 + S_1)S_3$$

3. Repeat Exercise 1 for the equation

$$L = S_1 S_2 + (S_1 + S_3)$$

4. Determine whether there is any operational difference between the equation in Exercise 3 and the equation

$$L = S_1 + S_3$$

Compare the circuits and the truth tables for the two equations.

5. Repeat Exercise 1 for the equation

$$L = S_1(S_2 + S_1 + S_3)$$

(See CD-ROM for Test 23-2.)

JOB 23-7

Switching Circuit Inverter

The previous developments have been with single-pole or double-pole switches. A switch has been considered either open or closed. To implement the invert function for the switching circuit concepts, a single-pole–double-throw switch is used as shown in Fig. 23-8. The switch position that previously would have been considered open is now the complement position. To use the complement of a variable

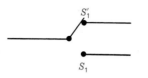

Figure 23-8   Switch arrangement for generation of a complement.

in a logic equation, a connection is made to the other pole of the switch. For example, to implement the logic equation $L = S_1'S_2$, a circuit such as that shown in Fig. 23-9 would be used.

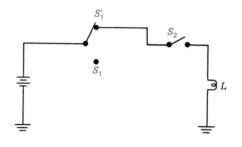

Figure 23-9 $L = S_1'S_2$.

An identity for the complement that is easy to see from a circuit point of view is $S + S' = 1$. Figure 23-10 shows the circuit for the equation $L = (S_1' + S_1)S_2$. The lamp in Fig. 23-10 will light no matter what position switch S_1 is in. The operation depends only on S_2. In logic arithmetic this is shown as:

$$L = (S_1 + S_1')S_2$$
$$= 1\,S_2 = S_2$$

from which the desired identity is obvious:

$$S + S' = 1$$

RULE   Whenever there is an OR combination of a variable and its complement, the result is always a 1.

Equations Using the Complement

Logic is often used in such common locations that we never realize it is being used. The EXCLUSIVE OR can be written in logic equation form as:

$$L = S_1 S_2' + S_1' S_2$$

This can be derived from the paths that give the truth table for the EXCLUSIVE OR. Another way to reason it is to express the EXCLUSIVE OR in words. The lamp will light only if one switch is on and the other is off. The circuit for the above equation is shown in Fig. 23-11. As soon as it is drawn out, you probably recognize this as the "three-way switch." The light can be turned on or off from the top of the stairs or from the bottom of the stairs.

By this time you should have a feeling for the close relationship between logic arithmetic and your everyday experiences. The purpose of this mathematics is to

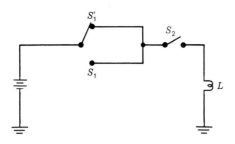

Figure 23-10 $L = (S_1' + S_1)S_2$.

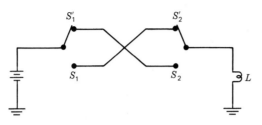

Figure 23-11 $L = S_1'S_2 + S_1S_2'$.

describe things that are familiar and common to you. The correctness of the fundamental rules and applications that we have seen is determined not by mysterious and obscure processes, but by what is common sense.

Exercises

Applications

1. Prepare a truth table for the function $L = S_1' (S_2 + S_3) + S_1$.
2. Make a truth table for the function $L_1 = S_1S_2 + S_1'S_2'$.
3. Make a circuit diagram for the equation in Exercise 2.
4. Construct a truth table for the two-way switch shown in Fig. 23-10.
5. What is the relationship between the equation for L in Exercise 4 and that for L_1 in Exercise 2? (The function L_1 is called an EXCLUSIVE NOR.) Is there any logical difference in the operation of the two circuits? Does this make any difference to the user? Could the user tell which circuit is being used?
6. Give a truth table for a "four-way switch"; that is, a light can be turned on or off from any of three possible locations. Start by assuming that for the switch positions S_1', S_2', S_3' the lamp is on. For the truth table, let S_1 be the LSD and S_3 the MSD. (*Hint:* Reason that from any given set of switch settings, if only one switch position is changed, the light will change state. This means that if the light was on, it will go off and if it was off, it will go on.)

(See CD-ROM for Test 23-3.)

JOB 23-8

Relays in Logic Circuits

Many control circuits use conventional or solid-state relays to control higher-power circuits. The contacts on a relay are treated as any other switch contacts

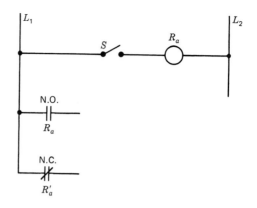

Figure 23-12   A switch-controlled relay showing relay-controlled contacts.

would be treated in a switching circuits problem. The primary difference is that the coil or actuator of the relay can be controlled by other logic variables; that is, the coil could be treated as the lamp was in the above problems. It would be considered as an output function and given a variable name such as R_a. The contacts that are associated with the relay are given the same name as the coil. This is shown in Fig. 23-12, where the relay coil is shown actuated by switch S. Two sets of contacts are shown associated with relay R_a. One set is normally open (NO), and the other set is normally closed (NC). If the conventions that we have chosen are used, the normally open contacts would represent the variable R_a. When R_a is turned on ($R_a = 1$), the contacts would be closed ($=1$). The normally closed contacts, on the other hand, represent the complement of R_a. When the relay is not actuated ($Ra = 0$), the normally closed contacts would be closed ($R_a = 1$). In relay control terminology, a relay that is energized is usually said to be picked up or picked. When the relay is turned off, it is said to be deenergized or dropped.

RULE   The NO contacts of a relay represent the relay variable, and the NC contacts represent the complement of the relay variable.

Figure 23-12 shows only one set of contacts that are normally opened and one set that are normally closed. Relays are made with many combinations of contacts. The identification system for the more complicated devices would be similar to the above. The identification of the contacts is associated with the coil that actuates them. The separate contacts may be further marked, but being able to associate contacts with coils makes the logic equations easier to understand.

To further explain the relationships between word descriptions of operation, hardware, and logic equations, let's take an example of a simple motor control. The motor operates a garage door. Pushing a button turns the motor on and raises the door as long as the button is depressed. Releasing the button stops the door. To prevent the door from going off the track or the motor from becoming overstressed, there is a limit switch that stops the motor when the door is fully opened. The door is very large, so in order to use a smaller switch, a relay is used to actuate the motor. This is a verbal description of the operation.

This description could be reduced to a logical description of the operation. We want the relay to be energized if the button is pushed and the limit switch has not been opened. This means that we want an AND function in the operation. We are assuming that the limit switch is a normally closed type. This gives us a logic equation for the relay operation.

$$R_a = L_S S$$

Since L_S is assumed NC under normal conditions, the value of the variable L_S is 1. This makes the above equation

$$R_a = 1\, S = S$$

The relay will actuate according to switch S as long as the limit switch is closed. If switch S is closed, the relay turns on. This is the desired operation.

The motor will actuate when the relay actuates; therefore

$$M = R_a$$

The logic equations say everything that the verbal description says but they are more compact and precise. The entire verbal description is exactly the same as the following set of logic equations:

$$R_a = L_S S$$
$$M = R_a$$

The above equations represent paths for the logic of the controls. The relay coil has a path $L_S S$ that goes through the limit switch, through the switch, and then through the coil of the relay. The motor is actuated by a path that goes through the NO set of contacts on the relay and then through the motor. These paths can be implemented as shown in Fig. 23-13. Notice how easy it was to take the equation, express it in paths, and then implement the logic.

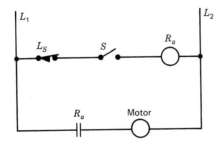

Figure 23-13 Diagram for relay-controlled motor.

Now modify the operation a little. We want the door to start to open as soon as the button is pushed, but when the button is released, we want the door to go to its fully opened position. What we are now saying logically is that we want the door to operate as before OR if it is once actuated to continue until the limit switch is reached. The logic equation is now:

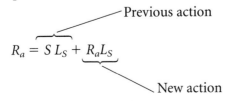

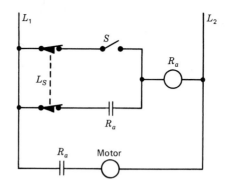

Figure 23-14 Alternate circuit for motor control.

The second term is the new condition for operation. It says that if the relay is actuated and the limit switch is still closed, the relay will remain actuated. To add this new operation to the system, all that was necessary was to add a term to the equation.

The new equation could be implemented by adding a path that goes through the relay coil, through the limit switch, and then through a set of NO contacts on the relay. This looks as shown in Fig. 23-14. The limit switch is now a little more complicated than it was before. It is now a double-pole–single-throw switch. If the logic equation is rearranged a little, just as you would with an algebraic equation, we get the following form:

$$R_a = L_S (S + R_a)$$

In this form we have the combination of an AND and an OR circuit that has removed the necessity of the more complicated limit switch. The limit switch is now back to being a single-pole–single-throw switch. This gives the simplified configuration shown in Fig. 23-15.

The operation of the circuit shown in Fig. 23-15 is called *latching* or *locking*. This means that once the relay is energized, it keeps itself energized. This comes about because the state of the relay depends on itself; in other words, the relay variable is on both sides of the logic equation:

$$R_a = L_S (S + R_a)$$

Using the rules of the OR function, it can be seen that when the relay is not actuated ($R_a = 0$), the equation for R_a is:

$$R_a = L_S (S + 0) = L_S S$$

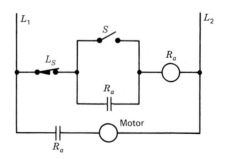

Figure 23-15 Alternate form for circuit shown in Fig. 23-14.

In this condition or "state," the relay depends only on the limit switch and the control switch.

Once the relay becomes actuated ($R_a = 1$), the logic equation is:

$$R_a = L_S(S + 1) = L_S$$

In this state the relay depends only on the limit switch.

RULE　If the state of a relay depends on its own state, that is, the relay variable is on both sides of the logic equation, the system has modes of operation.

This type of performance is very common in control circuits. Systems may have many "modes" of operation with the logic of operation changing depending on which mode the system is in. To the observer, the system looks like different systems, one in each mode of operation.

In the above example, moding or latching was desired. Sometimes in complicated systems this type of operation comes in by accident. A very unusual sequence of events may cause the system to get into the undesired mode, where unplanned operation can occur.

Another time that this can occur is when there is some type of equipment failure that puts the system into a form of operation other than what it was intended to be. Sometimes these faults can be very difficult to locate. One way is to look at the circuit diagram and make logic equations of the operation. After this is done, you can make different assumptions about parts that might be broken. Substitute 0s or 1s for variables according to the type of failure that you think they may have. Then look at the equations and see whether they describe the type of operation that you are seeing. If this does not work, the problem is that the circuit is not like the diagram. Either something has failed in a manner that changes the circuit, or it was built incorrectly.

Now consider some additional problems of this type. While doing these problems, try to improve your skills with the rules for the basic logic functions. Also try to learn the truth table for these functions so that you don't have to keep referring to them. Another purpose of these examples is to help improve your skills in going directly from circuit diagrams to logic equations so that you can better understand the operation of the circuit.

The next example of a latching circuit is shown in Fig. 23-16 on page 826. The circuit is supposed to be a self-latching relay that can be released with a release switch.

The contacts on the relay are drawn as NO contacts. This means that the function that the relay contacts represent is R_a. With this background on the intent of the circuit, determine the actual operation.

First list the paths that the network has. Start at the line L_1 and go through the relay coil. This now completes two important elements of the path requirement. We have gone from the source through the output. At the other side of the relay we can go directly through switch S_1 to line L_2. This completes a path. Therefore, the first path is just S_1.

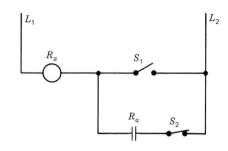

Figure 23-16   Self-latching circuit with release.

At the output of the coil the other path that we could have taken goes through the relay contacts and then through the NC switch S_2 to ground. This path has the value R_aS_2. Since this is all the paths that the circuit has, we can now write the logic equation for the relay operation as:

$$R_a = S_1 + R_a S_2$$

With R_a on both sides of the equation, the system has different modes of operation. Let's look at each of these modes. First let $R_a = 0$. This corresponds to the initial conditions before the relay is turned on. When this is done, then

$$R_a = S_1 + 0\ S_2$$
$$R_a = S_1$$

This means that when the relay is off ($R_a = 0$), only S_1 controls the operation of the circuit. As long as S_1 is open, the relay will remain off. When the switch closes, the relay will be energized. This will cause the system to go into the other mode of operation.

The second mode of operation is when the relay is energized and $R_a = 1$, the equation becomes

$$R_a = S_1 + 1\ S_2$$
$$R_a = S_1 + S_2$$

Switch S_2 is NC; therefore, under normal conditions the equation is

$$R_a = S_1 + 1 = 1$$

As long as S_2 is closed, it makes no difference what is done with S_1. The equation is independent of S_1; thus the relay remains energized or is "locked."

The relay can be "unlocked" and turned off if switch S_2 is opened. The switch S_2 then becomes ($S_2 = 0$), and the equation for the relay becomes

$$R_a = S_1 + 0 = S_1$$

If S_1 is open so that $S_1 = 0$, the relay will turn off. If S_1 is closed, the relay will not turn off. This may not be the type of operation that was desired. As the system is designed, the relay can be turned off only if S_1 and S_2 are both open. If S_2 is supposed to be an emergency switch, the operation is improper. Suppose that this were the door operation example that we discussed earlier and that S_2 were the limit switch. This could mean that when the operation limit is reached, the limit switch could be defeated by someone closing S_1.

On the other hand, if this were a ventilation fan that must be kept on, then turning it off would require two switches to be thrown. For this type of application, the operation provides additional safety. The logic of the circuit may be an advantage to one application and a disadvantage to another. The design and analysis of these circuits are complicated by the fact that what is proper operation depends on human decisions.

With a little experience you can just look at a circuit diagram and write the logic equation without going through all the above discussion. In your mind you can see all the paths that control a device. This is faster than following them with your finger or a pencil. Notice how fast and easy it is and how it implies all the subtle aspects of the operation.

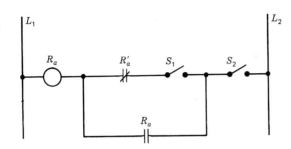

Figure 23-17 Self-latching circuit.

As another example look at the circuit shown in Fig. 23-17. This relay has two sets of contacts; one is NO and the other is NC. Find the logical operation of the system. First list the paths of the network:

Path 1—$R_a S_2$
Path 2—$R'_a S_1 S_2$

The equation for the operation of the relay is obtained as the sum of the paths:

$$R_a = R_a S_2 + R'_a S_1 S_2$$

This equation can be simplified a little if the S_2 is factored out to give

$$R_a = S_2(R_a + R'_a S_1)$$

Again the system has two modes of operation. To better understand these modes of operation, first let $R_a = 0$. This corresponds to the start of operation when the relay is not operated. The logic equation is then:

$$R_a = S_2(0 + 1\ S_1) = S_2 S_1$$

This shows that both S_1 and S_2 have to be closed so that the relay will be activated. Once it is activated, $R_a = 1$ and the equation becomes:

$$R_a = S_2 (1 + 0\ S_1) = S_2$$

This says that when the relay is actuated, S_2 has full control of its operation; that is, when the relay is actuated, only S_2 has to open so that the relay will be deenergized or "dropped."

Self-Test 23-6

1. The ——————— contacts on a relay represent the relay variable and the ——————— contacts represent the complement of the variable.
2. When a relay turns on and keeps itself on the relay, it is said to be ——————— or ———————.
3. When a relay is turned on it is said to be ——————— or ——————— or ———————.
4. When a relay is turned off it is said to be ——————— or ———————.
5. When a control system operates differently depending on the internal states of the relays, we say that the system has ——————— of operation.
6. Biomodal operation means that a system has ——————— modes of operation.
7. Biomodal operation of a relay is indicated when the logic equation for the relay has the relay variable on ——————— sides of the equation.

Exercises

Applications

For the following problems, use the motor control circuit shown in Fig. 23-18.

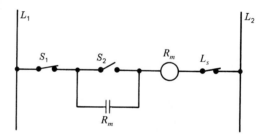

Figure 23-18   Control circuit for Exercises 1 to 5.

1. Show that for the motor control relay, the logic equation is given by:

$$R_m = S_1\, L_S(S_2 + R_m)$$

2. Assume that the motor is not operating and show that the control equation is:

$$R_m = S_1\, L_S\, S_2$$

3. Once the motor starts to operate, show that it can be stopped by either S_1 or L_S.
4. What would happen if the contacts broke off of relay R_m?
5. What would be the system operation if the contacts on relay R_m burned shut? If S_1 were an NC push-button switch, how could you stop the motor?

(See CD-ROM for Test 23-4.)

Types of switches found in electronic equipment include the toggle type, push button, rocker, slide, rotary, and dual-inline package. All switches have a current rating and a voltage rating. The current rating is the maximum allowable current that the switch can carry when it is closed. The current rating is based on the physical size of the switch contacts, as well as the type of metal used for the contacts. Many switches have gold- or silver-plated contacts to ensure a very low resistance when closed. The voltage rating is the maximum voltage that can safely be applied across the open contacts without internal arcing. The voltage rating does not apply when the switch is closed, because the voltage drop across the closed switch contacts is about zero.

Toggle switches.

A dual-inline package (DIP) switch.

A typical push-button switch.

A rotary switch.

Digital Logic Circuits

Starting in this section a slightly different type of circuit is discussed. All the basic concepts of logic that were learned previously will remain the same. The only difference is the relationship between the logic and the circuit form. The previous circuits were switching circuits, and the fundamental interest was whether there is a path for the current or not.

Digital logic circuits are usually voltage-level-operated. The most common of such devices is the transistor-transistor logic (TTL) family of devices. Many digital components that are not actually TTL are often designed to be "TTL-compatible." This means that the voltages are designed to be compatible as well as other aspects of power, supply voltages, and so on. First we shall look at the symbols that are used in this type of logic.

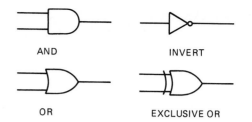

Figure 23-19   Elementary logic function symbols.

Symbols and Other Conventions

It is necessary that certain conventions be established so that communications in this field can be easier. If you find circuit diagrams that are 20 or more years old, you may find many different styles of symbols. To eliminate this confusion, some standards or conventions have been established. Most present-day diagrams will conform to the symbols and conventions described in this section. As stated earlier, this is the very early phase of a new field of mathematics. Because of the rapid changes in the technology of integrated circuits, the techniques of binary and logic circuits are changing very rapidly. At the present time a totally new set of standard symbols is evolving. These symbols are coming from the International Electrotechnical Commission, the Institute of Electrical and Electronic Engineers, and the American National Standards Institute.

The new symbols mostly represent new development in the technology. The symbols given there are likely to continue in use for the type of problems that we are analyzing. The symbols for the elementary logic functions are shown in Fig. 23-19.

There is one subtle convention in the symbols given in Fig. 23-19. The symbol for the inverter is really two symbols. The inverter is considered to be an amplifier followed by an inverter. For this reason the symbol has a triangle which represents an amplifier and an open circle which represents inversion of logic level. When we later review more complicated logic functions, it will be found that the open circle is used frequently on both inputs and outputs to logic devices (Fig. 23-20).

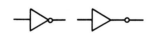

Figure 23-20   The inverter symbol.

The logic as we have used it in the truth tables given earlier is called *positive logic.* In positive logic a high or true logic value is represented by a 1 and a low or false value is represented by a 0. The opposite values can be used. When they are, it is called *negative logic.* For negative logic, a high or true value is represented by a 0 and a low or false value is represented by a 1.

When manufacturers specify a logic device, they generally state the logic convention that they are using. For example, they would say that a device is a "positive logic AND gate." This means that the device performs the logic function that we have described above as an AND function. In many newer control systems, particularly microprocessor-oriented systems, many control signals are negative logic. The logic performed on a negative logic signal by a positive logic device is not the same as it would be on a positive logic signal.

To better understand the characteristics of the digital logic devices, we shall look at the 54/74 families of TTL components a little more closely. The individual devices are generally numbered 5400, 74164, and so on. The leading 54/ or 74/ indicates that the device belongs to these families. The prefix 54- indicates a broader operating temperature range that is usually associated with military-type devices. The 7400 series is a commerical-grade device. Devices with comparable numbers other than the prefix are functionally interchangeable.

To conform to the standards, each device must give an output high voltage V_{oh} that is greater than 2 V and an output low voltage V_{ol} that is less than 0.4 V. At the input a device will treat any voltage less than 0.8 V as a low and any voltage greater than 2.4 V as a high. An additional characteristic is that if the input has nothing connected to it, then the device will treat it as an input high. The inputs are said to "float" high.

It is apparent that the voltages on which these devices operate are much less than most relays and house wiring. It is also apparent that the single-pole–single-throw switch configuration that we used at first in the switching networks section will not work for the control of inputs to a TTL device. If the switch were closed, the input would be a high. If the switch were open, the input would have nothing attached to it, so the device would again assume a high. The input logic level could not be changed.

The single-pole–double-throw configuration that was used will not work, either. In the switching circuits the complement was indicated by switching the voltage from one line to another. It did not change the voltage level, as is required by the TTL devices.

The manner in which a switch would be used in a digital logic circuit is shown in Fig. 23-21. To provide a high voltage to the input, the input is connected through a 1-kΩ resistor to the 5-V line. A low is provided by connecting the input to ground. If a low-voltage lamp such as a light-emitting diode (LED) is connected to the output of the AND gate, the logic operation is exactly the same as the switching circuit and function. Let $L = 1$ mean that the output is high and the lamp is on. An input high is represented by $S = 1$ and a low, by $S = 0$. The truth table would then be as shown in Table 23-11 on page 832.

There is no difference in the logic operation of the switching circuit AND function and the digital circuit AND function. There is a physical difference in the circuits. The digital circuit does not have the path configuration that the switching circuit had. This means only that different techniques are used when going back

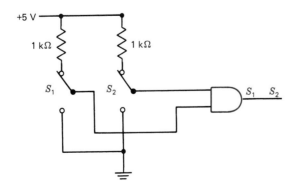

Figure 23-21 Input signals to digital logic using switches.

Table 23-11

AND Function

S_1	S_2	L
0	0	0
0	1	0
1	0	0
1	1	1

and forth between the circuits and the logic equations. We first consider how logic signals are traced through a logic circuit to determine the output for a given set of input conditions.

Assume the logic circuit shown in Fig. 23-22. Assume also that the input signals are $S_1 = 1$, $S_2 = 1$, and $S_3 = 0$. AND gate 1 has two 1s as inputs; therefore, the output is a 1 also. This means that the inputs to AND gate 2 are a 1 and a 0. The output from this gate is therefore 0; thus $L = 0$.

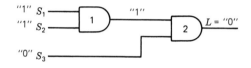

Figure 23-22

Example 23-7

For practice in following logic signals through a digital circuit, show that if the inputs to the circuit in Fig. 23-22 are all 1s, the output is $L = 1$. Show also that any other combination of inputs results in the output $L = 0$.

Solution

1. Look at gate 1. The inputs are 1 AND 1. From the truth table for an AND function (Table 23-11), the output is also a 1. This output is an input to gate 2.
2. Now look at the inputs to gate 2. They also are two 1s. The output is the desired result, which is a 1.
3. To show that any other combination of inputs will give a zero, prepare a truth table for the circuit. Use as inputs S_1, S_2, and S_3. Use L as the output.

For a second example, find the output of the circuit shown in Fig. 23-23 if the inputs are $S_1 = 1$, $S_2 = 1$, and $S_3 = 0$. The AND gate has the same inputs as in the previous example; therefore, the output is again 1. This 1 is now an input with S_3 to the EXCLUSIVE OR. Since only one input is a 1, the output of the EXCLUSIVE OR is 1. Thus $L = 1$ in this case.

Now that we have a better understanding of the concepts of digital circuits, let's make a comparison of the two techniques we have studied: switching circuits and digital circuits. We shall implement some of the same logic functions that we used in the switching circuits sections, but this time in digital circuits.

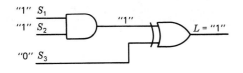

Figure 23-23

First recall Fig. 23-5, where we generated the logic function $S_1(S_2 + S_3) = L$. Figure 23-24 illustrates the equivalent function in digital logic. The path structure that was in the switching circuit is not present. To determine the logic function performed by this circuit, it is necessary to "trace" the signals through the circuit as shown in Fig. 23-24. The inputs to each gate are labeled with their variable names. The output is then labeled with the result of the function performed. This then becomes the input to the next gate. This process continues until the desired output is determined.

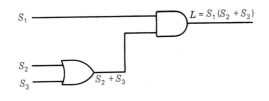

Figure 23-24

For one more example, construct the digital equivalent of Fig. 23-7. The first step is to generate the two OR functions and then to AND these together to achieve the desired result (Fig. 23-25).

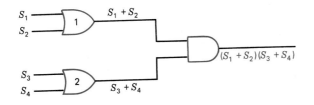

Figure 23-25

At this point you may ask why anyone would want to use such an apparently more complicated system. The reasons lie in cost, reliability, and size. Because of the development of the semiconductor fabrication industry what appears to be very complicated logic can be made on very small chips. These chips are not only small, but in large volume they become very inexpensive. Experience has shown that they are also very reliable and durable. Of course, they do not handle the power that switching circuits handle.

The result is that hybrid or combination circuits are likely to be around for quite a while. Switching circuits may be more cost-effective in units that handle higher power and are manufactured in small numbers. If the logic is complicated, it may be done digitally and then interfaced to switching circuits. This means that persons working at the technology level must be able to use both types of analysis.

We now do some problems that will help relate the two fields and to further develop skills in digital circuit analysis. Remember that once you are at the logic equation level, it makes no difference which system you are using. Truth tables, and so on, are exactly the same for each type of circuit.

Practice

1. Convert the circuit shown in Fig. 23-11 to digital logic.
2. Convert the circuit shown in Fig. 23-16 to digital logic.
3. Convert the circuit shown in Fig. 23-17 to digital logic.
4. Make a truth table for the circuit in Exercise 1.
5. Make a truth table for the circuit shown in Fig. 23-24.
6. Make a truth table for the circuit shown in Fig. 23-25.
7. In the circuit shown in Fig. 23-25, what is the output if the inputs are $S_1 = 0$, $S_2 = 1$, and $S_3 = 1$? Do you have to know the values of S_4 to determine the output? Why?
8. Using Fig. 23-25, determine the output if $S_1 = 0$ and $S_2 = 0$. Do you have to know the values of S_3 and S_4? Why?

(See CD-ROM for Test 23-5.)

1. Draw a wiring diagram for the following logic function: $LS_1 \cdot PS_1 \cdot FS_1 = 1$

2. What is the total number of combinations available for a group of 12 permissive contacts in series with a control relay?

3. Show the truth table for the following logic function: $LS_1 \cdot PS_1 \cdot FS_1 + HS_1 = 1$

4. On a truth table, how are the input variables arranged?

5. Draw a truth table for Fig. 23-26.

6. What is the complement of a switch function, and how is it drawn as a logic symbol?

7. For safety purposes, all permissive contacts (except motor overloads) are usually on the left side of the output device. Redraw Fig. 23-18 to comply with this requirement.

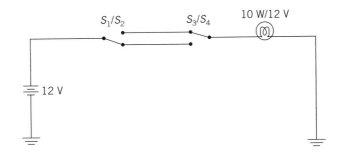

Figure 23-26

INTERNET
A C T I V I T I E S

Internet Activity A

View the following site to learn more about the EXCLUSIVE OR function. From the information provided at this site, give a verbal statement that describes the operation of the EXCLUSIVE OR, or XOR function.
http://www.play-hookey.com/digital/xor_function.html

Internet Activity B

At the site listed above, how many NAND gate ICs are needed to construct an XOR gate?

Chapter 23 Solutions to Self-Tests and Reviews

Self-Test 23-1

1. switching circuits, digital logic circuits
2. continuity, connectedness
3. voltage
4. Binary

Self-Test 23-2

1. 1, 0
2. 1
3. 0
4. may

5. independent
6. 0
7. that variable

Self-Test 23-3

1. independent variables
2. dependent variables
3. 2^N
4. binary number
5. most significant digit
6. least significant digit
7. LSD
8. MSD

Self-Test 23-4

1. inclusive, exclusive
2. independent
3. 1
4. that variable
5. 1
6. 0

Self-Test 23-5

1. segments, blocks
2. path
3. expanded
4. alternative

5. the same
6. path, paths

Self-Test 23-6

1. NO, NC
2. latched, locked
3. energized, picked up, picked
4. dropped, deenergized
5. modes
6. two
7. both

Signal Distribution

1. Use basic exponential formulas.
2. State the rule for multiplying exponents with a common base.
3. State the rule for any number raised to the zero power.
4. State the rule for dividing exponents with a common base.
5. Define a logarithm.
6. Show that the logarithm of a product of two numbers is equal to the sum of the logarithms of the two numbers.
7. Show that the logarithm of a quotient of two numbers is equal to the logarithm of the numerator minus the logarithm of the denominator.
8. Explain the meaning and use formulas associated with bels and decibels (dB).
9. Write the voltage and current forms of the dB equation.
10. Use the dB equations to determine the loss associated with cable distribution systems.

The high-technology world is seeing the integration of previously separate technologies. On the factory floor there is the implementation of computers to monitor, control, and collect information. This means that the cables that distribute the power to run the machines follow the same paths as the cables that distribute information and control signals.

At first there is a tendency to view the two technologies of power and signal distribution as being basically the same. Both systems use metal conductors to distribute electrical signals. During the construction and installation phases the established work crews are called upon to install the signal distribution cables along with the power cables. If you're going to pull a power cable, why can't you pull the signal cable along with it?

There are some very significant differences between the two technologies that lead to confusion and problems. In this chapter we shall learn some of the mathematics that is used to express the same concepts for signals as were discussed previously relative to power.

The concept of line drop was introduced in Job 6-4 and extended to distribution systems in Job 6-5. In Chap. 13 it was shown how wire size relates to the line resistance and hence line drop. The concept of efficiency was introduced in Job 11-4. In Job 11-6 it was shown that the efficiency of energy transfer depends on the impedance level of the source and the supply. These jobs provide the basic concepts of power distribution.

In signal distribution all these concepts are brought together in a single concept, the bel. The bel is based upon the ratio of two powers such as the power in and the power out, or a measured power and a standard power level.

In this chapter you will learn some new mathematics and some new concepts for the application of this mathematics. We will start by extending the mathematics and concepts of exponentials that have already been used in other areas. The concept of logarithms will next be introduced as an extension of the concept of exponentials. Next you will be introduced to the concept of a bel. The definition of bel will then be expanded to make the voltage and current relationships explicit. This will then be applied to the evaluation of coaxial cables for signal distribution.

JOB 24-1 — Exponentials and Logarithms

In this job we develop the mathematics of exponentials and logarithms. In the days prior to calculators and computers, logarithms were used in the scientific community to solve complex calculations. It is not the intention of this job to teach the use of logs for such purposes. The intention here is to learn to use the logarithm function itself. This is necessary for the applications that follow. It is also assumed that most persons will, when necessary, have access to some type of calculator or computer. In some cases it may be necessary to use a table of logarithms; therefore, this matter will be covered.

Exponentials

Exponentials are not entirely new to you, and you probably have more experience with them than you realize. In Jobs 3-5 and 3-6 the basic idea of exponents was covered. You may want to review some of this material before we generalize the use of exponents.

In general, we can raise any number to any power. The general form of an exponent is

$$Y = X^N \qquad \text{Formula 24-1}$$

This definition says take the number X and multiply it by itself N times and that will be the number Y that we want. We call X the base of the number and N is the exponent. Number systems are created using different bases. If the base is 10, we have the decimal number system that we use. If the base is 2, we have the binary number system that is used by computers. Other bases are used by persons who use computers. For example, the base 8 is the octal number system, and the base 16 is the hexadecimal number system. It is interesting to note that the common bases used in computers are themselves powers of 2: $2 = 2^1$, $8 = 2^3$, and $16 = 2^4$.

In some earlier jobs, in particular in Job 3-6, the base X was always 10. This is a very important case because the base for our number system is 10. That is why Job 3-7 on units of measure assigned names to various integer powers of 10. In this chapter we continue to look at the same situation, $X = 10$.

Some previous jobs have worked with special exponents and used any value for a base number. We used the case $N = 2$ extensively to get electrical power as V^2 or I^2. In those cases the current or the voltage is the base and could take on any value. The exponent was always 2. We saw $N = 2$ again when we studied the pythagorean theorem in Job 15-7. Here again, the sides of the triangle could be any value and we could square any number.

Let's review some of the rules of exponents. First is the rule for multiplying exponents that have a common base.

Hint for Success

A logarithm is an exponent. As an aid in remembering this, formula (24-1) can be rewritten as (base)$^{\log}$ = number.

$$X^{N_1} \times X^{N_2} = X^{(N_1 + N_2)} \qquad \text{Formula 24-2}$$

This rule says that to multiply numbers that have the same base X, just add the exponents N. This can be taken as the fundamental rule for exponents. The other rules can then be obtained from this one rule if we have one additional definition.

The definition covers the case X^0. In the discussion of the rule above, it was assumed that the exponents were all positive. What does it mean to have a negative exponent? To understand negative exponents and to have a consistent set of mathematics, we have to define what X^0 is. This sort of says do nothing, don't bother to multiply the number by anything. If you follow this idea, it says nothing will change—everything will stay the same. We can therefore make the definition:

$$X^0 = 1 \qquad \text{Formula 24-3}$$

This rule says that any number at all that is raised to the 0 power is equal to 1. This definition and the above rule help us to get the following rule for dividing two numbers:

$$\frac{X^{N_1}}{X^{N_2}} = X^{(N_1 - N_2)} \qquad \text{Formula 24-4}$$

We will now use these rules to expand our understanding of exponentials. In each of the situations that we have studied in earlier chapters the exponent has been an integer, V^2, 10^3, etc. In this job, as we generalize the values of N, we can have fractional powers of 10 such as

$$10^{\frac{1}{2}} \text{ or } 10^{\frac{1}{3}}$$

What does it mean to multiply 10 by itself half a time? To understand this use of exponentials, use the rule for multiplication of exponentials and suppose the following:

$$X^{\frac{1}{2}} \times X^{\frac{1}{2}} = X^{\frac{1}{2} + \frac{1}{2}} = X^1$$

This equation says take a number and multiply it by itself; the answer is X. This means that the numbers that were multiplied must have been the square root of X:

$$X^{\frac{1}{2}} = \sqrt{X}$$

In other words this is just another way to express square root, which was covered in Job 7-7 to Job 7-10. From this discussion we conclude that fractional exponents are roots of a number: $\frac{1}{2}$ is the square root, $\frac{1}{3}$ is the cube root, etc.

We can now interpret all exponents. Positive exponents and the sum of exponents are multiplication; negative exponents and the difference of exponents are division; and fractional exponents are roots.

Evaluating Exponentials

Exponentials are usually evaluated using a calculator or computer with a built-in power function. If these are not available, tables of exponentials are available. Table 24-1 on page 840 is a typical table of exponents. In the first column of Table 24-1 are values of X. The value of 10^x is given in the second column.

Light guide fibers used in telecommunications (fiber-optic cables) must be joined carefully so that minimal light escapes at the junction. Here a light guide fiber is readied to be spliced to another fiber. The fiber is held in place by the grooves in a pair of silicon chips. Once it is in this precisely fixed position, it can be merged with another fiber into the near-perfect alignment needed.

Example 24-1

Find the value of $10^{0.35}$

Solution

Using Table 24-1, proceed down column 1 until you come to the value 0.35. Adjacent to this value in column 2 is the number 2.23872. This is the answer.

The table can be expanded by using the rules for the exponential of products and quotients.

Table 24-1

Powers of 10

X	10^x	X	10^x
0.00	1.00000	0.50	3.16228
0.05	1.12202	0.55	3.54813
0.10	1.25893	0.60	3.98107
0.15	1.41254	0.65	4.46684
0.20	1.58489	0.70	5.01187
0.25	1.77828	0.75	5.62341
0.30	1.99526	0.80	6.30957
0.35	2.23872	0.85	7.07946
0.40	2.51189	0.90	7.94328
0.45	2.81838	0.95	8.91251

Example 24-2

Find the value of $10^{3.5}$.

Solution

Table 24-1 has values of X only up to 0.95. To find the exponential for 3.5, use the fact that $3.5 = 3 + 0.5$; then use the relationship

$$10^3 \times 10^{0.5} = 10^{(3.5)}$$
$$= 1000 \times 3.16228 = 3162.28$$

Logarithms

The concept of inverse operations such as addition versus subtraction, and multiplication versus division exists in mathematics. If you add a number and then subtract it, you are back where you started from. If you multiply by a number and then divide by the number, you are right back where you started from. For example, with addition and subtraction,

$$Y = X + N$$
$$N = Y - X$$

Addition and subtraction are inverse functions. You use the addition of X to get Y. When you have Y, you subtract X to get back to N. Addition does the opposite of subtraction.

The same concept applies to multiplication and division. If you multiply a number N by X, you get Y. If you have Y and want to get back to N, divide by X.

$$Y = X \times N$$
$$N = \frac{Y}{X}$$

Here again multiplication and division are inverse operations. When you use one operation, you can use the other operation to get back to where you started from. It is important to have inverse operations in mathematics so that you can freely move back and forth in equations.

For exponentials we have

$$Y = X^N \qquad \text{Formula 24-5}$$

This tells us how to get Y when we have X and N. We need a function that will help us go backward; when we have Y how do we get back to our number N?

This inverse function is defined as a logarithm or log.

$$N = \log_X(Y) \qquad \text{Formula 24-6}$$

The unknown X is often the base of the number system. The base of the decimal system is 10, and it is common to write $\log_{10}$ as just log.

The following are some important properties of the log function. The logarithm of the product of two numbers is the sum of the logarithms of the numbers. This is

$$\log(A \times B) = \log(A) + \log(B) \qquad \text{Formula 24-7}$$

The next important property is the logarithm of the quotient of two numbers. If the number A is divided by the number B, the logarithm of the quotient is the log of the numerator minus the log of the denominator.

$$\log\left(\frac{A}{B}\right) = \log(A) - \log(B) \qquad \text{Formula 24-8}$$

The last rule covers the logarithm of an exponential. If a number A is raised to a power N, the log of A^N is N times the log of A.

$$\log(A^N) = N \times \log(A) \qquad \text{Formula 24-9}$$

CALCULATOR HINT

Note that on many scientific calculators, the [LOG] key is tied to base 10. So $10^5 = 100\,000$ and $10{,}000$ [LOG] $= 4$.

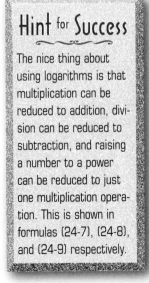

Hint for Success

The nice thing about using logarithms is that multiplication can be reduced to addition, division can be reduced to subtraction, and raising a number to a power can be reduced to just one multiplication operation. This is shown in formulas (24-7), (24-8), and (24-9) respectively.

Evaluating Logarithms

There are two common ways to evaluate logarithms, with computers or calculators and with log tables. Computers or calculators are the easiest and fastest way, but sometimes they are not available. If you do not have a calculator, you can use a table of logarithms. Table 24-2 is called a *two-place log table* because it supplies the logarithm for numbers with two digits. The table is laid out so that the first digit for the number is located in the column at the left of the table and the second digit is located in the row at the top of the table. To find the logarithm for a number, locate the first digit in the column at the left and then move across the table until you come to the column headed by the second digit. The number located in that table position is the logarithm for the number. The three formulas—(24-7), (24-8), and (24-9)—are important in extending the use of the table.

Table 24-2

Table of Two-Place Logarithms

	0	1	2	3	4
0		0.00000	0.30103	0.47712	0.60206
1	1.00000	1.04139	1.07918	1.11394	1.14613
2	1.30103	1.32222	1.34242	1.36173	1.38021
3	1.47712	1.49136	1.50515	1.51851	1.53148
4	1.60206	1.61278	1.62325	1.63347	1.64345
5	1.69897	1.70757	1.71600	1.72428	1.73239
6	1.77815	1.78533	1.79239	1.79934	1.80618
7	1.84510	1.85126	1.85733	1.86332	1.86923
8	1.90309	1.90849	1.91381	1.91908	1.92428
9	1.95424	1.95904	1.96379	1.96848	1.97313

	5	6	7	8	9
0	0.69897	0.77815	0.84510	0.90309	0.95424
1	1.17609	1.20412	1.23045	1.25527	1.27875
2	1.39794	1.41497	1.43136	1.44716	1.46240
3	1.54407	1.55630	1.56820	1.57978	1.59106
4	1.65321	1.66276	1.67210	1.68124	1.69020
5	1.74036	1.74819	1.75587	1.76343	1.77085
6	1.81291	1.81954	1.82607	1.83251	1.83885
7	1.87506	1.88081	1.88649	1.89209	1.89763
8	1.92942	1.93450	1.93952	1.94448	1.94939
9	1.97772	1.98227	1.98677	1.99123	1.99564

Example 24-3

Find the logarithm for 10.

Solution

Locate the first digit, 1, in the column at the left. Move across the table to the column headed by the second digit, 0. At that position read the number 1.00000. This is the logarithm of the number 10.

Example 24-4

Find the logarithm for 46.

Solution

Locate the first digit, 4, in the column at the left of the second part of the table. The second part is used because that is where the column headed by 6 is located. Move across the table to the column headed by the second digit, 6. At that position read the number 1.66276. This is the logaritm of the number 46.

Example 24-5

Find the logarithm for 460.

Solution

The number 460 has 3 digits. This is too large for the table; however, 460 = 46 × 10. We can use the rule for the log of the product of two numbers.

$$\log (460) = \log (46 \times 10) = \log (46) + \log (10)$$
$$= 1.66276 + 1.00000$$
$$= 2.66276$$

This example shows a very useful property of logarithms. Every time we multiply a number by 10, we add 1 to the logarithm of the number. This means that multiplication by 100 adds 2, and multiplication by 1000 adds 3, etc. The number we add is equal to the number of 0s in the multiplier.

CALCULATOR HINT

To use your calculator to find the logarithm of 460 in Example 24-5, just press the [LOG] key followed by the numbers [4] [6] [0]. On some calculators it may be necessary to press the [=] key. The calculator should display an answer of 2.662.

Exercises

1. Find the logarithm of the following numbers:
   a. 2
   b. 0.5
   c. $\dfrac{1}{2}$
   d. 50
   e. 500
   f. 0.004

2. Find the logarithm of the following numbers:
   a. 12 × 5
   b. 23 × 46
   c. 0.045 × 220
   d. 220 V × 5 A
   e. 42/12
   f. 0.046/55

Review of Exponentials and Logs

1. In the equation $Y = X^N$ the number X is the _____ and the number N is the _____.
2. In the equation $Y = X^N$ we say that X is raised to the Nth _____.
3. $X^{N_1} \times X^{N_2} = ?$
4. $X_0 = ?$
5. $\dfrac{X^{N_1}}{X^{N_2}} = ?$
6. $X^{1/2} = ?$
7. $\log (A \times B) = ?$
8. $\log (A/B) = ?$
9. $\log (A^N) = ?$

Bels and Decibels

Job 24-1 provided the mathematical basis that is needed to develop the applications in the remainder of this chapter. The new application concept is that of the bel or decibel. The decibel, which is usually abbreviated dB, is the more common. These concepts are very common in the field of electronics as opposed to electricity. The dB is used in many ways. First we shall look at signal distribution.

It is important to remember that a bel is defined in terms of the ratio of two powers. In the power electricity applications that were studied earlier, the focus was on voltage, current, or resistance. Power is more complex, and as we will see, this complexity makes the dB a more comprehensive concept.

The bel can be defined using the exponential form as

$$\frac{P_1}{P_2} = 10^{\text{bel}} \qquad \qquad \text{Formula 24-10}$$

This form of the definition emphasizes the power ratio concept of the bel. As we saw above, an equivalent way to state the definition is to use the log form of the inverse pair. The bel is defined using the base 10 for the logarithm. In the logarithm form the definition of the bel is

$$\text{bel} = \log_{10}\left(\frac{P_1}{P_2}\right) \qquad \qquad \text{Formula 24-11}$$

For many applications the bel is too small; therefore, we use a special variation of the definition that is 10 times larger, the decibel. The dB is defined as

$$\text{dB} = 10 \times \log_{10}\left(\frac{P_1}{P_2}\right) \qquad \qquad \text{Formula 24-12}$$

The dB can be expressed in the exponential form as

$$\frac{P_1}{P_2} = 10^{\text{dB}/10} \qquad \qquad \text{Formula 24-13}$$

Example 24-6

Find the number of dB for a device that has 1 W of power in and 2 W of power out.

Solution

Let P_1 be the power out and P_2 the power in.

$$dB = 10 \times \log_{10}\left(\frac{P_1}{P_2}\right)$$

$$= 10 \times \log\left(\frac{2}{1}\right) = 3.0103$$

If you use the log table, the log of 2 can be found as follows: The table is for two-place numbers. Write 2 as 02. The first digit is 0; therefore, enter the table at the row where 0 is in the first column.

Move to the right to the column headed by the number 2. Read the number 0.30103. Multiply this number by 10 to get dB.

Example 24-7

Find the number of dB for a device that has 2 W of power in and 1 W of power out.

Solution

Let P_1 be the power out and P_2 the power in. Use the relationship that log $(A/B) = \log(A) - \log(B)$.

$$dB = 10 \times \log_{10}\left(\frac{P_1}{P_2}\right)$$

$$= 10 \times \log_{10}\left(\frac{1}{2}\right)$$

$$= 10 \times (\log 1 - \log 2)$$

$$= -3.0103$$

These two examples show an important property of dB. If the power out is greater than the power in, there is a power gain and the dB is positive. On the other hand, if the power out is less than the power in, there is a loss and the dB is negative.

JOB 24-4

Voltage and Current Forms of the dB Equation

This job will continue to develop the concept of dB. We shall first substitute the voltage form of power into the dB equation. This will give a relationship that emphasizes voltage. Then the current form of the power equation will be used to develop a dB equation that emphasizes current. Resistance is in both of these power relationships. The resulting equations will therefore include resistance as a parameter.

Voltage Form of the dB Equation

The following familiar equation expresses the power dissipated in a resistor.

$$P = \frac{V^2}{R}$$

Whenever you measure a voltage, there must be some resistance that is creating this voltage. The power that would be measured would result from the measured voltage across the resistance. Let's use the dB equation but substitute the voltage measurements for the power measurements. If we do this for both powers, the dB equation becomes the following:

$$dB = 10 \times \log_{10}\left(\frac{V_1^2/R_1}{V_2^2/R_2}\right)$$

$$= 10 \times \log_{10}\left(\frac{V_1^2}{V_2^2}\frac{R_2}{R_1}\right)$$

$$= 10 \times \log_{10}\left[\left(\frac{V_1}{V_2}\right)^2 \times \frac{R_2}{R_1}\right]$$

In this equation V_1 and R_1 are the voltage and resistance associated with power P_1 and V_2 and R_2 are the voltage and resistance associated with power P_2. The form of the above equation could be used as it is, but it is useful to change the equation to another form. To do this, we use the rule for the logarithm of the product of two numbers. Using formula (24-7) the equation becomes

$$dB = 10 \times \log\left[\left(\frac{V_1}{V_2}\right)^2\right] + 10 \times \log\left(\frac{R_2}{R_1}\right) \qquad \text{Formula 24-14}$$

This form of the equation shows that if you do not take into consideration the two resistances, the measurement of dB will be off by a constant factor. This error is the second term:

$$\text{Error} = 10 \times \log\left(\frac{R_2}{R_1}\right) \qquad \text{Formula 24-15}$$

Many voltmeters have a dB scale. These scales assume a standard resistance, usually 600 Ω. The value 600 is used because it is the standard for many telephone lines. As we shall see, not all lines have the same impedance.

In some cases the error introduced by ignoring the resistance will not affect measurements or calculations. One such situation is when a meter is connected to a circuit and measurements are made under two different conditions such as two different frequencies. If it can be assumed that the resistance does not change with frequency, the error in the two measurements is the same. If you are interested only in the difference between the two readings, when you subtract them the error will cancel out.

Another important variation of the equation is obtained by using Eq. (24-14) for the logarithm of an exponent. When this is done, the equation for dB becomes

$$dB = 20 \times \log\left(\frac{V_1}{V_2}\right) + 10 \times \log\left(\frac{R_2}{R_1}\right) \qquad \text{Formula 24-16}$$

This equation includes the concepts of voltage drop, impedance levels, and power transfer all at once. The V_1/V_2 is just another way of looking at output voltage and input voltage or line drop. The coefficient 20 is used because we are interested in power. The second term involving R_2/R_1 reminds us that we cannot forget the impedance levels that we are working with.

Sometimes the resistances can be ignored as stated above. When this is done, the second term of the equation can be dropped. This gives the following voltage form of the dB equation:

$$db = 20 \times \log\left(\frac{V_1}{V_2}\right) \qquad \text{Formula 24-17}$$

It is often useful to have the above equation in the exponential form so that it is easier to solve for the voltages.

$$\frac{V_1}{V_2} = 10^{dB/20} \qquad \text{Formula 24-18}$$

Current Form of the dB Equation

The current form of the dB equation is obtained just as the voltage form was obtained. The difference is that the equation for power is

$$P - I^2 \times R$$

Substituting this equation into the dB equation gives

$$dB = 10 \times \log\left(\frac{I_1^2 \times R_1}{I_2^2 \times R_2}\right)$$

$$= 20 \times \log\left(\frac{I_1}{I_2}\right) + 10 \times \log\left(\frac{R_2}{R_1}\right) \qquad \text{Formula 24-19}$$

Notice that the current form of the equation is similar to the voltage form. The major difference to notice is the second term for resistance. The resistance ratio is the inverse of that for the voltage.

Exercises

Applications

1. A power transformer has 4000 W of power in and 3500 W out. What is the power loss in dB?
2. A power distribution system is 95% efficient. If 1000 W is put into the system, 950 W come out. What is the power loss in dB?
3. If the power increases to twice its original value, what is the dB gain in power?
4. An amplifier has 20 dB gain. What is the ratio of the power out to the power into the amplifier?
5. If you ignore the effects of resistance, what is the voltage gain of the amplifier with 20 dB gain?
6. If a voltmeter indicates 20 dB across a 12-Ω load, what is the measurement error if the meter was designed to indicate properly for a 600-Ω load?
7. An amplifier has an input voltage of 0.15 V across an input resistance of 100 kΩ and an output of 2.5 V across 12Ω. What is the power gain of the amplifier in dB?

Review of dB

1. Decibels are defined in terms of the ratio of two ―――.
2. The definition of dB is ―――.
3. The definition of dB in exponential form is ―――.
4. The dB definition in voltage ratio form is ―――.
5. The dB definition in current ratio form is ―――.
6. The voltage dB equation is ―――.
7. The exponential form of the voltage dB equation is ―――.

(See CD-ROM for Test 24-1.)

Cables and Cable Losses

In this job we shall apply the above mathematics to some real coaxial cables. Listed in Table 24-3 are some values for real cables. We are primarily interested in the cable loss in dB listed in the last column. As we have seen in the preceding jobs, it is also important to take into consideration the cable impedance that is listed in the second column.

Table 24-3

Coaxial Cable Loss, dB

RG/U type	Impedance, Ω	Diameter, in	dB/ 100 ft
RG8A/U	52	0.405	9.0
RG58C/U	50	0.195	17.5
RG108A/U	78	0.235	26.2
RG179B/U	75	0.100	25.0
RG188A/U	50	0.110	30.0
RG196A/U	50	0.080	45.0
RG213/U	50	0.405	9.0

It is common practice to match a cable to its characteristic impedance. In Job 11-6 it was shown that for maximum power transfer the impedance of the source and the load should be equal. When the load resistance is equal to the cable impedance, the load is said to be matched to the cable. In this case the resistance term of the dB equation can be ignored because $\log(1) = 0$.

Example 24-8

An RG58C/U cable 30 ft long is used to transmit data from a robot to a computer. If the cable is matched at both ends, find the loss in dB. Find the cable output voltage if the input voltage is 3.5 V.

Solution

First find the cable loss from Table 24-3. The table shows that an RG58C/U cable has a loss of 17.5 dB per 100 ft of cable. For 30 ft of cable this gives a total cable loss of 5.25 dB. Notice that we speak of the loss as positive, 5.25 dB, but for calculations we will have to use a negative value.

$$dB = 17.5 \times \frac{30}{100}$$

$$= 5.25 \text{ dB}$$

Because the cable is matched, there is no resistance effect. Therefore, we can use Eq. (24-18) to find the output voltage without any error.

$$\frac{V_1}{V_2} = 10^{dB/20}$$

$$\frac{V_{out}}{3.5} = 10^{-5.25/20}$$

$$V_1 = 3.5 \times 10^{-0.2625}$$

$$= 1.91 \text{ V}$$

Effect of Frequency

The loss of a cable is dependent upon the frequency of the signal that is applied to the cable. An RG58/U has a center conductor that is a No. 20 solid wire. As the frequency of the signal increases, the cable loss increases as shown in Table 24-4.

Table 24-4

dB Loss for RG58/U Cable vs. Frequency

Frequency, MHz	dB/ 100 ft
50	3.1
100	4.5
200	6.8
400	10.0
900	16.0

Example 24-9

An RG58/U cable 30 ft long operates at 50 MHz. If the frequency is increased to 100 MHz, how many additional dB will be lost at the higher frequency?

Solution

At 50 MHz the loss is 3.1 dB per 100 ft. This increases to 4.5 dB per 100 ft at 100 MHz. The additional loss is 1.4 dB per 100 ft. The total additional loss in the 30-ft cable is

$$dB = 1.4 \times \frac{30}{100}$$
$$= 0.42 \ dB$$

Fiber-Optic Cable

Fiber optics is becoming increasingly popular for data transmission. A fiber-optic cable transmits signals on light waves. Such cables are less subject to electrical noise, which is important in a factory environment. With fiber-optic cables we are interested in the light power into the cable and the light power out. Table 24-5 shows the loss in dB per km. Notice that the loss is much less than in the coaxial cables above. A coaxial cable's losses in 100 ft are as much as or more than a fiber-optic cable loses in 1 km.

Table 24-5

dB Loss for Fiber-Optic Cable

Cable	dB/km
GK-EFN050	5
GK-EFN100	6

Example 24-10

Compare the difference in loss between a 300-ft RG58C/U coaxial cable and a GK-EFN 100 fiber-optic cable. The losses for the two types of cables are expressed in different measures. The coaxial cable is in feet, and the fiber-optic cable is in kilometers. We can bring this into a common measure by using the conversion factor:

$$1 \text{ kilometer} = 3281 \text{ feet}$$

The loss for the RG58C/U cable is

$$dB = 17.5 \times \frac{300}{100}$$
$$= 52.5 \ dB$$

Next find the loss for the fiber-optic cable, remembering to convert kilometers to feet.

$$dB = 6 \times \frac{300}{3281}$$
$$= 0.55 \ dB$$

Applications

1. An RG179B/U is used to connect an automation station to a computer. The cable is 45 ft long. What is the loss in dB for the cable? If a 9-V signal is put into the cable, what is the voltage of the output assuming that the cable is terminated in a matched load?

2. An RG179B/U cable has an impedance of 75 Ω. If this cable is used to connect to a 50-Ω load, what is the error in dB caused by ignoring the impedance mismatch?

3. An amplifier has an output voltage of 7.2 V. If the receiver requires a minimum of 0.010 V to give adequate performance, what is the maximum allowable cable loss assuming matched conditions? What is the maximum cable length for an RG58C/U cable?

4. It is necessary to use as small a cable as possible for the amplifier installation in Exercise 3. An RG196A/U cable has the same impedance as the RG58C/U cable. If the RG196A/U cable is used, what is the maximum cable length?

5. A voltmeter is used to measure dB. The meter is calibrated for dB across a 600-Ω load. If the voltage is measured across a 270-Ω resistor, what is the error in the measurement?

6. If a cable has 1 W of output power and 2 W of input power how much dB loss is there in the cable? If the 1 W out of the cable is connected to a second cable with 3dB loss, what is the power out of the second cable? Why is the -3-dB level referred to as the "half-power point"?

7. Two cables are connected together. The power into the first cable is 800 W. The power at the output of the first cable is 400 W. The power out of the second cable is 200 W. How may dB are lost in the first cable? How many dB are lost in the second cable? What is the dB loss from the input of the first cable to the output of the second cable? Is the overall loss of the two cables in dB equal to the sum of the losses in each cable in dB?

8. An amplifier has a gain of 60 dB. What is the ratio of the power out of the amplifier to the power in? What is the ratio of the output voltage to the input voltage? What is the relationship between the power ratio and the voltage ratio? Why is this so?

9. Assume that the amplifier of Exercise 8 is a hi-fi amplifier. The output of the amplifier drives an 8-Ω speaker. The input impedance is 100 kΩ. If 1 mV in produces 1 V out, what is the voltage gain of the amplifier in dB? What is the dB gain due to the change in impedance from input to output? What is the power gain of the amplifier in dB?

10. An RG188A/V cable has an impedance of 50 Ω and a loss of 30.0 dB/100 ft. If the cable is 45 ft long, what is the ratio of the input voltage to the output voltage?

Assessment

1. Give a complete, well-worded definition of the term *logarithm*.
2. A motor pump system uses 25 hp. The actual work done amounts to 12.5 hp. What is the power loss in dB?

3. Refigure Example 24-8, using RG8A/U instead of the RG58C/U cable. Is there any difference in the power loss of the circuit?

INTERNET ACTIVITIES

Internet Activity A

View the following site to learn more about bels and decibels. What is the power level corresponding to 0 dBm?
http://www.cc.columbia.edu/~fuat/cuarc/dB.html

Internet Activity B

At the site listed above, determine the power level corresponding to −30 dBm.

Chapter 24 Solutions to Self-Tests and Reviews

Job 24-2 Review

1. base, exponent
2. power
3. $X^{(N_1 + N_2)}$
4. 1
5. $X^{(N_1 - N_2)}$
6. $\sqrt{X}$
7. $\log(A) + \log(B)$
8. $\log(A) - \log(B)$
9. $N \times \log(A)$

Jobs 24-3 and 24-4 Review

1. powers
2. $dB = 10 \times \log\left(\dfrac{P_1}{P_2}\right)$
3. $\dfrac{P_1}{P_2} = 10^{dB/10}$

4. $dB = 20 \times \log\left(\dfrac{V_1}{V_2}\right)$
   $+ 10 \times \log\left(\dfrac{R_2}{R_1}\right)$

5. $dB = 20 \times \log\left(\dfrac{I_1}{I_2}\right)$
   $+ 10 \times \log\left(\dfrac{R_1}{R_2}\right)$

6. $dB = 20 \times \log\left(\dfrac{V_1}{V_2}\right)$

7. $\dfrac{V_1}{V_2} = 10^{dB/20}$

Color Codes

Finding the Values of Fixed Resistors

Instead of the number of ohms of resistance being stamped on carbon-type resistors, the resistors are colored according to a definite system approved by the EIA (Electronics Industries Association). Each color represents a number according to the plan in Table A-1.

Table A-1			
Color*	**Number**	**Color**	**Number**
Black	0	Green	5
Brown	1	Blue	6
Red	2	Violet	7
Orange	3	Gray	8
Yellow	4	White	9

*Gold—multiply by 0.1; silver—multiply by 0.01.

The value of the resistor is obtained by reading the colors according to the following systems.

The three-band system. The first band represents the first number in the value. The second band represents the second number. The third band represents the number of zeros to be added after the first two numbers. If the third band is gold or silver, multiply the value indicated by the first two bands by 0.1 or 0.01, respectively, as indicated above.

Example A-1

Find the resistance of a resistor marked red, violet, yellow as shown in Fig. A-1.

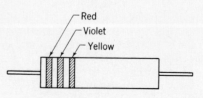

Figure A-1 The resistance value is indicated by three bands of color read in order from left to right.

Solution

First band	Second band	Third band
Red	Violet	Yellow
2	7	0000

$$\text{Resistance} = 270{,}000 \ \Omega \qquad \textit{Ans.}$$

Example A-2

A resistor is marked yellow, orange, black. What is its resistance?

Solution

First band	Second band	Third band
Yellow	Orange	Black
4	3	No zeros

$$\text{Resistance} = 43 \ \Omega \qquad \textit{Ans.}$$

Example A-3

A resistor is marked green, blue, gold. What is its resistance?

Solution

First band	Second band	Third band
Green	Blue	Gold
5	6	Multiply by 0.1

The first two bands indicate a value of 56 Ω. Therefore,

$$56 \times 0.1 = 5.6 \ \Omega \qquad \textit{Ans.}$$

The body-tip-dot system. The colors must be read in the following order: body, tip, and dot. The body color represents the first number, the right-hand tip represents the second number, and the center dot represents the number of zeros to be added after the first two numbers. If the dot is gold or silver, multiply the value indicated by the first two numbers by 0.1 or 0.01, respectively.

Example A-4

What is the resistance of the resistor shown in Fig. A-2?

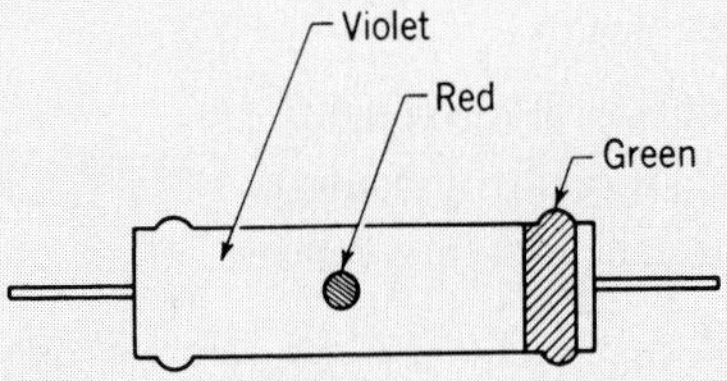

Figure A-2 The resistance value is indicated by the colors in order as follows: body, right-end tip, and center dot.

Solution

Body	Tip	Dot
Violet	Green	Red
7	5	00

$$\text{Resistance} = 7500 \ \Omega \quad \textit{Ans.}$$

Example A-5

A resistor has a brown body, a blue right end, and an orange dot in the center. What is its resistance?

Solution

Body	Tip	Dot
Brown	Blue	Orange
1	6	000

$$\text{Resistance} = 16{,}000 \ \Omega \quad \textit{Ans.}$$

Example A-6

A resistor has a gray body, a red right end, and a silver dot in the center. What is its resistance?

Solution

Body	Tip	Dot
Gray	Red	Silver
8	2	Multiply by 0.01

The first two colors indicate a value of 82 Ω. Therefore,

$$82 \times 0.01 = 0.82 \ \Omega \quad \textit{Ans.}$$

Example A-7

A mechanic needs a 510,000-Ω resistor. What combination of colors in the body-tip-dot system is needed?

Solution

The first digit is a 5, indicating green.
The second digit is a 1, indicating brown.
The four zeros that remain indicate yellow.
Therefore, the resistor will be color-coded as follows:

Body	Tip	Dot	
Green	Brown	Yellow	*Ans.*

Example A-8

What color combination is needed to indicate a 6.8-Ω resistor in the three-band system?

Solution

The first digit is a 6, indicating blue.
The second digit is an 8, indicating gray.

To obtain the number 6.8 from the number 68 it is necessary to multiply 68 by 0.1, which indicates gold. Therefore, the resistor will be color-coded as follows:

First band	Second band	Third band	
Blue	Gray	Gold	*Ans.*

Tolerance markings. A fourth band of color in the band system or a color on the left-hand side of the resistor in the body-tip-dot system is used to indicate how accurately the part is made to conform to the indicated markings. Gold means that the value is not more than 5% away from the indicated value. Silver indicates a tolerance of 10%, and black a tolerance of 20%. If the tolerance is not indicated by a color, it is assumed to be 20%.

Exercises

What value of resistance is indicated by each of the following color combinations?

1. Brown, black, yellow
2. Red, yellow, red
3. Gray, red, orange
4. Green, brown, red
5. Violet, green, black
6. Red, black, green
7. Yellow, violet, gold
8. Brown, green, yellow
9. Brown, red, gold
10. Brown, red, red
11. Brown, gray, green
12. Orange, white, silver
13. Blue, red, brown
14. Brown, gray, brown
15. White, brown, red
16. Orange, orange, gold
17. Yellow, violet, silver
18. Yellow, green, silver
19. Gray, red, black
20. Green, blue, gold

What color combination is needed to indicate each of the following resistances?

21. 240,000 Ω
22. 430,000 Ω
23. 51 Ω
24. 150 Ω
25. 10,000,000 Ω
26. 5.6 Ω
27. 3900 Ω
28. 0.47 Ω
29. 12 Ω
30. 750,000 Ω
31. 0.68 Ω
32. 1.2 Ω
33. 2.2 Ω
34. 100 Ω
35. 1.8 Ω
36. 1,000,000 Ω
37. 360 Ω
38. 6200 Ω
39. 1.5 Ω
40. 1 Ω

Preferred Values of Components

In both of the conventions shown there are two digits and a multiplier. The first two digits have only certain values according to industry and military standards. The number that they can have depends on the tolerance that the resistor has. For tolerances of 5, 10, and 20%, the standard values are as shown in Table A-2.

Table A-2

Tolerance, %

5	10	20
10	10	10
11		
12	12	
13		
15	15	15
16		
18	18	
20		
22	22	22
24		
27	27	
30		
33	33	33
36		
39	39	
43		
47	47	47
51		
56	56	
62		
68	68	68
75		
82	82	
91		

These preferred values are also used for some capacitors and zener diodes.

In the problems in the book no consideration was given to the standard resistor values. If a solution required a 65-Ω resistor, the answer stated a 65-Ω resistor. This is not a standard value, and it would have to be made specially for the application. According to the table, a choice would have to be made between 62 and 68 Ω; 65 Ω is 3 Ω greater than 62 and 3 Ω smaller than 68 Ω. There is no particular reason to select either 62 or 68 Ω over the other. Notice that 62 Ω is available only in 5% tolerance. The other choice, 68 Ω, can be obtained in 5, 10, or 20% tolerances. The higher-precision 5% resistor is usually more expensive and less readily available; therefore, you may decide to use 68-Ω 20% tolerance rather than 62-Ω 5% tolerance.

For problems 21 through 40, select the 10% tolerance resistor that is closest to the value given.

Trigonometric Functions of Decimal Degrees

Natural Trigonometric Functions

Degrees	Function	0.0°	0.1°	0.2°	0.3°	0.4°	0.5°	0.6°	0.7°	0.8°	0.9°
0	sin	0.0000	0.0017	0.0035	0.0052	0.0070	0.0087	0.0105	0.0122	0.0140	0.0157
	cos	1.0000	1.0000	1.0000	1.0000	1.0000	1.0000	0.9999	0.9999	0.9999	0.9999
	tan	0.0000	0.0017	0.0035	0.0052	0.0070	0.0087	0.0105	0.0122	0.0140	0.0157
1	sin	0.0175	0.0192	0.0209	0.0227	0.0244	0.0262	0.0279	0.0297	0.0314	0.0332
	cos	0.9998	0.9998	0.9998	0.9997	0.9997	0.9997	0.9996	0.9996	0.9995	0.9995
	tan	0.0175	0.0192	0.0209	0.0227	0.0224	0.0262	0.0279	0.0297	0.0314	0.0332
2	sin	0.0349	0.0366	0.0384	0.0401	0.0419	0.0436	0.0454	0.0471	0.0488	0.0506
	cos	0.9994	0.9993	0.9993	0.9992	0.9991	0.9990	0.9990	0.9989	0.9988	0.9987
	tan	0.0349	0.0367	0.0384	0.0402	0.0419	0.0437	0.0454	0.0472	0.0489	0.0507
3	sin	0.0523	0.0541	0.0558	0.0576	0.0593	0.0610	0.0628	0.0645	0.0663	0.0680
	cos	0.9986	0.9985	0.9984	0.9983	0.9982	0.9981	0.9980	0.9979	0.9978	0.9977
	tan	0.0524	0.0542	0.0559	0.0577	0.0594	0.0612	0.0629	0.0647	0.0664	0.0682
4	sin	0.0698	0.0715	0.0732	0.0750	0.0767	0.0785	0.0802	0.0819	0.0837	0.0854
	cos	0.9976	0.9974	0.9973	0.9972	0.9971	0.9969	0.9968	0.9966	0.9965	0.9963
	tan	0.0699	0.0717	0.0734	0.0752	0.0769	0.0787	0.0805	0.0822	0.0840	0.0857
5	sin	0.0872	0.0889	0.0906	0.0924	0.0941	0.0958	0.0976	0.0993	0.1011	0.1028
	cos	0.9962	0.9960	0.9959	0.9957	0.9956	0.9954	0.9952	0.9951	0.9949	0.9947
	tan	0.0875	0.0892	0.0910	0.0928	0.0945	0.0963	0.0981	0.0998	0.1016	0.1033
6	sin	0.1045	0.1063	0.1080	0.1097	0.1115	0.1132	0.1149	0.1167	0.1184	0.1201
	cos	0.9945	0.9943	0.9942	0.9940	0.9938	0.9936	0.9934	0.9932	0.9930	0.9928
	tan	0.1051	0.1069	0.1086	0.1104	0.1122	0.1139	0.1157	0.1175	0.1192	0.1210
7	sin	0.1219	0.1236	0.1253	0.1271	0.1288	0.1305	0.1323	0.1340	0.1357	0.1374
	cos	0.9925	0.9923	0.9921	0.9919	0.9917	0.9914	0.9912	0.9910	0.9907	0.9905
	tan	0.1228	0.1246	0.1263	0.1281	0.1299	0.1317	0.1334	0.1352	0.1370	0.1388
8	sin	0.1392	0.1409	0.1426	0.1444	0.1461	0.1478	0.1495	0.1513	0.1530	0.1547
	cos	0.9903	0.9900	0.9898	0.9895	0.9893	0.9890	0.9888	0.9885	0.9882	0.9880
	tan	0.1405	0.1423	0.1441	0.1459	0.1477	0.1495	0.1512	0.1530	0.1548	0.1566
9	sin	0.1564	0.1582	0.1599	0.1616	0.1633	0.1650	0.1668	0.1685	0.1702	0.1719
	cos	0.9877	0.9874	0.9871	0.9869	0.9866	0.9863	0.9860	0.9857	0.9854	0.9851
	tan	0.1584	0.1602	0.1620	0.1638	0.1655	0.1673	0.1691	0.1709	0.1727	0.1745
10	sin	0.1736	0.1754	0.1771	0.1778	0.1805	0.1822	0.1840	0.1857	0.1874	0.1891
	cos	0.9848	0.9845	0.9842	0.9839	0.9836	0.9833	0.9829	0.9826	0.9823	0.9820
	tan	0.1763	0.1781	0.1799	0.1817	0.1835	0.1853	0.1871	0.1890	0.1908	0.1926
11	sin	0.1908	0.1925	0.1942	0.1959	0.1977	0.1994	0.2011	0.2028	0.2045	0.2062
	cos	0.9816	0.9813	0.9810	0.9806	0.9803	0.9799	0.9796	0.9792	0.9789	0.9785
	tan	0.1944	0.1962	0.1980	0.1998	0.2016	0.2035	0.2053	0.2071	0.2089	0.2107
12	sin	0.2079	0.2096	0.2113	0.2130	0.2147	0.2164	0.2181	0.2198	0.2215	0.2232
	cos	0.9781	0.9778	0.9774	0.9770	0.9767	0.9763	0.9759	0.9755	0.9751	0.9748
	tan	0.2126	0.2144	0.2162	0.2180	0.2199	0.2217	0.2235	0.2254	0.2272	0.2290
13	sin	0.2250	0.2267	0.2284	0.2300	0.2318	0.2334	0.2351	0.2368	0.2385	0.2402
	cos	0.9744	0.9740	0.9736	0.9732	0.9728	0.9724	0.9720	0.9715	0.9711	0.9707
	tan	0.2309	0.2327	0.2345	0.2364	0.2382	0.2401	0.2419	0.2438	0.2456	0.2475
14	sin	0.2419	0.2436	0.2453	0.2470	0.2487	0.2504	0.2521	0.2538	0.2554	0.2571
	cos	0.9703	0.9699	0.9694	0.9690	0.9686	0.9681	0.9677	0.9673	0.9668	0.9664
	tan	0.2493	0.2512	0.2530	0.2549	0.2568	0.2586	0.2605	0.2623	0.2642	0.2661
15	sin	0.2588	0.2605	0.2622	0.2639	0.2656	0.2672	0.2689	0.2706	0.2723	0.2740
	cos	0.9659	0.9655	0.9650	0.9646	0.9641	0.9636	0.9632	0.9627	0.9622	0.9617
	tan	0.2679	0.2698	0.2717	0.2736	0.2754	0.2773	0.2792	0.2811	0.2830	0.2849
16	sin	0.2756	0.2773	0.2790	0.2807	0.2823	0.2840	0.2857	0.2874	0.2890	0.2907
	cos	0.9613	0.9608	0.9603	0.9598	0.9593	0.9588	0.9583	0.9578	0.9573	0.9568
	tan	0.2867	0.2886	0.2905	0.2924	0.2943	0.2962	0.2981	0.3000	0.3019	0.3038
17	sin	0.2924	0.2940	0.2957	0.2974	0.2990	0.3007	0.3024	0.3040	0.3057	0.3074
	cos	0.9563	0.9558	0.9553	0.9548	0.9542	0.9537	0.9532	0.9527	0.9521	0.9516
	tan	0.3057	0.3076	0.3096	0.3115	0.3134	0.3153	0.3172	0.3191	0.3211	0.3230
18	sin	0.3090	0.3107	0.3123	0.3140	0.3156	0.3173	0.3190	0.3206	0.3223	0.3239
	cos	0.9511	0.9505	0.9500	0.9494	0.9489	0.9483	0.9478	0.9472	0.9466	0.9461
	tan	0.3249	0.3269	0.3288	0.3307	0.3327	0.3346	0.3365	0.3385	0.3404	0.3424
19	sin	0.3256	0.3272	0.3289	0.3305	0.3322	0.3338	0.3355	0.3371	0.3387	0.3404
	cos	0.9455	0.9449	0.9444	0.9438	0.9432	0.9426	0.9421	0.9415	0.9409	0.9403
	tan	0.3443	0.3463	0.3482	0.3502	0.3522	0.3541	0.3561	0.3581	0.3600	0.3620
Degrees	Function	0′	6′	12′	18′	24′	30′	36′	42′	48′	54′

Natural Trigonometric Functions, Continued

Degrees	Function	0.0°	0.1°	0.2°	0.3°	0.4°	0.5°	0.6°	0.7°	0.8°	0.9°
20	sin	0.3420	0.3437	0.3453	0.3469	0.3486	0.3502	0.3518	0.3535	0.3551	0.3567
	cos	0.9397	0.9391	0.9385	0.9379	0.9373	0.9367	0.9361	0.9354	0.9348	0.9342
	tan	0.3640	0.3659	0.3679	0.3699	0.3719	0.3739	0.3759	0.3779	0.3799	0.3819
21	sin	0.3584	0.3600	0.3616	0.3633	0.3649	0.3665	0.3681	0.3697	0.3714	0.3730
	cos	0.9336	0.9330	0.9323	0.9317	0.9311	0.9304	0.9298	0.9291	0.9285	0.9278
	tan	0.3839	0.3859	0.3879	0.3899	0.3919	0.3939	0.3959	0.3979	0.4000	0.4020
22	sin	0.3746	0.3762	0.3778	0.3795	0.3811	0.3827	0.3843	0.3859	0.3875	0.3891
	cos	0.9272	0.9265	0.9259	0.9252	0.9245	0.9245	0.9239	0.9232	0.9225	0.9219
	tan	0.4040	0.4061	0.4081	0.4101	0.4122	0.4142	0.4263	0.4183	0.4204	0.4224
23	sin	0.3907	0.3923	0.3939	0.3955	0.3971	0.3987	0.4003	0.4019	0.4035	0.4051
	cos	0.9205	0.9198	0.9191	0.9184	0.9178	0.9171	0.9164	0.9157	0.9150	0.9143
	tan	0.4245	0.4265	0.4286	0.4307	0.4327	0.4348	0.4369	0.4390	0.4411	0.4431
24	sin	0.4067	0.4083	0.4099	0.4115	0.4131	0.4147	0.4163	0.4179	0.4195	0.4210
	cos	0.9135	0.9128	0.9121	0.9114	0.9107	0.9100	0.9092	0.9085	0.9078	0.9070
	tan	0.4452	0.4473	0.4494	0.4515	0.4536	0.4557	0.4578	0.4599	0.4621	0.4642
25	sin	0.4226	0.4242	0.4258	0.4274	0.4289	0.4305	0.4321	0.4337	0.4352	0.4368
	cos	0.9063	0.9056	0.9048	0.9041	0.9033	0.9026	0.9018	0.9011	0.9003	0.8996
	tan	0.4663	0.4684	0.4706	0.4727	0.4748	0.4770	0.4791	0.4813	0.4834	0.4856
26	sin	0.4384	0.4399	0.4415	0.4431	0.4446	0.4462	0.4478	0.4493	0.4509	0.4524
	cos	0.8988	0.8980	0.8973	0.8965	0.8957	0.8949	0.8942	0.8934	0.8926	0.8918
	tan	0.4877	0.4899	0.4921	0.4942	0.4964	0.4986	0.5008	0.5029	0.5051	0.5073
27	sin	0.4540	0.4555	0.4571	0.4586	0.4602	0.4617	0.4633	0.4648	0.4664	0.4679
	cos	0.8910	0.8902	0.8894	0.8886	0.8878	0.8870	0.8862	0.8854	0.8846	0.8838
	tan	0.5095	0.5117	0.5139	0.5161	0.5184	0.5206	0.5228	0.5250	0.5272	0.5295
28	sin	0.4695	0.4710	0.4726	0.4741	0.4756	0.4772	0.4787	0.4802	0.4818	0.4833
	cos	0.8829	0.8821	0.8813	0.8805	0.8796	0.8788	0.8780	0.8771	0.8763	0.8755
	tan	0.5317	0.5340	0.5362	0.5384	0.5407	0.5430	0.5452	0.5475	0.5498	0.5520
29	sin	0.4848	0.4863	0.4879	0.4894	0.4909	0.4924	0.4939	0.4955	0.4970	0.4985
	cos	0.8746	0.8738	0.8729	0.8721	0.8712	0.8704	0.8695	0.8686	0.8678	0.8669
	tan	0.5543	0.5566	0.5589	0.5612	0.5635	0.5658	0.5681	0.5704	0.5727	0.5750
30	sin	0.5000	0.5015	0.5030	0.5045	0.5060	0.5075	0.5090	0.5105	0.5120	0.5135
	cos	0.8660	0.8652	0.8643	0.8634	0.8625	0.8616	0.8607	0.8599	0.8590	0.8581
	tan	0.5774	0.5797	0.5820	0.5844	0.5967	0.5890	0.5914	0.5938	0.5961	0.5985
31	sin	0.5150	0.5165	0.5180	0.5195	0.5210	0.5225	0.5240	0.5255	0.5270	0.5284
	cos	0.8572	0.8563	0.8554	0.8545	0.8536	0.8526	0.8517	0.8508	0.8499	0.8490
	tan	0.6009	0.6032	0.6056	0.6080	0.6104	0.6128	0.6152	0.6176	0.6200	0.6224
32	sin	0.5299	0.5314	0.5329	0.5344	0.5358	0.5373	0.5388	0.5402	0.5417	0.5432
	cos	0.8480	0.8471	0.8462	0.8453	0.8443	0.8434	0.8425	0.8415	0.8406	0.8396
	tan	0.6249	0.6273	0.6297	0.6322	0.6346	0.6371	0.6395	0.6420	0.6445	0.6469
33	sin	0.5446	0.5461	0.5476	0.5490	0.5505	0.5519	0.5534	0.5548	0.5563	0.5577
	cos	0.8387	0.8377	0.8368	0.8358	0.8348	0.8339	0.8329	0.8320	0.8310	0.8300
	tan	0.6494	0.6519	0.6544	0.6569	0.6594	0.6619	0.6644	0.6669	0.6694	0.6720
34	sin	0.5592	0.5606	0.5621	0.5635	0.5650	0.5664	0.5678	0.5693	0.5707	0.5721
	cos	0.8290	0.8281	0.8271	0.8261	0.8251	0.8241	0.8231	0.8221	0.8211	0.8202
	tan	0.6745	0.6771	0.6796	0.6822	0.6847	0.6873	0.6899	0.6924	0.6950	0.6976
35	sin	0.5736	0.5750	0.5764	0.5779	0.5793	0.5807	0.5821	0.5835	0.5850	0.5864
	cos	0.8192	0.8181	0.8171	0.8161	0.8151	0.8141	0.8131	0.8121	0.8111	0.8100
	tan	0.7002	0.7028	0.7054	0.7080	0.7107	0.7133	0.7159	0.7186	0.7212	0.7239
36	sin	0.5878	0.5892	0.5906	0.5920	0.5934	0.5948	0.5962	0.5976	0.5990	0.6004
	cos	0.8090	0.8080	0.8070	0.8059	0.8049	0.8039	0.8028	0.8018	0.8007	0.7997
	tan	0.7265	0.7292	0.7319	0.7346	0.7373	0.7400	0.7427	0.7454	0.7481	0.7508
37	sin	0.6018	0.6032	0.6046	0.6060	0.6074	0.6088	0.6101	0.6115	0.6129	0.6143
	cos	0.7986	0.7976	0.7965	0.7955	0.7944	0.7934	0.7923	0.7912	0.7902	0.7891
	tan	0.7536	0.7563	0.7590	0.7618	0.7646	0.7673	0.7701	0.7729	0.7757	0.7785
38	sin	0.6157	0.6170	0.6184	0.6198	0.6211	0.6225	0.6239	0.6252	0.6266	0.6280
	cos	0.7880	0.7869	0.7859	0.7848	0.7837	0.7826	0.7815	0.7804	0.7793	0.7782
	tan	0.7813	0.7841	0.7869	0.7898	0.7926	0.7954	0.7983	0.8012	0.8040	0.8069
39	sin	0.6293	0.6307	0.6320	0.6334	0.6347	0.6361	0.6374	0.6388	0.6401	0.6414
	cos	0.7771	0.7760	0.7749	0.7738	0.7727	0.7716	0.7705	0.7694	0.7683	0.7672
	tan	0.8098	0.8127	0.8156	0.8185	0.8214	0.8243	0.8273	0.8302	0.8332	0.8361
40	sin	0.6428	0.6441	0.6455	0.6468	0.6481	0.6494	0.6508	0.6521	0.6534	0.6547
	cos	0.7660	0.7649	0.7638	0.7627	0.7615	0.7604	0.7593	0.7581	0.7570	0.7559
	tan	0.8391	0.8421	0.8451	0.8481	0.8511	0.8541	0.8571	0.8601	0.8632	0.8662
Degrees	Function	0′	6′	12′	18′	24′	30′	36′	42′	48′	54′

Natural Trigonometric Functions, Continued

Degrees	Function	0.0°	0.1°	0.2°	0.3°	0.4°	0.5°	0.6°	0.7°	0.8°	0.9°
41	sin	0.6561	0.6574	0.6587	0.6600	0.6613	0.6626	0.6639	0.6652	0.6665	0.6678
	cos	0.7547	0.7536	0.7524	0.7513	0.7501	0.7490	0.7478	0.7466	0.7455	0.7443
	tan	0.8693	0.8724	0.8754	0.8785	0.8816	0.8847	0.8878	0.8910	0.8941	0.8972
42	sin	0.6691	0.6704	0.6717	0.6730	0.6743	0.6756	0.6769	0.6782	0.6794	0.6807
	cos	0.7431	0.7420	0.7408	0.7396	0.7385	0.7373	0.7361	0.7349	0.7337	0.7325
	tan	0.9004	0.9036	0.9067	0.9099	0.9131	0.9163	0.9195	0.9228	0.9260	0.9293
43	sin	0.6820	0.6833	0.6845	0.6858	0.6871	0.6884	0.6896	0.6909	0.6921	0.6934
	cos	0.7314	0.7302	0.7290	0.7278	0.7266	0.7254	0.7242	0.7230	0.7218	0.7206
	tan	0.9325	0.9358	0.9391	0.9424	0.9457	0.9490	0.9523	0.9556	0.9590	0.9623
44	sin	0.6947	0.6959	0.6972	0.6984	0.6997	0.7009	0.7022	0.7034	0.7046	0.7059
	cos	0.7193	0.7181	0.7169	0.7157	0.7145	0.7133	0.7120	0.7108	0.7096	0.7083
	tan	0.9657	0.9691	0.9725	0.9759	0.9793	0.9827	0.9861	0.9896	0.9930	0.9965
45	sin	0.7071	0.7083	0.7096	0.7108	0.7120	0.7133	0.7145	0.7157	0.7169	0.7181
	cos	0.7071	0.7059	0.7046	0.7034	0.7022	0.7009	0.6997	0.6984	0.6972	0.6959
	tan	1.0000	1.0035	1.0070	1.0105	1.0141	1.0176	1.0212	1.0247	1.0283	1.0319
46	sin	0.7193	0.7206	0.7218	0.7230	0.7242	0.7254	0.7266	0.7278	0.7290	0.7302
	cos	0.6947	0.6934	0.6921	0.6909	0.6896	0.6884	0.6871	0.6858	0.6845	0.6833
	tan	1.0355	1.0392	1.0428	1.0464	1.0501	1.0538	1.0575	1.0612	1.0649	1.0686
47	sin	0.7314	0.7325	0.7337	0.7349	0.7361	0.7373	0.7385	0.7396	0.7408	0.7420
	cos	0.6820	0.6807	0.6794	0.6782	0.6769	0.6756	0.6743	0.6730	0.6717	0.6704
	tan	1.0724	1.0761	1.0799	1.0837	1.0875	1.0913	1.0951	1.0990	1.1028	1.1067
48	sin	0.7431	0.7443	0.7455	0.7466	0.7478	0.7490	0.7501	0.7513	0.7524	0.7536
	cos	0.6691	0.6678	0.6665	0.6652	0.6639	0.6626	0.6613	0.6600	0.6587	0.6574
	tan	1.1106	1.1145	1.1184	1.1224	1.1263	1.1303	1.1343	1.1383	1.1423	1.1463
49	sin	0.7547	0.7559	0.7570	0.7581	0.7593	0.7604	0.7615	0.7627	0.7638	0.7649
	cos	0.6561	0.6547	0.6534	0.6521	0.6508	0.6494	0.6481	0.6468	0.6455	0.6441
	tan	1.1504	1.1544	1.1585	1.1626	1.1667	1.1708	1.1750	1.1792	1.1833	1.1875
50	sin	0.7660	0.7672	0.7683	0.7694	0.7705	0.7716	0.7727	0.7738	0.7749	0.7760
	cos	0.6428	0.6414	0.6401	0.6388	0.6374	0.6361	0.6347	0.6334	0.6320	0.6307
	tan	1.1918	1.1960	1.2002	1.2045	1.2088	1.2131	1.2174	1.2218	1.2261	1.2305
51	sin	0.7771	0.7782	0.7793	0.7804	0.7815	0.7826	0.7837	0.7848	0.7859	0.7869
	cos	0.6293	0.6280	0.6266	0.6252	0.6239	0.6225	0.6211	0.6198	0.6184	0.6170
	tan	1.2349	1.2393	1.2437	1.2482	1.2527	1.2572	1.2617	1.2662	1.2708	1.2753
52	sin	0.7880	0.7891	0.7902	0.7912	0.7923	0.7934	0.7944	0.7955	0.7965	0.7976
	cos	0.6157	0.6143	0.6129	0.6115	0.6101	0.6088	0.6074	0.6060	0.6046	0.6032
	tan	1.2799	1.2846	1.2892	1.2938	1.2985	1.3032	1.3079	1.3127	1.3175	1.3222
53	sin	0.7986	0.7997	0.8007	0.8018	0.8028	0.8039	0.8049	0.8059	0.8070	0.8080
	cos	0.6018	0.6004	0.5990	0.5976	0.5962	0.5948	0.5934	0.5920	0.5906	0.5892
	tan	1.3270	1.3319	1.3367	1.3416	1.3465	1.3514	1.3564	1.3613	1.3663	1.3713
54	sin	0.8090	0.8100	0.8111	0.8121	0.8131	0.8141	0.8151	0.8161	0.8171	0.8181
	cos	0.5878	0.5864	0.5850	0.5835	0.5821	0.5807	0.5793	0.5779	0.5764	0.5750
	tan	1.3764	1.3814	1.3865	1.3916	1.3968	1.4019	1.4071	1.4124	1.4176	1.4229
55	sin	0.8192	0.8202	0.8211	0.8221	0.8231	0.8241	0.8251	0.8261	0.8271	0.8281
	cos	0.5736	0.5721	0.5707	0.5693	0.5678	0.5664	0.5650	0.5635	0.5621	0.5606
	tan	1.4281	1.4335	1.4388	1.4442	1.4496	1.4550	1.4605	1.4659	1.4715	1.4770
56	sin	0.8290	0.8300	0.8310	0.8320	0.8329	0.8339	0.8348	0.8358	0.8368	0.8377
	cos	0.5592	0.5577	0.5563	0.5548	0.5534	0.5519	0.5505	0.5490	0.5476	0.5461
	tan	1.4826	1.4882	1.4938	1.4994	1.5051	1.5108	1.5166	1.5224	1.5282	1.5340
57	sin	0.8387	0.8396	0.8406	0.8415	0.8425	0.8434	0.8443	0.8453	0.8462	0.8471
	cos	0.5446	0.5432	0.5417	0.5402	0.5388	0.5373	0.5358	0.5344	0.5329	0.5314
	tan	1.5399	1.5458	1.5517	1.5577	1.5637	1.5697	1.5757	1.5818	1.5880	1.5941
58	sin	0.8480	0.8490	0.8499	0.8508	0.8517	0.8526	0.8536	0.8545	0.8554	0.8563
	cos	0.5299	0.5284	0.5270	0.5255	0.5240	0.5225	0.5210	0.5195	0.5180	0.5165
	tan	1.6003	1.6066	1.6128	1.6191	1.6255	1.6319	1.6383	1.6447	1.6512	1.6577
59	sin	0.8572	0.8581	0.8590	0.8599	0.8607	0.8616	0.8625	0.8634	0.8643	0.8652
	cos	0.5150	0.5135	0.5120	0.5105	0.5090	0.5075	0.5060	0.5045	0.5030	0.5015
	tan	1.6643	1.6709	1.6775	1.6842	1.6909	1.6977	1.7045	1.7113	1.7182	1.7251
60	sin	0.8660	0.8669	0.8678	0.8686	0.8695	0.8704	0.8712	0.8721	0.8729	0.8738
	cos	0.5000	0.4985	0.4970	0.4955	0.4939	0.4924	0.4909	0.4894	0.4879	0.4863
	tan	1.7321	1.7391	1.7461	1.7532	1.7603	1.7675	1.7747	1.7820	1.7893	1.7966
61	sin	0.8746	0.8755	0.8763	0.8771	0.8780	0.8788	0.8796	0.8805	0.8813	0.8821
	cos	0.4848	0.4833	0.4818	0.4802	0.4787	0.4772	0.4756	0.4741	0.4726	0.4710
	tan	1.8040	1.8115	1.8190	1.8265	1.8341	1.8418	1.8495	1.8572	1.8650	1.8728

| Degrees | Function | 0′ | 6′ | 12′ | 18′ | 24′ | 30′ | 36′ | 42′ | 48′ | 54′ |

Degrees	Function	0.0°	0.1°	0.2°	0.3°	0.4°	0.5°	0.6°	0.7°	0.8°	0.9°
62	sin	0.8829	0.8838	0.8846	0.8854	0.8862	0.8870	0.8878	0.8886	0.8894	0.8902
	cos	0.4695	0.4679	0.4664	0.4648	0.4633	0.4617	0.4602	0.4586	0.4571	0.4555
	tan	1.8807	1.8887	1.8967	1.9047	1.9128	1.9210	1.9292	1.9375	1.9458	1.9542
63	sin	0.8910	0.8918	0.8926	0.8934	0.8942	0.8949	0.8957	0.8965	0.8973	0.8980
	cos	0.4540	0.4524	0.4509	0.4493	0.4478	0.4462	0.4446	0.4431	0.4415	0.4399
	tan	1.9626	1.9711	1.9797	1.9883	1.9970	2.0057	2.0145	2.0233	2.0323	2.0413
64	sin	0.8988	0.8996	0.9003	0.9011	0.9018	0.9026	0.9033	0.9041	0.9048	0.9056
	cos	0.4384	0.4368	0.4352	0.4337	0.4321	0.4305	0.4289	0.4274	0.4258	0.4242
	tan	2.0503	2.0594	2.0686	2.0778	2.0872	2.0965	2.1060	2.1155	2.1251	2.1348
65	sin	0.9063	0.9070	0.9078	0.9085	0.9092	0.9100	0.9107	0.9114	0.9121	0.9128
	cos	0.4226	0.4210	0.4195	0.4179	0.4163	0.4147	0.4131	0.4115	0.4099	0.4083
	tan	2.1445	2.1543	2.1642	2.1742	2.1842	2.1943	2.2045	2.2148	2.2251	2.2355
66	sin	0.9135	0.9143	0.9150	0.9157	0.9164	0.9171	0.9178	0.9184	0.9191	0.9198
	cos	0.4067	0.4051	0.4035	0.4019	0.4003	0.3987	0.3971	0.3955	0.3939	0.3923
	tan	2.2460	2.2566	2.2673	2.2781	2.2889	2.2998	2.3109	2.3220	2.3332	2.3445
67	sin	0.9205	0.9212	0.9219	0.9225	0.9232	0.9239	0.9245	0.9252	0.9259	0.9265
	cos	0.3907	0.3891	0.3875	0.3859	0.3843	0.3827	0.3811	0.3795	0.3778	0.3762
	tan	2.3559	2.3673	2.3789	2.3906	2.4023	2.4142	2.4262	2.4383	2.4504	2.4627
68	sin	0.9272	0.9278	0.9285	0.9291	0.9298	0.9304	0.9311	0.9317	0.9323	0.9330
	cos	0.3746	0.3730	0.3714	0.3697	0.3681	0.3665	0.3649	0.3633	0.3616	0.3600
	tan	2.4751	2.4876	2.5002	2.5129	2.5257	2.5386	2.5517	2.5649	2.5782	2.5916
69	sin	0.9336	0.9342	0.9348	0.9354	0.9361	0.9367	0.9373	0.9379	0.9385	0.9391
	cos	0.3584	0.3567	0.3551	0.3535	0.3518	0.3502	0.3486	0.3469	0.3453	0.3437
	tan	2.6051	2.6187	2.6325	2.6464	2.6605	2.6746	2.6889	2.7034	2.7179	2.7326
70	sin	0.9397	0.9403	0.9409	0.9415	0.9421	0.9426	0.9432	0.9438	0.9444	0.9449
	cos	0.3420	0.3404	0.3387	0.3371	0.3355	0.3338	0.3322	0.3305	0.3289	0.3272
	tan	2.7475	2.7625	2.7776	2.7929	2.8083	2.8239	2.8397	2.8556	2.8716	2.8878
71	sin	0.9455	0.9461	0.9466	0.9472	0.9478	0.9483	0.9489	0.9494	0.9500	0.9505
	cos	0.3256	0.3239	0.3223	0.3206	0.3190	0.3173	0.3156	0.3140	0.3123	0.3107
	tan	2.9042	2.9208	2.9375	2.9544	2.9714	2.9887	3.0061	3.0237	3.0415	3.0595
72	sin	0.9511	0.9516	0.9521	0.9527	0.9532	0.9537	0.9542	0.9548	0.9553	0.9558
	cos	0.3090	0.3074	0.3057	0.3040	0.3024	0.3007	0.2990	0.2974	0.2957	0.2940
	tan	3.0777	3.0961	3.1146	3.1334	3.1524	3.1716	3.1910	3.2106	3.2305	3.2506
73	sin	0.9563	0.9568	0.9573	0.9578	0.9583	0.9588	0.9593	0.9598	0.9603	0.9608
	cos	0.2924	0.2907	0.2890	0.2874	0.2857	0.2840	0.2823	0.2807	0.2790	0.2773
	tan	3.2709	3.2914	3.3122	3.3332	3.3544	3.3759	3.3977	3.4197	3.4420	3.4646
74	sin	0.9613	0.9617	0.9622	0.9627	0.9632	0.9636	0.9641	0.9646	0.9650	0.9655
	cos	0.2756	0.2740	0.2723	0.2706	0.2689	0.2672	0.2656	0.2639	0.2622	0.2605
	tan	3.4874	3.5105	3.5339	3.5576	3.5816	3.6059	3.6305	3.6554	3.6806	3.7062
75	sin	0.9659	0.9664	0.9668	0.9673	0.9677	0.9681	0.9686	0.9690	0.9694	0.9699
	cos	0.2588	0.2571	0.2554	0.2538	0.2521	0.2504	0.2487	0.2470	0.2453	0.2436
	tan	3.7321	3.7583	3.7848	3.8118	3.8391	3.8667	3.8947	3.9232	3.9520	3.9812
76	sin	0.9703	0.9707	0.9711	0.9715	0.9720	0.9724	0.9728	0.9732	0.9736	0.9740
	cos	0.2419	0.2402	0.2385	0.2368	0.2351	0.2334	0.2317	0.2300	0.2284	0.2267
	tan	4.0108	4.0408	4.0713	4.1022	4.1335	4.1653	4.1976	4.2303	4.2635	4.2972
77	sin	0.9744	0.9748	0.9751	0.9755	0.9759	0.9763	0.9767	0.9770	0.9774	0.9778
	cos	0.2250	0.2232	0.2215	0.2198	0.2181	0.2164	0.2147	0.2130	0.2113	0.2096
	tan	4.3315	4.3662	4.4015	4.4374	4.4737	4.5107	4.5483	4.5864	4.6252	4.6646
78	sin	0.9781	0.9785	0.9789	0.9792	0.9796	0.9799	0.9803	0.9806	0.9810	0.9813
	cos	0.2079	0.2062	0.2045	0.2028	0.2011	0.1994	0.1977	0.1959	0.1942	0.1925
	tan	4.7046	4.7453	4.7867	4.8288	4.8716	4.9152	4.9594	5.0045	5.0504	5.0970
79	sin	0.9816	0.9820	0.9823	0.9826	0.9829	0.9833	0.9836	0.9839	0.9842	0.9845
	cos	0.1908	0.1891	0.1874	0.1857	0.1840	0.1822	0.1805	0.1788	0.1771	0.1754
	tan	5.1446	5.1929	5.2422	5.2924	5.3435	5.3955	5.4486	5.5026	5.5578	5.6140
80	sin	0.9848	0.9851	0.9854	0.9857	0.9860	0.9863	0.9866	0.9869	0.9871	0.9874
	cos	0.1736	0.1719	0.1702	0.1685	0.1668	0.1650	0.1633	0.1616	0.1599	0.1582
	tan	5.6713	5.7297	5.7894	5.8502	5.9124	5.9758	6.0405	6.1066	6.1742	6.2432
81	sin	0.9877	0.9880	0.9882	0.9885	0.9888	0.9890	0.9893	0.9895	0.9898	0.9900
	cos	0.1564	0.1547	0.1530	0.1513	0.1495	0.1478	0.1461	0.1444	0.1426	0.1409
	tan	6.3138	6.3859	6.4596	6.5350	6.6122	6.6912	6.7720	6.8548	6.9395	7.0264
82	sin	0.9903	0.9905	0.9907	0.9910	0.9912	0.9914	0.9917	0.9919	0.9921	0.9923
	cos	0.1392	0.1374	0.1357	0.1340	0.1323	0.1305	0.1288	0.1271	0.1253	0.1236
	tan	7.1154	7.2066	7.3002	7.3962	7.4947	7.5958	7.6996	7.8062	7.9158	8.0285
Degrees	Function	0′	6′	12′	18′	24′	30′	36′	42′	48′	54′

Degrees	Function	0.0°	0.1°	0.2°	0.3°	0.4°	0.5°	0.6°	0.7°	0.8°	0.9°
83	sin	0.9925	0.9928	0.9930	0.9932	0.9934	0.9936	0.9938	0.9940	0.9942	0.9943
	cos	0.1219	0.1201	0.1184	0.1167	0.1149	0.1132	0.1115	0.1097	0.1080	0.1063
	tan	8.1443	8.2636	8.3863	8.5126	8.6427	8.7769	8.9152	9.0579	9.2052	9.3572
84	sin	0.9945	0.9947	0.9949	0.9951	0.9952	0.9954	0.9956	0.9957	0.9959	0.9960
	cos	0.1045	0.1028	0.1011	0.0993	0.0976	0.0958	0.0941	0.0924	0.0906	0.0889
	tan	9.5144	9.6768	9.8448	10.02	10.20	10.39	10.58	10.78	10.99	11.20
85	sin	0.9962	0.9963	0.9965	0.9966	0.9968	0.9969	0.9971	0.9972	0.9973	0.9974
	cos	0.0872	0.0854	0.0837	0.0819	0.0802	0.0785	0.0767	0.0750	0.0732	0.0715
	tan	11.43	11.66	11.91	12.16	12.43	12.71	13.00	13.30	13.62	13.95
86	sin	0.9976	0.9977	0.9978	0.9979	0.9980	0.9981	0.9982	0.9983	0.9984	0.9985
	cos	0.0698	0.0680	0.0663	0.0645	0.0628	0.0610	0.0593	0.0576	0.0558	0.0541
	tan	14.30	14.67	15.06	15.46	15.89	16.35	16.83	17.34	17.89	18.46
87	sin	0.9986	0.9987	0.9988	0.9989	0.9990	0.9990	0.9991	0.9992	0.9993	0.9993
	cos	0.0523	0.0506	0.0488	0.0471	0.0454	0.0436	0.0419	0.0401	0.0384	0.0366
	tan	19.08	19.74	20.45	21.20	22.02	22.90	23.86	24.90	26.03	27.27
88	sin	0.9994	0.9995	0.9995	0.9996	0.9996	0.9997	0.9997	0.9997	0.9998	0.9998
	cos	0.0349	0.0332	0.0314	0.0297	0.0279	0.0262	0.0244	0.0227	0.0209	0.0192
	tan	28.64	30.14	31.82	33.69	35.80	38.19	40.92	44.07	47.74	52.08
89	sin	0.9998	0.9999	0.9999	0.9999	0.9999	1.000	1.000	1.000	1.000	1.000
	cos	0.0175	0.0157	0.0140	0.0122	0.0105	0.0087	0.0070	0.0052	0.0035	0.0017
	tan	57.29	63.66	71.62	81.85	95.49	114.6	143.2	191.0	286.5	573.0
Degrees	Function	0′	6′	12′	18′	24′	30′	36′	42′	48′	54′

Symbols and Abbreviations

Term	Symbol	Abbreviation	Term	Symbol	Abbreviation
Alternating current		ac	Electromotive force	E	emf
Ampere (unit of current)	I	A	Emitter	E	
Milliampere		mA	Energy	W	
Microampere		μA	Farad (unit of capacitance)	C	F
American Wire Gage		AWG	Microfarad	C	μF
Apparent power	VA	VA	Picofarad	C	pF
Angle	$\angle$		Foot		ft
Area	A		Frequency	f	freq
Circular mils		cmil	Audio		AF
Square inches		in^2	Intermediate		IF
Base	B		High		HF
Candela		cd	Low		LF
Capacitance	C		Radio		RF
Collector	C		Resonant		f
Constant	K		Grid	G	
Continuous wave		cw	Ground		gnd
Cosine		cos	Henry (unit of inductance)	L	H
Coulomb	C	C	Millihenry	L	mH
Current	I		Microhenry	L	μH
Average value	I_{av}		Hertz		Hz
Change in current	ΔI		High pass		h-p
Effective value	I		Horsepower		hp
Instantaneous value	i		Hour	T	h
Maximum value	I_{max}		Impedance	Z	
Cycles			Inch		in
Cycles per second		Hz	Inductance	L	
Kilocycles per second		kHz	Kilo (1000)		k
Megacycles per second		MHz	Low pass		l-p
Decibel		dB	Maximum		max
Delta (Greek letter)	Δ		Mega (1,000,000)		M
Diameter	D, d	diam	Micro (one-millionth)		μ
Degree	$\circ$	deg	Milli (one-thousandth)		m
Diode	D		Minimum		min
Direct current		dc	Ohm (unit of resistance)	R	Ω
Efficiency		Eff	Pi (Greek letter)	π	

Term	Symbol	Abbreviation
Pico		p
Phase angle (theta)	θ	
Power	P	
Apparent power	VA	VA
Effective power	W	
Input power	P_i	
Output power	P_o	
Reactive power	VAR	var
Power factor		PF
Push-pull		pp
Quality	Q	
Radius	R, r	rad
Reactance	X	
Capacitive reactance	X_C	
Inductive reactance	X_L	
Resistance	R	
Root mean square		rms
Revolutions per minute		rpm
Second	T	s
Microsecond		μs
Sigma	Σ	
Sine		sin
Specific resistance	K	
Tangent		tan
Television		TV
Time	T	
Transistor	Q	
Dc base current	I_B	
Dc collector current	I_C	
Dc emitter current	I_E	
Alpha—the ratio of ΔI_C to ΔI_E	α	
Beta—the ratio of ΔI_C to ΔI_B	β	
Base bias resistor	R_B	
Collector-to-base resistance	R_{CB}	
Resistance in the collector leg; the load resistor	R_L	
Base-to-ground voltage	V_B	

Term	Symbol	Abbreviation
Base supply voltage	V_{BB}	
Base-to-emitter voltage drop	V_{BE}	
Collector-to-ground voltage	V_C	
Collector-to-base voltage	V_{CB}	
Collector supply voltage	V_{CC}	
Collector-to-emitter voltage	V_{CE}	
Emitter-to-ground voltage	V_E	
Emitter supply voltage	V_{EE}	
Turns ratio		TR
Vacuum tube		VT
Plate	P	
Plate current	I_p	
Plate resistance	R_p	
Plate voltage	E_p	
Screen current	I_S	
Volt	V	V
Kilovolt		kV
Millivolt		mV
Microvolt		μV
Voltage	V	
Average voltage	V_{av}	
Effective value	V	
Instantaneous value	ν	
Maximum value	V_{max}	
Voltampere	VA	VA
Kilovoltampere	KVA	kVA
Voltage ratio	VR	
Watt (unit of electric power)	P	W
Kilowatt		kW
Kilowatthour		kWh
Milliwatt		mW
Microwatt		μW
Watthour		Wh
Wavelength	λ	

Bibliography

Bartkiw, Walter L.: **Mathematics for Electricity and Electronics**, Glencoe/McGraw-Hill, Columbus, Ohio, 1996

Forster, Harry: **Mathematics for Electronics**, Glencoe/McGraw-Hill, Columbus, Ohio, 1997

Grob, Bernard: **Basic Electronics**, 8th ed., Glencoe/McGraw-Hill, Columbus, Ohio, 1997

Grob, Bernard: **Mathematics for Grob Basic Electronics**, 5th ed., Glencoe/McGraw-Hill, Columbus, Ohio, 1997

Malvino, Albert Paul: **Electronic Principles**, 6th ed., Glencoe/McGraw-Hill, Columbus, Ohio 1998

McIntyre, R. L., and Losee, Rex: **Industrial Motor Control Fundamentals**, 4th ed., McGraw-Hill, New York, 1990

Fowler, Richard J.: **Electricity: Principles and Applications**, 4th ed., Glencoe/McGraw-Hill, Columbus, Ohio, 1994

McPartland, Joseph, and McPartland, Brian J.: **National Electrical Code Handbook**, 23d ed., McGraw-Hill, New York, 1998

"National Electric Code 1993, National Fire Protection Association," Quincy, Maine

Quattro Pro 4.0 User's Manual, http://www.corel.com/support/suite8manuals/quattro, Borland International, Inc.

Answers to Selected Problems

Job 2-1: Page 14

1. 8
2. 8
3. 66
4. $\frac{1}{6}$
5. $\frac{1}{50}$
6. $\frac{1}{2}$
7. $45\frac{1}{3}$
8. 16
9. $25\frac{1}{2}$ ft
10. $3\frac{3}{4}$ hp
11. 90 V
12. 7 Ω
13. $2.81
14. 228 kWh
15. $37\frac{1}{2}$ h
16. $94\frac{1}{2}$ W
17. 36 lb
18. 20 ft 3 in

Job 2-2: Page 17

1. $\frac{1}{2}$
3. $\frac{1}{4}$
5. $\frac{3}{4}$
7. $\frac{1}{4}$
9. $\frac{1}{3}$
11. $\frac{3}{100}$
13. $\frac{3}{8}$
15. $\frac{7}{16}$
17. $\frac{1}{3}$
19. $\frac{3}{5}$
21. $\frac{2}{1}$
23. $\frac{49}{1}$
25. $\frac{2}{5}$
27. $\frac{3}{4}$

Job 2-2: Page 19

1. $\frac{3}{4}$
3. $\frac{20}{9}$

5. $\frac{13}{10}$
7. $\frac{25}{6}$
9. $\frac{37}{8}$
11. $\frac{10}{3}$

Job 2-2: Page 20

1. $1\frac{3}{4}$
3. $4\frac{1}{3}$
5. $3\frac{3}{4}$
7. $1\frac{3}{5}$
9. $3\frac{3}{10}$
11. 10
13. $9\frac{2}{3}$
15. 16
17. $3\frac{3}{5}$
19. 12
21. 7
23. $1\frac{7}{10}$
25. $5\frac{2}{9}$
27. 12
29. $12\frac{1}{4}$

Job 2-2: Page 22

1. 4
3. $\frac{3}{10}$
5. $2\frac{1}{2}$
7. $\frac{1}{15}$
9. 6
11. $5\frac{1}{4}$ hp
13. $\frac{1}{2}$ hp
15. 3200 W
17. $\frac{1}{3}$
19. $\frac{5}{7}$

Job 2-2: Page 23

1. $\frac{2}{3}$
3. $9\frac{1}{2}$
5. $\frac{1}{3}$
7. 5

9. $\frac{2}{3}$
11. $4\frac{1}{2}$ V
13. $7\frac{1}{8}$ h
15. $67\frac{1}{2}$ ft
17. $4\frac{1}{4}$ hp

Job 2-3: Page 27

1. 55.8 V
3. 44 V
5. 117.6 V
7. 3.72 V
9. 18.75 V
11. 0.05 V
13. $V_1 = 150$ V
$V_2 = 100$ V
$V_3 = 50$ V

Job 2-4: Page 29

1. 5.55 A
2. 0.015 25 μF
3. 0.0378 in
4. 0.35 A
5. 2.6 V
6. 1.185 in
7. 65.75 V
8. $0.60
9. 0.004 Ω
10. 12.55
11. 0.4 A
12. (a) Four and three tenths
(b) Three hundred fifty-nine thousandths
(c) Forty-one hundredths
13. 0.015
14. (a) 0.25
(b) 0.571
(c) 0.625

(d) 0.0625
15. 6.24
16. 1.1, 0.30, 0.050, 0.0070
17. 0.05
18. 2.25

Job 2-5: Page 32 (top)

1. 0.7
3. 0.114
5. 0.06
7. 0.018
9. 0.11
11. 0.013
13. 0.045
15. 0.6
17. 0.4, 0.16, 0.007
19. 0.8, 0.06, 0.040
21. 0.5, 0.18, 0.051
23. 0.4, 0.236, 0.1228
25. 0.19, 0.08, 0.004

Job 2-5: Page 32 (bottom)

1. 2.3
3. 3.144
5. 2.025
7. 2.020
9. 1.002
11. 9.145

Job 2-5: Page 35

1. 0.25
3. 0.625
5. 0.4
7. 0.15
9. 0.875
11. 0.444
13. 0.094
15. 0.625
17. 0.02

19. 0.192
21. 0.375 A

Job 2-5: Page 36

1. 0.625
3. 0.5625
5. 2.25
7. 1.140625
9. ⅜
11. ²⁹⁄₆₄
13. ¹⁄₃₂
15. ¹³⁄₃₂
17. ¾
19. ⁹⁄₆₄
21. 2¹³⁄₁₆
23. 1²⁷⁄₃₂
25. The lamp

Job 2-6: Page 39

1. 7.2
3. 3090
5. 3700
7. 0.6
9. 2700
11. 0.078
13. 15,400
15. 0.5
17. 80
19. 2340

Job 2-6: Page 40

1. 6.5
3. 880
5. 0.06
7. 0.835
9. 0.6538
11. 0.000 45
13. 0.0085
15. 0.0286
17. 0.002
19. 0.0398

Job 2-6: Page 43

1. 350 mm
3. 24 cm

5. 360 dm
7. 500 cm
9. 2450 mm
11. 12.5 cm
13. 42 dam
15. 750 hm
17. 390 m
19. 20,000 m
21. 150 dm
23. 345.6 mm

Job 2-7: Page 44

1. 13.492
3. 22.832
5. 11.967
7. 45.46 V
9. 0.1019 in
11. 0.0315 A
13. 25.38 mm
15. $44.90

Job 2-7: Page 45

1. 0.23
3. 0.34
5. 7.56
7. 0.317
9. 2.92
11. 5.9
13. 2.39
15. 0.262
17. 2.62
19. 0.515
21. (a) 0.304
    (b) 1.106
    (c) 0.39
    (d) 5.75
    (e) 1.98
23. 0.1446 in
25. 0.43 mm
27. 0.000 15 μF

Job 2-7: Page 47

1. 0.41
3. 0.463
5. 30.31
7. 0.221

9. 0.0186
11. 4.77
13. 4.55 A
15. (a) 250 mA
    (b) 25 mA
    (c) 2500 mA
17. 41.850 in
19. 69.6 A · h

Job 2-7: Page 49

1. 13
3. 161
5. 4.7
7. 786
9. 2.32
11. 200 ft
13. 60.7
15. 40

Job 2-7: Page 50

1. $A = 3.09$ in
   $B = 2.8475$ in
3. 0.082
5. (a) 735
   (b) 81,700,000
   (c) 3.5
   (d) 0.4
   (e) 0.0007
   (f) 0.006 28
   (g) 47.5
   (h) 0.000 397
7. $43.31
9. 22.62 mm
11. 1.875 A
13. 163.9 W
15. No. 10

Job 3-1: Page 54

1. 0.6 A
2. 300 Ω
3. 111 V
4. 4 A
5. $kW = \dfrac{I \times V}{1000}$
6. 2 A
7. 25

8. 0.125 A
9. 11 A
10. 100 V

Job 3-2: Page 58

1. $P = IV$
3. $Eff = \dfrac{P_o}{P_i}$
5. $X_C = \dfrac{159,000}{fC}$
7. $P = 2L + 2W$
9. $I_T = \dfrac{V_T}{R_1 + R_2}$

Job 3-2: Page 61

1. 14
3. 20
5. 18
7. 6¼
9. 7
11. 135 ft²
13. 10 A
15. 140.4 Ω
17. 20 Ω
19. ¼ A
21. 72.6 Ω

Job 3-3: Page 62

1. −$3
3. +23°
5. +2 dB
7. −10 mph
9. −5 blocks
11. −4 V

Job 3-3: Page 65

1. +12
3. +6
5. −17
7. +7
9. −17
11. −14
13. +16
15. −52
17. +1.9

19. $+\frac{1}{4}$
21. $-\frac{5}{6}$
23. $-5\frac{11}{16}$
25. $+3$
27. $+7$
29. -25
31. $+0.5$
33. -2.9

Job 3-3:
Page 66
(Multiplication)

1. $+18$
3. -8
5. $+30$
7. $+18R$
9. $-10R$
11. $+10R$
13. 0
15. -104
17. $-85R$
19. $-144R$
21. -6
23. $-2.4I$
25. $-3R$
27. -14.72
29. $-5.1R$

Job 3-3:
Page 66
(Division)

1. $+4$
3. -4
5. $+12$
7. $-6R$
9. $-\frac{1}{2}$
11. $+\frac{1}{2}$
13. 0
15. -7
17. $-7\frac{1}{2}$
19. -3
21. $+3\frac{1}{5}$
23. $+130$
25. -2
27. -3.2
29. -62.5

Job 3-4:
Page 67

1. 200
2. 360
3. 20 V
4. 65 Ω
5. 0.000 000 5 F

Job 3-5:
Page 69

1. 3600 cmil
3. 1024 ft
5. 0.45 A
7. 0.032 V
9. 132.25 W

Job 3-6:
Page 70

1. 7.2
3. 3090
5. 3700
7. 0.6
9. 2700
11. 0.078
13. 15,400
15. 1
17. 80
19. 234,000

Job 3-6:
Page 71

1. 65
3. 880
5. 0.06
7. 0.835
9. 0.6538
11. 0.0045
13. 0.0085
15. 0.0286
17. 0.000 02

Job 3-6:
Page 72

1. 25
3. 0.15
5. 6.25
7. 0.0025
9. 0.007 54

Job 3-6:
Page 73

1. 1920
3. 850
5. 7200
7. 0.000 045
9. 88,000,000
11. 3000
13. 3,000,000
15. 0.0006

Job 3-6:
Page 74

1. 6×10^3
3. 1.5×10^5
5. 2.35×10^5
7. 4.96×10^3
9. 9.8×10^2
11. 1.25×10
13. 4.82×10
15. 8.8×10^8
17. 3.83×10^4
19. 1.75×10^6
21. 4.83×10^5

Job 3-6:
Page 75

1. 6×10^{-3}
3. 3.5×10^{-3}
5. 4.56×10^{-1}
7. 7.85
9. 9.65×10^{-2}
11. 5×10^{-1}
13. 8.15×10^{-3}
15. 7.25×10
17. 6×10^{-1}
19. 3.6×10^{-2}

Job 3-6:
Page 76

1. 5
3. 6,400,000 or
   6.4×10^6
5. 30,000 or 3×10^4
7. 120
9. 0.03
11. 5×10^{-7}
13. 960

15. 628,000 or
    6.28×10^5

Job 3-6:
Page 78

1. 10^5
3. 120
5. 0.0002
7. 2×10^{-7}
9. 8
11. 5×10^{-3}
13. (a) 53 Ω
    (b) 63 Ω
    (c) 3.18 Ω

Job 3-7:
Page 82

1. 0.225 A
3. 3,500,000 Ω
5. 550,000 Hz
7. 0.07 MΩ
9. 65 mA
11. 0.075 V
13. 0.006 A
15. 0.0039 A
17. 5000 pF
19. 1,000,000 Hz
21. 8 mV
23. 60 pF
25. 0.000 000 15 F
27. 8 kW
29. 4 μA
31. 500,000 MHz
33. 0.015 GV

Job 3-8:
Page 83

1. 100 μV or 0.1 mV
3. 0.6 kW
5. 0.5 μA
7. 215 kΩ
9. 2625 pF
11. 4×10^{-5}
13. 14.1 V

Job 3-9:
Page 88

1. 5
3. 6

5. 9
7. 13⅔
9. 5.85
11. 40
13. 400
15. 200
17. 800
19. 390
21. ¹⁄₁₆

Job 3-9:
Page 90

1. $2\,\Omega$
3. $13\,\Omega$
5. $1\frac{1}{2}$ A
7. 20 A
9. $25\,\Omega$
11. $0.133\,\Omega$
13. $1200\,\Omega$
15. 2.25 A
17. 3.67 A
19. $0.0015\,\Omega$
21. $1.76\,\Omega$
23. 0.002 A
25. $21.43\,\Omega$

Job 3-11:
Page 94

1. 110 V
3. Yes
5. 18 V
7. $0.0007\,\Omega$
9. $10,000\,\Omega$
11. $0.004\,\Omega$
13. 10 A
15. 0.03 A
17. 0.0081 A

Job 4-1:
Page 102

1. (a) 24 V
   (b) 0.2 A
   (c) $120\,\Omega$
3. (a) 42.9 V
   (b) 0.3 A
   (c) $143\,\Omega$

Job 4-2:
Page 105

1. (a) 4 A
   (b) 74 V
   (c) $18.5\,\Omega$
3. (a) Dash = $2.5\,\Omega$,
      tail = $5\,\Omega$
   (b) $7.5\,\Omega$
   (c) 0.8 A
5. 0.0088 A
7. (a) 4.31 V
   (b) 7.7 V
   (c) 12.01 V

Job 4-3:
Page 111

1. 550 V
3. 4.8 V
5. 0.012 A
7. (a) 0.8 A
   (b) 1.6 V
   (c) $2\,\Omega$
9. 157.5 V

Job 4-6:
Page 118

1. 54 V
3. 6 V
5. (a) 45, 75, 150 V
   (b) 270 V
   (c) $18,000\,\Omega$
7. (a) 1.2 A
   (b) 132 V
   (c) $110\,\Omega$
9. $V_T = 32$ V
   $I_T = 0.4$ A
   $I_1 = I_2 = I_3$
      $= 0.4$ A
   $R_1 = 20\,\Omega$
   $R_2 = 25\,\Omega$
   $R_3 = 35\,\Omega$

Job 4-7:
Page 119

1. 29
3. 1.1 A
5. 7
7. 12
9. $25\,\Omega$

Job 4-8:
Page 121

1. 6
3. 70
5. 52
7. 25
9. 7.3
11. 28.2
13. 5¼
15. 6.1
17. 56.1
19. 120
21. 73
23. 0.000 08
25. 1.45

Job 4-8:
Page 123

1. 17
3. 22
5. 5
7. 12.5
9. 79
11. 10¾
13. 15¼
15. 80
17. 0.099
19. $21.83
21. 172
23. 17.05
25. 109

Job 4-8:
Page 126

1. 11
3. 6
5. 10
7. 12
9. 8
11. 1.54
13. 10
15. 11.5
17. 0.021
19. 11.48
21. 1½
23. 180
25. 30
27. 76.5
29. 0.000 25

Job 4-9:
Page 130

1. $25\,\Omega$
3. $8\,\Omega$
5. $84\,\Omega$
7. $90\,\Omega$
9. 202 V
11. $I_{max} = 11$ A
    $I_{min} = 2$ A
13. $0.025\,\Omega$

Job 4-10:
Page 132

1. $I_T = I_1 = I_3$
      $= 2$ A
   $V_T = 110$ V
   $R_T = 55\,\Omega$
   $R_1 = 12\,\Omega$
   $V_2 = 60$ V
   $V_3 = 26$ V
3. $12\,\Omega$
5. $28\,\Omega$, 2.67 A, 74.8 V
7. (a) 2.2 A
   (b) $10\,\Omega$
9. $30\,\Omega$

Job 5-1:
Page 140

1. (a) 5 A
   (b) $V_1 = V_2 = V_3$
         $= 110$ V
   (c) $22\,\Omega$
3. 4 A
5. 40 lamps; 30 lamps
7. 9.1 A; yes
9. 9.5 A, $1.26\,\Omega$

Job 5-2:
Page 142

1. (a) 18 V
   (b) $I_1 = 6$ A,
      $I_2 = 3$ A,
      $I_T = 9$ A
   (c) $2\,\Omega$
3. 2 A, 6 A; $I_T = 12$ A
5. (a) $V_T = V_1 = V_2$
         $= V_3 = 114$ V
   (b) $I_2 = 6, I_3 = 2$,
      $I_T = 20$ A

(c) $R_1 = 9.5\ \Omega$
  $R_T = 5.7\ \Omega$
7. (a) $30\ \Omega$ for 3 lamps;
    $15\ \Omega$ for fourth
    lamp
  (b) 0.25 A
  (c) $1\frac{1}{2}$ V
  (d) $6\ \Omega$
9. $I_T = 1.05$ A
  $V_T = 6.8$ V
  $R_T = 6.47\ \Omega$

Job 5-3:
Page 144

1. $\frac{17}{24}$ hp
3. $\frac{9}{64}$ in
5. $22\frac{7}{8}$ lb
7. $1\frac{13}{16}$
9. $1\frac{11}{18}$
11. $2\frac{1}{8}$
13. $4\frac{1}{4}$
15. $\frac{1}{60}$

Job 5-4:
Page 148

1. $1\frac{3}{8}$
3. $1\frac{11}{24}$
5. $\frac{1}{6}$
7. $\frac{2}{5}$
9. $\frac{1}{120}$
11. $\frac{57}{64}$
13. $\frac{3}{20}$
15. $1\frac{11}{32}$
17. $15\frac{7}{16}$
19. $\frac{15}{16}$
21. $\frac{11}{600}$ S
23. $29\frac{23}{24}$ ft

Job 5-5:
Page 152

1. $\frac{1}{2}$
3. $\frac{3}{8}$
5. $\frac{5}{16}$
7. $\frac{1}{12}$
9. $\frac{1}{6}$
11. $\frac{1}{12}$
13. $\frac{1}{80}$
15. $\frac{1}{500}$
17. $4\frac{3}{4}$

19. $7\frac{1}{3}$
21. $5\frac{7}{16}$
23. $\frac{1}{300}$
25. $76\frac{3}{4}$ ft
27. $\frac{7}{32}$ in
29. $\frac{7}{8}$ A

Job 5-6:
Page 153

1. 1800
3. 2 A
5. 6 V

Job 5-7:
Page 155

1. 24
3. 12
5. 1
7. 2
9. 8
11. 6750
13. 54
15. 4000
17. 30
19. 93.97

Job 5-7:
Page 157

1. 2
3. 4
5. $3\frac{1}{3}$
7. 15
9. 15
11. 1.2
13. 10
15. $3\frac{1}{2}$
17. 7.8
19. 315
21. 120 turns
23. $33\frac{1}{3}$
25. $60\ \Omega$
27. 500 lb
29. $1.02\ \Omega$

Job 5-7:
Page 159

1. 6
3. 80
5. 51

7. 27
9. 25
11. 10
13. 1.44
15. $5\frac{1}{3}$
17. 0.1
19. 1.84

Job 5-8:
Page 162

1. $1.5\ \Omega$
3. $1.64\ \Omega$
5. $1.9\ \Omega$
7. $2.5\ \Omega$
9. $R_A = 12.4\ \Omega$
  $R_B = 11.25\ \Omega$
  $R_C = 9.47\ \Omega$
11. $60\ \Omega$
13. $1200\ \Omega$
15. $20\ \Omega$

Job 5-8:
Page 164

1. $50\ \Omega$
3. $0.75\ \Omega$
5. (a) $7.33\ \Omega$
  (b) 45 A

Job 5-8:
Page 165

1. $20\ \Omega$
3. $24\ \Omega$
5. $7143\ \Omega$
7. $3.53\ k\Omega$
9. $28.6\ \Omega$

Job 5-9:
Page 166

1. 40 V
3. 4.8 V
5. 14.4 V
7. 18.5 V
9. 4.8 V

Job 5-10:
Page 171

1. $I_1 = I_2 = 4$ A
  (The current in a cir-
  cuit divides equally

between equal parallel
branches.)
3. $V_T = 6$ V
  $I_1 = I_2 = 4$ A
5. 9 A, 6 A, 4.5 A
7. $I_2 = 0.024$ A
  $I_1 = 0.006$ A
9. (a) $2.14\ \Omega$
  (b) 28 V
  (c) $V_1 = V_2 = V_3$
    $= 28$ V
  (d) $I_1 = 5.6$ A
    $I_2 = 4$ A
    $I_3 = 3.5$ A

Job 5-11:
Page 173

1. (a) 5.75 A
  (b) 110 V
  (c) $19.1\ \Omega$
3. $I_T = 47$ A
  $R_T = 9.36\ \Omega$
  $V_1 = V_2 = V_3$
    $= 440$ V
  $I_2 = 22$ A
  $R_1 = 44\ \Omega$
  $R_3 = 29.3\ \Omega$
5. $26.7\ \Omega$
7. (a) $10\ \Omega$
  (b) $8\ \Omega$
  (d) Series
  (e) $18\ \Omega$
  (f) 20 A
9. $600\ \Omega$
11. $85.7\ \Omega$
13. (a) $2550\ \Omega$
  (b) 102 V

Job 6-2:
Page 181

1. $R_T = 24\ \Omega$
  $I_T = 5$ A
  $I_1 = I_2 = 3$ A
  $I_3 = 2$ A
  $V_1 = 30$ V
  $V_2 = 90$ V
  $V_3 = 120$ V
3. $R_T = 60\ \Omega$
  $I_T = 2$ A

5. (a) 12,000 Ω
   (b) 0.03 A
   (c) 360 V
   (d) 0.012 A
   (e) 0.018 A
   (f) 54 V
   (g) 306 V
7. 8 Ω

Job 6-3:
Page 193

1. $I_T = 10$ A
   $I_1 = 10$ A
   $I_2 = 7.5$ A
   $I_3 = 2.5$ A
   $R_T = 13$ Ω
   $V_1 = 100$ V
3. 0.5 A
5. 10 mA
7. $R_T = 30$ Ω
9. (a) 60 kΩ
   (b) 90 kΩ
11. $R_T = 25$ Ω

Job 6-4:
Page 201

1. $V_l = 2.4$ V
   $V_L = 114.6$ V
3. 0.25 Ω
5. 6.048 V, No. 14
7. 113.3 V

Job 6-5:
Page 205

1. (a) $V_1 = 15.2$ V
      $V_2 = 4.8$ V
   (b) 101.8 V
   (c) 97 V
3. $V_A = 112$ V
   $V_B = 109.6$ V
5. $V_G = 126.4$ V
   $V_B = 111.1$ V
7. 125.8 V
9. $V_{M_1} = 109$ V
   $V_{M_2} = 99$ V

Job 7-1:
Page 215

1. 480 W

3. 22 W
5. 374 W
7. 2.2 W
9. 2.3 W
11. 900 W
13. 1 W
15. 2508 W

Job 7-2:
Page 217

1. 11.2 W
3. Yes. (0.135 W developed)
5. 58.8 W
7. 2246 W
9. $I_T = 1$ A
   $P_1 = 2$ W
   $P_2 = 0.64$ W
   $P_3 = 3.84$ W
   $P_4 = 19.2$ W
   $P_5 = 1.6$ W
   $P_6 = 3$ W
   $P_T = 13$ W

Job 7-3:
Page 218

1. 5 A
3. 50 A
5. 1.2 W
7. 0.5 A
9. 5.45 A
11. 20 A

Job 7-4:
Page 220

1. 720 W
3. 6.25 W, 13 W
5. 0.006 W
7. 3361 W
9. 661 W
11. (a) 60 Ω
    (b) 2 A
    (c) 240 W
13. (a) 100 Ω
    (b) 2 A
    (c) 400 W

Job 7-5:
Page 223

1. 10^{10}
3. 9×10^6
5. 10^{10}
7. 64×10^6
9. 121×10^4
11. 625×10^{-10}
13. 2.25×10^6
15. 1.44×10^{-6}

Job 7-6:
Page 225

1. 2.25 kW
3. 132.3 W
5. 6.5 W
7. 6.36 A; 16.2 W
9. 11.1 W; 28.9 W

Job 7-7:
Page 226

1. 8
2. 13
3. 4.2
4. 59
5. 112
6. 7.6
7. 3.74
8. 23.8
9. 276.5
10. 5.48
11. 0.807
12. 932.6

Job 7-8:
Page 233

1. 59
3. 3.9
5. 137
7. 6.32
9. 8.06

Job 7-9:
Page 236

1. 10^4
3. $4 \times 10^2 = 400$
5. $4 \times 10^2 = 400$
7. 0.005

9. 60
11. $6.32 \times 10^2 = 632$
13. 4
15. $2 \times 10^2 = 200$
17. $1.732 \times 10^3 = 1732$
19. 12×10^3
21. 1.5×10^{-2}
23. 250

Job 7-10:
Page 238

1. 10 A
3. 2 A
5. 20 V
7. 144 Ω
9. 38.7 A

Job 7-11:
Page 240

1. 660 W
3. 550 W
5. 43.2 W
7. 6.25 V
9. (a) $R_T = 16.8$ Ω
   (b) $P_T = 6.05$ W
11. 12.3 W
13. 90 Ω
15. $I_l = 20$ A
    $V_L = 110$ V
    $P_L = 2200$ V
    No. = 44 lamps
17. (a) $R_T = 200$ Ω
    (b) $I_T = 0.5$ A
    (c) $P_T = 50$ W

Job 8-1:
Page 245

1. $8x$
3. $6R$
5. $12x$
7. $5x$
9. $4.4R$
11. $3.5x$
13. $\frac{5}{6}T$
15. $9R$
17. $7.5x$
19. $4.5x$
21. $6.6R$
23. $2\frac{7}{8}x$

25. $3x$

27. $1.12x$

Job 8-2:
Page 247

1. $10R + 7I$

3. $2x + 9$

5. $6R + 8$

7. $3R + 60$

9. $6x + 5y$

11. $0.9I + 1.9R$

13. $8I_1 + 10I_2$

15. $4x + 5y + 10$

17. $1.2I_2 + 2$

19. $2.4 + 3.4I$

21. $6x + 55$

23. $0.45R$ V $+ 25$ V

25. $19I_3 + 3I_2$

Job 8-3:
Page 252

1. 7

3. 5

5. 5

7. 5

9. 12

11. 5

13. $5\frac{1}{3}$

15. 200

17. 30

19. 70

21. 5

23. 0.8

25. 5

27. 2

29. $24\frac{1}{2}$

31. 7

33. 3

35. 6

37. 6

39. $\frac{1}{4}$

41. 0.7

43. $6\frac{2}{3}$

45. $10\frac{1}{5}$

47. 2

49. 40

51. 6 in, 3 in, 12 in

53. $A = 13, B = 0.4,$
    $C = 21.6$

55. $A = 32, B = 16,$
    $C = 8$

57. (a) $2R + 150 = 200$
    (b) $R = 25 \ \Omega$

59. Motor $= 4$ A
    Lamp $= 2$ A
    Iron $= 7$ A

61. $V_1 = 30$ V
    $V_2 = 60$ V
    $V_3 = 30$ V

Job 8-4:
Page 255

1. 10

3. 30

5. 16

7. 15

9. 10

11. 13.2

13. 3

15. 28

17. 12

19. $12\frac{5}{8}$ ft

Job 8-5:
Page 258

1. $-2x + 7$

3. $-4I_1 - 2I_2$

5. $-2x - 4y$

7. $-3x - 7$

9. $-20 - 2x$

Job 8-6:
Page 260

1. 2

3. 9

5. 4

7. 10

9. 5

11. 7

13. $\frac{1}{3}$

15. 5

17. 4

19. 9

21. 7

23. 8

25. 5

27. $\frac{1}{3}$

29. 2.1

Job 8-7:
Page 261

1. -8

3. -5

5. $-\frac{1}{2}$

7. -8

9. $+7$

11. -4

13. -1

15. -5

17. -11

19. -3

21. -2

23. -7

Job 8-8:
Page 264

1. $6 - 8x$

3. $-3x + 12$

5. $3x - 5$

7. $6I - 24$

9. $1 - 2R$

11. $29 - 3x$

13. $20 + 2I_1 + 2I_2$

15. $7R - 12$

17. $5 - 2R$

19. $3I_1 + I_2 - 24$

Job 8-9:
Page 266

1. 1

3. -2

5. 2

7. -3

9. 3

11. -1

13. 3

15. -4

17. 7

19. 4

21. -21

23. $\frac{1}{2}$

25. 3

27. $5\frac{1}{2}$

29. 1.1

31. 25

33. 48 Ω

35. 650 W

Job 8-10:
Page 269

1. 2

3. 40

5. 30

7. 8

9. 7

11. 6

13. 300

15. 10

17. 2

19. $17\frac{1}{2}$

21. 108

Job 8-11:
Page 275

1. $x - 3, y = 2$

3. $x - 5, y = 2$

5. $I_1 = 6, I_2 = 2$

7. $V = 7, R = 2$

9. $I = 7, V = -2$

11. $a = 1, b = -3$

13. $x = 5, y = 1$

15. $a = 6, b = 10$

17. $I_2 = 7, I_3 = -3$

19. $I_2 = 5, I_3 = 2$

Job 8-12;
Page 279

1. $x = 5, y = 3$

3. $x = 5, y = 2$

5. $x = 2, y = 7$

7. $x = 5, y = 3$

9. $x = 1, y = 4$

11. $x = 1, y = 4$

Job 9-3:
Page 294

1. (a) $I_T = 4$ A
   (b) $V_2 = 20$ V
       $V_4 = 12$ V

3. (a) $I_T = 2.4$ A
   (b) $V_2 = 14.4$ V
       $V_3 = 9.6$ V
       $V_5 = 12$ V

5. (a) $I_T = 3$ A
   (b) $V_2 = 12$ V
       $V_4 = 15$ V
       $V_6 = 6$ V

7. 8.43 A

1. $V_T = V_1 = V_2$
   $= V_3 = 90$ V
   $I_T = 15$ A
   $I_2 = 4.5$ A
   $I_3 = 1.5$ A
   $R_2 = 20\ \Omega$
3. $V_T = V_1 = V_2$
   $= V_3 = 300$ V
   $I_T = 20$ A
   $I_2 = 2.5$ A
   $I_3 = 7.5$ A
   $R_2 = 120\ \Omega$
5. $V_T = 108$ V
   $R_3 = 6\ \Omega$
7. $I_1 = 1.95$ A
   $I_2 = 0.65$ A
   $V_T = 7.80$ V
9. Same as Prob. 7
11. (a) $I_{16} = 6$ A
    $I_{48} = 2$ A
    $I_{24} = 4$ A
    (b) $V_G = 96$ V

Job 9-5:
Page 313
Set No. 1

1. $x = 2$ A, $y = 3$ A,
   $z = 5$ A
3. $x = 3$ A, $y = 2$ A,
   $z = 1$ A
5. $v = 12$ A, $w = 3$ A,
   $x = 5$ A, $y = 10$ A,
   $z = 2$ A, $V_T = 104$ V
7. $x = 12$ A, $y = 4$ A,
   $z = 8$ A

Job 9-5:
Page 315 (top)
Set No. 2

1. $x = 4$ A, $y = 2$ A,
   $z = 6$ A
3. $x = 1$ A, $y = 0.5$ A,
   $z = 0.5$ A
5. $v = 6$ A, $w = 5$ A,
   $x = 3$ A, $y = 8$ A,

$V_T = 170$ V, $z = 2$ A
7. $x = 7$ A, $y = 3$ A,
   $z = 4$ A

Job 9-5:
Page 315 (bottom)
Set No. 3

1. $x = 0.5$ A,
   $y = 1.5$ A, $z = 2$ A
3. $x = 0.8$ A,
   $y = 0.6$ A, $z = 0.2$ A
5. $v = 1$ A, $w = 2.5$ A,
   $x = 0.5$ A, $y = 3$ A,
   $z = 2$ A, $V_T = 26$ V
7. $x = 5.5$ A, $y = 2$ A,
   $z = 3.5$ A

Job 9-7:
Page 325

1. $R_a = R_b = R_c$
   $= 20\ \Omega$
3. $R_a = 6\ \Omega$
   $R_b = 4\ \Omega$
   $R_c = 2.4\ \Omega$
5. $R_{AD} = 3.67\ \Omega$
7. $R_{AD} = 10.95\ \Omega$
9. (a) $R_T = 5\ \Omega$
   (b) $v = 5$ A,
   $w = 5$ A,
   $x = 4.5$ A,
   $y = 5.5$ A,
   $z = 0.5$ A
   (c) $V_T = 50$ V
11. (a) $R_T = 12.67\ \Omega$
    (b) $v = 4$ A,
    $w = 6$ A,
    $x = 7.87$ A,
    $y = 2.13$ A,
    $z = 1.87$ A
    (c) $V_T = 126.7$ V

Job 9-8:
Page 337

1. $I_L = 1$ A
   $V_L = 10$ V
3. $I_L = 12$ A
   $V_L = 43.2$ V
5. $I_L = 0.2$ A
   $V_L = 4.6$ V

7. $I_5 = 3$ A
   $V_5 = 18$ V
9. $I_L = 0.0025$ A
   $V_L = 25$ V

Job 10-1:
Page 348

1. 5.56
3. 3
5. 0.051
7. 0.747
9. 0.21

Job 10-2:
Page 350

1. 15 mA
3. 40.4 mA
5. 70.7 A

Job 10-3:
Page 353

1. (a) 290,000
   (b) 168 V
3. (a) 2995
   (b) 21.6 V
5. (a) 17,000
   (b) 51,000
7. (a) 45,000
   (b) 95,000
   (c) 145,000

Job 10-4:
Page 355

1. 0.182
3. 60.6 mA
5. 333 Ω/V
7. 50 μA
9. 1500 Ω
11. (a) 2250 Ω
    (b) 12.8 V
13. 0.56 Ω

Job 10-5:
Page 360

1. $R_1 = 50,000\ \Omega$
   $R_2 = 7143\ \Omega$
   $R_3 = 20,000\ \Omega$
3. $R_1 = 10,000\ \Omega$

$R_2 = 3333\ \Omega$
$R_3 = 1111\ \Omega$
5. $R_1 = 10,000\ \Omega$
   $R_2 = 7143\ \Omega$
   $R_3 = 12,500\ \Omega$

Job 10-6:
Page 365

1. 1250 Ω
3. 11.1 Ω
5. $V_k = 33.3$ V
   $V_x = 66.7$ V
7. (a) $V_1 = 25.6$ V
   (b) $V_2 = 58.1$ V
   (c) $V_3 = 116.3$ V
9. 130 V
11. 0.1 V

Job 10-7:
Page 367

1. 843 Ω
3. 6.04 Ω
5. 1289 Ω
7. 465 Ω
9. 0.16 Ω

Job 10-8:
Page 371

1. $z = 9$ A
3. $z = 3$ A
5. $I_5 = 2.25$ A

Job 10-9:
Page 378

1. (a) $I_{RL} = 1.44$ mA
   (b) $V_{RL} = 7.2$ V
   (c) $V_C = -2.8$ V
3. (a) $I_C = 0.25$ mA
   (b) $I_E = 0.25$ mA
   (c) $V_{RL} = 10$ V
   (d) $V_C = -3$ V
   (e) $V_{RE} = 0.5$ V
   (f) $V_{CE} = 2.5$ V

Job 10-10:
Page 381

1. (a) $I_C = 1$ mA
   (b) $V_{RL} = 1$ V

(c) $V_C = -9$ V

3. (a) $I_E = 3.67$ mA
   (b) $V_E = -0.73$ V
   (c) $I_C = 3.63$ mA
   (d) $V_{RL} = 3.63$ V
   (e) $V_C = -4.37$ V

Job 10-11:
Page 386

1. 4×10^{-6} A
3. $V_{BE} = 11.5$ mV

Job 10-12:
Page 391

1. $R_1 = 12$ Ω,
   $R_2 = 3.75$ Ω
3. $R_1 = 5$ Ω,
   $R_2 = 10$ Ω
5. $R_1 = 36$ Ω,
   $R_2 = 40$ Ω
7. $R_1 = 20$ Ω,
   $R_2 = 400$ Ω
9. $R_1 = 75$ Ω,
   $R_2 = 33.3$ Ω

Job 10-12:
Page 393

1. $R_1 = R_3 = 20$ Ω
   $R_2 = 80$ Ω
3. $R_1 = R_3 = 20$ Ω
   $R_2 = 80$ Ω
5. $R_1 = R_3 = 7.14$ Ω
   $R_2 = 3.43$ Ω
7. $R_1 = R_2 = 77.7$ Ω
   $R_2 = 25.3$ Ω

Job 10-13:
Page 395

1. $R_1 = 20$ Ω—10 W
   $R_2 = 40$ Ω—5 W
   $R_L = 40$ Ω—5 W
3. $R_1 = 10$ Ω—55 W
   $R_2 = 25$ Ω—113 W
   $R_3 = 20$ Ω—113 W
5. $R_1 = 3$ Ω, 7 W
   $R_2 = 4$ Ω, 7 W
   $R_3 = 12$ Ω, 2 W

Job 11-1:
Page 398

1. 100 Ω
2. 0.765 lb
3. 17.25 A
4. No
5. 70.5 V
6. 90%
7. 93.3%
8. 57.6 W
9. 117.6 V
10. $81.25
11. $79.63
12. 20%
13. $3.33
14. 12.8 oz

Job 11-2:
Page 399

1. 0.38
3. 0.06
5. 0.04
7. 0.036
9. 1.25
11. 0.167
13. 0.625
15. 0.125
17. 0.0225
19. 0.0425

Job 11-2:
Page 400

1. 50%
3. 20%
5. 145%
7. 100%
9. 62½%
11. 22.2%
13. 70%
15. 5½%

Job 11-2:
Page 401

1. 25%
3. 40%
5. 62½%
7. 30%
9. 50%
11. 65%

13. 66⅔%
15. 45.5%
17. 33⅓%
19. 22.2%

Job 11-2:
Page 403

1. 33
3. $7.50
5. $56.25
7. 17.25 A
9. 116.7 V
11. 5
13. 7.5 W
15. $19.18

Job 11-2:
Page 404

1. 25%
3. 40%
5. 2%
7. 150%
9. 33⅓%
11. 12½%
13. 40%
15. 2.25%
17. 20%
19. 5%

Job 11-2:
Page 406

1. 20
3. 40
5. 2000
7. $238.33
9. 15 A

Job 11-2:
Page 407

1. (a) 0.62
   (b) 0.03
   (c) 0.056
   (d) 0.008
   (e) 1.16
   (f) 0.045
   (g) 0.0625
3. (a) 75%
   (b) 60%
   (c) 42.9%

(d) 30%
(e) 23.1%
(f) 16⅔%
(g) 46.1%
5. $90
7. 40%
9. $48.75
11. 20.8%
13. 1.5%
15. 116.4 V
17. $22.80, $307.80
19. $88.80

Job 11-3:
Page 409

1. 6500 W
3. 2.3 kW
5. 50 W
7. ¾ kW
9. 1.69 kW
11. 0.094 kW
13. 5.33 hp
15. 13⅓ hp

Job 11-4:
Page 411

1. 83⅓%
3. 96%
5. 88.9%
7. 91.3%
9. 80.2%
   19.8%

Job 11-5:
Page 414

1. 3.6 hp
3. 0.587 hp
5. 4508 W
7. 2647 W
9. 2200 W, 3.26 hp
11. 220 V

Job 11-6:
Page 416

1.

R_L	I_L	P_{RL}	P_{Pth}	Eff, %
6	¾ A	5.63	3.38	38
8	⅔ A	4.44	3.56	45
10	⅗ A	3.60	3.60	50
12	⁶⁄₁₁ A	2.98	3.57	55
14	½ A	2.58	3.50	58

3. $R = 100\ \Omega$

Job 11-7:
Page 417

1. (a) 0.5 kW
   (b) ⅔ hp
   (c) 1500 W
   (d) 1125 W
   (e) 13.3 hp
   (f) 7.5 kW
3. 91.4%
5. 0.62 hp
7. 0.625 kW

Job 12-1:
Page 420

1. 2:5
2. 4:1
3. 9:1
4. 10:11
5. 3:4
6. 0.007 Ω
7. $23.75
8. 45 oz
9. 15 V
10. 104 Ω
11. $\dfrac{R_1}{R_2} = \dfrac{I_2}{I_1}$
12. 2 V

Job 12-2:
Page 424

1. 1:4
3. 2:3
5. 2:1
7. 4:1
9. (a) 3:2
   (b) 2:3

11. 1:4
13. 1:4
15. 83.3%
17. 171
19. 40
21. 10
23. 1:48
25. (a) 2 V
    (b) 1.7%
    (c) 98.3%

Job 12-2:
Page 427

1. 2750 lb
3. 4½ h
5. 71.4 ft³
7. 1000 Ω
9. 45 V

Job 12-2:
Page 431

1. 5 in
3. 66.7
5. 500
7. 0.08 A
9. 0.93 Ω
11. 0.25 A

Job 12-3:
Page 432

1. (a) 1:3
   (b) 1:6
   (c) 1:10
   (d) 5:8
   (e) 3:4
3. 36
5. 72
7. 0.6 A
9. 10 V

Job 12-4:
Page 434

1. No. 21, 0.0285 in
3. No. 2, 250 mil
5. No. 6, 162 mil
7. 64.08 mil, 0.0641 in
9. No. 18, 0.0403 in

Job 12-4:
Page 436

1. 0.010 in, 100 mil
3. 0.064 in, 64 mil
5. 32 mil, 1024 cmil
7. 0.0872 in, 7604 cmil
9. 0.173 in, 173 mil
11. 397,886 cmil

Job 12-5:
Page 438

1. 19.6 Ω
3. 12.5 Ω

Job 12-6:
Page 440

1. 3.68 Ω
3. 4.06 Ω
5. 1.205 Ω, 5 A
7. 32.82 Ω, 3.35 A
9. 33.7 Ω, 6.74 V

Job 12-7:
Page 442

1. 52.5 Ω
3. 508 V
5. 12.86 Ω

Job 12-8:
Page 444

1. 0.104 Ω
3. 0.0062 Ω
5. 4.46 Ω, 24.6 A, 2.71 kW
7. (a) 0.145 A
   (b) 0.87 W
9. 0.2 V

Job 12-9:
Page 446

1. 1.7 Ω
3. 96.1 ft
5. 3 ft
7. 370 ft
9. Nichrome

Job 12-10:
Page 448

1. 9.4 mil
3. 6.93 mil
5. 32 mil
7. No. 17
9. No. 12

Job 12-11:
Page 457

1. 1.93 Ω/kft
3. 3.25 Ω/kft
5. No. 8
7. No. 10
9. 45

Job 12-12:
Page 458

1. (a) 25 mil, No. 22
   (b) 102 mil, No. 10
   (c) 125 mil, No. 8
   (d) 128 mil, No. 8
3. 44.9 Ω
5. 1.4 Ω
7. 3.07 ft

Job 13-1:
Page 463

1. 55 A
3. No. 6
5. 34.5 V
7. 0.000 25 Ω/ft
9. No. 3

Job 13-2:
Page 465

1. No. 6
3. No. 4
5. No. 1
7. No. 10
9. RH = No. 0
    V = No. 1
   AF = No. 3

Job 13-3:
Page 470

1. No. 4
3. No. 0

5. No. 3
7. No. 12
9. (a) 5250 W
   (b) 250 W
   (c) 44 A
   (d) 5.7 V
   (e) No. 1
11. 450 ft

Job 13-4:
Page 472 (top)

1. 95
3. 30
5. 30
7. 135
9. 25

Job 13-4:
Page 472 (bottom)

1. 47.85
3. 110.7
5. 94
7. 22.75
9. 11.6

Job 13-5:
Page 472

1. 40 A
3. No. 3
5. No. 14
7. No. 10
9. No. 1
11. No. 14

Job 14-1:
Page 483

1. $\sin = 0.2462$
   $\cos = 0.9692$
   $\tan = 0.2540$
3. $\sin = 0.3243$
   $\cos = 0.9459$
   $\tan = 0.3428$
5. $\sin = 0.8823$
   $\cos = 0.4706$
   $\tan = 1.8750$

Job 14-2:
Page 485 (top)

1. 18°

3. 80°
5. 30°
7. 30°
9. 60°

Job 14-2:
Page 485 (bottom)

1. 15°
3. 58°
5. 65°
7. 16°
9. 46°

Job 14-3:
Page 488

	$\angle A$	$\angle B$
1.	53°	37°
3.	30°	60°
5.	55°	35°
7.	51°	39°
9.	11°	79°

11. 67°, 113°
13. 3°
15. 58°

Job 14-4:
Page 493

1. 10 in, 30°
3. 89.4 ft, 40°
5. 151 Ω, 60°
7. 940.4 ft, 62°
9. 123 Ω, 75°
11. 35.1 m, 33 m
13. 8.5 cm
15. 9.1 m
17. $I_x = 12.69$ A,
    $I_y = 5.92$ A
19. 2128 Ω

Job 14-5:
Page 495

1. 0.5878
3. 2.6051
5. 0.9063
7. 65°
9. 11°
11. 17°
13. 73°
15. 37°

17. 71°
19. 58°
21. 930 W
23. 38.5
25. 833 Ω
27. 2828 W
29. 236 V
31. 81°, 99°
33. 3°
35. 1.72 in
37. 23°
39. 233.3 Ω

Job 15-1:
Page 504

1. (a) 0.54 A
   (b) 0.63 A
   (c) 0.66 A
   (d) 0.762 A
   (e) 0.79 A
   (f) 0.87 A
   (g) 0.99 A
5. Max eff = 80%; 50 hp

Job 15-3:
Page 516

1. 10,000 m
3. 126.4 V
5. 17°
7. 43.3 A
9. (a) 30°
   (b) 37°
   (c) 60°

Job 15-4:
Page 519

	Max	Eff	Instant	Phase Angle
1.	—	24.7	17.5	—
3.	622	—	476	—
5.	—	109.6	77.5	—
7.	155.5	—	140.9	—
9.	35	24.7	—	—
11.	—	70.7	—	26°

13. 17.3 V
15. (a) 10.6 A
    (b) 900 w

17. 285 W

Job 15-5:
Page 522

1. 90 V
3. (a) 150 mV
   (b) 0.2 Hz
5. 600 V
7. 2.5 ms
9. ⅓ V

Job 15-6:
Page 525

1. 60 Ω
3. 110 V
5. 160 V, 3.2 W
7. 1.82 A, 60.4 Ω
9. 111 V, 24.7 Ω

Job 15-7:
Page 528

1. 10
3. 16
5. 26.5
7. 4.5
9. 72.11 m
11. 7.62 ft
13. 46 ft
15. 32 in

Job 15-8:
Page 531

1. 1 A
3. 25.12 V
5. 3000 Ω, 7.96 H

Job 16-3:
Page 547

1. (a) 439.6 Ω
   (b) 8792 Ω
3. (a) 7536 Ω
   (b) 0.02 A
5. 37.7 Ω
7. 5 H
9. 62,800 Ω
11. 2864 Ω
13. 2.7 mH
15. 255 μH

Job 16-4:
Page 555

1. (a) 13 Ω
   (b) 8 A
   (c) $V_R = 40$ V
      $V_L = 96$ V
   (d) 67°
   (e) 39.1%
   (f) 321 W
3. 0.92 A, 74°
5. (a) 150 Ω
   (b) 170 Ω
   (c) 85 V
7. 1.5 μA
9. 4.5 A
11. 3770 Ω, 0.01 A
13. 28.6 H
15. $X_L = 7.5$ Ω; 43°

Job 16-5:
Page 558

1. 4
3. 31.4
5. (a) 10
   (b) 1000 Ω
7. (a) 9420 Ω
   (b) 10,665 Ω
   (c) 3.8 mA
9. (a) 314 Ω
   (b) 330 Ω
11. 0.7 mA

Job 16-6:
Page 561

1. 3.18 H
3. 0.796 H
5. 0.51 H

Job 16-7:
Page 562

1. (a) 125.6 Ω
   (b) 1256 Ω
   (c) 12,560 Ω
   (d) 125,600 Ω
3. 3140 Ω
5. 60
7. 30 Ω
9. 0.61 H
11. 0.25 H

Job 16-8:
Page 566

1. 0.365 mH
3. (a) 32.30 mH
   (b) 117 V

Job 16-9:
Page 570

1. 360 V
3. (a) 48:1
   (b) 48:1
5. 600 turns
7. 6000 V
9. 1.5 V
11. 300 turns
13. 240 V
15. (a) 2 turns
    (b) 5¼ turns
    (c) 500 turns

Job 16-10:
Page 574

1. 2 A
3. (a) 0.16 A
   (b) 19.2 W
5. (a) 0.136 A
   (b) 7.5 V
7. (a) 0.086 A
   (b) 9.45 W
9. (a) 5 A
   (b) 1500 V
11. 511 A

Job 16-11:
Page 576

1. 20:1
3. 4:1
5. 1:5
7. 12,100 Ω
9. 8:1
11. 39:1
13. 1:3

Job 16-12:
Page 578

1. (a) 176 turns
   (b) 8:1
3. 1750 turns
5. (a) 100 turns
   (b) 22.7 turns
7. (a) 1.25 A
   (b) 150 W
9. (a) 14 A
   (b) 28.75 V
   (c) 402.5 W
11. (a) 135 W
    (b) 2.7 A
13. 45,000 Ω

Job 16-13:
Page 586

1. 1210 Ω
3. 0.392 H
5. 0.209 Ω

Job 16-14:
Page 590

1. 87.5 percent
3. (a) 180 W
   (b) 168 W
   (c) 93.3%
5. 80%
7. (a) 10,417 W
   (b) 5.21 A
9. (a) 77 W
   (b) 88 V
11. (a) $R_m = 8.52$ Ω
    (b) $R_1 = 0.02$ Ω

Job 16-15:
Page 602

1.

Cable Size	Resistance, R	cmil, A	K
4	0.0318	41,470	13.18746
2	0.0203	66,360	13.47108
1	0.0162	83,690	13.55778
0	0.0130	105,500	13.715

2.

Cable Size	Resistance, R	Reactance, X	Impedance, Z	Drop, V	%
4	0.053	0.00490	0.0532260275	12.77	11.1
2	0.0335	0.00457	0.033810278	8.11	7.1
1	0.0267	0.00440	0.0270601183	6.49	5.6
0	0.0212	0.00410	0.0215928229	5.18	4.5
00	0.017	0.00396	0.0174551311	4.19	3.6
000	0.0138	0.00386	0.0143296755	3.44	3
0000	0.01103	0.00381	0.01097	2.63	2.3

Job 17-2:
Page 609

1. 373 pF
3. 5065 pF
5. 30,085 pF
7. 150 pF

Job 17-3:
Page 612

1. 1.333 μF
3. 19 pF
5. 31 to 162 pF
7. 3 μF
9. 8 μF

Job 17-4:
Page 616

1. (a) 17,667 Ω
   (b) 5300 Ω
   (c) 662 Ω
3. 0.19 A

5. 400 pF; 3975 Ω
7. 1.67 μF
9. (a) 15.9 Ω
   (b) 3.18 Ω
   (c) 0.795 Ω
11. 13,250 Ω
13. 60 Hz
15. (a) 31.8 Ω
    (b) 0.636 Ω
17. 0.02 μF
19. 442 μF

Job 17-5:
Page 620

1. (a) 2.25 A
   (b) 0
   (c) 0 W
3. 124 Ω
5. 39.75 Ω, 0.1 A
7. 0.25 mA
9. 100 Ω, 0.001 59 μF

Job 17-6:
Page 621

1. 2.65 μF
3. 2.94 μF
5. 10 μF

Job 17-7:
Page 626

1. (a) 17 A
   (b) 7.06 Ω
   (c) 28°
   (d) 88.2%
   (e) 1799 W
3. (a) 2.23 A
   (b) 53.9 Ω
   (c) 63°
   (d) 44.8%
   (e) 120 W
5. 0.566 A
7. 44.8% AF, 0.05% RF;
   Yes, but not a good one.
9. 200 μA

Job 17-8:
Page 629

1. 0.0497 μF
3. 1.6 μF
5. 187 pF
7. (a) 26.5 Ω
   (b) 0.053 Ω
9. 44.7 Ω
11. 0.381 μF

Job 18-2:
Page 635

1. (a) 17 Ω
   (b) 7 A
   (c) V_R = 56 V
      V_C = 105 V
   (d) 62°
   (e) 47.1%
   (f) 392 W
3. (a) 141 Ω
   (b) 0.8 A
   (c) V_R = 80 V
      V_C = 80 V
   (d) 45°
   (e) 70.9%
   (f) 64 W

5. (a) 159,000 Ω
   (b) 18,800 Ω
7. (a) 500 Ω
   (b) 400 Ω
   (c) 6.63 μF
9. 113 Hz

Job 18-3:
Page 644

1. (a) 65 Ω
   (b) 2 A
   (c) V_R = 32 V
      V_L = 166 V
      V_C = 40 V
   (d) 76°
   (e) 24.6%
   (f) 64 W
3. (a) X_L = 6280 Ω
      X_C = 3180 Ω
   (b) 5060 Ω
   (c) 0.025 A
   (d) 37°
   (e) 79.1%
   (f) 2.47 W
5. 0.7 V
7. 1.5 A
9. 271 Ω
11. 0.01 H

Job 18-4:
Page 648

1. (a) 795 kHz
   (b) 50 Ω
3. (a) 2 kHz
   (b) 502 Ω
   (c) 502 Ω
   (d) 12 Ω
   (e) 0.5 A
   (f) 251 V
5. 1988 kHz
7. 1988 kHz
9. 1590 kHz

Job 18-5:
Page 651

1. 84.3 μH
3. 0.506 H
5. 35.1 μF
7. 72 μH
9. 2.53 mH

Job 18-6:
Page 654

1. 180 V
3. 5.3 H
5. (a) 6360 Ω
   (b) 7.9 mA
   (c) V_R = 7.9 V
      V_L = 49.6 V
   (d) 81°
   (e) 15.7%
   (f) 0.06 W
7. (a) 41.2 Ω
   (b) 4790 Ω
9. 71.3 kHz

Job 19-1:
Page 663

1. (a) 14.4 A
   (b) 8.33 Ω
   (c) 1728 W
3. (a) 1.5 A
   (b) 66.7 Ω
   (c) 0
   (d) 0 W
5. 0.11 A; 1636 Ω; 0 W
7. (a) 0.159 A, 0.064 A
   (b) 0.223 A
   (c) 538 Ω
   (d) 0
   (e) 0 W

Job 19-2:
Page 668

1. (a) 13 A
   (b) 9.23 Ω
   (c) 67°
   (d) 38.5%
   (e) 600 W
3. (a) 2.15 A
   (b) 46.5 Ω
   (c) 22°

(d) 93%
(e) 200 W
5. 70%, 100%;
   Yes, but not a good filter, since it
   also passes low frequencies.

Job 19-3:
Page 673

1. R = 7.2 Ω;
   C = 276 μF
2. R = 28.2 Ω;
   L = 16 mH
4. (a) 4.12 A
   (b) 29.1 Ω
   (c) 14° leading
   (d) 97.1%
   (e) 480 W
   (f) R = 28.2 Ω;
      C = 376 μF
6. (a) X_L = 7536 Ω
      X_C = 3312 Ω
   (b) I_L = 0.029 A
      I_C = 0.066 A
      I_R = 0.1 A
   (c) 0.106 A
   (d) 2075 Ω
   (e) 19° leading
   (f) 94.3%
   (g) 22 W
   (h) R = 1963 Ω;
      C = 3.9 μF

Job 19-4:
Page 678

1. I_x = 17.32 A
   I_y = 10 A
3. I_x = 10 A
   I_y = −17.32 A
5. I_x = 19.32 A
   I_y = 17.4 A
7. I_x = 64.3 mA
   I_y = −76.6 mA
9. I_x = 12.69 A
   I_y = 5.92 A

Job 19-5:
Page 688

1. EMI = electromagnetic interference;
   RFI = radio frequency interference
3. Radiated EMI; conducted EMI
5. L = open circuits;
   C = short circuits
7. Used to specify EMI filters
9. Measurements between two lines

Job 19-6:
Page 700

1. (a) 23.4 A
   (b) 5.13 Ω
   (c) 35° leading
   (d) 2305 W
3. (a) 14.1 A
   (b) 7.09 Ω
   (c) 0°
   (d) 1410 W
5. (a) 1.8 A
   (b) 55.5 Ω
   (c) 48° lagging
   (d) 120 W

Job 19-7:
Page 706

1. (a) Z = 21.1 Ω
   (b) I_T = 4.74 A
   (c) θ = 17° lagging
   (d) 95.6%
   (e) P = 454 W
3. (a) Z = 60.7 Ω
   (b) L_T = 1.98 A
   (c) θ = 31° lagging
   (d) 85.9%
   (e) P = 204 W
5. (a) Z = 30.9 Ω
   (b) I_T = 3.9 A
   (c) θ = 12° leading
   (d) 97.6%
   (e) P = 457 W

Job 19-8:
Page 708

1. 5300 kHz
3. 400 pF
5. 80 pF
7. 0.92 μH
9. 40.2 pF

Job 19-9:
Page 712

1. (a) 0.083 A
   (b) 301 Ω
   (c) 0 W
3. (a) I_R = 0.5 A
      I_L = 0.16 A
   (b) I_T = 0.52 A
   (c) 192 Ω
   (d) 16°
   (e) 96.2%
   (f) 50 W
5. (a) 128 mA
   (b) 1875 Ω
   (c) 51°
   (d) 62.5%
   (e) 19.2 W
7. (a) 17.1 A
   (b) 7 Ω
   (c) 31° leading
   (d) 85.6%
   (e) 1756 W
9. 120 μH
11. (a) 10 Ω
    (b) I_T = 10 A
    (c) 37° leading
    (d) 800 W

Job 20-1:
Page 720

1. 80 percent
3. 360 W
5. (a) 4400 VA
   (b) 90.9%
7. 2.5 A
9. 45.45 A
11. 3450 VA; 2933 W
13. 10,417 VA
15. 40 A
17. 800 var
19. 10 A

Job 20-2:
Page 727

1. (a) 1232 W
   (b) 80% leading
   (c) 1540 VA
   (d) 14 A
3. (a) 1650 W
   (b) 89.9%
   (c) 1835 VA
   (d) 16.7 A
5. (a) 27 kW
   (b) 54.5%
   (c) 49.5 kVA
7. (a) 13 kW
   (b) 73.1%
   (c) 17.8 kVA
   (d) 80.9 A
9. (a) 72 kW
   (b) 64.3%
   (c) 112 kVA

Job 20-3:
Page 733

1. (a) 950 W
   (b) 99.8%
   (c) 952 VA
3. (a) 5680 W
   (b) 92.1%
   (c) 6167 VA
   (d) 51.4 A
5. 98.8%
7. (a) 22 kW
   (b) 90.6%
   (c) 24.28 kVA
   (d) 105.6 A
9. (a) 1584 W
   (b) 94.6%
   (c) 1674 VA
   (d) 13.95 A

Job 20-4:
Page 744

1. 17.4%
3. 58.8%
5. 38 μF
7. 6.67 kVA
9. 109.6 kvar;
   Graph = 108;
   4 standard 25-C kvar

63.5 kVA released;
Graph = 64 kVA
11. (a) 111.6 kW
    (b) 93.3 kvar
    (c) 76.6%
    (d) 145.7 kVA
    (e) (1) 25 *C* kvar plus
        (1) 15 *C* kvar

Job 20-5:
Page 747

1. 28%, 4.8 A
3. (a) 1944 W
   (b) 90%
   (c) 2160 VA
   (d) 18 A
5. (a) 22.5 kW
   (b) 62.9%
   (c) 35.8 kVA
7. (a) 1936 W
   (b) 99.5%
   (c) 1946 VA
   (d) 16.6 A
9. 45.4%
11. (a) 80%
    (b) 900 var
    (c) 900 var
13. (a) 82.4 kVA
    (b) 32°
    (c) 43.7 kvar
    (d) 1-15 *C* kvar, plus
        1-25 *C* kvar
    (e) 99.9%
15. 22.6 *C* kvar;
    7.5 kVA released

Job 21-1:
Page 762

1. 2078 V
3. 40 A
5. (a) 460 V
   (b) 4505 W
   (c) 13.52 kW
7. (a) 9.01 kVA
   (b) 88%
9. (a) 4858 kVA
   (b) 0.823
   (c) 3810 V
11. (a) 12,000 V

(b) 361 A
(c) 7500 kVA
(d) 6000 kW
13. (a) 6.93 A
    (b) 6.93 A
    (c) 1440 W
15. (a) 480 V
    (b) 13 Ω
    (c) 21.3 A
    (d) 21.3 A
    (e) 0.923
    (f) 16.34 kW
17. (a) 4157 V
    (b) 127.4 kW
    (c) 20.8 A
    (d) 20.8 A
19. (a) 100 Ω
    (b) 138.6 V
    (c) 1.39 A
    (d) 0.866
    (e) 500 W
    (f) 577 VA

Job 21-2:
Page 769

1. 103.9 A
3. 440 V
5. (a) 5.77 A
   (b) 10 A
7. (a) 240 V
   (b) 10 A
   (c) 17.3 A
   (d) 7.2 kW
   (e) 7.2 kVA
9. (a) 0.82
   (b) 35°
   (c) 61 kVA
   (d) 35 kvar
11. (a) 72 kVA
    (b) 72 kVA

Job 21-3:
Page 774

1. (a) 17.3 A
   (b) 240 V
   (c) 12.5 kVA
3. 27.4 kW
5. (a) 350 A
   (b) 208 V

(c) 100.8 kW
7. 120 kW
9. (a) 520 kW
   (b) 166.7 A
   (c) 288.7 A
   (d) 1300 V
11. (a) 6 A
    (b) 1.73 kW
    (c) 3.46 A
    (d) $R = 16 \Omega$
        $X_L = 12 \Omega$
13. (a) 40.7 kW; 36.8 kvar
    (b) 20.0 kVA;
        12.0 kvar
    (c) 56.7 kW; 24.8 kvar
    (d) 0.914
    (e) 74.62 A

Job 22-3:
Page 779

1. (a) 5   (b) 8.7   (c) 15   (d) 8.7
3. All answers are $I = 14.4$ A
   because the input line voltage and
   the load power is fixed and the
   transformer is assumed lossless.
   Check all of the configurations to
   see that this is true.
5. Problem 3 showed that the line
   currents are all 14.4 A. The phase
   currents are determined by the
   delta and wye relationships
   between line and phase currents.
   (a) 14.4   (b) 8.3   (c) 8.3
   (d) 14.4

Job 23-2:
Page 807

1. (a) 0
   (b) 1
   (c) 0
3. $L = S_1 \cdot S_2$

Job 23-3:
Page 810

1. 8
3. 4

Job 23-4:
Page 813

1. 1.16

3.

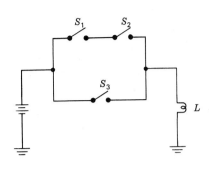

5.

S_1	S_2	$(S_1 S_2)'$	$S'_1 + S'_2$
0	0	1	1
0	1	1	1
1	0	1	1
1	1	0	0

Job 23-6:
Page 819

1.

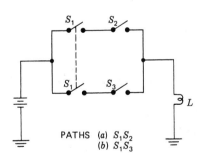

PATHS (a) $S_1 S_2$
     (b) $S_1 S_3$

3.

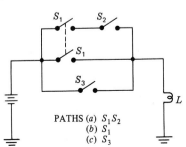

PATHS (a) $S_1 S_2$
     (b) S_1
     (c) S_3

5.

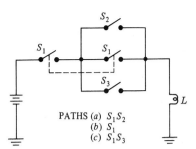

PATHS (a) $S_1 S_2$
     (b) S_1
     (c) $S_1 S_3$

Job 23-7:
Page 821

1.

S'_1	S_2	S_3	$(S_2 + S_3)$	S_1	L
0	0	0	0	1	1
0	0	1	1	1	1
0	1	0	1	1	1
0	1	1	1	1	1
1	0	0	0	0	0
1	0	1	1	0	1
1	1	0	1	0	1
1	1	1	1	0	1

3.

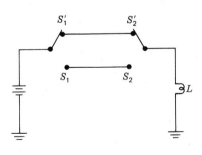

5. $L_1 = L'_1$

The user cannot tell the difference in the operation.

Job 23-8:
Page 828

1.

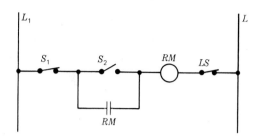

3. If $R_M = 1$, then
   $R_M = S_1 L_S$
5. (a) If the relay burned shut, then
   $R_M = S_1 L_S$.
   (b) Either push S_1 or let the motor run to the limit switch.

Job 23-9:
Page 834

1.

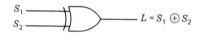

3.

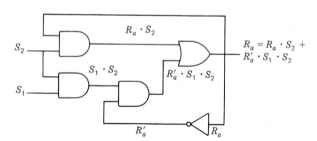

5. $L = S_1(S_2 + S_3)$

S_3	S_2	S_1	$S_2 + S_3$	L
0	0	0	0	0
0	0	1	0	0
0	1	0	1	0
0	1	1	1	1
1	0	0	1	0
1	0	1	1	1
1	1	0	1	0
1	1	1	1	1

7. 1; No; since $S_3 = 1$, $S_3 + S_4 = 1$. The output is independent of S_4.

Job 24-1:
Page 843

1. (a) 0.30103
   (b) −0.30103
   (c) −0.30103
   (d) 1.69897
   (e) 2.69897
   (f) −2.39794

Job 24-3:
Page 847

1. −0.58 dB
3. 3.01 dB
5. 10
7. 63.65 dB

Job 24-4:
Page 851

1. (a) 11.25 dB
   (b) 2.47 V
3. (a) 57.15 dB
   (b) 327 ft
5. 3.47 dB
7. (a) 3.01 dB
   (b) 3.01 dB
   (c) 6.02
   (d) yes
9. (a) 60 dB
   (b) 40.97 dB
   (c) 100.97 dB

Appendix A:
Color Codes:
Page 857

1. 100,000 Ω
3. 82,000 Ω
5. 75 Ω
7. 4.7 Ω
9. 1.2 Ω
11. 1,800,000 Ω
13. 620 Ω
15. 9100 Ω
17. 0.47 Ω
19. 82 Ω
21. Red, yellow, yellow
23. Green, brown, black
25. Brown, black, blue
27. Orange, white, red
29. Brown, red, black
31. Blue, gray, silver
33. Red, red, gold
35. Brown, gray, gold
37. Orange, blue, brown
39. Brown, green, gold

Appendix A:
Preferred Values:
Page 859

21. 220,000 Ω
23. 47 Ω
25. 10,000,000 Ω
27. 3900 Ω
29. 12 Ω
31. 0.68 Ω
33. 2.2 Ω
35. 1.8 Ω
37. 33 Ω or 39 Ω
39. 1.5 Ω

Answers to Internet Activities

Chapter 1: Page 12

A. Each resistor is marked with colored stripes to indicate its resistance value in ohms.

B. The speed of the electron flow in a metal wire is a few millimeters per second.

Chapter 2: page 52

A. Form 1: $DV = I \times R$ or $V = I \times R$, Form 2: $I = DV/R$ or $I = V/R$, Form 3: $R = DV/I$ or $R = V/I$.

B. When adding fractions such as $\frac{2}{4}$ and $\frac{1}{4}$, just add the numerators (you never add the denominators!). The denominators must be the same when you add fractions. Since in this case they are the same, just add the numerators. So $\frac{2}{4} + \frac{1}{4} = \frac{3}{4}$. When subtracting fractions, just look at the denominators. If they are the same proceed to subtract the second numerator from the first numerator, for example, $\frac{6}{5} - \frac{1}{5} = \frac{2}{5}$.

Chapter 3: page 96

A. (1) Whole number: A whole number is a positive number that has no fractions or decimals. Often referred to as natural numbers, these are the numbers we use when starting at 1 and counting up. (2) Integers: The family of numbers that includes whole numbers, negative values of whole numbers, and zero. (3) Examples of whole numbers: 1, 56, 127, 5639; examples of numbers that are not whole numbers: -3, 6.3, $4\frac{1}{2}$; ex-

amples of integers: -5, 0, and 146.

B. The minus symbol ($-$) means two different things in math. If it is between two numbers, it means subtraction; if it is in front of one number, it means the opposite (or negative) of that number.

Chapter 4: page 134

A. (1) Push the "test" button. The reset button should pop up. This indicates that power to the protected circuit has been discontinued. (2) If the reset button does not pop up when the test button is pushed, a loss of ground fault protection is indicated. DO NOT USE. Call a qualified electrician. (3) To restore power, push the reset button.

B. The term *effective resistance* is the resistance of a single resistor having the same effect as a combination of resistors. The effective resistance of series resistors is $R = R_1 + R_2 + R_3$. This states that for series resistors the effective resistance is the sum of the individual resistors.

Chapter 5: page 175

A. (1) Steps to finding GCF: Write the factors of all your numbers. Look for all the numbers that are the same. The highest common number is the GCF. (2) 12.

B. A node is a single point in a parallel circuit where the current divides among the parallel connected resistors.

Chapter 6: page 210

A. A *combination circuit* has a combination of series and parallel paths through which the electricity can flow. Its properties are a synthesis of the two. In the example shown, the parallel section of the circuit is like a subcircuit and actually is part of an overall series circuit.

B. (1) Loads 1 and 2. (2) Load 3.

Chapter 7: page 242

A. Cost = $1.26.

B. (1) The square roots of 9 are $+3$ and -3; the square roots of 16 are $+4$ and -4; the square roots of 25 are $+5$ and -5. (2) The principal root is the positive square root of a number.

Chapter 8: page 282

A. $x = 3.6$, $y = 0.4$.

B. The point where the lines cross ($x = 3.6$, $y = 4$) is the solution that satisfies both equations.

Chapter 9: page 343

A. No. The actual voltage at the location of the ground could be something other than 0 V, and this would not change any of the calculations of voltage and current in the circuit.

B. $V_{eq} = 36$ V, $V_1 = 24$ V, $I_1 = 2$ A, $I_2 = 1.2$ A, $I_3 = 0.8$ A, and $V_3 = 12$ V.

Chapter 10: page 396

A. A moving-coil galvanometer consists of a coil of wire on an iron core that is pivoted on bearings between the poles of a permanent magnet.

B. The scale of a galvanometer is commonly marked with intervals on both sides of a central zero. When the coil current is in one direction, the pointer needle is deflected to the right. If the current direction is reversed, the needle is deflected to the left.

Chapter 11: page 418

A. (hp × 746)/(E × eff × PF).

B. A switch-mode regulator overcomes the drawbacks of linear regulators. Switch-mode power supplies are more efficient and tend to have an efficiency of 80% or more. They can be packaged in a fraction of the size of linear regulators. Unlike linear regulators, switch-mode power supplies can step up or step down the input voltage.

Chapter 12: page 459

A. Southwire type TFFN or MTW or AWM can be used as fixture wire, machine tool wiring, or appliance wiring material as specified in the national Electrical Code 2. When used as type TFFN, the conductor is suitable for use at temperatures not to exceed 90°C. When used as type MTW, the conductor is suitable for use in wet locations or when exposed to oil or coolant at temperatures not to exceed 60°C or dry locations at temperatures not to exceed 90°C (with ampacity limited to that for 75°C conductor temperature per NFPA 79).

Conductor temperatures are not to exceed 105°C in dry locations when rated AWM and used as appliance wiring

material. Voltage for all applications is 600 V.

B. Alcatel Telecommunications, Belkin Components, C & M Inc., Eastman Wire & Cable, General Cable Corporation, and Northern Wire & Cable are firms that produce various types of cable in the United States.

Chapter 13: page 475

A. Current-carrying capacity is defined as the amperage a conductor can carry before melting either the conductor or the insulation. Heat caused by an electrical current flowing through the conductor will determine the amount of current a wire will handle. Theoretically, the amount of current that can be passed through a single bare copper wire can be increased until the heat generated reaches the melting temperature of the copper.

B. This table gives closest equivalent size cross-references between metric and American wire sizes. In Europe, wire sizes are expressed in cross-sectional area in square millimeters and also as the number of strands of wires of a diameter expressed in millimeters. For example, 7/0.2 means 7 strands of wire each 0.2 mm in diameter. This example has a cross-sectional area of 0.22 mm². In the United States, the most common system is the AWG numbering scheme, in which the numbers are applied not only to individual strands but also to equivalent size bunches of smaller strands. For example, 24 AWG could be made of 1 strand of 24 AWG wire (1/24) or 7 strands of 32 AWG wire (7/32). Because

standard metric wire sizes commonly used in manufacturing do not generally correspond exactly to American wire sizes, some stranding configurations do not have equivalents in practice.

Chapter 14: page 496

A. The answer can be found at http://www.hofstra.edu/~matscw/trig/trigans2.html

B. The word *trigonometry* comes from the Greek words *treis*, meaning "three," *gonia*, meaning "angle," and *metron*, meaning "measure." The early Greeks developed trigonometry by studying the relationship between the arc of the circle—the measure of the central angle—and the chord of the arc. Initially trigonometry was used in architecture, navigation, surveying, and engineering, but during the last two centuries it has been used more for mathematical analysis and for investigating repeating waves and periodic phenomena.

Chapter 15: page 539

A. *Hot* and *neutral* refer to wires irrespective of whether they happen to be carrying electrons toward or away from your lava lamp at any given ac half-cycle. A voltmeter placed between either hot and neutral would read 120 Vac rms. A voltmeter placed between either hot and ground would read 120 Vac rms. A voltmeter placed between both hots would read 240 Vac rms.

B. Some references for misconception research: Bill Beaty's www page: **http://www.eskimo.com/~billb/miscon/miscon.html.**

Proceedings of the 2d International Seminar—Misconceptions and Educational Strategies in *Science and Mathematics,* July 26–29, 1987. Mario Iona, *Why Johnny Can't Learn Physics from Textbooks I Have Known.* Millikan Award Lecture, *Am J. Phys.* 55 (4), April 1987, pp. 299–307. Mario Iona, *Would You Believe* Series of column in *The Physics Teacher* (AAPT Publication). Mario Iona, *How Should We Say It?* Series of columns in *The Science Teacher,* 1970–1972.

Chapter 16: page 603

A. (1) The charging of the transformer winding inductance. (2) The actual measurement of the winding resistance. (3) The safe dissipation of the energy stored in the transformer.

B. Raceway Wizard is a computer program that calculates electrical raceway size, overcurrent device size, voltage drop, fault current, and withstand ampacity. It also calculates energy savings for conductors that are larger than the required size. Raceway Wizard uses your input to calculate wire size. It uses a sophisticated algorithm—based on NEC rules and criteria—that repeatedly checks tables of electrical data. These tables contain nearly 100,000 pieces of information. When it's done working, the Wizard selects the largest suitable wire. Raceway Wizard is smart. It operates in the background, continually checking your work. If the Wizard detects an error, it displays a Stop Sign to alert you. You can see how Raceway Wizard performs its calculations by clicking the See Wizard Calculations button in the program.

Chapter 17: page 630

A. Capacitors store electrical energy and release it when the system demands more than your car's battery can provide. "Caps" are typically used in fully decked-out car stereo systems with power amps and subwoofers. They are used to provide extra impact and definition on heavy bass passages. Unlike a battery, which is designed to deliver moderate amounts of energy over a long period of time, a cap is designed to instantly release large amounts of energy on demand. When connected in parallel with the car battery, caps deliver instant energy to the amplifier. This instant energy is released during deep bass notes and loud crescendos so that the music isn't distorted when the system demands more than the battery can provide.

An electronics circuit board was being powered by an unregulated low-voltage power supply set to the nominal voltage required. The board was fitted with a tantalum electrolytic capacitor that "exploded, throwing out white-hot fragments which burned small holes in the surrounding furniture." The explosion was undoubtedly due to the rapid generation of gases within the capacitor together with overheating, both resulting from the passage of an alternating current greater than that which the capacitor was designed to handle.

B. Capacitance arises whenever electrical conductors are separated by a dielectric, or insulating, material. The fact that the conductors are separated by a dielectric material implies that electric charge is not transported through the capacitor. Although the application of a voltage to the terminals of the capacitor cannot cause a movement of charge through the dielectric, it can cause a displacement of charge within the dielectric. As the voltage varies with time, the displacement of charge within the dielectric varies with time, causing what is known as the *displacement current.* From the point of view of the terminals, the displacement current is indistinguishable from a conduction current.

Chapter 18: page 656

A. See Website.

B. This method requires the use of a signal generator and an analog or digital multimeter. (1) Measure the wire resistance of the inductor under measurement; call it r. (2) Assemble the following circuit using a known resistance R between 100 s and 1 ks. Adjust the generator to a frequency of 2 to 10 kHz, sine wave, and 1 V rms. The voltmeter must be connected across points A and B, that is, reading the voltage from the generator. (3) Move the voltmeter test leads across points C and B and record the reading; call it x. (4) Substitute your values into the following formula to determine the inductance:

$$L = (\sqrt{R/x})2 - (r + R)^2/(2f).$$

Chapter 19: page 714

A. Transients generally arise from two basic causes, from natural occurrence or from those generated from the use of equipment.

B. Every track in a PCB will radiate or receive unwanted harmonics to and from other parts of the circuit or even to other electronic equipment in the vicinity. Also, there may exist many subtle paths via parasitic elements that are difficult to observe, such as the stray capacitance between the heat sink of the switching transistor and the chassis ground. Finally, the victims. Noise generated by one part of the SMPS may cause the malfunctioning of other parts of the SMPS, such as the sensitive part of the PWM control circuit and the input line filter; this makes the design rather difficult. SMPS emissions affect the proper functioning of other electronic products. The effect may be some "snow" on the TV screen, noise on the audio system, or malfunction of other electronic appliances. When several pieces of electronic equipment are plugged to the same power source, if one piece of equipment generates interference, the noise signal will couple conductively to others through the polluted main supply. Besides the pollution of the main supply, a polluted output voltage will affect the functioning of the load circuit.

Chapter 20: page 752

A. Yes. Energy charge, demand charge, and power factor penalty charge.

B. Conventional rectifiers draw pulsed currents from the line, resulting in various problems such as the creation of harmonics and RF, high losses, reduced maximum power. All rectified ac sine wave signals with capacitive filtering draw high-amplitude current pulses from their source. Usually, the current peak value is in order of six times the current necessary for the same power on an ohmic load! Agencies such as the FCC and the VDE guard against pollution of power lines and RF emissions. Filtering, shielding, and other precautions are necessary to comply with their requirements.

Chapter 21: page 776

A. $P(\text{total}) = (3)^{\frac{1}{2}} \mid V\text{line} \mid \mid I\text{line} \mid \cos q$, $Q(\text{total}) = (3)^{\frac{1}{2}} \mid V\text{line} \mid \mid I\text{line} \mid \sin q$, $\mid S(\text{total}) \mid = (3)^{\frac{1}{2}} \mid V\text{line} \mid \mid I\text{line} \mid$, q is the angle between Vline and Iline.

B. Three-phase power has three "hot" wires, 120° out of phase with each other. These are usually used for large motors because it is more "efficient," provides a bit more starting torque, and because the motors are simpler and hence cheaper.

Bringing in a three-phase feed to a private residence is usually ridiculously expensive, or impossible. If the equipment has a standard motor mount, it is *much* cheaper to buy a new 110-V or 220-V motor for it. In some cases it is possible to run three-phase equipment on ordinary power if you have a "capacitor start" unit or use a larger motor as a (auto-)generator. These are tricky, but are a good solution if the motor is a non-standard size or too expensive or too big to replace. The Taunton Press book *The Small Shop* has an article on how to do this if you must.

Chapter 22: page 801

A. (1) Wye-wye: It should be avoided unless a neutral is provided. There are also problems with third harmonics. (2) Wye-delta: It is commonly used in stepping down from a high voltage to a medium or low voltage. One of the reasons is that a neutral is provided for grounding on the high-voltage side. (3) Delta-wye: It should be used for stepping up to high voltage. (4) Delta-delta: In a three-phase bank, this connection has the advantage that a transformer may be removed for repair while the remaining two continue to function; however, the rating is reduced to 58% of that of the original bank.

B. *Note:* The National Electrical Code (NEC) Report on Proposals and Report on Comments are works in progress. They are available for your information prior to their publication. The editing process, however, is ongoing, and these files may contain typographical or other errors or omissions.

Chapter 23: page 835

A. Either A or B but not both.
B. Only 1.

Chapter 24: page 852

A. 1 milliwatt (1 mW).
B. 1 microwatt (1 μW).

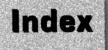

Index

AC (alternating current):
    cycle, 512
    frequency, 512
    generation, 506–514
    instantaneous value, 514–515
    measurement with oscilloscopes, 520–522
    power, 716–751
    sine waves, 511–512
    wavelength, 512–513
    (*See also* Parallel circuits, AC; Series circuits, AC)
Accuracy, degree of, 35
Addition, 120–121
    of decimals, 43–44
    defined, 63–64
    of fractions, 144–147
    of negative numbers, 63–64
    of phasors, 544–547, 689–690
    solving simultaneous equations with, 270–273
    symbols of, 56
Algebra, 244–281
    combining like terms in, 244–245
    combining unlike terms in, 246–247, 257–258
    cross multiplication in, 254–255
    fractions in, 267–269
    negative answers and, 260–261
    parentheses in, 262–265
    simultaneous equations in, 270–278
    solving for multiple unknowns in, 258–259
    solving for one unknown in, 248–250
    transposition in, 254–255
Ambient temperature, 454–456
American Wire Gage, 432–436
Ammeters:
    defined, 345
    extending range of, 345–349
    use of, 6
Ampacity, table, 462
Ampere, defined, 4–5
AND function, 805–807
Angles, 477–478, 485–487
Atoms, 1
Attenuators, 387–393
    L-pad, 388–390

Attenuators *(cont.):*
    T-type, 392–393
Autotransformers, 568

Balanced loads, 113–114
Batteries, 3
Bels, 844
    (*See also* Decibels)
Binary concepts, 803
Break frequency, 687

Cables:
    coaxial, 848–850
    fiber-optic, 850
    power distribution and, 591–601
Capacitors:
    coulombs, 4, 606–607
    defined, 605
    distributed capacitance in, 614–615
    farad, defined, 606–607
    leakage resistance in, 623
    measurement of, 620–621
    parallel, 607–609
    parallel AC, 661–663, 669–671
    reactance, 613–616
    real capacitors, 621–626
    series, 609–612
    series AC, 632–635, 637–643
    working voltage, 608, 610–612
Centimeters, 37
Circuits:
    analysis of
        delta, 319–321
        equivalent, 319–321, 368–371
        equivalent Norton, 338–340
        equivalent Thevenin, 327
        equivalent transistor, 381–386
        fixed-bias, 378–381
        H-parameter, 381–385
        ideal, 207–209
        Kirchhoff's current law, 284–286
        Kirchhoff's voltage law, 286–288
        load matching in, 414–415
        Norton's theorem of, 338–340

RMS (root mean square), 517

Safety:
    ground-fault protection, 115–116
    from shocks, 109–111
Scaler, defined, 523
Scientific notation, 91
SCR, 376–377
Series circuits:
    AC (alternating current), 632–655
        resistance, inductance, and capacitance in, 637–643
        resistance and capacitance in, 632–635
        series resonance in, 645–650
    advantages of, 177
    capacitors, 609–612
    DC (direct current), 98–133
        control of current in, 127–129
        Ohm's law in, 102–108
        power distribution in, 112–115
        safety measures, 109–111, 115–116
        total current in, 99–100, 107–108
        total resistance in, 100–101, 108
        total voltage in, 100, 106–107
    defined, 98
    Kirchhoff's law in, 289–292
    resistance and inductance in, 548–553
    (*See also* Combination circuits)
Series-parallel circuits, 178, 183–191, 701–705
Series resonance, 645–650
Shock safety, 109–111
Signed numbers (*see* Negative numbers)
Signs of operation, 55–56
Simultaneous equations:
    solving by addition, 270–273
    solving by subtraction, 276–278
Sines, 480, 484
Sine waves, 511–512
Square roots, 226–235
    power of 10, 233–235
Square waves, 533–534
Star circuits, 319–321
    (*See also* Wye connections)
Subtraction, 122
    of decimals, 45
    of fractions, 149–151
    solving simultaneous equations with, 276–278
    symbols of, 56
Symbols:
    circuit, 8, 9

Symbols (*cont.*)
    mathematical, 55–56
    TTL, 829–830

Tangents, 479–480, 484, 485
Temperature coefficient of resistance, 448–456, 471–472
Thevenin's theorem, 325–334, 414
Three-phase generation, 754
Three-phase power, 754–769
Transformers:
    auto-, 568
    copper loss in, 581
    core loss in, 581
    current ratio, 571–573
    defined, 566–567
    efficiency, 587–590
    impedance
        matched, 575
        referred, 575
    impedance transformation, 575–576
    primary, 567–568
    real, model of, 579–581
    secondary, 567–568
    test
        open-circuit, 581, 582–584
        short-circuit, 581, 585–586
    three-phase, 777–800
        balanced loads in, 781–794
        delta-delta, 781–793
        power from two transformers, 795–799
        wye-wye, 793–794
    turns ratio, 568
    voltage ratio, 568
Transistors, 372–375
    current gain, 374–375
    fixed-bias, 378–381
    operation of, 374
    self-biased, 372–378
Transposition, 123–125, 254–255
    (*See also* Formulas)
Triac, 377
Triangles, 478, 485–491
    Pythagorean theorem and, 526–527
Trigonometry, 477–495
    acute angles, 478, 485–487
    angles, 477–478, 485–487
    cosines, 480–482
    defined, 477
    Pythagorean theorem, 526–527

– A Word on Illustrations –

The horse is a beautiful animal, and descriptions of him and his activities are much enhanced by visual images. The author is most grateful to those photographers, artists, collectors and instititions who have made their works available to illustrate this book. These contributors include:

Ron Aira
Daphne Alcock
Howard Allen
Anita Baarns
Carey Beer and Adrienne Hewitt
Alf Baker
Terry Branham
Leonard "Sonny" Brown
Paul Brown Studios
The William Buseman Family
Dasher Chagna
The Chilcote Family
Anna de Roo
Foxcroft School
Juliet Graham
Sarah Greenhalgh
Marshall Hawkins –
     courtesy of McClannahan Camera
The Hermitage (Andrew Jackson's home)
Tanja Hess and the M.A.R.E. Center
Flora Hillman
Janet Hitchen
Lesley Howells
Harry Huberth
Kimberly Hurst
Mrs. Walter M. Jeffords
The Jockey Club
Kassie Kingsley
Matthew Klein

Raleigh Kraft
Caroline Leake
Karl Leck
Nancy Lee
Douglas Lees
Sandy Lerner
Sebastian Lezica
Lift Me Up!
The MacKay-Smith Family
Pat MacVeagh
Dan Marzani
Susan McHugh
Kelly Meister
Mischka Farm
Nat and Sherry Morison
National Sporting Library
Kitty Newman
Betsy Parker
"Boots" Parker
Donna Rogers
Dr. Garfield Royer
David L. Sally
Muffy Seaton
M.E. Smith
Steeplechase Times and Sean Clancy
Genie Stewart-Spears
Trinity Church
The Virginia Historical Society
Virginia Trails Association

To each of you, and to your subjects, many thanks.

Bruce Smart
Fall, 2003